Toward a Science of Consciousness
The First Tucson Discussions and Debates

Complex Adaptive Systems
John H. Holland, Christopher Langton, and Stewart W. Wilson, advisors

Adaptation in Natural and Artificial Systems: An Introductory Analysis with Applications to Biology, Control, and Artificial Intelligence, John H. Holland

Toward a Practice of Autonomous Systems: Proceedings of the First European Conference on Artificial Life, edited by Francisco J. Varela and Paul Bourgine

Genetic Programming: On the Programming of Computers by Means of Natural Selection, John R. Koza

From Animals to Animats 2: Proceedings of the Second International Conference on Simulation of Adaptive Behavior, edited by Jean-Arcady Meyer, Herbert L. Roitblat, and Stewart W. Wilson

Intelligent Behavior in Animals and Robots, David McFarland and Thomas Bösser

Advances in Genetic Programming, edited by Kenneth E. Kinnear, Jr.

Genetic Programming II: Automatic Discovery of Reusable Programs, John R. Koza

Turtles, Termites, and Traffic Jams: Explorations in Massively Parallel Microworlds, Mitchel Resnick

From Animals to Animats 3: Proceedings of the Third International Conference on Simulation of Adaptive Behavior, edited by Dave Cliff, Philip Husbands, Jean-Arcady Meyer, and Stewart W. Wilson

Artificial Life IV: Proceedings of the Fourth International Workshop on the Synthesis and Simulation of Living Systems, edited by Rodney A. Brooks and Pattie Maes

Comparative Approaches to Cognitive Science, edited by Herbert L. Roitblat and Jean-Arcady Meyer

Artificial Life: An Overview, edited by Christopher G. Langton

Evolutionary Programming IV: Proceedings of the Fourth Annual Conference on Evolutionary Programming, edited by John R. McDonnell, Robert G. Reynolds, and David B. Fogel

An Introduction to Genetic Algorithms, Melanie Mitchell

Catching Ourselves in the Act: Situated Activity, Interactive Emergence, and Human Thought, Horst Hendriks-Jansen

Toward a Science of Consciousness: The First Tucson Discussions and Debates, edited by Stuart R. Hameroff, Alfred W. Kaszniak, and Alwyn C. Scott

Toward a Science of Consciousness
The First Tucson Discussions and Debates

Edited by
Stuart R. Hameroff
Alfred W. Kaszniak
Alwyn C. Scott

A Bradford Book
The MIT Press
Cambridge, Massachusetts
London, England

© 1996 Massachusetts Institute of Technology

All rights reserved. No part of this book may be reproduced in any form by any electronic or mechanical means (including photocopying, recording, or information storage and retrieval) without permission in writing from the publisher.

This book was set in Palatino by Ruttle Graphics, Inc. and was printed and bound in the United States of America.

Library of Congress Cataloging-in-Publication Data

Toward a science of consciousness: the first Tucson discussions and debates / edited by Stuart R. Hameroff, Alfred W. Kaszniak, Alwyn C. Scott.
    p.   cm. — (Complex adaptive systems)
"A Bradford book."
Includes index.
ISBN 0-262-08249-7 (alk. paper)
    1. Consciousness—Congresses.   I. Hameroff, Stuart R.
II. Kaszniak, Alfred W., 1949–   III. Scott, Alwyn. 1931–
IV. Series.
QP411.T68   1996
612.8'2—dc20   95-26556
                                                                              CIP

Contents

Preface		xi
I	**Philosophy of Mind**	1
1	Facing Up to the Problem of Consciousness *David J. Chalmers*	5
2	Consciousness and the Introspective Link Principle *Güven Güzeldere*	29
3	The Place of Qualia in the World of Science *Leopold Stubenberg*	41
4	The Binding Problem and Neurobiological Oscillations *Valerie Gray Hardcastle*	51
5	Deconstructing Dreams: The Spandrels of Sleep *Owen Flanagan*	67
II	**Cognitive Science**	89
6	Unconscious Processes in Social Interaction *John F. Kihlstrom*	93
7	Efference and the Extension of Consciousness *Thaddeus M. Cowan*	105
8	Edelman's Biological Theory of Consciousness *John J. Boitano*	113
9	The Structure of Subjective Experience: Sharpen the Concepts and Terminology *David Galin*	121

10	The Varieties of Conscious Experience: Biological Roots and Social Usages	
	Karl H. Pribram	141
III	**Medicine**	**165**
11	Induction of Consciousness in the Ischemic Brain	
	James E. Whinnery	169
12	Conflicting Communicative Behavior in a Split-Brain Patient: Support for Dual Consciousness	
	Victor Mark	189
13	Left Brain Says Yes, Right Brain Says No: Normative Duality in the Split Brain	
	Marco Iacoboni, Jan Rayman, and Eran Zaidel	197
14	Inkblot Testing of Commissurotomy Subjects: Contrasting Modes of Organizing Reality	
	Polly Henninger	203
15	Evidence for Language Comprehension in a Severe "Sensory Aphasic"	
	Britt Anderson and Thomas Head	223
16	Self-Awareness of Deficit in Patients with Alzheimer's Disease	
	Alfred W. Kaszniak and Gina DiTraglia Christenson	227
IV	**Experimental Neuroscience**	**243**
17	Toward the Neuronal Substrate of Visual Consciousness	
	Christof Koch	247
18	Visual Perception and Phenomenal Consciousness	
	Petra Stoerig and Alan Cowey	259
19	Levels of Awareness and "Awareness Without Awareness": From Data to Theory	
	Gary E. Schwartz	279
20	Implicit Memory During Anesthesia	
	Randall C. Cork	295

21	Experimental Evidence for a Synchronization of Sensory Information to Conscious Experience *Mikael Bergenheim, Håkan Johansson, Brittmarie Granlund, and Jonas Pedersen*	303
22	Positron Emission Tomography, Emotion, and Consciousness *Eric M. Reiman, Richard D. Lane, Geoffrey L. Ahern, Gary E. Schwartz, and Richard J. Davidson*	311
23	Dimensional Complexity of Human EEG and Level of Consciousness *Richard C. Watt*	321
24	Collapse of a Quantum Field May Affect Brain Function *C. M. H. Nunn, C. J. S. Clarke, and B. H. Blott*	331
25	Neural Time Factors in Conscious and Unconscious Mental Functions *Benjamin Libet*	337
V	**Neural Networks**	349
26	Modeling What It Is Like To Be *John Taylor*	353
27	Artificial "Attention" in an Oscillatory Neural Network *Tokiko Yamanoue*	377
28	The Emergence of Memory: Categorization Far From Equilibrium *Andrew Wuensche*	383
VI	**Subneural Biology**	393
29	Water Clusters: Pixels of Life *John G. Watterson*	397
30	Microtubular Self-Organization and Information Processing Capabilities *J. A. Tuszyński, B. Trpisová, D. Sept, and M. V. Satarić*	407
31	Quantum Computation in the Neural Membrane: Implications for the Evolution of Consciousness *Ron Wallace*	419

32	Computer Simulation of Anesthetic Binding in Protein Hydrophobic Pockets *Dyan Louria and Stuart R. Hameroff*	425
VII	**Quantum Theory**	**435**
33	Consciousness and Bose-Einstein Condensates *Danah Zohar*	439
34	On the Quantum Mechanics of Dreams and the Emergence of Self-Awareness *Fred Alan Wolf*	451
35	Percolation and Collapse of Quantum Parallelism: A Model of Qualia and Choice *Michael Conrad*	469
36	Subcellular Quantum Optical Coherence: Implications for Consciousness *Mari Jibu, Scott Hagan, and Kunio Yasue*	493
37	Orchestrated Reduction of Quantum Coherence in Brain Microtubules: A Model for Consciousness *Stuart R. Hameroff and Roger Penrose*	507
VIII	**Nonlocal Space and Time**	**541**
38	Time and Consciousness: The Uneasy Bearing of Relativity Theory on the Mind-Body Problem *Avshalom C. Elitzur*	543
39	New Insights from Quantum Theory on Time, Consciousness, and Reality *Jeff Tollaksen*	551
40	Consciousness: A New Computational Paradigm *Douglas J. Matzke*	569
41	A Mathematical Strategy for a Theory of Consciousness *Saul-Paul Sirag*	579
42	Nonlocality on a Human Scale: Psi and Consciousness Research *Mario Varvoglis*	589
43	Synchronicity and Emergent Nonlocal Information in Quantum Systems *E. M. Insinna*	597

IX	**Hierarchical Organization**	609
44	Self-Referent Mechanisms as the Neuronal Basis of Consciousness *Erich Harth*	611
45	A Framework for Higher-Order Cognition and Consciousness *Nils A. Baas*	633
46	Bioenergetic Foundations of Consciousness *B. Raymond Fink*	649
47	The Hierarchical Emergence of Consciousness *Alwyn C. Scott*	659
X	**Phenomenology**	673
48	Pharmacology of Consciousness: A Narrative of Subjective Experience *Andrew Weil*	677
49	What Can Music Tell Us About the Nature of the Mind? A Platonic Model *Brian D. Josephson and Tethys Carpenter*	691
50	Intention, Self, and Spiritual Experience: A Functional Model of Consciousness *Arthur J. Deikman*	695
51	Enhanced Vigilance in Guided Meditation: Implications of Altered Consciousness *Richard P. Atkinson and Heath Earl*	707
52	The Stream Revisited: A Process Model of Phenomenological Consciousness *José-Luis Díaz*	713
XI	**Overview**	725
53	Three Kinds of Thinking *I. N. Marshall*	729
54	The Possibility of Empirical Test of Hypotheses About Consciousness *Jean E. Burns*	739

55	Toward a Science of Consciousness: Addressing Two Central Questions *Willis W. Harman*	743

Postscript	753
Contributors	755
Index	759

Preface

The role of consciousness in science has had its ups and downs over the past century. After dominating the stage throughout William James's Long Course in Principles of Psychology, the concept was banned from polite discussion for many decades by the behaviorists. Even after the emergence of cognitive science and neuroscience in the sixties and seventies—which undermined behaviorism by demonstrating the objective reality of inner states—the C-word remained slightly off-color in polite scientific discourse. Why discuss something that can't be measured?

In the past few years much of this has changed. Scientists with unimpeachable credentials in a wide range of fields, from psychology to molecular biology to mathematical physics, have begun to assert that understanding the nature of consciousness is an important scientific goal, perhaps the most important question that science faces at the present time. But how to begin? We take the view that the problem of consciousness transcends the traditional boundaries of scientific organization. Clearly, psychologists and psychiatrists have important contributions to make, but so do biochemists who study the various actions of mood-altering chemicals. In addition to philosophers, who have thought about the nature of consciousness for many centuries, computer scientists are now entering the discussion, as are neural network analysts, electrophysiologists, quantum physicists, and ethnologists. Such a diverse set of backgrounds might lead to the supposition that the discussion is spinning out of control, but we are more optimistic. It is a time of great intellectual excitement; old perspectives are changing and new concepts are emerging in this vast interdisciplinary area.

This is the background that led us to organize an international conference entitled: "Toward a Scientific Basis for Consciousness" at the University of Arizona in April of 1994. Our aim was not to predetermine the outcome of the discourse by favoring certain points of view, but to provide a forum in which many opinions and approaches could interact. Our diverse professional backgrounds—we are, respectively, an anesthesiologist, a psychologist, and an applied mathematician—have helped us to achieve a necessary breadth of view. This is not to suggest that we do not have individual opinions about the nature of consciousness. We most certainly do, and we have expressed our individual opinions in the proper place as contributors to the conference. As organizers of the conference, however, we have tried to act as honest brokers, to create a level playing field upon which a wide variety of ideas could interact. We were surprised and pleased by the enthusiastic response to the conference. Amid the warmth and beauty of the desert spring, some three hundred and fifty participants converged from many lands. Scientists, philosophers, and journalists engaged in vigorous and illuminating discussions from early in the morning until late at night, at formal sessions and over meals, teas, and drinks. There was a sense of excitement throughout the conference, a feeling that we were taking part in something new and needed, a whiff of Woodstock in the air. Favorable reports in *Scientific American, New Scientist,* the *Journal of Consciousness Studies,* and other periodicals documented the impact of the Conference, and as the papers collected in this book show, a substantial record of the views presented from the speaker's platform and posters has also been obtained.

What, then, was established about the nature of consciousness in Tucson in April of 1994? First of all—we must admit—the problem of consciousness was not solved. Nonetheless, several interesting lines of investigation were suggested. Perhaps the most important issue concerned the ontological basis for a theory of consciousness. Is consciousness based on a nonphysical aspect of reality as the dualists would suggest? Or does it emerge from the electrochemical interactions of tens of billions of neurons as is assumed in the field of cognitive neuroscience? Is classical neuroscience up to the task of explaining consciousness, or must one appeal to the mysteries of quantum mechanics to capture its essence? Is a new, and yet to be discovered, element of physical reality at the base of consciousness? In the papers presented here, you can read arguments for each of these positions, and more. What was accomplished in Tucson and from its aftermath is a clearer understanding of the important questions to be asked about consciousness. These include: How do traditional neuroscience and neural network theory address the question of qualia or the essential nature of our experience? If the brain is but a mechanism, why does it need an inner life? Could it not operate just as well without such a subjective phenomenon? What are the neural correlates of consciousness? How precisely can they be determined? How are conscious aspects of attention, vision, language, and memory mediated? What are the distinctions and mechanisms of transition among pre-, sub-, and nonconscious processes, and consciousness itself? What is the relation between consciousness and time? How does the brain provide for both simultaneity of events and a flow of time? Beyond being a physical theory that is as mysterious as is consciousness, how might quantum theory contribute to the understanding of consciousness? In what sense is this theory more than merely another reductive mechanism? Is there any experimental evidence to indicate that quantum effects play a role in the processes of the mind? Are there experimental observations that might help to decide among any of the various theories of mind? Can such experiments be suggested? In what ways can bridges be built among the insights of hitherto unrelated areas of scientific inquiry into the nature of consciousness? How can we better learn to learn from each other? Is there a useful analogy between the concepts of consciousness and life? Does consciousness emerge from the hierarchical levels of the brain as life emerges from the hierarchical levels of molecules, cells, and tissues? Can we expect to understand consciousness if we don't understand life? What progress has been made in understanding the ways in which anesthetic agents and other consciousness-altering drugs act on the biomolecular constituents of the brain? How do such developments contribute to an understanding of consciousness? What can we learn about the nature of consciousness from experiments in parapsychology? Are the data statistically reliable? How can evidence collected from spiritual practices (for example, meditation, shamanism, and mysticism) and nonstandard medical practices (mind/body medicine, hypnotism, and psychopharmacology) contribute to our scientific understanding of the nature of consciousness? Might insights from aesthetics enlarge our understanding? Is the sense of beauty—as it emerges in music, dance, and poetry—relevant to our discussion?

In preparing this book for publication, we faced a serious problem: more than twice the number of papers were submitted than we had space to include. We were thus forced to make some difficult editorial decisions. Several invited talks were not included, while many of the submitted poster presentations were. In general we tended to favor papers that reported the results of well-controlled experimental investigations over those proposing unsubstantiated theories. We also strove for balance, and when we found several papers that argued for the same point of view, we selected one or two that—in our imperfect wisdom—seemed to present the ideas most neatly. Were we successful? You the reader must decide.

The chapters in this volume are arranged in eleven sections ordered roughly as they were presented at the conference. This sequence, we feel, has some logical progression and should give the reader a sense of the way things went at "Tucson I." Each section is preceded by a brief introduction that summarizes and integrates contents of various chapters and will hopefully help traverse interdisciplinary barriers. Thus, the book begins with papers from the first morning on the philosophy of mind and the psychology of cognition. These are followed by observations from a variety of paradigms including effects on consciousness of brain injury, physical stress, and disease. On the second morning, the focus was on insights from the vast field of neuroscience, with emphasis on experimental results from a variety of studies in animals and human subjects. Papers presented the second afternoon examined the phenomenon of consciousness at a biochemical level within neurons: membranes, cell water, proteins, and the cytoskeleton. The third day of the conference was devoted to considerations of the role that quantum theory might play in the phenomenon of consciousness, and on possible means for quantum effects to arise at the biochemical or cytoskeletal levels. On the fourth morning, the phenomenon of emergence was considered, which is closely related to the hierarchical organization of biological systems and the properties of neural networks. This morning session was followed in the afternoon by an important session on the phenomenology of consciousness, with papers on pharmacological effects, meditation, and the relationship between consciousness and aesthetic perception. A section on nonlocal space and time effects in consciousness is included, gleaned mostly from submitted posters. The meaning of conscious now, and whether our perceptions are three-dimensional shadows projected from multidimensional hyperspace are among the issues raised. The last section of this book—from the final morning of the conference—is devoted to general overviews of the problem of consciousness. Can reductive science adequately explain consciousness? Where do we go from here? We invite all students of consciousness from every point of view to read through these papers with a critical eye, to ask what is missing, and to consider how she or he might contribute to a continuation of the discussion at "Tucson II" in the spring of 1996, and beyond.

Acknowledgments
The seeds of the Tucson Conference leading to this book were a series of annual weekend conferences on the topic of consciousness held in Sierra Vista, Arizona. They were sponsored by Gordon Olson, M.D., a physician in Sierra Vista, a small city approximately 90 miles southeast of Tucson. Dr. Olson's deep concern for the mechanisms of consciousness stemmed from his beloved daughter Maria, whose tragic coma the Olson family endured for a number of years prior to her death. Dr. Olson's courage and commitment inspired the conference.

Jim Laukes of Conference Services, University of Arizona Extended University, skillfully managed the conference and related events. Richard Hofstad helped with the organization and was an essential contributor to the preparation of the book. Harry Stanton of MIT Press provided sound guidance and encouragement, and David Chalmers gave useful and insightful advice. Professor Burnell Brown, Chairman of the Department of Anesthesiology, generously supported the endeavor, thereby extending the horizon of his specialty. We sincerely thank these people, conference participants, our families, friends and colleagues, and students of consciousness everywhere.

Stuart R. Hameroff
Alfred W. Kaszniak
Alwyn C. Scott

I Philosophy of Mind

Although philosophers have pondered consciousness for millennia, modern advances in reductionist brain sciences and computers have fueled contemporary interest (Churchland 1984, Dennett 1991, Searle 1992, Flanagan 1992, Scott 1995, Chalmers 1996). This volume begins with a series of philosophical chapters devoted mostly to the explanatory chasm between reductionist mechanisms and the subjective phenomenon of conscious experience. The chasm is so daunting that many support "dualism," the notion that the mind is distinct from the brain and merely interacts with it.

In the opening paragraph of his classic *Principles of Psychology*, William James—the dean of American philosophers—described dualism as the theory of "scholasticism and common sense." He then went on to suggest that "Another and a less obvious way of unifying the chaos is to seek common elements in the diverse mental facts rather than a common element behind them, and to explain them constructively by the various forms of arrangement of these elements, as one explains houses by stones and bricks."

More than a century later, although far better informed about the brain and benefiting from the insights of computers, we are faced with the same stark choice between dualism and reductionism. The task has not become easier, as David Chalmers points out in the opening chapter of this section. Any reductive explanation of consciousness, he argues, must fail because it could work as well without the "feel" and the sense of human experience and free will.

Chalmers divides the problems of consciousness into two classes, which he labels as the *easy problems* and the *hard problems*. The easy problems yield to usual scientific methods and include the differences between wakefulness and sleep, introspective access, and reportability. The hard problems involve explanations of the subjective experience of mental states. On this, reductionist science is still at sea. Why do we have an inner life experience, and what exactly is it? Within this framework Chalmers proposes a *double-aspect theory*, in which *information* (seen as the basic ingredient of the universe) has both physical and experiential aspects. He also sees structural coherence and organizational invariance as fundamental principles for a theory of consciousness.

Many philosophers and psychologists have taken the position that consciousness is not an ordinary mental state or process, but rather consists in the awareness of such. This perspective, conveyed by Locke (1689/1959): "Consciousness is the perception of what passes in a man's own mind," is referred to as *introspective consciousness* (Armstrong 1980). In Chapter 2, Güven Güzeldere considers higher-level representation of lower-level mental states and processes. He distinguishes *higher order perception*, as implied by Locke, from *higher order thought* and argues the latter more closely describes introspective consciousness. Güzeldere suggests the two are connected by an "introspective link."

In Chapter 3, Leopold Stubenberg examines the experiential aspects of consciousness, or *qualia*, in light of the "neutral monism" of philosopher/mathematician Bertrand Russell (1954). Russell argued that a common underlying entity that was neither physical nor mental gave rise to both the physical and mental. Stubenberg claims that *qualia* are such a common entity. Moving beyond reductionism, he sees the "stones and bricks" from which the mind's house is constructed as *qualia*—more fundamental than either physical particles or mental activity.

Although a philosopher, in Chapter 4 Valerie Gray Hardcastle takes a close look at some details of the brain's construction and considers the *"binding problem,"* which appeared insuperable to William James (1890/1950, pp. 158–162). How does the brain fuse together the many disparate features of a complex perception like Martin Luther King, Paris in the springtime, or your grandmother? Not in a "grandmother cell," certainly, because there aren't enough of them. Hardcastle judges other recently suggested neurobiological solutions to also be inadequate. In particular, she critically examines the evidence for 40-Hz oscillatory cortical firing patterns as a mechanism for unifying perceptual experiences, and finds that no conclusive evidence links phase–locked oscillations with the psychological phenomenon of binding. Hardcastle further asserts that limits in currently available methodologies presently prevent neuroscience from providing a link between bound perceptual unity and activities of neurons.

In Chapter 5, Owen Flanagan discusses the significance of dreams from a *natural method* stance, which treats phenomenology, cognitive science, and neuroscience with equal importance. While consciousness, according to Flanagan, has "depth, hidden structure, hidden and possibly multiple functions, and hidden natural and cultural history," he maintains that dreams are mere epiphenomenal side effects ("spandrels"—after an architectural detail) occurring during sleep. Arguing against the significance of dreams as representations of subconscious processes as espoused by Freud, Jung, and many others, Flanagan claims that dreams are only complex "inkblots" to which we attach *post hoc* importance. For a quite different view of dreams, see Chapter 34 by Fred Alan Wolf.

Valerie Gray Hardcastle illustrates the reductionist/dualist paradox: the closer one looks at the brain's biology, the more enigmatic consciousness appears. Several chapters articulate the paradox further, and offer general descriptions of the "mind-stuff" from which consciousness derives. In the view of David Chalmers, consciousness derives from the experiential aspect of information; in that of Güven Güzeldere, introspective consciousness is a hierarchical emergence; Leopold Stubenberg sees qualia as comprising the monistic entity underlying both physical and mental substance. In different ways these philosophical descriptions attempt to relate consciousness to the nature of underlying reality.

Owen Flanagan's chapter on dreams turns our attention to the question of subconscious processes, and how consciousness is organized within the brain. These topics are then taken up in Part 2.

REFERENCES

Armstrong, D. 1980. *The nature of mind and other essays.* Ithaca, NY: Cornell University Press.

Chalmers, D. J. 1996. *The conscious mind: In search of a fundamental theory.* New York: Oxford University Press.

Churchland, P. 1984. *Matter and consciousness.* Cambridge, MA: MIT Press.

Dennett, D. C. 1991. *Consciousness explained.* New York: Little-Brown.

Flanagan, O. 1992. *Consciousness reconsidered.* Cambridge, MA: MIT Press.

James, W. 1890/1950. *The varieties of religious experience.* New York: Dover.

Locke, J. 1689/1959. *An essay concerning human understanding,* Vol. 1. New York: Dover.

Russell, B. 1954. *The analysis of matter.* New York: Dover.

Scott, A. 1995. *Stairway to the mind.* New York: Springer-Verlag (Copernicus).

Searle, J. R. 1992. *The rediscovery of the mind.* Cambridge, MA: MIT Press.

1 Facing Up to the Problem of Consciousness

David J. Chalmers

INTRODUCTION

Consciousness poses the most baffling problems in the science of the mind. There is nothing that we know more intimately than conscious experience, but there is nothing that is harder to explain. All sorts of mental phenomena have yielded to scientific investigation in recent years, but consciousness has stubbornly resisted. Many have tried to explain it, but the explanations always seem to fall short of the target. Some have been led to suppose that the problem is intractable, and that no good explanation can be given.

To make progress on the problem of consciousness, we have to confront it directly. In this paper, I first isolate the truly hard part of the problem, separating it from more tractable parts and giving an account of why it is so difficult to explain. I critique some recent work that uses reductive methods to address consciousness, and argue that such methods inevitably fail to come to grips with the hardest part of the problem. Once this failure is recognized, the door to further progress is opened. In the second half of the paper, I argue that if we move to a new kind of nonreductive explanation, a naturalistic account of consciousness can be given. I put forward my own candidate for such an account: a nonreductive theory based on principles of structural coherence and organizational invariance, and a double-aspect theory of information.

THE EASY PROBLEMS AND THE HARD PROBLEMS

There is not just one problem of consciousness. "Consciousness" is an ambiguous term, referring to many different phenomena. Each of these phenomena needs to be explained, but some are easier to explain than others. At the start, it is useful to divide the associated problems of consciousness into "hard" and "easy" problems. The easy problems of consciousness are those that seem directly susceptible to the standard methods of cognitive science, whereby a phenomenon is explained in terms of computational or neural mechanisms. The hard problems are those that seem to resist those methods.

The easy problems of consciousness include those of explaining the following phenomena:

- the ability to discriminate, categorize, and react to environmental stimuli;
- the integration of information by a cognitive system;
- the reportability of mental states;
- the ability of a system to access its own internal states;
- the focus of attention;
- the deliberate control of behavior;
- the difference between wakefulness and sleep.

All of these phenomena are associated with the notion of consciousness. For example, one sometimes says that a mental state is conscious when it is verbally reportable, or when it is internally accessible. Sometimes a system is said to be conscious of some information when it has the ability to react on the basis of that information, or, more strongly, when it attends to that information, or when it can integrate that information and exploit it in the sophisticated control of behavior. We sometimes say that an action is conscious precisely when it is deliberate. Often, we say that an organism is conscious as another way of saying that it is awake.

There is no real issue about whether *these* phenomena can be explained scientifically. All of them are straightforwardly vulnerable to explanation in terms of computational or neural mechanisms. To explain access and reportability, for example, we need only specify the mechanism by which information about internal states is retrieved and made available for verbal report. To explain the integration of information, we need only exhibit mechanisms by which information is brought together and exploited by later processes. For an account of sleep and wakefulness, an appropriate neurophysiological account of the processes responsible for organisms' contrasting behavior in those states will suffice. In each case, an appropriate cognitive or neurophysiological model can clearly do the explanatory work.

If these phenomena were all there was to consciousness, then consciousness would not be much of a problem. Although we do not yet have anything close to a complete explanation of these phenomena, we have a clear idea of how we might go about explaining them. This is why I call these problems the easy problems. Of course, "easy" is a relative term. Getting the details right will probably take a century or two of difficult empirical work. Still, there is every reason to believe that the methods of cognitive science and neuroscience will succeed.

The really hard problem of consciousness is the problem of *experience*. When we think and perceive, there is a whir of information processing, but there is also a subjective aspect. As Nagel (1974) has put it, there is *something it is like* to be a conscious organism. This subjective aspect is experience. When we see, for example, we experience visual sensations: the felt quality of redness, the experience of dark and light, the quality of depth in a visual field. Other experiences go along with perception in different modalities: the sound of a clarinet, the smell of mothballs. Then there are bodily sensations, from pains to orgasms; mental images that are conjured up internally; the felt quality of emotion; and the experience of a stream of conscious thought. What unites all of these states is that there is something it is like to be in them. All of them are states of experience.

It is undeniable that some organisms are subjects of experience. But the question of how it is that these systems are subjects of experience is perplexing. Why is it that when our cognitive systems engage in visual and auditory information processing, we have visual or auditory experience: the quality of deep blue, the sensation of middle C? How can we explain why there is something it is like to entertain a mental image, or to experience an emotion? It is widely agreed that experience arises from a physical basis, but we have no good explanation of why and how it so arises. Why should physical processing give rise to a rich inner life at all? It seems objectively unreasonable that it should, and yet it does.

If any problem qualifies as *the* problem of consciousness, it is this one. In this central sense of "consciousness," an organism is conscious if there is something it is like to be that organism, and a mental state is conscious if there is something it is like to be in that state. Sometimes terms such as "phenomenal consciousness" and "qualia" are also used here, but I find it more natural to speak of "conscious experience" or simply "experience." Another useful way to avoid confusion (used by Newell 1990, Chalmers 1996) is to reserve the term "consciousness" for the phenomena of experience, using the less loaded term "awareness" for the more straightforward phenomena described earlier. If such a convention were widely adopted, communication would be much easier; as things stand, those who talk about "consciousness" are frequently talking past each other.

The ambiguity of the term "consciousness" is often exploited by both philosophers and scientists writing on the subject. It is common to see a paper on consciousness begin with an invocation of the mystery of consciousness, noting the strange intangibility and ineffability of subjectivity, and worrying that so far we have no theory of the phenomenon. Here, the topic is clearly the hard problem—the problem of experience. In the second half of the paper, the tone becomes more optimistic, and the author's own theory of consciousness is outlined. Upon examination, this theory turns out to be a theory of one of the more straightforward phenomena—of reportability, of introspective access, or whatever. At the close, the author declares that consciousness has turned out to be tractable after all, but the reader is left feeling like the victim of a bait-and-switch. The hard problem remains untouched.

FUNCTIONAL EXPLANATION

Why are the easy problems easy, and why is the hard problem hard? The easy problems are easy precisely because they concern the explanation of cognitive *abilities* and *functions*. To explain a cognitive function, we need only specify a mechanism that can perform the function. The methods of cognitive science are well-suited for this sort of explanation, and so are well-suited to the easy problems of consciousness. By contrast, the hard problem is hard precisely because it is not a problem about the performance of functions. The problem persists even when the performance of all the relevant functions is explained. (Here *function* is not used in the narrow teleological sense of something that a system is designed to do, but in the broader sense of any causal role in the production of behavior that a system might perform.)

To explain reportability, for instance, is just to explain how a system could perform the function of producing reports on internal states. To explain internal access, we need to explain how a system could be appropriately affected by its internal states and use information about those states in directing later processes. To explain integration and control, we need to explain how a system's central processes can bring information contents together and use them in the facilitation of various behaviors. These are all problems about the explanation of functions.

How do we explain the performance of a function? By specifying a *mechanism* that performs the function. Here, neurophysiological and cognitive modeling are perfect for the task. If we want a detailed low-level explanation, we can specify the neural mechanism that is responsible for the function. If we want a more abstract explanation, we can specify a mechanism in computational terms. Either way, a full and satisfying explanation will result. Once we have specified the neural or computational mechanism that performs the function of verbal report, for example, the bulk of our work in explaining reportability is over.

In a way, the point is trivial. It is a *conceptual* fact about these phenomena that their explanation only involves the explanation of various functions, as the phenomena are *functionally definable.* All it means for reportability to be instantiated in a system is that the system has the capacity for verbal reports of internal information. All it means for a system to be awake is for it to be appropriately receptive to information from the environment and for it to be able to use this information in directing behavior in an appropriate way. To see that this sort of thing is a conceptual fact, note that someone who says "you have explained the performance of the verbal report function, but you have not explained reportability" is making a trivial conceptual mistake about reportability. All it could *possibly* take to explain reportability is an explanation of how the relevant function is performed; the same goes for the other phenomena in question.

Throughout the higher-level sciences, reductive explanation works in just this way. To explain the gene, for instance, we needed to specify the mechanism that stores and transmits hereditary information from one generation to the next. It turns out that DNA performs this function; once we explain how the function is performed, we have explained the gene. To explain life, we ultimately need to explain how a system can reproduce, adapt to its environment, metabolize, and so on. All of these are questions about the performance of functions, and so are well-suited to reductive explanation. The same holds for most problems in cognitive science. To explain learning, we need to explain the way in which a system's behavioral capacities are modified in light of environmental information, and the way in which new information can be brought to bear in adapting a system's actions to its environment. If we show how a neural or computational mechanism does the job, we have explained learning. We can say the same for other cognitive phenomena, such as perception, memory, and language. Sometimes the relevant functions need to be characterized quite subtly, but it is clear that insofar as cognitive science explains these phenomena at all, it does so by explaining the performance of functions.

When it comes to conscious experience, this sort of explanation fails. What makes the hard problem hard and almost unique is that it goes beyond problems about the performance of functions. To see this, note that even when we have explained the performance of all the cognitive and behavioral functions in the vicinity of experience—perceptual discrimination, categorization, internal access, verbal report—there may still remain a further unanswered question: *Why is the performance of these functions accompanied by experience?* A simple explanation of the functions leaves this question open.

There is no analogous further question in the explanation of genes, or of life, or of learning. If someone says "I can see that you have explained how DNA stores and transmits hereditary information from one generation to the next, but you have not explained how it is a *gene*," then they are making a conceptual mistake. All it means to be a gene is to be an entity that performs the relevant storage and transmission function. But if someone says "I can see that you have explained how information is discriminated, integrated, and reported, but you have not explained how it is *experienced*," they are not making a conceptual mistake. This is a nontrivial further question.

This further question is the key question in the problem of consciousness. Why doesn't all this information processing go on "in the dark," free of any inner feel? Why is it that when electromagnetic waveforms impinge on a retina and are discriminated and categorized by a visual system, this discrimination and categorization is experienced as a sensation of vivid red? We know that conscious experience does arise when these functions are performed, but the very fact that it arises is the central mystery. There is an *explanatory gap* (a term due to Levine 1983) between the functions and experience, and we need an explanatory bridge to cross it. A mere account of the functions stays on one side of the gap, so the materials for the bridge must be found elsewhere.

This is not to say that experience has no function. Perhaps it will turn out to play an important cognitive role. But for any role it might play, there will be more to the explanation of experience than a simple explanation of the function. Perhaps it will even turn out that in the course of explaining a function, we will be led to the key insight that allows an explanation of experience. If this happens, though, the discovery will be an *extra* explanatory reward. There is no cognitive function such that we can say in advance that explanation of that function will *automatically* explain experience.

To explain experience, we need a new approach. The usual explanatory methods of cognitive science and neuroscience do not suffice. These methods have been developed precisely to explain the performance of cognitive functions, and they do a good job of it. But as these methods stand, they are *only* equipped to explain the performance of functions. When it comes to the hard problem, the standard approach has nothing to say.

SOME CASE STUDIES

In the last few years, a number of works have addressed the problems of consciousness within the framework of cognitive science and neuroscience. This might suggest that the analysis above is faulty, but in fact a close examination of the relevant work only lends the analysis further support. When we investigate just which aspects of consciousness these studies are aimed at, and which aspects they end up explaining, we find that the ultimate target of explanation is always one of the easy problems. I will illustrate this with two representative examples.

The first is the "neurobiological theory of consciousness" outlined by Crick and Koch (1990; see also Crick 1994). This theory centers on certain 35–75-hertz neural oscillations in the cerebral cortex; Crick and Koch hypothesize that these oscillations are the basis of consciousness. This is partly because the oscillations seem to be correlated with awareness in a number of different modalities—within the visual and olfactory systems, for example—and also because they suggest a mechanism by which the *binding* of information contents might be achieved. Binding is the process whereby separately represented pieces of information about a single entity are brought together to be used by later processing, as when information about the color and shape of a perceived object is integrated from separate visual pathways. Following others (for example, Eckhorn *et al*. 1988), Crick and Koch hypothesize that binding may be achieved by the synchronized oscillations of neuronal groups representing the relevant contents. When two pieces of information are to be bound together, the relevant neural groups will oscillate with the same frequency and phase.

The details of how this binding might be achieved are still poorly understood, but suppose that they can be worked out. What might the resulting theory explain? Clearly it might explain the binding of information contents, and perhaps it might yield a more general account of the integration of information in the brain. Crick and Koch also suggest that these oscillations activate the mechanisms of working memory, so that there may be an account of this and perhaps other forms of memory in the distance. The theory might eventually lead to a general account of how perceived information is bound and stored in memory, for use by later processing.

Such a theory would be valuable, but it would tell us nothing about why the relevant contents are experienced. Crick and Koch suggest that these oscillations are the neural *correlates* of experience. This claim is arguable—does not binding also take place in the processing of unconscious information?—but even if it is accepted, the *explanatory* question remains: Why do the oscillations give rise to experience? The only basis for an explanatory connection is the role they play in binding and storage, but the question of why binding and storage should themselves be accompanied by experience is never addressed. If we do not know why binding and storage should give rise to experience, telling a story about the oscillations cannot help us. Conversely, if we *knew* why binding and storage gave rise to experience, the neurophysiological details would be just the icing on the cake. Crick and Koch's theory gains its purchase by *assuming* a connection between binding and experience, and so can do nothing to explain that link.

I do not think that Crick and Koch are ultimately claiming to address the hard problem, although some have interpreted them otherwise. A published interview with Koch gives a clear statement of the limitations on the theory's ambitions.

Well, let's first forget about the really difficult aspects, like subjective feelings, for they may not have a scientific solution. The subjective state of play, of pain, of pleasure, of seeing blue, of smelling a rose—there seems to be a huge jump between the materialistic level, of explaining molecules and neurons, and the subjective level. Let's focus on things that are easier to study—like visual awareness. You're now talking to me, but you're not looking at me, you're looking at the cappuccino, and so you are aware of it. You can say, 'It's a cup and there's some liquid in it.' If I give it to you, you'll move your arm and you'll take it—you'll respond in a meaningful manner. That's what I call awareness." ("What is consciousness," *Discover* November 1992, p. 96.)

The second example is an approach at the level of cognitive psychology. This is Baars's global workspace theory of consciousness, presented in his book *A Cognitive Theory of Consciousness.* According to this theory, the contents of consciousness are contained in a *global workspace,* a central processor used to mediate communication between a host of specialized nonconscious processors. When these specialized processors need to broadcast information to the rest of the system, they do so by sending this information to the workspace, which acts as a kind of communal blackboard for the rest of the system, accessible to all the other processors.

Baars uses this model to address many aspects of human cognition, and to explain a number of contrasts between conscious and unconscious cognitive functioning. Ultimately, however, it is a theory of *cognitive accessibility,* explaining how it is that certain information contents are widely accessible within a system, as well as a theory of informational integration and reportability. The theory shows promise as a theory of awareness, the functional correlate of conscious experience, but an explanation of experience itself is not on offer.

One might suppose that according to this theory, the contents of experience are precisely the contents of the workspace. But even if this is so, nothing internal to the theory *explains* why the information within the global workspace is experienced. The best the theory can do is to say that the information is experienced because it is *globally accessible.* But now the question arises in a different form: why should global accessibility give rise to conscious experience? As always, this bridging question is unanswered.

Almost all work taking a cognitive or neuroscientific approach to consciousness in recent years could be subjected to a similar critique. The "neural Darwinism" model of Edelman (1989), for instance, addresses questions about perceptual awareness and the self-concept, but says nothing about why there should also be experience. The "multiple drafts" model of Dennett (1991) is largely directed at explaining the reportability of certain mental contents. The "intermediate level" theory of Jackendoff (1987) provides an account of some computational processes that underlie consciousness, but Jackendoff stresses that the question of how these "project" into conscious experience remains mysterious.

Researchers using these methods are often inexplicit about their attitudes to the problem of conscious experience, although sometimes they take a clear stand. Even among those who are clear about it, attitudes differ widely. In placing this sort of work with respect to the problem of experience, a number of different strategies are available. It would be useful if these strategic choices were more often made explicit.

The first strategy is simply to *explain something else.* Some researchers are explicit that the problem of experience is too difficult for now, and perhaps even outside the domain of science altogether. These researchers instead choose to address one of the more tractable problems such as reportability or the self-concept. Although I have called these problems the "easy" problems, they are among the most interesting unsolved problems in cognitive science, so this work is certainly worthwhile. The worst that can be said of this choice is that in the context of research on consciousness it is relatively unambitious, and the work can sometimes be misinterpreted.

The second choice is to take a harder line and *deny the phenomenon.* (Variations on this approach are taken by Allport 1988, Dennett 1991, and Wilkes 1988.) According to this line, once we have explained the functions such as accessibility, reportability, and the like, there is no further phenomenon called "experience" to explain. Some explicitly deny the phenomenon, holding for example that what is not externally verifiable cannot be real. Others achieve the same effect by allowing that experience exists, but only if we equate "experience" with something like the capacity to discriminate and report. These approaches lead to a simpler theory, but are ultimately unsatisfactory. Experience is the most central and manifest aspect of our mental lives, and indeed is perhaps the key explanandum in the science of the mind. Because of this status as an explanandum, experience cannot be discarded like the vital spirit when a new theory comes along. Rather, it is the central fact that any theory of consciousness must explain. A theory that denies the phenomenon "solves" the problem by ducking the question.

In a third option, some researchers *claim to be explaining experience* in the full sense. These researchers (unlike those above) wish to take experience very seriously; they lay out their functional model or theory, and claim that it explains the full subjective quality of experience (for example, Flohr 1992, Humphrey 1992). The relevant step in the explanation is usually passed over quickly, however, and usually ends up looking something like magic. After some details about information processing are given, experience suddenly enters the picture, but it is left obscure how these processes should suddenly give rise to experience. Perhaps it is simply taken for granted that it does, but then we have an incomplete explanation and a version of the fifth strategy below.

A fourth, more promising approach appeals to these methods to *explain the structure of experience*. For example, it is arguable that an account of the discriminations made by the visual system can account for the structural relations between different color experiences, as well as for the geometric structure of the visual field (see for example, Clark 1992 and Hardin 1992). In general, certain facts about structures found in processing will correspond to and arguably explain facts about the structure of experience. This strategy is plausible but limited. At best, it takes the existence of experience for granted and accounts for some facts about its structure, providing a sort of nonreductive explanation of the structural aspects of experience. (I will say more on this later.) This is useful for many purposes, but it tells us nothing about why there should be experience in the first place.

A fifth and reasonable strategy is to *isolate the substrate of experience*. After all, almost everyone allows that experience *arises* one way or another from brain processes, and it makes sense to identify the sort of process from which it arises. Crick and Koch put their work forward as isolating the neural correlate of consciousness, for example, and Edelman (1989) and Jackendoff (1987) make similar claims. Justification of these claims requires a careful theoretical analysis, especially as experience is not directly observable in experimental contexts, but when applied judiciously this strategy can shed indirect light on the problem of experience. Nevertheless, the strategy is clearly incomplete. For a satisfactory theory, we need to know more than *which* processes give rise to experience; we need an account of why and how. A full theory of consciousness must build an explanatory bridge.

THE EXTRA INGREDIENT

We have seen that there are systematic reasons why the usual methods of cognitive science and neuroscience fail to account for conscious experience. These are simply the wrong sort of methods; nothing that they give to us can yield an explanation. To account for conscious experience, we need an *extra ingredient* in the explanation. This makes for a challenge to those who are serious about the hard problem of consciousness: What is your extra ingredient, and why should *that* account for conscious experience?

There is no shortage of extra ingredients to be had. Some propose an injection of chaos and nonlinear dynamics. Some think that the key lies in nonalgorithmic processing. Some appeal to future discoveries in neurophysiology. Some suppose that the key to the mystery will lie at the level of quantum mechanics. It is easy to see why all these suggestions are put forward. None of the old methods work, so the solution must lie with something new. Unfortunately, these suggestions all suffer from the same old problems.

Nonalgorithmic processing, for example, is put forward by Penrose (1989, 1994) because of the role it might play in the process of conscious mathematical insight. The arguments about mathematics are controversial, but even if they succeed and an account of nonalgorithmic processing in the human brain is given, it will still only be an account of the *functions* involved in mathematical reasoning and the like. For a nonalgorithmic process as much as an algorithmic process, the question is left unanswered: why should this process give rise to experience? In answering *this* question, there is no special role for nonalgorithmic processing.

The same goes for nonlinear and chaotic dynamics. These might provide a novel account of the dynamics of cognitive functioning, quite different from that given by standard methods in cognitive science. But from dynamics, one only gets more dynamics. The question about experience here is as mysterious as ever. The point is even clearer for new discoveries in neurophysiology. These new discoveries may help us make significant progress in understanding brain function, but for any neural process we isolate, the same question will always arise. It is difficult to imagine what a proponent of new neurophysiology expects to happen, over and above the explanation of further cognitive functions. It is not as if we will suddenly discover a phenomenal glow inside a neuron!

Perhaps the most popular "extra ingredient" of all is quantum mechanics (for example, Hameroff 1994). The attractiveness of quantum theories of consciousness may stem from a Law of Minimization of Mystery: consciousness is mysterious and quantum mechanics is mysterious, so maybe the two mysteries have a common source. Nevertheless, quantum theories of consciousness suffer from the same difficulties as neural or computational theories. Quantum phenomena have some remarkable functional properties, such as nondeterminism and nonlocality. It is natural to speculate that these properties may play some role in the explanation of cognitive functions, such as random choice and the integration of information, and this hypothesis cannot be ruled out *a priori.* But when it comes to the explanation of experience, quantum processes are in the same boat as any other. The question of why these processes should give rise to experience is entirely unanswered.

(One special attraction of quantum theories is the fact that on some interpretations of quantum mechanics, consciousness plays an active role in "collapsing" the quantum wave function. Such interpretations are controversial, but in any case they offer no hope of *explaining* consciousness in terms of quantum processes. Rather, these theories *assume* the existence of consciousness, and use it in the explanation of quantum processes. At best, these theories tell us something about a physical role that consciousness may play. They tell us nothing about how it arises.)

At the end of the day, the same criticism applies to any purely physical account of consciousness. For any physical process we specify there will be an unanswered question: Why should this process give rise to experience? Given any such process, it is conceptually coherent that it could be instantiated in the absence of experience. It follows that no mere account of the physical process will tell us why experience arises. The emergence of experience goes beyond what can be derived from physical theory.

Purely physical explanation is well-suited to the explanation of physical *structures,* explaining macroscopic structures in terms of detailed microstructural constituents; and it provides a satisfying explanation of the performance of *functions,* accounting for these functions in terms of the physical mechanisms that perform them. This is because a physical account can *entail* the facts about structures and functions: once the internal details of the physical account are given, the structural and functional properties fall out as an automatic consequence. But the structure and dynamics of physical processes yield only more structure and dynamics, so structures and functions are all we can expect these processes to explain. The facts about experience cannot be an automatic consequence of any physical account, as it is conceptually coherent that any given process could exist without experience. Experience may arise from the physical, but it is not entailed by the physical.

The moral of all this is that *you can't explain conscious experience on the cheap.* It is a remarkable fact that reductive methods—methods that explain a high-level phenomenon wholly in terms of more basic physical processes—work well in so many domains. In a sense, one can explain most biological and cognitive phenomena on the cheap, in that these phenomena are seen as automatic consequences of more fundamental processes. It would be wonderful if reductive methods could explain experience, too; I hoped for a long time that they might. Unfortunately, there are systematic reasons why these methods must fail. Reductive methods are successful in most domains because what needs explaining in those domains are structures and functions, and these are the kind of thing that a physical account can entail. When it comes to a problem over and above the explanation of structures and functions, these methods are impotent.

This might seem reminiscent of the vitalist claim that no physical account could explain life, but the cases are disanalogous. What drove vitalist skepticism was doubt about whether physical mechanisms could perform the many remarkable functions associated with life, such as complex adaptive behavior and reproduction. The conceptual claim that explanation of functions is what is needed was implicitly accepted, but lacking detailed knowledge of biochemical mechanisms, vitalists doubted whether any physical process could do the job and put forward the hypothesis of the vital spirit as an alternative explanation. Once it turned out that physical processes could perform the relevant functions, vitalist doubts melted away.

With experience, on the other hand, physical explanation of the functions is not in question. The key is instead the *conceptual* point that the explanation of functions does not suffice for the explanation of experience. This basic conceptual point is not something that further neuroscientific investigation will affect. In a similar way, experience is disanalogous to the *élan vital*. The vital spirit was put forward as an explanatory posit, in order to explain the relevant functions, and could therefore be discarded when those functions were explained without it. Experience is not an explanatory posit but an explanandum in its own right, and so is not a candidate for this sort of elimination.

It is tempting to note that all sorts of puzzling phenomena have eventually turned out to be explainable in physical terms. But each of these were problems about the observable behavior of physical objects, coming down to problems in the explanation of structures and functions. Because of this, these phenomena have always been the kind of thing that a physical account *might* explain, even if at some points there have been good reasons to suspect that no such explanation would be forthcoming. The tempting induction from these cases fails in the case of consciousness, which is not a problem about physical structures and functions. The problem of consciousness is puzzling in an entirely different way. An analysis of the problem shows us that conscious experience is just not the kind of thing that a wholly reductive account could succeed in explaining.

NONREDUCTIVE EXPLANATION

At this point some are tempted to give up, holding that we will never have a theory of conscious experience. McGinn (1989), for example, argues that the problem is too hard for our limited minds; we are "cognitively closed" with respect to the phenomenon. Others have argued that conscious experience lies outside the domain of scientific theory altogether.

I think this pessimism is premature. This is not the place to give up; it is the place where things get interesting. When simple methods of explanation are ruled out, we need to investigate the alternatives. Given that reductive explanation fails, nonreductive explanation is the natural choice.

Although a remarkable number of phenomena have turned out to be explicable wholly in terms of entities simpler than themselves, this is not universal. In physics, it occasionally happens that an entity has to be taken as *fundamental*. Fundamental entities are not explained in terms of anything simpler. Instead, one takes them as basic, and gives a theory of how they relate to everything else in the world. For example, in the nineteenth century it turned out that electromagnetic processes could not be explained in terms of the wholly mechanical processes that previous physical theories appealed to, so Maxwell and others introduced electromagnetic charge and electromagnetic forces as new fundamental components of a physical theory. To explain electromagnetism, the ontology of physics had to be expanded. New basic properties and basic laws were needed to give a satisfactory account of the phenomena.

Other features that physical theory takes as fundamental include mass and space-time. No attempt is made to explain these features in terms of anything simpler. But this does not rule out the possibility of a theory of mass or of space-time. There is an intricate theory of how these features interrelate, and of the basic laws they enter into. These basic principles are used to explain many familiar phenomena concerning mass, space, and time at a higher level.

I suggest that a theory of consciousness should take experience as fundamental. We know that a theory of consciousness requires the addition of *something* fundamental to our ontology, as everything in physical theory is compatible with the absence of consciousness. We might add some entirely new nonphysical feature, from which experience can be derived, but it is hard to see what such a feature would be like. More likely, we will take experience itself as a fundamental feature of the world, alongside mass, charge, and space-time. If we take experience as fundamental, then we can go about the business of constructing a theory of experience.

Where there is a fundamental property, there are fundamental laws. A nonreductive theory of experience will add new principles to the furniture of the basic laws of nature. These basic principles will ultimately carry the explanatory burden in a theory of consciousness. Just as we explain familiar high-level phenomena involving mass in terms of more basic principles involving mass and other entities, we might explain familiar phenomena involving experience in terms of more basic principles involving experience and other entities.

In particular, a nonreductive theory of experience will specify basic principles telling us how experience depends on physical features of the world. These *psychophysical* principles will not interfere with physical laws, as it seems that physical laws already form a closed system. Rather, they will be a supplement to a physical theory. A physical theory gives a theory of physical processes, and a psychophysical theory tells us how those processes give rise to experience. We know that experience depends on physical processes, but we also know that this dependence cannot be derived from physical laws alone. The new basic principles postulated by a nonreductive theory give us the extra ingredient that we need to build an explanatory bridge.

Of course, by taking experience as fundamental, there is a sense in which this approach does not tell us why there is experience in the first place. But this is the same for any fundamental theory. Nothing in physics tells us why there is matter in the first place, but we do not count this against theories of matter. Certain features of the world need to be taken as fundamental by any scientific theory. A theory of matter can still explain all sorts of facts about matter, by showing how they are consequences of the basic laws. The same goes for a theory of experience.

This position qualifies as a variety of dualism, as it postulates basic properties over and above the properties invoked by physics. But it is an innocent version of dualism, entirely compatible with the scientific view of the world. Nothing in this approach contradicts anything in physical theory; we simply need to add further *bridging* principles to explain how experience arises from physical processes. There is nothing particularly spiritual or mystical about this theory—its overall shape is like that of a physical theory, with a few fundamental entities connected by fundamental laws. It expands the ontology slightly, to be sure, but Maxwell did the same thing. Indeed, the overall structure of this position is entirely naturalistic, allowing that ultimately the universe comes down to a network of basic entities obeying simple laws, and allowing that there may ultimately be a theory of consciousness cast in terms of such laws. If the position is to have a name, a good choice might be *naturalistic dualism.*

If this view is right, then in some ways a theory of consciousness will have more in common with a theory in physics than a theory in biology. Biological theories involve no principles that are fundamental in this way, so biological theory has a certain complexity and messiness to it; but theories in physics, insofar as they deal with fundamental principles, aspire to simplicity and elegance. The fundamental laws of nature are part of the basic furniture of the world, and physical theories are telling us that this basic furniture is remarkably simple. If a theory of consciousness also involves fundamental principles, then we should expect the same. The principles of simplicity, elegance, and even beauty that drive physicists' search for a fundamental theory will also apply to a theory of consciousness.

(A technical note: Some philosophers argue that even though there is a conceptual gap between physical processes and experience, there need be no metaphysical gap, so that experience might in a certain sense still be physical [Hill 1991, Levine 1983, Loar 1990]. Usually this line of argument is supported by an appeal to the notion of *a posteriori* necessity [Kripke 1980]. I think that this position rests on a misunderstanding of *a posteriori* necessity, however, or else requires an entirely new sort of necessity that we have no reason to believe in; see Chalmers 1996 [also Jackson 1994 and Lewis 1994] for details. In any case, this position still concedes an *explanatory* gap between physical processes and experience. For example, the principles connecting the physical and the experiential will not be derivable from the laws of physics, so such principles must be taken as *explanatorily* fundamental. So even on this sort of view, the explanatory structure of a theory of consciousness will be much as I have described.)

TOWARD A THEORY OF CONSCIOUSNESS

It is not too soon to begin work on a theory. We are already in a position to understand certain key facts about the relationship between physical processes and experience, and about the regularities that connect them. Once reductive explanation is set aside, we can lay those facts on the table so that they can play their proper role as the initial pieces in a nonreductive theory of consciousness, and as constraints on the basic laws that constitute an ultimate theory.

There is an obvious problem that plagues the development of a theory of consciousness, and that is the paucity of objective data. Conscious experience is not directly observable in an experimental context, so we cannot generate data about the relationship between physical processes and experience at will. Nevertheless, we all have access to a rich source of data in our own case. Many important regularities between experience and processing can be inferred from considerations about one's own experience. There are also good indirect sources of data from observable cases, as when one relies on the verbal report of a subject as an indication of experience. These methods have their limitations, but we have more than enough data to get a theory off the ground.

Philosophical analysis is also useful in getting value for money out of the data we have. This sort of analysis can yield a number of principles relating consciousness and cognition, thereby strongly constraining the shape of an ultimate theory. The method of thought experimentation can also yield significant rewards, as we will see. Finally, the fact that we are searching for a *fundamental* theory means that we can appeal to such nonempirical constraints as simplicity, homogeneity, and the like in developing a theory. We must seek to systematize the information we have, to extend it as far as possible by careful analysis, and then to make the inference to the simplest possible theory that explains the data while remaining a plausible candidate to be part of the fundamental furniture of the world.

Such theories will always retain an element of speculation that is not present in other scientific theories, because of the impossibility of conclusive intersubjective experimental tests. Still, we can certainly construct theories that are compatible with the data that we have, and evaluate them in comparison to each other. Even in the absence of intersubjective observation, there are numerous criteria available for the evaluation of such theories: simplicity, internal coherence, coherence with theories in other domains, the ability to reproduce the properties of experience that are familiar from our own case, and even an overall fit with the dictates of common sense. Perhaps there will be significant indeterminacies remaining even when all these constraints are applied, but we can at least develop plausible candidates. Only when candidate theories have been developed will we be able to evaluate them.

A nonreductive theory of consciousness will consist in a number of *psychophysical principles*, principles connecting the properties of physical processes to the properties of experience. We can think of these principles as encapsulating the way in which experience arises from the physical. Ultimately, these principles should tell us what sort of physical systems will have associated experiences, and for the systems that do, they should tell us what sort of physical properties are relevant to the emergence of experience, and just what sort of experience we should expect any given physical system to yield. This is a tall order, but there is no reason why we should not get started.

In what follows, I present my own candidates for the psychophysical principles that might go into a theory of consciousness. The first two of these are *nonbasic principles*—systematic connections between processing and experience at a relatively high level. These principles can play a significant role in developing and constraining a theory of consciousness, but they are not cast at a sufficiently fundamental level to qualify as truly basic laws. The final principle is my candidate for a *basic principle* that might form the cornerstone of a fundamental theory of consciousness. This final principle is particularly speculative, but it is the kind of speculation that is required if we are ever to have a satisfying theory of consciousness. I can present these principles only briefly here; I argue for them at much greater length in Chalmers (1996).

The Principle of Structural Coherence

This is a principle of coherence between the *structure of consciousness* and the *structure of awareness.* Recall that awareness was used earlier to refer to the various functional phenomena that are associated with consciousness. I am now using it to refer to a somewhat more specific process in the cognitive underpinnings of experience. In particular, the contents of awareness are to be understood as those information contents that are accessible to central systems, and brought to bear in a widespread way in the control of behavior. Briefly put, we can think of awareness as *direct availability for global control.* To a first approximation, the contents of awareness are the contents that are directly accessible and potentially reportable, at least in a language-using system.

Awareness is a purely functional notion, but it is nevertheless intimately linked to conscious experience. In familiar cases, wherever we find consciousness, we find awareness. Wherever there is conscious experience, there is some corresponding information in the cognitive system that is available in the control of behavior, and available for verbal report. Conversely, it seems that whenever information is available for report and for global control, there is a corresponding conscious experience. Thus, there is a direct correspondence between consciousness and awareness.

The correspondence can be taken further. It is a central fact about experience that it has a complex structure. The visual field has a complex geometry, for instance. There are also relations of similarity and difference between experiences, and relations in such things as relative intensity. Every subject's experience can be at least partly characterized and decomposed in terms of these structural properties: similarity and difference relations, perceived location, relative intensity, geometric structure, and so on. It is also a central fact that to each of these structural features, there is a corresponding feature in the information processing structure of awareness.

Take color sensations as an example. For every distinction between color experiences, there is a corresponding distinction in processing. The different phenomenal colors that we experience form a complex three-dimensional space, varying in hue, saturation, and intensity. The properties of this space can be recovered from information processing considerations: examination of the visual systems shows that waveforms of light are discriminated and analyzed along three different axes, and it is this three-dimensional information that is relevant to later processing. The three-dimensional structure of phenomenal color space therefore corresponds directly to the three-dimensional structure of visual awareness.

This is precisely what we would expect. After all, every color distinction corresponds to some reportable information, and therefore to a distinction that is represented in the structure of processing.

In a more straightforward way, the geometric structure of the visual field is directly reflected in a structure that can be recovered from visual processing. Every geometric relation corresponds to something that can be reported and is therefore cognitively represented. If we were given only the story about information processing in an agent's visual and cognitive system, we could not directly observe that agent's visual experiences, but we could nevertheless infer those experiences' structural properties.

In general, any information that is consciously experienced will also be cognitively represented. The fine-grained structure of the visual field will correspond to some fine-grained structure in visual processing. The same goes for experiences in other modalities, and even for nonsensory experiences. Internal mental images have geometric properties that are represented in processing. Even emotions have structural properties, such as relative intensity, that correspond directly to a structural property of processing; where there is greater intensity, we find a greater effect on later processes. In general, precisely because the structural properties of experience are accessible and reportable, those properties will be directly represented in the structure of awareness.

It is this isomorphism between the structures of consciousness and awareness that constitutes the principle of structural coherence. This principle reflects the central fact that even though cognitive processes do not conceptually entail facts about conscious experience, consciousness and cognition do not float free of one another but cohere in an intimate way.

This principle has its limits. It allows us to recover structural properties of experience from information processing properties, but not all properties of experience are structural properties. There are properties of experience, such as the intrinsic nature of a sensation of red, that cannot be fully captured in a structural description. The very intelligibility of inverted spectrum scenarios, where experiences of red and green are inverted but all structural properties remain the same, show that structural properties constrain experience without exhausting it. Nevertheless, the very fact that we feel compelled to leave structural properties unaltered when we imagine experiences inverted between functionally identical systems shows how central the principle of structural coherence is to our conception of our mental lives. It is not a *logically* necessary principle, as after all we can imagine all the information processing occurring without any experience at all, but it is nevertheless a strong and familiar constraint on the psychophysical connection.

The principle of structural coherence allows for a very useful kind of indirect explanation of experience in terms of physical processes. For example, we can use facts about neural processing of visual information to indirectly explain the structure of color space. The facts about neural processing can entail and explain the structure of awareness; if we take the coherence principle for granted, the structure of experience will also be explained.

Empirical investigation might even lead us to better understand the structure of awareness within a bat, shedding indirect light on Nagel's vexing question of what it is like to be a bat. This principle provides a natural interpretation of much existing work on the explanation of consciousness (Clark 1992 and Hardin 1992 on colors, and Akins 1993 on bats, for example), although it is often appealed to inexplicitly. It is so familiar that it is taken for granted by almost everybody, and is a central plank in the cognitive explanation of consciousness.

The coherence between consciousness and awareness also allows a natural interpretation of work in neuroscience directed at isolating the *substrate* (or the *neural correlate*) of consciousness. Various specific hypotheses have been put forward. For example, Crick and Koch (1990) suggest that 40-Hz oscillations may be the neural correlate of consciousness, whereas Libet (1993) suggests that temporally extended neural activity is central. If we accept the principle of coherence, the most direct physical correlate of consciousness is awareness: the process whereby information is made directly available for global control. The different specific hypotheses can be interpreted as empirical suggestions about how awareness might be achieved. For example, Crick and Koch suggest that 40-Hz oscillations are the gateway by which information is integrated into working memory and thereby made available to later processes. Similarly, it is natural to suppose that Libet's temporally extended activity is relevant precisely because only that sort of activity achieves global availability. The same applies to other suggested correlates such as the "global workspace" of Baars (1988), the "high-quality representations" of Farah (1994), and the "selector inputs to action systems" of Shallice (1972). All these can be seen as hypotheses about the *mechanisms of awareness*: the mechanisms that perform the function of making information directly available for global control.

Given the coherence between consciousness and awareness, it follows that a mechanism of awareness will itself be a correlate of conscious experience. The question of just which mechanisms in the brain govern global availability is an empirical one; perhaps there are many such mechanisms. But if we accept the coherence principle, we have reason to believe that the processes that *explain* awareness will at the same time be part of the *basis* of consciousness.

The Principle of Organizational Invariance

This principle states that any two systems with the same fine-grained functional organization will have qualitatively identical experiences. If the causal patterns of neural organization were duplicated in silicon, for example, with a silicon chip for every neuron and the same patterns of interaction, then the same experiences would arise. According to this principle, what matters for the emergence of experience is not the specific physical makeup of a system, but the abstract pattern of causal interaction between its components. This principle is controversial, of course. Some (Searle

1980, for example) have thought that consciousness is tied to a specific biology, so that a silicon isomorph of a human need not be conscious. I believe that the principle can be given significant support by the analysis of thought-experiments, however.

Very briefly: suppose (for the purposes of a *reductio ad absurdum*) that the principle is false, and that there could be two functionally isomorphic systems with different experiences. Perhaps only one of the systems is conscious, or perhaps both are conscious but they have different experiences. For the purposes of illustration, let us say that one system is made of neurons and the other of silicon, and that one experiences red where the other experiences blue. The two systems have the same organization, so we can imagine gradually transforming one into the other, perhaps replacing neurons one at a time by silicon chips with the same local function. We thus gain a spectrum of intermediate cases, each with the same organization, but with a slightly different physical makeup and slightly different experiences. Along this spectrum, there must be two systems A and B between which we replace less than one tenth of the system, but whose experiences differ. These two systems are physically identical, except that a small neural circuit in A has been replaced by a silicon circuit in B.

The key step in the thought experiment is to take the relevant neural circuit in A, and install alongside it a causally isomorphic silicon circuit, with a switch between the two. What happens when we flip the switch? By hypothesis, the system's conscious experiences will change; from red to blue, say, for the purposes of illustration. This follows from the fact that the system after the change is essentially a version of B, whereas before the change it is just A.

But given the assumptions, there is no way for the system to *notice* the changes! Its causal organization stays constant, so that all of its functional states and behavioral dispositions stay fixed. As far as the system is concerned, nothing unusual has happened. There is no room for the thought, "Hmm! Something strange just happened!" In general, the structure of any such thought must be reflected in processing, but the structure of processing remains constant here. If there were to be such a thought it must float entirely free of the system and would be utterly impotent to affect later processing. (If it affected later processing, the systems would be functionally distinct, contrary to hypothesis.) We might even flip the switch a number of times, so that experiences of red and blue dance back and forth before the system's "inner eye." According to hypothesis, the system can never notice these "dancing qualia."

This I take to be a *reductio* of the original assumption. It is a central fact about experience, very familiar from our own case, that whenever experiences change significantly and we are paying attention, we can notice the change; if this were not to be the case, we would be led to the skeptical possibility that our experiences are dancing before our eyes all the time. This hypothesis has the same status as the possibility that the world was created five minutes ago: perhaps it is logically coherent, but it is not plausible. Given the extremely plausible assumption that changes in experience correspond to changes in processing, we are led to the conclusion that the original hypothesis is impossible, and that any two functionally isomorphic systems must have the same sort of experiences. To put it in technical terms, the philosophical hypotheses of "absent qualia" and "inverted qualia," while logically possible, are empirically and nomologically impossible.

(Some may worry that a silicon isomorph of a neural system might be impossible for technical reasons. That question is open. The invariance principle says only that *if* an isomorph is possible, then it will have the same sort of conscious experience.)

There is more to be said here, but this gives the basic flavor. Once again, this thought experiment draws on familiar facts about the coherence between consciousness and cognitive processing to yield a strong conclusion about the relation between physical structure and experience. If the argument goes through, we know that the only physical properties directly relevant to the emergence of experience are *organizational* properties. This acts as a further strong constraint on a theory of consciousness.

The Double-Aspect Theory of Information

The two preceding principles have been nonbasic principles. They involve high-level notions such as "awareness" and "organization," and therefore lie at the wrong level to constitute the fundamental laws in a theory of consciousness. Nevertheless, they act as strong constraints. What is further needed are *basic* principles that fit these constraints and that might ultimately explain them.

The basic principle that I suggest centrally involves the notion of *information*. I understand information in more or less the sense of Shannon (1948). Where there is information, there are information states embedded in an information space. An information space has a basic structure of *difference* relations between its elements, characterizing the ways in which different elements in a space are similar or different, possibly in complex ways. An information space is an abstract object, but following Shannon we can see information as physically realized when there is a space of distinct physical states, the differences between which can be transmitted down some causal pathway. The states that are transmitted can be seen as themselves constituting an information space. To borrow a phrase from Bateson (1972), physical information is a *difference that makes a difference*.

The double-aspect principle stems from the observation that there is a direct isomorphism between certain physically realized information spaces and certain *phenomenal* (or experiential) information spaces. From the same sort of observations that went into the principle of structural coherence, we can note that the differences between phenomenal states have a structure that corresponds directly to the differences embedded in physical processes; in particular, to those differences that make a difference down certain causal pathways implicated in global availability and control. That is, we can find the same abstract information space embedded in physical processing and in conscious experience.

This leads to a natural hypothesis: that information (or at least some information) has two basic aspects, a physical aspect and a phenomenal aspect. This has the status of a basic principle that might underlie and explain the emergence of experience from the physical. Experience arises by virtue of its status of one aspect of information, when the other aspect is found embodied in physical processing.

This principle is lent support by a number of considerations, which I can only outline briefly here. First, consideration of the sort of physical changes that correspond to changes in conscious experience suggests that such changes are always relevant by virtue of their role in constituting *informational changes*—differences within an abstract space of states that are divided up precisely according to their causal differences along certain causal pathways. Second, if the principle of organizational invariance is to hold, then we need to find some fundamental organizational property for experience to be linked to, and information is an organizational property *par excellence.* Third, this principle offers some hope of explaining the principle of structural coherence in terms of the structure present within information spaces. Fourth, analysis of the cognitive explanation of our *judgments* and *claims* about conscious experience—judgments that are functionally explainable but nevertheless deeply tied to experience itself—suggests that explanation centrally involves the information states embedded in cognitive processing. It follows that a theory based on information allows a deep coherence between the explanation of experience and the explanation of our judgments and claims about it.

Wheeler (1990) has suggested that information is fundamental to the physics of the universe. According to this "it from bit" doctrine, the laws of physics can be cast in terms of information, postulating different states that give rise to different effects without actually saying what those states *are.* It is only their position in an information space that counts. If so, then information is a natural candidate to also play a role in a fundamental theory of consciousness. We are led to a conception of the world on which information is truly fundamental, and on which it has two basic aspects, corresponding to the physical and the phenomenal features of the world.

Of course, the double-aspect principle is extremely speculative and is also underdetermined, leaving a number of key questions unanswered. An obvious question is whether *all* information has a phenomenal aspect. One possibility is that we need a further constraint on the fundamental theory, indicating just what sort of information has a phenomenal aspect. The other possibility is that there is no such constraint. If not, then experience is much more widespread than we might have believed, as information is everywhere. This is counterintuitive at first, but on reflection I think the position gains a certain plausibility and elegance. Where there is simple information processing, there is simple experience, and where there is complex information processing, there is complex experience. A mouse has a simpler information processing structure than a human, and has correspondingly simpler experience; perhaps a thermostat, a maximally simple information processing structure, might have maximally simple experience? Indeed, if experience is truly a fundamental property, it would be surprising for it to arise only every now and then; most fundamental properties are more evenly spread. In any case, this is very much an open question, but I believe that the position is not as implausible as it is often thought to be.

Once a fundamental link between information and experience is on the table, the door is opened to some grander metaphysical speculation concerning the nature of the world. For example, it is often noted that physics characterizes its basic entities only *extrinsically*, in terms of their relations to other entities, which are themselves characterized extrinsically, and so on. The intrinsic nature of physical entities is left aside. Some argue that no such intrinsic properties exist, but then one is left with a world that is pure causal flux (a pure flow of information) with no properties for the causation to relate. If one allows that intrinsic properties exist, a natural speculation given is that the intrinsic properties of the physical—the properties that causation ultimately relates—are themselves phenomenal properties. We might say that phenomenal properties are the internal aspect of information. This could answer a concern about the causal relevance of experience—a natural worry, given a picture on which the physical domain is causally closed, and on which experience is supplementary to the physical. The informational view allows us to understand how experience might have a subtle kind of causal relevance in virtue of its status as the intrinsic nature of the physical. This metaphysical speculation is probably best ignored for the purposes of developing a scientific theory, but in addressing some philosophical issues it is quite suggestive.

CONCLUSION

The theory I have presented is speculative, but it is a candidate theory. I suspect that the principles of structural coherence and organizational invariance will be planks in any satisfactory theory of consciousness; the status of the double-aspect theory of information is less certain. Indeed, right now it is more of an idea than a theory. To have any hope of eventual explanatory success, it will have to be specified more fully and fleshed out into a more powerful form. Still, reflection on just what is plausible and implausible about it, on where it works and where it fails, can only lead to a better theory.

Most existing theories of consciousness either deny the phenomenon, explain something else, or elevate the problem to an eternal mystery. I hope to have shown that it is possible to make progress on the problem even while taking it seriously. To make further progress, we will need further investigation, more refined theories, and more careful analysis. The hard problem is a hard problem, but there is no reason to believe that it will remain permanently unsolved.

ACKNOWLEDGMENTS

The arguments in this paper are presented in greater depth in my book *The Conscious Mind*. Thanks to Francis Crick, Peggy DesAutels, Matthew Elton, Liane Gabora, Christof Koch, Paul Rhodes, Gregg Rosenberg, and Sharon Wahl for their comments.

REFERENCES

Akins, K. 1993. "What is it like to be boring and myopic?" In *Dennett and his critics*, edited by B. Dahlbom. London: Blackwell.

Allport, A. 1988. "What concept of consciousness?" In *Consciousness in contemporary science*, edited by A. Marcel and E. Bisiach. Oxford: Oxford University Press.

Baars, B. J. 1988. *A cognitive theory of consciousness*. Cambridge, England: Cambridge University Press.

Bateson, G. 1972. *Steps to an ecology of mind*. Chandler Publishing.

Chalmers, D. J. 1996. *The conscious mind: In search of a fundamental theory*. New York: Oxford University Press.

Clark, A. 1992. *Sensory qualities*. Oxford, England: Clarendon.

Crick, F., and C. Koch. 1990. Toward a neurobiological theory of consciousness. *Seminars in the Neurosciences* 2:263–275.

Crick, F. 1994. *The astonishing hypothesis: The scientific search for the soul*. New York: Scribners.

Dennett, D. C. 1991. *Consciousness explained*. Boston: Little, Brown.

Edelman, G. 1989. *The remembered present: A biological theory of consciousness*. New York: Basic Books.

Farah, M. J. 1994. "Visual perception and visual awareness after brain damage: A tutorial overview." In *Consciousness and unconscious information processing: Attention and performance 15*, edited by C. Umilta and M. Moscovitch. Cambridge, MA: MIT Press.

Flohr, H. 1992. "Qualia and brain processes." In *Emergence or reduction?: Prospects for nonreductive physicalism*, edited by A. Beckermann, H. Flohr, and J. Kim. Berlin: De Gruyter.

Hameroff, S. R. 1994. Quantum coherence in microtubules: A neural basis for emergent consciousness? *Journal of Consciousness Studies* 1:91–118.

Hardin, C. L. 1992. "Physiology, phenomenology, and Spinoza's true colors." In *Emergence or reduction?: Prospects for nonreductive physicalism*, edited by A. Beckermann, H. Flohr, and J. Kim. Berlin: De Gruyter.

Hill, C. S. 1991. *Sensations: A defense of type materialism*. Cambridge, England: Cambridge University Press.

Humphrey, N. 1992. *A history of the mind*. New York: Simon & Schuster.

Jackendoff, R. 1987. *Consciousness and the computational mind*. Cambridge, MA: MIT Press.

Jackson, F. 1994. "Finding the mind in the natural world." In *Philosophy and the Cognitive Sciences,* edited by R. Casati, B. Smith, and S. White. Vienna: Holder-Pichler-Tempsky.

Kripke, S. 1980. *Naming and necessity.* Cambridge, MA: Harvard University Press.

Levine, J. 1983. Materialism and qualia: The explanatory gap. *Pacific Philosophical Quarterly* 64:354–61.

Lewis, D. 1994. "Reduction of mind." In *A companion to the philosophy of mind,* edited by S. Guttenplan. Oxford, England: Blackwell.

Libet, B. 1993. "The neural time factor in conscious and unconscious events." In *Experimental and Theoretical Studies of Consciousness* (Ciba Foundation Symposium 174), edited by G. R. Block and J. Marsh. New York: John Wiley and Sons.

Loar, B. 1990. Phenomenal states. *Philosophical Perspectives* 4:81–108.

McGinn, C. 1989. Can we solve the mind-body problem? *Mind* 98:349–66.

Nagel, T. 1974. What is it like to be a bat? *Philosophical Review* 4:435–50.

Newell, A. 1990. *Unified theories of cognition.* Cambridge, MA: Harvard University Press.

Penrose, R. 1989. *The emperor's new mind.* Oxford, England: Oxford University Press.

Penrose, R. 1994. *Shadows of the mind.* Oxford, England: Oxford University Press.

Searle, J. R. 1980. Minds, brains and programs. *Behavioral and Brain Sciences* 3:417–57.

Shallice, T. 1972. Dual functions of consciousness. *Psychological Review* 79:383–93.

Shannon, C. E. 1948. A mathematical theory of communication. *Bell Systems Technical Journal* 27: 379–423.

Wheeler, J. A. 1990. "Information, physics, quantum: The search for links." In *Complexity, entropy, and the physics of information,* edited by W. Zurek. Reading, MA: Addison-Wesley.

Wilkes, K. V. 1988. "—, Yishi, Duh, Um and consciousness." In *Consciousness in contemporary science,* edited by A. Marcel and E. Bisiach. Oxford, England: Oxford University Press.

2 Consciousness and the Introspective Link Principle

Güven Güzeldere

INTRODUCTION

There is a strong intuition, which dates back several centuries, that consciousness is not, or does not consist in, an ordinary mental state or process itself, but it is, or it consists in, the awareness of such states and processes. Locke (1689/1959) epitomized this intuition in his celebrated statement: "Consciousness is the perception of what passes in a man's own mind" (p. 138). Various versions of this maxim have appeared in the writings of philosophers and psychologists from William James (1890/1950) to Franz Brentano (1924/1973) and more recently received endorsement by David Armstrong (1980), Paul Churchland (1984), David Rosenthal (1986, 1990), Peter Carruthers (1989), and William Lycan (1996), as well as (though rather indirectly) by Daniel Dennett (1991). Armstrong calls this form of consciousness the "perception of the mental," or "introspective consciousness," and promotes it as "consciousness in the most interesting sense of the word":

Introspective consciousness is a perception-like awareness of current states and activities in our own mind. The current activities will include sense-perception: which latter is the awareness of current states and activities of our environment and our body (p. 61).

Churchland echoes Armstrong's conviction in calling introspective consciousness "just a species of perception: self-perception " (p. 74). According to Rosenthal (1990), the awareness of mental states and activities comes in the form of a cognitive, rather than a perceptual, state: [A] mental state is conscious just in case it is accompanied by a higher-order, nondispositional, assertoric thought to the effect that one is in that very state" (p. 7). And Lycan claims that "consciousness is the functioning of internal attention mechanisms directed upon lower-order psychological states and events" (p. 2).

These assertions constitute the core of a substantial body of work among the philosophical theories of consciousness today. What is common to all these theories is the claim that consciousness is, or consists in, some sort of higher-level representing of lower-level mental states and processes. This representing may be perception-like (as Armstrong, Churchland, and Lycan claim, after Locke), or thought-like (as Rosenthal, and to some extent, Carruthers and Dennett, claim).

The psychological plausibility and philosophical merit of such accounts depend on how well the details and mechanics of such representings are spelled out—whether it is the perception or the thinking about of "what passes in one's own mind." Those details are what I will try to lay out in this essay. In doing so, I hope to show, in particular, that the Lockean "higher-order perception theories" of consciousness face a serious dilemma: they either necessarily commit themselves to an "intentional fallacy," or turn into a species of another, competing type of theory. The first horn of the dilemma stems from a confusion between properties of what is represented and properties of that which represents what is represented, in their account of second-order, introspective awareness. Avoiding this fallacy is possible only under a particular interpretation of the phrase "perception of what passes in one's mind," which, in turn, forces the "higher-order perception theories" to transform into a species of "higher-order thought" theories. That is the second horn of the dilemma.[1]

TWO SENSES OF CONSCIOUSNESS AND THE INTROSPECTIVE LINK PRINCIPLE

Let me start with a simple set of distinctions with respect to consciousness (which can be traced back to James [1890/1950], Brentano [1874/1973], and more recently in Rosenthal [1990]). Among the things that the term "conscious" can be used to predicate upon, the two most important kinds are individual beings and mental states that belong to, or occur in, these individuals. There is a difference in my being conscious (as opposed to, say, in a coma) and my having a mental state which is conscious (as opposed to, say, subconscious). One can be conscious in the sense of being awake and alert, and furthermore one's consciousness can be directed upon something, for example, a tune coming from the radio, or a lingering thought about a past conversation. This is the individual sense of consciousness.

On the other hand, it makes sense to talk about whether a particular mental state is conscious or not. This is not quite the same as someone's being conscious. The individual sense of consciousness denotes an overall state one is in; the other one classifies one's (mental) states as of one type or another. Following Rosenthal (1990), I will call this sense of consciousness, which functions as a type-identifier for mental states, state consciousness.[2]

Given this distinction, the natural next step is to investigate the relation between these two senses of consciousness. One way of doing this is to formulate a tighter relation between the (state-)consciousness of some particular mental state and one's (individual-)consciousness of that state.[3] So far, I haven't specified the domain of objects towards which individual consciousness can be directed. For the purposes of this paper, I will adopt a fairly nondetailed position: One can be individual conscious of physical things (cups, telephones), abstract objects (relations, theorems), facts, states of affairs, and events. These are all "external," in the sense that they can be objects of other people's consciousnesses as well. Alternatively, one can be conscious of the proprioceptive states of one's own body (the position of one's leg under the desk), or, as it is generally accepted, of one's own mental states (a desire for a chocolate bar). Broadly speaking, both of these are "internal"; only I have access to my proprioceptive states or desires, at least in the way that I do (see for example, Lyons [1986]).

This classification roughly corresponds to a common sense understanding of perception (seeing a cup on the counter, seeing that the cup is on the counter), proprioception (feeling one's foot getting numb), and introspection (dwelling on one's thoughts, desires, and motives). That is to say, I want to spell out individual consciousness in terms of familiar mental phenomena, and remain neutral about further details at this stage. This is not to say, however, that this usage is unproblematic. In particular, the claim that mental states can be the object of one's individual consciousness, which itself is a mental state, is multiply ambiguous, and it cries out for further analysis.

What does "being conscious of one's belief" mean? Being conscious of the content of that belief, being conscious of having that belief with the content that it has, or being conscious of the belief state qua the mental state that it is (that is, the representing vehicle itself, not what it represents)? The discussion of these issues will constitute the core of my assessment of the "higher-order perception theories" of consciousness in section 4 later in the chapter.

The Higher-Order Representation (HOR) Theories of Consciousness

Let me now state a canonical formulation of the relation between the state and the individual senses of consciousness. This formulation is meant to capture the thesis that underlies both the perception-like and the thought-like higher-order theories of consciousness:

The Introspective Link Principle
A mental state M in a subject S is a state-conscious state if and only if S is individual-conscious of M.

I would like to give the consciousness accounts that are built around this general linking principle a common name: the theories of Higher-Order (Mental) Representation, or in short, HOR. Implicit in the Introspective Link Principle, and hence in all the HOR theories, is the employment of a metalevel mental state, which is responsible for the consciousness of the lower-level states. S's (individual-) consciousness of M is itself a mental state, M', in S, presumably directed upon M. M', then, endows M with state-consciousness, by virtue of being a mental state in S itself, directed upon M. In other words, M' is a second-order state that represents M, an ordinary first-order mental state. M may, for instance, just be a visual perceptual state whose content is a scene S is looking at, or a thought S is entertaining about her dinner plans.

HOR comes in two main flavors, depending on the nature of the postulated higher order mental representation. This is only natural since mental representation is possible via both perceptual and cognitive states, and indeed there is a sufficient number of philosophers taking each route in spelling out the "R" of HOR. If M is regarded as a perception-like state, where M's relation to its object M is somehow like the relation between a perceptual state and its object, I call this version of HOR the thesis of Higher-Order Perception, or HOP. As I noted earlier, Armstrong, Churchland, and Lycan, following Locke, all defend this view.

Alternatively, M can be taken to be a type of cognitive mental representation, a higher-level thought-like state, the content of which is either the content of M, or (roughly speaking) a fact involving M; for example, that S is having M. This is the thesis of Higher-Order Thought, or HOT. Rosenthal and Carruthers both subscribe to the HOT thesis, and Dennett comes close, though they all defend different variations of it.

In either case, some form of hierarchical mental structure and of higher-order representing is the key idea for explaining consciousness in these accounts. For that reason, I call them "double-tiered" theories. The HOR accounts constitute the orthodox line with regard to theories of "introspective consciousness" in analytic philosophy today. The only account which explicitly criticizes these accounts and hints at the sketch of a counter thesis is Dretske (1993).[4]

WHAT'S WRONG WITH HIGHER-ORDER PERCEPTION: SPELLING OUT THE "P" IN "HOP"

Locke, Armstrong, Churchland, and Lycan all talk about consciousness as the awareness, or perception, or monitoring, or scanning of one's mental states. There is an ambiguity inherent in all these statements concerning the nature of the proper object of such internal, perception-like, awareness. What exactly is being perceived in the "perception of what passes in one's own mind"—the content of the mental state that happens to be "passing through" one's mind at the time, or the mental state itself? Or the content of another thought to the effect that one is having such a mental state?

There is surprisingly little attention paid to spelling out the answer to this question in any detail. Perhaps it is because the answer seems obvious or self-evident to everyone. Or perhaps not much is thought to depend on it. I will argue that there is actually nothing quite obvious or self-evident about such an answer, especially when the implicit ontological assumptions underlying it are made explicit. Nor is it true that not much turns on providing such details—what is at stake is simply the plausibility of the whole class of HOP style theories.

Three Options for the HOP Theorist

So far, one of the predominant distinctions on which the discussion has been resting has been perceptual versus cognitive mental states; for example, one's seeing a cup on the desk versus one's believing that the cup is on the desk (seeing that the cup is on the desk). Since vision is paradigmatic of perception, and beliefs are paradigmatic of cognitive states, let me proceed as such, assuming no loss of generality.

Assume that our subject S is sitting at her desk and looking at her cup right in front of her, under all the "normal circumstances"—perfect light, veridical experience, no demons or "evil neurosurgeons" playing tricks on S, so on and so forth. And despite the immense complexity and dynamical nature of perception, let's assume that we can capture a single instant of S's perceptual processing as she continues to eye the cup, freezing time and thus securing a momentary state of her bodily/mental conditions—"a polaroid snapshot of S's mind," if you will. Call the visual state S is in at this moment, the state which is in her (as well as for her) a representation of the cup, V. In other words, V is the state in virtue of which S sees the cup in the way she does. Let me, in all simplicity, denote the content of S's visual state as: [cup in front].

Now, according to the HOP theorists, S's consciousness consists in her "perception of V." Call this S's second-order "perception-like awareness" of V, her "monitoring" or "scanning" of her first-order mental state V. Whatever we may call it (and let me henceforth use the term *awareness* to cover all these), this is a junction in the conceptual landscape where some unpacking is necessary.

I can come up with at least three different readings of "S is aware of her visual state V, which has the content: [cup in front]." Let me enumerate:

1. S is aware of the cup in front of her, simpliciter.

2. S is aware of having a visual state V, which has the content: [cup in front].

3. S is aware of the visual state V, itself. That is, she is aware of the (cup-representing) internal state in her qua the internal state (as the representing vehicle, not what it represents).

Which one of these three readings truly reflects what the HOP theorists have in mind?

Option 1: Collapsing the Double-Tiered Structure

Option 1 will not do: an (introspective) awareness of one's mental state is not, by the HOP theorists' definition, nothing but just having that mental state. This option would no longer leave room for higher-order representing, and hence not allow for a double-tiered account, collapsing the (first) level of an ordinary mental state and the (second) level of an introspective mental state into one. Such an option may be available to those (single-tiered theorists of consciousness) who would like to claim that introspection is but an illusion, and that every time one tries to become introspectively aware of one's mental state, one finds oneself simply in that very mental state. But this would mean doing away with second-order representing altogether, something central to HOP accounts. Option 1 is therefore really a nonoption for the HOP theorist.

Option 2: HOP as HOT

Option 2 has more plausibility. In ordinary circumstances, we are generally aware of having a variety of mental states, and can easily report the presence of "what is passing in our minds."[5]

At least, it seems commonplace to be able to make judgments of the sort: "I am now looking at a coffee cup with bird motifs on it that sits right in front of me, on my desk." But notice that this way of construing "introspective awareness" really boils down to the ability to think a thought or entertain a belief about the mental state one is currently having. And as it happens, this is precisely the thesis of Higher-Order Thought (HOT), not Higher-Order Perception (HOP). The HOP theorist cannot opt for this route, while defending an account separate from and independent of the HOT account. Thus, option 2 is really not an option for HOP accounts, either.[6]

Option 3: Introspecting the Representation Itself

Finally, what about option 3? The idea that introspective consciousness may be the direct awareness of mental states qua states (as carriers of content) rather than of what those states represent (their content) is an interesting one. It can also be read in one of two ways, depending on the accompanying theory of mental representations. I will look at them both.

The contemporary materialist theories of mental representations, details aside, have this much in common: Mental representations are states of the nervous system that represent to the subject in whom they occur various objects and facts in the subject's environment, as well as the subject's bodily states. Now, these representations, being neuronal structures interacting in various electrochemical ways, have themselves such properties as spike frequencies and electric charge. In turn, they represent the world to the subject as being a certain way, via (at least, seemingly) different properties—those properties we are all familiar with in our phenomenology: colors, textures, temperature, and so on. So, at least on the face of it, it seems that the properties of the representers and those of the represented do not always coincide—in fact, they often seem radically different.

On the other hand, there used to be a time when the relation between "ideas" in the mind and objects in the world external to the mind was taken to be one of resemblance. Traces of this view, though not exactly in this form, can be found in Aristotle, and more recently in Berkeley's conception of mental representations as images of sorts (Cummins 1991). Even though no one I know explicitly defends the "resemblance theory of mental representations" anymore, I think that it is still operant, albeit implicitly, in many discussions regarding the philosophy of perception. In fact, I will argue that the HOP theorists, too, come close to embracing such a resemblance account of mental representations.

Let me start with the former reading: This may be the reading that Churchland (1984, 1985) has in mind, though I don't think any of the other authors I have been discussing subscribe to it. Churchland's materialist position oscillates between that of an eliminativist and of a reductionist. In either case, he claims that an account of "human subjective consciousness" can be given in the language of an advanced neuroscience, with a promissory note that this may enable, when applied to our own case, the "direct introspection of brain states." Intertwined with Churchland's claim is the assertion that not only are mental states identical to brain states, but also the properties of the former will turn out to be identical to the properties of the latter. That is to say, Churchland does bite the bullet and claim that one day we will indeed be able to find out about ourselves by "directly introspecting our brain states," via such properties as spike frequencies of neuronal structures, serotonin secretions, and the like.

Churchland's case, for all its worth, has been, and for the time being, remains to be, promissory. Furthermore, the claim about the direct introspection of brain states as a method that will reveal not only of which type those states are, but also what they are about (for example, a visual state, about a cup in front of the subject) seems to involve a conflation of properties of the representer and those of the represented. It is almost never the case, except in iconic representation and the like, that one can find out about the nature of what is being represented via the properties of whatever it is that does the representing. Why should we expect, then, to find out all about the cup a subject sees by examining the neural properties of the visual state that subject happens to have in looking at that cup?[7]

Well, what about "indirect access" to one's brain states? I can certainly look at real-time images of my brain produced by magnetic resonance imaging (MRI), and thus not only have various ongoing first-order experiences, but also have simultaneous second-order viewing of my brain states. I doubt that is what Locke had in mind when he talked about "the perception of what passes in one's own mind," and nor do Armstrong and Lycan. Insisting on option 3, it seems to me, could stem from but a confusion of the representer and the represented. Mental states qua brain states (in a materialist ontology) are the vehicles that represent for us the world around, as well as in, us. What we thus become aware of is what those states represent as being a certain way, via the properties of what is represented. What this reading of option 3 does is to try again to replace what is being represented with those that do the representing. I think that conflating the two, and trying to make a case using the properties of the representers, when what is called for is how the representers represent what they represent, that is, the properties of the represented, is to commit a fallacy worthy of a name: I will call it the "intentional fallacy."

The latter reading of option 3 also involves a variation of this fallacy, even when it is already a dead-end street, considering the implausibility of the theory of mental representations it is based on. Given the recent advances in neuroscience and computational modeling, as well as the philosophical understanding of intentionality, the "picture theory" of mental representations is now considered hopelessly defunct.

The only way to make sense out of it would be by embracing a "mind's eye" model. Here is how it could go: By visual perception, S forms, in her mind, a representation of the cup in front of her—this representation is her visual state, V. At the same time, she goes on to "perceive" this representation, in her "mind's eye." And as a result, she perceives the cup.

Such a story doesn't make sense unless we commit ourselves literally to a "mind's eye" and to mental representations that bear a resemblance relation to those they represent. If S had, in her mind, a little replica of what she saw in front of her—a cup—and with the aid of a "third eye" she were able to view this replica, and as a result had a conscious visual experience of the real cup, only then would she have constituted a true example of the fourth reading of the HOP account. This is all fairy tales, of course, and this reading of option 3, with its unacceptable ontological baggage, is no option for the HOP theorist either.

Finally, notice how both cases, the Churchland position as well as the "resemblance theory of mental representations," involve a conflation of the properties of what represents and of what is represented, though in slightly different ways. In Churchland, the properties of the representer are taken to be the properties of the represented, identifying what is in the head with the properties of what is perceived (something outside the head). In the "resemblance theory," the properties of the represented are taken to be the properties of the representer, importing the properties of what are outside the head as the properties of the representers inside the head. Under either reading, option 3 is committed to the "intentional fallacy."

CONCLUSION

Let me recapitulate: The HOP theorists are at a crossroads, and neither direction looks promising. In fact, they face a serious dilemma: They either have to commit themselves to the "intentional fallacy," or give up the HOP account altogether (by embracing either the unilevel or the HOT interpretation).

Option 3 presents the first horn of this dilemma. Both of the two readings of option 3 involve a conflation of the properties of representations and those of what is being represented. The result for the HOP theorist is having to either bet on a programmatic and not-so-promising Churchlandesque promise, or accept an outdated and not-so-acceptable resemblance theory of mental representations. Facing the other horn would involve opting for either option 1 or option 2. Either disjunct means giving up the HOP account altogether, in favor of either a single-tiered theory, or a version of the HOT account, respectively. This, of course, is not quite a desirable route for the HOP theorist, either.

In other words, there is no single construal of a HOP theory of consciousness that does not look substantially problematic. Furthermore, it is not clear that the four major proponents of HOP theories will all opt for the same interpretation. In fact, everyone may be forced to take a different path, given their prior theoretical commitments. If the Locke-Armstrong-Churchland-Lycan pact could thus simply disintegrate under this sort of an argumentative pressure, it would be hard to say whether there was any genuine theoretical backbone in the HOP paradigm to begin with.

In closing, let me note that I intend this analysis to be exhaustive of all possible readings, but of course there can be no saying on my part that such is the case. Nonetheless, I hope to have hereby posed a challenge to the HOP theorists, and shifted the burden of spelling out "perception of what passes in one's own mind" onto them. Given that the HOP accounts do lack any such analysis, it is up to the defenders of this view to show that their account is alive and well.

NOTES

1. I also have a second half to this story: that the "higher-order thought theories" of consciousness, though they fare better than their cousin, provide only a partial account of consciousness, unless complemented by a nonintrospective, unilevel account based on the idea of "attention as a limited cognitive resource necessary for consciousness." That half lies outside the scope of this paper, however. I try to present the whole case in Güzeldere (forthcoming).

2. For a more comprehensive attempt to sort out definitional and conceptual issues in the study of consciousness, see my two-part essay (Güzeldere 1995b, 1995c).

3. Here, I take the sense in which a mental state is said to be conscious at least conceptually distinct from the subject's consciousness of it. Not everyone thinks so. Lycan (forthcoming), for instance, states: "[I] cannot myself hear a natural sense of the phrase 'conscious state' other than as meaning 'state one is conscious of being in.'" So much so that he quotes a statement from Dretske (1993) that "an experience can be conscious without anyone—including the person having it—being conscious of having it" and labels it "oxymoronic." However, he also adds that "the philosophical use of `conscious' is by now well and truly up for grabs, and the best one can do is to be as clear as possible in one's technical specification" (p. 10).

Although generally in agreement with Lycan's latter observation (but not the former claim), let me nonetheless note that there are various established uses that one ought to pay attention to—if not the folk psychological use, then, for instance, the use grounded in the Freudian tradition. In any case, "giving a clear technical specification" (albeit alternative to Lycan's) is precisely what I aim at doing here. Furthermore, as evidenced from contemporary psychology literature, research on type-identifying mental states as conscious versus nonconscious, and research on the nature of consciousness of the subjects who have such states can be, and actually are being, pursued on independent conceptual grounds.

4. There are of course other accounts which say interesting things about consciousness (Dennett 1991, Flanagan 1992, Searle 1992, Block 1995) but they do not explicitly address the HOR thesis. It is instructive, however, to examine these accounts in the light of HOR, for in some cases, they turn out to be not orthogonal to it, but rather tacitly in agreement. Hence, the discrediting of the HOR thesis may very well undermine the plausibility of these accounts.

5. Easily, but not necessarily accurately. Above and beyond some rudimentary ability to report what we are experiencing, human subjects are notoriously bad in making correct and accurate judgments about the causes of their reasoning and

behavior. For many "surprising" results along these lines, see Nisbett and Wilson (1977), Wilson and Dunn (1986), Lyons (1986).

6. Let me note a point internal to the HOP camp, here. Even though Armstrong explicitly talks about (introspective) consciousness as a "perception-like awareness of current states and activities of one's own mind," and that Lycan locates Armstrong in the same lineage with Locke and himself, Armstrong's position may ultimately be closer to those of the HOT theorists, such as Rosenthal. For Armstrong thinks that "perception is the acquiring of beliefs." From the writings of these authors, it is hard to comparatively judge just how conceptually loaded Armstrong wants to view perception vis Lycan. But it is at least a possibility that Armstrong may turn out to be on the same side of the fence with Rosenthal, not with Lycan, despite the prima facie dissimilarity of their respective vocabulary.

7. Churchland's case may have more plausibility regarding brain states that do not have any obvious representational content, for example, emotions and moods. Perhaps we can indeed learn to individualize such states, which do not (if such is the case) represent anything "outside the head" via their intrinsic qualities. For the purposes of this paper, I choose to leave this issue open, and make my point in terms of straightforwardly representational states, such as visual states and beliefs.

8. A substantially expanded version of this article appears in Metzinger (1995) under the title "Is consciousness the perception of what passes in one's own mind?"

ACKNOWLEDGMENTS

I would like to thank Fred Dretske, John Perry, Owen Flanagan, and Murat Aydede for valuable comments on this essay, and David Chalmers, Stefano Franchi, and Brian C. Smith for helpful discussion on related issues.

REFERENCES

Armstrong, D. 1980. *The nature of mind and other essays*. Ithaca, NY: Cornell University Press.

Block, N. 1995. On a confusion about a function of consciousness. *Behavioral and Brain Sciences* 19:227–87.

Block, N., O. Flanagan, and G. Güzeldere. 1996. *The nature of consciousness: Philosophical and scientific debates*. Cambridge, MA: MIT Press.

Brentano, F. 1924/1973. *Psychology from an empirical standpoint*, translated by A. Rancurello, D. B. Terrell, and L. McAlister. New York: Humanities Press.

Carruthers, P. 1989. Brute experience. *The Journal of Philosophy* 86:258–69.

Churchland, P. 1985. Reduction, qualia, and the direct introspection of brain states. *The Journal of Philosophy* 82(1): 2–28.

Churchland, P. 1984. *Matter and consciousness*. Rev. ed. Cambridge, MA: MIT Press (1988).

Cummins, R. 1989. *Meaning and mental representation*. (paperback edition, 1991). Cambridge, MA: MIT Press.

Dennett, D. 1991. *Consciousness explained*. Boston, MA: Little, Brown Co.

Dretske, F. 1993. Conscious experience. *Mind* 102(406): 263–83.

Flanagan, O. 1992. *Consciousness reconsidered.* Cambridge, MA: MIT Press.

Güzeldere, G. 1995b. Consciousness: What it is, how to study it, what to learn from its history. *Journal of Consciousness Studies* 2:30–51.

Güzeldere, G. 1995c. Problems of consciousness: A perspective on contemporary issues, current debates. *Journal of Consciousness Studies* 2.

Güzeldere, G. 1995a. *The many faces of consciousness.* Stanford, CA: Stanford University Press (Ph.D. dissertation).

James, W. 1890/1950. *The principles of psychology,* Vol. 1. New York: Dover Publications.

Locke, J. 1689/1959. *An essay concerning human understanding,* Vol. 1. New York: Dover Publications.

Lycan, W. 1996. "Consciousness as internal monitoring." In *The nature of consciousness: Philosophical and scientific debates,* edited by N. Block, O. Flanagan, and G. Güzeldere. Cambridge, MA: MIT Press.

Lyons, W. 1986. *The disappearance of introspection.* Cambridge, MA: MIT Press.

Metzinger, T. 1995. *Conscious experience.* Paderborn, Germany: Schoningh-Verlag.

Nisbett, R., Wilson, T. 1977. Telling more than we can know: Verbal reports on mental processes. *Psychological Review* 84:231–58.

Rosenthal, D. 1990. Explaining consciousness. (Manuscript) CUNY, Graduate School.

Rosenthal, D. 1986. Two concepts of consciousness. *Philosophical Studies* 94:329–59.

Searle, J. 1992. *The rediscovery of the mind.* Cambridge, MA: MIT Press.

Wilson, T., Dunn, D. 1986. Effects of introspection on attitude-behavior consistency: Analyzing reasons versus focusing on feelings. *Journal of Experimental and Social Psychology* 22:249–63.

3 The Place of Qualia in the World of Science

Leopold Stubenberg

INTRODUCTION

The most difficult aspect of the problem of consciousness is the qualitative aspect of consciousness. I assume that a philosophical account of qualitative consciousness (or sentience, raw sensation, the subjective character of experience, the phenomenology of awareness) has to make use of the notion of *qualia*. Qualia are the phenomenal properties—the colors, sounds, smells, tastes, tickles, pains—we are immersed in at every moment of our waking and dreaming lives.

I assume that color qualia have the following higher order properties: they are monadic, simple, homogeneous, and irreducible. That is, the nature of qualia is independent of their relations to other things. They lack inner structure or complexity. And this lack manifests itself in their homogeneity (or grainlessness) and in their irreducibility (or unanalyzability). I also assume that qualia are neutral, in the sense that it is an open question whether they qualify mental particulars (like sense–data) or nonmental particulars (like physical objects).

Given these assumptions, a philosophical account of qualitative consciousness can be compatible with science only if qualia are compatible with science. To give this investigation more focus I shall concentrate on color qualia. This, then, is the question I want to ask and answer: *Is the existence of color qualia compatible with science?* The compatibility thesis answers "Yes." The incompatibility thesis answers "No." A defense of compatibilism must show how one can make sense of the idea that qualia are properties of the objects or processes or states that populate the scientific world. I shall argue that incompatibilism is false and that most versions of compatibilism are implausible. I end by tentatively endorsing a currently unpopular version of compatibilism, namely, neutral monism.

THREE INTERPRETATIONS OF SCIENCE

Whether the existence of qualia is compatible with science depends on how science is interpreted. Different interpretations are suggested by the three answers to the following question: What does it take for science to be true?

1. That it correctly predicts our experiences.
2. That it correctly predicts our experiences and that the entities involved in scientific explanations have all and only the properties ascribed to them in these explanations.
3. That it correctly predicts our experiences and that the entities involved in scientific explanations have all the properties ascribed to them in these explanations.

Answer No. 1

Antirealism Answer no. 1 provides a minimal interpretation of science. Even a solipsist or an instrumentalist can accept it. On this reading the domain of scientific discourse is limited to experiences; that is, to episodes in which we become acquainted with qualia. Thus, the question whether the existence of qualia is compatible with the deliverances of science does not arise. Though this is an attractive result of answer no. 1, it has been widely rejected because of its inability to give a satisfactory account of the scientific enterprise as a whole.

Answer No. 2

Science and the Elimination of Qualia This answer enshrines what always has been and still is the most popular reason for holding the incompatibility thesis. Ever since Democritus announced that "by convention color exists, by convention bitter, by convention sweet but in reality atoms and the void," there has been a tradition of sweeping color qualia out of the physical world and into the "philosophical dustbin of the mind" (Armstrong 1980, p. 176). This relocation of phenomenal color is carried out in the name of science. We find it in Galileo, Hobbes, Descartes, Locke, and Leibniz; and the tradition is still alive and well today. Thus, we read in a recent paper that "the projectivist account of color experience (that is, the account that takes color to be an intrinsic, sensational quality of the visual field that is projected upon external objects) is . . . the one that occurs naturally to anyone who learns the rudimentary facts about light and vision" (Boghossian and Velleman 1989, p. 97). Examples could be multiplied. Note that the scientific broom sweeps all parts of the scientific world clean of qualia. The surface of the tomato, the light it reflects into the eye, the eye itself, the optic nerve, and the brain—none of these can bear the red quale we experience. The world of science has no place for qualia. Barring dualism this means that science tells us that there are no qualia. This is the conclusion that the qualia eliminativists have drawn. Hence you cannot have a qualia–based account of consciousness that is compatible with science.

Science vs. Physicalism I want to reject this conclusion by insisting that science is much more tolerant of qualia than the champions of a certain "scientific world view" pretend. Not physics but physicalism has made qualia into homeless properties. But we must not confuse physics with physicalism. Physics does not entail physicalism. Therefore, a respect for physics (and science in general) does not commit one to physicalism. It is necessary to insist on this distinction if we are to have a qualia-based account of consciousness that is compatible with science.

A full characterization of physicalism is difficult (Poland 1994). But the rough idea is simple. Physicalism holds that "all entities, properties, relations, and facts are those which are studied by physics or other physical sciences. . . . physical science [enjoys] a unique ontological authority: the authority to tell us what there is" (Crane and Mellor 1990, p. 185). Or as Wilfrid Sellars has so memorably put it: "in the dimension of describing and explaining the world, science is the measure of all things, of what is that it is, and of what is not that it is not (Sellars 1963, p. 173). Notice that this is not a scientific claim; it is an enormously strong philosophical claim about science. It states, in dramatic fashion, what has been called the principle of scientific realism (PSR): Only properties whose attribution helps explain something scientifically are admissible in our world view as truly descriptive or reality (Sosa 1990, p. 211). This principle makes the difference between the scientific attitude and physicalism. By entailing the PSR, answer no. 2 makes the truth of physicalism into a condition of the truth of science, thereby sanctioning the systematic confusion of the claims of science with the claims of a physicalist metaphysics. Answer no. 3 severs this mistaken union.

In deciding to accept the PSR, you choose an ontologically sparse world—just how sparse will depend on how broadly you construe the term "science." A hard-core physicalist will only admit the attributes that feature in the explanations of physics. But no matter how inclusive your notion of the sciences is—your world is going to lose an enormous amount of furniture, only a tiny part of which will be the qualia.

Physicalism Without Eliminativism Many physicalists reject the claim that physicalism entails qualia elimination. Physicalists have employed two strategies to render qualia physicalistically respectable. Reductive physicalism attempts to show that qualia are reducible to physicalistically respectable properties. Nonreductive physicalism attempts to show that qualia—though not strictly reducible to physical properties—depend on physical properties in such a way that the hegemony of physics is not threatened. Physics remains basic in the important sense that everything that does not strictly belong to physics is still made sense in terms of it. Thus, reductive as well as nonreductive physicalism can acknowledge the existence of qualia without jeopardizing the basic tenets of physicalism. Qualia eliminativism is not obligatory for the physicalist.

Objections to Noneliminative Physicalism But the friend of qualia must resist both of these physicalist strategies. First, reductive physicalism: by assumption qualia are irreducible; therefore reductive physicalism is out. The reasonableness of this assumption is strongly suggested by the unglamorous record of reductive physicalism. An awareness of this record has persuaded most physicalists to opt for nonreductive physicalism.

Second, nonreductive physicalism: nonreductive physicalism exists in many forms, all of which seem to face the same problem. A satisfactory version of physicalism must not simply state that but also explain why and how a given quale gets instantiated in virtue of the instantiation of certain physical properties. Summing up a detailed survey of all extant versions of physicalism (excluding only his own), Poland writes that they all fail to "engage the deepest and most important motivations of the physicalist programme: the drive to develop an understanding of how the non–physical aspects of the worlds are embedded in the physical fabric that, according to the physicalist, exhausts, determines, and realizes all that there is" (Poland 1994, p. 105). Schiffer put the matter more bluntly when he wrote that the appeal to the supervenience relation—the favorite relation of most nonreductive physicalists—"is obscurantist in the extreme" (Schiffer 1987, 154). Kim, himself a lapsed nonreductive physicalist, has sharpened this disenchantment with nonreductive physicalism into a dilemma: "a physicalist has only two genuine options, eliminativism and reductionism. That is, if you have already made your commitment to a version of physicalism worthy of the name, you must accept the reducibility of the psychological to the physical, or failing that, you must consider the psychological as falling outside your physicalistically respectable ontology" (Kim 1993, p. 267).

Nonreductive physicalism is a blind alley. Reductive physicalism is no option for the friend of qualia. Hence one is led back to the view that physicalism leads to qualia eliminativism. And if one mistakes physicalism for science one will think that science leads to qualia eliminativism. Thus you will conclude that you have to choose between accepting science and accepting a qualia based philosophy of consciousness. Consideration of answer no. 3 shows that this alternative is mistaken.

Answer No. 3

Answer no. 3 combines an unwavering descriptive realism[1] about science with an awareness that science may not afford an exclusive or exhaustive description of reality. A true description of x need not be a complete description of x. Answer no. 3 embodies this simple insight and therefore does not entail the PSR. Thus physicalism is no longer a condition for the truth of science. This opens the door for a number of ways of coherently conjoining a respect for science with a belief in irreducible color qualia.

Simple Compatibilism This is how Broad states the case for compatibilism:

It is sometimes thought that the physical theories of light and heat positively *disprove* the common–sense view that physical objects are literally coloured or hot. This is a sheer logical blunder. The physical theory of light, e.g., asserts that, whenever we sense a red sensum, vibrations of a certain period are striking our retina. This does not prove that bodies which emit vibrations of that period are not literally red, for it might well be that only bodies which are literally red can emit just these vibrations. The vibrations might simply be the means of stimulating us to sense the red colour, which is literally in the body, whether we happen to sense it or not. (Broad 1969, p. 280)

As Broad points out, this consideration does not establish that physical objects bear qualia; it merely points out that descriptive scientific realism is compatible with this possibility. Whether this simpleminded form of compatibilism is a reasonable view to adopt is a different matter altogether—Broad himself takes it to be unlikely in the extreme.

Critique of the Simple View It is liberating to find out that a sober form of scientific realism does not, automatically, rule out the phenomenologically obvious view that experienced color qualia are borne by the perceived physical object. But that must not blind one to the many objections to this way of integrating qualia into the world of science. Here is a sampling of them. (1) Illusions, hallucinations, dreams, and the facts of perceptual relativity make it very difficult to hold that experienced qualia are always situated on the surface of the experienced object. In some of these cases there is no object; in others the surface of the object appears to be overcrowded by incompatible qualia. (2) The causal facts involved in perception militate against the view that experienced qualia are part of the experienced object rather than part of the experiencing subject's mind/brain. Bertrand Russell, for example, argued that the experiences "come at the end of a causal chain of physical events leading, spatially, from the object to the brain of the percipient. We cannot suppose that, at the end of this process, the last effect suddenly jumps back to the starting–point, like a stretched rope when it snaps" (Russell 1954, p. 320). (3) Given a complete scientific account of perception, the externally located qualia would appear to become epistemically superfluous and inaccessible. Qualia might be out there but they would be unknowable. (4) If qualia play no causal role in perception or anything else, they are mere epiphenomena. To many philosophers epiphenomena are anathema. (5) If qualia are assigned a causal role, then our current science is incomplete—for it assigns no such role to qualia. Thus the price of acknowledging causally active qualia is the rejection of current science—a price too high to pay. (6) Such lawlike connections as there may be between scientific properties and qualia pose problems of their own. Such laws are "nomological danglers" that threaten the coherence of the scientific view of the world. (7) The final problem that I want to mention is Sellars's famous "grain problem." How could groups of discrete, colorless scientific objects (that is, ordinary things) instantiate homogeneous phenomenal color properties? It will not do to simply assert that all the scientific properties and all the qualia are harmoniously coinstantiated. If the object really has the properties that science ascribes to it then it is difficult to see how it could also exemplify color qualia. The sort of thing that can have all the scientific properties does not seem like the sort of thing that can have the relevant phenomenal properties.

These objections are a very mixed bag. I take the last objection—Sellars's grain argument—very seriously. It persuades me that neither external objects of perception—like tomatoes—nor inner physical states of the perceiver can serve as bearers of color qualia. That is, I think that there is no physical object anywhere that instantiates the phenomenal redness (of the sort that we experience upon viewing a ripe tomato) by means of having this redness spread out over its surface.

But this negative conclusion is not universally accepted. John Searle, for example, has recently proposed a view—biological naturalism—that does not see any difficulty in the idea that physical and phenomenal properties are instantiated by the very same physical object, namely, the brain. Searle starts out by reiterating the core of the simple view: the facts of physics are not inconsistent with the fact that our experiences have "*irreducible* phenomenological properties" (Searle 1992, p. 28). And Searle feels that the coinstantiation of these two sorts of properties in the human brain is no more problematical than the fact that a given body of water instantiates both microphysical properties and the property of liquidity. If you cannot see that it is that simple then—so Searle's diagnosis—you are still suffering from sort of Cartesian hangover. But once you have seen that all the properties of water and all the properties of the brain are natural properties, the question how these diverse properties could be had by one thing no longer poses itself.

I too am convinced that color qualia are natural if all this means is that they aren't supernatural. Of course they aren't! But my belief in the naturalness of qualia does not help me one bit in understanding how qualia can be instantiated by a physical object, be it a brain, a ripe tomato, or whatever. Though I find myself incapable of adopting Searle's simple view, I am still unwilling to plead guilty of Cartesianism. And there is a way out of Searle's dilemma; for Searle's alternative—either biological naturalism or Cartesianism of some sort or other—does not exhaust all the viable possibilities. I want to close this exploration of the compatibility thesis by looking at a version of a reductive doctrine of a somewhat unusual nature.

THE PRIMACY OF QUALIA: NEUTRAL MONISM

The simple view exploited the fact that the descriptive realism embodied in answer no. 3 allows for things to have irreducible, nonscientific properties in addition to their scientific properties. The simple view insists on the consistency of such a double attribution of properties to a given physical object and leaves it at that. It does not volunteer an account of how such a remarkable coinstantiation of heterogeneous properties is possible. For the simple view these properties are, so to speak, on a par—neither one is treated as more fundamental than the other; both properties coexist in what appears to be a miraculous harmony.

But the descriptive realism of answer no. 3 allows to combine these two sorts of properties in a different manner: one group of properties can be treated as more fundamental than the other group. The various failed versions of physicalistic reductionism treated the scientific properties as more fundamental than the phenomenal properties. We might, however, try to invert the order of fundamentalness. This sort of approach places qualia at the basis. Nonqualitative properties are then understood as in some way constructed from or derivative of qualia. Berkeley and the later phenomenalists accepted this kind of approach. So did the neutral monists: Hume, James, Mach, Russell, and Ayer to name just a few. The demarcation between phenomenalism and neutral monism is a source for unending discussions; so are the details of and the differences between all the various neutral monisms that these authors developed. I shall not enter into any of this, and simply present Russell's version of this doctrine.

Russell's neutral monism is a monism because it recognizes a single basic category. It is a monism that is neutral as between materialistic monism and mentalistic monism inasmuch as the elements of its single basic category are neither material nor mental.

Russell's basic elements are events, paradigms of which are percepts: the immediately known events, examples of which are "patches of color, noises, smells, hardnesses, etc., as well as perceived spatial relations" (Russell 1954, p. 257). We can think of Russell's percepts as instances of phenomenal properties (or qualia). The red you experience upon seeing a particular ripe tomato is a Russellian percept.

The causal account of perception persuaded Russell that our percepts are located in our heads. When I see a tomato my tomato-percepts are in my head—the terminus of the causal chain—not on or in the external, physical tomato—the origin of the causal chain.

That the basic "building blocks" of the universe are events is something that Russell took to be a lesson of relativistic physics. As he sees it, physics showed that the traditional materialistic "billiard-ball" model of matter is untenable. Therefore all things—medium–sized drygoods as well as the particles of physics—have to be understood as processes ultimately constituted by events. That is, the microscopic particles into which physics analyzes macroscopic objects are themselves subject to further analysis:

Electrons and protons, however, are not the stuff of the physical world: they are elaborate logical structures composed of events . . . As to what the events are that compose the physical world, they are, in the first place, percepts, and then whatever can be inferred from percepts . . . (Russell 1954, p. 386).

The above quote has already given away the startling conclusion of neutral monism. The events that compose the elementary particles of physics are of the nature of patches of red, pancake smells, and chocolate tastes. Russell feels most comfortable in asserting this view about the particles that make up the brains of living, experiencing beings. The events that are the percepts of these beings occur in a region where, physically speaking, the brain of these beings is located. Physically speaking again, their brain is made exclusively of elementary particles. Those elementary particles are, metaphysically speaking, made of events. Now, we already know what sort of events are going on in the region occupied by the brain: percepts. So the obvious way to bring together the physically elementary building blocks of these living, experiencing brains with the percepts that make up this lived experience is this: the percepts of a living, experiencing brain are the metaphysically basic building blocks of the elementary particles that make up this living experiencing brain. That is, the physically elementary constituents of living experiencing brains are made of (constituted by) the percepts of this brain.

The preceding is merely a sketch of one aspect of Russell's neutral monism. But it should suffice to adumbrate the neutral monist way of integrating qualia—conceived of as nonmental, nonphysical events—into the world of science. Compared to the other integrative programs presented here neutral monism has much going for it. It is compatible with descriptive realism and it does, I think, allow for fairly satisfactory answers to some of the more worrisome objections that have been leveled against the simple view. Instead of making qualia into high-level properties of large groups of scientific objects, neutral monism makes qualia into the basic building blocks from which individual elementary particles of the scientific world are construed. By thus standing the relationship between the world of science and qualia on its head, neutral monism indicates a promising way of upholding the compatibility thesis and integrating qualia into the world of science.

But I shall not attempt to hide what appears to me to be the most disheartening fact about neutral monism. Despite its not so short history, this metaphysical "theory" has never been more than a bare sketch of a metaphysical program. And even these sketches prove to be rather problematical. So it must be said, I think, that neutral monism is not a solution but, at best, a speculative pointer into the direction of a possible solution to the problem of integrating qualia into the scientific world.

NOTE

1. I borrow this term from Nicholas Rescher. Descriptive realism is the "doctrine that *science describes the real world:* that the world actually is as science takes it to be and that its furnishings are as science envisages them to be.... On this realistic construction of scientific theorizing, the declarations of science are factually true generalizations about the actual behavior of objects that exist in the world" (Rescher 1982, p. 284). Rescher contrasts descriptive realism with scientific instrumentalism. Neither one of these positions has anything to do with the contrast between metaphysical (or ontological) realism and idealism. One more terminological comment. I suggest that the term "scientific realism" is best understood as meaning descriptive realism. The realism captured by the PSR would more fittingly deserve a label like "scientific exclusivism" or "scientific imperialism."

REFERENCES

Armstrong, D. M. 1980. *The nature of mind.* Brighton, England: Harvester Press.

Boghossian, P. A., and D. J. Velleman 1989. Color as secondary quality. *Mind* 98:81–103.

Broad, C. D. 1969. *Scientific thought.* (First published in 1923) New York: Humanities Press.

Crane, T., and D. H. Mellor 1990. There is no question of physicalism. *Mind* 90:185–206.

Kim, J. 1993. *Supervenience and the mind. Selected philosophical essays.* Cambridge, England: Cambridge University Press.

Poland, J. 1994. *Physicalism. The philosophical foundations.* Oxford: Oxford University Press.

Rescher, N. 1982. *Empirical inquiry.* Totowa, NJ: Rowman and Littlefield.

Russell, B. 1954. *The analysis of matter.* (First published in 1927) New York: Dover.

Schiffer, S. 1987. *Remnants of meaning.* Cambridge, MA: MIT Press.

Searle, J.R. 1992. *The rediscovery of the mind.* Cambridge, MA: MIT Press.

Sellars, W. 1963. *Science, perception and reality.* London: Routledge & Kegan.

Sosa, E. 1990. "Perception and reality." In *Information, semantics and epistemology,* edited by E. Villanueva. Oxford, England: Basil Blackwell.

4 The Binding Problem and Neurobiological Oscillations

Valerie Gray Hardcastle

INTRODUCTION

It is well known that the brain processes visual data in segregated, specialized cortical areas (Hubel and Livingstone 1987, Livingstone and Hubel 1987). For example, (roughly speaking) the interblob region computes the orientation of lines and edges, the intralaminar pathway responds to color, and the magnocellular stream calculates movement (Ramachandran 1990, Ramachandran and Anstis 1986, Zeki 1992). Nevertheless, since our conscious experiences are unified, the separately computed features must somehow come together after processing. That is, our conscious experiences must reflect the various features and subcontents as joined together in a single interpretive unit or as a single percept after being processed in initially distinct areas.

However, given what we know about the segregated nature of the brain and the relative absence of multimodal association areas in the cortex (Kandel and Schwartz 1987 for relevant neurophysiology, see also Damasio 1989, Goldman and Rosvold 1970, Goldman-Rakic *et al.* 1983, Goldman-Rakic 1987, Sereno 1990), how conscious percepts become unified into single perceptual units is not clear. If we lack true association areas, then the more popular (and intuitive) neurophysiological solutions to psychological binding, such as "grandmother" neurons and convergence zones, cannot be correct. But, if we could figure out how and where the brain joins together segregated outputs, we would have a start in localizing the neuronal processes that correlate with perceptual experiences.

This difficulty is known as the "binding problem." It arises in modern psychology in the construction of visual percepts. Generally speaking, in order for us to interpret incoming stimuli semantically (that is, in order for us to perceive some object as that object), our brains first must extract constant features from an incoming array of light patterns, then construct temporary (pre-)representations from the patterns, before they begin to associate the object representation with its meaning (Treisman 1986). It also surfaces in neuroscience as scientists realize there is no central cortical "informational exchange," given the dearth of "association" areas. Psychology's question of how disparate perceptual features can yield single interpretations for objects becomes for neurophysiologists: How do brains link together the various outputs from different processing modules such that some system activates unified percepts? As we come to know more about the brain and how it is organized, it becomes clear that there is going to be no simple solution to the binding problem.

Here I critically examine data apparently relevant for understanding the neurophysiological underpinnings of perception. In particular, I examine the possibility that 40-Hz oscillatory firing patterns in cortex are important lower level neuronal events related to perceptual experiences (Blakeslee 1992, Crick and Koch 1990, Flanagan 1992, Koch 1992 this volume, Stryker 1989). In what follows, I discuss ongoing research in visual perception in which the scientists involved are deliberately and conscientiously making explanatory connections between perceptual awareness and neural oscillatory firing patterns. I conclude however that the scientists involved are making a mistake and that the binding problem, a fundamental difficulty in understanding consciousness, remains unsolved.

TIME AND THE BINDING PROBLEM

The biggest obstacle to solving the problem stems from accepting parallel distributed processing as the fundamental computational mechanism of the brain. Since we understand ourselves as growing, developing organisms, we must hypothesize that the different mental "symbols" that we use to represent objects or aspects of objects in the real world have to be able to coexist within the same physical hardware. Most importantly, new symbols cannot require new pieces of hardware, for otherwise our brain would have to grow each time we think an original thought or have an original perceptual experience. However, if the brain is in fact fundamentally and massively parallel, we are left with what von der Malsburg calls the "superposition catastrophe" (von der Malsburg 1987). That is, if we take mental "symbols" to be different subsets of coactive neurons within the same brain structure, as they are in the classical framework of neurobiology, then if more than one symbol becomes active at a time (as they surely must, given what we know about feature extraction and later binding), they become superposed by coactivation and any information carried in the original differentiated subsets is lost. The lack of internal structure of neuronal assemblies in the classical framework prevents computational combinations, which, in turn, rules out most interesting mental calculations. (See also Legéndy 1970, for discussion.)[1]

Fodor and Pylyshyn (1988) argue that this coactivation conundrum entails that the parallel distributed architecture of the brain is merely a lower–level implementation of a higher–level discrete symbol manipulating device. Segregate each cell assembly unit to physically distinct areas of the brain, for example, the different parts of the topologically organized somatosensory system in the postcentral gyrus of the parietal cortex. The firing patterns of the different cell assemblies can "speak" to one another through select connections among areas. The symbols can then interact across the areas as unstructured units since all the lower level detail within cell assemblies is lost at the higher level of connected groups of segregated areas. In this way, the parallel architecture masquerades as a von Neumann-style computing device.

However, as von der Malsburg argues, this solution can only work in severely restricted environments with a specific set of patterns, since slicing a system up into so many subsystems destroys most of a machine's flexibility. Requiring a different subsystem for each interpretation would not permit the machine to respond to new and different stimuli very efficiently. Moreover, given the sophistication of the human mind, it also places too heavy a burden on either phylogenetic development or ontogenetic organization because each distinct subsystem would have to be either already programmed in or would have to develop quickly in response to each distinct but representable stimulus. Such a solution probably would allow only incomplete and abbreviated interpretations of the rapidly changing external world. Finally, since the solution assumes a different dedicated physical subsystem for each concept, it entails that symbol patterns for new input or newly learned concepts require new and different areas of the brain, and this violates what we know about how brains develop. The basic problem is that most physical connections within the head (that we know about) just develop too slowly to allow for the rapid conceptual learning we continually exhibit if changes in the physical connections were the only mechanism for representing new ideas.

The Addition of Time

But if using the physical connections studied thus far among neurons to tie different subsets of firing patterns together cannot work, obviously the brain must find additional degrees of freedom elsewhere. Time is a likely candidate, for mental and biological functions operate on two different time scales. There is the psychological time scale, characterizing mental processes and ordered on tenths of a second; there is also the faster time scale ordered on only thousandths of a second. The mean unit activity of neurons evolves on the longer psychological time scale, but the activity of a single cell fluctuates around the mean on the faster time scale (compare von der Malsburg 1987, p. 424). It is possible that correlating the activity fluctuations of a set of coactivated neurons could thereby bind that firing pattern into a single unit. Several such units could coexist if their activity is desynchronized relative to one another, thus solving the superposition catastrophe (see also König, Engel, and Singer 1992, Legéndy 1970, Sejnowski 1981).

Here then could be the procedure our parallel and distributed perceptual systems use to synthesize disparate feature contents into a single unified interpretation. Given the complexity of what we represent in the environment and the flexibility and computational demands on our representational machinery, it seems the simple answer to the binding problem—that all binding requires is a serial architecture—must fail. Something occurring in a lower level of neural organization must be importantly relevant. That is, single-cell activity (which would underlie and support higher level averaged activity) may be the appropriate place to look for the causal mechanisms in order to account for the "higher level" perceptual phenomena (compare von der Malsburg 1982, 1987).

However, it is difficult to translate questions dealing with how populations of neurons interact to questions concerning how a single cell behaves. Since each area of the brain is composed of a staggeringly enormous number of neurons, recording the firing pattern of a single cell may not capture any relevant higher-level patterns or processes. Indeed, since there are tens of thousands of cells in each visual column, any one cell from which a scientist records could have a function orthogonal to the task at hand. And until neuroscientists could make the leap from studying single cells to studying populations of cells, many interesting questions concerning visual perceptual experience in the nervous system were simply unapproachable.

But within the last ten years or so, two sorts of techniques have emerged as ways to start studying the behavior of larger numbers of neurons: multiple simultaneous recordings from single cells and multiple–channel electroencephalograph recordings from relatively small populations of cells. As means by which to connect neurophysiological structures with larger neurophysiological or psychological functional units, these procedures allow scientists to examine the intrinsic dynamic operations of circumscribed cortical areas and to start assessing their relevance for visual perception. In particular, when coupled with computer technology (which permits simultaneous visual stimulation of different areas in the visual field that are keyed to the cells whose spike trains are being recorded), they gave scientists a way to start determining the principles of visual processing within single cortical areas. The function of the distributed systems connecting the various visual subsystems can then be studied using mathematical correlation methods to assess the cooperative firing across groups of neurons in different areas (compare Eckhorn and Reitbock 1988).

One Possible Neurobiological "Solution"

Though it is still largely unknown how the visual system links features of a visual scene together, neuroscientists are beginning to give a few answers to some of the smaller questions and are starting to speculate on what principles might underwrite psychology's larger story. For example, Eckhorn and his research group at Philipps University in Germany took simultaneous multiple recordings of single unit activity, multiple unit activity, and local slow wave field potentials from areas 17 and 18 with independently drivable microelectrodes in lightly anesthetized cats. They evaluated their data for receptive field properties, orientation and direction tuning, and short-epoch cross-correlations between various combinations of the different types of recordings and discovered that the neurons of an assembly partly synchronize their outputs through a transition to a phase-locked oscillatory state. Assemblies relatively far from one another that have similar receptive field properties—including assemblies in different areas of the brain—synchronize their activities as well if they are stimulated at the same time. Operating at a frequency between 40 and 80 Hz (the gamma range), these oscillations occur in all three types of recordings (see Figure 4.1). Eckhorn speculates that this sort of oscillatory pattern may serve as a general

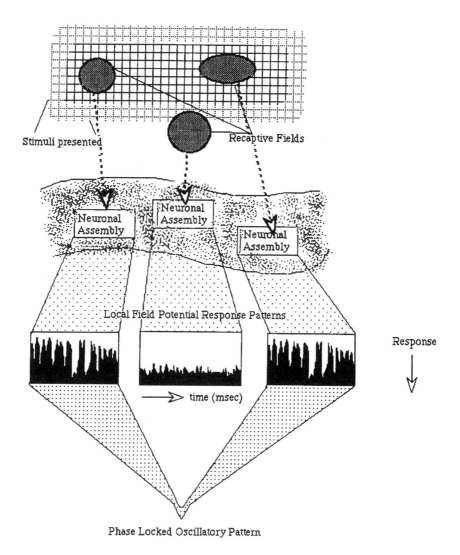

Figure 4.1 A sketch of Eckhorn's synchronized neuronal assemblies. Spatially separated neural assemblies with similar receptive fields synchronize outputs through a transition to a phase-locked oscillatory state for simple stimuli (the grid pattern shown here). Scientists can record the phase-locked transitions using local field potentials, as well as by simultaneous multiple recordings from single unit activity and multiple unit activity. Neural assemblies that do not have similar receptive fields, even though they are spatially contiguous, do not synchronize their outputs.

mechanism for binding of stimuli features in the primary visual cortex by linking excitatorially connected neurons with similar receptive fields. In this way, the dynamic assemblies would "define" an object in virtue of resemblence among features (Eckhorn, Bauer and Reitbock 1988, Eckhorn *et al.* 1988a, 1988b, Eckhorn *et al.* 1989, see also Engel, Kreiter and Singer 1992, Schwartz, Aertsen and Bolz 1992, Young, Tanaka and Yame 1992).

There are three possible sources for the oscillations in area 17 (Jagadeesh, Gray, and Ferster 1992): intrinsic membrane properties in presynaptic cells (for example, the interneurons, Llinas, Grace, and Yarom 1991; or the pyramidal cells Silva, Amitai, and Conners 1991; see also Rhodes 1992), oscillations in thalamic input (Ghose and Freeman 1990, Steriade *et al.* 1991), or intracortical feedback pathways (Freeman 1975).

However, it appears that intrinsic membrane properties have little effect (Jagadeesh, Gray, and Ferster 1992, Gray 1992), so some rhythmic synaptic input probably leads to the oscillatory behavior. Using single and multi-unit recordings, Gray and Singer have shown that the dorsal lateral geniculate nucleus of the thalamus in anesthetized cats shows no general synchronized oscillation pattern in the gamma range, suggesting that the synchronous oscillations are probably cortical phenomena alone (see also Engel *et al.* 1991). These results strongly suggest that the brain uses these temporal patterns for some peculiarly cortical activity (Gray and Singer 1989; Gray *et al.* 1992).

Moreover, Gray and Singer have shown that phase-locked synchronization across spatially separate columns for cells with similar orientation preferences are influenced by global properties of the stimulus. (Indeed, such synchronization has been found between neurons in area 17 of left and right cerebral hemispheres [Engel *et al.* 1991].) When two short light bars move in opposite directions over two similar receptive fields, the responses of the cells show no phaselocking. When they move in the same direction, the cells are marginally synchronized. When the cells are shown a single long bar of light, their responses are strongly phaselocked (see also Kretier, Engel, and Singer 1992) (See Figure 4.2).

These effects indicate that phase-locked oscillations depend on various large-scale aspects of the stimuli, such as form or motion, which local responses cannot reflect when taken individually. Therefore, the sychronization may also serve to represent the higher-order features in a pattern. For example, it may help in figure-ground segregation (Gray *et al.* 1989, see also Engel *et al.* 1991a, Engel *et al.* 1991b, von der Malsburg and Buhman 1992).[2]

Although the phase-locked oscillations may yet turn out be an artifact of some other, more fundamental, process,[3] it may be that "[stimulus-evoked] resonances are a general phenomena, forming the basis of a correlation code which is used within and between different sensory systems and perhaps even throughout the entire brain" (Eckhorn *et al.* 1988, p. 129; see also König, Engel, and Singer 1992). And as modelers in artificial intelligence pick up on the mechanism of phaselocked oscillations and explore the

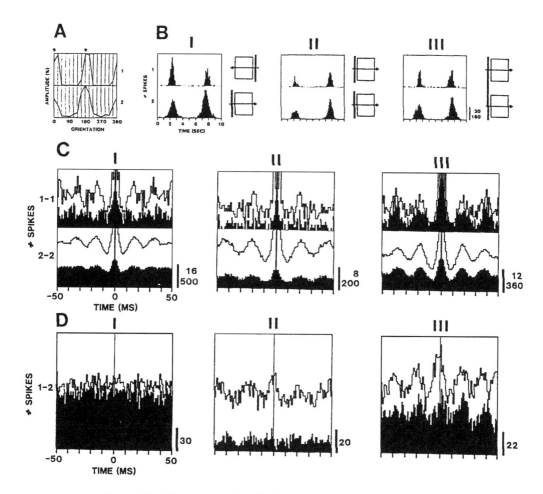

Figure 4.2 Phase-locked oscillations depend on global properties of the stimuli. Long-range oscillatory correlations reflect global stimulus properties. (A) Orientation tuning curves of neuronal responses recorded from two electrodes (1, 2) separated by 7 mm show a preference for vertical light bars (0° and 180°) at both recording sites. (B) Post-stimulus time histograms of the neuronal responses recorded at each site for each of the three different stimulus conditions: (I) two light bars moved in opposite directions, (II) two light bars moved in the same direction, and (III) one long light bar moved across both receptive fields. A schematic diagram of the RF locations and the stimulus configuration used is displayed to the right of each post-stimulus time histogram. (C, D) Auto-correlograms (C, 1–1, 2–2) and cross-correlograms (D, 1–2) computed for the neuronal responses at both sites (1 and 2 in A and B) for each of the three stimulus conditions (I, II, III) displayed in B. For each pair of correlograms except the two displayed in C (I, 1–1) and D (I) the second direction of stimulus movement is displayed with unfilled bars. (Reprinted with permission from *Nature* and the author [Gray et al., 1989] Copyright © 1989) Macmillan Magazines Limited)

principles underlying this sort of binding solution (for example, Cotterill and Nielsen 1991, Desmond and Moore 1991, Finkel and Edelman 1989, Humel and Biederman 1992, Sporns *et al.* 1989, Sporns, Tonomi, and Edelman 1991, 1992, Tonomi, Sporns, and Edelman 1992, Wilson and Bower 1991), it does appear that this hypothesis provides the ideal sort of answer to psychology's binding problem.

In fact, it is congenial to both Triesman's hypothesis of a master map of locations in which attention is used to tie features together (Treisman 1986, Treisman and Gelade 1980, Treisman and Schmidt 1982) and Prinzmetal's notion that organizational assumptions in perception plus feedback from higher-level processes accounts for perceptual binding (Prinzmetal 1981, Prinzmetal and Keysar 1989, Prinzmetal, Presti, and Posner 1986). For example, Kammen, Koch, and Holmes (1991) have demonstrated that some nonlocal feedback must play a fundamental role in the initial synchronization and the dynamic stability of the oscillation. And this higher-level comparator could exist either as Treisman's central feature-locator attentional mechanism or as Prinzmetal's lower level, distributed recurrent feedback loops (see also Lumer and Huberman, 1991). Moreover, in investigating the architecture required to segment and bind more than one object at a time, Horn, Sagi, and Usher (1991) have stumbled across something like the illusory conjunction effect in their computer simulations of oscillatory binding. It is no wonder then that phase-locked oscillations are being taken seriously as a new and exciting solution to superposition catastrophe and the binding problem.

Problems with Neuroscience's "Solutions"

However, the neurophysiologists who suggest that synchronized oscillations are the key to solving the binding problem are guilty of confusions about stages in processing. The sort of underlying model of mental function which neuroscientists assume divides visual perception into three different stages: segmentation, binding, and association, each of which the brain actively pursues using some computational means or other (see Figure 4.3). (See also Horn, Sagi and Usher 1991, Baird 1991 for discussion.)

Early stages in the visual system must be dedicated to parsing the incoming two-dimensional pattern of retinal stimulation into cohesive features. They must somehow figure out how to indicate that this neuron over here with this particular receptive field is signalling a bit of the same feature as that neuron over there with that particular receptive field. Differential response properties of cells, as well as through the axonal and dendritic connections among individual neurons, could carry out some of these required "computations." However, since our perception of single features must remain constant across such disruptive events as blinks, saccades, head movements, and changes in lighting, the architecture of the neural net probably will not be able to handle the task alone. The brain must also segment incoming data actively. And only after the brain determines which

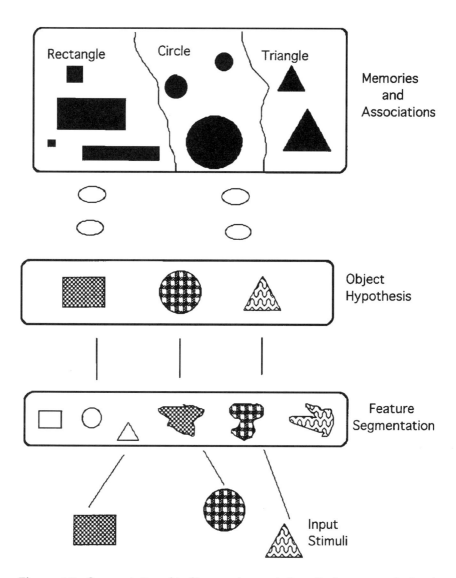

Figure 4.3 Segmentation, binding, and association. Early stages of visual processing parse the incoming two-dimensional stimuli patterns into simple cohesive features (feature segmentation). The next stage binds these features into perceptual objects, in conjunction with some top-down expectations of which features and objects are actually there (object hypothesis). Finally, these "guesses" about which objects are present are associated with memories of previous experiences with similar objects (memory and association).

features are present can it begin to bind them into unified objects (though top-down expectations most likely do play some role in determining which features the brain decides are there). Some of this binding may be architecturally constrained as well. For example, the connections between topological maps in different areas can preserve the approximate spatial relationships of objects in the external world. However, since our world is in a constant flux and we are constantly being exposed to novel visual stimuli, purely "structural" architectural solutions again will fall short. Finally, only after this initial processing can the brain assign meanings to the bundles of features through associating the temporary object hypotheses with previously encountered stimuli and their related associations.

What Eckhorn and his group and Gray and his colleagues have indexed is segmentation, not binding. They presented their cats with very simple stimuli, a bar or a grid pattern, and then tested the response of cells with similar receptive fields for single features. If two cells are sensitive to similar orientations, for example, then their experiments predict that the cells will fire in phase-locked synchrony if the visual system is shown a bar in that orientation. This is no trivial result, but it does not show what would happen if two cells, one color responsive and one orientation responsive, were both activated by an appropriately colored and oriented bar. In fact, in cells they tested, Gray and Singer found that if neurons in different hypercolumns were sensitive to different features, but still responded to the same particular input, then these cells were not phase-locked (1987, 1989).[4]

Moreover, the difficulty with generalization to a population which plagued single cell recordings affects the multiple unit recordings as well. Scientists currently can record from such a small number of neurons at a time that we have little assurance that the behavior we are witnessing reflects the general trend, even within a particular column. (In this particular case, the difficulty is compounded by the fact that not all the neurons probed became phaselocked, even though they had the proper receptive fields [Gray et al. 1989]).[5] Without additional converging evidence that the brain uses phaselocked oscillations to tie together disparate features into a single object, it is premature to conclude that these oscillations form a binding mechanism. Given current state-of-the-art technology, we are simply unable to devise single cell (or local field potential) experiments that would be able to test the oscillation hypothesis definitively.

I must conclude that those who see phase-locked oscillations in individual neurons as the additional temporal process that overcomes superposition woes celebrate prematurely. Though oscillations may be the correct answer, we do not have anything near conclusive evidence linking phase-locked oscillations with the psychological phenomena of binding, and perhaps more importantly, multiple-unit recording technology itself falls short of being able to support our higher-level functional analyses. Methodological limitations appear to halt any sort of actual explanatory connection between individual neural firing patterns and psychological effects right here. Neuroscience cannot (now) provide the link that psychology requires between bound perceptual unity and individual neurons simply because neuroscience lacks the tools to do so.

NOTES

1. More sophisticated versions of this problem exist in which a few symbols can be active at a time in some system without collapse.

2. These results have now been extended to include interareal synchronization of area 17 and the posteromedial lateral suprasylvian area (PMLS), which is a visual association area specialized for motion.

3. There is some concern that stimulus–induced nonoscillatory synchronizations may be more fundamental for feature binding (Kruse *et al.* 1992).

4. An apparent exception to this finding is Engel *et al.* 1991b. They found synchronization between areas specialized for spatial continuity (area 17) and coherence of motion (PMLS area). Area 17 engages in a high resolution, fine-grained analysis of local features, while PMLS has large receptive fields, is strongly directionally selective, but poorly trained for orientation. This interareal synchronization serves to bind local with global features. However, from a higher-level psychological perspective, it is not clear that these two areas do in fact code different psychologically defined features. Taken together, both areas could still be concerned with shape alone in that they (possibly) differentiate among objects by uniting different orientation of outlines with coherent motion. There is still no evidence that 40-Hz oscillations or other synchronous firing patterns tie radically different features domain together.

5. Until recently, about the only place where phase-locked, 40-Hz oscillations were reliably found in unanesthetized animals was motor cortex (see Flament, Fortier, and Fetz 1992, Gaal, Sanes, and Donoghue 1992, Murthy and Fetz 1992, Murthy, Chen, and Fetz 1992, Smith and Fetz 1989). However, even though the oscillation results remain difficult to reproduce (though see Engel *et al.* 1992, Kreiter, Engel, and Singer 1992, Schwartz, Aertsen, and Bolz 1992), they have now been found in areas 17 and 18 of alert cats. Unfortunately, the discharges were also found in a wide range of states not connected to stimulus interpretation, including saccadic eye movements and slow wave sleep (Lee *et al.* 1992). Hence, synchronous oscillations cannot be sufficient for feature binding (though they may still be necessary).

REFERENCES

Baird, B. 1991. Associative memory in a simple model of oscillating cortex. In *Advances in neural information processing systems 2,* edited by D. S. Touretzky. San Mateo, CA: Morgan Kaufman.

Blakeslee, S. 1992. Nerve cell rhythm may be key to consciousness. *The New York Times,* October 27, C1.

Cotterill, R. M., and C. Nielsen. 1991. A model for cortical 40 hz oscillations invokes interarea interactions. *Neuroreport* 2:289–92.

Crick, F. H. C., and C. Koch. 1990. Towards a neurobiological theory of consciousness. *Seminars in Neuroscience* 2:263–75.

Damasio, A. R. 1989. The brain binds entities and events by multiregional activation from convergence zones. *Neural Computation* 1:123–32.

Desmond, J. E., and J. W. Moore. 1991. Altering the synchrony of stimulus trace processes: tests of a neural-network model. *Biological Cybernetics* 65:161–69.

Eckhorn, R., and H.J. Reitbock. 1988. "Assessment of cooperative firing in groups of neurons: Special concepts for multiunit recordings from the visual system." In *Brain dynamics,* edited by E. Basar. New York: Springer-Verlag.

Eckhorn, R., R. Bauer, and H. J. Reitbock. 1988. "Discontinuities in visual cortex and possible functional implications." In *Dynamics of sensory and cognitive processing by the brain,* edited by E. Basar and T. H. Bullock. Springer Series in Brain Dynamics, Vol. 2. New York: Springer.

Eckhorn, R., R. Bauer, W. Jordan, M. Brosch, W. Kruse, and M. Munk. 1988. Functionally related modules of cat visual cortex show stimulus–evoked coherent oscillations: A multiple electrode study. *Investigations in Opthamological Visual Science* 29:12.

Eckhorn, R., R. Bauer, W. Jordan, M. Brosch, W. Kruse, M. Munk, and H. J. Reitbock. 1988. Coherent oscillations: A mechanism of feature linking in the visual cortex. *Biological Cybernetics* 60:121–30.

Eckhorn, R., H. J. Reitbock, M. Arndt, and P. Dicke. 1989. "A neural network for feature linking via synchronous activity: Results from cat visual cortex and from simulations." In *Models of brain function,* edited by R. M. J. Cotteril. Cambridge, England: Cambridge University Press.

Engel, A. K., P. König, and W. Singer. 1991. Direct physiological evidence for scene segmentation by temporal coding. *Proceedings of the National Academy of Sciences of the United States of America* 88:9136–40.

Engel, A. K., P. König., A. K. Kreiter, and W. Singer. 1991. Interhemispheric synchronization of oscillatory neuronal responses in cat visual cortex. *Science* 252:1177–79.

Engel, A. K., A. K. Kreiter, P. Konig, and W. Singer. 1991. Synchronization of oscillatory neuronal responses between striate and extrastriate cortical areas of the cat. *Proceedings of the National Academy of Sciences of the United States of America* 88:6048–52.

Engel, A. K., A. K. Kreiter, and W. Singer. 1992. Oscillatory responses in the superior temporal sulcus of anaesthetized macaque monkeys. *Society for Neuroscience Abstracts* 18:12.

Finkel, L. H., and G. M. Edelman. 1989. Integration of distributed cortical systems by reentry: A computer simulation of interactive functionally segregated visual areas. *Journal of Neuroscience* 9:3188–208.

Flament, D., P. A. Fortier, and E. E. Fetz. 1992. Coherent 25- to 35-Hz oscillations in the sensorimotor cortex of awake behaving monkeys. *Proceedings of the National Academy of Sciences of the United States of America* 89:5670–74.

Flanagan, O. 1992. *Consciousness reconsidered.* Cambridge, MA: MIT Press.

Fodor, J. A., and Z. Pylyshyn. 1988. Connectionism and cognitive architecture: A critical analysis. *Cognition* 28:3–71.

Freeman, W. J. 1975. *Mass action in the nervous system.* New York: Academic Press.

Gaal, G., J. N. Sanes, and J. P. Donoghue. 1992. Motor cortex oscillatory neural activity during voluntary movement in *Macaca fasicularis. Society for Neuroscience Abstracts* 18:848.

Ghose, G. M., and R. D. Freeman. 1990. *Society of Neuroscience Abstracts* 16: 523.

Goldman, P. S., and H. E. Rosvold. 1970. Localization of function within the dorsolateral prefrontal cortex of the rhesus monkey. *Experimental Neurology* 27(2):291–304.

Goldman-Rakic, P., I. Isseroff, M. Schwartz, and N. Bugbee. 1983. "The neurobiology of cognitive development." In *The handbook of child development,* edited by F. Plum and V. Mountcastle. Bethesda, MD: American Physiological Society.

Goldman-Rakic, P. S. 1987. Circuitry of primate prefrontal cortex and regulation of behavior by representational memory. In *Handbook of physiology—The nervous system V.* Bethesda, MD: American Physiological Society.

Gray, C. M. 1992. Bursting cells in visual cortex exhibit properties characteristic of intrinsically bursting cells in sensorimotor cortex. *Society for Neuroscience Abstracts* 18:292.

Gray, C. M., and W. Singer. 1987. Stimulus dependent neuronal oscillations in the cat visual cortex area 17. *Neuroscience [Suppl]* 22:1301P.

Gray, C. M., and W. Singer. 1989. Stimulus–specific neuronal oscillations in orientation columns of cat visual cortex. *Proceedings of the National Academy of Sciences of the United States of America* 86:1698–702.

Gray, C. M., A. K. Engel, P. König, and W. Singer. 1992. Synchronization of oscillatory neuronal responses in cat striate cortex: Temporal properties. *Visual Neuroscience* 8:337–47.

Gray, C. M., P. König, A. K. Engel, and W. Singer. 1989. Oscillatory responses in cat visual cortex exhibit inter-columnar synchronization which reflects global stimulus properties. *Nature* 338:334–37.

Horn, D., D. Sagi, and M. Usher. 1991. Segmentation, binding, and illusory conjunctions. Technical Report CS91–07, Weismann Institute of Science.

Hubel, D. H., and M. S. Livingstone. 1987. Segregation of form, color, and stereopsis in primate area 18. *The Journal of Neuroscience* 7:3378–415.

Humel, J. E., and I. Biederman. 1992. Dynamic binding in a neural network for shape recognition. *Psychological Review* 99:480–517.

Jagadeesh, B., C. M. Gray, and D. Ferster. 1992. Visually evoked oscillations of membrane potential in cells of cat visual cortex. *Science* 257:552–4.

Kammen, D., C. Koch, and P. J. Holmes. 1991. Collective oscillations in the visual cortex. In *Advances in neural information processing systems 2*, edited by D. S. Touretzky. San Mateo, CA: Morgan Kaufman.

Kandel, E. R., and J. H. Schwartz, eds. 1987. *Principles of neural science, 2nd ed.* New York: Elsvier.

Koch, C. 1992. The connected brain. *Discover* 96–8.

König, P., A. E. Engel, and W. Singer. 1992. Gamma–oscillations as a vehicle for synchronization. *Society for Neuroscience Abstracts* 18:12.

Kreiter, A. K., A. K. Engel, and W. Singer. 1992. Stimulus dependent synchronization in the caudal superior temporal sulcus of macaque monkeys. *Society for Neuroscience Abstracts* 18:12.

Kruse, W., R. Eckhorn, T. Schanze, and H. J. Reitbock. 1992. Stimulus-induced oscillatory sychronization is inhibited by stimulus-locked non-oscillatory synchronization in cat visual cortex: Two modes that might support feature linking. *Society for Neuroscience Abstracts* 18:292.

Lee, C., J. Kim, J. Park, and S. Chung. 1992. Oscillatory discharges of the visual cortex in the behaving cats. *Society for Neuroscience Abstracts* 18:292.

Legéndy, C. R. 1970. The brain and its information trapping device In *Progress in cybernetics*, edited by J. Rose, Vol. 1. New York: Gordon and Beach.

Livingstone, M. S., and D. H. Hubel. 1987. Psychophysical evidence for separate channels for the perception of form, color, movement, and depth. *The Journal of Neuroscience* 7:3416–68.

Llinas, R. R., A. A. Grace, and Y. Yarom. 1991. In vitro neurons in mammalian cortical layer 4 exhibit intrinsic oscillatory activity in the 10- to 50-hz frequency range.

Proceedings of the National Academy of Sciences of the United States of America 88:897–901.

Lumer, E., and B. A. Huberman. 1991. Binding hierarchies: A basis for dynamic perceptual grouping. Manuscript.

Murthy, V. N., and E. E. Fetz. 1992. Coherent 25- to 35-Hz oscillations in the sensorimotor cortex of awake behaving monkeys. *Proceedings of the National Academy of Sciences of the United States of America* 89:5670–74.

Murthy, V. N., D. F. Chen, and E. E. Fetz. 1992. Spatial extent and behavioral dependence of coherence of 25–35-hz oscillations in primate sensorimotor cortex. *Society for Neuroscience Abstracts* 18:847.

Prinzmetal, W. 1981. Principles of feature integration in visual perception. *Perception and Psychophysics* 30:330–40.

Prinzmetal, W., and B. Keysar. 1989. Functional theory of illusory conjunctions and neon colors. *Journal of Experimental Psychology: General.* 118:165–89.

Prinzmetal, W., D. E. Presti, and M. I. Posner. 1986. Does attention affect visual feature integration? *Journal of Experimental Psychology: Human Perception and Performance* 12:361–69.

Ramachandran, V. S., 1990. "Visual perception in people and machines." In *AI and the Eye,* edited by A. Balke and T. Troscianko. New York: John Wiley and Sons.

Ramachandran, V. S., and S. M. Anstis. 1986. The perception of apparent motion. *Scientific American* 254:101–09.

Rhodes, P. A. 1992. The long open time of the NMDA channel facilitates the self–organization of invariant object responses in cortex. *Society for Neuroscience Abstracts* 18:740.

Schwartz, C., A. Aertsen, and J. Bolz. 1992. Dynamics of coherent firing in cat visual cortex. *Society for Neuroscience Abstracts* 18:292.

Sejnowski, T. 1981. "Skeleton fibers in the brain." In *Parallel models of associative memory,* edited by G.E. Hinton and J. A. Anderson. Hillsdale, NJ: Lawrence Erlbaum.

Sereno, M. I. 1990. Language and the primate brain. *CRL Newsletter* 4:4.

Silva, L. R., Y. Amitai, and B. W. Conners. 1991. Intrinsic oscillations of neocortex generated by layer 5 pyramidal cells. *Science* 251:432.

Smith, W. S., and E. E. Fetz. 1989. Effects of synchrony between primate corticomotoneuronal cells on post–spike facilitation of muscles and motor units. *Neuroscience Letters* 96:76–81.

Sporns, O., J. A. Gally, G. N. Reeke, Jr., and G. M. Edelman. 1989. Reentrant signalling among simulated neuronal groups leads to coherency in their oscillatory activity. *Proceedings of the National Academy of Sciences of the United States of America* 86: 7265–69.

Sporns, O., G. Tonomi, and G. M. Edelman. 1991. Modeling perceptual groupings and figure–ground segregation by means of active reentrant connections. *Proceedings of the National Academy of Sciences of the United States of America* 88:129–33.

Sporns, O., G. Tonomi, and G. M. Edelman. 1992. Constructive and correlative reentry in the visual system: Computer simulations and psychophysics. *Society for Neuroscience Abstracts* 18:741.

Steriade, M., R. C. Dossi, D. Pare, and G. Oakson. 1991. Fast oscillations 20–40 hz in thalamocortical systems and their potentiation by mesopontine cholinergic nuclei in the cat. *Proceedings of the National Academy of Sciences of the United States of America* 88:4396–400.

Stryker, M. 1989. Is grandmother an oscillation? *Nature* 338:297–98.

Tononi, G., O. Sporns, and G. M. Edelman. 1992. Modeling integration in the visual cortex. *Society for Neuroscience Abstracts* 18:741.

Treisman, A. 1986. Features and objects in visual processing. *Scientific American* 255(5):B114.

Treisman, A., and G. Gelade. 1980. A feature integration theory of attention. *Cognitive Psychology* 12:97–136.

Treisman, A., and H. Schmidt. 1982. Illusory conjunctions in the perception of objects. *Cognitive Psychology* 14:107–41.

Van Essen, D. C., C. H. Anderson, and D. J. Felleman. 1992. Information processing in the primate visual system: An integrated systems perspective. *Science* 255:419–23.

von der Malsburg, C. 1982. *The correlation theory of brain function.* Göttengen, Germany: Max-Planck Institute for Biophysical Chemistry.

von der Malsburg, C. 1987. Synaptic plasticity as basis of brain organization. In *The Neural and Molecular Bases of Learning,* edited by J.–P. Changeux and M. Konishi. New York: John Wiley and Sons.

von der Malsburg, C., and J. Buhman. 1992. Sensory segmentation with coupled neural oscillators. *Biological Cybernetics* 67:233–42.

Wilson, M. A., and J. M. Bower. 1991. "Computer simulation of oscillatory behavior in cerebral cortical networks." In *Advances in Neural Information Processing Systems 2,* edited by D. S. Touretzky. San Mateo, CA: Morgan Kaufman.

Young, M. P., K. Tanaka, and S. Yame. 1992. On oscillating neuronal responses in the visual cortex of the monkey. *Journal of Neurophysiology* 67:1464–74.

Zeki, S. 1992. The visual image in mind and brain. *Scientific American* 267:68–77.

5 Deconstructing Dreams: The Spandrels of Sleep

Owen Flanagan

FOUR PHILOSOPHICAL PROBLEMS ABOUT DREAMS

There are two famous philosophical problems about dreams.[1]

1. How can I be sure I am not always dreaming? (Descartes's problem; also Plato's and Cicero's)

2. Can I be immoral in dreams? (Augustine's problem)

After his transformation from philandering pagan to an ascetic Christian, poor Augustine proposed a theory for how dreams might contain sinful content without being sins. The proposal in modern terms was that dreams are happenings not actions. Augustine wrote,

> You commanded me not to commit fornication . . . But when I dream [thoughts of fornication] not only give me pleasure but are very much like acquiescence to the act . . . Yet the difference between waking and sleeping is so great [that] I return to a clear conscience when I wake and realize that, because of this difference, I am not responsible for the act, although I am sorry that by some means or other it happened in me.[2]

A third problem emerged in the twentieth century in the hands of Norman Malcolm[3] and Daniel Dennett,[4] a natural sequel to the prominence of verificationist ideas generally. The question was this:

3. Are dreams experiences? Or are so-called "dreams" just reports of experiences we think we had while sleeping but which in fact are, in so far as they are experiences at all, constituted by certain thoughts we have while waking or experiences we have while giving reports upon waking?

The answers to the first three dream problems are these: (1) Rest assured, you're not always dreaming; (2) Augustine's right that committing adultery, murder, and so on in dreams is not sinful; and (3) Dreams are experiences that take place during sleep.[5]

Given that dreams are experiences that take place while sleeping, a fourth dream problem suggests itself:

4. Is dreaming functional?

Now there are many senses of functional and my reflections on this, the fourth dream problem, which take up the remainder of the paper, will try to sort out the way dreams look using different senses of function. But I want to state my general answer to the fourth dream problem up front so that the reader can understand better the use I am making of various sorts of empirical evidence along the way. The answer is this: Although there are credible adaptationist accounts for sleep and the phases of the sleep cycle itself, there is reason to think that the mentation—the phenomenal mentation—that occurs during sleep is a *bona fide* example of a byproduct of what the system was designed to do during sleep and sleep-cycling. If this is right then there is a sense in which dreaming, phenomenally speaking, is an "automatic sequelae,"[6] a spandrel,[7] and exaptation.[8]

CONSTRUCTIVE NATURALISM AND THE NATURAL METHOD

I raise the epiphenomenalist suspicion about dreams not as another skeptical philosophical exercise. It is intended as a serious proposal I've been lead to in my work on consciousness within the framework of two general assumptions. (1) Consciousness has depth, hidden structure, hidden and possibly multiple functions, and hidden natural and cultural history. (2) Consciousness is heterogeneous in kind.

Regarding the first assumption, consciousness has a first-personal phenomenal surface structure. But from a naturalistic point of view, the subjective aspects of consciousness (call these *p-aspects*—the 'p' for phenomenal) do not exhaust the properties of consciousness. Part of the hidden structure of conscious mental states involves their neural realization. Conscious mental states supervene on brain states (call the neural realization the *b-aspect*(s)—'b' for brain). These brain states are essential aspects or constituents of the conscious states, as are the phenomenal aspects (the *p-aspects*) of these states. But of course nothing about neural realization is revealed at the phenomenal surface, not even that there is such realization. The phenomenal surface often hints at or self–intimates the causal role of conscious states. But the phenomenology leaves us clueless as to how conscious thoughts and intentions actually get the system doing what it does; and of course, experience intimates nothing about the causal origins and evolutionary function, if there is any, for the different kinds of consciousness.

From the first-person point of view consciousness only has *p-aspects*. So with respect to dreaming all we know about first personally is p-dreaming. The story of the brain side—of *b-dreaming*—as it were, will need to be provided by the neuroscientists, and the functional-causal role(s) of dreaming (now taking both the phenomenal and brain sides together) will need to be nested in a general psychological cum evolutionary (both natural and cultural) account.

The idea here is to deploy what I call the natural method.[9] Start by treating three different lines of analysis with equal respect. Give phenomenology its due. Listen carefully to what individuals have to say about how things seem. Also, let the psychologists and cognitive scientists have their say. Listen carefully to their descriptions about how mental life works, and what jobs if any consciousness has in its overall economy.[10] Third, listen carefully to what the neuroscientists say about how conscious mental events of different sorts are realized, and examine the fit between their stories and the phenomenological and psychological stories. Now this triangulation procedure will, I claim, yield success in understanding consciousness if anything will. The expectation that success is in store using this method is what makes my kind of naturalism constructive rather than anticonstructive, as is the naturalism of philosophers like Colin McGinn[11] who thinks that although consciousness is a natural phenomena, we will never be able to understand it.

I need to emphasize that the troika of phenomenology, psychology, and neuroscience despite playing the initial and central role in the triangulation procedure I favor are not enough. Evolutionary biology and cultural and psychological anthropology will also be crucial players as the case of dreams will make especially clear. Embedding the story of consciousness into theories of evolution (biological and cultural), thinking about different forms of consciousness in terms of their ecological niche, and in terms of the mechanisms of drift, adaptive selection, and free-riding will be an important part of understanding what consciousness is, how it works, and what, if anything, it is good for. As consilience and reflective equilibrium emerge, if they do emerge, from these different informational sources we will

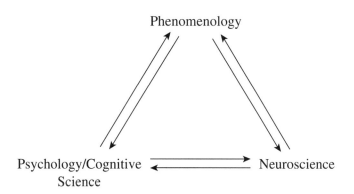

Also: Evolutionary Biology
Cultural and Psychological Anthropology
Social Psychology

Figure 5.1 The Natural Method: Knowledge by triangulation.

understand the nature of consciousness more deeply. Even claims about how things seem can change as our views about how things are change.[12]

DREAMS: A DOUBLE ASPECT MODEL[13]

I've said that p-dreaming is a good example of one of the heterogeneous kinds of conscious experience, and it is at the same time, given neuroscientific evidence and evolutionary considerations, a likely candidate for being given epiphenomenalist status from an evolutionary point of view. P-dreaming is an interesting side effect of what the brain is doing, the function(s) it is performing during sleep.[14]

To put it in slightly different terms: p-dreams, despite being experiences, have no interesting biological function—no evolutionary proper function. The claim is that p-dreams (and possibly even rapid eye movements [REM] after development of the visual system is secured) are likely candidates of epiphenomena. Since I think that all mental phenomena supervene on neural events, I don't mean that p-dreams are nonphysical side effects of certain brain processes. I mean in the first instance the p-dreaming was probably not selected for, that p-dreaming is neither functional nor dysfunctional in and of itself, and thus that whether p-dreaming has a function depends not on Mother Nature's work as does, for example, the phenomenal side of sensation and perception. It depends entirely on what we as a matter of cultural inventiveness—*memetic selection*,[15] one might say—do with p-dreams and p-dream reports. We can, in effect, create or invent functions for dreams. Indeed, we have done this. But as temporally significant aspects of conscious mental life, they are a good example, the flip side say of awake perceptual consciousness which is neither an evolutionary adaptation nor ontogenetically functional or dysfunctional until we do certain things with "our dreams"—for example use them as sources of information about "what's on our mind," utilize dream mentation in artistic expression, and the like.

Despite being epiphenomena from an evolutionary perspective, the way the brain operates during sleep guarantees that the noise of p-dreams is revealing and potentially useful in the project of self understanding. Thus, many things stay the same on the view I am staking out. But there is a paradox: p-dreams are evolutionary epiphenomena, noise the system creates while it is doing what it was designed to do, but because the cerebral cortex is designed to make sense out of stimuli, it tries half successfully to put dreams into narrative structures already in place, structures which involve modes of self-representation, present concerns, and so on. But the cortex isn't designed to do this for sleep stimuli, it is designed to do it for stimuli period and it is ever vigilant. The idea is that it did us a lot of good to develop a cortex that makes sense out of experience while awake, and the design is such that there are no costs to this sense-maker always being ready to do its job. So it works during the chaotic neuronal cascades of part of the sleep-cycle that activate certain sensations and thoughts. So p-dreams despite their bizarreness, and epiphenomenal status, are meaningful and interpretable up to a point.

I've been using p-dreaming so far to refer to any mentation that occurs during sleep. But the term 'p-dreaming' despite being useful for present purposes ultimately won't carve things in a sufficiently fine-grained way. An example will help see why: Until I started working on dreams a year or two ago, I often woke up remembering and reporting dreams like this: "A late tenure letter (ten years late) was just received by the Provost, it was negative, and my tenure has been taken away." Now to say that this is a dream is to use the term "dream" as it is commonly used to refer to any mentation occurring during sleep.[16] But research has shown that this sort of perseverative fearful thought is most likely to occur during non-REM (NREM) sleep, the sleep standardly divided into four stages which occupies about 75% of the night. Sleep-walking, sleep-talking, and tooth-grinding are also NREM phenomena—and no one knows for certain whether we should say persons walking and talking in sleep are p-conscious or not.

Night terrors, a common affliction of young children (my daughter had severe ones for a time and still does at 10 if she is feverish), are very puzzling since the child seems totally awake, eyes wide open, running about speaking alternately sense and nonsense, but almost impossible to comfort and wake up entirely and, on most every view, suffering terrifying hallucinations (which even if the child is finally awakened are remembered much less well than hallucinatory REM dreams). But, and here's the anomaly, the terrorized child is almost certainly in stage III or IV NREM sleep.

The first point is that mentation occurs during NREM sleep as well as during REM sleep, and we report mentation occurring in both states as "dreams." Now since the discovery of REM sleep, and its close association with reports of vivid fantastic dreaming, some have simply identified dreaming with REM-ing or with mentation occurring during REM sleep.[17] But this goes against the grain of our folk psychological usages of the term "dream."

So much the worse for the folk psychological term one might say. But if one wants to regiment language in this way, the stipulation must be made explicitly and there are costs with the explicit stipulation. To the best of my knowledge, only one dream research team has made any explicit sort of definitional maneuver along these lines. Allan Hobson's group at Harvard defines "dreams" as the bizarre, fantastic, image rich mentation that occurs during REM sleep. Hobson's group leaves the logically perseverative tenure dream, worries about tomorrow's agenda, that the car needs gas first thing in the morning, and the like, on the side of *conscious but nondreaming mentation* associated with NREM sleep[18]. This definitional maneuver cleans things up and helps in general to draw a helpful distinction between different kinds of sleep-mentation. We can imagine people—I now do this—reporting as real dreams only the really weird sleep mentation, and thinking of the recurring thought of failing the exam as NREM mentation. But the definitional maneuver has costs for it doesn't deal well with the NREM states, like night terrors, that (probably) involve hallucinations and bizarre mentation. These will turn out to be nondreams because they occur during NREM sleep, but, at the same time, dreams because they are bizarre and hallucinatory. And there are, of course, daydreams which, at least, phenomenally may be closer to the old wish fulfillment model than the mentation of NREM or REM sleep. So everything doesn't become neat and tidy once we use the natural method to make principled distinctions. One reason is that terms that have their roots in folk psychology, although often helpful, are not designed to specify scientific kinds or natural kinds, if there are such kinds in the science of the mind.

Having recognized the benefits and costs of the definitional maneuver, it will do no harm for present purposes if I continue to use p-dreams to refer to any mentation occurring during sleep recognizing full well that since mentation occurs in all stages of NREM and REM sleep, p-dreaming isn't precise enough ultimately to type mentation during sleep from either a phenomenological or neuroscientific point of view.

Now some of the essential questions that any good theory of sleep and dreams will need to explain are these:

1. Why (and how) despite involving vivid experiences do p-dreams involve shut-downs of the attentional, motor, and memory systems and insensitivity to disturbance by external stimuli?

2. Why do the phenomenology of non-REM and REM mentation differ in the ways they do?

3. What function(s) does sleep serve and how do the clocklike cycling of NREM and REM sleep contribute to these functions?

There are numerous additional questions that need addressing but these are the ones that are both somewhat tractable at the present time and most useful for pressing the epiphenomenalist suspicion.

The short answers to questions 1 and 2 are these: Sleeping in general is controlled by a clock in the suprachiasmatic nucleus of the hypothalamus—the hypothalamus is an area importantly implicated in the manufacture of hormones and thermoregulation. This clock gets us into non-REM sleep, a hypometabolic form of sleep, and moves us through its four stages. There appears to be a second clock in the pons (the pontine brainstem) (P) that sets off REM movements and its accompanying mentation (see Figure 5.2).

In REM sleep, pulsing signals originate in the brainstem and reach the lateral geniculate (G) body of the thalamus. When awake this area G is a relay between the retina—on certain views part of the brain itself—and visual processing areas. Other pulses go to the occipital cortex (O)—the main visual processing area of the brain. So *PGO waves* (see Figure 5.3) are the prime movers of REMing. This much accounts for the saliency of visual imagery in the dreams of sighted people. But the PGO noise is going to lots of different places and reverberating every which way. This is why people who work at remembering dreams will report loads of auditory, olfactory, tactile kinesthetic, and motor imagery as well as visual imagery. There is nice convergence of neuroscientific and phenomenological data here.

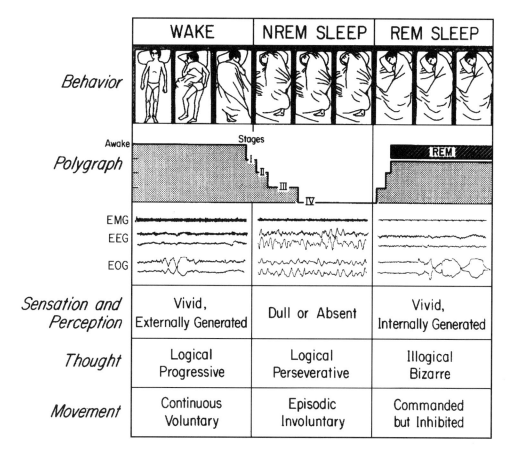

Figure 5.2 Stages of Sleep. Reprinted with permission from Hobson (1989).

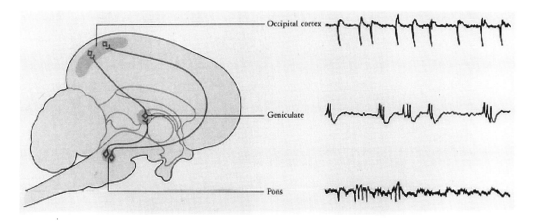

Figure 5.3 The visual brain stimulates itself in REM sleep via a mechanism reflected in EEG recording as PGO waves. Originating in the pons (P) from the neurons that move the eyes, these signals are conducted both to the lateral geniculate (G) body in the thalamus and to the occipital cortex (O). Roffward has proposed that this autostimulation system facilitates brain development. Reprinted with permission from Hobson (1989).

Recent studies have shown that the parts of the brain that reveal robust activity on PETs, MIRs or magneto–encephalographs indicate that "mentation during dreaming operates on the same anatomical substrate as does perception during the waking state."[19] But the main point is that PGO waves are dominant during REM sleep and quiescent during NREM sleep and this explains (by inference to the best available explanation) a good deal about why the mentation of REM sleep involves vivid, bizarre, and multimodal imagery.[20]

The answer to another piece of the puzzle, namely, why don't we in fact get up and do or try to do the things we dream about doing has to do with the fact that a certain area in the brainstem containing the bulbar reticular formation neurons sends hyperpolarizing signals to the spinal cord blocking external sensory input and motor output. People with certain sorts of brainstem lesions do get up in the middle of the night and play linebacker to their dresser—presumably imagined to be an oncoming fullback. (Figure 5.4 gives two sorts of pictures of how REM-dreaming works).

**THE FOURTH DREAM PROBLEM
AND FUNCTIONAL EXPLANATION**

So far I've tried to answer questions 1 and 2 about the differences between REM sleep and NREM sleep and the accompanying mentation. The answer to question 3—the question to the function(s) of sleep and sleep-cycling—is not as well understood as some of the matters just discussed. Based on my reading, I have a list of over 50 distinct functions that have been attributed to sleep and dreams in the last decade alone! Using the best theory in town principle: here is how things look to me regarding the function question.

First, some facts. (1) Fish and Amphibia rest but do not sleep at all. The most ancient reptiles have only NREM sleep, while more recent reptiles and birds have robust NREM sleep and some REM sleep. All mammals save one, the egg-laying marsupial echidna of Australia, have REM sleep. (2) In creatures that REM, REMing is universally more frequent at the earliest stages of development. So for humans, newborns are REMing during half the sleep cycle, this drops to 33% at 3 months, and at puberty REM sleep comprises about 25% of all sleep. It decreases in relative amount as we age, as does stage III and IV NREM sleep.

The fact that NREM is the oldest form of sleep and is hypometabolic suggests the following hypothesis: It was selected for to serve restorative and/or energy conservation and/or body building functions. Now some people find this hypothesis empty—akin to saying sleep is for rest which although true is thought to be uninformative. But things are not so gloomy if we can specify some of the actual restorative/conservatory/building mechanisms and processes in detail. And we can. The endocrine system readjusts all its levels during sleep. For example, testosterone levels in males are depleted while awake regardless of whether any sexual or aggressive behavior has occurred, and are restored during sleep—indeed

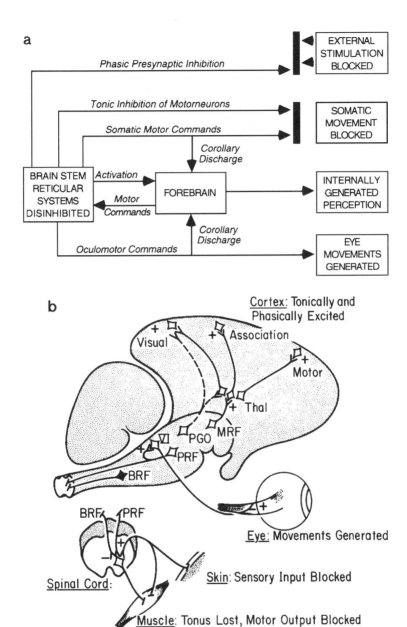

Figure 5.4 The Activation-Synthesis Hypothesis. Reprinted with permission from Hobson and Stickgold (1993).

levels peak at dawn. Pituitary growth hormone does its work in NREM sleep. Growth hormone promotes protein synthesis throughout the body—new cell growth helps with tissue repair—for example, cell repair of the skin is well–studied and known to be much greater while sleeping than awake. Protein synthesis in the cerebral cortex and the retina follow the same pattern of having a faster rate in sleep than while awake. And, of course, the amount of food needed for survival is lowered insofar as metabolic rate is. To be sure, much more needs to be said, and can be found in medical textbooks about the restorative/conservatory/building processes that are fitness enhancing and associated with NREM sleep.[21]

Regarding REM sleep, two functions suggest themselves. First, the much larger percentage of REMing in development across mammals suggests that it is important in helping build and strengthen brain connections, particularly ones in the visual system, that are not finished being built in utero. On the other side, the prominence of significantly greater time spent in REM sleep as an infant—where one doesn't know or care about right and wrong—than as a adolescent bubbling over with vivid and new socially acceptable wishes should go the other way, one would think, if anything like the orthodox Freudian view of dream function were true. What instinctual impulses, what sexual and aggressive fantasies are being released by a newborn, or even less credibly by 30-week old fetuses which, according to some experts, go through phases of REMing 24 hours a day?

Now the biggest difference between waking and NREM sleep and REM sleep has to do with the ratios of different types of neurochemicals, modulators, and transmitters in the soup. In particular, the ratios of cholinergic and aminergic neurochemicals flip-flop. Neurons known to release serotonin and norepinephrine shut off in the brainstem during REM and neurons secreting acetylcholine are on (see Figures 5.5 and 5.6).

What good could this do? Here's one possible answer. The best theory of attention, namely Posner's and Peterson's[22] says that norepinephrine is crucial in getting the frontal and posterior cortical subsystems to do a good job of attending. Furthermore, both norepinephrine and serotonin are implicated in thermoregulation as well as in learning, memory, and attention; and dopamine has been shown to play an essential role in learning at least in sea slugs. Now what happens in REM sleep that is distinctive in addition to the dream–mentation part is that there is a complete shift in the ratios of certain classes of neurochemicals. In particular, in waking serotonin is working hard as are dopamine and norepinephrine. The aminergic neurons that release these neurochemicals quiet down in NREM sleep and turn off during REM sleep—this helps explain why memory for dreams is degraded. Meanwhile, cholinergic neurons, for example, those releasing acetylcholine turn on. Here is a credible hypothesis for why this might be: By a massive reduction in firing during REM sleep the neurons releasing the neurochemicals most directly involved in attention, memory, and learning get a rest. While resting they can synthesize new neurotransmitters. The evidence points to a major function of REM sleep as involving

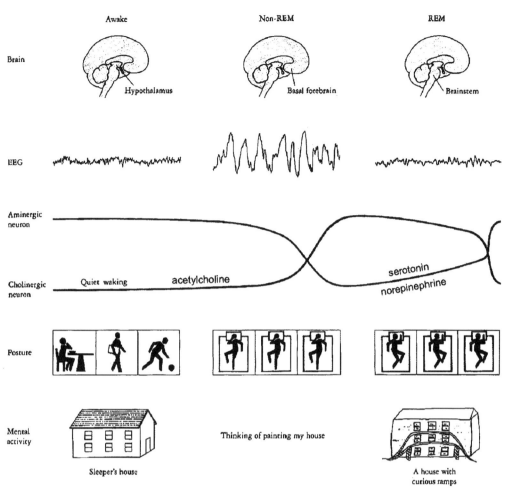

Figure 5.5 The hypothalamus, the basal forebrain, and the pontine brainstem are believed to control the states of waking, REM sleep, and NREM sleep. As we go from one state to another, a series of coordinated changes occur in EEG signals, neurotransmitter level of activity, posture, and mental activity. Posture shifts occur during the transition to and from REM sleep. The vivid perceptions of reality in waking shift to thoughtlike, nonvisual, nonvisual cognition in NREM sleep and then to the bizarre visual imagery of dreams. Reprinted with permission from Hobson (1989).

"stockpiling" the neurotransmitters that the brain will need in the morning for the day's work.[23]

Another hypothesized function of sleep and of REM sleep in particular, that I haven't yet mentioned, is that something like disk maintenance, compression, trash disposal, and memory consolidation take place.[24] These seem like good things for the system to do. But it's pie in the sky hypothesizing until some mechanism is postulated that could do the job. How could such memory consolidation or junkyard functions work? What sort of mechanism could govern such a process or processes? One idea is this: for memories to be retained they must be converted from storage in the

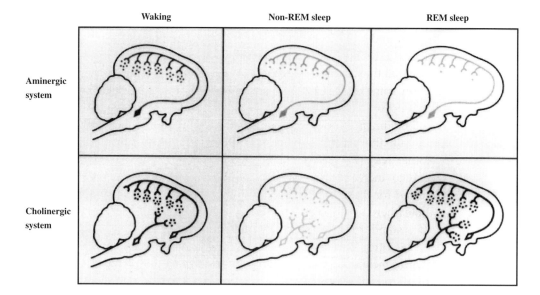

Figure 5.6 During waking both the aminergic and cholinergic neurons are active, and the brain is bathed in high and balanced levels of both kinds of neurotransmitter (dots). In NREM sleep, the balance is retained, but the levels of both decline (fewer dots). Alertness also declines, but the logical and thoughtlike nature of mental activity does not change. In REM sleep, the chemical balance of the brain shifts radically as aminergic neurotransmitter levels plummet to their nadir and cholinergic neurotransmitter levels rise again to waking levels. The effects of this shift are seen in the unique formal features of dreaming, including bizarre events, illogical explanations, and poor recall. Reprinted with permission from Hobson (1989).

halfway house of distributed electrical patterns into stable protein structures within neurons, in particular at the synapses. To get a feel for the need here: imagine your computer crashing and the difference Saves make. The idea is that memory reactivation involves the reactivation of the neural networks whose synaptic strengths have been altered. What happens during REM sleep is that the cholinergic neurons that are on and releasing acetylcholine interact with the temporary but connected electrical synaptic hot spots constituting a memory from the day, and change those hot spots to a more stable form—to some sort of protein structure.[25]

NATURAL FUNCTIONS

Enough theory and data are now on the table to see how I intend the argument for the hypothesis I floated at the beginning of the paper to go. The hypothesis can be formulated somewhat more precisely given what has been said so far. It is that sleep and the phases of the sleep cycle—NREM and REM sleep—were selected for and are maintained by selective pressures. They are adaptations in the biological sense.[26] However, the mental aspects of sleep, the thoughts that occur during NREM sleep, as well as the dreams, and lucid dreams (dreams which contain the awareness that one is dreaming) that occur during REM sleep are probably epiphenomena in the sense that they are serendipitous accompaniments of what sleep is for.

Now some things that were originally selected for to serve a certain function, end up being able—with some engineering modifications—to serve another function. Selection pressures then work as it were to select and maintain the adaptation because it serves both purposes, or to put it another way both the original phenotypic feature and the extended one, serve to increase fitness. For example, feathers were almost certainly selected for thermoregulation, but now selective pressures work to maintain feathered wings because they enable flight.

It is standard in evolutionary biology to say of some "automatic sequelae," pleiotropic or secondary characteristic, that it is a nonadaptation only if it is a concomitant of a trait that was selected for and if in addition no concurrent positive selection or independent modification operate on the trait.[27] So the capacity to fly may have been a sequelae of selection pressures to design efficient thermoregulation, but feathered wings are an adaptation because despite being a secondary characteristic they were (and are) subject to positive selection and modification pressures. But the color of blood and the human chin are common examples of sequelae that are nonadaptations.

The biological notion of an adaptation and even a nonadaptation needs to be marked off from the concept of adaptativeness or functionality. The biological notion is tied to selection pressures which contribute to reproductive success in a particular environment or set of environments.[28] But we also say of mechanical devices, intentional human acts or act-types, and of cultural institutions that they are adaptive or functional. Here we mean that the device, act, act-type, institution does what it is designed to do.[29]

We need to draw one further distinction within the nest of meanings of the terms 'function' and 'functional': this between (1) a causal contribution sense of function and (2) a functional versus dysfunctional sense. So to use Kitcher's example, mutant DNA causing tumor growth is functioning as it is supposed to; it is making the causal contribution we expect, but it is dysfunctional—bad biologically and psychologically for the organism in which the tumor is growing.

Now my argument is this: sleep and sleep-cycling is an adaptation for reasons given above—it restores, conserves, and builds and we can specify some of the specific things it does and the mechanisms these are done by.[30] There is some reason to wonder whether REMing and NREMing, that is the moving or nonmoving of eyes is an adaptation. And there is very good reason to be positively dubious about the adaptive significance of the phenomenal experiences that supervene on REM and NREM–sleep. Dreaming, broadly construed, is pleiotropic, an automatic sequelae, a spandrel. It is doubtful that dream–consciousness once in play as a sequelae of causal processes originating in the brainstem that tickle the visual areas producing REMs was subjected to positive selection pressures and modification. Actually I should put it this way: for reasons discussed earlier the brainstem is designed to activate the visual system to finish building it during the first year of life. Once the system is built the continuation of the activation of the visual system serves no obvious further developmental function. Furthermore, whereas the PGO waves of REM sleep are implicated in the processes of stockpiling neurochemicals for the next day's work, for making what is learned more stable so that it can be remembered, and possibly for trash disposal, there is no reason to believe that these jobs require mentation of any sort.

Assuming, tentatively, that the stabilizing idea is right, there is no phenomenological evidence that as electrical patterns are transformed into protein structures that the associated mentation involves the activation of the thoughts worth remembering. People remember nonsense syllables better after sleep than if tested right after learning but before sleep. But, to the best of my knowledge, people never report p-dreaming about nonsense syllables. Nor do students of mathematics work through the proofs of the previous day in dreams. It may well be that the proof of the Pythagorean theorem would go in one ear and out the other if we didn't sleep in between. But I would place large bets that one will have trouble getting any phenomenological reports of sophomore geometry students dreaming through the steps of the theorem in REM sleep. The point is the PGO waves are causally implicated in the neurochemical stockpiling of amines (serotonin, norepinephrine, etc.) and in setting acetylcholine and its friends to the task of bringing stability to what has been learned. But there is no reason, so far as I can see, to think that the mentation caused by the PGO waves is causally relevant to these processes. The right circuits need to be worked on, but no mentation about the information that those circuits contain is needed—and typically such mentation does not occur. The visual, auditory, propositional, and sensory–motor mentation that occurs is mostly noise. One might be drawn to a different conclusion if the mentation was, as it were, about exactly those things one needs to stabilize for memory storage, but phenomenologically that seems not to be the case. It can't be the actual thoughts that occur during the bizarre mentation associated with REMing which the system is trying to stabilize, remember, or store—most of that is weird. Sure some is not weird, and of course the so-called day's residue makes occasional appearances in dreams. It would be surprising if it didn't—it's on your mind. The incorporation of external stimuli is also easily explained—the system is designed to be relatively insensitive to outside noise, but it would be a pathetic survival design if it was completely oblivious to outside noise. So dripping faucets, cars passing on the street outside, are being noticed but in a degraded way, they won't wake you, but a growling predator at your campsite will.

P-dreams are a special but by no means unique case where the epiphenomenalist suspicion has a basis. P-dreaming is to be contrasted with cases where phenomenal awareness was almost certainly selected for. Take normal vision, for example. It is, I think, a biological adaptation. Blindsighted persons who have damage to area V1 in the visual cortex get visual information but report no phenomenal awareness of what is in the blindfield. They behave in degraded ways towards what's there if asked to guess what is there, or reach for it, which is why we say they are getting some information. But the evidence suggests that the damage to V1, which is essentially implicated in phenomenal visual awareness, explains why the performance is degraded.[31] And this suggests that the phenomenal side of vision is to be given an adaptationist account along with, and as part of, an adaptationist account of visual processing generally. This is not so with p-dreaming.

INVENTED FUNCTION

The phenomenal aspects associated with sleeping are nonadaptations in the biological sense. The question remains does p-dreaming serve a function. If it does it is a derivative psychological function constructed via mechanisms of cultural imagination, and utilization of the fact that despite not serving a direct biological or psychological function,[32] the content of dreams are not totally meaningless (this has to do with what the system is trying to do during sleep and sleep-cycling) and thus dreams can be used to shed light on mental life, on well-being, and on identity. What I mean by the last remark is this: the cortex's job is to make sense out of experience and it doesn't turn off during sleep. The logically perseverative thoughts that occur during NREM sleep are easy for the cortex to handle since they involve real, but possibly "unrealistic" ideation about hopes, worries, and so on. Indeed, from both a phenomenological and neuroscientific perspective, awake mentation and NREM sleep mentation differ more in degree than in kind.

REM mentation is a different story. It differs in kind. Phenomenologically and brainwise it is a radically different state—closer to psychosis than to any other mental state types. Still the cortex takes what it gets during REM sleep and tries to fit it into the narrative, scriptlike structures it has in place about how my life goes and how situations, for example, restaurant scenes, visits to amusement parks, to the beach, and so on, go. Indeed, Hobson's group[33] is now studying the underlying grammar of dream mentation and the evidence is fascinating. Settings and scenes are fairly unconstrained, plot is intermediate, and characters and objects tend to be transformed in the most gradual ways. So one's true love might be in China one second and Brazil the next, there will be some work to make the plot coherent which will be hard, so what she's doing in China and Brazil might involve an odd plot shift but not a completely incoherent one. But you may also find that she has turned into an old true love, but probably not into a sofa, in the transcontinental scene shift.

Now mentation about one's current true love and one's old true love might be informative about what's on your mind, or it might be uninformative, just the best story line the cortex can bring to the materials offered up, but this could be true and such a dream could still be a good place to start from in conversation with a psychiatrist if your love life is going badly, or if you are suffering from any psychosocial malady. This may be because the content itself is revealing—remember there is a top-down/bottom-up struggle going on between the noise from below and the cortex which is trying to interpret the noise in terms of narrative structures, scripts, and self-models it utilizes in making sense of things while awake. It could be that the dream is uninterpretable, or is meaningless as an intentional narrative construct, or has nothing like the meaning we are inclined to assign, but is nonetheless a good conversation starter for someone trying to figure out the shape of his life. Obviously, from what I have said, the cortex is expressing what's on your mind, how you see things. Your dreams are expressions of the way you uniquely cast noise that someone else would cast differently. So things remain the same, phenomenal–dreams make a difference to your life. They may get you thinking in a certain way upon waking. You may find yourself in a hard to shrug-off mood despite learning that the imagined losses of loved-ones causing that mood were dreamed and didn't really happen.[34] You may be inspired to write a poem or a mystical text, and you may work at the project of interpretation. This is not silly. What you think while awake or while asleep is identity-expressive. The project of self-knowledge is important enough to us that we have learned to use the serendipitous mentation produced by a cortex working with the noise the system produces to further the project of self-knowledge and identity location. This is resourceful of us.

Spandrels serve functions even though they are sequelae of the design the architect is focused on putting in place. Being a spandrel doesn't make something nonfunctional. Spandrels are beautiful accompaniments of dome and arch designs. They are probably nice enough to look at as a matter of structural geometry if left plain but putting frescoes on them makes the world more beautiful. So it is with dreams. They are mental spandrels, but we can work them into all sorts of useful, creative, and fun things we have learned to do in our lives. One final point: purely informational. After the hormonal poisoning accompanying adolescence returns to normal levels, only about 6 in 100 dreams have romantic or sexual content and involve feelings of affection or eroticism. I thought I'd tell you that because you were probably wondering.

ACKNOWLEDGMENTS

This paper was written as the Presidential Address for the Society for Philosophy and Psychology, 1994, June. It received a test run as the kickoff speech for the Tucson, Arizona conference on "Consciousness," April, 1994, and again in the Philosophy-Cognitive Science seminar at East Carolina University. Deborah Stahlkopf has been a great help listening to my ideas and helping me sort through the literature on the function question. Thanks also to Allan Hobson, Greg Cooper, Robert Brandon, Patricia Churchland, Gail Marsh, and especially to David Sanford who has patiently listened to each day's new discoveries about sleep and dreams.

NOTES

1. See Matthews, G. B. 1981. On being immoral in a dream, *Philosophy* 56:47–54.

2. St. Augustine. *Confessions*, R. S. Pine-Coffin (trans.) New York: Penguin, pp 233–34 (10.30).

3. Malcolm, N. 1959. *Dreaming*. London: Routledge and Kegan Paul.

4. Dennett, D. 1976. Are dreams experiences? *Philosophical Review* 151–71.

5. But Malcolm and Dennett are right to express the worry that dream reports are putrid evidence that this is so. We need a neuroscientific account to shore up the phenomenological confidence that dreams are experiences that take place while asleep. We now, I claim, have such a theory, although I won't set out the argument here (I have in Flanagan, O. "Towards a unified theory of consciousness, or what dreams are made of," forthcoming in *Scientific Approaches to the Question of Consciousness: 25th Carnegie Symposium on Cognition*, edited by J. Cohen and J. Schooler, Hillsdale, NJ: Lawrence Erlbaum; also see Hobson, J. A. 1988. *The dreaming brain* New York: Basic Books and *Sleep* (New York: Scientific American Library, 1989; and unpublished papers by Hobson, J. A., et al., Dreaming: A Neurocognitive Approach, 1993, for the theory—AIM—I depend on). Some of this research has just been published in *Consciousness and Cognition*, vol. 3, no. 1, 1994. Dreams turn out to be experiences on that theory and thus to belong to the heterogenous set of experiences we call 'conscious.'

6. Gould, S. J. 1984. Covariance sets and ordered geographic variation in *Cerion* from Aruba, Bonaire and Curacao: A way of studying nonadaptation. *Systematic Zoology* 33(2):217–37.

7. Gould, S. J., and R. C. Lewontin. 1979. The spandrels of San Marco and the Panglossian paradigm. *Proceedings of the Royal Society of London, series B*, 205:217–37.

8. Gould, S. J., and E. Vrba. 1982. Exaptation: A missing term in the science of form. *Paleobiology* 8:4–15.

9. Flanagan, O. 1992. *Consciousness reconsidered* Cambridge, MA: MIT Press.

10. When I say let the psychologists and cognitive scientists have their say I mean also to include amateurs—let folk wisdom be put out on the table along with everything else.

11. McGinn, C. 1989. "Can we solve the mind–body problem?" *Mind* 98:349–66, reprinted in McGinn 1991. *The problem of consciousness* Oxford, England: Blackwell.

12. The case of dreams is a case in point. If we ever came to have really good theoretical reasons for thinking that dreams were not experiences, they might well seem less like experiences. It is hard to imagine giving up the idea that there are perceptual experiences since such experiences take place in the specious present (or so it strongly seems); but even dreamers will admit that they are remembering both the alleged experience and the content of the alleged experience. At the Carnegie-Mellon Conference on Consciousness in the Spring of 1993, Clark Glymour proposed that theory should be as biologically constrained as possible; and he expressed the worry that cognitive information-processing models often fail to attend to biological realism. I quite agree with the normative point. But our present knowledge of the brain is thin and often hard to interpret. Sometimes it is hard to know what the neuroscientific data are or mean and thus hard to know how they should constrain our theories. For example, many effective antidepressants work by affecting dopamine or serotonin levels, absorption rates, and so on. But in many cases neither the FDA, nor the pharmaceutical companies know exactly how these

drugs work. Judgments about what they do at the psychological level, what some of their phenomenological and physiological side effects are, as well as assessments about overall safety and effectiveness, are made without anything approaching complete understanding at the level of brain chemistry.

13. Crick and Koch have suggested that subjective awareness is linked to oscillation patterns in the 40-Hz range in the relevant groups of neurons, that is, neurons involved in a certain decoding task "synchronize their spikes in 40-Hz oscillations" (Crick, F. and C. Koch. 1990. Towards a neurobiological theory of consciousness. *Seminars in the Neurosciences* 2 263–75). 40-Hz oscillations have been found in single neurons and neural nets in the retina, olfactory bulb, in the thalamus, and neocortex. Recently Llinás, R. and D. Paré. 1991. Of dreaming and wakefulness. *Neuroscience* 44(3):521–35, have produced strong evidence that such oscillation patterns characterize REM sleep. Llinás, R. and Ribary. 1993. Coherent 40-Hz oscillation characterizes dream state in humans. *Proceedings of the National Academy of Science* 90:2078–81, report that "during the period corresponding to REM sleep (in which a subject if awakened reports dreaming), 40-Hz oscillation patterns similar in distribution phase and amplitude to that observed during wakefulness is observed." The second finding of significance they express this way: "during dreaming 40-Hz oscillations are not reset by sensory input . . . We may consider the dreaming condition a state of hyperattentiveness in which sensory input cannot address the machinery that generates conscious experience." Within Hobson's theory the 40-Hz oscillations pertain to (A) activation level, while the tuning out of external stimuli is explained by the mechanisms of input–output gating (**I**). The main point for present purposes is that the reason dreams seem like conscious experiences is because they are conscious experiences and they are like awake conscious experiences in certain crucial respects.

Llinás and Ribary suggest this unifying hypothesis: 40-Hz activity in the nonspecific system comprised of the thalamo-cortical loop provides the temporal binding of contentful states that involve 40-Hz oscillations in the areas devoted to particular modalities. That is, the neural system subserving a sensory modality provides the *content* of an experience and the nonspecific system consisting of resonating activity in the thalamus and cortex provides "the *temporal binding* of such content into a single cognitive experience evoked either by external stimuli or, intrinsically during dreaming" (1993, p. 2081). Llinás and Paré write "it is the dialogue between the thalamus and the cortex that generates subjectivity" (1991, p. 532).

These data and hypotheses, in light of other data and theory, increase the credibility of the claim linking REM sleep with vivid experiences. Whether it is really true that dreams are experiences depends, of course, on whether it is true that 40-Hz oscillations turn out to be a marker, a constituent component, or a cause (which one they are is, of course, very important) of vivid conscious experiences. The point is that the neuroscientific data push credible theory in a certain direction. If these data bring us closer to the answer to one question they open a host of others, suggesting occasional answers—a sure sign of a progressive research program.

One might wonder for example, whether 40-Hz oscillation patterns will turn out to be necessary or sufficient for experience or enable us to differentiate different kinds of experiences. There is the possibility that alive human beings might always be in some experiential state or other, that is, that we are never wholly unconscious—if, that is, 40-Hz patterns are sufficient for experience. If this sounds like an incredible prospect, remember that persons awakened from NREM sleep often report having experiences—albeit experiences lacking the vivacity of post-REM reports. And sleep talking and sleep walking are well known to take place during NREM sleep (postural muscles are turned off during REM sleep) and it is obscure whether, or in what precise sense, sleep walkers and talkers are experiential blanks. Globality of 40-Hz activity may turn out to be the relevant feature of robust conscious experiences, not the mere presence or absence of some 40-Hz activity (Llinás and Paré 1991 p. 527). Now one worry for the 40-Hz necessary condition hypothesis is this: the mentation occurring during NREM when measured by EEG doesn't appear to involve the 40-Hz oscillations despite involving mentation, that is, 'p-dreaming.' So 40-Hz oscillations may be a reliable marker of certain kinds of conscious mentation but not necessary for all mentation (see Steriade, M., D. A. McCormick, and T. J. Sejnowski. 1993. Thalamocortical oscillations in the sleeping and aroused brain. *Science* 262:679–85). On the other hand, when measured with MEG, one does find 40-Hz oscillations, but much attenuated in amplitude and we don't pick up much in the way of amplitude modulations.

14. See the exchange between Kathleen Emmett, 1979. Oneiric experiences, *Philosophical Studies* 35:315–18). Dreams are fodder for skeptics about the prospects for a theory of consciousness for a number of reasons. One route to skepticism is simple and straightforward. Common sense says that "conscious" involves, among other things, being awake. But dreaming takes place during sleep, and, thus, by the distinguished deliverances of conceptual analysis, dreams cannot be conscious experiences. But common wisdom also says that dreams are experiences that take place during sleep, so our common sense taxonomy of "consciousness" is worse than a hodge podge; it's riddled with inconsistency from the start.

15. Larry Rosenwald pointed out that examples of obsession about whether the car lights are off, or about soon to come conversations with the boss might be more familiar.

16. The exciting work of Jouvet, M. 1962. Récherches sur les structures nerveuses et les mécanismes résponsables des différentes phases du sommeil physiologique, *Archives Italiennes de Biologie* 100:125–206), Aserinsky, E. and N. Kleitman. 1955. Two types of ocular motility occurring in sleep. *Journal of Applied Physiology* 8:1–10, Dement, W. 1958. The occurrence of low voltage, fast, electroencephalogram patterns during behavioral sleep in the cat. *Electroenceph. clin. Neurophyslol.* 10:291–96 in the late 1950s led to the identification of "dream mentation" with REM sleep. But this it appears is a mistake: NREM sleep is also associated with reports of mentation, and although the phenomenological content of such mentation is mundane and fairly nonbizarre, involving such things as worries about something that needs to be done the next day, subjects do not say they were just thinking about what they needed to do the next day, but that they were "dreaming" about it (Foulkes, D. 1962. Récherches les structures nerveuses et les mécanismes résponsables des différentes phases du sommeil physiologique. *Archives Italiennes de Biologie*, 100:125–206; Herman, J. H., S. J. Ellman, and H. P. Roffward. 1978. The problem of NREM dream recall re-examined. In *The Mind in Sleep*, edited by Arkin, A., Antrobus, J. and Ellman, S. Hillsdale, NJ: Lawrence Erlbraum. Indeed if one thinks that mentation is dreaming only if it occurs during REM sleep then people are even more disastrously bad at giving dream reports than most have thought: for many reported dreams are of the "I was supposed to do x, but didn't" sort and the evidence points to greater likelihood that such mentation occurs during NREM sleep than during REM sleep. It was Foulkes who led the credible movement, not yet won, to disassociate the virtually analytic connection that had been drawn and continues to be drawn by most researchers between REM and dreaming. Indeed, someone—I can't remember who it was—founded "The Society for the prevention of cruelty to NREM sleep." The idea was to let dreaming be, as a first pass, any mentation that takes place during sleep, and go from there. The frequency of NREM mentation casts doubt on the idea that dreaming is a natural kind, although we may well be able to discern differences between NREM mentation and REM mentation. Nonetheless, some researchers (Herman, J. H., S. J. Ellman, and H. P. Roffward, *op. cit.*) suggest that the conclusion to be gained from expanding the concept of dreaming is this "The hope that one stage of sleep, or a given physiological

marker, will serve as the sole magic key for vivid dream mentation has all but faded from view." (1978, p. 92). The overall point is this: P-dreams as we normally think of them include both the perseverative, thoughtlike mentation of NREM sleep and the bizarre and fantastic mentation of REM sleep, but the foregoing scientific considerations suggest reasons for restricting the use of the term "dreams" for certain scientific and philosophical purposes only to REM–mentation, but there are some reasons against doing this—for example the hallucinatory night terrors of Stage III and IV NREM sleep. A further issue is this: since we are always in either NREM or REM sleep or awake it is possible that we are never unconscious. Alternatively it is possible that there are times in both NREM and REM sleep when virtually nothing thoughtlike is happening—or perhaps we are informationally sensitive (this could explain how sleep walking without awareness could be possible—similar to the way blindsight patients process some information about what's in the blind field without being p-aware of, or experientially sensitive to what's in that field). No one knows for sure.

17. Hobson, J. and R. Stickgold. 1994. Dreaming: A neurocognitive approach. *Consciousness and Cognition* 3:1–15.

18. Llinás, R. R. and D. Paré. 1991. Of dreaming and wakefulness. *Neuroscience,* 44(3):521–35. This helps explain why prosopagnosiacs don't report dreaming of faces and why people with right parietal lobe lesions who can't see the left side of the visual field report related deficits in their dream imagery (p. 524). On the other hand, it tells us something about memory that visual imagery sustains itself better in both the dreams and the awake experiences of people who develop various kinds of blindness in later life.

19. Once such imagery is overrated, dreaming is equated with REMing and the sensorily dull but thoughtlike mentation of NREM sleep is ignored and the foolish inference that NREM sleep especially Stage IV NREM sleep is a period of unconsciousness.

20. See Kryger, M. H., T. Roth, and W. Dement. 1994. *Principles and practice of sleep medicine*, 2nd ed. London: W. B. Saunders.

21. Posner, M. I., and S. E. Petersen. 1990. The attention system of the human brain. *Annual Review of Neuroscience* 13:25–42.

22. Hobson, J. A. *Sleep, op. cit.* footnote 6.

23. See, for example, Crick, F. and G. Mitchison. 1983. The function of dream sleep. *Nature* 304. Hopfield, J. J., D. I. Feinstein, and R. G. Palmer. 1993. "Unlearning" has a stabilizing effect in collective memories, *Nature* 304. Steriade, M., D. A. McCormick, and J. Sejnowski (*op. cit.*, footnote 14).

24. See Hobson, J. A., *Sleep, op. cit.* footnote 6.

25. See West-Eberhard, M. J. 1992. "Adaptation: Current usages" and Burian, R. 1992. "Adaptation: Historical perspectives," in *Keywords in Evolutionary Biology*, edited by E. F. Keller and E. Lloyd, Cambridge, MA: Harvard University Press pp. 13–18 and pp. 7–12. Also see Kitcher, P. (forthcoming). *Function and Design* (forthcoming) and Godfrey-Smith, P. (forthcoming). *A modern history theory of functions*.

26. West-Eberhard, M. J. *op.cit.*, footnote 26.

27. Brandon, R. 1990.*Adaptation and environment.* Princeton, NJ: Princeton University Press.

28. Kitcher argues that this idea unifies attributions of *function* across biological and nonbiological contexts. See his forthcoming paper "Function and Design."

29. I haven't even mentioned some of the other systems that are being worked on in sleep, for example, the immune system. People who are kept from sleeping die, not from lack of sleep as such but from blood diseases. Without sleep the immune system appears to breakdown.

30. See Block, N. (forthcoming). *Brain and behavioral sciences,* for an objection to this sort of reasoning; and see my response in my 1992, *op. cit.* pp.145–52.

31. I need to be clear here the very same processes that produce p-dreaming as an effect produce the switch in neurotransmitter ratios that do serve an important psychological function. But p-dreams don't serve this function.

32. Stickgold, R., C. D. Rittenhouse, and J. A. Hobson. 1994. Constraint on the transformation of characters, objects, and settings in dream reports. *Consciousness and Cognition.* 3:100–13, and Dream splicing: A new technique for assessing thematic coherence in subjective reports of mental activity. *Consciousness and Cognition* 3:114–28.

33. I have talked very little about the activation of limbic areas during dreams. Hobson's group finds that emotions and moods are coherently coordinated with dream content. This, it seems to me, is good evidence of cortical domination of the plot line and of limbic cooperation.

REFERENCES

Armstrong, D. 1980. *The nature of mind and other essays.* Ithaca, NY: Cornell University Press.

Block, N. 1995. Forthcoming. On a confusion about a function of consciousness. *Behavioral and Brain Sciences.*

Brentano, F. 1924/1973. *Psychology from an empirical standpoint.* Translated by A. Rancurello, D. B. Terrell, and L. McAlister. New York: Humanities Press.

Carruthers, P. 1989. Brute experience. *The Journal of Philosophy* 86(5):258–69.

Churchland, P. 1985. Reduction, qualia, and the direct introspection of brain states. *The Journal of Philosophy* 82(1):2–28.

Churchland, P. 1988. *Matter and consciousness.* Revised edition. Cambridge, MA: MIT Press.

Cummins, R. 1991. *Meaning and mental representation.* (paperback edition.) Cambridge, MA: MIT Press.

Dennett, D. 1991. *Consciousness explained.* Boston: Little, Brown Co.

Dretske, F. 1993. Conscious experience. *Mind* 102(406):263–83.

Flanagan, O. 1992. *Consciousness reconsidered.* Cambridge, MA: MIT Press.

Güzeldere, G. 1995. (Forthcoming) *The many faces of consciousness.* Stanford University, Ph.D. dissertation.

James, W. 1890/1950. *The principles of psychology,* Vol. 1. New York: Dover Publications.

Locke, J. 1689/1959. *An essay concerning human understanding* Vol. 1. New York: Dover Publications.

Lycan, W. 1995. (Forthcoming) "Consciousness as internal monitoring." In *Philosophical Perspectives,* edited by J. Tomberlin.

Lyons, W. 1986. *The disappearance of introspection.* Cambridge, MA: MIT Press.

Nisbett, R, and T. Wilson. 1977. Telling more than we can know: Verbal reports on mental processes. *Psychological Review* 84(3):231–58.

Rosenthal, D. 1990. *Explaining consciousness.* (Manuscript) CUNY, Graduate School.

Rosenthal, D. 1986. Two concepts of consciousness. *Philosophical Studies* 94(3):329–59.

Searle, J. 1992. *The rediscovery of the mind.* Cambridge, MA: MIT Press.

Wilson, T., and D. Dunn. 1986. Effects of introspection on attitude-behavior consistency: Analyzing reasons versus focusing on feelings. *Journal of Experimental and Social Psychology* 22:249–63.

II Cognitive Science

Cognitive science is carried on by a diverse assembly of scholars and scientists from among the fields of philosophy, psychology, linguistics, anthropology, computer science, and neuroscience who share a mutual interest in the nature of the human mind (for example, Gardner 1985, Posner 1989, Stillings *et al.* 1987). Recent symposia and books (Bock and Marsh 1993, Revuonso and Kamppinen 1994) illustrate cognitive approaches to consciousness along issues of mental representation, the varying demands of different cognitive tasks, and theoretical constructs of information processing. Hirst (1995) notes that most empirical cognitive studies on consciousness attempt to answer one or more of six questions:

1. Can individuals be aware of their cognitive processes as well as the product of these processes?
2. What are the limits on conscious experience?
3. Do individuals go beyond the information provided in a stimulus array?
4. What and how much information can be processed unconsciously?
5. What guides unconscious information processing?
6. How does unconscious processing affect conscious behavior?

Chapters in this section attempt to answer some of these questions.

In Chapter 6, John Kihlstrom blends theory and empirical data regarding unconscious processes in social interaction. He presents the general features of a cyclic social interaction, illustrating how interpersonal behavior is cognitively mediated, but not entirely by processes accessible to awareness. Kihlstrom classifies the cognitive unconscious into domains of procedural knowledge (strictly unconscious), declarative knowledge which is preconscious, and declarative knowledge which is subconscious. Citing recent social psychology research on both healthy and brain–damaged subjects, Kihlstrom concludes preconscious processes are extensively involved in various aspects of social cognition (for example, attitudes, impressions) and behavior (compliance, aggression). He closes with an examination of the analytical limits on preconscious processing of socially relevant information.

Thaddeus Cowan addresses in Chapter 7 how conscious images are projected into space. He considers perceptual phenomena as sequences of events: stimulus → neurological representation → projection of the conscious image. Cowan cites four examples that highlight the conscious projection: effects of movements on the appearance and duration of a prolonged visual afterimage, "sensory capture" in the conscious experience of taste, location of the conscious experience of touch, and the conscious experience of projection of visual objects into space (as opposed to being experienced at the receptor site). The perception sequences are modeled topologically using a branch of mathematics called "braid theory." Cowan suggests that the efferent projection mechanism itself is close to consciousness.

In Chapter 8, John Boitano explores Gerald Edelman's (1989, 1992) biological account of the different psychological functions necessary for the development of consciousness. Edelman's proposals have captured scientific and popular interest, generating both praise and controversy. Boitano's review focuses on some central features of Edelman's theory of consciousness: (1) The evolutionary principle of natural selection through survival of favorable variants applied to neural firing patterns ("neural Darwinism"); (2) Edelman's global mapping construct, in which multiple reentrant local maps connecting sensory and motor areas are also interconnected with nonmapped brain areas (cerebellum, hippocampus, basal ganglia), which order sensory and motor activity; and (3) Edelman's distinction between primary consciousness (immediate, ongoing, present-time) and higher-order consciousness (awareness of past and future, self concept, language, and awareness of one's own consciousness). Boitano concludes by reviewing some relevant empirical evidence, and suggesting where modifications of the theory are indicated.

Cognitive psychology has, in general, focused on neurological and cognitive–behavioral processes and structures supporting experience, while overlooking direct examination of subjective experience itself. In Chapter 9 David Galin attempts to redress this neglect, and proposes a taxonomy for components of subjective experience based on the type of information each carries. Providing examples from domains of action, emotion, and metacognition, Galin argues that subjective experience of an object, event, or idea is a complex that includes feature awareness, informational aspects (for example, meaning), evaluative or metacognitive components, and nonconscious background knowledge related to other topics or goals. Complementing the prevalent "moving spotlight" metaphor for consciousness, Galin offers a precise and useful characterization of the experiential structure being illuminated.

In the final chapter of this section, Karl Pribram examines the biological roots and social usages of the varieties of conscious experience. He begins by reviewing various historical developments in the present century that link aspects of consciousness with neuronal operations. Pribram argues that the "language" used by the brain in operations relevant to consciousness is not exclusively that of neuronal switching, but relies heavily on microprocesses occurring in synaptodendritic fields. As sources for his argument, Pribram cites case histories of patients with neurological disorders, theories of feedback and feedforward mechanisms in brain organization, his own famous "holographic" model of neural image processing (and his response to criticisms of this model), and from recent research concerning both states and contents of consciousness.

Chapters in this section are concerned with organization, representation, and qualities of conscious and nonconscious cognitive processes. John Kihlstrom shows how unconscious processing can affect conscious behavior, and Thaddeus Cowan describes how our perception of spatial reality is consciously projected and suggests a relationship between the projected perception and reality itself. John Boitano presents Gerald Edelman's theories for organizational strategies for consciousness, and David Galin portrays the various components of subjective experience. Karl Pribram gives an historical account of neurological representations of consciousness, and nominates subneural synaptodendritic field dynamics as the primary "language" of consciousness.

In the next section, some abnormal states of consciousness are examined.

REFERENCES

Bock, G. R., and J. Marsh eds. 1993. *Experimental and theoretical studies of consciousness (Ciba Foundation Symposium 174)*. Chichester, England: John Wiley & Sons.

Edelman, G. M. 1989. *The remembered present*. New York: Basic Books.

Edelman, G. M. 1992. *Bright air, brilliant fire*. New York: Basic Books.

Gardner, H. 1985. *The mind's new science: A history of the cognitive revolution*. New York: Basic Books.

Hirst, W. 1995. Cognitive aspects of consciousness. In *The cognitive neurosciences*, edited by M. S. Gazzaniga. Cambridge, MA: MIT Press, pp 1307–19.

Posner, M. I. 1989. *Foundations of cognitive science*. Cambridge, MA: MIT Press.

Revuonso, A., and M. Kamppinen, eds. 1994. *Consciousness in philosophy and cognitive neuroscience*. Hillsdale, NJ: Lawrence Erlbaum Associates.

Stillings, N. A., M. H. Feinstein, J. L. Garfield, E. L. Rissland, D. A. Rosenbaum, S. E. Weisler, and L. Baker-Ward. 1987. *Cognitive science: An introduction*. Cambridge MA: MIT Press.

6 Unconscious Processes in Social Interaction

John F. Kihlstrom

It is a cardinal principle of social psychology that an individual's experience, thought, and action is highly sensitive to details of the social situation in which it occurs. This principle accounts for the extraordinary amount of cross-situational variability that is characteristic of human social behavior: people appear to be quite inconsistent in what they do and say from one context to the next. However, this does not mean that human behavior is inherently unpredictable; nor, as the behaviorists might have had it, that persons are at the mercy of environmental stimuli and reinforcement schedules. Rather, we now know that the effects of the environment are largely mediated by the cognitive processes by which the individual gives meaning to each situation which he or she encounters; a meaning that can differ widely depending on the precise set of cues available in the setting and the cognitive resources that the individual brings to it.

Thus, our behavior is not affected by the situation as it would be objectively described by a third-person observer, but rather as it is subjectively construed by the actor him- or herself. Social behavior, then, is not inconsistent, nor is the lawfulness of human behavior given by tracking the functional relationships between stimulus and response. Rather, the consistencies in social behavior are to be found at the cognitive level: people behave similarly across situations they construe as similar; their behavior changes when their construal changes. For this reason, cognitive social psychologists have long been interested in how social knowledge is represented and processed in the mind: how we form impressions of ourselves, others, and the social context in which we interact; how we explain, and give meaning to, others' and our own behaviors; how we predict what is going to happen next as an interaction unfolds; and how we plan our responses to events that actually transpire, and the actions that we hope will achieve our short- and long-term goals.

THE GENERAL SOCIAL-INTERACTION CYCLE

The cognitive perspective on social interaction is illustrated in Figure 6.1, which depicts what might be called the *general social-interaction cycle*. The scheme is also known as the general social interaction sequence (see Darley and Fazio 1980), but I prefer to emphasize its cyclical aspects. This scheme can be used to represent any dyadic (two-person) interaction, from something as mundane to buying a toothbrush to something as monumental as proposing marriage. The two participants have been assigned to the respective roles of actor and target. In some respects this assignment is arbitrary, because—as will be clear in a moment—both participants are simultaneously actor and target of the other's action; but for purposes of this illustration, the actor role is assigned to the person who initiates the interaction.

In the first phase, the actor enters the situation, which may be defined as the immediate context in which he or she physically encounters the target. At this point, the actor has a goal, or something that she intends to accomplish—like asking the target for a date Friday night; but she also carries with her a fund of social knowledge concerning herself, the target, and other more generic factual information that is relevant to her current goals; and she also carries a repertoire of cognitive and motor skills that she can use in the course of the interaction. She needs to know, for example, whether the target is currently dating someone else, and if he has expressed any interest in her. She also needs to know how to start a conversation and how to bring it around to the subject of Friday night. This fund of declarative and procedural social knowledge may be called *social intelligence* (Cantor and Kihlstrom 1987).

In the second phase, the actor forms an impression of the target in context. Does he still seem interested? Is this a good time to ask? In doing so, she combines her preexisting knowledge (retrieved from memory) with information derived from the immediate situation (acquired through perception). On the basis of this impression, she either approaches the target or shies away, and she either pops the question or not. In short, she acts on the basis of her impression of the situation. Let's assume that she asks him for a date.

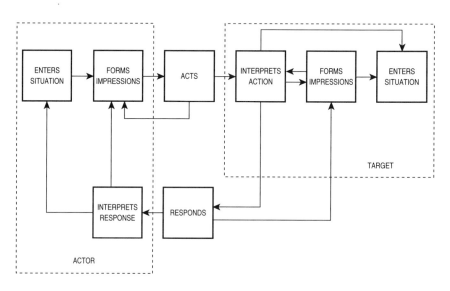

Figure 6.1 The general social interaction cycle (adapted from Darley and Fazio 1980).

Once she has acted, our attention turns to the target. He too, has entered the situation, either actively or passively, and he too comes with a fund of preexisting knowledge. And he, too, has been forming an impression of the situation in which he finds himself. Now that she has actually asked him for a date, the situation is clarified. As it happens, he is free Friday night, but he stops to wonder whether he should play "hard to get," or perhaps wait for a better offer from someone else. In the end, he decides to hold his options open for Friday night; but he doesn't want to spurn the actor entirely. Therefore, on the basis of his impression of the total situation, he responds to her initial salvo: he says he can't see her Friday, but proposes Saturday instead.

Now attention shifts back to the actor: she has to revise her impressions. Perhaps he is seeing someone else after all, which is why he is busy Friday night; but he is obviously not uninterested in her, or else why would he offer to substitute Saturday night? As it happens, she's free Saturday night, but if she says yes, she immediately communicates that she doesn't presently have a date for either Friday or Saturday night. Is this something he should know? If she agrees to Saturday night, does she make herself a pawn in whatever game he might be playing with the woman he's dating Friday night? Or is the Friday-night woman a pawn in a game he's playing with *her*? In the clinch, she takes a chance and accepts. Now the ball is back in the target's court. And so it goes, in a cycle of exchanges in which each participant is trying to make sense of what the other one is doing, and planning and executing behaviors in accordance with this understanding.

Of course, in addition to the cycle going on between actor and target, there are also cycles going on within each of these participants, as their own behavior feeds back to them and changes their impressions of themselves. Thus, for example, the actor may have wondered if she had the nerve, and the social skill, required to ask this man for a date: now she knows she does (this is self-efficacy information which she will now use to guide other social interactions in the future; Bandura 1986). Similarly, the target may never before have been in a position where he had to negotiate two overlapping dating relationships, and now he knows that this particular skill is in his repertoire. In any event, the point is that each participant behaves in accordance with his or her construal of the other, and of him- or herself; each participant's construal is modified by the other's behavior; and by his or her own behavior as well—a situation reflecting what Bandura (1986) has called *reciprocal determinism.*

Of course, not all of the mediators of social interaction are cognitive in nature. The participants' feelings and motives play important roles as well. Each participant enters the situation with a particular goal in mind, and other goals arise very quickly as the interaction unfolds. Furthermore, the participants' behaviors are determined by the attitudes that each has toward the other and the emotions that each arouses in the other. This situation fits nicely with the general doctrine of mentalism that underlies contemporary psychology: what we do is caused by what we think, feel, and want.

THE PSYCHOLOGICAL UNCONSCIOUS

In analyzing the cognitive, emotional, and motivational determinants of social interaction, and other forms of behavior, most research has focused on conscious mental states. This emphasis parallels that of psychology in general, which in its beginnings, and again since the outbreak of the cognitive revolution, has focused on conscious percepts, memories, and thoughts. However, in recent years it has become increasingly clear that behavior, including social behavior, is also influenced by mental structures and processes that operate outside of phenomenal awareness and voluntary control. These unconscious aspects of cognition, emotion, and motivation may be called the *psychological unconscious* (Kihlstrom 1990).

Elsewhere, I have distinguished among three different aspects of the psychological unconscious (Kihlstrom 1984, 1990, 1994b). There are, first of all, procedural knowledge structures that operate automatically, without conscious intent, and whose execution consumes little or no attentional capacity. Procedural knowledge is unavailable to introspective phenomenal awareness, and can be known only indirectly, by inference. It is unconscious in the strict sense of the term. Second, there are cases of preconscious processing, which involve declarative rather than procedural knowledge. Most demonstrations of implicit memory (Schacter 1987) and implicit perception (Kihlstrom, Barnhardt, and Tataryn 1992) fall into this category, because they involve stimuli or memory traces that have been degraded to such an extent that they are not accessible to conscious awareness. But there is a third category of unconscious influence, observed in hypnosis and states of pathological dissociation, where the percepts and memories are not degraded in any sense—yet they are still inaccessible to awareness. Following the tradition of James, Janet, and Prince, I refer to them as *subconscious* or *coconscious*. Subconscious processing is theoretically important, because it indicates that consciousness is not merely a matter of attention or activation, or the engagement of particular brain structures or neural patterns. In my view, subconscious processing indicates that consciousness requires that a connection be made, and preserved, between two activated mental representations: of an event, and of the self as the agent or experiencer of that event (Kihlstrom 1994a).

The literature documenting the role of strictly unconscious processes in social cognition and interpersonal behavior is huge (for reviews see Bargh 1994, Uleman and Bargh 1989). It is by now generally accepted that attitudes, impressions, and other social judgments, as well as aggression, compliance, and other social behaviors, are often mediated by automatic processes that generally operate outside our awareness and voluntary control. Most of this literature, however, assumes that these automatic processes operate on cognitive contents—mental representations in perception and memory—that are themselves accessible to consciousness. This is a more controversial matter in social psychology, just as it is in cognitive psychology (Greenwald 1992, Kihlstrom 1990, 1994b). Accordingly, in what follows I will sample studies in which implicit percepts and memories mediate the cognitive, affective, or behavioral aspects of social interaction.

SOCIAL JUDGMENT AS IMPLICIT MEMORY

Consider, for example, the brain–damaged patient Boswell, studied by Damasio and his colleagues (Damasio, Tranel, and Damasio 1989). By virtue of bilateral damage to temporal and basal forebrain structures resulting from herpes encephalitis, Boswell is densely amnesic: he is completely disoriented as to time and place, and fails to recognize health professionals who have worked with him for as long as 13 years. Nevertheless, he is able to make accurate social judgments about the people around him. In an experiment, three staff members were instructed to behave in a consistently positive, neutral, or negative manner toward Boswell. Later, Boswell was presented with pictures of the staff members' faces, paired with the faces of people he had never met, and asked which he liked the best, whom he would approach for rewards, treats, and favors.

Although Boswell never recognized the faces as familiar, he consistently chose the individual who had treated him well, and rejected the person who had treated him badly. These choices are clearly related to his past experiences with these individuals. Boswell's behavior illustrates the dissociation between explicit and implicit memory (Schacter 1987): his choices are clearly affected by past experiences, in the absence of conscious recollection of those experiences. Put another way, Boswell's past social interactions are unconsciously influencing his present ones. It's very much the situation portrayed by the Roman epigrammatist Martial, as freely translated by the seventeenth-century English poet Tom Brown (see Howell 1980):

> I do not love you, Dr. Fell, but why I cannot tell;
> But this I know full well, I do not love you, Dr. Fell.

Boswell, too, knows whom he likes; but he doesn't know why.

In Boswell's case, social judgment is an expression of implicit memory. Johnson and her colleagues (Johnson, Kim, and Risse 1985) observed a similar effect in a group of patients with Korsakoff's syndrome. These patients suffer from bilateral damage to the diencephalon, including the mammillary bodies, which renders them densely amnesic for events that have occurred since the onset of their illness. In the experiment, the patients viewed pictures of faces paired with fictional biographies that characterized the target person in either positive or negative terms. Korsakoff patients show a gross anterograde amnesia, so they had no memory for the exposures on subsequent testing. However, when asked to indicate which faces they liked, they generally preferred the "good" faces (which had been paired with positive characteristics) to the "bad" ones (which had been paired with the negative characteristics).

IMPLICIT PERCEPTION IN EVALUATION AND BEHAVIOR

Social impressions can be influenced even when there is no substantive contact between perceiver and target. A salient example is based on a phenomenon known as the *mere exposure effect* (Zajonc 1968). The general idea behind mere exposure is that familiarity breeds admiration, not contempt: repeated exposure to unfamiliar stimuli increases likability ratings of those stimuli. The most prominent explanation of this effect is that we prefer the familiar to the unfamiliar; however, the effect occurs even when subjects do not consciously recognize the targets as familiar.

The mere exposure effect was demonstrated in a now–classic experiment by Kunst-Wilson and Zajonc (1980), who presented subjects with five tachistoscopic exposures of ten irregular polygons—presentations that were too brief to be consciously perceived; another ten polygons were never presented at all, and served as controls. Proof that the stimuli had not been perceived consciously was provided by a recognition test, in which the subjects were unable to distinguish critical from neutral stimuli. However, when presented with the same stimuli, and asked to make preference rather than recognition judgments, there was a clear difference between targets and lures. Thus, mere exposure, in the absence of any substantive contact between subject and stimulus, had an impact on affective judgments, even though the exposures themselves were not consciously perceived. Again, the subjects knew what they liked, but they did not know why.

Technically, the influence of prior exposures on subsequent evaluations counts as evidence of implicit memory. But in this case, we have more than a failure of conscious recollection. The critical events were never consciously perceived in the first place. Therefore, in this case evidence of implicit memory also counts as evidence of implicit perception (Kihlstrom *et al.* 1992). At this point, the subliminal mere exposure effect has been replicated many times—and as it happens, the mere exposure effects induced by subliminal stimuli appear to be greater than those induced by supraliminal ones (Bornstein 1989). The likeliest explanation for this difference is that when subjects consciously recognize the critical stimuli as having been presented before, they attribute their affective judgments to their prior exposures, and discount them. But subliminal exposures do not leave consciously accessible memory traces, so the proper attribution, and discounting, isn't possible. Conscious awareness is the logical prerequisite to conscious control: we cannot cope with something about which we do not know.

Subliminal stimulation can affect preferences, but can it affect actual behavior as well? One would like to think so. The Kunst-Wilson and Zajonc study (1980) bears on social interaction because social psychologists believe that an important determinant of behavior is the actor's attitude toward the target. If mere exposure induces more favorable attitudes, then interactions should go better. This hypothesis was put to the test by Bornstein and his colleagues (Bornstein, Leone, and Galley 1987), who presented college students with subliminal exposures of other students' faces. After four exposures, recognition was at chance levels, but subjects showed a clear preference for those faces that had been presented previously—another demonstration of the subliminal mere exposure effect.

In a follow-up experiment (Bornstein *et al.* 1987), male subjects received subliminal exposures of the face of one of two other men. Then they were actually brought into contact with these targets, who were in fact confederates of the experimenter, under the guise of a poetry-rating task. The three subjects were instructed to read a series of poems, and then guess whether the poet was male or female. On predetermined trials the two confederates disagreed, forcing the lone subject to take sides. On these trials, subjects who had been exposed to confederate 1 tended to agree with confederate 1, while those who had been exposed to confederate 2 tended to agree with confederate 2; those who had not been exposed to either confederate divided their agreement evenly between them. So, in this case, subliminal exposure, which inculcates positive social judgments, also produced more positive behavior toward the objects of those evaluations.

The fact that subliminal stimuli can elicit emotional responses, even in the absence of repeated exposure, has been demonstrated by Niedenthal (1990). In one experiment, subjects studied a series of target slides depicting a cartoon character; these slides were preceded by a priming slide depicting a human face expressing either joy or disgust. By virtue of the tachistoscopically brief presentation of the face, and metacontrast produced by the cartoon, the primes were not consciously perceived. In the next phase of the experiment, the subjects performed a recognition task, distinguishing old from new cartoon figures. On this test trial, the targets were preceded by preconscious primes that were the same as, or different from, those presented on the study trial.

Considering the response latencies for targets, there were several results of interest. First, response latencies are longer for cartoons that were studied, or tested, in the context of disgust compared to the context of joy. Second, regardless of emotional valence, response latencies are shorter when the affective context at test is congruent with the affective context at study. A second experiment, using both emotional faces and emotional scenes as primes, yielded similar effects. So, the emotional valence of the primes had a palpable effect on recognition, even though the primes themselves were never consciously seen.

PRECONSCIOUS PROCESSING IN IMPRESSION FORMATION

Preconscious processing affects the content of social impressions, not just the speed of recognition. This fact was demonstrated by an experiment performed by Bargh and his colleagues. (Bargh *et al.* 1986). These investigators asked subjects to read a paragraph describing 12 behaviors displayed by a fictional person—what has come to be known in the trade as "The Donald Story" (Wyer and Srull 1979). Depending on the experimental condition, five of these twelve items could be interpreted as acts of kindness or shyness, while the remaining seven items were neutral with respect to these traits. Moreover, the five kind or shy behaviors were themselves ambiguous, in that they could be attributed either to dispositional or situational causes. For example, one of the items read, "One of Donald's colleagues asked him to donate $2 to the Red Cross, and he agreed." From this information, we do not know whether Donald himself is charitable, or whether he succumbed to social pressure. However, the manifest task of the subjects was to rate Donald's personality in terms of his kindness or shyness.

Prior to reading the paragraph, the subjects had engaged in an ostensible vigilance task, in which they indicated whether targets appeared on the left or right of a fixation point in the center of a computer screen. Unknown to the subjects, the targets were actually words, presented for 60 milliseconds and followed by a central pattern mask—and thus too fast to be consciously perceived. In two conditions, 80 percent of the masked words were adjectives related to kindness or shyness; in a third condition, the words were unrelated to personality. Memory for these words, assessed on a subsequent recognition test, was very poor, confirming that the items had not been consciously perceived.

Obviously, the purpose of the vigilance task was to prime subjects' ratings of Donald as kind or shy. And, in general it succeeded. When Donald was presented as ambiguously kind, subjects who were primed with kind adjectives rating him as significantly more kind than those who were primed with neutral words. Similarly, when Donald was presented as ambiguously shy, subjects who were primed with shy adjectives rated him as significantly more shy, compared to those who were primed with neutral words.

Another study gave similar results (Bargh and Pietromonaco 1982). Here, masked adjectives relating to hostility were presented in the vigilance task, and the narrative contained a set of behaviors that were ambiguous with respect to that trait. Again, there was no recognition of the trait adjectives. However, the subjects who got hostile primes gave higher ratings of hostility to the target, compared to a control condition in which the primes were neutral. In both cases, then, the trait adjectives presented preconsciously during the vigilance task clearly influenced the subjects' subsequent impressions of the target Donald. This occurred despite the fact that the adjectives themselves were masked and thus not consciously processed by the subjects at the time of presentation. Thus, the effect qualifies as another instance of implicit perception.

LIMITS ON PRECONSCIOUS PROCESSING

The experiments by Niedenthal (1990; see also Niedenthal, Setterlund, and Jones 1994) and Bargh (Bargh and Pietromonaco 1982, Bargh *et al.* 1986, see also Bargh 1994) bear on the vexatious question of the degree to which information processing can occur outside of conscious awareness. In general, the field has been divided between those who think that preconscious processing is limited to the physical and structural features of the stimulus, and those who believe that semantic features can be processed as well. Niedenthal's (1990) experiment, in which the emotional context affected face recognition, is somewhat ambiguous in view of the claim by some emotion theorists that emotional expressions can be read directly off the face, without need for any kind of complex cognitive analysis. The implications of the Bargh experiments (Bargh and Pietromonaco 1982, Bargh *et al.* 1986) are clearer: the fact that the content of the adjectives—whether they pertained to kindness or shyness—influenced subjects' later impressions of the target obviously indicates that semantic, not just perceptual, processing occurred outside of awareness.

It should be understood, however, that even if preconscious semantic processing is possible (as I believe the evidence now indicates), it is also clearly limited. These limits are clearly indicated by the work of Greenwald and his colleagues, who examined the semantic priming of evaluative judgments of words (Greenwald 1992, Greenwald, Klinger, and Liu 1989). (This psycholinguistic work is actually relevant to social interaction, in that social psychologists have long been interested in attitudes as determinants of behavior, and positive and negative evaluations are expressions of these attitudes.) In these experiments, subjects are asked to judge whether a target word is positive or negative in connotative meaning. The target itself is preceded by another word, a prime, which is clearly positive or negative. Thus, a prime like ENEMY might precede a target word like LOSE. The prime is separated from the target by a central pattern mask, such that it is not consciously perceived by the subjects.

Nevertheless, the emotional connotations of the prime influence judgments of the targets: over several experiments, Greenwald has shown that evaluative judgments are speeded when a positive prime precedes a positive word, or a negative prime precedes a negative word (Greenwald *et al.* 1989). And, as in the case of mere exposure, there is more priming when the prime is subliminal, than when it is supraliminal.

In other experiments, however, Greenwald has attempted to prime evaluative judgments with two-word phrases. The critical conditions are where a prime composed of two negative words has a positive meaning, such as ENEMY LOSES—which is a good thing. If ENEMY LOSES is a good thing, then it should prime evaluations of other good things, such as words like GAME. But no: It turns out that phrases like ENEMY LOSES facilitate judgments of negative, not positive words. Apparently, preconscious processing can extract the meaning of a single word, but it cannot construct the meaning of a two-word phrase. So, as Greenwald (1992) concludes, preconscious semantic processing is possible, but it is also analytically limited.

In my view, preconscious processing is also limited in at least two other ways (Kihlstrom 1993). First, we have to adopt the distinction drawn by Merikle and his colleagues between subjective and objective thresholds (for example, Merikle and Reingold 1993). The subjective threshold is that point at which the subject reports no awareness of the stimulus; the objective threshold is that point at which forced-choice guesses of presence or absence fall to chance levels. Semantic processing is possible when stimuli are presented between the objective and subjective threshold (and the closer to the subjective threshold, the better), but not when stimuli are presented below the objective threshold. Second, we have to adopt the distinction, drawn by many attention theorists, between automatic and controlled processing (for example, Logan 1989). Preconscious stimuli can be analyzed by automatic, but not controlled, processes. After all, people cannot consciously deal with something of which they are not consciously aware.

Even though preconscious processing is limited, it can still have effects on experience, thought, and action in the social domain. A rapidly developing body of research shows that our feelings, social judgments, and interpersonal behaviors can be influenced by cues in the environment that are so subtle that we are not even aware of them, and pay them no conscious attention—something to conjure with the next time we meet Dr. Fell.

ACKNOWLEDGMENTS

This paper was presented at a conference, "Toward a Scientific Basis for Consciousness," sponsored by the University of Arizona, Tucson, April 1994. The point of view represented here is based on research supported by Grant no. MH-35856 from the National Institute of Mental Health. I thank Melissa Berren, Lawrence Couture, Elizabeth Glisky, Martha Glisky, Heather Law, Chad Marsolek, Victor Shames, Susan Valdiserri, and Michael Valdiserri for their comments.

REFERENCES

Bandura, A. 1986. *Social foundations of thought and action: A social cognitive theory.* Englewood Cliffs, NJ: Prentice–Hall.

Bargh, J. A. 1994. "The four horsemen of automaticity: Awareness, intention, efficiency, and control in social cognition." In *Handbook of social cognition,* 2nd ed., Vol. 1, edited by R. S. Wyer and T. K. Srull. Hillsdale, NJ: Lawrence Erlbaum, pp. 1–40.

Bargh, J. A., and P. Pietromonaco. 1982. Automatic information processing and social perception: The influence of trait information presented outside of conscious awareness on impression formation. *Journal of Personality and Social Psychology* 43:437–49.

Bargh, J. A., R. N. Bond, W. J. Lombardi, and M. E. Tota. 1986. The additive nature of chronic and temporary sources of construct accessibility. *Journal of Personality and Social Psychology* 50:869–78.

Bornstein, R. F. 1989. Exposure and affect: Overview and meta–analysis of research, 1968–1987. *Psychological Bulletin* 106:265–89.

Bornstein, R. F., D. R. Leone, and D. J. Galley. 1987. The generalizability of subliminal mere exposure effects: Influence of stimuli perceived without awareness on social behavior. *Journal of Personality and Social Psychology* 53:1070–79.

Cantor, N., and J. F. Kihlstrom. 1987. *Personality and social intelligence.* Englewood Cliffs, NJ: Prentice-Hall.

Damasio, A. R., D. Tranel, and H. Damasio. 1989. "Amnesia caused by herpes simplex encephalitis, infarctions in basal forebrain, Alzheimer's disease, and anoxia/ischemia." In *Handbook of neuropsychology,* edited by F. Boller and J. Graffman. Amsterdam: Elsevier, pp. 149–66.

Darley, J. M., and R. H. Fazio. 1980. Expectancy confirmation processes arising in the social interaction sequence. *American Psychologist* 35:867–81.

Greenwald, A. G. 1992. New look 3: Unconscious cognition reclaimed. *American Psychologist* 47:766–79.

Greenwald, A. G., M. R. Klinger, and T. J. Liu. 1989. Unconscious processing of dichoptically masked words. *Memory and Cognition* 17:35–47.

Howell, P. 1980. *A commentary on Book I of the Epigrams of Martial.* London: Athlone Press.

Johnson, M. K., J. K. Kim, and G. Risse. 1985. Do alcoholic Korsakoff's syndrome patients acquire affective reactions? *Journal of Experimental Psychology: Learning, Memory, & Cognition* 11:22–36.

Kihlstrom, J. F. 1984. "Conscious, subconscious, unconscious: A cognitive perspective." In *The unconscious reconsidered,* edited by K. S. Bowers and D. Meichenbaum. New York: John Wiley & Sons, pp. 149–211.

Kihlstrom, J. F. 1990. "The psychological unconscious." In *Handbook of personality: Theory and research,* edited by L. Pervin. New York: Guilford, pp. 445–64.

Kihlstrom, J.F. 1993. The continuum of consciousness. *Consciousness and Cognition* 2:334–54.

Kihlstrom, J. F. 1994a. "Consciousness and me–ness." In *Scientific approaches to the question of consciousness,* edited by J. Cohen and J. Schooler. Hillsdale, NJ: Lawrence Erlbaum (in press).

Kihlstrom, J. F. 1994b. "The rediscovery of the unconscious." In *The mind, the brain, and complex adaptive systems,* edited by H. Morowitz and J. Singer. Santa Fe Institute Series. Reading, MA: Addison-Wesley, pp. 123–43.

Kihlstrom, J. F., T. M. Barnhardt, and D. J. Tataryn. 1992. "Implicit perception." In *Perception without awareness,* edited by R. F. Bornstein and T. S. Pittman. New York: Guilford, pp. 17–54.

Kunst-Wilson, W. R., and R. B. Zajonc. 1980. Affective discrimination of stimuli that cannot be recognized. *Science* 207:557–58.

Logan, G. D. 1989. "Automaticity and cognitive control." In *Unintended thought,* edited by J. S. Uleman and J. A. Bargh. New York: Guilford, pp. 52–74.

Merikle, P. M., and E. M. Reingold. 1992. "Measuring unconscious perceptual processes." In *Perception without awareness,* edited by R. F. Bornstein and T. S. Pittman. New York: Guilford, pp. 55–80.

Niedenthal, P. M. 1990. Implicit perception of affective information. *Journal of Experimental Social Psychology* 26:505–27.

Niedenthal, P. M., M. B. Setterlund, and D. E. Jones. 1994. Emotional organization of perceptual memory. In *The heart's eye: Emotional influences in perception and attention,* edited by P. M. Niedenthal and S. Kitayama. San Diego, CA: Academic Press, pp. 87–113.

Schacter, D. L. 1987. Implicit memory: History and current status. *Journal of Experimental Psychology: Learning, Memory, and Cognition* 13:501–18.

Uleman, J. S., and J. A. Bargh. 1989. *Unintended thought.* New York: Guilford.

Wyer, R. S., and T. K. Srull, 1979. *Memory and cognition in its social context.* Hillsdale, NJ: Lawrence Erlbaum.

Zajonc, R. B. 1968. Attitudinal effects of mere exposure. *Journal of Personality and Social Psychology Monograph* 9 (2, Pt. 2):1–27.

7 Efference and the Extension of Consciousness

Thaddeus M. Cowan

Consider the following experiment: An observer enters a light sealed room with a patterned wallpaper design on some wall. With hand outstretched, palm parallel with the frontal plane occluding the wallpaper design, the observer turns off the lights and then pops a flashbulb. A very strong, long lasting (five to ten seconds), highly detailed positive afterimage image of the room, wallpaper pattern, and outstretched arm and hand will follow. When the detail is at its peak, the hand is moved downward. The afterimage of the hand will appear to shift leaving a black hole in the shape of the hand in its place. Remember that this takes place in a completely darkened room, and what the observer sees is not the hand but the image of the hand impressed on the retina; it is this image that appears to move.

This affect was reported by Davies (1973). Weintraub included it at the end of his *Annual Review* chapter (1975) in what appeared to be a bewildering afterthought. "Any hypotheses?" asked Weintraub. This paper offers steps toward a possible explanation that relates to the issue of conscious awareness of images.

THE EXTENSION OF EXPERIENCE ACROSS THE VARIOUS SENSES

Rather than attempt to answer questions like "What is consciousness?" this paper will examine the components in which consciousness manifests itself. When we talk about "conscious image" we use too broad a brush stroke. Think of an acoustical signal of which we are aware as an example. It has components of frequency, amplitude, and complexity, and each of these can be described in terms of our conscious experience of pitch, loudness, and timbre. We can also be aware of direction and distance (or, collectively, extension) of the stimulus from our egocenters. When we speak of the extent of consciousness, we are isolating the part of experience that answers the question, "Where is it?"

Taste and Smell

The conscious experience of taste is clearly located in the mouth on the tongue and soft palate. This occurs even when odors interplay with it. For example, peppermint molecules interact with the olfactory epithelium, not the taste buds, and in spite of the distance of peppermint's neural trigger from the mouth, we experience the after–dinner mint in the mouth as flavor, not in the nose as essence. Its conscious experience is transmitted to the mouth by a kind of sensory ventriloquism psychologists call "sensory capture" seen when information coming from two different sensory sources appears to be coming from only one of them.

Taste, being internal, has no directionality or distance. Compare this with smell. Like taste, we experience essences internally but in the nose. Unlike taste, we can gather information about the direction of smell. We catch a whiff of apple pie cooling on the window sill; we turn toward it guided by the strength of its vapors. Kinesthesis (orientation) is obviously responsible for the awareness of direction.

Tactile Senses

The tactile senses are multifaceted, but only touch will be addressed here. The conscious experience of touch is clearly at the frontier between the skin (fingertip) and the external world; it protrudes no further than one's reach. It obviously has both direction and distance (extent). The location of the experience of touch surely involves kinesthesis since muscular movement directs our finger in a particular direction over a particular distance.

Vision (Audition)

The conscious experiences of taste, smell, and touch occur at the receptor site. With vision and audition, something seemingly miraculous occurs: the conscious event breaks free and is projected into the space that surrounds us. Only vision will be examined since the experience of vision is exceptionally rich in primates with intact systems of sight; the conclusions that can be drawn here apply to audition.

PROXIMAL AND DISTAL EVENTS

What we see obviously has direction and distance. In order to motivate the understanding of the conscious extension of vision we will employ the following observation of Aleister Crowley (as quoted by Hughes and Brecht 1976): "A red rose absorbs all colours but red; red, therefore, is the one colour the rose is not." [p. 66]

The red wavelength reflected from the rose enters the eye and triggers a cortical response (Figure 7.1a, top left). The physical rose is a distal event, and the cortical response is a proximal event (not to be confused with distal and proximal stimuli which refer to the physical object and the retinal image respectively). The neurological response causes the red image to appear projected to the location of the real rose, and a mental distal event is produced (Figure 7.1a, bottom left). The sequence of events is distal physical

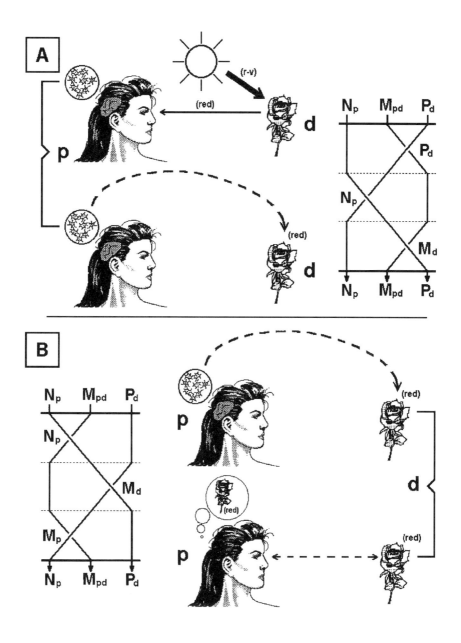

Figure 7.1 (A) Crowley's paradox presented pictorially (left) and by its braid representation, *dpd* (right). (B) A homeomorphism, *pdp*, of the braid in (A) left, and its visual portrayal in terms of conscious perception.

object to proximal neurological translation to distal mental image or *dpd*. Sequence *dpd* is an intuitive assumption about the nature of conscious images; furthermore, it occurs in mathematics, and it can serve us here.

A FORMAL MODEL

Sequence *dpd* can be represented by the nodes of a set of woven strings. The nodes of interwoven or twisted strings are defined by under (s_k^{-1}) and overcrossings (s_k) of the kth string from the left. Figure 7.1a, right, shows a three node topological "braid," $s_2^{-1}s_1s_2$ (Artin 1927). Note that the "k^{th} string from the left" refers to the position of the string along any line parallel to the frames and not merely the position at the top or base frame. The "cornered" form (Figure 7.1a, right) elucidates this definition; the sequence of nodes is read from the top down in keeping with Figure 7.1a, left.

Let the right-most string position represent any physical event (P); let the left-most string position be a neurological event (N), and let the middle string position portray some mental event (M). Obviously, physical events (P_d) are always distal, and neurological events (N_p) are always proximal. Mental events ($M_{p,d}$), given by the middle string position, are ambiguous with regard to distal and proximal placement. Proximal and distal nodes are seen to the left and right of the midline respectively.

Let us proceed from top to bottom and examine each node separately to see how closely the twisted form ($P_dN_pM_d$) in Figure 7.1a, right, conforms to the situation shown in Figure 7.1a, left; each node reflects a dominance relationship at a distal or proximal location. The first node encountered is formed by two strings (P_d and M_d). String P dominates, and this tells us that the first event is a distal physical event. The second node met is a confrontation between two proximal states (N_p and M_p); N dominates so the second event is a neurological process presumably triggered by the reflection of the light off the rose. Finally, the third node is formed by a crossing of P_d and M_d (again), but this time M dominates, obviously the projected image of the rose. The node positions mimic these events as they appear in Figure 7.1a, left. Sequence $P_dN_pM_p$ is an intuitive interpretation of *dpd*.

MANIPULATION OF THE MODEL AND ITS INTERPRETATION

The deepest string can flex to the left. The lateral positions of the nodes change, but the number of nodes and their relative vertical positions remain invariant. This produces the sequence $s_1s_2s_1^{-1}$ (Figure 7.1b, left), and the homotopy of the two braids, the smooth distortion of one into the other, is represented by

$$s_2^{-1}s_1s_2 = s_1s_2s_1^{-1},$$

or

$$dpd = pdp.$$

In Figure 7.1b, left, N_p dominates M_p, M_d dominates P_d, and $M_{p'}$ dominates N_p as shown by the descending nodes. Sequence $N_p M_d M_p$ is a prediction generated by the model; it is unexpected, interpretable, and has empirical support. Furthermore it has something profound to say about an established approach to the problem of consciousness.

Interpretation of *pdp*

First the independent flexing of the deepest string of the model suggests that *pdp* is separable from *dpd*, yet, because of their equivalence, they both play an integral role in the conscious experience of the image. Since *dpd* precedes *pdp*, it follows that *dpd* sets a stage (by creating extent, it is assumed) and *pdp* activates the conscious experience.

Interpretation of $N_p M_d M_p$

The neurological response (N_p) to the physical stimulus projects the image of the rose (M_d) to the location of the physical rose. We do not see the physical rose, but rather we observe, proximally (M_p), the mental projection of it. This is not an unreasonable supposition since the rose we "see" in space is truly mental; it is red, or "... the one colour the rose is not." We could easily assume that this observation keeps the neural response going, which continually projects the image that we continually observe. As long as this loop continues, we are conscious of the rose.

The problem with the loop is that it is entirely composed of mental and neurological processes; the physical event (P_d) seems to have been lost or rendered irrelevant. This is patently absurd of course; remove the physical rose, and the perception stops. Topological homotopy between *dpd* and *pdp*, seems to keep the two forms together in some way making P_d a viable event, yet the two forms appear to possess a kind of independence. The relation *dpd = pdp* and its two components, *dpd* and *pdp*, are specific to three definable empirical situations.

EMPIRICAL SUPPORT

The relation *pdp = dpd*, or $P_d N_p M_d = N_p M_d M_p$, describes normal perception from the presentation of a real object, followed by transduction to a cortical representation, ending with the projection and subsequent observation of the mental image.

The last half of the relation, *dpd* or $N_p M_d M_p$, determines the actual conscious perception. Alone, it is stripped of the physical object. Hallucinations seem to fit this description, as do certain nondrug-induced experiences such as afterimages (vision) and the missing fundamental (audition). In each of these cases, the effects are private, seemingly external to the individual hence real, yet lacking physical support.

As for *dpd* or $P_d N_p M_d$ alone (the first half of the relation), the implication is that this does not produce the conscious experience, but rather it sets the stage for it by projecting the image to be observed. That is, no conscious experience is felt, but a reaction to the stimulus is observed. The obvious behavior befitting these conditions is automatism or automaticity. One drives the interstate from here to there pondering some pressing problem without awareness of what transpires en route, yet this segment of the longer journey is negotiated without incident.

Note that the separation of *pdp* and *dpd* is very much in step with the findings of Libet and his coworkers (1987) who discovered that blocking certain accompanying waveforms of a sensory signal blocks the conscious experience in spite of the fact that the patient responds to the sensory input. The motor response to the sensory signal does not produce a conscious experience, but a later event does.

EFFERENCE THEORY

What produces the change from $P_d N_p M_d$ to $N_p M_d M_p$? That is, what makes the model flex? Efference theory (von Helmholtz 1962, Münsterberg 1899, Washburn 1916, Taylor 1962, and Festinger *et al.* 1967) states that the conscious experience of perception is created by efferent commands. This paper offers a modification of this view which suggests that efference is responsible only for the conscious perception of the location of a pattern and not the perception of the pattern itself.

There are three phases to efferent activity: (1) the command that initiates the movement, (2) the active movement itself, and (3) a quantum potential well suggested by Sperling (1970) which locks a limb or the eyes into its resting position. It is the first of these that Festinger *et al.* had in mind when they spoke of "efferent command." Here it is proposed that this command places the image in space (*dpd*); the final position lock accompanies the second phase (*pdp*), the conscious experience, and the movement is responsible for the transition or in terms of the model, the shift of the middle string.

Let us return to the experiment described at the beginning. Shifting the arm downward causes us to see the afterimage shift even though we do not see the arm actually move and the afterimage is fixed relative to its surrounds. Whether this is due to efference or afference cannot be determined here, but investigations with succinylcholine chloride dihydrate, which blocks the connections between the motor neurones and the musculature (Campbell *et al.* 1964), suggests that efference is the cause (movement is felt by the subject, but not observed by the experimenter).

Finally, the proposition offered in this paper, that the placement of the image in space is separable from the conscious awareness of it, is testable using the afterimage procedure: The image of the pattern of the hand is the afterimage, location is created by efference. One can create efference without the pattern by first obtaining an afterimage of the surround with the hand out of the line of sight (on the lap), then moving the hand into an outstretched position. Efference is produced in the presence of a visual surround, but no pattern of arm or hand should be seen. Preliminary experiments in our laboratories tend to verify these results, which supports the theory put forth here that efference is responsible only for the image of movement and location and not the image of the other components of the pattern.

REFERENCES

Artin, E. 1925. Theorie der Zöpfe. *Hamburger Abhandlungen* 4:47–72.

Campbell, D., R. E. Sanderson, and S. G. Laverty. 1964. Characteristics of a conditioned response in human subjects during extinction trials following a simple traumatic conditioning trial. *Journal of Abnormal and Social Psychology* 68:627–39.

Davies, P. 1973. Effects of movements upon the appearance and duration of a prolonged visual afterimage: I. Changes arising from the movement of a portion of the body incorporated in the afterimage scene. *Perception* 2:147–53.

Festinger, L., H. Ono, C. A. Burnham, and D. Bamber. 1967. Efference and the conscious experience of perception. *Journal of Experimental Psychology Monograph* 74:637.

von Helmholtz, H. 1962. *Physiological optics,* Translated by J. P. C. Southall. New York: Dover.

Hughes, P. and G. Brecht. 1976. *Viscious circles and infinity.* London: Jonathan Cape.

Libet, B. 1987. "Consciousness: Conscious, subjective experience." In *Encyclopedia of Neuroscience,* edited by G. Adelman. Boston: Birkhäuser.

Münsterberg, H. 1899. The physiological basis of mental life. *Science* 9:442–7.

Sperling, G. 1970. Binocular vision: A physical and neural theory. *American Journal of Psychology* 83:461–534.

Taylor, J. G. 1962. *The behavioral basis of perception.* New Haven, CT: Yale University Press.

Washburn, M. F. 1916. *Movement and mental imagery.* Boston: Houghton-Mifflin.

Weintraub, D. 1975. Perception. *Annual Review of Psychology* 26:263–89.

8 Edelman's Biological Theory of Consciousness

John J. Boitano

Edelman's (1989) theory is a biological account of the different psychological functions necessary for the development of consciousness and is based on the principles of Darwin's evolution and those of population dynamics. Instead of referring to the workings of individual neurons, his view expounds on enormous collections of neurons interacting with each other. Note the similarity to the modern version of mind as suggested by Uttal (1987): "... mind ... is a process that is embodied within the interaction of the extremely complex, yet fundamentally discrete neuronal network. It is the ebb and flow of signals among the vast number of neurons in the brain ..." (p.673).

NEURAL DARWINISM

The evolutionary principle of natural selection through survival of favorable variants is fundamental to Edelman's Theory of Neuronal Group Selection (TNGS) upon which rests his theory of consciousness. He postulates three basic tenets: (1) The *Primary Repertoire* is a group of neurons in a given brain region forming local circuits and neural networks that have arisen out of huge number of individual neurons. The number of neurons originally generated at birth greatly exceeds the number of neurons surviving beyond the developmental period. The surviving neurons exhibit remarkable variation in their local connectivities due to the dynamic regulation of the cellular adhesion molecules and the substrate adhesion molecules, and the random fluctuation of cell movements, as neurites are extended and retracted. Note that this is a large and diverse population in which many of its elements have been selected, and in principle is similar to evolutionary concepts. (2) From the primary repertoire, a *Secondary Repertoire* is established involving further selection by the strengthening/weakening of a population of synapses acquired through the experience of the behaving organism. "Experiential selection ... occur(s) as a result of differential amplification of certain synaptic populations" (Edelman 1989, p. 46). Behavior "selects out" functioning circuits. The type of information conveyed between groups involves large-scale signalling rather than coded signals over individual neurons. (3) *Reentrant* (reciprocal) connections between the primary and secondary repertoires selects groups of neurons in maps. Mapping is an important feature of brain organization. Any point in a receptor sheet is represented in the cortex by a population of cells connected to sensory fibers coming from that particular receptor. When that point is stimulated, that population of cortical neurons will be excited. In the mammalian visual system there are over 25 different maps outside the primary visual cortex, vertically arrayed and reacting to similar stimulus properties. While this columnar organization responding to one-dimensional stimulus parameters like contour orientation or eye dominance has been known for some time (Hubel and Wiesel 1962), it has recently been found that neuronal groups in the inferotemporal area of the monkey cortex, a higher association area, respond to multiple stimulus features (like horizontally elongated overall shape, upper part darker than lower, and darkest in the center), and that the organization is indeed columnar (Fujita *et al.* 1992). The maps that are formed from the primary and secondary repertoires are connected by massively parallel and two-way connections. For example, Map 1 could refer to the tactile and proprioceptive/kinesthetic feedback from a moving arm throwing a baseball guided by vision projecting onto Map 2. Over time and with multiple encounters of the stimuli, the synapses of the reentrant connections become strengthened, thereby allowing patterns in Map 1 to excite patterns in Map 2. This provides a neurological basis for generalizations and perceptual categorization which is defined as "the selective discrimination of an object or event from other objects or events (in the environment) for adaptive purposes" (Edelman 1992, p. 87).

GLOBAL MAPPING

Edelman postulates a supramap called *Global Mapping* which contains multiple reentrant local maps connecting sensory and motor areas that are also interconnected with such nonmapped brain areas as the hippocampus, the basal ganglia, and the cerebellum. It is these structures that are involved in the ordering of both sensory phenomena and patterns of motor activity. The cerebellum regulates the execution of movement by comparing the internal information about plans for movement with the external feedback from ongoing motor performance. The basal ganglia are involved in the cognitive aspects of motor control; that is, the long–range planning and execution of complex motor strategies. These structures aid in the correlation of an entire sequence of gestures into a motor plan. The hippocampus is not only involved in the transfer of information from short-term to long-term memory, but because of its convergent and divergent reentrant connections with the cortex and other limbic structures, notably the hypothalamus by way of the septum, it receives inputs from both hedonic structures subserving value and from reentry patterns between maps in global mappings. Thus, the hippocampus assists in ordering "events that have been immediately categorized by the cortex and then ensures that these categorized events effect further synaptic changes in the cortex to enable long term memory" (Edelman 1992, p. 107). With these brain areas interacting with cortical structures, Edelman postulates that:

> ... global mapping ensures the creation of a dynamic loop that continually matches the animal's gestures and postures to the sampling of several kinds of sensory signals. Selection of neuronal groups within the local maps of a global mapping then results in particular categorical responses (Edelman 1992, p. 89).

Global mapping changes over time and with appropriate behavior, which is determined by the organism's internal criteria of value already specified by evolutionary constraints. These are the motivational systems and include the brain areas concerned with regulating homeostasis such as: respiration, cardiac output, sexual responses, eating, drinking, endocrine, and autonomic responses. The level of each of these systems is changed by behavior which is a result of perceptual categorization/classification selecting neuronal groups, thereby rearranging global mapping.

PRIMARY CONSCIOUSNESS

Primary consciousness is immediate, ongoing, present–time consciousness. It is an awareness of objects and events in the world; a mental scene depicting the here and now. Edelman's schema of primary consciousness starts with a description of two neural systems subserving different functions. The brain stem/limbic system is the hedonic system mediating internal values based on evolutionary selection for homeostatic/autonomic and endocrine functions. It facilitates the expression of appetitive, consummatory, and defensive behaviors, such as food or water procurement, sexual/reproductive behavior, aggressivity. It includes the brain stem reticular formation, the hypothalamus, the septum, amygdala, and hippocampus. It receives many interoceptive inputs from many different organ systems and the autonomic nervous system. It was designed to take care of internal body functions and developed early in evolution. The second system, the thalamocortical system, evolved to receive inputs from external sensory receptors, and to transcript these signals into movement when appropriate. The thalamus projects these external world inputs to the various sensory maps located in the cortex. As mentioned earlier, these cortical maps have massive reentrant connections arranged in local circuits within and between columns. The cortex evolved to receive multitudinous parallel inputs quickly over different sensory paths simultaneously. It evolved later than the brain stem/limbic system, and was designed to facilitate increasingly complex adaptive behaviors involving diverse perceptual categorization. Both systems are connected through learning as manifested by adaptive responding. The learning or matching between the two systems may take place in a number of loci including the fornix and septal systems, the hippocampus, the temporal cortex, the forebrain, and cingulate gyrus.

Emerging as a result of the increasingly sensitive interplay between these two systems is a new kind of memory, which Edelman labels "value–category" memory. It is localized in the cortex of the frontal, temporal, and parietal areas. Its major function is to recategorize and store the interactions between the simple perceptual categorization system and the internal value system. It allows conceptual responses to occur and be remembered between the categorized exteroceptive stimuli and the interoceptive signals reflecting homeostatic needs.

The third component is the development of reentrant signalling in each sensory modality between value–category memory and ongoing exteroceptive global mappings located in the primary and secondary cortices before these new perceptual signals enter or are stored in (value-category) memory. The reentrant circuitry provides an immediate feedback mechanism allowing classification and interpretation of ongoing multiple sensations (perceptions), and it is this recursive interaction that permits the construction of a complex scene or mental image. This is primary consciousness. The phenomenal outcome of such activities appears as a representation of current external world happenings.

It is the discriminative comparison between a value-dominated memory involving the conceptual system and current ongoing perceptual categorization that generates primary consciousness of objects and events . . . In effect, primary consciousness results from the interaction in real time between memories of past value-category correlations and present-world input as it is categorized by global mappings (but before the components of these mappings are altered by internal states) (Edelman 1989, p. 155).

Primary consciousness connects the features of the perceptual scene with their importance as determined by the organism's prior history and values. It is constrained to a small time span around a present happening. In Edelman's terminology, it is a kind of "remembered present." It does not contain notions of the past or the future, nor has it the concept of the self. For these we turn to higher-order consciousness.

HIGHER–ORDER CONSCIOUSNESS (HOC)

HOC involves the recognition by the organism of one's own feelings and behavior. It consists of a direct awareness of mental episodes having past and future characteristic; a concept of the personal self as distinguished from other objects in the world; the use of language; and being conscious of one's own consciousness. Edelman's concept of HOC imposes two new cortical areas to the schema of primary consciousness; namely, Broca's area and Wernicke's area. Both have to do with the acquisition of language. Broca's area is specialized for the production/expression of speech while Wernicke's area is involved in the recognition/comprehension of articulated speech sounds. The acquisition of speech, whether it be in an infant or through evolution, follows three basic steps: (1) The production of speech sounds became associated with concepts and gestures through learning. Nouns became correlated with objects/actions through phonology, and this led to the development of semantics/meanings. (2) A lexicon/vocabulary accrued as a result of semantics. (3) Syntax, the way in which words are juxtaposed to form phrases and sentences, and stored in memory, emerged with the correlation of preexisting conceptual learning and the learning of a large vocabulary. Thus, the brain gives meaning to utterances and then generates syntax by "rules developing in memory as objects for conceptual manipulation" (Edelman 1992, p. 130); for example, the rule that the noun/subject precedes the verb/predicate. The proposed anatomical substrate for this process is Broca's and Wernicke's area in the frontal and temporal/parietal cortices. These are connected via reentrant paths to other cortical areas subserving value-category memory and conceptual memory (where concepts are stored with their respective meanings) also located in the frontal, temporal, and parietal lobes. Since verbalizations symbolically represent concepts, the result is an enhancement of the ability to create, improve, remember, and produce many new ideas. Thus, the acquisition of language leads to the development of new memory for concepts and imagery, autonomous of direct sensory stimulation.

Now, how does all of this relate to being aware of one's own awareness? The answer resides in being able to distinguish and remember concepts of the self from nonself; of discerning the true self acting on the environment. For the development of a self-concept, it is necessary to store symbolic (social) relations acquired and affectively reinforced through (social) interactions with other conspecifics. This is done linguistically through the categorization of sentences relating self and nonself "through verbs of various acts" (Edelman 1992, p. 132) depicting self-nonself distinctions. The interplay between the speech areas (where verbal symbols are stored) and value-category memory (where concepts are stored) provides an internal representation (symbolic model) of the world and of the self in that world. It is also the essential element in liberating the organism from the constraints of the present because the self–nonself distinction must reflect some amount of temporal autonomy. The ability to differentiate conceptual-symbolic models in long-term memory from what is happening in the immediate perceptual present permits a concept of the past to be acquired. "Higher-order consciousness, with its self-nonself distinction and freedom from immediate time constraints and with its increased riches of social communication eventually led to capabilities allowing the anticipation of future states and planned behavior" (Edelman 1989, p. 192). Thus, HOC provides an adaptive flexibility characterized by a concept of one's self and therefore, one's own consciousness; the ability to model the world and plan for the future (free of present real time); the ability to compare outcomes on the basis of personal values and past histories; the capability of reorganizing memories and plans; and the development of linguistic communication.

EVALUATION

It is premature to offer valuative comments at the present time. Edelman's brain-based theory needs experimental verification if it is to survive, or definitive evidence to indicate its fallacies. Already some evidence is accumulating in support of reentrant connections in his TNGS. Gray *et al.* (1989) have demonstrated correlated firings from the orientation columns in two separate visual maps. Synchronous 40-Hz oscillations were observed from both areas in response to the presentation of a bar of light. Is this finding suggestive of a solution to the binding problem? The evidence derived from somatosensory plasticity (Clark *et al.* 1988), demonstrate a rearrangement of two separate cortical maps originally corresponding to two separate digits into one map as a result of surgically fusing digits 3 and 4 of the owl monkey. These findings support his notions regarding interactive competition between neuronal groups with subsequent strengthening of synaptic connections. Furthermore, patients with brain damage either by stroke, accident, drug overdose, or direct surgical intervention always involve a disruption of reentrant connections that can lead to alterations in consciousness. Examples of a partial loss of conscious functioning include hippocampal deficiencies in such publicized cases as HM (Corkin 1984) and Clive (Restak 1988); blindsight victims (Weiskrantz *et al.* 1974); and the contralateral neglect syndrome (Sacks 1985). A complete loss of cognition and awareness but with intact arousal was recently attributed to hypoxia-ischemia in the remarkable case of Karen Ann Quinlan who suffered cardiopulmonary arrest after drug overdose. She nonetheless lived in a vegetative state for more than ten years following the accident without ever regaining consciousness. The neuropathological findings (Kinney *et al.* 1994) revealed extensive bilateral and symmetric damage to the thalamus with less severe damage observed in the cortex, cerebellum, and basal ganglia. With this new information, the role of the thalamus in Edelman's theory may have to be reevaluated.

Nevertheless, Edelman has marshalled an impressive array of scientific data to support his incisive insights, both in the realms of neural acumen and contemplative speculations. The theory rests on principles of evolution, supported by the concepts of developmental neurobiology, and augmented by the psychological effects of daily environmental change. The ideas of altered reentrant connectivities within the global maps resulting in greater linguistic capability, and therefore, enhanced conscious adaptive flexibility, all seem immanently feasible.

ACKNOWLEDGMENTS

The following sources were quoted or summarized:
The Remembered Present: A Biological Theory of Consciousness, by Gerald M. Edelman, © 1990 by Basic Books, Inc., and *Bright Air, Brilliant Fire,* by Gerald M. Edelman, © 1992 by Basic Books, Inc. Courtesy of HarperCollins Publishers, Inc.

REFERENCES

Clark, S. A., T. Allard, W. M. Jenkins, and M. Merzenich. 1988. Receptive fields in the body-surface map in adult cortex defined by temporally correlated inputs. *Nature* 332:444–45.

Corkin, S. 1984. Lasting consequences of bilateral medial temporal lobectomy: Clinical course and experimental findings in H. M. *Seminars in Neurology* 4:249–59.

Edelman, G. M. 1990. *The remembered present.* New York: Basic Books.

Edelman, G. M. 1992. *Bright air, brilliant fire.* New York: Basic Books.

Fujita, I., K. Tanaka, I. Minami, and K. Cheng. 1992. Columns for visual features of objects in monkey inferotemporal cortex. *Nature* 360:343–46.

Gray, C. M., P. Konig, A. K. Engel, and W. Singer. 1989. Oscillatory responses in cat visual cortex exhibit inter-column synchronization which reflects global stimulus properties *Nature* 338:334–37.

Hubel, D. H., and T. N. Wiesel. 1962. Receptive fields, binocular interaction and functional architecture in the cat's visual cortex. *Journal of Physiology (London)* 160:106–54.

Kinney, H. C., J. Korein, A. Panigrahy, P. Dikkes, and R. Goode. 1994. Neuro–pathological findings in the brain of Karen Ann Quinlan. *New England Journal of Medicine* 330:1469–75.

Restak, R. M. 1988. *The mind.* New York: Bantam Books.

Sacks, O. 1985. *The man who mistook his wife for a hat.* New York: Harper and Row.

Uttal, W. R. 1987. The psychobiology of mind. In *Encyclopedia of neuroscience,* edited by George Adelman. Boston: Birkhauser.

Weiskrantz, L., E. K. Warrington, M. D. Sanders, and J. Marshall. 1974. Visual capacity in the hemianopic field following a restricted occipital ablation. *Brain* 97:709–28.

9 The Structure of Subjective Experience: Sharpen the Concepts and Terminology

David Galin

INTRODUCTION

There are two big problems strangling research in consciousness. Problem #1 is that there is very little agreement about what needs to be studied. Problem #2 is that there is almost no agreement about language with which to discuss Problem #1. The two are deeply entwined.

Take Problem #2 first. Our situation regarding terminology is a disaster zone: each of us is perfectly clear what we mean but nobody else uses the same words in the same way. It is said that an academic would rather use someone else's toothbrush than their terminology. Some speak of consciousness as a special stuff, or as a state of regular stuff, or as a process (like respiration), or as a whole system (like the cardiopulmonary-vascular system) with parts and functions and controls. Is it fundamentally a noun, or a verb, or an adjective? Some say that we should not speak of consciousness at all because it is an illusion, or that it is a real but hopelessly heterogeneous hodge-podge, or that it is a wastebasket category.[1]

It is argued by some thinkers that the terminology can only coevolve as the concepts are sharpened, and we should try to "avoid the heartbreak of premature definitions" (Churchland 1988). This argument points back to Problem # 1, What-to-Study. (I have addressed Problem #1 more fully in an essay on defective questions, "What is the difference between a duck?" [Galin 1995].) To some extent researchers are guided (or constrained) by philosophers as well as by the agendas of the funding agencies. Philosophers are expert in examining assumptions, methods, and their implications. Can X be studied? Is it science? If so, what sort of data is admissible? What classes of questions are or are not permissible? (See the admirable book *Consciousness and Contemporary Science* [Marcel and Bisiach 1988], and Wimsatt 1976a, 1976b).

But there is another more personal basis for deciding what to study—what is it that interests us? What is it that we want explained (Galin 1995)? Most current work on consciousness focuses on the mechanisms (neural or cognitive) that underlie and control it, or on the levels of behavioral achievement. I assert that what is most interesting about mental life for most ordinary people is not mechanism, not performance, not information processing; it is what it feels like! Subjective experience! It is possible that you, dear reader, may be different; if you think so try this test on yourself: Imagine that you can get a treatment that will augment your perception, memory, and problem-solving by 1000%. Imagine also that you will be able to achieve your goals 1000% more effectively. The only catch is that you would no longer be aware. Would you want the treatment? (Some people claim that this is what is offered by higher education).

I believe that a relatively complete theory of consciousness will include accounts at the subjective level as well at the neurological and cognitive levels of analysis (Flanagan 1995, Jackendoff 1987). Not all researchers are naive reductionists trying to reduce consciousness away or to show that it is "nothing but." Nevertheless, in the rush to account for awareness in terms of information processing or neurophysiology, one may not sufficiently examine just what it is that needs to be explained.

In this paper, as a first step toward a complete theory, I call attention to key descriptive aspects of subjective experience that have been rather neglected. Anyone who examines her subjective experience finds a great variety of qualitatively different components: for example, sensory feelings, feelings of intention, feelings of expectancy, the feeling of knowing. I propose a taxonomy for the components based on the kind of information they carry. The descriptive analysis is then set in the framework of self-monitoring functions, including the domains of emotion, action, and metacognition.

My proposals are built in part on the insights of William James who analyzed the structure of awareness a century ago in his revered but now nearly unread *Principles of Psychology* (1890/1950). James's conception that awareness consists of two qualitatively different parts is quite distinct from current models, and it addressed phenomena that still must be accounted for. His model has nearly been forgotten; it was called to my attention by Mangan (1991), apparently the only contemporary psychologist who recognized the potential of James' idea and built upon it.

James's Model of Awareness

In his famous Chapter IX, "The Stream of Thought" (James 1890/1950), James railed against the "traditional psychology" of his day which held that definite images were the sole building blocks of mental life. He made three key points:

1. Awareness has two parts. He called them the definite and the vague, or the nucleus and the fringe. He also used the metaphor of a halo or penumbra spreading out around a distinct image, or water surrounding a rock in the stream.[2]
2. The nucleus and the fringe present qualitatively different kinds of information. What is presented in fringe experiences are not just dim or preliminary or fleeting or defective versions of nucleus experience (see Galin 1994 for a full discussion of this point).

3. The experiences of the fringe are critically important in their own right. They represent the context and web of relations that give meaning to the particularized contents in the nucleus. James said:

> The definite images of our traditional psychology form but the very smallest part of our minds as they actually live. . . . The significance, the value, of the image is all in this halo or penumbra that surrounds and escorts it . . . (pp. 254–255)

For James, meaning is not intrinsic to a thing but is given by the network of other knowledge and relations in which it is embedded. This applies to any sort of thing, such as an object, event, word, idea, or bit of knowledge. For example, an object's meaning is given by the totality of its aspects such as its name, physical properties, hedonic value, uses, and history (similar to Quillian 1972). Whereas a few particular features of an object are presented in the nucleus, some indications of its extended connections are presented in awareness in the form of a variety of fringe experiences (James 1890/1950, pp. 254–255, 265, 269, James 1904/1976, pp. 28–29, McDermott 1976). Note that both parts of James's distinction "fringe versus nucleus" refer to what is present in awareness. Do not confuse the fringe with the unconscious or the preconscious of Freud, or Polanyi's tacit knowledge (1969), or Schacter's implicit knowledge (1987), which are not in awareness. Neither do nucleus and fringe correspond to the focus of attention versus unattended contents of awareness, nor to Neisser's attended versus pre-attended contents (1967), nor to Gestalt psychology's figure/ground (Koffka 1935). Although it is common to say that there are two kinds of mind, or consciousness, or modes of thought, James's two parts do not correspond to any of the familiar dichotomies (for lists of such pairs see Bogen 1969). It is difficult to grasp a new idea if we say too quickly, "Oh, that is just like so-and-so."

Types of Fringe Experience

James indicated a great many fringe experiences, but he did not attempt an exhaustive list, or a systematic analysis of their relations to each other, or to other mental phenomena. He offered a few examples: feelings of familiarity; feelings of knowing; feelings of action tendency (for example, the intention to say so-and-so); expectant feelings (the commands, "wait," "look," "hark," elicit distinct feelings of the domain from which a new impression is to come). [Note that after struggling to find a single term to denote conscious states (pp. 185–186), James settled on "thought" and "feeling" depending on context, used synonymously with experience and awareness.] James considered the fringe feeling of "rightness" or "on-the-right-track-ness" to be one of the most important, because he believed that it guides thought. This is a feeling that the content currently in the nucleus of awareness is congruent in some global way with our current goal structure (what James calls the "topic" of our thought). He uses a number of synonyms for rightness such as "harmony," or "fittingness."

There is another large group of experiences that seem to belong to the fringe: the subjective component of emotions. But this group is conspicuously missing from James' examples. I will return to emotion later.

The Tip-of-the-Tongue Experience

Several important features of fringe experience are illustrated in James' account of trying to recall a forgotten name, the ubiquitous tip-of-the-tongue phenomena (for contemporary treatments see Baars 1988, p. 225, Brown 1991, Hart 1965, Mangan 1991). James declares that the feeling of an absence is different from (and much more than) the absence of a feeling. I have extended his analysis to clarify what is occurring. Three fringe experiences are active simultaneously: the feelings of knowing, meaning, and mismatch (the inverse of "rightness"). In addition to the intense feeling of knowing that you know, there is an experience in a global way of the web of connections of the missing word, and it is this which gives the experience its feeling of specificity and "singularly definite" quality. "If wrong names are proposed to us, this singularly definite gap acts immediately to negate them . . . the gap of one word does not feel like the gap of another, all empty of content as both might seem . . ." (James 1890/1950, pp. 251–252). Individual features from this web of connections may appear in the nucleus, such as the name's initial sound, or its rhythm (p. 252), but isolated features do not fill enough of the constraints of the web of connections to elicit the complete feeling of "rightness." James's use of the term gap expresses the mismatch between the fringe experience of global meaning for which an adequate token is being sought and the nucleus experience of explicit contents being tried out as candidate tokens.

The tip-of-the-tongue phenomenon illustrates that several fringe experiences can occur together, that they can be intense and specific, and that they may carry directional or control information. Mangan (1991, 1993) has stressed the control aspects, for example, that the feeling of knowing has the effect of sustaining memory search when retrieval of an item is temporarily unsuccessful (Hart 1965, Nelson *et al.* 1984, 1986). The feeling of being on-the-right-track in some way keeps the search in the neighborhood currently being examined, rather than switching to another region of knowledge.

I have presented James's views in some detail because he pointed precisely and gracefully to what is still missing in contemporary accounts of awareness: the qualitative differences among the elements that constitute subjective experience, and the relations among these elements. It must be noted that James himself never developed the notion of fringe and nucleus systematically, and did not make further use of it in *Principles of Psychology* in his extensive treatment of attention, his theory of emotion, or his conception of the self, nor in *Varieties of Religious Experience*.

RELATION OF JAMES'S MODEL TO MODERN CONCEPTS

Contemporary Models Undervalue or Ignore James's Fringe

In another paper (Galin 1994) I have contrasted James's views with four very prominent modern researchers representing different theoretical and methodological perspectives (Neisser 1967, Rumelhardt *et al.* 1986, Crick and Koch 1990, Baars 1988). I contended that they do not adequately deal with the simple introspective data that concerned James in 1890. In summary, this sample of modern theorists, each of whom I value highly for other reasons, have not heeded James's exhortation that "the reinstatement of the vague to its proper place in our mental life" is paramount for scientific psychology. These exemplars were selected because they explicitly include something they call vague awareness in their models, but not as carrying a special class of information. Rather, such feelings are said to be vague because they are either preliminary, defective, or fleeting. Other theorists do not even consider the vague. Because these components of awareness have been categorized as preliminary and/or defective, they have been ignored in most research and theory.[3]

James's Model Confused with the Spotlight Metaphor for Awareness

The moving spotlight is a powerful metaphor that guides much of current research and theory in awareness, implicitly or explicitly. Because a spotlight has a clear focus and a fuzzy periphery, this metaphor is easily confused with James's model, given his choice of terms with similar spatial connotations (vague fringe, halo, or penumbra contrasted with the clear, definite nucleus). It is important to distinguish the quite fundamental differences between James's nucleus/fringe model and the spotlight model.

In the spotlight metaphor the concept "attention" is used to indicate the adjustments of the beam (Crick 1984, Crick and Koch 1990, Jung 1954, Kahneman and Treisman 1984). Attention may denote either the beam's present locus, its stability, intensity, breadth, degree of focus; or the control mechanisms that change these parameters. Sometimes the term attention is used synonymously with awareness (Mandler 1975, pp. 236–238). Some people confuse James's nucleus with the "focus of attention" (Mangan 1993). This is a serious error. That which is attended typically includes both nucleus and fringe components. For example, in James's terms, in the experiencing of a word, whether it is being attended or incidental, the phonological image (nucleus) usually comes along with one or more fringe elements—such as the feeling of its meaning and the feeling of its degree of rightness for the present context. The relative dominance or foreground quality of the nucleus and fringe components varies; in the tip-of-the-tongue experience the fringe elements of the meaning of the desired word and the feeling of gap are more prominent and more persistent than the nucleus images of candidate words or word fragments that rapidly succeed one another.

Neither should James's concept of fringe experience be confused with the unattended, or the dim, fuzzy, fringe of the spotlight beam metaphor. James's fringe presents a separate class of information than the nucleus, not just the same kind of information at a lower resolution. Therefore, whereas dim information at the edge of the spotlight can be brightened (brought into awareness) by recentering the beam on it, in James's model the contents of the fringe as such cannot be brought into the nucleus. They are "attended" in their form as fringe experiences, not converted to the form of nucleus experiences (Jackendoff 1987, Chapter 4, presents compelling arguments on the importance of form).

Mangan's Rehabilitation of Fringe Awareness

To my knowledge the only contemporary psychologist who has explicitly responded to James' distinction between the nucleus and the fringe is Mangan (1991, 1993).[4] Mangan's important insight was that fringe experiences are an overlooked major way to finesse the limited capacity of awareness. He proposed that a fringe experience is a radical condensation of a very large web of nonconscious information relevant to the current topic, which would otherwise exceed the limited processing capacity. The fringe experience is a way to present *a summary form* of the contexts or relations that give meaning to the discrete items presented in the nucleus of awareness.

Note that the condensation or summary provided by the fringe experience is not the same as the summary provided by Miller's chunking (1956, 1962). A new chunk simply replaces many separate items; the fringe experience does not replace features but rather annotates, amplifies, and explicates them.

Mangan focused primarily on the feeling of "rightness," which he characterized as signaling the fit or compatibility between the small number of articulated features currently in the nucleus of awareness and the *nonconscious* information structure that makes up the current "topic" or mental context. For example, imagine that while visiting in an unfamiliar home you are seeking the bathroom. You look into a room and see a stove. The mismatch between the stove and the topic (finding the bathroom) is presented in your awareness as a fringe feeling of "wrongness," accompanying the nucleus awareness of "stove." In the next room you see a bathtub. The fit between this item and the topic is presented in your awareness as the feeling of rightness, or on-the-right-track-ness. For the interested reader unfamiliar with these aspects of cognitive psychology, a full treatment of how rooms and their identifying features are handled in terms of "schemata" and connectionist networks is given in Rumelhart *et al.* 1986, pp. 22–38.

The critical point here is that the fringe awareness of rightness carries a different class of information than the nucleus awareness of the stove or tub. The fringe information in this case has to do with *the relation* of the nucleus items to the topic, the bathroom schema; this could not be given by just another nucleus item. It is providing evaluation or relational information, not just more content information.

Mangan strongly advocates and expands James's position that experiences such as rightness or familiarity have a control function, directing the stream of thought. He links the concept of fringe awareness to important connectionist concepts, arguing (p. 190) that the feeling of rightness carries the same sort of information as the formal connectionist metric "goodness-of-fit" or "harmony." This is an abstract mathematical expression that provides a score for how well the pattern of activation over a PDP network as a whole has matched its input (Rumelhart *et al.* 1986, Smolensky 1986). Because it is a global measure, reflecting information distributed over the whole network, goodness-of-fit could act as a summary or condensation. As far as I know, no connectionist models actually compute and use a value for goodness-of-fit; it is only used by the researcher to describe the global state of the system and to compare states. Mangan suggests that in human minds something like it may actually be computed, and that it shows up in subjective experience as the feeling of rightness.

Connectionist ideas have been predominantly concerned with "the microstructure of cognition" (Rumelhart *et al.* 1986). Mangan's hypothesis makes a bridge from the microcognitive to the level of awareness. This is important because, in my view, one of the purposes served by the emergence of awareness as a level of organization beyond the neural and the cognitive levels (Galin 1992b) is to provide a frame of reference in which to compare and evaluate elements that appear as unwieldy global properties at the microcognitive level.

PROBLEMS WITH JAMES'S MODEL

What Is Meant by "Vague" and "Definite"

James's characterization of the nucleus and the fringe as definite versus vague, also emphasized by Mangan, introduces several sources of confusion, and furthermore, it obscures the really important functional and phenomenal characteristics of these experiences.

First, the terms have several senses. Vague has many synonyms with overlapping but distinguishable meanings: imprecise, undefinable, indistinct, hazy, ineffable. In common usage we may describe a visual image as vague if it is dim, or if the boundaries are not sharp, or if parts are missing, or if the resolution of the parts is poor, or its class membership is uncertain. But certainly an image in nuclear awareness may be both dim and easily classifiable, or both incomplete and sharply resolved. When should we call it vague?

Just as the term vague may be used to describe aspects of nuclear awareness, so too fringe experiences could be described as "definite." There need be no uncertainty about detecting their occurrence or distinguishing one from another. For example, the feeling of knowing in the tip-of-the-tongue experience is very intense, not dim, and it is also very specific in that even a closely related word that is suggested will be rejected. Neither is it fleeting. It is misleading to call it vague.

Second, sometimes the vagueness can be attributed to the objects of awareness rather than to the awareness itself (for example, a puff of smoke or a rippled reflection in a pond).

Third, although an experience was quite definite, the report of it may be vague due to the reporter's inarticulateness or fleeting memory. On the other hand the report may be more definite than the experience, as with the confabulations of a Korsakoff's syndrome patient, or the testimony of an overly eager eyewitness.

Thus, James's and Mangan's rather informal characterization of the fringe as vague and the nucleus as definite needs revising. The ambiguities resolve when we see that what makes the term vague appropriate or not depends crucially on the user's purposes. We speak of something (an image, a thought, a report, or more generally, a representation) as vague if it does not give us all the information we seek from it. For example, sharp boundaries may not be important if our purpose is determining class membership, but would be critical if we were interested in exact size. And in addition, the resolution needed must only match the scale of our question: if you asked the distance to Paris an answer in miles would not be considered vague, but it would be if you had asked the distance between goalposts. Thus, *whether or not we consider an experience vague depends on our purposes for the information it presents.*

This analysis brings us to see that the difference between fringe and nucleus awareness is not simply that one is intrinsically vague and the other intrinsically definite. Rather, their difference is that the information they carry is suitable for different purposes. Therefore I propose that we categorize these different types of experience by the kinds of information they provide, and I will suggest some possibilities below.

Beyond the Dichotomy

I find a second problem in James's formulation. The power of his idea was in showing that the fringe and nucleus experiences were qualitatively different, and in calling attention to the overemphasis on the nucleus and the richness of what was left out. But neither James (nor Mangan) attempted to sort the very heterogeneous array of fringe awarenesses into categories, either phenomenally or functionally. Thus they implicitly gave us the dichotomization of awareness into nuclear versus everything else. We must explicitly reject this parsing of awareness. It is important to distinguish more than two basic varieties. Fringe experiences seem to *differ among themselves functionally and phenomenally as much as they each differ from the nucleus.* For example, the feeling of rightness is as subjectively distinct from the feeling of intention as they both are from the experience of seeing.

This difficulty is compounded further because Mangan has not clarified James's concept of nucleus awareness. What sort of information does the nucleus present? Mangan has accepted by default a loose equation of the nucleus with the concept of focus of attention as it is used in the long-lived "searchlight" model of consciousness discussed above.

Because the kinds of awareness are so numerous, some sort of subgrouping, however provisional, would be useful.[5] In the following sections I suggest an abstract framework and some new nomenclature that will allow us to reformulate what James was pointing at in terms more congenial to cognitive psychology. This is plainly speculative; I propose it merely as heuristic, as what mathematicians or physicists call a conjecture; its purpose is to focus and stimulate thinking before rigorous theorems or experimental data can be produced.

VARIETIES OF AWARENESS

Reconceptualization of James's Nucleus as "Feature Awareness"

The sum of our conscious and nonconscious knowledge of objects can be thought of as a very large set of properties: size, shape, color, history, uses, cost, hedonic value (this applies to events and ideas as well as objects). Consider these general properties as dimensions defining an abstract multidimensional "space." Particular properties (for example, blueness, or price, or pleasantness) can be thought of as specific values arrayed along these dimensions (blue is a particular value along the color dimension). A value may be specified with more precision (a particular shade of blue) or with less precision (any blue between green and violet). Such values are what we commonly call "features." Our nonconscious knowledge of a particular object (or an event or idea) can be described by the whole set of values on the dimensions of its property space. In general, all of our nonconscious knowledge can be thought of as a set of representations in such spaces, which might be separate, or linked, or overlapped, or nested. Many authors have proposed detailed architectures of concept spaces and levels of representation (Jackendoff 1987, Rosch and Lloyd 1978), but this oversimplified sketch will do for the present purpose.

When we become aware of objects, events, or ideas, they are typically experienced in the context of a topic or set of goals. Some dimensions of the object or event are more relevant to the current goal than others; they maximally discriminate among items or choices. For example, when shopping we may be guided sometimes by the dimension of color, and on another occasion by cost. Only the small subset of the object's dimensions most relevant to the current topic need be selected.[6] I propose that when we become aware of an object, event, or idea, the values on these selected dimensions are presented as one part of our awareness. This part corresponds to James's nucleus. I call it "feature-awareness," in order to be descriptive of function, to get away from the spatial and status connotations of "nucleus," and to distinguish it from the contemporary "focus of attention" which can include nucleus, fringe, or both.

Other Varieties of Awareness: Reconceptualizing James's Fringe

But there is much more in awareness than just the most topic-relevant features. This is what James was trying to tell us with his concept of the fringe. I have already stressed that many types of awareness have been conflated under this label. What is different about the information they carry?

One type provides evaluation or relational information, explicating the items in feature-awareness. This type was illustrated above in Mangan's analysis of the feeling of rightness and my example of looking for the bathroom. I emphasized that the information carried by the feeling of rightness concerned the *relation* of the items in feature-awareness (stove or tub) to the topic, the nonconscious bathroom schema. Such information could not be given by just adding more features. Other examples of this type are the feelings of source that tell us vividly whether current feature-awareness is derived from current perception or from memory, or the feelings of agency that tell us whether a proprioceptive image of movement was self-initiated, autonomous, or passive.

Another category of fringe experience presents information about dimensions of the current input *not selected* for feature-awareness. For example, our experience of an object's global meaning can be thought of as a condensation or summary over a relatively large part of its multidimensional property space, in contrast to the few most topic-relevant features. It is this, in addition to the feature-awareness, that fleshes out our experience of an object as more than an assemblage of parts. It is this which is missing in patients with associative agnosia following brain injuries (Damasio 1990, Rubens 1985, Sacks 1985). Such patients may be able to see an object, describe it, draw it, but not recognize what it is or how it is used, hence the title story of Sacks' popular book, *The Man who Mistook his Wife for a Hat* (Sacks 1985). I call this type of experience "current topic summary awareness."

The reader who is not sure of what meaning feels like in awareness may be helped by the contrasting experience of loss of meaning. If you repeat a word aloud twenty times (try "fascination"), your original awareness of its meaning will fade and disappear. What remains of it (or perhaps, what replaces it), is the representation of a few phonemes in feature-awareness, with fringe representation of only a few fragments of its usual extensive connections, or new meanings related to new segmentations of the phoneme string. The experience of the fading of word meaning with repetition is the opposite of the tip-of-the-tongue experience, in which the condensation of the word's meaning is vividly present in the absence of feature-awareness of its key phonemic features. This awareness of a condensed specific meaning is distinct from the awareness of generic meaningfulness (rightness, relevance, significance) that Mangan has emphasized.

A third very important type of fringe information is not related to current input, current feature-awareness, or the immediate topic. Rather, it concerns nonconscious knowledge related to other topics or goals further in the background of goal hierarchies (for example, Baars' nonconscious "contexts," 1988). I call this type "competing topic summary awareness." One example is the feeling of having left something undone, with no awareness of what it is (call home, or file taxes, or spouse's birthday). Another example is a persistent blue or angry mood relating to a previous or anticipated social encounter, or so-called free-floating anxiety that may actually relate to something like a pending exam or surgery.

To summarize thus far, I have proposed that (1) one part of awareness (feature-awareness, née James's nucleus) presents the few most discriminating features of a knowledge structure related to our most immediate topic, and (2) other parts of awareness (James's fringe) present evaluational or relational information that explicates the bare features, or present condensations, or global indices of other nonconscious knowledge, both declarative and procedural, which are related to or competing with the current topic.

The term "fringe" should be replaced. These awarenesses are too varied to be captured by a one-word descriptor, and fringe has a misleading connotation of relative unimportance. Therefore, until we find a less awkward nomenclature, I will use specific labels like "relational awareness" and "summary awareness," and use "nonfeature-awareness" for the collective.

Obviously there are a great many categories beyond the few examples I have offered. It would be useful if we could formulate a broader frame of reference within which to order them. I will present one such frame here: the concept of self-monitoring, which has been attracting attention recently in widely different areas of psychology (Frith 1987, Johnson and Hirst 1992, Prigatano and Schacter 1991, Zaidel 1987).

Categories of Awareness Related to Self-Monitoring

The Self-Monitor[7] Many complex self-regulating systems that adapt to their environment use a regularly up-dated map of their own state and the adaptations made. The processes that keep track of the current state of the self in its environment are likely to be hierarchically organized and distributed, but for simplicity I use the term self-monitor in the singular to refer to all of them collectively. There are some empirical data available to constrain hypotheses on the details of self-monitoring although it has not often been considered as a separate process or general capacity apart from the specific performance being studied, and no general theory of monitoring has been offered (Johnson 1991, Kihlstrom and Tobias 1991, Landis, Graves, and Goodglass 1981, Prigatano and Schacter 1991, Zaidel

1987). Recently I discussed self and self-monitoring in the context of unawareness of deficits following certain brain injuries (Galin 1992b).

We infer the existence of a self-monitor because we know a lot about our present "mode" of organization, including such things as the level and quality of our awareness and our cognition, and our status as agent. For example, we know how aroused we are (drowsy or alert or drunk); we can distinguish imagining from remembering, we can even sometimes realize that we are dreaming and not awake. We have information about our goals and actions; when the doctor taps our knee and elicits a reflex knee-jerk, we can say "I didn't do that." I believe it is more than a figure of speech when a person says "I don't feel like myself today." It is certainly more than that in pathological conditions, for example, when patients with multiple personality disorder experience a radical shift between discrete selves, or when a "split brain" patient watches his left hand (right hemisphere) make a response and says, "I know it wasn't me that did that!" (Galin 1974).

It is important to distinguish self-monitoring from self-awareness. Self-awareness logically means simply awareness of information about the self (Galin 1992b). Numerous experiments have demonstrated that even very complex information processing can go on without our being aware of it (Kihlstrom 1987, Velmans 1991). Therefore it should not be surprising that much or all of what a self-monitor does could be done without awareness. I propose that *when self-monitoring information does enter awareness, it is largely in the form of a wide variety of evaluative, relational, and explicating experiences, with key details as usual in the form of feature-awareness*. The nonfeature awarenesses can be thought of as condensed summaries of the state of certain aspects of our overall system. A summary is needed because the whole map is too big for the limited representational capacity of awareness.

The self-concept is another hyphenated term that needs to be considered because it is often conflated with what I have called self-monitoring and self-awareness. In my set of definitions, the self-concept is a knowledge structure (a body of information) consisting of knowledge, beliefs, and attitudes *about* "who one is" as an object in the world. It is similar in form to our concepts of any other objects, for example, our next-door neighbor or the Washington Monument. Like any other concept we have, it can be incomplete and in some respects incorrect. Some of this knowledge may be inference from earlier reports from the self-monitor, but much of it comes from other sources, such as the opinions of our relatives. It is clearly distinguishable from self-monitoring as defined above. I hypothesize that when information from the self-concept is brought into awareness, it is represented in the same fashion as other concepts.

Taxonomy for Experiences Related to Self-Monitoring

Awarenesses derived from self-monitoring can be provisionally sorted according to the domains to be monitored. I suggest three to start: knowledge, action, and goals. I do not mean this to be an exhaustive list.

The group of awarenesses related to monitoring the domain of *knowledge* includes metacognitive experiences, such as feelings of knowing, familiarity, and source identification (distinguishing memory from perception or imagination; see Johnson 1991). The feeling of knowing determines how long a person persists in memory search; when it is inaccurate one may stop too soon or obsess too long (discussed by Mangan 1991 citing: Gruneberg and Sykes 1978, Hart 1965, Nelson et al. 1984).

The second group, related to the monitoring of *action,* includes experiences such as intentions, sense of effort, and sense of agency. Feelings of intention can be thought of as summary representations of action plans currently readied. This view is complementary with the theoretical framework for the role of attention and will in the control of action proposed by Norman and Shallice (1986). These authors give particular emphasis to the different ways in which an action is subjectively experienced, and this guides the development of their theory. Libet and his colleagues, in studies of cerebral processes related to awarenesses in the action domain, have also emphasized the importance of fine subjective distinctions, for example, between the feeling of preparation (intending to move "soon") and the feeling of intending (wishing to move now). These have distinct electrophysiological correlates (Libet 1985.)

Emotion and the Domain of Goals   The experiences related to monitoring the domain of goals includes a major group conspicuously missing from James' examples: the subjective component of emotions. James treated emotional awareness in a separate theory (1884), with no regard to his categories of fringe and nuclear awareness. In fact, it seems that in his emphasis on the core role of somatic sensations in emotional awareness he was making the error he warned of, that is, emphasizing the nucleus and ignoring the fringe (for a modern treatment clarifying misstatements and misunderstandings of James's theory of emotion, see Papanicolaou 1989).

Emotions are complex events, including subjective, physiologic, behavioral, and cognitive components. I am calling attention here only to the subjective component. These components are organized in coherent patterns (Tomkins 1962). Emotional responses (as distinct from moods) can be conceptualized as brief reactions to events that have significance for our personal goals (Ekman 1977, Lazarus 1991). For example, if progress to the goal is externally blocked we may feel angry; if the event renders the goal irretrievably lost we may feel sad; if the event signals movement closer to the goal we feel happy. I propose that *the subjective component of emotion is a summary representation of where we stand with respect to our current personal goals*. In addition, a few key details of the event, of the goal, or of our physical state may or may not be present in feature-awareness. In this formulation the subjective component of emotion is seen as part of a family of self-monitoring events, and not of a wholly different nature from other events in awareness involved in metacognition, evaluation, and control. This is consistent with a number of modern theories of emotion that take an information processing approach (Carver and Scheier 1990, Oatley and Johnson-Laird 1987, Ortony, Clore, and Collins 1988).

These are just initial steps toward a taxonomy of nonfeature awarenesses, organized in relation to self-monitoring. The few categories suggested do not capture numerous other awarenesses, vivid to most people, that also seem related to self-monitoring. One omitted domain includes the feelings of health, energy, and vitality, as contrasted with feelings of illness or fatigue. Another group concerns sense of social distance and quality, such as the feelings of being isolated or connected, "at home," loving or indifferent. However, the scheme I have sketched indicates a direction for further development.

GENERAL CONCLUSIONS

The test of any new descriptive framework is whether it helps us to group together things whose connections we had not noticed, or to focus our attention on discriminative aspects that we had previously ignored. Although there has been a fair amount of empirical work on some kinds of evaluative, summary, or explicating experiences, in general such research has not been identified as awareness research, and therefore more effort has been spent on assessing performance or mechanism than on phenomenal details and how they vary. Attending explicitly to subjective experience and distinguishing among the varieties of awareness that I have suggested might be helpful with current problems. One example is the study of the feeling of knowing. This is usually classed as memory research, and focuses on correlations of a simple estimate of the feeling's intensity with subsequent memory performance. Few studies have examined how feelings of knowing are related to feelings of familiarity and to feelings of global meaning, and what aspects of the underlying knowledge structure the feeling of knowing represents or omits (see Schacter 1989, 1991, and Brown's 1991 review for exceptions). The details of questions or instructions to subjects are particularly critical because, as Schacter has observed concerning tests of implicit memory, how you test and what instructions you give determines what you find (1991, pp. 140–141). The importance of this self-monitoring awareness is shown in the phenomenon of blindsight in patients with occipital cortical lesions (Cowey and Stoerig 1992, Weiskrantz 1986). Unless forced to "guess" the patients make no attempt to use "visual" information from their blind field. Even though they have obviously processed the information at some level, they have no subjective experience indicating that it is available. There is no associated feeling of knowing.

A complete theory of awareness will have to include accounts of phenomena at the subjective level as well as at the neurological, cognitive, and behavioral levels (Flanagan 1994, Jackendoff 1987). For instance, most work on associative agnosia has only matched up what is missing at the neurological level (for example, lesions of occipital-parietal cortex) with what is missing at the cognitive-behavioral level (inability to identify an object, its use, or its class membership [Rubens 1985]). In this paper I have identified a specific component that is missing in these patients at the phenomenal level, the feeling of meaning, and I have hypothesized that its relation to the cognitive level is that of a condensation or global summary of the knowledge structure that makes up that object's property space. This is not an explanation; it is a better specification of what needs to be explained.

Although contemporary psychologists hold various convictions on the causal potency of awareness (Edelman 1989, Jackendoff 1987, Norman 1986, Sperry 1976, Velmans 1991), all can benefit from James's exhortations to attend to subjective experience; even if these distinct types of awareness do not serve control functions in their own right, they must be indicators of important underlying structures.

An important part of scientific psychology is to describe phenomena and explicate their functions. This is distinct from reductive mapping onto other levels of organization and complements it. Theory-building is an interactive process in which description precedes attempts at reduction and, in turn, is modified by them (Wimsatt 1976a, 1976b). In this paper I have called attention to key descriptive aspects of awareness and extended the distinctions James drew between nucleus and fringe; not as a dichotomy of two types, but as a much larger set. I have emphasized that the important differences among them are not their degrees of vagueness, but rather the types of information they convey. This descriptive analysis was set in the functional framework of self-monitoring, which is relevant to many subdisciplines in psychology. The categories I have suggested, feature-awareness and the variety of evaluative, summary, and explicating awarenesses, give us a more differentiated vocabulary and a more precisely characterized set of variables with which to work. This should be useful for psychologists interested in awareness whether their focus is in computer simulations, neuroscience, or clinical interventions.

NOTES

An extended version of many of the issues discussed in this paper has been presented in Galin 1994 and 1995.

1. The situation is similarly muddy for the terms *awareness, will, voluntary,* and *self*. For a discussion of the acute problem of terminology, see Galin 1992a, 1992b, p. 153. By "awareness" I mean first-person subjective experience, all that James included in the "stream of consciousness." Unless otherwise specified, I am not referring to the underlying mechanisms and control systems that generate it, sustain it, and modify it. I try to use only *aware* and *awareness* (or subjective experience) instead of the more common terms *conscious* and *consciousness* which are often used in more inclusive senses. I find it useful to think of awareness as a medium in which information is presented. The code and what is encoded are conceptually distinguishable from the medium. This concept may be clarified if we consider more familiar examples of media and information. Paper-and-ink is a medium in which information at the level of alphabetic characters can be

presented. Alphabetic characters, in turn, are a medium in which words can be presented. Words are a medium in which concepts can be presented. In this series, each succeeding medium is an emergent property of the information at the previous level, but no rigid dependence is implied; clearly words can be represented in media other than alphabetic characters. A helpful discussion of levels and emergence is given by Wimsatt (1976a).

2. James often changed metaphors to evoke complementary aspects of subjective experience. Sometimes he spoke of awareness as a bird's trajectory, alternating flights and perchings, corresponding to the "transitive" (transitory) and "substantive" (stable) aspects (p. 243). This is different from his nucleus versus fringe distinction. The bird's trajectory metaphor refers to the dynamic aspects of awareness—its duration, direction of change, and rate of change—whereas the nucleus and fringe refer to the form of thought contents at a particular moment. Note also that James spoke of "the stream of thought," flowing seamlessly like a river. He asserted strongly that experience is subjectively unitary and continuous; his division of awareness into functionally and phenomenally distinct parts was only for analytic purposes.

3. I must note one more contemporary psychologist, Jackendoff (1987), who has incorporated some features quite similar to James's two-part structure in his theory, although he developed it without being aware of James's ideas (personal communication, 1992). His model is quite thoroughly worked out with respect to aspects of awareness that would correspond to James's nucleus. But important phenomena were still unaccounted for, and in the penultimate chapter he sketched a second qualitatively different aspect that carried the same sorts of information as the fringe: source information (which distinguishes percept from imagination), feelings of agency, of novelty, of meaningfulness, and a simple emotional liking versus disliking. Unfortunately he has not developed this part of his extremely interesting and useful theory further, and to my knowledge, it has not been picked up by others. Like James, Jackendoff puts great weight on paying attention to subjective experience as a guide to what any cognitive theory must explain. This is the more noteworthy because he considers himself an epiphenomenalist and holds firmly to his belief that consciousness can have no effects on the computational mind or the brain.

4. Mangan's highly original work focuses on the feeling of "rightness" or "on-track-ness." He argues that it plays a central role (and therefore can be used as a bridging concept) in three areas of psychology usually seen as completely disparate: cognition (in attention, problem-solving, and awareness), aesthetics, and religious experience. I sketch only a few of his ideas here.

5. In a very thoughtful essay, Shallice (1988) also commented on the great variety of types of awareness, and noted that there is nothing like an exhaustive account in the cognitive literature. He offered what he called a rough-and-ready list, but did not make use of anything like James's distinctions.

6. What governs the selection, and how "relevance to topic" is defined are complex issues, deliberately glossed over here. More or less detailed models are proposed by Norman and Shallice (1986), Jackendoff (1987), Baars (1988). In my view there are always several topics, some competing and some nested within each other, related to longer or shorter time frames, and more general or more specific goals. Sudden intense stimuli can presumably always gain access to feature awareness; the current topic is then shifted to "what is it?"

7. In considering the role of nonfeature-awarenesses in self-monitoring, it is useful to first clarify the stem concept "self" and separate self-monitoring from the other hyphenated derivatives with which it is often confused: self-concept and self-awareness. Self is rarely explicitly defined even in technical psychology and

philosophy, and can carry with it a great deal of unacknowledged conceptual baggage. In a recent paper I proposed a new definition: the term *self* should denote the overall organization that makes a person an entity (Galin 1992b). A person is made up of component subsystems which can be more or less tightly integrated or autonomous. If we define self as the organization, not just another subsystem, we will have a clear referent for common phrases such as "a more integrated self," or "a development of the self," or "a loss of self." The concept of self as organization, varying dynamically in degree and quality, works at the neurological level of description as well as at the psychological level, and across levels. However, this definition is not necessary for the arguments in the rest of the paper.

REFERENCES

Baars, B. 1988. *A cognitive theory of consciousness.* Cambridge, England: Cambridge University Press.

Bogen, J. E. 1969. The other side of the brain: II. An appositional mind. *Bulletin of the Los Angeles Neurological Society* 34:135–62.

Brown, A. S. 1991. A review of the tip-of-the-tongue experience. *Psychological Bulletin* 109:204–23.

Carver, C. S., and M. F. Scheier. 1990. Origins and functions of positive and negative affects: A control-process view. *Psychological Review* 97:19–35.

Churchland, P. 1988. "Reduction and the neurobiological basis of consciousness." In *Consciousness in contemporary science,* edited by A. J. Marcel and E. Bisiach. Oxford, England: Clarendon Press.

Cowey, A., and P. Stoerig. 1992. "Reflections on blindsight." In *The neuropsychology of consciousness,* edited by A. D. Milner and M. D. Rugg. London: Academic Press, pp. 11–38.

Crick, F. 1984. Function of the thalamic reticular complex: The searchlight hypothesis. *Proceedings of the National Academy of Science, USA* 81:4586–90.

Crick, F., and C. Koch. 1990. Towards a neurobiological theory of consciousness. *Seminars in the Neurosciences* 2:263–75.

Damasio, A. R. 1990. Category-related recognition defects as a clue to the neural substrates of knowledge. *TINS* 13:95–9.

Edelman, G. 1989. *The remembered present.* New York: Basic Books.

Ekman, P. 1977. "Biological and cultural contributions to body and facial movement." In *Anthropology of the body,* edited by J. Blacking. London: Academic Press, pp. 58–84.

Flanagan, O. 1995. "Prospects for a unified theory of consciousness." In *Scientific approaches to consciousness,* edited by J. D. Cohen and J. W. Schooler. Hillsdale, NJ: Lawrence Erlbaum (in press).

Frith, C. D. 1987. The positive and negative symptoms of schizophrenia reflect impairments in the perception and initiation of action. *Psychological Medicine* 17:631–48.

Galin, D. 1974. Implications for psychiatry of left and right hemisphere specialization. *Archives of General Psychiatry* 31:572–83.

Galin, D. 1992a. The blind wise men and the elephant of consciousness. *Consciousness and Cognition* 1:8–11.

Galin, D. 1992b. Theoretical reflection on awareness, monitoring, and self in relation to anosognosia. *Consciousness and Cognition* 1:152–62.

Galin, D. 1994. The structure of awareness: contemporary applications of William James' forgotten concept of "the fringe." *Journal of Mind and Behavior* 15(4):375–400.

Galin, D. 1995. "What is the difference between a duck?" In *Scientific approaches to the question of consciousness: 25th Carnegie Symposium on Cognition,* edited by J. D. Cohen and J. W. Schooler. Hillsdale, NJ: Lawrence Erlbaum (in press).

Gruneberg, M. M., and R. N. Sykes, 1978. "Knowledge and retention: The feeling of knowing and reminiscence." In *Practical aspects of memory,* edited by M. M. Gruneberg, P. E. Morris, and R. N. Sykes. New York: Academic Press, pp. 189–96.

Hart, T. 1965. Memory and the feeling-of-knowing experience. *Journal of Educational Psychology* 56:208–16.

Jackendoff, R. 1987. *Consciousness and the computational mind.* Cambridge, MA: MIT Press.

James, W. 1884. What is an emotion? *Mind* 9:118–205.

James, W. 1950. *The principles of psychology.* New York: Dover. (Originally published 1890)

James, W. 1976. *Essays in radical empiricism.* Cambridge, MA: Harvard University Press. (Originally published 1904)

Johnson, M. K. 1991. "Reality monitoring: Evidence from confabulation in organic brain disease patients." In *Unawareness of deficit after brain injury,* edited by G. P. Prigatano and D. S. Schacter. New York: Oxford University Press, pp. 176–97.

Johnson, M. K., and W. Hirst. 1992. "Processing subsystems of memory." In *Perspectives in cognitive neuroscience,* edited by R. G. Lister and H. J. Weingartner. New York: Oxford University Press, pp. 197–217.

Jung, R. 1954. "Correlation of bioelectric and autonomic phenomena with alterations of consciousness and arousal in man." In *Brain mechanisms and consciousness,* edited by J. F. Delafresnaye. Springfield, IL: Charles C Thomas, pp. 310–44.

Kahneman, D., and A. Treisman. 1984. "Changing views of attention and automaticity." In *Varieties of attention,* edited by R. Parasuraman and D. R. Davies. Orlando, FL: Academic Press, pp. 29–62.

Kihlstrom, J. F. 1987. The cognitive unconscious. *Science* 237:1445–52.

Kihlstrom, J. F., and B. A. Tobias. 1991. "Anosognosia, consciousness, and self." In *Unawareness of deficit after brain injury,* edited by G. P. Prigatano and D. S. Schacter. New York: Oxford University Press, pp. 198–222.

Koffka, K. 1935. *Principles of Gestalt psychology.* New York: Harcourt Brace.

Landis, T., R. Graves, and H. Goodglass. 1981. Dissociated awareness of manual performance on two different visual associative tasks: A "split-brain" phenomenon in normal subjects? *Cortex* 17:435–40.

Lazarus, R. 1991. *Emotion and adaptation.* New York: Oxford University Press.

Libet, B. 1985. Unconscious cerebral initiative and the role of conscious will in voluntary action. *Behavioral and Brain Sciences* 8:529–66.

Mandler, G. 1975. "Consciousness: Respectable, useful, probably necessary." In *Information processing and cognition,* edited by R. Solso. Hillsdale, NJ: Lawrence Erlbaum, pp. 229–254.

Mangan, B. 1991. Meaning and the structure of consciousness: An essay in psychoaesthetics. University of California Berkeley, Doctoral dissertation. University Microfilms No. 92033636.

Mangan, B. 1993. Taking phenomenology seriously: The "fringe" and its implications for cognitive research. *Consciousness and Cognition* 2:89–108.

Marcel, A., and E. Bisiach. Eds. 1988. *Consciousness in contemporary science.* Oxford: Clarendon Press.

McDermott, J. J. 1976. Introduction In W. James, *Essays in radical empiricism.* Cambridge, MA: Harvard University Press.

Miller, J. 1956. The magical number seven, plus or minus two. *Psychological Review* 63:81–97.

Miller, J. 1962. *Psychology: The science of mental life.* New York: Harper and Row.

Neisser, U. 1967. *Cognitive psychology.* San Francisco: W. H. Freeman.

Nelson, T. O., D. Gerler, and L. Narens. 1984. Accuracy of feeling-of-knowing judgements for predicting perceptual identification and relearning. *Journal of Experimental Psychology: General* 113:282–300.

Nelson, T. O., R. J. Leonesio, R. S. Laandwehr, and L. Narens. 1986. A comparison of three predictors of an individual's memory performance. *Journal of Experimental Psychology: Learning, Memory, and Cognition* 12:279–87.

Norman, D. A. 1986. "Reflections on cognition and parallel distributed processing." In *Parallel distributed processing,* Vol. 2, edited by J. L. McClelland and D. E. Rumelhard. Cambridge, MA: MIT Press, pp. 531–46.

Norman, D. A., and T. Shallice. 1986. "Attention to action." In *Consciousness and self-regulation,* edited by R. J. Davidson, G. E. Schwartz and D. Shapiro. New York: Plenum Press, pp. 1–18.

Oatley, K., and P. N. Johnson-Laird. 1987. Towards a cognitive theory of emotion. *Cognition and Emotion* 1:29–50.

Ortony, A., G. L. Clore and A. Collins. 1988. *The cognitive structure of emotions.* New York: Cambridge University Press.

Papanicolaou, A. C. 1989. *Emotion: A reconsideration of the somatic theory.* New York: Gordon and Breach.

Polanyi, M. 1969. *Knowing and being.* Chicago: University of Chicago Press.

Prigatano, G. P., and D. S. Schacter. 1991. *Unawareness of deficit after brain injury.* New York: Oxford University Press.

Quillian, M. R. 1972. "How to make a language user." In *Organization of memory,* edited by E. Tulving and W. Donaldson. New York: Academic Press, pp. 310–54.

Rosch, E., and B. B. Lloyd, eds. 1978. *Cognition and categorization.* Hillsdale, NJ: Lawrence Erlbaum.

Rubens, A. 1985. "Agnosia." In *Clinical neuropsychology,* 2nd ed, edited by K. Heilman and E. Valenstein. New York: Oxford University Press, pp. 187–242.

Rumelhart, D. E., and J. L. McClelland, eds. 1986. *Parallel distributed processing,* Vol. 1: *Foundations.* Cambridge, MA: MIT Press.

Rumelhart, D. E., P. Smolensky, J. L. McClelland, and G. E. Hinton. 1986. "Schemata and sequential thought processes in PDP models." In *Parallel distributed processing,*

Vol. 2, edited by J. L. McClelland and D. E. Rumelhardt. Cambridge, MA: MIT Press, pp. 7–57.

Sacks, O. 1985. *The man who mistook his wife for a hat.* New York: Summit Books.

Schacter, D. L. 1987. Implicit memory: History and current status. *Journal of Experimental Psychology Learning, Memory, and Cognition* 13:501–18.

Schacter, D. S. 1989. "On the relation of memory and consciousness." In *Varieties of memory and consciousness: Essays in honor of Endel Tulving,* edited by H. L. Roediger and F. I. M. Craik. Hillsdale, NJ: Lawrence Erlbaum, pp. 355–90.

Schacter, D. S. 1991. "Unawareness of deficit and unawareness of knowledge in patients with memory disorders." In *Unawareness of deficit after brain injury,* edited by G. P. Prigatano and D. S. Schacter. New York: Oxford University Press, pp. 127–51.

Shallice, T. 1988. "Information-processing models of consciousness: possibilities and problems." In *Consciousness in contemporary science,* edited by A. Marcel and E. Bisiach. Oxford, England: Clarendon Press, pp. 305–33.

Smolensky, P. 1986. "Information processing in dynamical systems: foundations of harmony theory." In *Parallel distributed processing,* edited by D. E. Rumelhart and J. L. McClelland, Vol. 1: *Foundations.* Cambridge, MA: MIT Press, pp. 194–281.

Sperry, R. W. 1976. "Mental phenomena as causal determinants in brain function." In *Consciousness and the brain,* edited by G. Globus, G. Maxwell, and I. Savodnic. New York: Plenum, pp. 163–77.

Tomkins, S. 1962. *Affect, imagery, and consciousness,* Vol. 1. NY: Springer.

Velmans, M. 1991. Is human information processing conscious? *Behavioral and Brain Sciences* 14:651–726.

Weiskrantz, L. 1986. *Blindsight.* Oxford, England: Clarendon Press.

Wimsatt, W. C. 1976a. "Reductionism, levels of organization, and the mind-body problem." In *Consciousness and the brain,* edited by G. G. Globus, G. Maxwell, and I. Savodnic. New York: Plenum, pp. 205–66.

Wimsatt, W. C. 1976b. "Reductive explanation: a functional account." In *Proceedings of the meetings of the Philosophy of Science Association, (1974),* edited by C. A. Hooker, G. Pearse, A. C. Michealos, and J. W. van Evra. Dordrecht, Netherlands: Reidel.

Zaidel, E. 1987. "Hemispheric monitoring." In *Duality and unity of the brain,* edited by D. Ottoson. Hampshire: Macmillan, pp. 77–97.

10 The Varieties of Conscious Experience: Biological Roots and Social Usages

Karl H. Pribram

> Let us ourselves look at the matter in the largest possible way. Modern psychology, finding definite psycho-physical connections to hold good, assumes as a convenient hypothesis that the dependence of mental states upon bodily conditions must be thoroughgoing and complete... According to the general postulate of psychology just referred to, there is not a single one of our states of mind, high or low, healthy or morbid, that has not some organic process as its condition. Scientific theories are organically conditioned just as much as religious emotions are; ... so of all our raptures and our drynesses, our longings and pantings, our questions and beliefs. They are equally organically founded, be they religious or of non–religious content.
>
> [However,] to plead the organic causation of a religious state of mind, then, in *refutation* of its claim to possess superior spiritual value, is quite illogical and arbitrary ... None of our thoughts and feelings, not even our scientific doctrines, not even our dis-beliefs, could retain any value as revelations of the truth, for every one of them without exception flows from the state of its possessor's body at the time.
>
> William James, *The Varieties of Religious Experience*, 1902/1929

INTRODUCTION: FORMS OF CONSCIOUS EXPERIENCE

Over the past two centuries, since the pioneering observations of Frances Gall and G. Spurtzheim (1809/1969), it has become clear that there is a special relation between brain tissue and the varieties of conscious experience. Gall initiated the procedure of comparing the locus of brain pathology with aberrations of behaviors of the patients whose brains he examined—a procedure that is continued today in the active field of clinical neuropsychology.

Though, on the whole, we today accept this special relation between brain and conscious experience, we are not at all agreed on the basic nature of the relationship nor on the consequences our understanding of this nature might have on our understanding of ourselves and our relation to others. In this respect we have apparently come no further than philosophers of the past two millennia. I believe the time is ripe for an advance in understanding because of the amount of detailed experimental and observational data now available which, when properly interpreted, can provide specific transfer functions that bind the mind-matter relation into a coherent science.

To begin with, the advent of computer science has provided the transparency that was lacking in an earlier period. I use my computer as a word processor by typing English words and sentences. The word processing system, by virtue of an operating system, assembler, ASCII, octal or hexadecimal, converts the keyboard input to binary which is the "language" of the computer. There is nothing in the description of English and that of binary machine language that appears to be similar. Despite this, by virtue of the various transformations produced in the encoding and decoding operations of the various stages leading from typescript to binary, the information of the typescript is preserved in the binary language of the operation of the computing machine.

In a similar fashion, there is little in conscious experience that resembles the operations of the neural apparatus with which it has such a special relation. However, when the various transformations, the transfer functions, the codes that intervene between experience and neural operations are sufficiently detailed, a level of description can be reached in which the transformations of experience are homomorphic with the "language" used by the brain. As will be reviewed in this essay, this language is the language of the operations of a microprocess taking place in synaptodendritic fields, a mathematical language similar to that which describes processes in micro (that is, subatomic) physics. Thus, the relation between brain and conscious experience becomes implemented at the microprocessing level. At neuronal and neural system levels the microprocesses become variously configured to produce the variety of conscious experiences.

First, there are *electrochemical synaptodendritic states* which are coordinate with *states of consciousness*. The operation of anesthetics and the very active field of psychoneuropharmacology attest to this relationship. Anesthetics work on a level that produces a quantitative rather than a qualitative change in consciousness; whereas catechol and indole amines act at specified brain sites to produce a variety of states of consciousness such as wakefulness and sleep; depression and elation; and probably even dissociated states such as those seen in schizophrenia. Relative concentrations of blood glucose and osmolarity produce hunger and thirst; sex hormones produce sexual feelings; and peptides such as the endorphins and enkephalins are related to the experiences of pain and stress and their converse, well being.

Second, there are detailed descriptions of the relations between the *sensory systems of the brain* and the sensory aspects of perception: these are responsible for the organization of the *contents* of consciousness (see for example, Pribram 1991, for a detailed account of the neural systems and neuronal functions involved in figural vision).

Third, states of consciousness often determine contents and, as often are determined by them. When hungry one tends to see restaurant signs (Zeigarnik 1972); walking past the fresh aromas emanating from a bakery whets the appetite. This connection between states and the contents of consciousness is mediated by a process—a process ordinarily called *attention*.

STATES OF CONSCIOUSNESS: THE SYNAPTODENDRITIC MICROPROCESS

The Mutual Exclusiveness of Conscious States

We ordinarily distinguish different states of consciousness much as does the physician and surgeon: when someone responds to prodding (for example, by grumbling "Oh leave me alone! Can't you see I'm trying to get some sleep!") we attribute to him a conscious state. When, on the other hand his response is an incoherent thrashing about, we say he is stuporous and if there is no response at all, we declare him comatose.

The interesting thing about such states is their mutual exclusiveness regarding experience: what is experienced in one state is not readily available to experience in another. Such state exclusiveness emerges in all sorts of observations: state-dependent learning in animal and human experiments; the fact that salmon spawning pay no attention to food, while when they are in their feeding state sexual stimuli are ignored; the observation in hypnosis that a person can be made unaware posthypnotically of suggestions made during hypnosis (although he carries out these suggestions); and the dissociation between experiences (and behavior) taking place during "automatisms" in temporal lobe epileptics and their ordinary state.

The evidence obtained in all of these situations suggests that the same basic synaptodendritic electrochemical substrate becomes variously organized to produce one or another conscious state. Hilgard (1977) has conceptualized these various organizations as a more or less "vertical" rearrangement of the substrate. One might picture such arrangements to resemble those that take place in a kaleidoscope: a slight rotation and an entirely new configuration presents itself (Pribram and Gill 1976, Chapter 5). Slight changes in relative concentrations of chemicals in the organization of synaptodendritic microprocesses in specific locations could, in similar fashion, result in totally different conscious states.

The Cortical Microprocess

To demostrate the basis upon which the organization of synaptodendritic processing domains operate and to portray it in a novel and realistic fashion we performed the following experiments on the rat somatosensory system. This system is convenient and the relation between whisker stimulation and central neural pathways has been extensively studied (see review by Gustafson and Felbain-Kermadias, 1977). Whiskers were stimulated by a set of rotating cylinders, each grooved with equally spaced steps, the step width and adjacent grooves subtending equal angles. Three cylinders were used with their steps measuring 30 degrees, 15 degrees, and 7.5 degrees, respectively. The cylinders were rotated at eight different speeds, varying from 22.5 degrees/second to 360 degrees/second. (The rotating cylinders were meant to mimic the drifting of gratings across the retinal receptors in vision.)

In most of our experiments an entire array of whiskers was subjected to contact with the rotating cylinders. This was done in order to bring the results of these somatosensory experiments into register with those performed in the visual system where an entire array of receptors is stimulated by the drifting grating.

Axonal spike trains recorded from single electrodes can be attributed to three separable processes: (1) those due to the sensory input per se, (2) those that are intrinsic to the operations of the synaptodendritic field potentials, and (3) those that reflect the output of the axon hillock (Pribram *et al.* 1981, Berger and Pribram 1992). In our experiments, sensory influences are generated by the frequency (spectrum) of the stimulus as modulated by the spacings of the grooves on the cylinders and the speed with which the cylinders are rotated. The results thus provide maps of the number of bursts or spikes generated at each spectral location as determined by the spatial and temporal parameters of the sensory input. (Figures 10.1a–d and 10.2a–f). The activity above or below baseline that resulted from whisker stimulation is plotted as a manifold describing total number of bursts (or spikes) per 100 seconds of stimulation. Spatial frequencies are scaled in

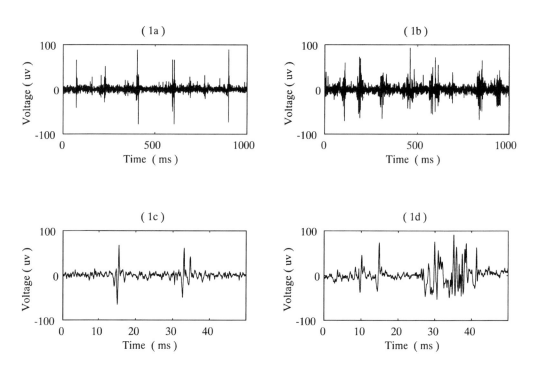

Figure 10.1 (1a) One second of a typical recording with no whisker stimulation (baseline). (1b) Data from the same location during one second of whisker stimulation (spatial frequency=24 grooves/revolution; temporal frequency=0.125 rps). Figures (1c) and (1d) show individual units during 50 mscs of baseline (1c) and the superposition of units during whisker stimulation.

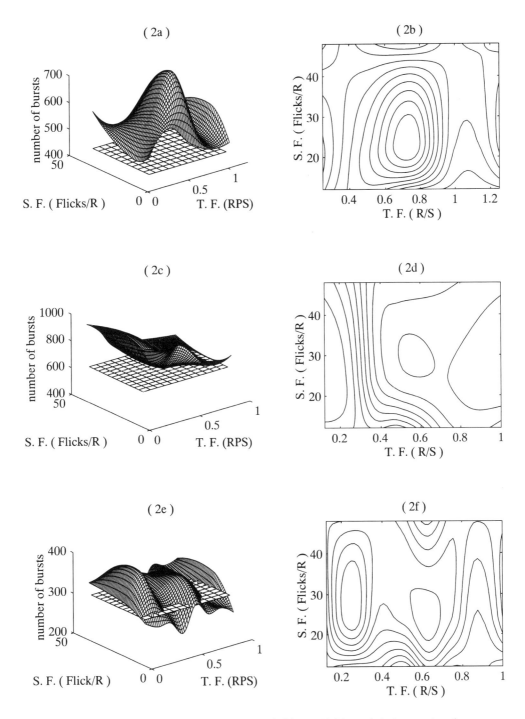

Figure 10.2 Examples of receptive field manifolds and their associated contour maps derived by an interpolation (spline) procedure from recorded whisker stimulation. The contour map was abstracted from the manifold by plotting contours in terms of equal numbers of bursts per recording interval (100 seconds). Each figure shows baseline activity (no whisker stimulation) at a given electrode location as a gr–plane located in terms of number of bursts per 100 seconds.

terms of grooves per revolution, while temporal frequencies are scaled in terms of revolutions per second. Thus, the density of stimulation of a whisker (or set of whiskers) is a function of both the spacings of the cylinder grooves and the speed with which the cylinder rotates. It is this density per se that composes the spectral domain.

In 27 experiments single whiskers were isolated and stimulated. Whiskers were identified according to accepted nomenclature as described by Simons (1978). The receptive field potential manifolds derived from such stimulations were irregular and broadly tuned to both spatial or temporal frequency. The intrinsic operations governing the configuration of the synaptodendritic field potentials are constrained by parameters such as the anatomical extent of each receptive field and the functional inhibitory and excitatory relationships among such fields. Our analyses were derived from both bursts of unit activity and from single units. We therefore sought to determine the relationships between the manifolds derived from bursts and those derived from single units composing the bursts.

A manifold (Figure 10.3) constructed from bursts is shown to encompass those of the individual units composing the bursts: Figures 10.4a–d illustrate manifolds from the four single units which compose the bursts. These units were identified using a template constructed from a spike sorting procedure that discriminated the shape of the action potential (spike) on the basis of spike amplitude and recovery slope. The four single unit manifolds show a gradual change in shape corresponding to slight changes in location within the burst manifold: Figures 10.4a and 10.4b illustrate two peaks which progressively become combined into a single broad peak in the manifolds of Figures 10.4c and 10.4d. This demonstration of continuity between two levels of analysis (bursts and single units) strongly supports the view proposed by Pribram (1991) that extended networks of synaptodendritic

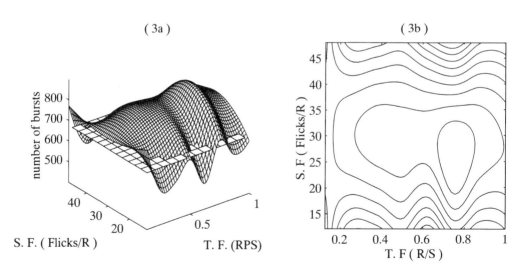

Figure 10.3 Lateral (3a) view of an empirically derived burst manifold and its associated contour map (3b).

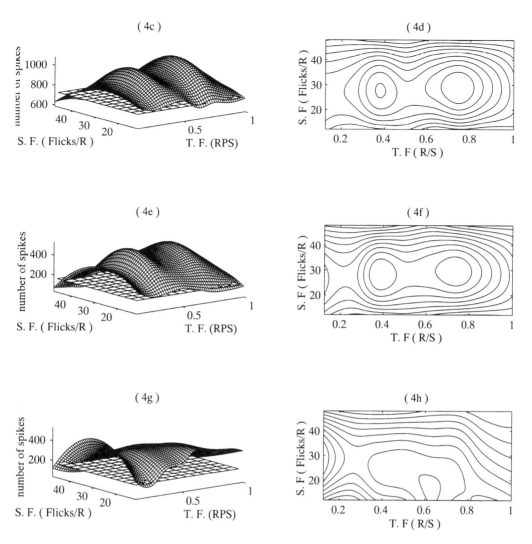

Figure 10.4 These figures illustrate manifolds and their associated contour maps from four single units which compose the bursts used to construct the manifold in Figure 10.3.

fields serve as the processing medium, and therefore, single neurons are sampling from overlapping areas of the synaptodendritic network.

The similarity of these manifolds obtained from recordings made from the somatosensory cortex to the receptive field characteristics demonstrated in the primary visual cortex (DeValois and DeValois 1988, Pollen and Taylor 1974, Pribram and Carlton 1986, Daugman 1990) suggests that this processing medium is ubiquitous in the cortical synaptodendritic network.

The manifolds derived from our data are constructed of two orthogonal dimensions: one dimension reflects the spatial frequency of the stimulus and the other its temporal frequency. Because spatial and temporal variables constrain the spectral density response, a Gabor-like rather than a simple Fourier representation describes our results. Thus the results of our experiments can be interpreted in terms of an information field composed of Gabor–like elementary functions, that is, of truncated two dimensional sinusoids.

THE CONTENTS OF CONSCIOUSNESS

Objective Consciousness: The Posterior Cerebral Convexity

The Projection of Sensory Experience—The Perifissural Cortex Surrounding the major fissures of the primate brain lie the terminations of the sensory and motor projection systems. Rose and Woolsey (1949) and Pribram (1960) have labeled these systems extrinsic because of their close ties (by way of a few synapses) with peripheral structures. The sensory surface and muscle arrangements are mapped more or less isomorphically onto the perifissural cortical surface by way of discrete, practically parallel lines of connecting fibre tracts. When a local injury occurs within these systems a sensory scotoma, or a scotoma of action, ensues. A scotoma is a spatially circumscribed hole in the "field" of interaction of organism and environment: a blind spot, a hearing defect limited to a frequency range, a location of the skin where tactile stimuli fail to be responded to. These are the systems where what Henry Head (1920) called epicritic processing takes place. These extrinsic sensory-motor projection systems are so organized that movement allows the organism to project the results of processing away from the sensory (and muscular) surfaces where the interactions take place, out into the world external to the organism. Thus processing within these extrinsic systems constructs an objective reality for the organism.

In between the perifissural extrinsic regions of cortex lie other regions of cortex variously named association cortex (Fleschig 1900), uncommitted cortex (Penfield 1969), or intrinsic cortex (Pribram 1960). These names reflect the fact that there is no apparent direct connection between peripheral structures and these regions of cortex that make up most of the convexity of the cerebrum.

The Personal/Extrapersonal Distinction—The Basal Ganglia and the Right Hemisphere Lesions of the intrinsic cortex of the posterior cerebral convexity result in sensory-specific agnosias in both monkey and man. Research on monkeys has shown that these agnosias are not due to failure to distinguish cues from one another, but due to making use of those distinctions in making choices among alternatives (Pribram and Mishkin 1955, Pribram 1969). This ability is the essence of information processing in the sense of uncertainty reduction (Shannon and Weaver 1949), and the posterior intrinsic cortex determines the range of alternatives, the sample size which a particular informative element must address. A patient with agnosia can tell the difference between two objects but does not know what the difference means. As Peirce (1934) once noted, what we mean by something and what we mean to do with it are synonymous. In short, alternatives, sample size, choice, cognition, information in the Shannon sense, and meaning are closely interwoven concepts. Finally, when agnosia is severe it is often accompanied by what is termed "neglect." The patient appears *not* only not to know that he doesn't know but to actively deny the agnosia. Typical is a patient I once had who repeatedly had difficulty in sitting up in bed. I pointed out to her that her arm had become entangled in the bedclothes—she would acknowledge this momentarily, only to "lose" that arm once more in a tangled environment. Part of the perception of her body, her personal consciousness seems to have become extinguished.

These results can readily be conceptualized in terms of extrapersonal and personal objective consciousness. For a time it was thought that personal body space depended on the integrity of the frontal intrinsic cortex and that the posterior convexal cortex was critical to the construction of extrapersonal reality (see for example, Pohl 1973). This scheme was brought to test in my laboratory in experiments with monkeys (Brody and Pribram 1978) and patients (Hersh 1980, Ruff, Hersh, and Pribram 1981) and found wanting. In fact, the personal/extrapersonal distinction involves the parietal cortex; in humans, most often the right parietal cortex and related structures. Studies by Mountcastle and his group (Mountcastle *et al.* 1975) in monkeys demonstrate how cells in the convexal intrinsic cortex respond when an object is within view, but only when it is also within reach. In short, our studies on patients and those of others have been unable to clearly separate the brain locations that produce agnosia from those that produce neglect. Furthermore, the studies on monkeys as well as those on humans (McCarthy and Warrington 1990, Chapter 2) indicate that agnosia is related to meaning as defined by corporeal use.

In monkeys the disturbances produced by restricted lesions of the convexal intrinsic cortex are also produced by lesions of the parts of the basal ganglia to which those parts of the cortex project. This finding takes on special meaning from the fact that lesions of the thalamus (which controls the relaying of sensory input to cortex) fail to produce such effects. Further, recent experiments have shown that the neglect syndrome can be produced in monkeys by lesions of the dopaminergic nigrostriatal system (Wright 1980). This special connection between intrinsic (recall that this is also called association) cortex and the basal ganglia further supports the conception that these systems make possible, on the basis of use, the distinction between an objective personal self (the "me") and, an extrapersonal reality. (See Pribram 1991, Lecture 6 for a detailed exposition of how this process operates.) An excellent review of the history of differentiating this objective "me" from a historical "I" can be found in Hermans, Kempen, and van Loon (1992). The next sections develop the relation between brain processing and the "I."

Episodic Consciousness: The Limbic Forebrain

Instinct as a Species–Shared Propensity   When resections were restricted to the amygdala and adjacent pole of the temporal lobe, the marked taming of the monkeys that had followed resection of the entire temporal lobe (Sanger–Brown and Schäefer 1888, Klüver and Bucy 1937) are reproduced (Pribram and Bagshaw 1953). Just what might this behavioral change signify?

First it was determined that not only were the monkeys tamed, but they also put everything in their mouths, gained weight, and increased their sexual behavior—all effects that had also followed the total temporal lobectomy. These changes in behavior were summarized under the rubric of the "four Fs": fighting, fleeing, feeding, and sex (Pribram 1960).

Historically these apparently disparate behaviors were classified together as "instinct" (a term still used to describe the processes underlying such behaviors in the psychoanalytic literature). More recently this concept came into disfavor (see for example, Beach 1955) and ethologists substituted the category "species specific" behaviors for instinct because these behaviors can be shown to have a common genetic component. But this substitution loses much of the meaning of the older terminology: Human language is species-spccific but not instinctive in the earlier sense. My preference is to retain the concept of instinct as descriptive of the four F's: What these behaviors have in common is the fact that their patterns are shared by practically all species. What makes the study of geese and other birds so interesting is that we recognize our own behavior patterns in the descriptions provided by ethologists (see for example, Lorenz 1969). It is therefore *species-shared* behavior-patterns that are of interest in tracking the effects of amygdalectomy.

The Boundaries of an Episode   The apparently disparate behaviors that characterize the four Fs were shown by careful analysis to be influenced by a common process. It is worth summarizing the highlights of this analysis because identifying a common process operating on apparently disparate behaviors is a recurring problem in behavioral neuroscience. In behavioral genetics the same problem entails identifying genotypes from phenotypical behaviors. Thus, qualitative and quantitative determinations were made in each of the four Fs with the following results.

In a social hierarchy fighting and fleeing were both diminished provided there was a sufficiently skillful antagonist (Rosvold, Mirsky, and Pribram 1954). In the study reported by Sanger–Brown and Schäefer (1888), when a monkey was returned to the social colony after amygdalectomy, he "voluntarily approaches all persons—and fellow monkeys indifferently." Also, having just interacted with his fellow monkey, and perhaps having been trounced, "he will go through the same process, as if he had entirely forgotten his previous experience."

This change in processing what would ordinarily be a deterrent outcome of behavior was dramatically demonstrated by displaying a lighted match to such monkeys. They would invariably grab the match, put it into their mouth, dousing the flame, only to repeat the grab when the next lit match was presented. This behavior could be elicited for a hundred consecutive trials unless either the monkey or the experimenter became bored before the session was ended (Fulton et al. 1949).

The increases in feeding that follow amygdalectomy were also shown to be due to a failure placing limits on actions. For instance, as reported by Sanger-Brown and Schäefer, monkeys with such resections appear to be indiscriminate in what they pick up, put in their mouths, and swallow. But when tests were performed and a record was kept of the order in which the food and nonfood objects were chosen, it turned out that the order of preference was undisturbed by the brain operation; only now the monkeys would continue to pick up additional objects beyond those that they had chosen first (Wilson 1959). In fact amygdalectomized animals may be a bit slow to start eating but continue eating far past the point when their controls stop eating (Fuller, Rosvold, and Pribram 1957).

These disturbances in feeding after amygdalectomy were shown to be due to connections with the satiety mechanism centered in the ventromedial region of the hypothalamus. For instance, a precise relationship was established between the amount of carbachol injected into the amygdala and amount of feeding (or drinking) once these behaviors had been initiated (Russel et al. 1968). Injections into the ventromedial hypothalamic region simply terminate feeding.

Modulation of a stop process was also shown responsible for changes in fighting behavior. Fall in a dominance hierarchy after amygdalectomy, when it occurred, was related to the amount of aggressive interaction between the dominant and submissive animals of the group. After amygdalectomy such interactions were overly prolonged leading to a reorganization of the dominance hierarchy. It was as if the amygdalectomized monkeys approached each interaction as novel. Prior experiences, which modulated the behavior of the control subjects, seemed to have little influence after amygdalectomy. This finding characterizes many of the experimental results to be described shortly.

Analyses of the effects of amygdalectomy and electrical stimulations of the amygdala on avoidance (fleeing) behavior brought a similar conclusion. Escape behavior is unaffected and sensitivity to shock is not diminished (Bagshaw and Pribram 1968). Nor is there a change in the generalization gradient to aversive stimulation (Hearst and Pribram 1964a, 1964b). What appears to be affected primarily is the memory aspect of avoidance—the expectation based on familiarity with the situation that aversive stimulation will occur. Such expectations are ordinarily referred to as feelings of fears that constrain behavior.

The theme recurs when the effects of amygdalectomy on sexual behavior are analyzed. The hypersexuality produced by the resections is found to be due to an increased territory and range of situations over which the behavior is manifest: Ordinarily cats perceive unfamiliar territory as inappropriate for such behavior (see Pribram 1960, for review). Sexual behavior is limited to familiar situations and situations become familiar as a consequence of rewarding sexual encounters.

The importance of the amygdala in more generally determining the spatial and temporal boundaries of a series of experiences or a behavioral routine—in short, an episode—is attested by the results of another set of experiments. Kesner and DiMattia (1987) presented a series of cues to animals to allow them to become familiar and then paired the initial, intermediate, and final cues of the series with novel cues in a discrimination. When similar tasks are administered to humans, they recall the initial and final cues of the series more readily than they recall the intermediate ones. These are termed the primacy and recency effects. Unoperated monkeys showed both effects in Kesner's experiments. However, after amygdalectomy, monkeys failed to show either a recency or a primacy effect. If the series is taken to be an episode, the effects of amygdalectomy can be considered to impair the demarcation of an episode. As described shortly, after resections of the far frontal cortex, ordering within an episode becomes deficient.

Familiarization: Episode as Context The demonstration of an episode is effected by an orienting reaction. What is oriented to, the novel, depends on the familiar, which serves as the context within which an event becomes appreciated as novel.

Familiarization is fragile. The process is readily disrupted by head injury or distraction. Some of the factors governing distractibility such as pro– and retroactive interference are well known. Amygdalectomy and resections of forebrain systems related to the amygdala have been shown to increase susceptibility to distraction (Douglas and Pribram 1969, Grueninger and Pribram 1969). Resistance to distraction is furnished by a visceroautonomic "booster" that places a value on the experience and thus leads to a *feeling* of familiarity. It is this booster process in which the amygdala is involved (Pribram, Douglas, and Pribram 1969).

Familiarity is a conscious feeling regarding an experience. In the clinic, patients who have a lesion in the region of the amygdala (and the adjacent horn of the hippocampus) describe experiences that are called "jamais vu" and "déjà vu"—the patient will enter a place such as his living room and experience a "jamais vu," a feeling of "never having seen," of complete unfamiliarity. Others will come into a place they have never been and feel that they have "already seen," are already, "déjà," completely familiar with it.

In the laboratory, familiarity has been shown to be related to reinforcement history. Monkeys were trained to select one of two cues on the basis of a 70 percent reinforcement schedule: that is, selection of one cue was rewarded on 70 percent of the trials; selection of the other cue was rewarded on 30 percent of the trials. Then the cue that had been most rewarded was paired with a novel cue. Control monkeys selected the previously rewarded cue. Monkeys who had their amygdalas removed selected the novel cue. Familiarization by virtue of previous reinforcing experience had little effect on monkeys who lacked the amygdala (Douglas and Pribram 1966). These monkeys were performing in a "jamais vu mode."

The process by which the history of reinforced episodes leads to conscious emotional and motivational states can be described as follows:

Consequences are the outcomes of behavior. In the tradition of the experimental analysis of behavior, consequences are reinforcers or deterrents that influence the recurrence of the behavior of which they are the consequences. Con-sequences are thus a series of events (Latin *ex-venire*, out-come), outcomes that guide action and thereby attain predictive value (confidence estimates). Such con-sequences, that is, sequence of events that form their own confidential context become, in humans, envisioned eventualities (Pribram 1963, 1971, 1991, Lecture 10 and Appendix G).

Confidence implies familiarity. Experiments with monkeys (Pribram *et al.* 1979) and humans (Luria, Pribram, and Homskaya 1964) have shown that repeated exposure to a stimulus habituates, that is, the orienting reaction gives way to familiarization. Familiarization is disrupted by limbic (amygdala) and frontal lesions (Pribram *et al.* 1979, Luria, Pribram, and Homskaya 1964). Disruption leads to repeated distraction and thus the outcomes-of-behaviors, events, become inconsequential. When intact, familiarization is segmented by orienting reactions into episodes within which confidence values can become established.

In such an episodic process the development of confidence is a function of coherences and correlations among the events being processed. When coherence and correlation spans multiple episodes, the organism becomes *committed* to a course of action (a prior intention, a strategy) which then guides further action and is resistant to perturbation by particular orienting reactions (arousals). The organism is now *competent* to carry out the action (intention-in-action; tactic). Particular outcomes now guide competent performance, they no longer produce orienting reactions (Brooks 1986, Pribram 1980). This cascade which characterizes episodic processing leads ultimately to considerable autonomy (an emotional or motivational state) of the *committed* competence. This state is not some vague feeling, however: Contents specific to the reinforcement history are addressed.

NARRATIVE CONSCIOUSNESS: THE FAR FRONTAL CORTEX AND THE LEFT HEMISPHERE

As is well known, frontal lesions were produced for a period of time in order to relieve intractable suffering, compulsions, obsessions, and endogenous depressions. When effective in pain and depression, these psychosurgical procedures portrayed in man the now well-established functional relationship between frontal intrinsic cortex and the limbic forebrain which was undertaken in nonhuman primates as a result of this clinical experience (Pribram 1950, 1954, 1958a, 1958b). Further, frontal lesions can lead either to perseverative, compulsive behavior or to distractibility in monkeys, and this is also true of humans (Pribram *et al.* 1964, Oscar-Berman 1975). Extreme forms of distractibility and obsession are due to a lack of "sensitivity" to feedback from consequences. Both the results of experiments with monkeys (Pribram 1960, 1962) and clinical observations attest to the fact that subjects with frontal lesions, whether surgical, traumatic, or neoplastic, fail to be guided by consequences (Luria, Pribram, and Homskaya 1964, Konow and Pribram 1970).

In a continually changing situation where episodic demarcation of consequences becomes difficult, or when transfer among the eventualities that comprise contexts is blocked, other resources must be mobilized. Such situations demand executive intervention if action is to be consequential. This part of the essay addresses the issue of an executive processor, a brain system that directs and allocates the resources of the rest of the brain. Ordinarily, input from sensory or internal receptors preempts allocation (for discussion see Miller, Galanter, and Pribram 1960) by creating a "temporary dominant focus" of activation within one or another brain system (for review, see Pribram 1971a, pp. 78–80). However when extra demands are placed on the routine operations of allocation, coherences among proprieties and priorities must be organized, and practical inference initiated. Proprieties must structure competencies, priorities must be ordered and practicalities assessed.

Proprieties, Priorities, and Practicalities The far frontal cortex is surrounded by systems that, when electrically excited, produce movement and visceroautonomic effects. On the lateral surface of the frontal lobe lies the classical precentral motor cortex (for review see Bucy 1944, Pribram 1991, Lecture 6). On the mediobasal surface of the lobe lie the more recently discovered "limbic" motor areas of the orbital, medial frontal, and cingulate cortex (Kaada, Pribram, and Epstein 1949, Pribram 1961). It is therefore likely that the functions of the far frontal cortex are, in some basic sense, related to these somatomotor and visceroautonomic effects.

At the same time, the far frontal cortex derives an input from the medial portion of the thalamus, the *n. medialis dorsalis*. This part of the diencephalon shares with those from anterior and midline nuclei (the origins of the input to the limbic cortex) an organization different from that of the projections from the ventrolateral group of nuclei to the cortex of the convexity of the hemisphere (see Chow and Pribram 1956, Pribram 1991 for review).

The close anatomical relationship of the far frontal cortex to the limbic medial forebrain is also shown by comparative anatomical data. In cats and other nonprimates, the gyrus proreus is the homologue of the far frontal cortex of primates. This gyrus receives its projection from the midline magnocellular portion of the *n. medialis dorsalis*. This projection covers a good share of the anterior portion of the medial frontal cortex; gyrus proreus on the lateral surface is limited to a narrow sliver. There appears to have been a rotation of the medial frontal cortex laterally (just as there appears to have occurred a rotation medially of the occipital cortex—especially between monkey and man) during the evolution of primates.

From these physiological and anatomical considerations it appears likely that the far frontal cortex is concerned with relating the motor functions of the limbic to those of the dorsolateral convexity. This relationship has been expressed by Deecke *et al.* (1985) in terms of the what, when, and how of action.

Deecke *et al.* (1985) concluded in an extensive review of their studies using electrical recordings made in humans that: The orbital cortex becomes involved when the question is what to do; the lateral cortex becomes active when the question is how something is to be done, and the dorsal portions of the lobe mediate when to do it. According to the anatomical connections of the far–frontal portions of lobe, described below, "what" can be translated into propriety, "how" into practicality, and "when" into priority.

On an anatomical basis, the far frontal systems have been shown to comprise three major divisions (see Pribram 1987, 1990 for review): One, an orbital, is derived from the same phylogenetic pool as, and is reciprocally connected with, the amygdala (and other parts of the basal ganglia such as the *n. accumbens,* which have been shown to be involved in limbic processing). As might be predicted from the role of the amygdala in familiarizing, in déjà and jamais vu phenomena, this orbital system augments and enhances sensitivities as to what to do, to propriety based on episodic processing (see below).

The second, a dorsal system, is derived from the same root as, and has connections with, the hippocampal system which includes the limbic medial frontal-cingulate cortex. As might be expected from the involvement of the hippocampus in recombinant processing—in innovation—the dorsal far frontal system controls flexibility when actions are to be engaged, in ordering priorities to ensure effective action.

The third, a laterally located system has strong reciprocal connections with the posterior cerebral convexity. It is this system that involves the far frontal cortex in a variety of sensory-motor modalities when sensory input from the consequences of action incompletely specifies the situation. In such situations practical inference becomes necessary.

Organizing Coherence In addition to its demarcation by successive orienting reactions, a defining attribute of an episode is that what is being processed coheres—processing must deal with covariation in terms of familiarity, equivalence, and novelty. Covariation can lead to interference, thus resulting in the inability to order the processing of events. Recall that primacy and recency effects were impaired after amygdala and hippocampal damage. With far frontal damage, monkeys show impairment in processing the latter part of the middle of a series. This impairment is attributed to increased pro- and retroactive interference among items in the series (Malmo and Amsel 1948).

The impairment is also shown by patients with damage to their frontal cortex. These patients fail to remember the place in a sequence in which an item occurs: The patients lose the ability to "temporally tag" events, that is, to place them within the episode. With such patients, Milner (1971, see also Petrides and Milner 1982) performed a series of experiments demonstrating how the processing impairment affects the middle portions of an episode. In her studies, it is *relative recency*, the *serial position* of covarying experiences, that becomes muddled. Other patients with fronto–limbic damage are described by Kinsbourne and Wood (1975). In keeping with the proposals put forward in this essay, they interpret the impairment in processing serial position as due to a derangement of the context that structures an episode.

Fuster (1988) conceptualized the far frontal processing of context in terms of cross-temporal contingencies. Relative recency, for instance, implies that a temporal context exists within which recencies can be relative to one another. However, as indicated by experimental results in which spatial context is manipulated, as in variants of object constancy tasks (Anderson *et al.* 1976) the contextual influence can be spatiotemporal as well as temporotemporal. In fact, in other experiments (Brody and Pribram 1978, Pribram, Spinelli, and Kamback 1967) data were obtained indicating far frontal involvement whenever processing is influenced by two or more distinct sets of covarying contextual contingencies, even when both are spatial.

The computation of this covariation demands that cross-temporal, spatiotemporal, and cross-spatial contingencies be processed. In classical and operant conditioning, the consequences of behavior are contiguous in time and place with the stimulus conditions that initiate the behavior. Contiguity determines the episode or conditioning "trial." When contiguity is loosened, stimulation that intervenes between initiation and consequence has the potential to distract and thus to prevent the processing of covariation. Processing is destabilized. Perturbation is controlled only if a stable state, established coherence, instructs and directs the process.

By virtue of the processes mediated by the middle part of the far frontal cortex covarying episodes are woven into a story expressed in language when the left hemisphere becomes engaged. This story can become a narrative, the myth by which "I" live. This narrative composes and is composed of an intention, a strategy that works for the individual in practice, a practical guide to action in achieving (temporary) stability in the face of a staggering range of variations of events (Pribram 1991, 1992). When the narrative becomes dysfunctional, as during endogenous depression or the persistence of obsessions or compulsions, alleviation of symptoms can be produced by changing the neural substrate of such persistence, that is changing the neural substrate of the "I."

Consciousness is manifest (by verbal report) when familiarization is perturbed, an episode is updated and incorporated into the narrative (Pribram 1991, Appendices C and D). Consciousness becomes attenuated when actions and their guides cohere—the actions become skilled, graceful and automatic (Miller, Galanter, and Pribram 1960).

TRANSCENDENTAL CONSCIOUSNESS: THE FAR FRONTAL CORTEX AND THE RIGHT HEMISPHERE

As noted in the epigram that introduces this essay, James did not limit his exploration to ordinary states of consciousness. The esoteric tradition in Western culture and the mystical traditions of the Far East are replete with instances of uncommon states that produce uncommon contents. These states are achieved by a variety of techniques such as meditation, Yoga, or Zen. The contents resulting from processing in such states appear to differ from ordinary feelings or perceptions. Among others, experiences such as the following are described (Morse, Venecia and Milstein 1989 and Stevenson 1974, for review): (1) oceanic, that is, a merging of corporeal and extracorporeal reality, and out-of-body, that is, corporeal and extracorporeal realities continue to be clearly distinguished but are experienced by still another reality: "a meta-me," or (2) the "I" becomes a transparent throughput experiencing everything everywhere, there is no longer a segmentation of experience into episodes; nor do events become enmeshed in a narrative structure. All of these experiences can be induced by drugs including anesthetics that are known to act on the neural cytoskeleton and the synaptodendritic microprocess. As to contents, they are often attributed to some encompassing structure. It is these transcendental contents which address domains that are ordinarily termed "spiritual." As will be developed below, these spiritual aspects of consciousness can be accounted for by assuming that there is an overriding effect of excitation of the frontolimbic forebrain (especially that of the right hemisphere) on the dendritic microprocess that characterizes cortical synaptodendritic domains in the sensory extrinsic systems (involved in the construction of objective reality).

In addition to the gross correspondence between cortical synaptodendritic domains and the organization of sensory surfaces which gives rise to the overall characteristics of processing in the extrinsic systems, a microprocess which depends on the internal organization of each domain comes into play. As noted in Part I, this internal organization of domains embodies, among other characteristics, a spectral dimension: Synaptodendritic domains are tuned to limited bandwidths of frequencies of radiant energy (vision), sound and tactile vibration. I have reviewed this evidence extensively on a number of occasions (Pribram 1966, 1971a, 1982, 1991, Pribram, Nuwer, and Barron 1974).

As late as the 1950s, how the brain operated to make perception possible and how the effects of perceived experiences could be stored remained enigmatic. For the most part, this was due to the fact that no one could inquire how distributed processes could operate to produce a palpable reality. Thus Karl Lashley (1950) exclaimed that his lifelong search for an encoded memory trace had been in vain, and Gary Boring (1929) indicated in his *History of Experimental Psychology* that little was to be gained, at this stage of knowledge, by psychologists studying brain function.

As noted in the Introduction to this essay, all this was dramatically changed when engineers, in the early 1960s, found ways to produce optical holograms using the mathematical formulation proposed by Dennis Gabor (1948). The mathematics of holography and physical properties of holograms provided a palpable instantiation of distributed memory and how percepts (images) could be retrieved from such a distributed store. Engineers (Van Heerden 1963), psychophysicists (Julez and Pennington 1965), and neuroscientists (Pribram 1966, and Pollen, Lee, and Taylor 1971) saw the relevance of holography to the hitherto intractable issues of brain function in memory and perception (Barrett 1969a, 1969b, Campbell and Robson 1968, and Pribram, Nuwer, and Barron 1974).

This timeless/spaceless/causeless aspect of processing is instigated by frontolimbic excitation that practically eliminates the inhibitory surrounds of receptive fields in the sensory systems (Spinelli and Pribram 1975) allowing these systems to function holistically. It is this holistic type of processing that is responsible for the apparently extrasensory dimensions of experience which characterize the esoteric traditions: Because of their enfolded property these processes tend to swamp the ordinary distinctions such as the difference between corporeal and extracorporeal reality.

The ordinary distinctions result from an enhancement of the inhibitory surrounds of the receptive fields when the systems of the posterior cortical convexity become activated (Pribram, Lassonde, and Ptito 1981). As a consequence the sensory system becomes an information processing system in Shannon's sense: intentional choices among alternatives become possible. This is comparable to the process called the "collapse of the wave function" in quantum physics. By contrast, in the esoteric traditions, consciousness is not limited to choices among alternatives and is therefore "intuitive."

An intriguing and related development (because it deals with the specification of a more encompassing, "cosmic" order) has occurred in quantum physics. Over the past 50 years it has become clear that there is a limit to the accuracy with which certain measurements can be made when others are being taken. This limit is expressed as an indeterminacy. Gabor, in his description of a quantum of information, showed that a similar indeterminacy describes telecommunication: this leads to a unit of minimum uncertainty, a unit describing the maximum amount of information that can be packed for processing.

These contributions have resulted in a convergence of our understanding of the microstructure of communication—and therefore of observation—and the microstructure of matter. The necessity of specifying the observations that lead to inferring the properties of matter has led noted physicists to write a representation of the observer into the description of the observable. Some of these physicists have noted the similarity of this specification to the esoteric descriptions of consciousness. Books with such titles as *The Tao of Physics* (Capra 1975) and *The Dance of the Wu Li Masters* (Zukav 1971) have resulted.

There is therefore in the making a real revolution in Western thought. The scientific and esoteric traditions have been clearly at odds since the time of Galileo. Each new scientific discovery and the theory developed from it has, up until now, resulted in the widening of the rift between objective science and the subjective spiritual aspects of man's nature. The rift reached a maximum toward the end of the nineteenth century: mankind was asked to choose between God and Darwin; heaven and hell were shown by Freud to reside within us and not in our relationship to the natural universe. The discoveries of twentieth-century science briefly noted here, but reviewed extensively elsewhere (Pribram 1986, 1991) do not fit this mold. For once the recent findings of science and the spiritual experiences of mankind are consonant. This augurs well for the upcoming new millennium—a science which comes to terms with the spiritual nature of mankind may well outstrip the technological science of the immediate past in its contribution to human welfare.

SUMMARY AND CONCLUSION

The varieties of conscious experience reviewed can be accounted for within the purview of science. Brain science has contributed sizably to this accounting.

States of consciousness are linked to electrochemical configurations of synaptodendritic domains; conscious processing to attention, volition and thought. The contents of consciousness fall into three major categories: (1) the construction of personal and extrapersonal objective reality by virtue of processing by systems of the posterior cerebral convexity; (2) the construction of narrative composed of episodes and eventualities as processed by the frontolimbic forebrain. Finally, (3) we distinguish a transcendental variety of consciousness that goes beyond narrative by freeing the cortical dendritic microprocess entirely from the spatiotemporal constraints so essential to the construction of personal and extrapersonal reality.

As with so many categorical constructions that we take to be unitary (for example, time, complexity, mind, brain), understanding demands deconstruction into components that appear to be only loosely connected. The puzzle is why these components are grouped together in the first place. There is some intuition that fails to be addressed by analysis. With regard to consciousness, this intuition may rest simply on the unity of experiencing—not what is experienced or how experience is generated. Rather it is *that* experiencing occurs. The dictionary definition of experiencing is to try, to test. This definition accounts for the affinity between consciousness and conscience which in some languages are not distinguished. In a deep sense, therefore, consciousness occurs when some destabilizing "trying" event or series of events occurs: An interrupted action or plan; an earthquake or other unanticipated environmental challenge; a neurochemical disequilibration of a homeostatic process. Thus, to be conscious is to experience, to be tried and to try. To paraphrase Descartes: I am tried and I try, therefore I am conscious.

REFERENCES

Anderson, R. M., S. C. Hunt, A. Vander Stoep, and K. H. Pribram, 1976. Object permancy and delayed response as spatial context in monkeys with frontal lesions. *Neuropsychologia* 14:481–90.

Bagshaw, M. H., and K. H. Pribram. 1968. Effect of amygdalectomy on stimulus threshold of the monkey. *Experimental Neurology* 20:197–202.

Barrett, T. W. 1969a. The cortex as interferometer: The transmission of amplitude, frequency, and phase in cortical structures. *Neuropsychologia* 7:135–48.

Barrett, T. W. 1969b. The cerebral cortex as a diffractive medium. *Mathematical Biosciences* 4:311–50.

Beach, F. A. 1955. The descent of instinct. *Psychological Review* 62:401–10.

Bekesy Von, G. 1967. *Sensory inhibition.* Princeton, NJ: Princeton University Press.

Berger, D. H. and K. H. Pribram. 1992. The relationship between the Gabor Elementary Function and a stochastic model of the inter-spike interval distribution in the responses of visual cortex neurons. *Biological Cybernetics* 67:191–4.

Boring, E. G. 1929. *A history of experimental psychology.* New York: The Century Co.

Brody, B. A., and K. H. Pribram. 1978. The role of frontal and parietal cortex in cognitive processing: Tests of spatial and sequence functions. *Brain* 101:607–33.

Brooks, C. V. 1986. How does the limbic system assist motor learning? A limbic comparator hypothesis. *Brain and Behavioral Evolution* 29:29–53.

Bucy, P. C. 1944. *The precentral motor cortex.* Chicago: University of Illinois Press.

Campbell, F. W., and J. G. Robson. 1968. Application of Fourier analysis to the visibility of gratings. *Journal of Physiology* 197:551–66.

Capra, F. 1975. *The Tao of physics.* Boulder, CO: Shambhala.

Chow, K. L., and K. H. Pribram. 1956. Cortical projection of the thalmic ventrolateral nuclear group in monkeys. *J. Comp. Neurol.* 104:37–75.

Daugman, J. G. 1990. "An inormation-theoretic view of analog representation in striate cortex." *Computational neuroscience,* edited by E. Schwartz. Cambridge, MA: MIT Press.

Deecke, L., H. H. Kornhuber, M. Long, and H. Schreiber. 1985. Timing function of the frontal cortex in sequential motor and learning tasks. *Human Neurobiology* 4:143–54.

DeValois, R. and K. K. DeValois. 1988. *Spatial vision* (Oxford psychology series No. 14). New York: Oxford University Press.

Douglas, R. J., and K. H. Pribram. 1969. Distraction and habituation in monkeys with limbic lesions. *Journal of Comparative and Physiological Psychology* 69:473–80.

Fleschig, P. 1900. Les centres de projection et d'association de cerveau humain. *XIII Congress International de Medecine (Sect. Neurologie):* 115–21. Paris.

Fuller, J. L., H. E. Rosvold, and K. H. Pribram. 1957. The effect of affective and cognitive behavior in the dog of lesions of the pyriform-amygdala-hippocampal complex. *Journal of Comparative and Physiological Psychology.* 50:89–96.

Fulton, J. F., K. H. Pribram, J. A. F. Stevenson, and P. Wall. 1949. Interrelations between orbital gyrus, insula, temporal tip and anterior cingulate gyrus. *Transactions of the American Neurological Association* 175–79.

Fuster, J. M. 1988. *The prefrontal cortex.* Anatomy, physiology and neuropsychology of the frontal lobe (2nd ed.). New York: Raven.

Gabor, D. 1948. A new microscopic principle. *Nature* 61:777–78.

Gall, F. J., and G. Spurtzheim. 1809/1969. "Research on the nervous system in general and on that of the brain in particular." In *Brain and Behavior*, edited by K. H. Pribram. Middlesex, England: Penguin, pp. 20–26.

Grueninger, W. E., and K. H. Pribram. 1969. Effects of spatial and nonspatial distractors on performance latency of monkeys with frontal lesions. *Journal of Comparative and Physiological Psychology* 68:203–09.

Gustafson, J. W., and S. L. Felbain-Kermadias. 1977. Behavioral and neural approaches to the function of the mystacial vibrissae. *Psychol. Bull.* 84:477–88.

Head, H. 1920. *Studies in neurology.* London: Oxford University Press.

Hearst, E., and K. H. Pribram. 1964a. Facilitation of avoidance behavior by unavoidable shocks in normal and amygdalectomized monkeys. *Psychological Reports* 14:39–42.

Hearst, E., and K. H. Pribram. 1964b. Appetitive and aversive generalization gradients in amygdalectomized monkeys. *Journal of Comparative and Physiological Psychology* 58:296–98.

Hermans, H. J. M., H. J. G. Kempen, and R. J. P. van Loon. 1992. The dialogical self: Beyond individualism and rationalism. *American Psychologist* 47(1):23–33.

Hersh, N. A. 1980. *Spatial disorientation in brain injured patients.* Unpublished dissertation, Department of Psychology, Stanford University.

Hilgard, E. R. 1977. *Divided consciousness: Multiple controls in human thought and action.* New York: Wiley.

James, W. 1902/1929. *The varieties of religious experience.* New York: Random House.

Julez, B., and D. S. Pennington. 1965. Equidistributed information mapping: An analogy to holograms and memory. *Journal of the Optical Society of America* 55:605.

Kaada, B. R., K. H. Pribram, and J. A. Epstein. 1949. Respiratory and vascular responses in monkeys from temporal pole, insular, orbital surface and cingulate gyrus. *Journal of Neurophysiology* 12:347–56.

Kesner, R. P., and B. V. DiMattia. 1987. "Neurobiology of an attribute model of memory." In *Progress in psychobiology and physiological psychology,* edited by A. N. Epstein and A. Morrison. Vol. 12. New York: Academic Press. 12:207–77.

Klüver, H., and P. C. Bucy. 1937. "Psychic blindness" and other symptoms following bilateral temporal lobectomy in Rhesus monkeys. *American Journal of Physiology* 119:352–53.

Konow, A., and K. H. Pribram. 1970. Error recognition and utilization produced by injury to the frontal cortex in man. *Neuropsychologia* 8:489–91.

Lashley, K. S. 1960. "The thalamus and emotion." In *The neuropsychology of Lashley*, edited by F. A. Beach, D. O. Hebb, C. T. Morgan, and H. W. Nissen. New York: McGraw-Hill, pp. 345–60.

Lorenz, K. 1969. Innate bases of learning. In *On the biology of learning*, edited by K. H. Pribram. New York: Harcourt Brace and World, pp. 13–94.

Luria, A. R., K. H. Pribram, and E. D. Homskaya. 1964. An experimental analysis of the behavioral disturbance produced by a left frontal arachnoidal endothelioma meningioma. *Neuropsychologia* 2:257–80.

Malmo, R. B., and A. Amsel. 1948. Anxiety–produced interference in serial rote learning with observations on rote learning after partial frontal lobectomy. *JEP* 38:440–54.

McCarthy, R. A., and E. K. Warrington. 1990. *Cognitive neuropsychology: A clinical introduction.* London: Academic Press.

Miller, G. A., A. Galanter, and K. H. Pribram. 1960. *Plans and the structure of behavior.* New York: Henry Holt and Company.

Milner, B. 1971. Interhemispheric differences in the localization of psychological processes in man (Review). *British Medical Bulletin* 27(3):272–77.

Morse, M., D. Venecia, and J. Milstein. 1989. Near-death experiences: A neurophysiological explanatory model. *Journal of Near-Death Studies* 8:45–53.

Mountcastle, V. B., J. C. Lynch, A. Georgopoulos, H. Sakata, and C. Acuna. 1975. Posterior parietal association cortex of the monkey: Command functions for operations within extrapersonal space. *Journal of Neurophysiology* 38:871–908.

Oscar-Berman, M. 1975. The effects of dorso-lateral-frontal and ventrolateral-orbito-frontal lesions on spatial discrimination learning and delayed response in two modalities. *Neuropsychologia* 13:237–46.

Peirce, C. S. 1934. *Collected papers.* Cambridge, MA: Harvard University Press.

Penfield, W. 1969. "Consciousness, memory and man's conditioned reflexes." In *On the biology of learning*, edited by K. H. Pribram. New York: Harcourt, Brace and World, pp. 127–68.

Petrides, M., and B. Milner. 1982. Deficits on subject-ordered tasks after frontal and temporal-lobe lesions in man. *Neuropsychologia* 20:249–62.

Pohl, W. G. 1973. Dissociation of spatial and discrimination deficits following frontal and parietal lesions in monkeys. *Journal of Comparative Physiological Psychology* 82:227–39.

Pollen, D. A., J. R. Lee, and J. H. Taylor. 1971. How does the striate cortex begin reconstruction of the visual world? *Science* 173:74–77.

Pollen, D. A., and J. H. Taylor. 1974. "The striate cortex and the spatial analysis of visual space." In *The neurosciences third study program*, edited by F. O. Schmitt and F. G. Worden. Cambridge, MA: MIT Press, pp. 239–47.

Pribram, K. H. 1950. Psychosurgery in midcentury. *Surgery, Gynecology and Obstetrics* 91:364–67.

Pribram, K. H. 1954. "Toward a science of neuropsychology (method and data)." In *Current trends in psychology and the behavior sciences*, edited by R. A. Patton. Pittsburgh, PA: University of Pittsburgh Press, pp. 115–42.

Pribram, K. H. 1958a. "Comparative neurology and the evolution of behavior." In *Evolution and behavior*, edited by G. G. Simpson. New Haven, CT: Yale University Press, pp. 140–64.

Pribram, K. H. 1958b. "Neocortical functions in behavior." In *Biological and biochemical bases of behavior,* edited by H. F. Harlow and C. N. Woolsey. Madison, WI: University of Wisconsin Press, pp. 151–72.

Pribram, K. H. 1960. "The intrinsic systems of the forebrain." In *Handbook of physiology, neurophysiology, II,* edited by J. Field, H. W. Magoun, and V. E. Hall. Washington, DC: American Psychological Society, pp. 1323–44.

Pribram, K. H. 1961. "Limbic system." In *Electrical stimulation of the brain,* edited by D. E. Sheer. Austin, TX: University of Texas Press, pp. 563–74.

Pribram, K. H. 1959/1962. Interrelations of psychology and the neurological disciplines. In *Psychology: A study of a science,* edited by S. Koch. New York: McGraw-Hill, pp. 119–57.

Pribram, K. H. 1963. Reinforcement revisited: A structural view. In *Nebraska symposium on motivation,* edited by M. Jones. Lincoln, NE: University of Nebraska Press, pp. 113–59.

Pribram, K. H. 1966. "Some dimensions of remembering: Steps toward a neuropsychological model of memory." In *Macromolecules and behavior,* edited by J. Gaito. New York: Academic Press, pp. 165–87.

Pribram, K. H. 1969. On the neurology of thinking. *Behavioral Science* 4:265–87.

Pribram, K. H. 1971a. *Languages of the Brain: Experimental paradoxes and principles in neuropsychology.* Englewood Cliffs, NJ: Prentice–Hall; Monterey CA: Brooks/Cole, 1977; New York: Brandon House, 1982. (Translations in Russian, Japanese, Italian, Spanish).

Pribram, K. H. 1971b. The realization of mind. *Synthese* 22:313–22.

Pribram, K. H. 1980. "The orienting reaction: Key to brain representational mechanisms." In *The orienting reflex in humans,* edited by H. D. Kimmel. Hillsdale, NJ: Lawrence Erlbaum, pp. 3–20.

Pribram, K. H. 1982. "Localization and distribution of function in the brain." In *Neuropsychology after Lashley,* edited by J. Orbach. New York: Lawrence Erlbaum, pp. 273–96.

Pribram, K. H. 1986. The cognitive revolution and the mind/brain issue. *American Psychologist* 41:507–20.

Pribram, K. H. 1987. Subdivisions of the frontal cortex revisited. In *The frontal lobes revisited,* edited by E. Brown and E. Perecman. IRBN Press, pp. 11–39.

Pribram, K. H. 1990. "Frontal cortex—Luria/Pribram rapprochement." In *Contemporary neuropsychology and the legacy of Luria,* edited by E. Goldberg. Hillsdale, NJ: Lawrence Erlbaum, pp. 77–97.

Pribram, K. H. 1991. *Brain and perception: Holonomy and structure in figural processing.* Hillsdale, NJ: Lawrence Erlbaum.

Pribram, K. H., A. Ahumada, J. Hartog, and L. Roos. 1964. "A progress report on the neurological process disturbed by frontal lesions in primates." In *The frontal granular cortex and behavior,* edited by S. M. Warren and K. Akart. New York: McGraw Hill, pp. 28–55.

Pribram, K. H., and M. H. Bagshaw. 1953. Further analysis of the temporal lobe syndrome utilizing frontotemporal ablations in monkeys. *Journal of Comparative Neurology* 99:347–75.

Pribram, K. H., and E. H. Carlton. 1986. Holonomic brain theory in imaging and object perception. *Acta Psychologica* 63:175–210.

Pribram, K. H., R. J. Douglas, and B. J. Pribram. 1969. The nature of nonlimbic learning. *Journal of Comparative and Physiological Psychology* 69:765–72.

Pribram, K. H., and M. Gill. 1976. *Freud's "Project" Reassessed.* New York: Basic Books.

Pribram, K. H., M. C. Lassonde, and M. Ptito. 1981. Classification of receptive field properties, *Experimental Brain Research* 43:119–30.

Pribram, K. H. and D. McGuinness. 1975. Arousal, activation and effort in the control of attention. *Psychological Review* 82(2):116–49.

Pribram K. H., M. Nuwer, and R. Barron. 1974. "The holographic hypothesis of memory structure in brain function and perception." In *Contemporary developments in mathematical psychology,* edited by R. C. Atkinson, D. H. Krantz, R. C. Luce, and P. Suppes. San Francisco: W. H. Freeman, pp. 416–67.

Pribram, K. H., S. Reitz, M. McNeil, and A. A. Spevack. 1979. The effect of amygdalectomy on orienting and classical conditioning in monkeys. *Pavlovian Journal* 14(4):203–21.

Pribram, K. H., D. N. Spinelli, and M. C. Kamback. 1967. Electrocortical correlates of stimulus response and reinforcement. *Science* 57:94–96.

Rose, J. E., and C. N. Woolsey. 1949. Organization of the mammalian thalamus and its relationship to the cerebral cortex. *EEG Clinical Neurophysiology.* 1:391–404.

Rosvold, H. E., A. F. Mirsky, and K. H. Pribram. 1954. Influence of amygdalectomy on social interaction in a monkey group. *Journal of Comparative and Physiological Psychology* 47:173–78.

Ruff, R. M., N. A. Hersh, and K. H. Pribram. 1981. Auditory spatial deficits in the personal and extrapersonal frames of reference due to cortical lesions. *Neuropsychologia* 19(3):435–43.

Russell, R. W., G. Singer, F. Flanagan, M. Stone, and J. W. Russell. 1968. Quantitative relations in amygdala modulation of drinking. *Physiology and Behavior* 3:871–75.

Sanger-Brown, and E. A. Schäefer. 1888. An investigation into the functions of the occipital and temporal lobes of the monkey's brain. *Philosphical Transactions of the Royal Society of London* 179:303–27.

Shannon, C. E., and W. Weaver. 1949. *The mathematical theory of communications.* Urbana, IL: The University of Illinois Press.

Simons, D. J. 1978. Response properties of vibrissa units in rat SI somatosensory neocortex. *Journal of Neurophysiology,* 41(3):798–820.

Van Heerden, P. J. 1963. A new method of storing and retrieving information. *Applied Optics* 2:387–92.

Wilson, W. H. 1959. The role of learning, perception and reward in monkey's choice of food. *American Journal of Psychology* 72:560–65.

Wright, J. J., M. D. Craggs, and A. A. Sergejew. 1979. Visual evoked response in lateral hypothalamic neglect. *Experimental Neurology* 65(1):178–85.

Zeigarnik, B. V. 1972. *Experimental abnormal psychology.* New York: Plenum.

Zukav, G. 1971. *The dancing Wu Li masters.* New York: Morrow.

III Medicine

Although the study of consciousness is an old subject, there is a feeling that significant advances are soon to be made. One reason for this optimism, clearly, is that much more is now known about the structure and dynamics of the brain and the nature of the mind than was available to William James. This section explores the mind's behavior under various types of stress and neuropathology, in a tradition that goes back to Paul Broca's studies of the effects of brain injuries well over a century ago.

In Chapter 11, James Whinnery, an aeromedical scientist, describes transitions between conscious and unconscious states under the influence of acceleration–induced, high gravitational fields. He has studied numerous healthy subjects (fighter-pilot trainees) whirling around in a centrifuge with acceleration leading to inadequate brain blood flow. At sufficiently high acceleration–induced gravitational fields, the subjects lose consciousness, and subsequently regain consciousness in a sequence of steps that include: myoclonic convulsions, dreamlets, return of memory, vision, some awareness, motor function, clearing of confusion/disorientation, and finally normal consciousness. At the conference Whinnery showed a striking video; here he describes the sequence and compares it to reported "near death experiences" (including common elements such as tunnel vision/bright light, euphoric floating, "out of body sense"), and suggests the clinical symptoms indicate a thermodynamic framework for neurologic states.

The next three chapters concern "split-brain" patients, whose corpus callosum (the brain structure that connects left and right hemispheres) had been surgically severed ("commissurotomy") as treatment for intractable seizures. This work was pioneered by Roger Sperry (1977), who first suggested that both hemispheres are independently conscious following commissurotomy.

Victor Mark presented startling videotapes during the conference showing behavior of a split-brain patient with strong disagreements between the conscious judgments by the two sides of her brain. In Chapter 12, Mark describes the patient's uncontrollable self-contradictions in her speech and limb gestures, and how, although aware of her behavioral conflict, she had no insight to its origins. Agreeing with Sperry, Mark concludes that each hemisphere can be independently conscious, and that the commissurotomy lesion prevents resolution of their disparate interpretations and desires.

In Chapter 13, systematic studies by Marco Iacoboni, Jan Rayman, and Eran Zaidel agree with this conclusion. In their work, a complete commissurotomy patient was asked to distinguish between words and nonwords presented separately to both (isolated) visual fields, and hence separately to both hemispheres. Distinctly different interpretations were given.

Polly Henninger describes a comparable study in Chapter 14 isolating hemispheres in two commissurotomy subjects. She presented "inkblots" (Rorschach and those of her own design) to the subjects' left and right visual half-fields, and asked them to draw what they had seen with their left and right hands, respectively. Drawings by the left hand (right hemisphere) generally consisted of simplified, blot-like shapes; right hand (left hemisphere) drawings tended to be unidentifiable, or when verbalized, to resemble the verbal description. Henninger concludes that the right hemisphere organizes perception in terms of spatial templates, whereas the left hemisphere has difficulty with ambiguous shapes unless they can be verbally labeled.

Moving away from split-brain subjects in Chapter 15, Britt Anderson and Thomas Head studied patients with severe sensory aphasia, a condition that appears to preclude comprehension. The patients were told jokes that elicited no apparent conscious reaction; however, recorded skin potential measurements demonstrated autonomic responses. Anderson and Head's findings suggest subconscious appreciation of humor, a phenomenon that may be related to subconscious appreciation of vision and smell discussed in the next section.

Finally, in Chapter 16 Alfred Kaszniak and Gina DiTraglia Christenson report on impairments in self-awareness of deficits ("anosognosia") that can occur in a variety of of neurological disorders. They review empirical research on deficit self-awareness in patients with Alzheimer's disease, emphasizing how different experimental methods track different facets of the dementing disorder. Kaszniak and Christenson conclude that it is the self-monitoring aspect of awareness that deteriorates in Alzheimer's patients.

In this section, perturbations of brain structure and function reveal interesting features of consciousness. James Whinnery shows physiological correlates of transitions between conscious and unconscious states, Britt Anderson and Thomas Head suggest that humor may be appreciated at subconscious levels, and Alfred Kaszniak and Gina DiTraglia Christenson describe defects in self-monitoring of awareness in Alzheimer's patients. Victor Mark, Marco Iacoboni and colleagues, and Polly Henninger expand on Roger Sperry's work and explore separation of consciousness ("unbinding") in split-brain patients. The dueling, dual consciousness in each of these unfortunate commissurotomy patients is startling, but perhaps not surprising. Numerous models of the organization of consciousness (see Part 5 on Neural Networks and Part 9 on Hierarchical Organization, and, for example, Minsky 1986, and Gazzaniga 1985) involve multiple, competing modules, or agents, capable of occupying center stage.

The next section more closely pursues neural correlates of consciousness.

REFERENCES

Gazzaniga, M. S. 1985. *The social brain—Discovering the networks of the mind.* New York: Basic Books.

Minsky, M. 1986. *The society of mind.* New York: Simon & Schuster.

Sperry, R. W. 1977. Forebrain commissurotomy and conscious awareness. *Journal of Medicine and Philosophy* 2:101–26.

11 Induction of Consciousness in the Ischemic Brain

James E. Whinnery

INTRODUCTION

The opportunity to have as experimental preparations completely healthy humans without consciousness and then to observe what is required to establish normal consciousness in them is unique. Inducing both unconsciousness and consciousness in these healthy humans was achieved by globally altering blood flow to the nervous system through exposure to high, sustained +Gz-stress using a human centrifuge. The risk associated with such research involving healthy humans is balanced by the benefit gained through the progress made in reducing the fighter aviation mishaps resulting from +Gz-induced loss of consciousness (G-LOC).

The emphasis of this manuscript will be to (1) briefly describe the fighter aviation research into the neurologic state of consciousness and its alterations, and (2) outline the resulting neurophysiologic theory of consciousness and unconsciousness based on experimental data from healthy humans. The references included in this manuscript are restricted to those involving fighter aviation medicine and G-LOC. This literature may not be familiar to many researchers interested in consciousness and its neurologic basis.

FIGHTER AVIATION AND +GZ–STRESS

It is probable that most researchers working toward an understanding of consciousness are unfamiliar with the fighter aviation environment and the nature of in-flight acceleration (+Gz) stress that can cause loss of consciousness. Before discussing the primary information that should be of interest to those endeavoring to understand consciousness, it is important to briefly describe the fighter aviation medicine and +Gz-stress aspects of the research. Modern fighter type aircraft are highly maneuverable and capable of rapid-onset, sustained, high levels of acceleration (+Gz). Aerial combat maneuvering (ACM) frequently requires full utilization of aircraft maneuverability which exceeds the unprotected G-tolerance of the pilot. The net effect of the +Gz-stress induced by ACM on the human body is illustrated in Figure 11.1. The human centrifuge closely simulates the in-flight ACM acceleration environment of fighter aircraft. When the

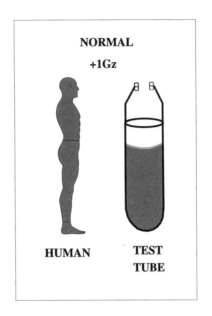
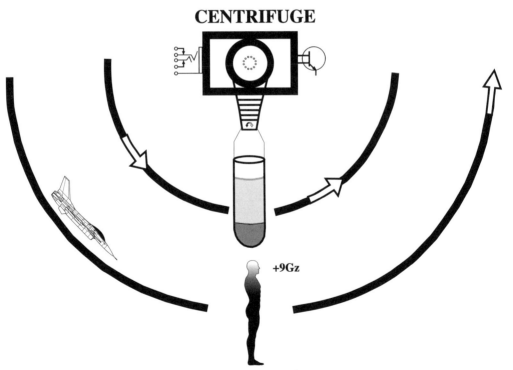

Figure 11.1 Illustration of the effects of +Gz acceleration on the human body in comparison to what happens in a test tube of blood during centrifugation.

fighter aircraft executes a rapid turn the pilot is pressed down into his seat. Although the entire body is pushed downward in the cockpit seat, it is held in place by the seat and harness. A 170-pound man in the earth's +1Gz gravitational field will weigh 1530 pounds at +9Gz. During ACM with rapid-onset acceleration this change from 170 to 1530 pounds can occur within 1 second. The vasculature within the body is distensible and the blood therefore pools in the dependent areas below the heart (abdomen and extremities). At high levels of +Gz, the heart is unable to generate adequate pressure to maintain critical perfusion to portions of the cephalic nervous system (CPNS), including the brain. When blood pressure is reduced to a critical level by the +Gz-stress, G-LOC results from the ischemia. When this occurs on the centrifuge, the +Gz-stress is reduced and blood flow to the CPNS is restored. In a fighter aircraft when the pilot loses consciousness, he loses motor control and therefore loss of aircraft control. Blood flow to the CPNS is also restored in the aircraft as back pressure on the flight control stick is released when G-LOC occurs. The fighter pilot is therefore at minimal risk for injury to the nervous system from this transient ischemia, however the G-LOC incapacitation for periods of 15 to 30 seconds is long enough to allow a high speed aircraft to descend from as high as 24,000 feet to ground impact. Numerous aircraft and aircrew have been lost as a result of G-LOC mishaps. Therein lies the challenge to fighter aviation medicine specialists, to achieve a solution for the G-LOC problem. In clinical medicine loss of consciousness (syncope) is merely a symptom of some underlying problem. If the underlying abnormality is corrected, the loss of consciousness is solved. In aviation medicine G-LOC is a primary problem in completely healthy fighter pilots.

EXPERIMENTAL APPROACH

Problems with G-LOC have been present from almost the very beginning of aviation. Significant research efforts to enhance aviator tolerance to +Gz-stress occurred during the period surrounding World War II. The thrust at that time was to prevent "blackout" and the approach was cardiovascular in nature. Undoubtedly G-LOC was recognized as a threat at that time and was described; however, a concentrated effort to approach the neurophysiologic aspects was not launched. In the late 1970s a new generation of high performance fighter aircraft was developed including the F-15, F-16, and F-18. The G-LOC threat was greatly enhanced. We therefore endeavored to approach G-LOC systematically from a neurophysiologic standpoint as shown in Figure 11.2. The first step was to qualitatively describe the G-LOC phenomenon. Qualitative analysis required acquisition of a large number of G-LOC episodes. It took many years to acquire sufficient data. As this was transpiring, we refined the methods and techniques to measure and analyze G-LOC in healthy humans (Whinnery 1989a). Once adequate data was available we began the second step, to develop a quantitative description of G-LOC (Whinnery 1990b). The entire

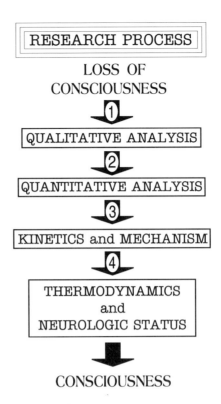

Figure 11.2 The experimental approach taken toward developing a comprehensive theory of consciousness and unconsciousness in the normal, completely healthy human.

symptom complex associated with the loss and recovery of consciousness could then be described as the G-LOC syndrome (Whinnery 1990a). The third step was to establish the kinetic relationships of the events of the G-LOC syndrome as shown in Figure 11.3. The kinetics of G-LOC provided an opportunity to develop a mechanistic theory of the G-LOC syndrome (Whinnery 1989b, 1991a). Experiments based on the theory of G-LOC were initiated and completed (Cammarota 1991). The fourth step was to develop a thermodynamic description of consciousness and unconsciousness. Thermodynamic treatment of the G-LOC syndrome leads to a description of consciousness and unconsciousness as energy dependent states of the nervous system (Whinnery 1994). The majority of the symptoms of the G-LOC syndrome become part of the neurologic state of subconsciousness, which is traversed during the loss of and recovery of consciousness. The experimental approach described above is essentially the same pathway that is followed for other phenomenon of the chemical, physical, and biologic world as outlined in Figure 11.2. This suggests that at least a general description of consciousness and other neurologic states can be approached using classical techniques. Even if such an approach ultimately proves unsuccessful, it should be extremely instructive relative to where

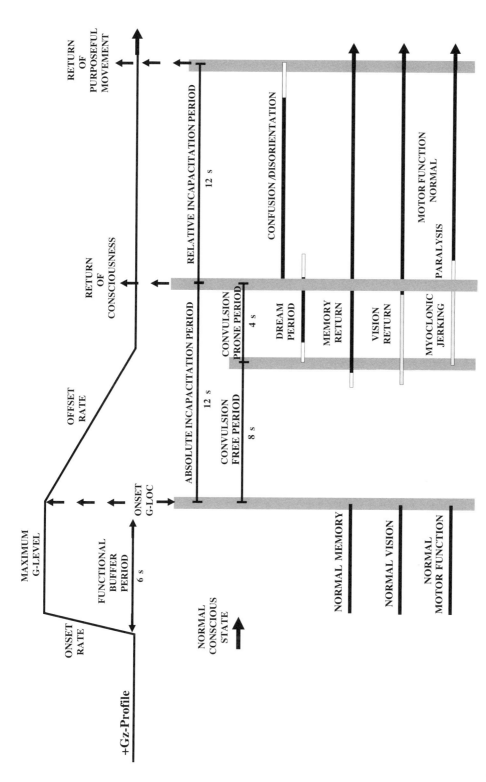

Figure 11.3 Quantitative and kinetic description of the G-LOC syndrome resulting from rapid alteration of blood flow to the cephalic nervous system in a normal human induced by +Gz-stress.

the classical inconsistencies develop and, if necessary, the best approach to pursue for ultimately describing consciousness in an appropriate manner. It should be recalled that this was exactly what happened when classical physics gave way to quantum physics, which indeed was born when irreconcilable difficulties arose between experimental evidence and classical theory.

The G-LOC experiments in completely healthy humans represent a unique vantage point from which the neurologic state of consciousness can be manipulated and understood. In comparison to other approaches toward understanding consciousness, for instance using vision as a model for consciousness, these techniques involve the entire spectrum of substrates and faculties that actually are consciousness. To ultimately understand consciousness, it is necessary to have a method to investigate consciousness, full and complete consciousness, in the completely normal human being. The G-LOC experiments and resulting theoretical developments represent one of the few opportunities to do just that.

THE G-LOC SYNDROME

The entire symptom complex and associated psychophysiologic alterations that are produced by +Gz stress in the normal human have been cumulatively classified as the G-LOC syndrome (Forster and Whinnery 1992). The G-LOC syndrome constitutes the normal response to +Gz-induced alterations in CPNS blood flow. From a fighter aviation medicine viewpoint, the importance of describing the phenomenon in this fashion is to emphasize that the loss of consciousness, electroencephalographic (EEG) changes, and myoclonic jerking, for instance, are not pathologic processes and are completely normal. If +Gz-stress is high enough and sustained long enough, everyone would experience the G-LOC syndrome. Fighter pilots are no longer automatically restricted or disqualified from flying duties when they experience the G-LOC syndrome as they were up until the 1980s. Implicit in this position is that there is no permanent residual from the G-LOC syndrome (Whinnery 1989c, 1991b). Recovery is complete and without pathologic consequences.

The characteristic nervous system response to rapid–onset +Gz-stress up to a sustained level above human tolerance, that is to a magnitude of +Gz-stress where CPNS ischemia is rapidly produced, results in the symptoms shown in Figure 11.3. It should be noted that the symptoms produced are dependent not only on the level of the +Gz-stress but also the rate of onset and offset of the +Gz-stress. This is equivalent to stating that it is not only the magnitude of the ischemia but also the rate of the ischemic change that determines the symptomatic response. Figure 11.3 represents the response to one type of ischemic profile, rapid-onset and rapid-offset ischemia.

For most routinely investigated +Gz-profiles there are few recognizable symptoms during the loss of consciousness. The majority of the symptoms occur during the recovery (induction) of consciousness. G-LOC theory leads to the prediction that if the ischemic insult is carefully titrated most of the same symptoms that are observed during recovery of consciousness should be producible during the loss of consciousness. Experiments have successfully confirmed this, by measuring loss of memory and producing myoclonic convulsions prior to loss of consciousness (Cammarota 1991). In addition, by manipulating the rate of return of blood flow to the CPNS it is possible to modify the kinetic characteristics of the G-LOC syndrome (Whinnery and Whinnery 1990, Whinnery 1990d). Such manipulations are not only of theoretical importance, by establishing experimental confirmation of the validity of G-LOC theory, they also have the operational significance of reducing the period of incapacitation that a fighter pilot would experience should G-LOC occur in flight (Whinnery, Hamilton, and Cammarota 1991, Whinnery and Burton 1986). The symptom complex sequence of the G-LOC syndrome is listed in Table 11.1. Each of the individual symptoms corresponds to a specific functional configuration of the nervous system. The specific configurations are produced by alteration of energy (blood flow) to a nervous system that has a normal anatomic structure.

The eyes are somewhat unique with respect to the effect +Gz-stress has on them. This uniqueness stems from the increased pressure (intraocular) within the eye in comparison to the remainder of the nervous system. As the perfusion pressure generated by the cardiovascular system becomes inadequate, because of the +Gz-stress, the eyes begin to lose perfusion before many other areas of the nervous system at an equivalent level within the +Gz field or at equivalent perfusion pressure levels. The visual symptoms begin with loss of peripheral vision, progress to tunnel vision, and end with complete blackout. Blackout can exist without loss of consciousness if the +Gz-stress is carefully manipulated. The visual symptoms result

Table 11.1

Symptom complex of the G-LOC syndrome resulting from exposure to +Gz-stress and recovery to baseline +1-Gz environment.

1. Loss of peripheral vision
2. Tunnel vision
3. Blackout
4. Loss of consciousness
5. Loss of motor control/function
6. Loss of sensory input
7. Loss of memory capability
8. Unconsciousness
9. Myoclonic convulsions
10. Recovery of sensory input
11. Recovery of memory capability
12. Dreamlets
13. Return of vision
14. Return of consciousness
15. Confusion/disorientation
16. Return of purposeful movement

from regional ischemic differential within the retina. Distal perfusion farthest from the central retinal artery is affected first. The areas nearest the central retinal artery are the last to be affected. If the eyes are located at the same level within the +Gz-field and the vascular system supplying each eye is symmetrical, then vision is bilaterally affected similarly and bilateral blackout results. If the eyes are positioned at different levels within the +Gz-field, the highest eye will be affected first, resulting in unilateral visual symptoms. As would be expected, if sustained +Gz-stress is very rapidly applied to a level well above tolerance, loss of consciousness and loss of vision all occur at once. Blood flow in this case is concurrently compromised to the eyes and critical areas of the CPNS to a degree that simultaneously compromises function. There are no recognizable visual symptoms prior to loss of consciousness in this situation.

A major question relative to loss of consciousness is exactly what the perfusion pattern is within the CPNS when consciousness is compromised. Just as we observe with visual symptoms, it is very likely that regional ischemic differential within the CPNS is responsible for the different symptoms (6,14). Our current hypothesis suggests that it is the combination of the effects of (1) +Gz-stress that affects areas most superior within the +Gz-field, (2) the vascular system architecture in which areas most distal to the blood supply would be affected first, (3) the differential sensitivity of the various regions within the CPNS to ischemia, and (4) the characteristics of the specific +Gz-profile. Exactly which one of these factors plays the major role or if each has a particular role is unknown. Since our research involves +Gz-stress, we have assumed it plays at least a partial causal role in the G-LOC ischemic pattern. It should be noted that our analysis of the available data involving strangulation in physiologically and anatomically normal prisoners reveals very similar characteristics between G-LOC and strangulation–induced loss of consciousness (Forster and Whinnery 1992). Strangulation and +Gz-stress involve slightly different modes of inducing CPNS ischemia, yet they produce strikingly similar symptom complexes and analogous kinetic profiles. Such similarities may reflect that +Gz-stress simply produces global CPNS ischemia. Whatever the exact hemodynamic mechanism involved, it remains most probable that it is regional ischemic differential that produces the functional configurations of the CPNS which correspond to the symptoms that are observed.

The G-LOC syndrome symptoms do not arise from a CPNS configuration that simply reflects the areas that continue to function as they do during periods of normal energy supply and those that have lost function because of inadequate energy supply. When specific areas have ischemically induced compromise of normal conscious function, this has a resultant effect on the remainder of the nervous system. When any portion of the nervous system is altered, there are reciprocal changes in regional areas that remain functionally uncompromised. A considerable portion of these interactions are apparently mediated by mutual reciprocal activation-inhibition relationships. The net effect is such that one must consider which areas are primarily altered by ischemia and become dysfunctional, which areas are not affected by ischemia, and finally the resultant symptoms that would be manifest by the altered but still functioning areas of the nervous system. Certainly, the highest, most sophisticated reasoning areas of the CPNS are compromised while the more primitive areas remain functional. This contrast in functional versus dysfunctional areas within the nervous system can be observed by the normal functioning of the cardiorespiratory control centers in the brain stem (as evidenced by normal cardiorespiratory function during G-LOC episodes) and altered functioning of the neurologic substrate supporting memory. The transitions between conscious and unconscious states reflect these rapidly changing processes (Whinnery 1989b, 1991a). The rapidity of the changes makes it difficult to absolutely pinpoint the exact kinetic relationships of all the changes over the short 12-second period on the average.

LOSS OF CONSCIOUSNESS: A PROTECTIVE MECHANISM

Life in the terrestrial environment appears to have always involved a continuous struggle for survival within a particular organism's environmental niche. Successful survival requires the organism to have the capability of sensing adverse environmental conditions, hopefully so they can be avoided. In effect, the sensory mechanism of an organism requires specific segments to be functionally altered in response to changes in the environment. Sensory segments appropriately connected with segments for locomotion provide the structure necessary to have the ability to take evasive action to avoid or escape from adverse environments. Without segments for locomotion or the overall capability to escape from an adverse environment, an alternative strategy might be for the organism to alter its activities or configuration until favorable conditions are established and sensed so that normal activities can be resumed.

The earth's gravitational field has been continuously present over the entire duration of the existence of life on earth. Organismal existence in the different mediums on earth, water versus air for instance, can alter the affect of gravitation on the organism. In addition, the orientation the organism assumes within the gravitational field also can alter the effects gravity has on the organism. If indeed evolution involved a transition from the ocean to the land and, for humans, a subsequent transition from a horizontal to an upright orientation within the gravitational field, then considerable influence on evolutionary development has been exerted by the subtle, ever-present gravitational field of the earth. The ascent of man would have been accomplished by transition from a zero +Gz-environment in the oceans to a −1Gx (posterior to anterior force) environment horizontally on land and finally to a +1Gz environment when living upright on land. The success of our adaptation to earth's +1Gz gravitational environment serves to make such an evolutionary transition appear almost trivial. Healthy humans rarely, if ever, have problems coping with normal gravitational stress. The real problems with increased levels of gravitational stress were realized when man left terra firma and entered the aerospace environment. It should be noted that adaptation to any G-environment other than what the organism has evolved in is usually difficult. This is amply illustrated by the difficulty astronauts have in adapting to the microgravity environment of space. The hypergravity environment of fighter aviation presents even more acute difficulty for the human. To even enter the modern aviation and space environments is life-threatening, requiring life support equipment to prevent compromise of consciousness in completely healthy humans. Altitude-induced hypoxia and acceleration (+Gz)-induced ischemia/hypoxia result from exposure to the ACM environment and remain constant threats to a pilot's conscious function. Problems with the earth's (+1Gz) gravitational environment do become evident in patients whose illness may compromise their tolerance threshold.

The results of centrifuge-based acceleration research reveal the responses of the organism to the high +Gz environment. All of the responses of the body are attempts to preserve normal function. They are oriented to protect the organism as a whole along with the individual systems that are component parts of the organism. As the cardiovascular system's protective responses that support delivery of the main energy source (blood flow) to the nervous system are overcome by +Gz-stress, the nervous system must also initiate its own protective responses. The G-LOC syndrome responses appear to constitute a continuation of the organismal and nervous system protective mechanisms. The protective sequence is shown schematically in Figure 11.4. Each of the steps listed and described below serves to enhance the survival of organism and concurrently protect the organism and its component systems.

Anatomy and physiology have combined to provide protection of the nervous system. The central nervous system is anatomically encased within a bony structure with minimal openings for sensory input and responsive output. The first lines of physiologic defense are provided by the cardiovascular system. Neurologic function remains normal over a range of vascular pressures and perfusion. This constitutes a cardiovascular reserve over which conscious function is maintained. When inadequate cardiovascular reserve is sensed, cardiovascular reflexes respond to maintain conscious function. Neurovascular reserve and reflexes respond when cardiovascular reserve and reflexes are inadequate to maintain an adequate internal energy supply. The CPNS has the ability to maintain conscious function for a short period of time in the face of ischemic/hypoxic embarrassment. This built-in functional buffer period lasts about 6 seconds during which normal function is retained. The nervous system does not have pain response to serve as a warning to recognize ischemic challenge. Warning symptoms do exist however, for the usual types of ischemia/hypoxia that the nervous system has evolved in. This includes the symptoms of loss of peripheral vision through blackout and sensations of dizziness, lightheadedness, and nausea. If the warning symptoms are not heeded during the functional buffer period and evasive action taken then the nervous system initiates the next protective step which is to try

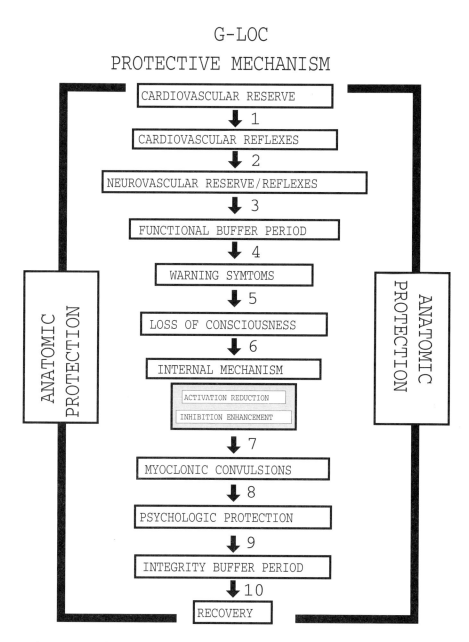

Figure 11.4 The protective sequence evoked in response to increasing magnitude of +Gz-induced nervous system ischemia.

and remove the nervous system from the +Gz-environment. This is accomplished by loss of postural tone and placing the heart and brain at the same level within the +Gz-field (horizontal) which is heralded by loss of consciousness. The internal neurologic mechanisms that conserve energy involve reducing conscious functioning so that the remaining available energy can be used to maintain basic cellular integrity. The nervous system also has a period of time over which an ischemic/hypoxic result can be tolerated without permanent pathologic alteration. This period of time over which the nervous system has altered function but does not suffer injury is the integrity buffer period. As long as blood flow returns to adequate levels within this time period, nervous system recovery is complete. When the insult is severe enough, enhancement of the return of blood flow from the peripheral circulation is facilitated by contraction of the skeletal muscle and myoclonic convulsions occur. Because loss of consciousness is a significant threat to organismal survival in a hostile external environment, it is extremely important to avoid its occurrence if at all possible. Since recognition of a loss of consciousness episode is frequently very different, it would be advantageous to have some type of mechanism to enhance recognition of a loss of consciousness episode. The dreamlet and psychophysiologic alterations serve to enhance the ability to recognize the episode and hopefully avoid subsequent exposure to such an environment. The final step in the protective mechanism is to continue to rapidly reduce the confusion/disorientation. Once again, if the insult is severe enough the nervous system will initiate a self–touch response to reorient the individual. The nervous system seems to ask where the individual is and by initiating an orientational motor response to touch oneself (especially the facial area), is able to answer the question, by sensing through touch the input being interpreted as "Here I am."

Considering loss of consciousness and the neurologic state of unconsciousness as sequential aspects of an evolved protective mechanism is a powerful theoretical tool. The search for the more detailed aspects of the neurophysiology of the normal responses of the central nervous system is facilitated by approaching the problems from the perspective of the mechanisms being protective in nature. Such a perspective also has implications concerning the neurologic state of consciousness. If the neurologic states of the nervous system are based on protective mechanisms, then consciousness must also be considered protective in nature. In contrast to loss of consciousness and unconsciousness representing primary protection of the nervous system in the face of the internal ischemic threat, consciousness represents primary protection of the organism as a whole in the face of external threats that arise. That is, consciousness represents the neurologic mechanism that has evolved to enhance survival in the organism's earthly environment. A delicate balance must exist between consciousness and unconsciousness. The extreme success of the human to survive in its environment is primarily due to the existence of powerful nervous system capabilities. As compared to more primitive nervous systems, which provide only the ability to survive temporarily in a hostile environment or to avoid and escape from hostile environments, the human nervous system has also provided the capability to modify its environment to ensure survival. These capabilities are based on a fully functional conscious state of the nervous system. It is only when the nervous system itself is severely threatened internally that consciousness is compromised. If consciousness was lost too frequently, at low internal ischemic threat levels, the organism would be at increased risk for survival in the external environment. On the other hand, if consciousness were maintained at too severe a magnitude of internal ischemic threat, such that the nervous system would suffer damage, then the primary system responsible for organismal survival would be compromised. The states of the nervous system, consciousness and unconsciousness, are maintained based on the available neurologic energy. Consciousness, by allowing optimal organismal function for survival in the environment, requires a continuous supply of energy. Unconsciousness, by reducing organismal and nervous system activities, conserves energy in a nervous system that is critically threatened with potential injury from inadequate blood flow (internal energy).

THERMODYNAMIC CONSIDERATIONS: NEUROLOGIC STATES

Considering the completely healthy human taken to unconsciousness by +Gz-induced reduction of the internal energy (blood flow), we are then afforded the opportunity to determine exactly what is required to develop consciousness in the human. Experimentally the induction of consciousness would appear to be rather simplistic and uneventful, simply add energy and consciousness will rapidly be observed to return (as long as the ischemic insult has not persisted for an excessive duration). Careful examination of the event sequence of the G-LOC syndrome, however, reveals the rather complex buildup of neurologic faculties that are characteristic of the state of full consciousness. Between the unconscious and conscious states, when only portions of the nervous system have adequate energy for normal function, a state of subconsciousness exists. Subconsciousness is a neurologic state where portions of the nervous system become functional as internal energy (blood flow) is partially restored, which, in turn, allows the nervous system to begin to absorb (sense) external energy, such as light and sound. The final transition from subconsciousness to consciousness is evidenced by the ability of the nervous system to emit energy in the form of responses. It becomes evident that the state of normal consciousness can be measured (observed) only when the normal nervous system structure is capable of absorbing internal and external energy, converting them into neurologic energy, and emitting energy to the nervous system's environment. A complete thermodynamic analysis of the sequential buildup process required to establish consciousness has been developed (Whinnery 1994). Careful attention must be paid to the specific environment in which the system exists. The organism (body) exists in the external environment. The nervous system exists within the body, an internal environment.

The most important result of establishing a thermodynamic description of loss of consciousness, unconsciousness, recovery of consciousness, and consciousness is that it leads to the development of specifically observable states of the nervous system as shown in Figure 11.5. For a given nervous system structure, the normal human nervous system, these neurologic states are dependent on the neurologic energy. Overall, the thermodynamic description of the neurologic states in the normal human nervous system

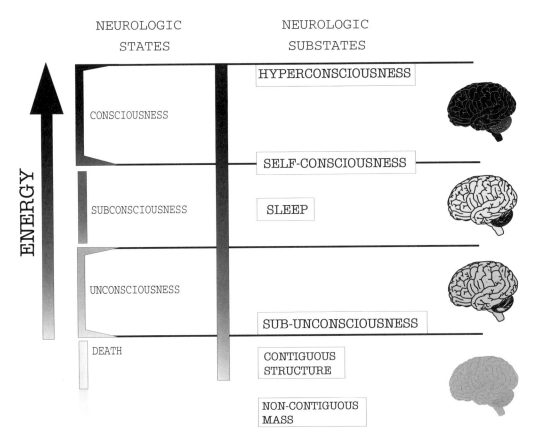

Figure 11.5 Neurologic states of the nervous system developed from a thermodynamic analysis of the G-LOC syndrome.

produced by alterations of the neurologic energy induced by +Gz-stress can be expanded to describe nervous systems in general. In the simplest way, the neurologic state (C) of a nervous system can be described as a function of neurologic mass (m), energy (E), and structure (S): $C \alpha ES/m$. The thermodynamic description of neurologic states becomes independent of the specific alteration mechanism (ischemia, hypoxia, trauma, disease). A unified theory of the loss of consciousness is the result. The manipulation of the neurologic states of the normal human nervous system using +Gz-stress becomes a manipulation of energy to a (nervous) system having a constant (and normal) mass and structure. Manipulation of mass and structure also alter the neurologic states, but are not a part of G-LOC experimentation.

Each neurologic state represents a range of energy within the nervous system and therefore a functional range. An individual neurologic state merges as a continuum with adjacent states since the amount of neurologic energy contained within a nervous system, at least appears from an observational level to be continuous. Although the neurologic energy content appears to be continuous, this is probably a result of our poor measurement (observation) capability. Energy absorption, emission, and content within the nervous system are more likely to all be subject to the same quantum effects of all matter. As measurement capability improves, it may be resolved that even in the complex nervous system, energy content is quantized.

The current measurement capability requires an arbitrary establishment for the normal ranges of the neurologic states. We are required to define the neurologic states representing specific capabilities that are observable. Definition of each neurologic state can be accomplished mathematically utilizing a thermodynamic equation of state based on the functions that can be observed, such as memory, response, and sensing. Starting with the concept of a brain in a black box, that is a viable, normal nervous system devoid of energy, or energy exchange with its environment, it is possible to sequentially supply the energy and structural components needed to establish the state of consciousness. The sequential buildup of energy requirements is guided by observations of recovering consciousness from the G-LOC syndrome previously described.

Although the process described above for thermodynamically defining neurologic equations of state becomes complex and at the same time, an oversimplification of the true nature of the nervous system, it does begin to systematically reveal what is required for consciousness. It is a classical approach to the problem of consciousness, based on experimental data where consciousness is lost by a nervous system and what is required for consciousness to be established. If there are indeed incompatibilities of experimental data and classical theory, they should become evident. In lieu of other models of the neurologic processes associated with the complete state of conscious function in healthy humans, it would appear that this approach could be useful to a variety of other scientists investigating consciousness.

THE NEUROLOGIC STATE OF DEATH

Everyone should have an interest in loss of consciousness, since they will experience it at least once, albeit only a terminal one perhaps. Death is covered by an unconscious veneer. To die one must traverse the unconscious state. If a sojourn into the unconscious state is made and followed by complete recovery, then from a physiologic perspective it was only a loss of consciousness episode. As compared to the conscious and subconscious states, unconsciousness represents a neurologic state nearer to death. The recent increase in interest in what have been classified as near-death experiences (NDEs) provides an opportunity to utilize and apply G-LOC theory.

As seen in the thermodynamic treatment of neurologic states, it was necessary to include death as the state which has a neurologic energy content even less than the state of unconsciousness. It remains unknown what minimum neurologic energy content can exist within a nervous system and still be compatible with life. It is known that such a question involves not only thermodynamics but also kinetics. Pathologic change in a nervous system depends not only on the extent of the energy supply decrement, it also depends on the duration that the perfusion decrement exists. Without an almost continuous supply of energy, the nervous system becomes dependent on any intrinsic reserve it has. The nervous system evidently has minimal reserve upon which to maintain its integrity. As evidenced by the G-LOC kinetics when +Gz-induced ischemia is induced, it is 6 seconds or less. Death, in a clinical sense, becomes a state from which higher neurologic states, specifically the state of normal consciousness, cannot be attained. Since the entire nervous system is not normally affected in a global sense simultaneously by the insults that may cause death and further the nervous system is not ideal (different regions have different tolerances), a spectrum of neurologic states can arise from various ischemic (energy) insults. If, however, a neurologic state that falls within the normal range of consciousness is regained after a NDE, then it would appear logical that at least some of the symptoms of the NDE and a G-LOC episode would be similar. This would be particularly true for an NDE reported in association with cardiac arrest, for example, since G-LOC and cardiac arrest have more similar causative mechanisms.

Comparison of the G-LOC syndrome symptoms with the NDE does reveal striking similarities. As previously mentioned, if full recovery from the NDE occurs it can at least physiologically be considered a loss of consciousness (and recovery of consciousness) episode, granted a really significant loss of consciousness episode. It is probable that the NDE more frequently involves a much greater magnitude of ischemic/hypoxic insult than that of the G-LOC episode. As shown in Figure 11.6, this suggests that if the NDE and G-LOC episode symptoms are compared, then similar symptoms are the result of the physiologic alterations associated with loss/recovery of consciousness. If there is truly anything unique about the NDE, then it should show up only in the NDE and be absent from the G-LOC episode. This approach would serve to focus NDE research on what, if anything, occurs uniquely in association with NDEs. The specific psychophysiologic symptoms of the overall experience need further description. This includes the characteristics of the G-LOC dreamlets. We refer to them as dreamlets since they meet the criteria for being classified as dreams and are essentially indistinguishable from the dreams experienced during sleep, save their recognizably brief duration. The dream content of G-LOC episodes has many similarities to the NDE. A summary of the dream content and psychophysiologic alterations similar to NDEs following recovery from a G-LOC episode are listed in Table 11.2.

CONCLUSION

The investigation of G-LOC in healthy humans provides the opportunity to experimentally manipulate the neurologic state of the nervous system. The standard scientific approach used to describe other chemical, physical, and

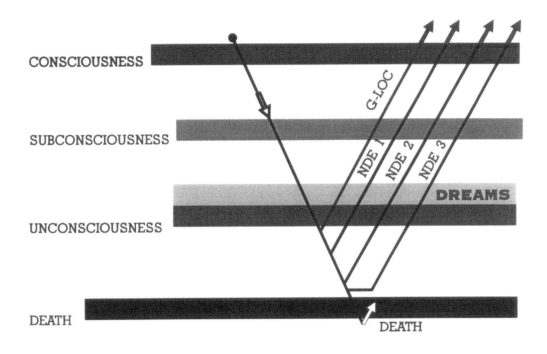
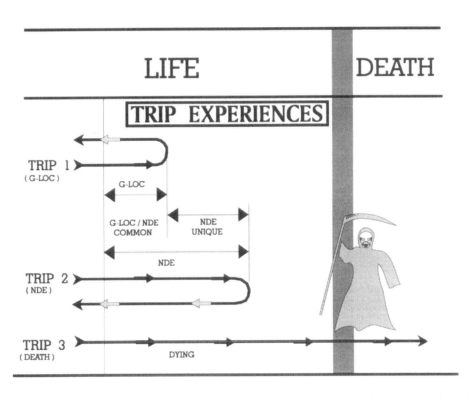

Figure 11.6 Comparison of the neurologic states that would be entered and traversed for G-LOC experiences and near death experiences (NDE).

Table 11.2

The major psychophysiologic characteristics of G-LOC experiences that have commonality with near death experiences.

1. Tunnel vision/bright light
2. Floating
3. Automatic movement
4. Autoscopy
5. Out-of-body experience
6. Not wanting to be disturbed
7. Paralysis
8. Vivid dreamlet/beautiful places
9. Pleasurable experience
10. Psychologic state alteration
    a. Euphoria
    b. Dissociation
11. Friends/family inclusion
12. Prior memories/thought inclusion
13. Very memorable (when remembered)
14. Confabulation
15. Strong urge to understand experience

biological phenomenon has proven useful in developing an understanding of neurologic states. Qualitative, quantitative, and kinetic analysis of the G-LOC syndrome allowed establishment of mechanistic and thermodynamic theory of the loss and recovery of consciousness. Understanding and predicting neurologic function is facilitated by considering the neurologic states of consciousness and unconsciousness to be evolved processes to ensure survival and protection of the organism and its component systems. A balance between the two states provides the mechanism for successful survival in threatening environments.

Utilizing the +Gz-stress technique to safely modify consciousness along with the development of a sound theory for describing neurologic function together serve as a useful probe into understanding consciousness. This is one of the ways to experimentally investigate the concept of normal consciousness. As the experimental evidence and theory of consciousness become further developed, if there are any inconsistencies that become evident, they should be recognizable. Recognition of experimental and theoretical inconsistencies should provide insight into the potential requirement to employ more sophisticated techniques and approaches to explain consciousness and other neurologic states.

The scope of G-LOC research involves a broad range of information applicable to evolutionary development in the ever-present terrestrial gravitational field, the mechanisms which evolved to ensure survival, the neurologic states present during life, and the ultimate pathway followed at the end of life. Fighter aviation medicine endeavors to understand the acute alterations of normal consciousness that result from inadequate supply of internal energy to the nervous system and how to preserve consciousness in fighter pilots. The unique experimental data obtained in the course of this aviation research should be of significant interest to others involved in the study of consciousness.

REFERENCES

Cammarota, J. P. 1991. Symptoms of +Gz-induced incapacitation during simulated aerial combat maneuvering. *Aviation and Space Environmental Medicine* 62:485.

Forster, E. M., and J. E. Whinnery. 1992. Statistical analysis of the human strangulation experiments: Comparison to +Gz-induced loss of consciousness. Naval Air Warfare Center, NAWCADWAR Report No. –92026–60. Warminster, PA: 18974, 18 Feb 1992.

Whinnery, J. E. 1989a. Methods for describing and quantifying +Gz-induced loss of consciousness. *Aviation and Space Environmental Medicine* 60:798–802.

Whinnery, J. E. 1989b. Observations on the neurophysiologic theory of acceleration (+Gz) induced loss of consciousness. *Aviation and Space Environmental Medicine* 60:589–93.

Whinnery, J. E. 1989c. Defining risk in aerospace medical unconsciousness research. *Aviation and Space Environmental Medicine* 60:688–94.

Whinnery, J. E. 1990a. The G-LOC syndrome. Naval Air Development Center Report No. NADC–91042–60. Warminster, PA: 18974, 31 Oct 1990.

Whinnery, J. E. 1990b. Recognizing +Gz-induced loss of consciousness and subject recovery from unconsciousness on a human centrifuge. *Aviation and Space Environmental Medicine* 61:406–11.

Whinnery, J. E. 1990c. Physiologic "transection" of the brain stem in healthy humans and the neurologic basis of consciousness. Naval Air Development Center Report No. NADC– 91053–60. Warminster, PA: 18974, 10 Feb 1990.

Whinnery, J. E. 1991a. Theoretical analysis of acceleration induced central nervous system ischemia. *IEEE Eng. Med. Bio.* 10:41–5.

Whinnery, J. E. 1991b. Medical considerations for human exposure to acceleration–induced loss of consciousness. *Aviation and Space Environmental Medicine* 62:618–23.

Whinnery, J. E. 1994. Thermodynamic considerations of the conscious state. Naval Air Warfare Center (Report in press).

Whinnery, J. E., and R. R. Burton. 1986. +Gz-induced loss of consciousness: A case for training exposure to unconsciousness. *Aviation and Space Environmental Medicine* 58:468–72.

Whinnery, J. E., R. J. Hamilton, and J. P. Cammarota. 1991. Techniques to enhance safety in acceleration research and fighter aircrew-training. *Aviation and Space Environmental Medicine* 62:989–93.

Whinnery, C. C. M., and J. E. Whinnery. 1990a. The effect of +Gz offset rate on recovery from acceleration-induced loss of consciousness. *Aviation and Space Environmental Medicine* 61:929–34.

Whinnery, J. E., and A. M. Whinnery. 1990b. Acceleration-induced loss of consciousness. *Archives of Neurology* 47:764–76.

12 Conflicting Communicative Behavior in a Split-Brain Patient: Support for Dual Consciousness

Victor Mark

INTRODUCTION

The presentations at this conference chiefly concerned either the definition of consciousness or its physiological or physical basis. In contrast, this report presents observations from a single person to suggest that consciousness may be dual within an individual. That is, awareness, reflection, and response preparation may occur in distinct brain regions both *simultaneously* as well as *differently*.

If the inference of parallel consciousness is correct, then we not only have a valuable perspective of the mind's organization, but we also recognize that processes that constitute consciousness may be physically separated, producing qualitative changes in the individual's behavior, and yet still maintain consciousness in their separation. Therefore this report, along with a few others presented at the conference on patients with anatomically defined cerebral disorders, may benefit our understanding of the brain regions that are required for consciousness and to what extent they may be isolated but still sustain conscious activity.

CASE REPORT

A young woman developed increasingly frequent epileptic seizures from around eight years of age that included sudden attacks of postural collapse (atonic seizures) and momentary lapses of attention (petit mal, or absence epilepsy). There was no report of abnormal pregnancy in her mother. However, the patient has indications of unusual prenatal brain development. First, she has always strongly favored her left hand for writing and drawing and her right hand for all other activities, while all of her other family members are exclusively right handed. Second, magnetic resonance imaging (MRI) of her brain indicates cortical heterotopia, the accumulation of neuronal cell bodies in the fibrous white matter of the cerebral hemispheres (Sarnat 1991) due to interference in the outward migration of neurons from central brain regions to form the external grey matter (cortex) of the brain during fetal development (Cowan 1979). Such disorders are commonly associated with mental retardation and severe epilepsy (Huttenlocher,

Taravath, and Mojtahedi 1994, Palmini *et al.* 1991). Her highest total score on intelligence testing has been 75, which possibly was affected by concurrent antiepileptic medication. Nonetheless she achieved a two-year college degree and performed clerical work.

By age 33 her seizures were occurring several times a day despite antiepileptic medication, and so her doctors proposed to divide her corpus callosum along its midline. The corpus callosum is the main fiber bundle that connects the two cerebral hemispheres (Figure 12.1), and division along its midline (corpus callosotomy, also known as the split–brain procedure) is often effective when seizures are severely disabling despite medication (Mamelak *et al.* 1993). Initially only the anterior portion of her callosum was divided, but because the frequency of her seizures did not change, the remainder of her callosum was cut three months later. Her seizure frequency declined markedly thereafter. Stearns *et al.* (1989) detail her epilepsy and surgery; they believe that she was the first person with cortical heterotopia ever to have undergone callosotomy.

The patient came under my care at a rehabilitation hospital one month after her second operation to treat the gait incoordination and speech dysfluency that accompanied her surgery. During her two-month stay she showed behaviors that were very unusual for our rehabilitation service. At times her left hand disrupted actions that were voluntarily performed by her right, such as when she learned to set the controls on a microwave oven. The left hand behaved in a seemingly purposeful manner, using carefully coordinated movements, and yet the patient claimed not to understand why she behaved this way and was often frustrated.

Conflicting Communicative Behavior

Even more striking was that the patient contradicted herself in certain circumstances. When answering questions about her epilepsy or performing challenging psychologic tests such as matching spoken sentences with pic-

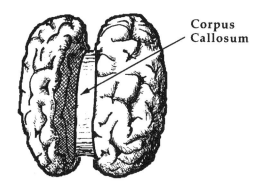

Figure 12.1 Schematic overhead view showing the relationship of the corpus callosum to the cerebral hemispheres. The distance between the hemispheres has been expanded to reveal the callosum.

tures that were minimally different, she tended to alternate rapidly and vigorously among the responses "Yes!," "No!," and "That's not right!" Despite attempts to correct herself, she had great difficulty accepting what she had just said or indicated, and sometimes she withdrew from the conflict by sighing deeply and declining to answer further. She often became emotionally upset and either pounded the table or slapped herself on the arm or leg. Her distress ceased when the topic or task were changed.

During these conflicts the patient never appeared to change in her personality or to be inattentive toward me, and she did not adopt any consistent body or gaze orientation. During these episodes she often asked, "Why do I lie to you?" My reply that her surgery was in some way responsible never completely reassured her and did not prevent further conflict.

Opposition between her hands was associated with some of these outbursts. Two instances occurred during interviews that were videotaped with her permission. Excerpts from these sessions were presented at the conference and are described here.

On one occasion she mentioned that she did not have feelings in her left hand. When I echoed the statement, she said that she was *not* numb, and then the torrent of alternating "Yes!" and "No!" replies ensued, followed by a despairing "I don't know!" I asked what she meant by not having feelings. She replied that she had been unable to identify numbers that I had traced onto her left palm with her eyes closed during formal neurologic examination. She also could not name objects held by her left hand while blindfolded, but she indicated by head nodding that she was fully aware of being touched on her left hand. In contrast, with her right hand she could readily name tracings and objects without seeing them.

Her surgery probably had caused her hemispheres to have opposing views on her left hand's ability to feel. Inability to name tactile stimuli with the left hand is common following corpus callosotomy (Bogen 1993), since skin sensations from each side of the body are relayed primarily to the opposite cerebral hemisphere (Carpenter 1985), while naming requires only the left hemisphere in most people (Benson 1979). Thus her right hemisphere, which received sensations mainly from her left side, could neither label the tracings nor send them to the other hemisphere for verbal interpretation. Accordingly the left hemisphere was unaware of her left hand's feelings.

It also seems that each hemisphere could organize speech. This is supported by the patient's presurgical evaluation, when she could still speak following the injection of sodium amobarbital (a barbiturate) into either carotid artery separately. This procedure, commonly known as the Wada test after its developer (Wada and Rasmussen 1960), temporarily inactivates most of the hemisphere on the same side as the artery that is injected (Biersack et al. 1987). It is often performed prior to epilepsy surgery to determine which hemisphere chiefly participates in language, since surgeons try to avoid injuring brain regions essential for basic language functions (Kupfermann 1991). Typically, right handers become language impaired (aphasic) after only left hemisphere inactivation, while non-right handers are considerably more variable, indicating left, right, or bilateral hemispheric control of language (Oxbury and Oxbury 1984, Serafetinides, Hoare, and Driver 1965).

Therefore the patient's mixed hand preference and indications of bilateral hemispheric control of speech agree with previous studies. Nonetheless she seemed able to name only with her left hemisphere, which suggests that her hemispheres' language abilities were not identical. Other authors have reported hemispheric differences in basic language functions within the same individual (Daffner *et al.* 1991, Kurthen *et al.* 1992).

Further indication that each hemisphere held a different opinion about her left hand's sensation appeared when I next had her use a sheet of paper marked with the words "YES" and "NO" to indicate her answer. When I asked "Is your left hand numb?" the left hand jabbed at the word "NO" and the right pointed to "YES." The patient became emotionally upset by the lack of unanimity and furiously and repeatedly tried to indicate the "correct" answer, but with the same results. Ultimately the left hand forced aside the right and covered the word "YES"! The patient by this point was obviously upset by the lack of resolution, and I removed the paper.

The uniform association of her right hand with "YES" and her left hand with "NO" suggests that her right hemisphere was communicating through her left hand and vice versa. This is consistent with animal research studies and evaluations of callostomized patients indicating that distal limb movements are controlled by the opposite hemisphere (Mack, Rothi, and Heilman 1993). The following occurrence in which distinct gestures from each hand were associated with conflicting statements further supports that her hemispheres held opposing views.

I asked the patient whether she were still having seizures. Again she gave contradictory responses and appeared frustrated. She then tried to enumerate her seizures. The right hand held up two fingers, but then the left hand reached over and tried to force it down. After trying several times to tally her seizures—and failing each time—she paused, softly counted to ten, and then simultaneously displayed three fingers with her right hand and one with her left. She appeared astonished and confused and asked why she behaved this way.

The patient's left hemisphere apparently tried to indicate a larger number of seizures than did her right; also the right hemisphere tried to suppress the left hemisphere's manual response. It is not clear why this difference of opinion occurred. However, epileptic patients often face conflicting interests when they review their illnesses with their doctors. If they conform to societal expectations and faithfully report their seizures, certain cherished activities such as becoming pregnant and driving a car may be medically forbidden. If they honor their heartfelt wishes, they risk lying. It is noteworthy that the patient had emphatically told me at other times that she had both desires. These desires must have been especially strong because she had declared them spontaneously. Moreover, she had expected that the surgery would control her seizure disorder. Thus her hemispheres may have had competing interests when she discussed her epilepsy.

Support For Dual Hemispheric Consciousness

The foregoing observations suggest that each cerebral hemisphere in this patient maintained opinions and desires different from the other in certain circumstances, comprehended speech and print at least on a basic level, and could announce its views through speaking, pointing, and distinct gesture (counting with fingers). Although the defining characteristics of consciousness are controversial, the appearance of two entities within the same person, each capable of understanding conversation and expressing unambiguously in words and gesture, suggests that each hemisphere could be considered "conscious" in commonly understood terms. Furthermore, during these manifestations of dual hemispheric consciousness it appeared that neither hemisphere was aware of the thoughts of the other prior to overt expression.

It is unlikely that these behavioral conflicts resulted from epileptic fits since they were not uniform, the patient's speech and limb movements appeared to be voluntary (although self-contradictory), she remained attentive to her environment, and she was aware of her incongruity. Indeed, she often complained of her lack of resolution and failed to understand its origin. The patient and family members maintain that this behavior surfaced only after her surgery. Therefore the demonstrations of dual consciousness appear to have been consequential to her corpus callosotomy.

It is also unlikely that the patient had multiple personality disorder, since there was no alteration in identity or amnesia for her behavior (Kluft 1991). Putnam (1991) indicates that hemispheric disconnection is unlikely to account for multiple personality disorder.

Split-brain patients generally behave in a unified manner and do not show mutual conflict to the extent described here. Special clinical testing is usually used to examine dissociated hemispheric function in split-brain patients (Bogen 1993). For example, at the conference Iacoboni, Rayman, and Zaidel (1994) reported a callosotomized patient who tended to identify briefly visible letter strings as words when she used her right hand to signal and as nonwords with her left hand, regardless of the sequence of letters. They suggested that the hand used to respond determined which hemisphere interpreted the letter strings.

In split-brain patients the distinct impressions of each hemisphere may be integrated by relying on smaller fiber bundles not surgically severed that link the hemispheres (for example, anterior commissure, brain stem connections) and by using compensatory eye movements and auditory cues to make each hemisphere aware of the same environmental stimuli (Sergent 1987, Sidtis 1986). In this way behavioral unity may be preserved. Indeed, the patient in this study showed intrapersonal conflict mostly during challenging tests and when discussing the sensitive topic of her epilepsy, although conflict between her limbs sometimes occurred on daily living activities. Thus while each hemisphere was vying independently for control of her body for at least some of the time, her hemispheres usually cooperated and circumvented conflict. This patient's unusual brain development may have interfered with integration to a greater extent than usually seen in split-brain patients.

The concept of dual hemispheric consciousness is not new, and Bogen (1993) comprehensively reviews supporting and dissenting claims. Briefly, the refutation of dual consciousness is based on the observation of behavioral unity in split-brain patients in routine activities, which suggests a unitary basis for consciousness, and the contention that the right hemisphere should not be considered conscious without its possessing language. Space does not allow examination of the latter point, which is well discussed by Bogen. Based on my review of published reports of split-brain patients, the present patient demonstrated the most extreme degree of hemispheric *dis*-integration described, and furthermore, her right hemisphere *could* be interviewed. Thus despite the usual presentation of the split-brain individual, the present patient appears to have had dual, independent hemispheric modules that simultaneously interpreted stimuli, formulated opinions, and communicated unambiguously in a symbolic manner. Such hemispheric independence may not have occurred all of the time, but it was sufficiently apparent to reflect dual consciousness.

These observations suggest that consciousness may be divided by severing the major interhemispheric connection. Each hemisphere has a distinct, parallel arrangement of executive, receptive, expressive, analytic, self–monitoring, and information storage and reconstitution functions that are fundamental for consciousness (Zaidel 1987). However, hemispheric connection to the brain stem ascending reticular activating system appears to be necessary for cerebral cortical arousal (Plum and Posner 1980). To judge from a relevant case report (Edinger and Fischer 1913), a human being with a brain stem but who lacks cerebral hemispheres does not behave in a recognizably conscious manner.

Thus dual consciousness seems possible after the left and right hemispheres are disconnected, but not so upon separating the superior and inferior regions of the brain. Whether an individual could be considered "conscious" after transection in a third way, between front and back parts of the hemispheres, is theoretically interesting but for which there exists no suitable example. Kinsbourne (1982) speculates that such an individual would be "inconsequential"—acting without the influence of perception and perceiving without effecting action. However, the networks that have been proposed to mediate fundamental faculties such as spatial attention and maintenance of object awareness range extensively within the hemispheres (Mesulam 1990, Stuss and Benson 1986), so that a familiar kind of consciousness probably would not occur after such disconnection, to say the least.

It therefore seems that the essential structure for consciousness is one intact hemisphere connected to the brain stem. However, callosotomy is associated with a lack of vigilance (Trevarthen 1990). Thus while the disconnected hemispheres may be independently conscious, they are not fully alert; the callosum facilitates the sustenance of attention, among its other functions.

CONCLUSION

The overt dissension in the patient presented here stemmed from her inability following hemispheric disconnection to mediate hemispheric disparity internally. The different viewpoints of each hemisphere arose either from sensory input restricted to one hemisphere or from cognitive processing differences that distinguish the hemispheres' responses to the same stimuli (Hellige 1993). However, much of the time her hemispheres collaborated. Similarly, while normal individuals may feel turmoil when trying to address competing desires or interpretations, they generally act without uncontrolled, self-directed spoken or manual combat because of inner resolution. In normal humans each hemisphere may be capable of being conscious without the influence of the other, but a variety of mechanisms promote internal interaction, so that the person is of one mind. Despite the potential for hemispheric independence, its occurrence is most unusual and follows impaired compensation for interhemispheric dissociation.

ACKNOWLEDGMENT

Sandra L. Smith made valuable comments on the manuscript.

REFERENCES

Benson, D. F. 1979. "Neurologic correlates of anomia." In *Studies in neurolinguistics*, edited by H. Whitaker and H. A. Whitaker. Vol. 4. New York: Academic Press.

Biersack, H. J., D. Linke, F. Brassel, K. Reichmann, M. Kurthen, H. F. Durwen, B. M. Reuter, J. Wappenschmidt, and H. Stefan. 1987. Technetium-99m HM-PAO brain SPECT in epileptic patients before and during unilateral hemispheric anesthesia (Wada test): Report of three cases. *Journal of Nuclear Medicine* 28:1763–67.

Bogen, J. E. 1993. "The callosal syndromes." In *Clinical neuropsychology*, edited by K. M. Heilman and E. Valenstein. 3rd Ed. New York: Oxford University Press.

Carpenter, M. B. 1985. *Core text of neuroanatomy*, 3rd Ed. Baltimore: Williams and Wilkins.

Cowan, M. W. 1979. The development of the brain. *Scientific American,* September, 112–133.

Daffner, K. R., D. L. Schomer, G. R. Cosgrove, N. Rubin, and M. M. Mesulam. 1991. Broca's aphasia following damage to Wernicke's area. *Archives of Neurology* 48:766–68.

Edinger, L., and B. Fischer. 1913. Ein Mensch ohne Grosshirn. *Pflüger's Archiv für die gesamte Physiologie* 152:535–61.

Hellige, J .B. 1993. *Hemispheric asymmetry: What's right and what's left.* Cambridge, MA: Harvard University Press.

Huttenlocher, P. R., S. Taravath, and S. Mojtahedi. 1994. Periventricular heterotopia and epilepsy. *Neurology* 44:51–5.

Iacoboni, M., J. Rayman, and E. Zaidel. 1996. "Left brain says yes, right brain says no: Normative duality in the split brain". In *Toward a science of consciousness: The first Tucson discussions and debates,* edited by S. R. Hameroff, A. W. Kaszniak, and A. C. Scott. Cambridge, MA: MIT Press, pp. 197–202.

Kinsbourne, M. 1982. Hemispheric specialization and the growth of human understanding. *American Psychologist* 37:411–20.

Kluft, R. P. 1991. "Multiple personality disorder." In *Review of psychiatry,* edited by A. Tasman and S. M. Goldfinger, Vol. 10. Washington, DC: American Psychiatric Press.

Kupfermann, I. 1991. "Localization of higher cognitive and affective functions: The association cortices." In *Principles of neural science,* edited by E. R. Kandel, J. H. Schwartz, and T. M. Jessell, 3rd Ed. New York: Elsevier.

Kurthen, M., C. Helmstaedter, D. B. Linke, L. Solymosi, C. E. Elger, and J. Schramm. 1992. Interhemispheric dissociation of expressive and receptive language functions in patients with complex–partial seizures: an amobarbital study. *Brain and Language* 43:694–712.

Mack, L., L. J. Gonzalez Rothi, and K. M. Heilman. 1993. Hemispheric specialization for handwriting in right handers. *Brain and Cognition* 21:80–6.

Mamelak, A. N., N. M. Barbaro, J. A. Walker, and K. D. Laxer. 1993. Corpus callosotomy: A quantitative study of the extent of resection, seizure control, and neuropsychological outcome. *Journal of Neurosurgery* 79:688–95.

Mesulam, M. M. 1990. Large-scale neurocognitive networks and distributed processing for attention, language, and memory. *Annals of Neurology* 28:597–613.

Oxbury, S. M., and J. M. Oxbury. 1984. Intracarotid amytal test in the assessment of language dominance. *Advances in Neurology* 42:115–23.

Palmini, A., F. Andermann, J. Aicardi, O. Dulac, F. Chaves, G. Ponsot, J. M. Pinard, F. Goutières, J. Livingston, D. Tampieri, E. Andermann, and Y. Robitaille. 1993. Diffuse cortical dysplasia, or the 'double cortex' syndrome. *Neurology* 41:1656–62.

Plum, F., and J. B. Posner. 1980. *The diagnosis of stupor and coma.* 3rd Ed. Philadelphia: F. A. Davis.

Putnam, F. W. 1991. "Dissociative phenomena." In *Review of psychiatry,* edited by A. Tasman and S. M. Goldfinger, Vol. 10. Washington, DC: American Psychiatric Press.

Sarnat, H. B. 1991. "Developmental disorders of the nervous system." In *Neurology in clinical practice,* edited by W. G. Bradley, R. B. Daroff, G. M. Fenichel, and C. D. Marsden, Vol. II. Boston: Butterworth-Heinemann.

Serafetinides, E. A., R. D. Hoare, and M. V. Driver. 1965. Intracarotid sodium amylobarbitone and cerebral dominance for speech and consciousness. *Brain* 88:107–30.

Sergent, J. 1987. A new look at the human split brain. *Brain* 110:1375–92.

Sidtis, J. J. 1986. "Can neurological disconnection account for psychiatric dissociation?" In *Split minds/split brains,* edited by J. M. Quen. New York: New York University Press.

Stearns, M., A. L. Wolf, E. Barry, G. Bergey, and F. Gellad. 1989. Corpus callosotomy for refractory seizures in a patient with cortical heterotopia: Case report. *Neurosurgery* 25:633–36.

Stuss, D. T., and D. F. Benson. 1986. *The frontal lobes.* New York: Raven Press.

Trevarthen, C. 1990. "Integrative functions of the cerebral commissures." In *Handbook of neuropsychology,* edited by F. Boller and J. Grafman, Vol. 4. Amsterdam: Elsevier.

Wada, J., and T. Rasmussen. 1960. Intracarotid injection of sodium amytal for the lateralization of cerebral speech dominance. *Journal of Neurosurgery* 17:266–82.

Zaidel, E. 1987. "Hemispheric monitoring." In *Duality and unity of the brain,* edited by D. Ottoson. New York: Plenum Press.

13 Left Brain Says Yes, Right Brain Says No: Normative Duality in the Split Brain

Marco Iacoboni, Jan Rayman, and Eran Zaidel

INTRODUCTION

In order to address the controversial issue of unity or duality of mind after surgical commissurotomy, we investigated the capacity of a split-brain patient to coordinate decisions about the lexical status (word or nonword) of letter strings flashed to one hemisphere alone or simultaneously to both hemispheres. The simultaneous presentation of different letter strings to each visual field, even though only one of them was cued as the target, introduced some ambiguity into the lexical decision task. This ambiguity was used in hope of revealing some aspects of the mechanism of hemispheric control in the split-brain (Iacoboni and Zaidel, in press). The patient was NG, a 60-year-old woman who had a complete cerebral commissurotomy in 1963. The completeness of her callosal section was confirmed by MRI (Bogen, Schultz, and Vogel 1988).

METHODS

The patient performed a lexical decision task. She decided whether a target letter string tachistoscopically presented to the left visual field (LVF) or to the right visual field (RVF) was a word or a nonword. NG responded by pressing one button to indicate a word decision and another to indicate a nonword decision. She used her left hand (Lh) for one block of trials and her right hand (Rh) for another block.

In the ipsilateral conditions the target appeared on the same side as the response hand (LVF-Lh, RVF-Rh); in the contralateral conditions the target and response hand were on opposite sides (LVF-Rh, RVF-Lh). Note that in these conditions the hemisphere receiving the target is different from the hemisphere controlling the response hand and a potential conflict is created.

A fixation cross was displayed during the entire experiment. A warning tone sounded 750 milliseconds before the presentation of stimuli. Displays of horizontal lowercase letter strings were presented for 150 milliseconds. The innermost edge of the letter string appeared 1.5 degrees to the right and/or to the left of the fixation cross. In the unilateral condition, a target string appeared alone in one visual field. In the bilateral condition, the target string was accompanied by a distractor string which appeared in the opposite visual field. An arrow presented at fixation indicated which item was the target. Half of these distractors were of the same type as the target (both words or both nonwords) and half were of opposite types (one word and one nonword).

RESULTS

NG's performance was significantly above chance only when the target was flashed in the RVF and the patient responded with the Rh. There was no difference between unilateral and bilateral presentations either in accuracy or in latency. Words (724 milliseconds) were recognized significantly faster than nonwords (964 milliseconds) (P<0.003).

The patient exhibited an unusually strong bias for word decisions whenever a stimulus (whether target or distractor, word or nonword) occurred in the RVF and responses were made with the Rh (Figure 13.1). Conversely, the patient exhibited an unusually strong bias for nonword decisions whenever a stimulus (whether target or distractor, word or nonword) occurred in the LVF and responses were made with the Lh (Figure 13.2). In the two unilateral conditions when the target and response were processed

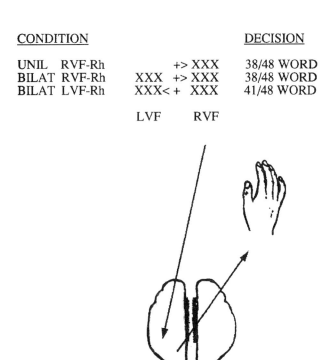

CONDITION		DECISION
UNIL RVF-Rh	+> XXX	38/48 WORD
BILAT RVF-Rh	XXX +> XXX	38/48 WORD
BILAT LVF-Rh	XXX< + XXX	41/48 WORD

LVF   RVF

Figure 13.1 The patient exhibited an unusually strong bias for word decisions whenever a stimulus (whether target or distractor) occurred in the right visual field and responses were made with the right hand.

CONDITION		DECISION
UNIL LVF-Lh	XXX < +	45/48 NONWD
BILAT LVF-Lh	XXX < + XXX	40/48 NONWD
BILAT RVF-Lh	XXX +> XXX	36/48 NONWD

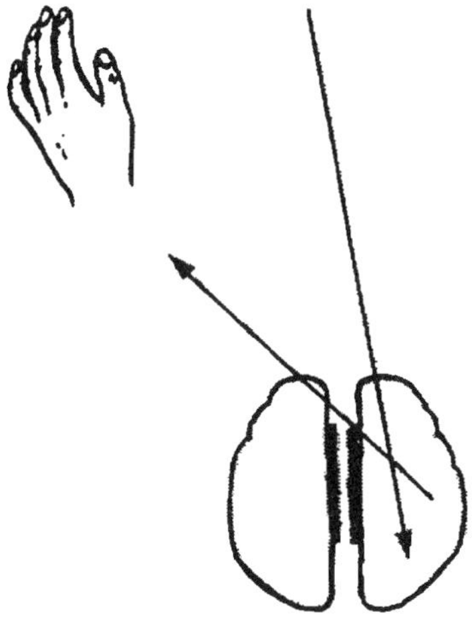

                      LVF   RVF

Figure 13.2 The patient exhibited an unusually strong bias for nonword decisions whenever a stimulus (whether target or distractor) occurred in the left visual field and responses were made with the left hand.

by opposite hemispheres (LVF-Rh and RVF-Lh), the hemisphere that received the target dominated (Figures 13.3 and 13.4).

DISCUSSION

These data reflect the effects of both lexical processing and decision mechanisms in the two hemispheres of NG. Her RVF-Rh accuracy is statistically above chance. Despite her strong word bias in the RVF-Rh condition, NG's nonword responses were correct in 19 of 20 responses. This indicates that lexical processing was taking place and that her performance was not due only to bias. The RVF-Rh condition, therefore, seems to reflect a successful balance between lexical access and decision bias (words) in the left hemisphere. By contrast, the unilateral RVF-Lh condition reflects complete dominance by the decision bias in the left hemisphere. Similarly, NG's LVF-Lh performance, though not different from chance in overall accuracy, does not show a random distribution of "word" and "nonword" responses, but rather a nonword bias. Finally, the unilateral LVF-Rh condition reflects only weak dominance by the decision bias of the right hemisphere. Thus, the left hemisphere is stronger than the right hemisphere in establishing decision bias dominance.

During bilateral presentations, the decision processes in NG's two disconnected hemispheres are in direct conflict. The same letter string can

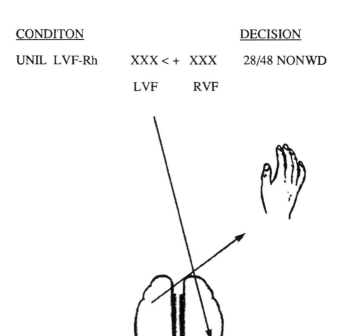

Figure 13.3 When the target and response were processed by opposite hemispheres (the right hemisphere receiving the stimulus and the left hemisphere controlling the right hand response), the right hemisphere dominated.

produce opposite bias depending on which hemisphere is in control. During bilateral presentations, the response hand determines which hemisphere will dominate. Certainly, decision bias changed dynamically from trial to trial, as between LVF-Lh and RVF-Lh trials in the same block. We may speculate that any stimulus in a visual field (even unattended) triggered automatic lexical access and a subsequent decision in the contralateral hemisphere, and that there was some competition for dominance between the lexical access and response bias in the same hemisphere as well as between the decision bias in the two hemispheres.

CONCLUSIONS

These findings suggest the existence of two independent, parallel, and sometimes opposing "decision makers" in the two disconnected hemispheres of NG. In lexical decision by normal subjects, there is often a small, but reliable, difference in hemispheric response criteria (Chiarello 1988), with subjects showing a nonword bias in the LVF and word bias in the RVF. In NG, on the other hand, we have observed a most radical separation of response criteria, where the left hemisphere responds "word" and the right hemisphere responds "nonword" almost whenever they "see" a stimulus and control the response hand.

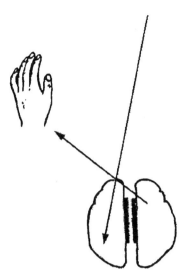

Figure 13.4 When the target and response were processed by opposite hemispheres (the left hemisphere receiving the stimulus and the right hemisphere controlling the left hand response), the left hemisphere dominated.

These findings are inconsistent with the hypothesis that NG has one undivided control system for supervisory evaluation of alternative choices (MacKay and MacKay 1982), and suggest instead the existence of two independent, parallel, and sometimes opposing "decision makers" in the two disconnected hemispheres. These decision makers have separate access to the two hands and the decision criterion can change when the patient changes her response hand. These data also show a dissociation between the lexical processing and the decision mechanisms in the two hemispheres of NG. Any stimulus in a visual field (even unattended) triggers automatic lexical access and a subsequent decision in the contralateral hemisphere, and there is some competition for dominance between the lexical access and decision processes in the same hemisphere, as well as between the decision processes in the two hemispheres. The coexistence of these decision processes suggests normative duality in the split brain. This duality is resolved by motor action plans determined by response factors (here, response hand).

REFERENCES

Bogen, J. E., D. H. Schultz, and J. P. Vogel. 1988. Completeness of callosotomy shown by magnetic resonance imaging in the long term. *Archives of Neurology* 45:1203–05.

Chiarello, C. 1988. "Lateralization of lexical processes in the normal brain: A review of visual-half field research." In *Contemporary reviews of neuropsychology*, edited by H. A. Whitaker, New York: Springer Verlag, pp. 36–76.

Iacoboni, M., and E. Zaidel. In press. Hemispheric independence in word recognition: Evidence from unilateral and bilateral presentations. *Brain and Language*

MacKay, D. M., and V. MacKay. 1982. Explicit dialogue between left and right half-systems of split-brains. *Nature* 295:690–91.

14 Inkblot Testing of Commissurotomy Subjects: Contrasting Modes of Organizing Reality

Polly Henninger

One of the problems in the study of consciousness is relating subjective experience to the underlying brain anatomy and physiology. Although subjective awareness often appears to be unified, research with commissurotomy ("split-brain") patients has shown that each cerebral hemisphere can function independently, seemingly unaware of the other's conscious contents. The commissurotomy subject acts or speaks in a manner suggesting that two consciousnesses are operating. For example, commissurotomy patient A was asked to write whatever came to mind with his right hand. (Because of the clinical nature of some of the material described, subjects from the Vogel-Bogen series described herein will be identified only by a randomly assigned letter.) He wrote "Wha lime is it?" (apparently forgetting to include one t and to cross another) and then read aloud "What time is it?" When asked to write the same thing with his left hand, he wrote "PUCK" and then two amorphous shapes that he claimed were supposed to be "time" and "is" (Figure 14.1).

He said, in an angry and frustrated tone, "I get so disgusted with myself. I know what I want to do but the hand won't co-operate with the mind." When asked what he had written, he replied without hesitation in a tone that seemed to indicate that it was obvious what he had written, "Puck, hockey puck." The intensity of his response raised doubt as to whether the word were indeed puck or if the "p" had been intended to be an "f," but in any case, the word bore no resemblance to "What time is it?" This example clearly shows the presence of two independent sources of control, but one is out of the awareness of the other.

Commissurotomy provides a means of investigating the nature of each hemisphere's contribution to consciousness and thus offers a means by which we may better understand our subjective sense of unity. It also offers a possible means of understanding those instances when it is absent, for example, chronically in certain dissociative disorders (Henninger 1992), and at certain times, in everyone.

Cerebral commissurotomy, commonly known as "split-brain" operation, applied to the human, denotes complete callosal section, with or without anterior commissurotomy, which is usually performed for medically intractable, multifocal epilepsy (Bogen 1993). This procedure largely

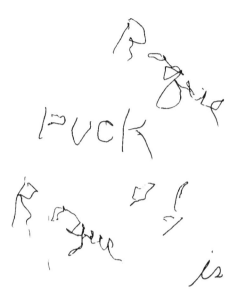

Figure 14.1 Writing by right (upper right-hand corner, upper signature, "is" as lower right- hand corner) and left hand (lettering "puck," two amorphous shapes, and lower signature) of commissurotomy patient A.

isolates the two hemispheres from each other, allowing each to receive stimuli unavailable to the other and to be tested independently.

Although research with subjects with commissurotomy is generally designed to better understand the functions of each hemisphere, the left hemisphere often dominates in the testing situation (Zaidel and Sperry 1973). The right hemisphere may begin a task but frequently the left hemisphere takes control and completes it. For example, in a study by Nebes and Sperry (1971), commissurotomy patient B was shown pictures of common objects in the left visual half field, one of which was a cat similar to his family pet. He was unable to name this object, indicative of the left hemisphere's lack of awareness of the picture, but began to write his answer with his left hand. He wrote "ca," stopped, said "No, that's wrong," added two loops and then said "bottle." It appeared that the right hemisphere knew what the object was, but because of its limited verbal abilities could only spell the first two letters. The left hemisphere, unaware of the object, took control of the writing hand and confabulated in order to produce a response.

The tendency of the left hemisphere to take control in the testing situation of the commissurotomy patient may be related to our tendency to equate consciousness with verbal consciousness. We commonly consider consciousness to be the content of awareness which we can describe verbally or our inner verbal stream of thought. It is possible but difficult to communicate content that we are unable to verbalize and methodological constraints often limit studies to verbal consciousness. It is easy to overlook that consciousness itself has both verbal and nonverbal elements. Research with commissurotomy subjects suggests a secondary parallel perceptual system (Levy, Trevarthen, and Sperry 1972) which is primarily nonverbal, but research with aphasics with and without semantic paralexias, suggests that the verbally dominant left hemisphere inhibits its expression in the callosally-intact person (Landis, et al. 1983).

Study of the right hemisphere processing of the split–brain subject can inform us about the nonverbal aspects of consciousnesss. However, since the left hemisphere frequently takes control of processing in the experimental situation, a method is needed for keeping the right hemisphere active and undominated. This paper describes efforts to develop such a method.

BACKGROUND TO THE RESEARCH

Research with commissurotomy patients has shown that the right hemisphere is preferentially engaged in form perception and visual-spatial processing (Bogen 1969a, Nebes 1973, 1974). Research with both brain-damaged and normal subjects is consistent with this finding (Bogen 1969b). One of the commonly used techniques for studying visual perception clinically is to show inkblots to a subject and ask what he or she sees in them. Inkblots were employed as part of an experimental approach to the study of imagination and creativity at the end of the last century by Dearborn at Harvard University (1897, 1898 cited by Exner 1993). The Rorschach Inkblot Test, the first major clinical and research instrument to be classified as a projective technique (Rabin 1981), was developed by Hermann Rorschach in 1921 for use as a perceptual-diagnostic tool. The fact that inkblots are not representational and thus not quickly verbalized and that the right hemisphere is preferentially engaged in visuospatial processing suggests that the Rorschach inkblot methodology might be a viable means of eliciting sustained right hemisphere processing and avoiding left hemisphere dominance. Moreover, numerous studies have found evidence indicating that the right hemisphere is specialized for global processing and the left hemisphere is specialized for local processing of detail (Cronin-Golomb 1986, Delis, Robertson, and Efron 1986, Robertson and Lamb 1991; Robertson, Lamb, and Knight 1991, Robertson, Lamb, and Zaidel 1993), suggesting that inkblots, which are likely to elicit global processing, will be preferentially processed by the right hemisphere.

Patients with right-sided lesions give a lower percentage of responses with good form on the Rorschach test than patients with left-sided lesions (Belyi 1988). Apparently tumors localized in the right hemisphere disrupt visual gestalts and in this way interfere with the process of comparing new visual impressions with previous ones. Although Belyi's findings suggest that the right hemisphere preferentially processes the inkblots, no study has tested this hypothesis directly. Studies with complete commissurotomy subjects either have not tried to lateralize the Rorschach Inkblot Test (Lewis 1979), have combined manual and verbal responding (Regard 1991), or have used it as a test of perception (Yates 1994). Although these studies have yielded valuable information, they do not address the issue of dominance.

In Yates' study (1994), a male (patient C in this study) and female split–brain subject from the Vogel–Bogen series were asked to identify which Rorschach blot they had seen tachistoscopically by pointing to an array of blot miniatures in free vision or giving a verbal association first and then pointing. When pointing, the female subject showed right hemisphere superiority and the male subject did not show a superiority. Both subjects showed left hemisphere superiority when a verbal association was made with a substantial decrement in right hemisphere performance. The female subject's performance supports the hypothesis that the right hemisphere is specialized for processing the inkblots. Other investigators also have found that dominance over the motor response switches sharply from right to left hemisphere in these subjects when instructions require a verbal instead of a manual response (Levy, Trevarthen, and Sperry 1972). It appears that to elicit and maintain right hemisphere responding in the commissurotomy subject, a nonverbal test such as the Rorschach with a nonverbal means of responding must be employed.

A study by Levy, Nebes, and Sperry (1971) suggested that drawing with the left hand may be an effective means of eliciting and maintaining right hemisphere processing. In that study commissurotomy patient B was presented with household objects to his left hand and asked to write the names of the test items. After a smoking pipe had been presented to the left hand, he first printed "PI," pressing down very hard with the pencil. After completing the first two letters, he stopped, delayed and then in a more relaxed writing manner, completed the word "Pencil." After another pause, he scratched out the last four letters and stated that he did not know what the object was. He was then asked to draw a picture of the test item with his left hand out of sight behind a screen and drew a facsmile of a pipe. This study suggests that drawing with the left hand may be an effective means of lateralizing the response to the right hemisphere.

The present study investigates the hypothesis that inkblots are preferentially processed by the right hemisphere by presenting inkblots to the left and right visual fields of two commissurotomy patients and having them draw what they see with their left and right hands respectively. In this way both the presentation of the stimuli and the response are lateralized. It is hypothesized that the right hemisphere will be more adept than the left at extracting sufficient information from the inkblot to copy it. It is predicted that drawings of the blots shown to the left visual field, executed by the left hand, will be matched correctly to the blot by judges matching the drawings to the blots significantly more often than drawings of the blots shown to the right visual field, executed by the right hand.

The "don't know" response allows a judge to indicate that a drawing does not look like a blot. This response provides a means of potentially identifying the occasions when the hemisphere to which the stimulus has been presented is not in control of responding. Differentiating among blots that cannot be identified correctly offers more opportunity for qualitative analysis (Kaplan 1988). It is predicted that drawings executed by the right hand will be more difficult to match than drawings executed by the left hand, resulting in more "don't know" responses for right–handed drawings than left–handed drawings.

EXPERIMENT 1

Method

Subjects Two complete commissurotomy patients of the Vogel-Bogen series, patients B and C, were tested. Both are right-handed males who underwent the commissurotomy procedure for the treatment of intractible epilepsy. These patients have been studied extensively psychologically (Sperry, Gazzaniga, and Bogen 1969, Sperry 1974), neurologically (Bogen and Vogel 1975), and with MRI (Bogen, Schultz, and Vogel 1988).

Materials Four inkblots were created by dropping black permanent ink on good quality bond paper and folding the paper in half along the side of the ink (Henninger inkblots). Twenty–three common household articles were used to lateralize responding in preliminary testing prior to inkblot presentation, including such items as a comb, spoon, walnut, marker pen lid, and clothespin.

Procedure Subjects were presented common objects behind a screen in the left or right hand and then asked to draw what they felt with that hand to train the subject (Ramachandran, Cronin-Golomb, and Myers 1986) to use the hand on the same side as stimulated for responding and to prime and engage the hemisphere before it was given the inkblot test. Inkblots were presented individually twice for a total of eight presentations in both the left and right visual half–fields by use of the lateral limits technique developed by Myers and Sperry (1982) for testing these subjects (Figure 14.2). This technique has been used extensively in this laboratory (Cronin-Golomb 1986a, 1986b, Myers 1985) and has been validated elsewhere with a callosotomy patient (Trope, Rozin, and Gur 1988, Trope *et al.* 1992).

With the head held in position by use of a standard bite bar, the subject moves the eyes horizontally until the limit of rotation in that direction is reached (at approximately 45 to 55 degrees off the vertical midline for most

Figure 14.2 Lateral limits technique (See text for description.)

subjects) (Cronin-Golomb 1986a). All stimuli that appear in the space lateral to this limit fall on the temporal half of each retina which projects only to the contralateral hemisphere. With this procedure, stimuli may be restricted for prolonged durations to the left or right visual half-fields while remaining in central vision. Limited vertical scanning is also possible. To minimize head-turning, a nose guard which triggered a light every time the patient moved the head was also used.

The first time each inkblot was shown in each visual field it was followed by a multiple choice presentation of three blots and the subject had to point to the blot he had just seen (Figure 14.2). The inkblot was then presented again individually, removed from sight, and the patient was instructed to draw what it might be with the ipsilateral hand.

Drawings were numbered, intermixed and presented with the original four inkblots to three judges naive to the subjects, the Rorschach, and the nature of the study. The judges were told that the drawings were attempts to draw the inkblots and instructed to indicate in writing which inkblot they believed to be the blot copied or to indicate "don't know."

Results

Patient B successfully discriminated each inkblot from three choices. His drawings of the blots presented in the left visual field generally looked like simplified versions of the blots (See the drawing in the lower left hand corner of Figure 14.3). His responses to blots in the right visual field commonly

Figure 14.3 Henninger inkblot and subjects' drawings of it, in the order they were obtained. Patient C, first drawing in first testing of left visual half-field, upper right-hand corner; second testing of left visual half-field, upper left-hand corner; drawing of same stimulus in testing of right visual half-field, upper center. Patient B, drawing in left visual half-field, lower left-hand corner; drawing in right visual half-field, lower right-hand corner.

were elaborated and often had wiggly lines suggesting detail. (See the drawing in the lower right hand corner of Figure 14.3). Many were unidentifiable and required filling in to form a gestalt. On half of the trials presented to patient B's right visual half field, he showed evidence suggesting that he wanted to reject the task. Once he did not want to go on the bite bar, once he quietly put down the pencil, slipped his left hand over, picked up the pencil and tried to use his left hand instead, and twice he commented that it was hard. On no left-sided trial did he attempt to reject the task.

Patient C accurately discriminated the blots presented in the left visual field from competing ones. His response to the first blot in the left visual field was to verbalize, "the Rorschach, you're giving me the Rorschach. I'm not nuts." His drawing of the blot presented in the left visual field was very stylized, did not look like a blot, and was drawn in the extreme upper righthand corner of the page. After drawing, he reported "kite," which his drawing did indeed resemble and which did not resemble the inkblot, suggesting that despite the extensive priming with the left hand, the left hemisphere had taken control of processing and motor output. (See blot in center and drawing in upper righthand corner of Figure 14.3). He then said "eel," "man-a-ray eel, long with a long tail at back (which did resemble the blot). That's the second thing I think of."

After the third trial, he requested a break which he was given after the fourth trial. Examination of the placement of his drawings showed that the first three were on the right side of the page but that each successive drawing was closer to the midline. From the fourth drawing on, all drawings were on the left side of the midline.

After the fourth blot the subject said, "the faces ... this always looks like two people kissing," an unlikely response given that he had never been shown this original blot before. In trials 4–7 he verbalized prior to drawing and in trial 8, he drew first, saying "it doesn't remind me of anything," and then labelled it "hands?" writing the question mark backwards. He wrote the question mark again, in correct orientation, and said "ah," in a tone suggesting that now he knew what the blot was. Apparently, despite right hemisphere activation, information was being transferred to the left hemisphere for output. In case the instructions contributed to the initial left hemisphere dominance, they were changed to "draw what you see." The eight presentations were repeated in the left visual field for patient C, making a total of 24 presentations in the left visual field and 16 in the right.

In the second testing of patient C in the left visual field, he was completely silent throughout testing. His first drawing with his left hand is shown at the upper lefthand corner of Figure 14.3. His drawings were more fluid and less stylized and he showed some perseveration of figures kissing.

When the blots were presented in the right visual field, patient C misidentified two of the four. He drew four drawings that lacked structure and showed no resemblance to the blots (see Figure 14.3, drawing of the pictured blot is shown in the upper center), one which showed vague resemblance, and three which were stylized drawings, two of which he labelled "elephants sitting down" and "elephants dancing" and one which looked like his first drawing of the same stimulus which he had labeled "two people with an obelisk between them." On one trial during presentations to the right visual field he attempted to use his left hand instead of his right. In the testing sessions of both men it appeared that the right hemisphere tried to control processing.

Both subjects easily described the four stimuli verbally when presented in free vision and asked what they might be. Patient B gave short, single answers such as two rhinos or a cat. Patient C gave more complex responses such as "It looks like two dogs standing back to back, yapping off," with multiple responses on two cards.

Judges' ratings showed adequate reliability of judgment. Two of the three judges gave the same rating on 90% of the trials; all three judges gave the same rating on 50% of the trials). Left-sided blots were identified correctly 61% of the time; 44% of the right-sided blots were identified correctly. A "don't know" response was given to 15% of the left-sided blots and to

38% of the right-sided blots. The distributions of responses to each hand were statistically different ($\chi^2 = 7.8$, $P<0.05$; see Table 14.1).

Further analysis showed that the distribution of judges' responses for each hand did not differ between the two subjects (left hand = 4.24, P>0.1; right hand = 1.43, P>0.1). The overall distribution of responses did not differ between the two subjects either ($\chi^2 = 1.2$, not significant). All three judges marked more right-sided blots "don't know" than left-sided blots. Indeed half of patient C's right-sided responses were marked "don't know." An analysis of the judges' responses to patient C's four drawings before and after the break revealed that six of twelve judgments of the drawings prior to the break were correct whereas eleven of twelve judgments after the break were correct ($\chi^2 = 5.04$, df = 1, $P<0.05$), a statistically significant difference that supports the observation that a change in strategy, and presumably hemispheric control, had occurred. (Judgments of the first four trials were removed from the left hand, right hemisphere, score and added to the right hand, left hemisphere score. This did not change the distribution of judgments between hands appreciably; the difference was still statistically significant.)

EXPERIMENT 2

Method

Subjects Patients B and C again participated.

Materials The ten cards from the Rorschach Inkblot Test were used for the experimental procedure. A comparison test comprising two line drawings of representational objects (a bird, a plate with two eggs on it, and fork), two additional inkblots created in the same manner as those in the

Table 1 Frequency of Judges' Responses by Hand

Type of Inkblot	Percent Response Type		
	Correct	Incorrect	Don't Know
Henninger Inkblots			
Left Hand	61	23.5	15
Right Hand	44	19	37.5*
Rorschach Inkblots			
Left Hand	22	48	30
Right Hand	20	27	53
Line Drawings			
Left Hand	92	0	8
Right Hand	92	0	8

Note: For the Henninger inkblots, $\chi^2 = 7.8$, $P<0.05$. For the Rorschach inkblots, $\chi^2 = 7.72$, $P<0.05$. For the line drawings χ^2 was not significant.

first experiment with the addition of red ink, and the original blots inverted was also used.

Procedure Within ten days of the first testing session, each subject was retested. He was given objects to palpate and draw, to prime the respective hemisphere. Each subject was given a graphite pencil and ten colored pencils and told to use whichever pencils he wished. This procedure was followed by two practice trials from the Henninger blots and then the Rorschach cards, and then the comparison test, presented with the lateral limits technique. All stimuli were presented initially to the right hemisphere (left hand, left visual half field) and then to the left hemisphere (right hand, right visual half field). Prior to each Rorschach drawing trial, an identification trial was given in which the subject was asked to point to the card he had just seen from among three choices. For Rorschach card I, which is black and white with additional red ink, a "fooler" in black and white, created by photocopying the original version, competed with it. After all drawing activity, the subjects were administered the Rorschach Test in the standard manner in which the cards were given serially to the subject and he was asked to say what each might be.

Three judges, one of the judges from the original experiment, an artist, and a therapist familiar with the Rorschach, attempted to match the drawings to the Rorschach cards, line drawings, and original blots.

Results

It was more difficult in the second (Rorschach) session of both patients to engage the right hemisphere. More priming trials were needed for patient B (12 versus 20) and patient C (21) to get them to stop talking. Indeed, patient C apologized for speaking after drawing object 12, and said, "the word just popped out." Patient B was given an additional priming task (selecting an object palpated from a basket of objects) and patient C was given another 20 trials after he failed to identify the blot among three in his initial practice trial. On trial six in the left visual half field in which Rorschach card VI was presented, patient B responded twice, once with each hand (see Figure 14.4).

After he copied the blot with his left hand (outline of a shape with five appendages on the left side) he reached over with his right hand, picked up the pencil and continued drawing (first the rectangle at the bottom and vertical lines to the right of the figure, additional lines superimposed on the original figure). Initially the additional lines were not synchronized with the original drawing and the style was different, resulting in the impression of a double exposure of two different contents. He then began filling in the initial figure (keeping within its boundary) as though he had become aware of its existence suggesting that at the outset he had not been. When the blot was presented later to the right visual half field, the drawing done by the right hand looked dissimilar to the one it had drawn when the blot was

Figure 14.4 Drawings of Rorschach card VI by left and right hands of patient B when presented in the left visual half-field on left; drawing by right hand when presented in the right visual half-field on right. (See text for further description.)

shown in the left visual half field, confirming that the left hemisphere had not had access to the visual image when the blot was presented in the left visual field. The drawing done by the left hand and the drawing done by the right hand when the subject could see the stimulus were relatively adequate portrayals of the blot but what appeared to be eyes had been added to the latter, suggesting that it had been conceptualized as a humanlike figure. All drawings by patient B were done with the graphite pencil. Drawings of the blots done with his left hand showed what appeared to be considerable perseveration. Each drawing appeared to be a variant on a basic pattern or template (the left-sided figure in Figure 14.4). Drawings executed with his right hand lacked structure and had more detail.

Patient C, when identifying Rorschach card II from three choices in the LVF, selected the correct version with red in it. When doing so in the RVF, he selected the black and white fooler. Similarly, when drawing with the left hand, copied Rorschach cards I–VII with the graphite pencil and the color cards, VIII–IX, in color. All drawings executed by the right hand were done with the graphite pencil only. Drawings either appeared representational (Blots I–V, VII) or were unidentifiable (VI, VIII, IX, X). Patient C labelled five blots in the right visual half field (I, II, IV, V, VI) and two in the left visual field (cards VI and VII). His label for VI in the right visual field was "splat cat" and in the left visual field was "splat cat. Kitty in the mac truck, didn't make it," spoken in a rhythmic voice. He did not label any color card (VIII, IX, and X).

The Rorschach judgment task with ten choices was more difficult than the judgment required in the original study in which there were only four different blots. The judges were able to identify the pictures in about one fifth of the drawings of both hands. Statistical analysis again showed a significant difference in the distribution of their responses ($\chi^2 = 7.72$, $P<0.05$). When the judges were unable to match correctly, they were more likely to select "don't know" when the blot was shown to the right side and to make an incorrect match when the blot was shown to the left side.

Copies of the two line drawings of common objects (intermixed with additional blots and the inverted original ones) were identified by the judges in 92% of both the right and left hand renderings (see Figure 14.5). One "don't know" response was elicited by drawings from each hand (see Table 14.1).

To evaluate the consistency and reliability of the styles reflected in the drawings, a clinical psychologist (N. Thurston), trained and experienced in interpreting Rorschach responses and naive to these patients, was given the

Figure 14.5 Line drawing of common objects (upper right-hand corner) and subjects' drawings of it in the order in which they were obtained. Patient C, drawing when presented in the left visual half-field, upper, left-hand corner, drawing when presented in the right visual half-field, upper, center. Patient B, drawing of figure in left visual half-field, lower left-hand figure, drawing of figure presented in the right visual half-field, lower right-hand figure.

40 renderings and asked to allocate each to one of four groups according to which she believed had been executed by the same hand (hemisphere) and subject without concern for the total number in each group. She accurately grouped 35 of the 40 drawings (binomial probability, $P<0.05$). An independent researcher familiar with these patients (Myers) was similarly asked to group the drawings and also was able to do so (36/40 drawings).

Examination of patient C's drawings suggested that once a blot was labelled verbally, the remaining drawings of it were more likely to be labeled, to look like the label, and to be identifiable. To investigate these observations further, patient C's drawings of the Rorschach cards were returned to the three original judges and the clinical psychologist, making a total of four judges, two who were unfamiliar with the Rorschach test and two who were experts with the test, with the six verbalizations he spontaneously emitted during testing. Judges were asked to review each drawing and indicate whether it matched one of the verbalizations. These judgments were compared to their earlier judgments of the blots.

Expert judges correctly matched the drawing to the verbalization in most trials (right-sided = 70%; left-sided = 75%) and identified the blots corresponding to those drawings in more than half the trials (right-sided = 60%, left-sided = 70%). Naive judges matched the drawing to the verbalization in few trials (right-sided = 30% and left-sided = 0%), but identified the blots from the drawings on the verbalized trials in 50% of the right-sided trials and 90% of the left-sided trials. When the blot wasn't verbalized, expert and naive judges performed similarly. They identified 50% of the right-sided drawings, and 40% and 30%, respectively of the left-sided drawings. On one occasion when the blot was presented in the right visual field and not verbalized (card VII), both experts correctly identified the drawing as matching the verbalization that had been given when this card was shown to the left visual field ("Hiawatha was an indian"), but did not match the drawing correctly to the blot, suggesting that the drawing looked more like the label than the blot. Patient C did not verbalize any of the color cards (VIII, IX, X); no judge matched any of the color cards to a verbalization. One expert spontaneously characterized two of the verbalizations as similar to nursery rhymes or children's books. These two verbalizations were the only ones given when the blots were presented in the left visual half-field.

DISCUSSION

The significant difference in the first experiment in the ease of identifying the blots made by the left and right hands and the rejecting behavior observed when the blots were presented to the right visual field but not the left strongly suggest that drawing the inkblots was successful in facilitating right hemispheric control of processing and output. The significant difference in distribution of the "don't know" responses in both experiments suggests that the left hemisphere at times was unable to do the task at all, either because the right hemisphere had control and could not see the stimuli when they were presented in the right visual field, or because the left hemisphere had trouble copying an ambiguous visual image and could only do so successfully once it had labelled it, or because it was drawing some unidentifiable mental construction. These results indicate a right hemisphere advantage for this task and suggest that copying the inkblot cards was a successful methodology for inducing right hemisphere control of processing. These findings are consistent with the view that the right hemisphere is specialized for global processing.

That the ability to identify copied line drawings of representational objects did not differ by hand indicates that the right hemisphere advantage observed in the inkblot task was not for drawing per se, but for drawing an image that was not easily verbalized. Apparently once a verbal label was achieved, the left hemisphere was able to do the task.

That the evaluators were able to successfully differentiate the drawings into four groups indicates that distinct styles characterize each hemisphere of each person. These reflect some combination of differences in manual dexterity, hemispheric specialization, and perhaps personality. Given that both men are right-handed, the fact that the drawings that both drew with their left hands were better recognized is strong support for the hypothesis. Taken in concert with findings by Bogen (1969b) that commissurotomy patients copied two-dimensional cubes and geometric crosses better with their left than their right hands, and research that has shown deficits of spatial articulation in patients with right but not left hemisphere parietal lobe disease (Warrington, Merle, and Kinsbourne 1966), these results suggest that the left hemisphere's drawings may not so much reflect copying the spatial component of the blot but may be more a rendition of the mental construction arising from the verbal label given.

That only drawings of the Rorschach color cards done with the left hand were drawn in color and that no right-sided drawings were further supports the conclusion that the right hemisphere is more perceptually accurate and the left hemisphere is more likely to construct meaning. That left-sided drawings of the color cards were done in color offers support to Exner's contention (1991) that these cards are emotionally activating, that these cards tap emotional processing, and the view that the right hemisphere is more involved in emotional processing. That the right-sided version with color of card II was rejected and a black and white version chosen instead suggests that the left hemisphere may avoid emotionally activating material.

Placement in extrapersonal space has been inferred to indicate functioning of the contralateral hemisphere (Plourde and Sperry 1984). The movement of patient C's placement of the figures from the right to the left side of the paper suggests that the right hemisphere became increasingly activated as the task progressed (for a discussion of activation theory of hemispheric functioning, see Kinsbourne 1970). His complete silence in the second testing in the left visual field suggests the right hemisphere was fully activated and the left hemisphere was suppressed. This silence, which had a still, noiseless quality that deepened as the session progressed, has been observed in this subject before and other commissurotomy subjects when the right hemisphere is fully engaged.

When the subjects were asked to describe the blots verbally, both subjects easily did so. Their behavior indicated that they did not have difficulty assigning verbal labels to the stimuli and suggested that the left hemisphere did not have difficulty doing the task when the blots were presented in free vision and the response was oral.

Subcortical transfer was observed in patient C in his second verbalization to the first blot presented in the left visual field, "eel." This more accurate description of the blot appeared to be the result of an association transferred from the right to the left hemisphere. This behavior is consistent with other studies that have shown that subcortical transfer of some contextual and connotative associations is possible in these patients, indicating that the product of one hemisphere still has some (highly restricted) access to what is happening in the other (Cronin-Golomb 1986b, Myers 1985). Studies have also shown the presence of a second parallel percept in the nonresponding hemisphere (Levy, Trevarthen, and Sperry 1972). The reversed question mark was similar to previous incidents of letter or object reversals seen in these patients when copying a word or object presented to the left visual field, strongly suggesting that the reversals are evidence of right hemisphere output. Patient C's comment after being shown a blot he had never seen before implies that what he reported was based on previous testing with the Rorschach. His rhythmical, rhyme-like verbalizations of two left-sided blots similarly suggested transfer of associations from the right to the left hemisphere.

Despite the right hemispheric control observed in the first session of both subjects, it was difficult to reestablish in the second session. This difficulty may have been due to the verbal descriptions requested at the end of the preceding session suggesting that a label would be wanted. Alternatively, both subjects had been tested with the Rorschach stimuli before by other investigators and may have felt anxious and thus reluctant to allow the nondominant right hemisphere to take control, the subjects may have recalled labelling the blots, or the finer gradations of ink in the Rorschach cards than the original blots may be less likely to elicit a global response.

These findings suggest that utilizing drawing of noneasily verbalized material facilitates right-hemispheric control of processing in commissurotomy subjects and inhibits dominance by the left hemisphere. However, the trial in which patient B first drew a copy with his left hand and then totally different material with his right hand dramatically demonstrated the presence of two independent sources of control. That the two right-handed drawings were dissimilar showed that the left hemisphere had attempted to take control despite its inability to see the stimulus. Apparently, even after the right hemisphere was engaged, the left hemisphere attempted to take control of the experimental situation.

In sum, it appears that each hemisphere brings its own abilities to the perceptual task. The left hemisphere attempts to convert the blot to something meaningful that can be verbally labelled and projects that on to reality. The right hemisphere attempts to relate the blot to a basic visuospatial pattern or structure inherent within its way of perceiving and misses detail. Neither hemisphere brings a blank slate to the situation. Neither perceives the blot totally without bias, but the bias of the left hemisphere is an error of commission and that of the right hemisphere is an error of omission. These results are consistent with findings that patients with right-hemisphere lesions confabulate, presumably relying on their intact left hemispheres, but patients with left-hemisphere lesions, presumably utilizing their intact right hemispheres, do not (Belyi 1988), and findings from other split-brain patients (Phelps and Gazzaniga 1992) that abilities to make inferences and interpret events are stronger in the left hemisphere than in the right hemisphere.

To what extent do these results generalize to the normal brain? Pilot testing of three normal subjects found that all three subjects experienced greater difficulty in drawing inkblots presented in the right visual field than in the left visual field, supporting the findings observed here and suggesting that further study with normal subjects would be valuable. The fact that expert judges better matched drawings to the verbal labels than naive judges but were no better in matching drawings to blots, and that expert judges were better at matching verbal labels than blots suggests that the verbal label does indeed influence perception and perhaps prevent people from perceiving reality directly.

These findings support the view that drawing is an effective means of accessing emotional material. Asking normal subjects to draw what they see in inkblots might be a fruitful methodology for accessing emotional material. The fact that none of the color cards were verbalized, and that drawings of these cards by the right hand were unidentifiable, and that a black and white photocopy version was selected over the correct card with red ink in the RF, suggests that emotionally activating material suppresses the left hemisphere. Comparing drawings with verbal responses might be useful for better understanding the accessing of, and suppression of, emotional material.

The commissurotomy subject provides a means of investigating the disconnection syndrome in light of distributed neural networks. The changes in the placement on the page of drawings done by patient C suggests that integrating disconnection phenomena with distributed changes in multiple neural systems might be productive for localizing consciousness. Findings that callosal efficiency is related to sustained attention (Rueckert, Sorenson, and Levy 1994) suggest that the attentional shifts in the split-brain subject, uncontrolled by callosally balanced activation, reflect shifts in hemispheric control. The fact that patient C's responses moved progressively across the page rather than discretely from one side to the other suggests a large, underlying source of activation that the two hemispheres share. This interpretation is consistent with the view that the locus of conscious awareness is subcortical.

The data reported here strongly suggest that awareness is greatly influenced by changes in hemispheric control. Study of nonverbal, particularly emotional processes, in the commissurotomy subject will further our understanding of right-hemispheric contributions to awareness and may assist in better understanding how the products of processing in the two independent hemispheres interrelate to create a sense of unified consciousness.

ACKNOWLEDGMENTS

I thank the subjects for their participation; R. W. Sperry for suggesting that I participate in this conference; G. Andrews, A. Ford, K. M. Henninger, N. Thurston, S. Perez, and S. Sabo, for judging the drawings; and J.E. Bogen, W. Brown, A. Cronin-Golomb, L. Ellenberg, M. Kinsbourne, J. MacCuish, J.J. Myers, and N. Thurston for their helpful comments and suggestions.

REFERENCES

Belyi, B. I. 1988. The role of the right hemisphere in form perception and visual gnosis organization. *International Journal of Neuroscience* 40 (3–4):167–80.

Bogen, J. E. 1969a. The other side of the brain I: Dysgraphia and dyscopia following cerebral commissurotomy. *Bulletin of the Los Angeles Neurological Society* 34:73–105.

Bogen, J. E. 1969b. The other side of the brain. II: An appositional mind. *Bulletin of the Los Angeles Neurological Society* 34:135–62.

Bogen, J. E. 1993. "The callosal syndromes." In *Clinical neuropsychology,* edited by Heilman and Valenstein, 3rd ed. New York: Oxford University Press, pp. 337–407.

Bogen, J. E., D. H. Schultz, and P. J. Vogel. 1988. Completeness of callosotomy shown by magnetic resonance imaging in the long term. *Archives of Neurology* 45:1203–05.

Bogen, J. E. and P. J. Vogel. 1975. "Neurologic status in the long term following complete cerebral commissurotomy." In *Les syndromes de disconnection calleuse chez l'homme,* edited by F. Michel and B. Schott. Lyon: Hospital Neurologigue.

Cronin-Golomb, A. 1986a. Comprehension of abstract concepts in right and left hemispheres of complete commissurotomy subjects. *Neuropsychologia* 24:881–87.

Cronin-Golomb, A. 1986b. Subcortical transfer of cognitive information in subjects with complete forebrain commissurotomy. *Cortex* 22:499–519.

Delis, D. D., L. C. Robertson, and R. Efron. 1986. Hemispheric specialization of memory for visual hierarchical stimuli. *Neuropsychologia* 205–14.

Exner, J. E., Jr. 1993. *The Rorschach: A comprehensive system,* Vol. 1, Basic Foundations. New York: John Wiley & Sons.

Exner, J. E., Jr. 1991. *The Rorschach: A comprehensive system,* Vol. 2, Interpretation (2nd Ed.). New York: John Wilcy & Sons.

Henninger, P. 1992. Conditional handedness: Handedness changes in multiple personality disordered subject reflect shift in hemispheric dominance. *Consciousness and Cognition* 1: 265–87.

Kaplan, E. 1988. "A process approach to neuropsychological assessment." In *Clinical neuropsychology and brain function: research, measurement and practice,* edited by T. Boll and B. K. Bryant. Washington, DC: American Psychological Association, pp. 129–67.

Kinsbourne, M. 1970. "The cerebral basis of lateral asymmetries in attention." In *Acta Psychologica 33 Attention and Performance III*, edited by A. F. Sanders. Amsterdam: North Holland, pp. 193–201.

Landis, T., M. Regard, R. Graves, and H. Goodglass. 1983. Semantic paralexia: A release of right hemispheric function from left hemispheric control? *Neuropsychologia* 21: 359–64.

Levy, J., R. Nebes, and R. Sperry. 1971. Expressive language in the surgically separated minor hemisphere. *Cortex* 7:49–58.

Levy, J., C. Trevarthen, and R. W. Sperry. 1972. Perception of bilateral chimeric figures following hemispheric deconnexion. *Brain* 95:61–78.

Lewis, R. T. 1979. Organic signs, creativity, and personality characteristics of patients following cerebral commissurotomy. *Clinical Neuropsychology* 1:29–33.

Myers, J. J. 1985. Interhemispheric communication after section of the forebrain commissure. *Cortex* 21:249–60.

Myers, J. J., and R. Sperry. 1982. A simple technique for lateralizing visual input that allows prolonged viewing. *Behavior Research Methods and Instrumentation* 14(3):305–08.

Nebes, R. D. 1973. Perception of spatial relationships by the right and left hemispheres in commissurotomized man. *Neuropsychologia* 11:285–89.

Nebes, R. D. 1974. Hemispheric specialization in commissurotomized man. *Psychological Bulletin* 81:1–14.

Nebes, R. D., and R. Sperry. 1971. Hemispheric deconnection syndrome with cerebral birth injury in the dominant arm area. *Neuropsychologia* 9:247–59.

Phelps, E. A., and M. S. Gazzaniga. 1992. Hemispheric differences in mnemonic processing: The effects of left hemisphere interpretation. *Neuropsychologia* 3:293–97.

Plourde, G., and R. W. Sperry. 1984. Left hemisphere involvement in left spatial neglect from right-sided lesions: A commissurotomy study. *Brain* 107:95–106.

Rabin, A. I. 1981. *Assessment with projective techniques*. New York: Springer-Verlag.

Ramachandran, V. S., A. Cronin-Golomb, and J. J. Myers. 1986. Perception of apparent motion by commissurotomy patients. *Nature* 320:358–59.

Regard, M. 1991. The perception and control of emotion: Hemispheric differences and the role of the frontal lobes. In Habilitationsschrift, pp. 7–19.

Robertson, L. C., and M. R. Lamb. 1991. Neuropsychological contributions to theories of /art/whole organization. *Cognitive Psychology* 23:299–330.

Robertson, L.C., M. R. Lamb, and R. T. Knight. 1991. Normal global-local analysis in patients with dorsolateral frontal lobe lesions. *Neuropsychologia* 29:959–68.

Robertson, L. C., M. R. Lamb, and E. Zaidel. 1993. Interhemispheric relations in processing hierarchical patterns: Evidence from normal and commissurotomized subjects. *Neuropsychology* 7:325–42.

Rueckhert, L., L. Sorensen, and J. Levy. 1994. Callosal efficiency is related to sustained attention. *Neuropsychologia* 32:159–74.

Sperry, R. W. 1974. "Lateral specialization in the surgically separated hemispheres." In *The neurosciences third study program*, edited by F. O. Schmitt and F. G. Worden. Cambridge, MA: MIT Press, pp. 5–19.

Sperry, R. W., M. S. Gazzaniga, and J. E. Bogen. 1969. "Role of the neocortical commissures." In *Handbook of clinical neurology,* edited by P. J. Vinken and G. W. Bruyn. Vol. 4. Amsterdam: North Holland.

Trope, I., P. Rozin, and R. C. Gur. 1988. Validation of the lateral limits technique with a callosotomy patient. *Neuropsychologia* 26:673–84.

Trope, I., P. Rozin, D. K. Nelson, and R. C. Gur. 1992. Information processing in the separated hemispheres of callosotomy patients: Does the analytic-holistic dichotomy hold? *Brain and Cognition* 19:123–47.

Warrington, E. K., J. Merle, and M. Kinsbourne. 1966. Drawing disability in relation to laterality of cerebral lesion. *Brain* 89:53–82.

Yates, J. L. 1994. "Rorschach and the split-brain perception of emotion." In *Psyche and the split-brain.* Lanham MD: University Press of America, pp. 60–65.

Zaidel, D., and R. W. Sperry. 1973. Performance on the Raven's colored progressive matrices test by subjects with cerebral commissurotomy. *Cortex* 9:34–39.

15 Evidence for Language Comprehension in a Severe "Sensory Aphasic"

Britt Anderson and Thomas Head

INTRODUCTION

Not all knowledge is accessible to conscious recall. In certain situations a person can be shown to "know" things that overtly they have no knowledge of. Blindsight is the prototypical example of this phenomenon, but it exists for other cognitive domains. Prosopagnosics may be able to access knowledge of persons whose faces they cannot identify (Young and Haan 1991) or show skin conductance responses to familar faces (Bauer 1984). Some subjects with hemianesthesia will show skin conductance responses to sensory stimuli they do not "feel" (Vallar *et al.* 1991). Neglect subjects may not detect the left side of objects but can be shown indirectly to know this information (Marshall and Halligan 1988). Amnesic subjects and controls may show priming (Mayes 1991). Does a similar situation exist for language?

METHODOLOGY

A 66-year-old aphasic man was read a humor inventory of funny and not funny riddles while skin potentials were recorded. The humor inventory consisted of 22 items: 2 blanks to fatigue the skin potential response to novel material, 10 children's riddles, and 10 like-worded unfunny phrases—the last 20 mixed in an unpredictable sequence. The humor inventory was first validated by showing that normal men could reliably discriminate the funny and not funny items, both by their verbal report in a post-test done after recording and via skin potential recording (Figure 15.1a). Skin potentials were measured from the left hand with the active electrode over hypothenar eminence and the reference electrode on the radial side of the forearm near the elbow.

RESULTS

The subject's skin potential responses to the 20 test items, divided funny or not funny, are shown in Figure 15.1b. Most skin potentials above the mean occurred to funny items (7/9) and most potentials below the mean

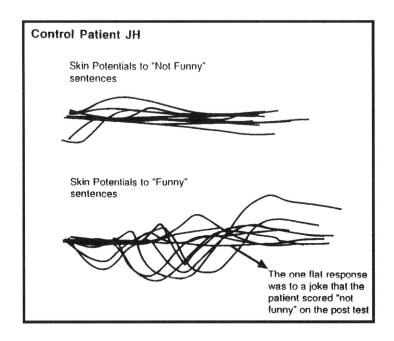
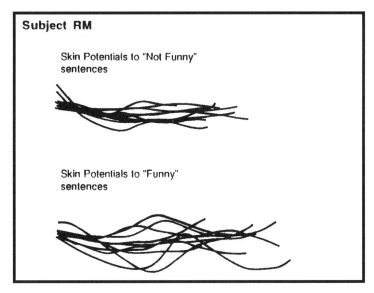

Figure 15.1 Superimposed skin potential tracings are shown for a control subject and the test subject. (A) The responses for the control subject. Note that the skin potential response for the one "funny" item the subject failed to find humorous (as reported in the post test) is flat. (B) The same division of responses for the test subject.

to not funny items (8/11). There was significant agreement between the examiner designated, control validated, categories of funny versus not funny and the subject's skin potential designation (Kappa = 0.5, $P<0.05$ by use of a Z statistic).

DISCUSSION

This study demonstrated that humorous material elicits a change in listener autonomic activity that can be detected as a change in skin potential. While there are numerous steps required to perceive a spoken joke as funny, a necessary requirement is the capacity to understand the complex spoken utterance. Thus, humorous material, with its attendant involuntary change in skin potential, allows an assessment of the comprehension of complex language in subjects who may not be able to comprehend and obey instructions.

Further, we have shown in a single subject covert comprehension of verbal material in advance of his overt capacity. Other researchers (Milberg and Blumstein 1981, Tyler 1991) have shown intact language modules in aphasic subjects, such as detection of verb agreement violation, but none has made a case for full comprehension of a complex spoken utterance in the absence of overt comprehension capacity as we demonstrate in our subject. If further work confirms this effect it will have direct implications for theories of language comprehension and conscious access.

REFERENCES

Bauer, R. 1984. Autonomic recognition of names and faces in prosopagnosia: A neuropsychological application of the guilty knowledge test. *Neuropsychologia* 22:457–69.

Marshall, J., and P. Halligan. 1988. Blindsight and insight in visuo-spatial neglect. *Nature* 336:766–67.

Mayes, A. 1991. "Automatic memory processes in amnesia: How are they mediated?" In *The neuropsychology of consciousness,* edited by A. Milner and R. Rugg. London: Academic Press, pp. 235–61.

Milberg, W., and S. Blumstein. 1981. Lexical decision and aphasia. *Brain and Language* 14:371–85.

Tyler, L. 1991. "The distinction between implicit and explicit language function; evidence from aphasia." In *The neuropsychology of consciousness,* edited by A. Milner and M. Rugg. London: Academic Press, pp. 159–78.

Vallar, G., G. Bottini, R. Sterzi, D. Passerini, and M. Rusconi. 1991. Hemianesthesia, sensory neglect, and defective access to conscious experience. *Neurology* 41:650–52.

Young, A., and E. D. Haan. 1991. "Face recognition and awareness after brain injury." In *The neuropsychology of consciousness,* edited by A. Milner and M. Rugg. London: Academic Press, pp. 69–90.

16 Self-Awareness of Deficit in Patients with Alzheimer's Disease

Alfred W. Kaszniak and Gina DiTraglia Christenson

In discussing the concept of consciousness, Farthing (1992) has drawn a distinction between primary consciousness and reflective consciousness. Primary consciousness is defined as ". . . the direct experience of percepts and feelings, and thoughts and memories arising in direct response to them" (Farthing 1992, p. 12). Reflective consciousness is held to consist ". . . of thoughts about one's own conscious experiences per se" (Farthing 1992, p. 13). Reflective consciousness, by this definition, is that which enables us to interpret our experience, evaluate our behavior and plan future actions, and judge our own knowledge. Reflective consciousness is thus necessary for self-awareness. Within cognitive psychology, the term metacognition (originally coined to characterize changes in self-awareness during childhood development; Flavell and Wellman 1977) is used to describe our own knowledge about how we perceive, think, remember, and behave. Research concerning metacognition has increased considerably in recent years (for a review, see Metcalfe and Shimamura 1994), with empirical research and theory focusing on clarifying the nature of metacognition, its psychological mechanisms and its neurobiological correlates. Much of this research and theory has addressed the question of whether metacognition actually represents privileged access to our own knowledge and mental states (as opposed to being inferences made without actual knowledge of the target information).

Flanagan (1992, pp. 11–14) has argued that the study of consciousness can best move forward through the simultaneous pursuit of three different research directions: the investigation of phenomenological reports, behavioral observations, and their neurobiological (particularly brain) correlates. The neuropsychological investigation of patients demonstrating disruptions of metacognition has pursued each of these three research directions and has provided important insights into the nature of self-awareness and its neurobiological correlates. As noted by Shimamura (1994), "Disruptions of metacognition can take many forms, depending on the aspect of knowledge that is inaccessible or unavailable to conscious awareness" (p. 253). The neuropsychological study of different varieties of impaired metacognition (for example, visual discrimination without awareness in blindsight, preserved implicit memory, and impaired explicit memory in organic amnesia or during surgical anesthesia), as discussed in other chapters within the present volume, has provided strong evidence that many cognitive processes can operate outside of awareness. Further, this research has suggested that different metacognitive impairments are due to damage of different neural structures and circuits, depending upon the type of cognitive function that is disrupted.

The present chapter provides a review and discussion of research concerned with metamemory failure (particularly impaired self-awareness of memory deficits) in persons suffering from Alzheimer's disease (AD). This review begins with a brief history of scientific interest in this phenomenon, and then moves to a discussion of the methods and results of relevant empirical research. The chapter concludes with an overview of the theoretical and practical implications of this body of research.

ANOSOGNOSIA

At the turn of the century, Babinksi (1914) observed curious behavior in several of his patients who suffered left-sided paralysis due to right-hemisphere cerebral damage. Specifically, he noted that these patients had "no idea of their paralysis," and "did not complain at all about their disability." Although related observations had been made earlier by others (for a review see Bisiach and Geminiani 1991), it was Babinski (1914) who coined the term *anosognosia* to describe the apparent failure of such patients to acknowledge their acquired deficits. Since that description, numerous reports have emerged confirming Babinski's original observations and extending application of the term anosognosia to include similar phenomena observed in patients suffering from a wide range of neurological syndromes (for a review see McGlynn and Schacter 1989, Prigatano and Schacter 1991).

Although anosognosia has been well documented in individuals who exhibit impaired sensory or motor functions as a result of cerebrovascular accidents or traumatic brain injury, only recently have reports of anosognosia in neurodegenerative disorders emerged. Clinical accounts of anosognosia in dementia syndromes such as AD and multi-infarct dementia (MID) began appearing in the literature over ten years ago (for example, Danielcyzk 1983, Gustafson and Nilsson 1982, Neary *et al.* 1986); however, empirical studies of the phenomenon have only more recently appeared. The majority of these studies have focussed on issues related to unawareness of deficits in AD, the most common dementing illness in older age.

A comprehensive review of anosognosia in various neurological syndromes (McGlynn and Schacter 1989) emphasized the specificity and dissociations found in different forms of unawareness of deficits. In many patients, awareness is preserved for some, but not other, deficits. Such reports of fractionation and dissociation in anosognosia have important implications for the study of unawareness in dementia because AD patients generally manifest impairments in a wide range of cognitive, behavioral, emotional, and functional areas. Thus, the study of self-awareness of deficits in AD provides an ideal opportunity for examining the degree to which impaired awareness is general or limited to particular aspects of cognitive deficit. Further, the clarification (through particular experimental manipulations) of the nature and neurobiological correlates of impaired awareness in AD will likely contribute to our understanding of specific brain processes important for different theoretically distinct aspects of self-awareness. As Bisiach and Geminiani (1991) have observed, ". . . the term anosognosia, rather than referring to a truly distinct symptom, may be (and indeed has been) used to denote aspects of patients' behavior in relation to their illness that are heterogeneous in appearance and unlikely to depend on a specific set of causes exclusively related to them." (p. 19)

UNAWARENESS OF COGNITIVE DEFICITS IN AD

AD is considered to be a progressive, cortical dementia (that is, the neuropathology of AD is most abundant in the cerebral cortex, and the primary signs and symptoms of AD appear due to cortical dysfunction). Of all the areas of disability associated with the clinical course of AD, the most striking deficits and the earliest to be recognized are cognitive in nature. It is not surprising, therefore, that the majority of studies investigating impaired awareness in AD have focussed on unawareness of cognitive deficits.

Impaired self-awareness of cognitive deficits has frequently been included in clinical descriptions of dementing illness, including AD, Pick's, and Huntington's diseases (for a review, see McGlynn and Kaszniak 1991a). Clinical reports have varied, however, in their conclusions regarding questions such as whether awareness of deficits is impaired in a particular dementing illness, and whether awareness is impaired early (for example, Frederiks 1985) or late (Schneck, Reisberg, and Ferris 1982) in the course of a progressive dementia. Within clinical reports, impaired awareness of deficits is most frequently reported as a feature of those dementias with predominant cortical degeneration, such as AD and Pick's disease. Some observers (Gustafson and Nilsson 1982) have suggested that loss of insight concerning symptoms occurs earlier in the course of illness for Pick's disease than for AD. Both of these kinds of dementia are typically associated with signs of frontal lobe pathology (see Cummings and Benson 1992, Kaszniak 1986), but frontal pathology is generally more severe in the early stages of Pick's disease than AD. This is of interest, given the association between frontal lobe pathology and impaired self-awareness of deficits in a variety of neuropsychological syndromes (see Beatty and Monson 1991, Janowsky, Shimamura, and Squire 1989, McGlynn and Schacter 1989), and speculations regarding the role of the frontal lobes in aspects of self-awareness (Stuss 1991, Grattan *et al.* 1994). In this respect, it is interesting to note that there have been recent empirical studies of the correlates of impaired self-awareness of cognitive deficit in AD that have also suggested a link to the frontal lobes. For example, Loebel *et al.* (1990) found an inverse relationship between fluency of speech (generally assumed to be dependent upon left frontal cortex) and self-awareness of memory impairment in AD. More recently, Reed, Jagust, and Coulter (1993) reported that anosognosia for memory loss was associated with decreased perfusion of the right dorsolateral frontal brain region (as determined by single photon emission computed tomography) of AD patients. Unfortunately, the assessment of anosognosia in the Loebel *et al.* (1990) and Reed *et al.* (1993) studies was based on clinical ratings, rather than on more objective empirical methodologies.

As discussed by Kaszniak and colleagues (Kaszniak 1992, Trosset and Kaszniak in press) investigators employing objective empirical approaches have typically used one of the following methodologies to study the phenomenon: (1) comparisons of patient self-report and caregiver ratings of patient disability; (2) examination of the concordance between the patient's subjective description of abilities and objective measures of the patient's cognitive abilities; (3) evaluation of the accuracy of patients' performance predictions on specific cognitive tasks; and (4) some combination of these three techniques.

Conclusions concerning the presence and degree of impaired awareness of deficits may be dependent upon the particular method employed. For example, although both Anderson and Tranel (1989) and DeBettignies, Mahurin, and Pirozzolo (1990) report evidence for impaired awareness of deficits in AD, Anderson and Tranel find the degree of deficit in awareness to be strongly associated with degree of intellectual impairment, while DeBettignies *et al.* find no relationship between awareness and general mental status. The Anderson and Tranel study defined impaired awareness as a discrepancy between the subject's description of abilities during a standardized interview and measurement of those abilities with neuropsychological and neurological evaluations. In the study by DeBettignies and colleagues, impaired awareness was defined as discrepancy between patient and caregiver report of patients' capacity for independent living, as assessed by response on standard activities of daily living (ADL) scales. Because results may be dependent on the particular method of assessing awareness, the following review is organized according to methodologic approach.

COMPARISONS OF PATIENT SELF-REPORT AND CAREGIVER RATINGS

Reisberg *et al.* (1985) published the first empirical study of unawareness of cognitive deficits in AD. Participants in their study included ten cognitively normal individuals, five with presumed benign senescent forgetfulness (age-related mild memory decline that is assumed not to indicate a dementing illness), and twenty patients with AD. Each subject and his/her spouse were interviewed separately and questioned about their own and their spouse's difficulties in four different areas (memory, emotion, social, and independent functioning). A comparison of the subjects' and spouses' responses revealed that subjects with mild or no cognitive deficit were in general agreement with their spouse regarding their abilities. However, the discrepancy between patient and spouse ratings increased with the severity of the patient's cognitive impairment. Specifically, compared to their spouses' assessments, individuals with moderate to moderately severe cognitive impairment tended to rate their memory difficulties, emotional problems, and inability to perform basic self-care and household chores as less severe. Patients' and spouses' appraisals of their ability to communicate with each other; however, were closely matched. Since the patients' and spouses' ratings of the spouses' functional abilities were also in close agreement, Reisberg *et al.* (1985) interpreted these findings as evidence that the "lack of insight" in AD patients is selective for processes affecting evaluations of self.

DeBettignies, Mahurin, and Pirozzolo (1989) examined unawareness of impairment in independent living skills (ILS) in dementia. In their study, 12 AD patients, 12 multi-infarct dementia (MID; a dementia syndrome associated with cerebrovascular disease) patients, and 12 normal elderly controls (NC) were asked to complete two ratings scales that assessed level of independence in performing activities of daily living. An informant also rated each subject on both scales. Loss of insight was operationally defined as the discrepancy between informant report and patient self-report on the combined total of the two scales (the Insight Discrepancy score). Results showed that, compared to the MID and normal control subjects, the AD patients demonstrated significantly greater loss of insight for ILS impairment. There was no significant difference between the MID and control groups' Insight Discrepancy scores. Across groups, the degree of loss of insight was unrelated to age, education, general intellectual function, or level of depression.

Freidenberg, Huber, and Dreskin (1990) examined factors related to "loss of insight" in AD. Twenty-eight probable AD patients and their spouses were asked to rate the patient on ten features of disability (for example, memory). A discrepancy score was computed based on the difference between the spouses' and patients' total ratings. Results showed that AD patients rated their overall function as better than or equal to ratings provided by their spouses. In contrast to the findings of DeBettignies, Mahurin, and Pirozzolo (1989) the extent of diminished awareness was found by Freidenberg and colleagues to be significantly related to greater dementia severity and to depressive and psychotic features.

Mangone *et al.* (1991) also investigated the psychiatric and neuropsychological correlates of "impaired insight" in 41 individuals with probable AD. Each subject and a caregiver rated the patient's difficulty on a standard instrumental activities of daily living questionnaire and an "insight score" was calculated to reflect the discrepancy between patient and caregiver reports on the measure. Mangone and colleagues reported that impaired patient insight was associated with greater severity of dementia and paranoid delusions. Further, patient performance on neuropsychological tests of sustained attention (Continuous Performance Test) and visual memory (Visual Reproduction) were the best predictors of "impaired insight."

In another study investigating the correlates of impaired awareness, Auchus *et al.* (1994) compared 13 AD patients who were clinically rated as unaware of their cognitive impairment to 13 AD patients who were aware of their deficits. Although the two groups did not differ in age, education, estimated premorbid intelligence, dementia severity, memory function, or language ability, the unaware group showed significantly poorer visuoconstructive ability. The authors suggest that this may reflect an association between extensive right cerebral hemisphere degenerative damage (inferred from the visuoconstructive impairment) and impaired awareness of deficit in AD. However, since damage to the frontal lobes, via disruption of intentional motor behavior (see Heilman and Watson 1991), can impair performance on visuoconstructive tasks, such inferences must be viewed skeptically.

McGlone *et al.* (1990) developed the Memory Observation Questionnaire to assess and compare patient self-report and informant ratings of the patient's current memory functioning and change in memory over time. Three groups of elderly individuals were included in the study: 29 "demeanting" individuals (possible AD), 28 "nondementing" elderly individuals with memory/cognitive complaints (including persons with diagnoses of depression, anxiety, and benign senescent forgetfulness), and 35 elderly normal controls. Results showed that the normal controls reported significantly fewer current memory complaints compared to the dementing and nondementing groups, which did not differ in their subjective ratings. There were no significant differences across groups on ratings of perceived change in memory functioning over the past few months. Informants' ratings of subjects' current memory abilities also revealed similar and significant group differences. Informants of the normal group rated subjects' current memory abilities as significantly better than did informants of the dementing and nondementing subjects. In addition, the informants' ratings of the dementing subjects' change in memory functioning over time were significantly lower (indicating more deterioration in memory functions over time), than were informants' ratings of normal controls and nondementing subjects. Informant ratings of patient's memory did not correlate significantly with patient self-report; however, they did correspond with objective measures of the patients' memory functioning. The patients' self-ratings of memory functioning were correlated only with their depression score, and not with the objective memory measures.

Verhey *et al.* (1993) examined 103 AD patients, 43 vascular dementia patients, and 24 persons with various other causes of dementia, in order to determine whether depressive symptoms are more likely to occur in dementia patients who have some degree of awareness of their cognitive impairment. Awareness was rated on a scale that assessed discrepancies between the patient's and the caregiver's reports concerning the patient's deficits. This measure of awareness was significantly related to the severity of dementia but not to presence or severity of depression.

Feher *et al.* (1991) computed discrepancy scores to quantify the extent to which patients' and spouses' assessments of patients' deficits differ. Feher *et al.* asked 38 individuals with probable AD and their caregivers to rate the patients' memory abilities on a 49-item memory questionnaire. A "Denial

Score" was computed for each patient by subtracting the caregiver's mean score from the patient's mean score. The results showed the AD patients to generally rate their memory abilities more positively than did their caregivers. In fact, 17 (45%) of the 38 AD patients rated their memory as above-average. Only the caregivers' ratings of the patients' memory abilities correlated significantly with the patients' actual performance on three standardized neuropsychological memory tests.

Weak relationships were also found between the patients' "Denial Scores" and objective measures of dementia severity, memory impairment, and depressive symptoms.

Kaszniak, DiTraglia and Trosset (1993) examined unawareness of cognitive and emotional difficulties in AD by assessing patient-caregiver discrepancies in ratings of the frequency and seriousness of patients' memory failures, problems in other areas of cognitive functioning, and emotional problems (for example, depression, anxiety, anger). A total of 19 mildly to moderately impaired AD patients and their caregivers participated in the study. Analysis of patient-caregiver difference scores revealed that, compared to their caregiver's assessments, AD patients underestimated both the frequency and seriousness of their difficulties with memory, language, praxis, and attention. However, patients and their caregivers were in agreement about the frequency and seriousness of patients' emotional difficulties.

Green *et al.* (1993) also compared awareness of memory deficits with awareness of other aspects of cognition and behavior. A group of 20 mildly impaired AD patients and their family members completed a questionnaire requiring them to rate the patients' abilities in recent memory, remote memory, attention, and everyday functional activities. The AD patients rated themselves as significantly more able than they were rated by their family members. However, there was significant variability in magnitude of patient-family member discrepancy across the different abilities rated. Discrepancies were largest for ratings of recent memory and everyday activity abilities, were smaller for attention, and were minimal for remote memory. The investigators concluded that accuracy of awareness in AD varies for different aspects of cognition and behavior.

Studies which have operationally defined self-awareness of deficits in AD by the discrepancy between patient self-report and caregiver ratings of patient disability are difficult to interpret for several reasons: (1) the contributions of patient's underestimation versus caregiver's overestimation of patient disability cannot be disentangled; (2) it is not clear whether the patient is unable to make judgments regarding the degree to which he/she experiences cognitive difficulties due to general intellectual impairment or general problems in cognitive estimation, versus a specific impairment in self-awareness; and (3) patient underestimation of abilities may reflect a defensive denial of deficits as opposed to a failure of self-awareness of deficit (Kaszniak 1992, Trosset and Kaszniak in press).

Some of these limitations to unambiguous interpretation can be at least partially addressed by data contained within the studies reviewed above. For example, McGlone *et al.* (1990) and Feher *et al.* (1991) found that only caregiver ratings correlated with objective measures of patient cognitive functioning. This suggests that the discrepancy between patient and caregiver ratings is probably not due to caregiver overestimation of patient's disability. Second, since patients' and spouses' judgments regarding the spouses' abilities have been found to be in general concordance (Reisberg *et al.* 1985), and severity of dementia has not always been found to be highly correlated with discrepancy measures (Feher *et al.* 1991), it is unlikely that general intellectual decline can solely account for patients' failure to accurately assess their cognitive abilities. Finally, Kaszniak *et al.* (1993) provide additional evidence, based upon comparisons of patient-caregiver discrepancies for frequency versus seriousness of forgetting, that patient-caregiver discrepancies cannot likely be accounted for by patient "defensive denial."

Unfortunately, the evidence contradicting alternative explanations of patient unawareness of deficit are scattered across different studies. To date, no single study employing this methodology has successively addressed the various limitations noted above. Thus, it appears that the discrepancy between patient and caregiver reports, in and of itself, is an insufficient measure of anosognosia in AD. Nonetheless, results of these studies are consistent with the hypothesis that AD patients lack awareness of the severity of their daily functioning, memory, and other cognitive deficits. They also provide some evidence for selectivity of unawareness, since patients were in closer agreement with their spouses on their ratings of some, but not other deficits (Kaszniak *et al.* 1993, Reisberg *et al.* 1985).

CONCORDANCE BETWEEN PATIENT'S SUBJECTIVE DESCRIPTION OF ABILITIES AND OBJECTIVE MEASURES

Feher *et al.* (1994) investigated correlates of awareness of memory abilities in 83 AD patients, 200 individuals with age-associated memory impairment (AAMI), and 64 healthy elderly individuals. Each subject and a family member rated the subject's abilities on a standardized memory questionnaire. The AD patients self-reported worse memory than the healthy elderly subjects, but many AD patients reported themselves as having normal memory, despite poor memory task performance. Multiple regression analyses showed that self-ratings of worse memory functioning were associated with greater dementia severity and higher scores on a depression scale. In the AAMI group, memory self-ratings were predicted by family ratings of patient memory ability, but not by memory test scores. Interestingly, despite the significantly lower memory test scores of the AD group as compared to the AAMI group, these two groups did not differ in self-reported memory ability. The authors interpreted these findings as indicating that the majority of AD patients are unaware of their memory deficits and that memory complaint is not useful diagnostically in differentiating between AD and AAMI.

Anderson and Tranel (1989) operationally defined unawareness as discrepancy between the subject's description of their abilities during interview and clinician judgments, based on neuropsychological and neurologic examinations, regarding impairment of these abilities. Thirty-two patients who had suffered cerebrovascular accidents, 49 with progressive dementia syndromes (29 AD, 5 MID, 15 mixed) and 19 head trauma victims were interviewed and asked to answer questions about their cognitive and motor impairments. Subjects' responses were coded to reflect the degree to which they verbalized awareness of deficits. Each subject was then administered a battery of neuropsychological tests that assessed the same five areas of cognitive functioning covered in the interview (intellect, orientation, memory, speech/language, and visuoperception). A deviation score was computed for each of the five areas of cognition by subtracting an "impairment score" (based on actual neuropsychological performance) from the patient's self-report of difficulties (reflected in the interview score). Through an analysis of the deviation scores, the authors found that 61 percent of the dementia group exhibited unawareness of cognitive deficits in at least one of the five areas of cognitive function. There were no differences across groups in the percentage of subjects showing unawareness of cognitive deficits, but unawareness of intellectual and memory impairments were most frequently observed in all three groups. Unawareness of deficits was associated with the overall degree of verbal intellectual impairment and temporal disorientation.

Although the deviation score was intended to represent a quantitative measure of unawareness in AD, there are several problems with Anderson and Tranel's (1989) methodology. Given the highly significant correlation of unawareness with verbal IQ and disorientation in the dementia group, it is possible that the "Awareness Index" simply reflects general intellectual decompensation in the AD patients. Second, since patients only rated their own impairments, the possible contributions of general problems in cognitive estimation versus a specific problem in self-awareness of deficit cannot be disentangled. Third, patients' responses to very general questions regarding their memory (for example, "Are you having any trouble with your memory?") and motor impairment were being compared with observations from neurological examinations and performance on neuropsychological tests, which assess highly specific functions. Finally, since a heterogeneous collection of etiologies were included in the dementia group, it is not clear whether the AD patients manifested the same severity and pattern of unawareness of cognitive deficits as did the dementia group as a whole.

ACCURACY OF PATIENTS' PERFORMANCE PREDICTIONS ON SPECIFIC COGNITIVE TASKS

Schacter *et al.* (1986, more fully described in Schacter, 1991) employed yet another approach in the study of metamemory and dementia. These researchers asked three groups of memory-disordered patients (due to head injury, ruptured anterior communicating artery aneurysms, and AD, respectively) to study two 20-item categorized lists and then to predict how many of the items they would be able to recall without cuing. Two trials of the task were administered. In the unblocked trial, the items were presented in a random order. This was followed by a blocked trial in which the items were grouped by semantic category and presented in successive order. Schacter *et al.* found that the head-injured and ruptured anterior communicating artery aneurism patients were nearly as accurate in their performance predictions as were their matched controls. However, compared to their controls, the AD patients consistently overpredicted their performance by about seven items on both the blocked and unblocked tasks. The observed differences in performance prediction accuracy between the three groups is of importance, since the degree of memory impairment (as determined by objective memory test performance) was fairly equivalent between the groups. Thus, the unawareness of memory deficit in the AD patients cannot be accounted for simply by assuming that patients are unable to remember that they have a deficit. Schacter (1991) concluded that the AD patients' inability to accurately predict their recall performance was consistent with an unawareness of deficit explanation. Unfortunately, alternative explanations of patient inaccuracy, such as nonspecific problems in cognitive estimation, task comprehension, or response bias, cannot be ruled out (Kaszniak 1992).

STUDIES EMPLOYING COMBINED METHODS

Correa and Graves (1993) examined metamemory processes in AD by employing a combination of methodological techniques. In their study, 20 AD patients, 18 elderly memory-impaired individuals, and 18 elderly normal controls were administered select subtests of the Metamemory in Adulthood (MIA) questionnaire. A relative was also asked to rate each subject on the MIA scale which assesses change in memory abilities. Although the three subject groups did not significantly differ in their responses on any of the self-report MIA scales, the informants for the AD group reported more change in the memory functioning of their family members than did the informants of the memory-impaired subjects or normal controls. A discrepancy score was calculated for each subject by subtracting the informant-reported Change score from the self-reported Change score. Based on an analysis of the discrepancy scores, it was found that AD patients reported less change in memory functions compared to reports obtained from their informants, whereas the memory-impaired and normal control groups actually reported more change.

Correa and Graves (1993) also found that self-monitoring of actual performance, as measured by performance postdiction accuracy, intrusion errors, and self-corrections, was impaired in the AD group relative to the memory-impaired and normal control subjects. The authors interpreted these findings as supporting and extending previous reports of diminished memory awareness in AD. However, the question of whether this reflects impaired self-awareness versus more general difficulties in judgment or cognitive estimation cannot be answered, because the AD patients were only required to rate their own abilities.

In an attempt to address the limitations and criticisms of the various methodologies described above, McGlynn and Kaszniak (1991b) developed two different measures to quantitatively assess the degree to which AD patients are unaware of their memory difficulties. In their study, eight individuals with AD and their relatives were asked to rate their own and their relative's memory-related difficulties encountered in everyday life. Each subject was also asked to make predictions regarding their own and their relative's performance on a number of objective memory tasks. The results showed that AD patients rated their own cognitive difficulties significantly lower than relatives rated the patients' problems. The discrepancy between patient and caregiver ratings increased with the patient's increasing level of cognitive impairment. Given that there were no significant differences between patients' and relatives' ratings of the relatives' cognitive difficulties, McGlynn and Kaszniak interpreted their findings as indicating that AD patients do not appreciate the severity of their own memory deficit, and that the data cannot be accounted for by any general impairment in cognitive estimation (which would have been expected to produce errors in patients' estimations of their caregivers' problems). Similarly, the results of the performance prediction task revealed that, in general, AD patients overestimated their own performance on objective tests of memory functions, but were more accurate when predicting their relatives performance. Interestingly, the AD patients were as sensitive in their predictions as were their relatives to expected effects of retention interval and the different demands of recall versus recognition memory tasks (that is, they predicted that they would perform better in recognition than in free recall memory tasks, and predicted that they would remember less in delayed than in immediate recall testing). Thus, the AD patients did not appear to have experienced a deterioration of general knowledge about memory processes (one aspect, in addition to that of memory self-monitoring, typically included in definitions of metamemory). Based on the several findings of their study, McGlynn and Kaszniak (1991b) proposed that particular metamemory functions, specifically the self-monitoring of memory functioning, breaks down as disease progresses in AD.

Kaszniak and Christenson (1995) recently replicated the general methodology of the McGlynn and Kaszniak (1991b) study, and added longitudinal observations. Twenty-one AD patients and their caregivers completed a questionnaire rating patients' memory and other cognitive difficulties. Patients and their caregivers also made predictions concerning each other's performance on various memory tasks. Results showed AD patients to underestimate (in comparison to caregiver ratings) both the frequency and seriousness of their memory and other cognitive deficits. Patient-caregiver discrepancy was greater for recent than for remote memory deficits, and increased with greater interval of comparison between present and past memory functioning. Analyses of performance prediction accuracy data supported the conclusion that patient-caregiver discrepancies reflected patients' impaired awareness of deficits rather than any general patient or caregiver response bias. Impaired awareness was not significantly correlated with either patient, caregiver, or clinician ratings of patients' or caregivers' depression-related symptoms. Finally, one-year reassessment data was generally consistent with the conclusion that self-awareness of cognitive deficit decreases with dementia progression.

McGlynn and Kaszniak (1991b) and Kaszniak and Christenson (1995) based their experimental protocols on modifications of the techniques introduced by Reisberg et al. (1985) and Schacter et al. (1986). In this way they were able to address many of the limitations inherent in using either approach in isolation. However, interpretation of their results has some clear limitations. Specifically, AD patients were only required to rate the cognitive abilities of their relatives, who have not likely experienced significant decline. It is thus possible that the apparently preserved general estimation abilities AD patients demonstrated when rating their caregiver's memory difficulties, or predicting their memory task performance, was based on recall of their caregiver's abilities of several years ago (prior to the beginning of the patient's own memory deterioration). The AD patients in the McGlynn and Kaszniak study may have failed to "update" mental representations about the current cognitive abilities of others, as well as of self. Given the probability that their caregivers' memory functioning was not deteriorating, this would account for patients underestimating their own abilities, while appearing to accurately assess their caregivers' abilities.

SUMMARY AND CONCLUSIONS

Despite the variety of methodologies employed, available studies are in agreement that persons with AD are impaired in their self-awareness of deficits. A majority of studies have found associations between various measures of deficit self-awareness and the severity of dementia, but a few have not. However, impaired awareness of deficit in AD does not appear to reflect some more general deficit in task comprehension or cognitive estimation. Available evidence is consistent with the interpretation that it is the memory self-monitoring aspect, rather than knowledge of how memory operates, that breaks down in AD. However, the question of whether this reflects a specific deficit in self-awareness, rather than a failure to update representations about one's own memory functioning, awaits an answer from future research. Specific neuropsychological deficits, particularly those thought related to frontal and right parietal cortical functioning, have been found to correlate with patient awareness. Further, there is some evidence to indicate that frontal lobe (particularly right hemisphere) cerebral blood flow is correlated with clinician ratings of impaired awareness in AD patients. Variation exists in reported findings regarding the relationship between unawareness of deficits and specific other symptoms. For example, several investigators have found depressive symptoms to be associated with the degree of impaired awareness, but others have found no such relation. The question of relationship to depressive symptoms has been of interest because there is evidence that, in healthy individuals, self-report of poor memory functioning is more related to negative affect than to actual memory performance (for example, Larrabee and Levin 1986, Seidenberg, Taylor, and Haltiner 1994).

Further research is needed to address the limitations in research methodology enumerated throughout this review. Specifically, future studies will need to employ more quantitative measures to directly compare the relative accuracy of patient and caregiver assessments/predictions of disability/performance, and should include an assessment of patient ability to evaluate the functional deficits of other individuals who are also experiencing progressive cognitive decline. Evidence of some specificity in impaired awareness has been reported, but further exploration of possible dissociations of anosognosia phenomenon in AD patients is needed. In this regard future research will need to simultaneously investigate unawareness of deficits in AD across functional domains, within and across subjects. Similarly, the increased use of longitudinal designs may elucidate how awareness of deficits and its correlates change over time in AD. Since premorbid patient characteristics may be related to anosognosia, prospective studies are also needed to address this issue. With attention to the methodologic issues described above, future research on impaired self-awareness of deficit in AD should provide data with important implications for models of brain processes contributing to self-awareness.

REFERENCES

Anderson, S. W., and D. Tranel. 1989. Awareness of disease states following cerebral infarction, dementia, and head trauma: Standardized assessment. *The Clinical Neuropsychologist* 3:327–39.

Auchus, A. P., F. C. Goldstein, J. Green, and R. C. Green. 1994. Unawareness of cognitive impairments in Alzheimer's disease. *Neuropsychiatry, Neuropsychology, and Behavioral Neurology* 7:25–9.

Babinski, M. J. 1914. Contribution a l'etude des troubles mentaux dans l'hemiplegie organique cerebral (Anosognosie). [Contribution to the study of mental disturbance in organic cerebral hemiplegia. (Anosognosia).] *Revue Neurologique* 12:845–48.

Beatty, W. W., and N. Monson. 1991. Metamemory in multiple sclerosis. *Journal of Clinical and Experimental Neuropsychology* 13:309–27.

Bisiach, E., and G. Geminiani. 1991. "Anosognosia related to hemiplegia and hemianopia." In *Awareness of deficit after brain injury: Clinical and theoretical issues*, edited by G. P. Prigatano and D. L. Schacter. New York: Oxford University Press, pp. 17–39.

Correa, D. D., and R. E. Graves. 1993. Awareness of memory deficit in Alzheimer's disease patient and memory-impaired older adults. *The Journal of Clinical and Experimental Neuropsychology* 15:30 (Abstract).

Cummings, J. L., and D. F. Benson. 1992. *Dementia: A clinical approach,* 2nd ed. Boston: Butterworth-Heinemann.

Danielczyk, W. 1983. Various mental behavioral disorders in Parkinson's disease, primary degenerative senile dementia, and multiple infarction dementia. *Journal of Neural Transmission* 56:161–76.

DeBettignies, B. H., R. K. Mahurin, and F. J. Pirozzolo. 1989. Insight for impairment in independent living skills in Alzheimer's disease and multi-infarct dementia. *Journal of Clinical and Experimental Neuropsychology* 12:355–63.

Farthing, G. W. 1992. *The psychology of consciousness.* Englewood Cliffs, NJ: Prentice Hall.

Feher, E. P., R. K. Mahurin, S. B. Inbody, T. H. Crook, and F. Pirozzolo. 1991. Anosognosia in Alzheimer's disease. *Neuropsychiatry, Neuropsychology, and Behavioral Neurology* 4:136–46.

Feher, E. P., G. J. Larrabee, A. Sudilovsky, and T. H. Crook. 1994. Memory self-report in Alzheimer's disease and in age-associated memory impairment. *Journal of Geriatric Psychiatry and Neurology* 7:58–65.

Flanagan, O. 1992. *Consciousness reconsidered.* Cambridge, MA: MIT Press.

Flavell, J. H., and H. M. Wellman. 1977. "Metamemory." In *Perspectives on the development of memory and cognition,* edited by R. V. Kail and J. W. Hagen. Hillsdale, NJ: Lawrence Erlbaum.

Frederiks, J. A. M. 1985. "The neurology of aging and dementia." In *Handbook of clinical neurology,* edited by J. A. M. Frederiks. Vol. 2. Amsterdam: Elsevier, pp. 199–219.

Freidenberg, D. L., S. J. Huber, and M. Dreskin. 1990. Loss of insight in Alzheimer's disease. *Neurology* 40(Suppl. 1): 240. (Abstract)

Grattan, L. M., P. J. Eslinger, K. E. Mattson, D. Rigamonti, and T. Price. 1994. Evidence for a specialized role of the frontal lobes in social self awareness. Paper presented at the Twenty Second Annual Meeting of the International Neuropsychological Society, Cincinnati, OH.

Green, J., F. C. Goldstein, B. E. Sirockman, and R. C. Green, 1993. Variable awareness of deficits in Alzheimer's disease. *Neuropsychiatry, Neuropsychology, and Behavioral Neurology* 6:159–65.

Gustafson, L., and L. Nilsson. 1982. Differential diagnosis of presenile dementia on clinical grounds. *Acta Psychiatr. Scand.* 65:194–209.

Heilman, K. M., and R. T. Watson. 1991. "Intentional motor disorders." In *Frontal lobe function and dysfunction,* edited by H. S. Levin, H. M. Eisenberg, and A. L. Benton. New York: Oxford University Press, pp. 199–213.

Janowsky, J. S., A. P. Shimamura, and L. R. Squire. 1989. Memory and metamemory: Comparisons between patients with frontal lobe lesions and amnesic patients. *Psychobiology* 17:3–11.

Kaszniak, A. W. 1992. Awareness of cognitive and behavioral deficit in Alzheimer's dementia. Address given at the Centennial Convention of the American Psychological Association, Washington, DC.

Kaszniak, A. W., and G. D. Christenson. 1995. One-year longitudinal changes in the metamemory impairment of Alzheimer's disease. *Journal of the International Neuropsychological Society* 1:145 (Abstract).

Kaszniak, A. W., G. DiTraglia, and M. W. Trosset. 1993. Self-awareness of cognitive deficit in patients with probable Alzheimer's disease. *The Journal of Clinical and Experimental Neuropsychology* 15:30. (Abstract).

Larrabee, G. J., and H. S. Levin. 1986. Memory self-ratings and objective test performance in a normal elderly sample. *Journal of Clinical and Experimental Neuropsychology* 8:275–84.

Loebel, J. P., S. R. Dager, G. Berg, and T. S. Hyde. 1990. Fluency of speech and self-awareness of memory deficit in Alzheimer's disease. *International Journal of Geriatric Psychiatry* 5:41–5.

Mangone, C. A., D. B. Hier, P. B. Gorelick, and R. J. Ganellen. 1991. Impaired insight in Alzheimer's disease. *Journal of Geriatric Psychiatry and Neurology* 4:189–93.

McGlone, J., S. Gupta, D. Humphrey, S. Oppenheimer, T. Mirsen, and D. R. Evans. 1990. Screening for early dementia using memory complaints from patients and relatives. *Archives of Neurology* 47:1189–93.

McGlynn, S. M., and A. W. Kaszniak. 1991a. "Unawareness of deficits in dementia and schizophrenia." In *Awareness of deficit after brain injury: Clinical and theoretical issues*, edited by G. P. Prigatano and D. L. Schacter. New York: Oxford University Press, pp. 84–110.

McGlynn, S. M., and A. W. Kaszniak. 1991b. When metacognition fails: Impaired awareness of deficit in Alzheimer's disease. *Journal of Cognitive Neuroscience* 3:183–9.

McGlynn, S. M., and D. L. Schacter. 1989. Unawareness of deficits in neuropsychological syndromes. *Journal of Clinical and Experimental Neuropsychology* 11:143–205.

Metcalfe, J., and A. P. Shimamura. 1994. *Metacognition: Knowing about knowing*, edited by J. Metcalfe and A. P. Shimamura. Cambridge, MA: MIT Press.

Neary, D., J. S. Snowden, D. M. Bowen, N. R. Sims, D. M. A. Mann, J. S. Benton, B. Northen, P. O. Yates, and A. N. Davison. 1986. Neuropsychological syndromes in presenile dementia due to cerebral atrophy. *Journal of Neurology, Neurosurgery, and Psychiatry* 49:163–74.

Prigatano, G. P., and D. L. Schacter, eds. 1991. *Awareness of deficit after brain injury: Clinical and theoretical issues*, New York: Oxford University Press.

Reed, B. R., W. J. Jagust, and L. Coulter. 1993. Anosognosia in Alzheimer's disease: Relationship to depression, cognitive function, and cerebral perfusion. *The Journal of Clinical and Experimental Neuropsychology* 15:231–44.

Reisberg, B., B. Gordon, M. McCarthy, S. H. Ferris, and M. J. deLeon. 1985. "Insight and denial accompanying progressive cognitive decline in normal aging and Alzheimer's disease." In *Geriatric psychiatry: Clinical, ethical, and legal issues*, edited by B. Stanley. Washington, DC: American Psychiatric Press.

Schacter, D. L. 1991. "Unawareness of deficit and unawareness of knowledge in patients with memory disorder." In *Awareness of deficit after brain injury: Clinical and theoretical issues*, edited by G. P. Prigatano and D. L. Schacter. New York: Oxford University Press, pp. 127–51.

Schacter, D. L., D. R. McLachlan, M. Moscovitch, and E. Tulving. 1986. Monitoring of recall performance by memory-disordered patients. *The Journal of Clinical and Experimental Neuropsychology* 8:130. (Abstract)

Schneck, M. K., B. Reisberg, and S. H. Ferris. 1982. An overview of current concepts of Alzheimer's disease. *American Journal of Psychiatry* 139:165–73.

Seidenberg, M., M. A. Taylor, and A. Haltiner. 1994. Personality and self-report of cognitive functioning. *Archives of Clinical Neuropsychology* 9:353–61.

Shimamura, A. P. 1994. "The neuropsychology of metacognition." In *Metacognition: Knowing about knowing,* edited by J. Metcalfe and A. P. Shimamura. Cambridge, MA: MIT Press, pp. 253–76.

Stuss, D. T. 1991. "Disturbance of self-awareness after frontal system damage." In *Awareness of deficit after brain injury: Clinical and theoretical issues,* edited by G. P. Prigatano and D. L. Schacter. New York: Oxford University Press, pp. 63–83.

Teri, L., and A. W. Wagner. 1991. Assessment of depression in patients with Alzheimer's disease: Concordance among informants. *Psychology and Aging* 6:280–85.

Trosset, M. W., and A. W. Kaszniak. In press. Measures of deficit unawareness in predicted performance experiments. *Journal of the International Neuropsychological Society*

Verhey, F. R., N. Rozendaal, R. W. Ponds, and J. Jolles. 1993. Dementia, awareness and depression. *International Journal of Geriatric Psychiatry* 8:851–56.

IV Experimental Neuroscience

Experimental studies of nervous systems from molluscs to man have unearthed enormous detail about brain function. Electrical recording, biochemical analysis, psychological testing, metabolic imaging, and many other techniques have encouraged reductive materialists to study consciousness. A widely held view is that consciousness emerges from "connectionist" behavior in which synaptic strengths regulate dynamical behavior of neural networks. Firing patterns of networks are taken as *neural correlates of consciousness.*

Whether the connectionist approach is complete, or asymptotic, is a matter of dispute. If consciousness does emerge purely from the dynamical complexity of connectionist networks, then it may be simulated and reproduced in computers. Indeed, many proponents of "strong artificial intelligence" believe that consciousness will one day exist in silicon, or other nonbiological media (for example, Moravec 1988). Skeptics of this view contend the evidence from neuroscience may tell us where consciousness occurs, but not precisely what it is (Dreyfus 1992). The belief that there is no knowable scientific explanation for what consciousness is leads to dualism. To begin to reconcile reductive materialism and dualism, research must aim at the difficult issues regarding consciousness. Papers in this section do so.

In Chapter 17, Christof Koch summarizes six years of collaboration with Francis Crick on the neuronal substrate of visual consciousness in humans and primates. Activity in primary visual cortex, the end-target of the visual system, does not correlate with awareness, and Koch concludes that most brain neurons in the visual system compute in a preprocessing mode. This preprocessing results in firing of a small neuronal subset for 100 to 200 milliseconds (msec) which correlates with awareness, according to Koch. Thus, consciousness is seen as the tip of an iceberg of pre- and subconscious activities. Koch adds that all events within the conscious 100 to 200 msec are somehow treated by the brain as being simultaneous.

Petra Stoerig and Alan Cowey further explore visual processing in Chapter 18 through studies of both human patients and monkeys with lesions of primary visual (striate) cortex causing "blindsight": reflexive visual processing with lack of phenomenal and consciously accessible vision. These patients and monkeys have preprocessing knowledge, but lack conscious awareness of visualized information. Stoerig and Cowey differentiate three stages of visual processing: *reflexive, phenomenal*, and *consciously accessible*. Blindsight patients and animals lack both phenomenal and consciously accessible visual processing. The authors conclude that phenomenal vision—the formation of an imagelike representation—occurs in striate cortex, and while not in itself conscious, is a necessary prelude to conscious visualization. Stoerig and Cowey also infer that conscious access, the highest stage, is "supramodal," combining various modalities, and hence probably involves "(ento- and perirhinal) temporal and prefrontal lobes." Conscious access to phenomenal information is functional, according to Stoerig and Cowey, permitting "thinking and planning and scheming and day-dreaming."

An occurrence comparable to blindsight in the sense of smell is discussed in Chapter 19 by Gary Schwartz. From EEG recordings of human volunteers exposed to odorant molecules, Schwartz reports that some odorants elicit robust EEG responses without conscious awareness ("blindsmell"). Schwartz comments on preconscious "awareness without awareness," which, he argues, can apparently occur in the same neurons as does conscious awareness. Schwartz's work illuminates a key question in consciousness research: By what mechanisms or criteria do pre-, sub-, or nonconscious processes become conscious?

Probing sub- and unconscious processes, Randall Cork in Chapter 20 examines implicit memory in anesthetized surgical patients. By playing audio tapes to patients during anesthesia and later testing the patients on tape content while awake, Cork and his colleagues find that (depending on the type of anesthetic) patients may have implicit memory: they can show knowledge of the data presented to them during anesthesia even though they lack conscious recall of the information presented, or of any other events.

Mikael Bergenheim, Håkan Johansson, Brittmarie Granlund, and Jonas Pedersen describe in Chapter 21 a study in human volunteers directly aimed at the issue of timing and apparent simultaneity in conscious awareness. Various sensory modalities reporting on the same event would appear to suffer time delay jitter due to differing pathway lengths, latencies, and neural conduction velocities along their separate pathways. However, a conscious subject experiences the various components simultaneously. Therefore subjective experience must occur either in the order of arrival, or via active brain synchronization to insure simultaneity. Bergenheim and colleagues report evidence for partial synchronization and discuss its implications.

In Chapter 22, Eric Reiman, Richard Lane, Geoffrey Ahern, Gary Schwartz, and Richard Davidson use positron emission tomography (PET) scanning, which images and localizes sites of brain metabolic activity during conscious activity. They elicited emotional responses from human volunteers and graphically visualized their associated brain activity. Reiman *et al.* also point to other recent PET scan studies dissecting and localizing various conscious functions. We can expect many more such developments in imaging brain activity correlated with consciousness.

As general anesthesia is the inverse of consciousness, Richard Watt reports in Chapter 23 on correlations of depth of anesthesia with electroencephalogram (EEG) parameters. Overall, EEG is a poor indicator of anesthetic depth, but Watt plots anesthetized patients' EEG data in phase space and analyzes the signals using dimensional analysis from chaos theory. He finds that as patients become more deeply anesthetized, the dimensionality of their EEGs decreases.

Employing EEG in a different context, Chris Nunn, C. J. S. Clarke, and B. H. Blott studied task performance in human volunteers in Chapter 24. They sought specifically to address Roger Penrose's (1989, 1994) linkage of quantum gravity and consciousness. Using EEG with movable pen recorders, Nunn and colleagues intermittently and without subjects' knowledge interrupted recording from cortical hemisphere relevant to the task being performed. If Penrose's quantum gravity idea is correct, they reasoned, observation/environmental entanglement by the EEG and movement of the recording pens should affect conscious thought and task performance. Thus whether or not the EEG was being recorded should alter task performance, even though the subjects could not tell the difference. Much to the authors surprise, observation of recording did alter task performance, which they believe supports Penrose's contention.

In Chapter 25, Benjamin Libet summarizes his widely known and highly acclaimed studies on awake humans undergoing brain surgery under local anesthesia. Libet was able to evoke conscious sensations by direct electrical stimulation of sensory cortex, and compare these to conscious sensations elicited by peripheral stimulation. He concludes that an essential feature in the preconscious to conscious transition is the duration of preconscious processing (up to 500 msec) and that conscious awareness of an event occurs after a delay on the order of hundreds of msec. To insure simultaneity and time consistency, Libet concludes that some aspects of conscious awareness may be referred backwards in time.

Several chapters in this section (Koch, Stoerig and Cowey, Schwartz, Libet, and Cork) differentiate pre-, sub- and nonconscious processing from consciousness itself. Others (Watt, Nunn et al., and Reiman *et al.*) use technology to represent or study various aspects of consciousness. Three chapters (Koch, Bergenheim *et al.*, and Libet) discuss peculiar aspects of time in relation to conscious experience, and indicate that temporal relations among perceived events are somehow actively processed to result in apparent simultaneity in consciousness. As we shall see in a later section, the concept of time in physics is quite different from our conscious sense of time, and may provide an important clue.

REFERENCES

Dreyfus, H. L. 1992. *What computers still can't do—a critique of artificial reason.* Cambridge, MA: MIT Press.

Moravec, H. 1988. *Mind children: The future of robot and human intelligence.* Cambridge, MA: Harvard University Press.

Penrose, R. 1989. *The emperor's new mind.* England: Oxford University Press.

Penrose, R. 1994. *Shadows of the mind.* England: Oxford University Press.

17 Toward the Neuronal Substrate of Visual Consciousness

Christof Koch

BASIC ASSUMPTIONS

What is the relationship between "awareness," in particular visual awareness, and neuronal activity in the nervous system? Within a larger context, this question is sometimes known as the "Mind–Body" problem, and has been asked since antiquity.

Over the last six years, Francis Crick and I have attempted to understand this relationship by focusing on the neural correlate of consciousness, the NCC. This chapter provides an overview of the framework we advocate and provides some of our more recent speculations. We have previously (Crick and Koch 1990a, 1990b, 1992, Koch and Crick 1995) described our general approach to the problem of visual awareness. In brief, we believe the next important step is to find experimentally the *neural correlates* of various aspects of visual awareness; that is, how best to explain our subjective mental experience in terms of the behavior of large groups of nerve cells. At this early stage in our investigation we will not worry too much about many fascinating but at the moment experimentally unaccessible aspects of the problem, such as the exact function of visual awareness, which species do and which species do not have awareness, different forms of awareness (such as dreams, visual imagination, etc.) and the deep problem of qualia. We here restrict our attention mainly to results on humans and on the macaque monkey, since their visual systems appear to be somewhat similar and, at the moment, we cannot obtain all the information we need from either of them separately.

Our main assumption is that, at any moment, the firing of some but not all the neurons in what we call the visual cortical system (which includes the neocortex and the hippocampus as well as a number of directly associated structures, such as the visual parts of the thalamus and possibly the claustrum) correlates with visual awareness. Yet, visual awareness is highly unlikely to be caused by the firing of all neurons in this system that happen to respond above their background rate at any particular moment. If at any given point in time only 1 percent of all the neurons in cortex fire significantly, about one billion cells in sensory, motor, and association cortices would be active and we would never be able to distinguish any particular event within this vast sea of active nerve cells. We strongly expect that the majority of neurons will be involved in carrying out computations, while only a much smaller number will express the results of these computations. It is probably only the latter that we become aware of. There is already preliminary evidence from the study of the firing of neurons during binocular rivalry that in area MT of the macaque monkey only a fraction of neurons follow the monkey's percept (Logothetis and Schall 1989). We can thus usefully ask the question: What are the essential differences between those neurons whose firing does correlate with the visual percept and those whose firing does not (in the above study, one difference is that the neurons that correlate with the monkey's perception are in the lower layers). Are these "awareness" neurons of any particular cell type? Exactly where are they located, how are they connected and is there anything special about their patterns of firing?

At this point, it may be useful to state our fundamental assumptions. They are:

1. To be aware of an object or an event, the brain has to construct an explicit, multilevel, symbolic interpretation of part of the visual scent.

By explicit, we mean that one such neuron (or a few closely associated ones) must be firing above background at that particular time in response to the feature they symbolize. The pattern of color dots on a television screen, for instance, contains an "implicit" representation of, say, a person's face, but only the dots and their locations are made explicit here; an explicit face representation would correspond to a light that is wired up in such a manner that it responds whenever a face appears somewhere on the television screen. By multilevel we mean, in psychological terms, different levels such as those that correspond, for example, to lines or to eyes or to faces. In neurological terms we mean, loosely, the different levels in the visual hierarchy (see Felleman and Van Essen 1991).

By symbolic, as applied to a neuron, we mean that a neuron's firing is strongly correlated with some "feature" of the visual world and thus symbolizes it (this use of the word "symbol" should not be taken to imply the existence of a homunculus who is looking at the symbol). The meaning of such a symbol depends not only on the neuron's receptive field (that is, what visual features the neuron responds to) but also to what other neurons it projects to (its projective field). Whether a neural symbol is best thought of as a scalar (one neuron) or a vector (a group of closely associated neurons as in population coding in the superior colliculus; Lee, Rohrer, and Sparks 1988) is a difficult question that we shall not discuss here.

2. Awareness results from the firing of a coordinated subset of cortical (and possible thalamic) neurons that fire in some special manner for a certain length of time, probably for at least 100 or 200 milliseconds (msec). This firing needs to activate some type of short–term memory by either strengthening certain synapses or maintaining an elevated firing rate or both. Experimental studies involving short–term memory tasks in the temporal lobe of the monkey (Fuster and Jervey 1981) have provided evidence

of elevated firing rates for the duration of the interval during which an item needs to be remembered. It is at present not possible to assess empirically to what extent synapses undergo a short–term change during a memory task in the animal.

We are assuming that the semiglobal activity that corresponds to awareness has to last for some minimum time (of the order of 100 msec) and that events within that time window are treated by the brain as approximately simultaneous. An example would be the flashing for 20 msec of a red light followed immediately by 20 msec of a green light in the same position. The observer sees a transient yellow light (corresponding to the mixture red and green) and not a red light changing into a green light (Efron 1973). Other psychophysical evidence shows that visual stimuli of less than 120–130 msec produce perceptions having a subjective duration identical of those produced by stimuli of 120–130 msec (Efron 1970a,b).

3. Unless a neuron has an elevated firing rate and unless it fires as a member of such an (usually temporary) assembly, its firing will not directly symbolize some feature of awareness at that moment.

These ideas, taken together, place restrictions on what sort of changes can reach awareness. An example would be the awareness of movement in the visual scene. Both physiological and psychophysical studies have shown that movement is extracted early in the visual system as a primitive (by the so-called short-range motion system; Braddick 1980). We can be aware that something has moved (but not what has moved) because there are neurons whose firing symbolizes movement as such, being activated by certain changes in luminance. To know what has moved (as opposed to a mere change of luminance) there must be active neurons somewhere in the brain that symbolize, by their firing, that there has been a change of that particular character.

3.1. As a corollary, we formulate our activity principle: Underlying every direct perception is a group of neurons strongly firing in response to that stimulus that come to symbolize it. An example is the "Kanizsa triangle" illusion, in which three Pacmen are situated at the corners of a triangle, with their open mouths facing each other. Human observers see a white triangle with illusory lines, even though the intensity is constant between the Pacmen. As reported by von der Heydt, Peterhans, and Baumgartner (1984), cells in V2 of the awake monkey strongly respond to such illusory lines. Another case is the filling-in of the blind spot in the retina (Fiorani *et al.* 1992). Since we do not have neurons that explicitly represent the blind spot and events within it, we are not aware of small objects whose image projects onto them and can only infer such object indirectly.

A semiglobal activity that corresponds to awareness does not itself symbolize a change within that short period of awareness unless such a change is made explicitly by some neurons whose firing makes up the semiglobal activity (for what else but another group of neurons can express the notion that a change has occurred?). These ideas are very counterintuitive and are not easy to grasp on first reading, since the "fallacy of the homunculus" slips in all too easily if one doesn't watch out for it.

3.2. It follows that active neurons in the cortical system that do not take part in the semiglobal activity at the moment can still lead to behavioral changes but without being associated with awareness. These neurons are responsible for the large class of phenomena that bypass awareness in normal subjects, such as automatic processes, priming, subliminal perception, learning without awareness, and others (Tulving and Schacter 1990, Kihlstrom 1987) or take part in the computations leading up to awareness. In fact, we suspect that the majority of neurons in the cortical system at any given time are not directly associated with awareness!

The elevated firing activity of these neurons also, of course, explains blindsight and similar clinical phenomena where patients with cortical blindness can point fairly accurately to the position of objects in their blind visual field (or detect motion or color) while strenuously denying that they see anything (Weiskrantz 1986, Storig and Cowey 1991).

We have argued (from the experiments on binocular rivalry) that the firing of some cortical neurons does not correlate with the percept. It is conceivable that all cortical neurons may be capable of participating in the representation of one percept or another, though not necessarily doing so for all percepts. The secret of visual awareness would then be the type of activity of a temporary subset of them, consisting of all those cortical neurons that represent that particular percept at that moment. An alternative hypothesis is that there are special sets of "awareness" neurons somewhere in cortex (for instance, layer 5 bursting cells; see below). Awareness would then result from the activity of these special neurons.

4. Such neurons must project directly to some part of the front of the cortex, in particular, to those areas in front of the primary motor area (M1, also called area 4). Such areas include both "premotor" as well as "prefrontal" areas but, unfortunately, the terminology is usually not very precise. We have only recently (Crick and Koch 1995) discussed the need for projections from the visual system to the front of the brain. Our basic argument assumes that in going from one visual area to another further up in the visual hierarchy—that is, further away from the retina (Zeki and Shipp 1988, Felleman and Van Essen 1991, Young 1992), the information is recoded at each step. This is certainly broadly compatible with the known fact that the "features" to which a neuron responds become more complex in going from the primary visual cortex, V1 (also called striate cortex, or area 17), to the higher levels in the visual hierarchy, such as the inferotemporal cortical areas (Ungerleider and Mishkin 1982, Maunsell and Newsome 1987; see Figure 17.1).

PREFRONTAL BRAIN AREAS AND PLANNING

This last assumption is based on the broad idea of the biological usefulness of visual awareness (or, strictly, of its neural correlate). This idea is to produce

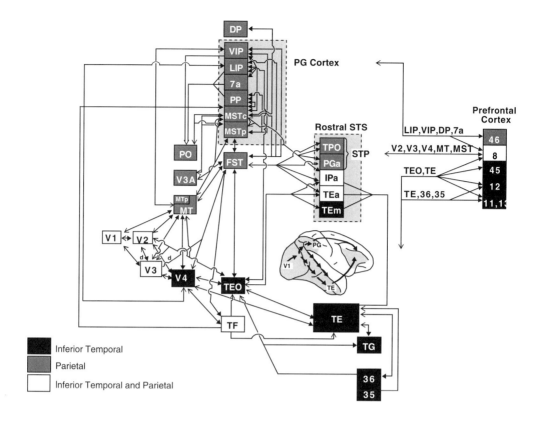

Figure 17.1 A summary diagram of the visual cortical hierarchy in the macaque monkey (adapted and updated from Distler *et al.* 1993). Only cortical connections are shown. The primary visual cortex (V1) receives a strong projection from both eyes via the lateral geniculate nucleus (not shown). Solid lines indicate connecting pathways originating from both central and peripheral parts of the visual scene, while dotted lines indicate connections restricted to the periphery. Solid arrowheads indicate feedforward connections; open arrows indicate feedback connections. Almost all connections among cortical areas are reciprocal. Black indicates areas that are part of the inferior temporal (the so–called "object" vision) processing stream, while gray indicates areas belonging to the parietal (the so–called "spatial" vision) processing stream. Areas shown in white receive input or project their outputs into both of these functional distinct streams. (From Leslie Ungerleider, personal communication.)

the best current interpretation of the visual scene, in the light of past experience either of ourselves or of our ancestors (embodied in our genes), and to make it available, for a sufficient time, to the parts of the brain that contemplate, plan, and execute voluntary motor outputs (of one sort or another).

Exactly how these prefrontal and premotor cortical areas operate is currently unknown, though there is now fragmentary evidence about the behavior of some of them. Even in the macaque, the details of neuroanatomical connections between all these areas have not yet been worked out in as much detail as they have for most of the visual areas of the macaque (Fuster 1989, 1993, Barbas 1992, Gerfen 1994).

It is probably a general rule that the further—connectionwise—a prefrontal area is from the primary motor area M1, the longer the time scale of the planning it is engaged in (Fuster 1989, Birbaumer *et al.* 1990). Moreover, these cortical areas are all heavily involved with the basal ganglia (which include the neostriatum, the globus pallidus, and the substantia nigra), whose main function, we speculate, is to provide a bias back to these areas (as well as to the superior colliculus in the midbrain) to influence the next step in their processing; that is, to assist some behaviors that involve a sequence of activities. The subject is additionally complicated for humans because of our highly developed language system and its usefulness for expressing our "thoughts" (in silent speech, for example).

Fortunately, at this stage, the details of the behavior of these "frontal" areas need not concern us. All we need to postulate is that unless a visual area has a direct projection to at least one of them, the activities in that particular visual area will not enter visual awareness directly.

PRIMARY VISUAL CORTEX AND ITS CONNECTIONS

This lack of any such projection appears to be true for area V1 of the macaque monkey, the almost exclusive recipient of the output of the lateral geniculate nucleus (LGN). V1 has no direct projections to the frontal eye fields (part of area 8), nor to the broad prefrontal region surrounding and including the principal sulcus (see Table 3 in Felleman and Van Essen 1991 and Figure 17.1); nor, as far as we know, to any other "frontal" area. Nor, for that matter, does primary visual cortex in the monkey project to the caudate nucleus of the basal ganglia (Saint-Cyr *et al.* 1990), to the intralaminar nuclei of the thalamus (L. G. Ungerleider personal communication), to the claustrum (Sherk 1986), or to the brainstem (with the exception of a small projection from peripheral V1 to the pons (Fries 1990). V1 does, of course, provide the dominant visual input to most of the posterior visual cortical areas, including V2, V3, V4, and area MT (see Figure 17.1). Among subcortical targets, the lower layers of V1 strongly project to the superficial layers of the superior colliculus (Sparks 1986), the lateral geniculate nucleus, and to the inferior and lateral pulvinar nuclei of the thalamus (Ungerleider *et al.* 1984, Robinson and Petersen 1992).

We think it unlikely that information sent along the pathway from V1 to the superior colliculus, responsible for controlling and initiating eye movements, can produce visual awareness. There is a multistage pathway from V1 to the colliculus, from there to the (inferior) pulvinar and thence to higher visual areas. This pathway may be involved in visual attention (Robinson and Petersen 1992); but according to our arguments, it is not sufficiently direct or strong to produce, by itself, vivid visual awareness of the neural activities in V1.

The pathway from V1 to the colliculus might possibly be used to produce involuntary eye movements so that psychophysical tests, using eye movements as the response, might show a form of blindsight in which subjects respond above chance while denying that they see anything. It is also possible that this or other pathways can produce vague feelings of some sort of awareness.

PRIMARY VISUAL CORTEX AND AWARENESS

Our hypothesis is too speculative to be convincing as it stands, since we are not yet confident as to how to think correctly about most of the operations of the brain, and especially about the detailed function of the so–called "back pathways." Many readers will find these suggestions counterintuitive. We would ask them: Do you believe that you are directly aware of the activity in your retina? Of course, without your retinas, you cannot see anything. If you do not believe this, what is the argument that you are directly aware of the neural activity in V1?

To avoid misunderstanding, let us underline what our hypothesis does not say. We are not suggesting that the neural activity in V1 is unimportant. On the contrary, we believe the detailed processing in V1 is crucial for normal vision, though recent work (Barbur *et al.* 1993) has shown that V1 in at least one patient is not essential for some limited form of visual awareness related to motion perception. All we are hypothesizing is that the activity in V1 does not directly enter awareness. What does enter awareness, we believe, is some form of the neural activity in certain higher visual areas, since they do project directly to prefrontal areas. This seems well established for cortical areas in the fifth tier of the visual hierarchy, such as MT and V4. For areas in the intervening tiers, such as V2, V3, V3A, VP, and PIP, we prefer to leave the matter open for the moment (see Table 3 in Felleman and Van Essen 1991).

Our hypothesis was suggested by neuroanatomical data from the macaque monkey. For humans, we are less certain, due to the present miserable state of human neuroanatomy (Crick and Jones 1993), but we surmise that our hypothesis, if true for the macaque monkey, is also likely to be true for apes and humans. To be established as correct, it also needs to fit with all the neurophysiological and psychological data. What kind of evidence would support it?

PHYSIOLOGICAL AND PSYCHOPHYSICAL EVIDENCE

A possible example may make this clearer. It is well known that the color we perceive at one particular visual location is influenced by the wavelength of the light entering the eye from surrounding regions in the visual field (Land and McCann 1971, Blackwell and Buchsbaum 1988). This mechanism acts to partially compensate for the effects of differently colored illumination. A white patch surrounded by patches of many colors still looks fairly white even when illuminated by pink light. This form of (partial) color constancy is often called the Land effect.

It has been shown in the anesthetized monkey (Zeki 1983, Schein and Desimone 1990) that neurons in V4, but not in V1, exhibit the Land effect. As far as we know, the corresponding information is lacking for alert monkeys. Since we cannot voluntarily turn off the Land effect, it would follow—if the same results could be obtained in a behaving monkey—that it would not be directly aware of the "color" neurons in V1. Notice that if neurons in both V1 and V4 in the alert monkey did turn out to show the full Land effect, this would not, by itself, disprove our hypothesis, as we do believe that we are visually aware of certain neural activity in V4 that could be triggered by activity in V1.

Psychophysical experiments would support our hypothesis if they demonstrate that we are not aware of neuronal activity that is highly likely to occur in V1. Such experiments have been done recently by Don MacLeod and Sheng He (personal communication). In brief, they have shown that exposure to high-contrast gratings that are so finely spaced that they cannot be seen (that is, cannot be distinguished from a uniformly gray surface) can produce an orientation-selective loss in sensitivity of human subjects to slightly less finely spaced gratings that can be visually perceived. Due to neuronal convergence in higher cortical areas and the associated reduction in receptive field size, neurons sensitive to the very fine conditioning grating appear to be restricted to V1 (DeValois, Albrecht, and Thorell 1982, Levitt, Kiper, and Movshon 1994). These psychophysical experiments are therefore compatible with the idea that certain neurons in V1 respond to very high spatial frequencies that we are not visually aware of. The support for our ideas would be greater if it were shown (by imaging methods such as PET or functional MRI) that these invisible gratings produced significant activity in V1 in humans. For an alert macaque, it might be possible to show experimentally that very finely spaced gratings, which activated certain neurons in V1, could not be reported by the monkey, although the animal could report less finely spaced ones.

There are neurons in V1 whose firing depends on which eye the visual signal is coming through. Neurons higher in the visual hierarchy do not make this distinction; that is, they are typically binocular. We are certainly not vividly and directly aware of which eye we are seeing with (unless we close or obstruct one eye), though whether we have some very weak awareness of the eye of origin is more controversial (Pickersgill 1961, Blake and Cormack 1979). These well-known facts suggest that we are not vividly aware of much of what goes on in V1 (for further direct support of this, see Kolb and Braun 1995).

These ideas would not be disproved if it were shown convincingly that (for some people) V1 is activated during visual imagery tasks (see the debate among P. E. Roland and others in Roland *et al.* 1994). There is no obvious reason why such top-down effects should not reach V1. Such V1 activity would not by itself, prove that we are directly aware of it, any more than the V1 activity produced there when our eyes are open proves this. This hypothesis (Crick and Koch 1995) then is a somewhat subtle one, though we believe that if it turns out to be true, it will eventually come to be regarded as completely obvious. We hope that further neuroanatomical work will make it plausible for humans, and further neurophysiological studies will show it to be true for most primates. We have yet to track down the location and nature of the neural correlates of visual awareness. (Kolb and Braun 1995). Our hypothesis, if correct, would narrow the search to areas of the brain further removed from the sensory periphery.

ACKNOWLEDGMENTS

All of the ideas discussed here were jointly developed with Dr. Francis Crick. The research reported here is supported by the Air Force Office of Scientific Research, the Office of Naval Research, and the National Science Foundation.

REFERENCES

Barbas, H. 1992. "Architecture and cortical connections of the prefrontal cortex in the rhesus monkey." In *Advances in neurology*, Vol. 57, edited by P. Chauvel, A. V. Delgado–Escueta, E. Halgren, and J. Bancaud. New York: Raven Press, pp. 91–115.

Barbur, J. L., J. D. G. Watson, R. S. J. Frackowiak, and S. Zeki. 1993. Conscious visual perception without V1. *Brain* 116:1293–1302.

Birbaumer, N., T. Elbert, A.G.M. Canavan, and B. Rockstroh. 1990. Slow potentials of the cerebral cortex and behavior. *Physiology Review* 70:1–41.

Blackwell, S. T., and G. Buchsbaum. 1988. Quantitative studies of color constancy. *J. Opt. Soc. Am. A.* 5:1772–80.

Blake, R., and R. H. Cormack. 1979. On utrocular discrimination. *Perception & Psychophysics* 26:53–68.

Braddick, O. J. 1980. Low-level and high-level processes in apparent motion. *Phil. Trans. R. Soc. B* 290:137–51.

Crick, F., and E. Jones. 1993. The backwardness of human neuroanatomy. *Nature* 361:109–10.

Crick, F., and C. Koch. 1990a. Towards a neurobiological theory of consciousness. *Seminar in the Neurosciences* 2:263–75.

Crick, F., and C. Koch. 1990b. Some reflections on visual awareness. *Cold Spring Harbour Symp. Quant. Biol.* 55:953–62.

Crick, F., and C. Koch. 1992. The problem of consciousness. *Scientific American* 267:153–59.

Crick, F., and C. Koch. 1995. Are we aware of neural activity in primary visual cortex? *Nature* 375:121–3.

DeValois, R., D. G. Albrecht, and L. G. Thorell. 1982. Spatial-frequency selectivity of cells in macaque visual cortex. *Vision Res.* 22:545–59.

Distler, C., D. Boussaoud, R. Desimone, and L. G. Ungerleider, 1993. Cortical connections of inferior temporal area IEO in macaque monkeys. *J. Comp. Neurology* 334:125 –50.

Efron, R. 1970a. The relationship between the duration of a stimulus and the duration of a perception. *Neuropsychologia* 8:37–55.

Efron, R. 1970b. The minimum duration of a perception. *Neurophysiologia* 8:57–63.

Efron, R. 1973. Conservation of temporal information by perceptual systems. *Percept. Psychophys.* 14:518–30.

Felleman, D. J., and D. C. Van Essen. 1991. Distributed hierarchical processing in the primate visual cortex. *Cerebral Cortex* 1:1–47.

Fiorani, M. Jr., M. G. P. Rosa, R. Gattass, and C. E. Rocha–Miranda. 1992. Dynamic surrounds of receptive fields in primate striate cortex: A physiological basis for perceptual completion. *Proc. Natl. Acad. Sci. USA* 89:854–51.

Fries, W. 1990. Pontine projection from striate and prestriate visual cortex in the macaque monkey: An anterograde study. *Visual Neurosci.* 4:205–16.

Fuster, J. M. 1989. *The prefrontal cortex*. 2nd ed. New York: Raven Press.

Fuster, J. M. 1993. Frontal lobes. *Current Opinion Neurobiol.* 3:160–65.

Fuster, J. M., and J. P. Jervey. 1981. Inferotemporal neurons distinguish and retain behaviorally relevant features of visual stimuli. *Science* 212:952–55.

Gerfen, C. R. 1994. "Relations between cortical and basal ganglia compartments." In *Motor and cognitive functions of the prefrontal cortex*, edited by A. M. Thierry, J. Glowinski, P. S. Goldman-Rakic, and Y. Christen. Berlin: Springer-Verlag, pp. 78–92.

Kihlstrom, J. F. 1987. The cognitive unconscious. *Science* 237:1445–52.

Koch, C., and F. Crick. 1994. "Some further ideas regarding the neuronal basis of awareness." In *Large-scale neuronal theories of the brain*, edited by C. Koch and J. Davis. Cambridge, MA: MIT Press, pp. 93–110.

Kolb, F. C., and J. Braun. 1995. Blindsight in normal observers. *Nature* 377:336–39.

Land, E. H., and J. J. McCann. 1971. Lightness and retinex theory. *J. Opt. Soc. Am.* 61:1–11.

Lee, C., W. H. Rohrer, and D. L. Sparks. 1988. Population coding of saccadic eye movements by neurons in the superior colliculus. *Nature* 332:357–60.

Levitt, J. B., D. C. Kiper, and J. A. Movshon. 1994. Receptive fields and functional architecture of macaque V2. *J. Neurophysiol.* 71:2517–42.

Logothetis, N. K., and J. D. Schall. 1989. Neuronal correlates of subjective visual perception. *Science* 245:761–63.

Maunsell, J. H. R., and W. T. Newsome. 1987. Visual processing in monkey extrastriate cortex. *Ann. Rev. Neurosci.* 10:363–401.

Pickersgill, M. J. 1961. On knowing with which eye one is seeing. *Quart. J. Exp. Psychol.* 11:168–72.

Robinson, D. L., and S. E. Petersen. 1992. The pulvinar and visual salience. *Trends Neurosci.* 15:127–32.

Roland, P. E. *et al.* 1994. Visual imagery: A debate. *Trends Neurosci.* 17:281–97.

Saint-Cyr, J. A., L. G. Ungerleider, and R. Desimone. 1990. Organization of visual cortex inputs to the striatum and subsequent outputs to the pallidonigral complex in the monkey. *J. Comp. Neurol.* 298:129–56.

Schein, S. J., and R. Desimone. 1990. Spectral properties of V4 neurons in the macaque. *J. Neurosci.* 10:3369–89.

Sherk, H. 1986. "The claustrum and the cerebral cortex." In *Cerebral cortex*, Vol. 5, edited by E. G. Jones and A. Peters. New York: Plenum Press, pp. 467–99.

Sparks, D. L. 1986. Translation of sensory signals into commands for control of saccadic eye movements: Role of primate superior colliculus. *Physiol. Rev.* 66:1181–72.

Stoerig, P., and A. Cowey. 1991. Wavelength sensitivity in blindsight. *Nature* 342:916–18.

Tulving, E., and D. L. Schacter. 1990. Priming and human memory systems. *Science* 247:301–06.

Ungerleider, L. G., and M. Mishkin. 1982. "Two cortical visual systems." In *Analysis of visual behavior,* edited by D. J. Ingle, M. A. Goodale, and R. J. W. Mansfield. Cambridge, MA: MIT Press, pp. 549–86.

Ungerleider, L. G., R. Desimone, T. W. Galkin, and M. Mishkin. 1984. Subcortical projections of area MT in the macaque. *J. Comp. Neurol.* 223:368–86.

von der Heydt, R., E. Peterhans, and G. Baumgartner. 1984. Illusory contours and cortical neuron responses. *Science* 224:1260–62.

Weiskrantz, L. 1986. *Blindsight.* Oxford, England: Oxford University Press.

Young, M.P. 1992. Objective analysis of the topological organization of the primate cortical visual system. *Nature* 358:152–55.

Zeki, S.M. 1983. Colour coding in the cerebral cortex: The reaction of cells in monkey visual cortex to wavelengths and colors. *Neuroscience* 9:741–65.

Zeki, S., and S. Shipp. 1988. The functional logic or cortical connections. *Nature* 335:311–17.

18 Visual Perception and Phenomenal Consciousness*

Petra Stoerig and Alan Cowey

INTRODUCTION

Consciousness, exiled by the behaviorist movement to become the forbidden fruit of scientific inquiry for the past decades, is once again a touchstone for philosophical and neuroscientific research. This development is witnessed by numerous publications and recent meetings devoted to its discussion (Dennett 1991, Edelman 1989, Flanagan 1992, Marcel and Bisiach 1992, McGinn 1991, Penrose 1989, 1994, and Searle 1992). The theoretical considerations focus on the question of qualia, their irreducibility, and their possible or impossible functional role (Dennett 1991); on the conglomerate nature of consciousness (Block 1995, Nelkin 1993, 1995, and in press); and the multiple uses we have for the term (Wilkes 1992). Empirical approaches use a diversity of methods to get a hold on the neuronal processes involved in conscious as opposed to unconscious behavior. They include neuropsychological studies of global or selective disturbances of consciousness through brain lesions, anesthetics, toxins, or electrical stimulation; they include functional imaging of normal and lesioned brains in overt and covert states of mental activation; and they include behavioral neurophysiology to correlate neuronal and perceptual processes in animals. In this wide field of inquiry, three basic questions are especially intriguing to us: Who has consciousness? What is its neuronal basis? and What is its function? (Stoerig 1995).

The second of these questions, directed at the neuronal correlate of consciousness, is presently the one most vigorously tackled by neuroscientific and psychological inquiry. However, the three questions are clearly connected. Knowing who possesses consciousness would delimit the possible neuronal basis; knowing its neuronal basis would delimit the organisms that possess it; knowing its function would delimit the number of candidates aspiring to its assignment. Although it is going to be very demanding, we believe that it will eventually be possible to get satisfactory answers to these questions, provided one constrains one's expectations sufficiently to accept as satisfactory explanations of a correlative nature. If conscious

---

*Previously printed in *Behavioural Brain Research* Volume 71. Reprinted by permission of Elsevier Science Publishers.

behavior was possible only if a certain part of the brain was intact, if assembly formation was possible, if NMDA-receptors were in working order, if synchronized activity occurred in a certain frequency range, if prefrontal cortex was functional and receiving its inputs from sensory cortical areas, or if whatever else turned out to be essentially involved in the distinction of conscious versus unconscious representation was functional at a given moment, one could put forward detailed hypotheses about the neuronal processes involved. If, however, one was unwilling to accept such hypotheses as satisfactory explanations, because they fail to explain at a deep level why these processes evoke consciousness, or how physiological processes of any kind could do any such thing, one would have to follow in DuBois-Reymond's (1872) footsteps and declare the impossibility of a satisfactory answer in any foreseeable future, and possibly in principle. Accordingly adopting a pragmatically low level of expectation/explanation throughout, we shall first briefly address the first question.

WHO HAS CONSCIOUSNESS?

In his *Contingencies of Reinforcement*, Skinner (1969, p. 288) poignantly stated that

> The real question is not whether machines think, but whether men do.

Every naturalist, that is, everyone who assumes that mind and consciousness are natural biological phenomena found in living beings, and that their functions are, in very general terms, to promote survival and adaptation, will agree with Skinner. The naturalist thesis is at odds with the modern functionalist view, defended for instance by Fodor (1975) and Dennett (1991), which claims that mind and consciousness are sets of algorithms that can in principle be implemented in any kind of set-up, be it computers, soda cans, the Paris sewers, or a brain, provided that the system performs this as-yet-to-be-defined set of operations. We disagree, not just because aspects of consciousness are likely to be noncomputable (Penrose 1989, 1994), but because none of these systems other than the brain—and we should stress that it is not the brain per se that is different, but the brain as-part-of-a-living-organism—is alive; only the living organism is concerned with survival, is a being with needs. Contrary to computers and sewers (and disregarding pathological states for the moment), living beings, even very simple ones, behave to ensure their survival and well-being; a paramecium paddles itself away from noxious stimuli, seeks nice things to ingest, searches for the company of conspecifics. To do that, the living being must be able to distinguish itself from its environment, must distinguish self from nonself. We suggest that this selfishness, this basic distinction, is the node of the mind that reflects the environment in relation to the organism itself, and that without it, there is neither mind nor consciousness.

Leaving aside speculations about attempts at creating artificial life, we can ask where consciousness begins, once this node of a self is there. Some (for example, Hameroff 1994) think it is there even in organisms as simple as paramecium, in the microtubules of the cytoskeleton. Others (like us) think it may not be the same, because many of the basic needs of more complex organisms with specialized nervous systems are regulated by neurophysiological processes that are not consciously accessible. The brain regulates the body's production of hormones, bowel activity, breathing, blood pressure; as the word signifies, the autonomous or vegetative nervous system is not a conscious and voluntary system, although its activity feeds back on conscious mentality—provided you have that.

It is not only these autonomous functions that (at least in us) are not consciously represented; neither are reflexes that link a sensory input—something hot on the skin, light in the eye, a knock to the knee cap—swiftly to a behavioral response—withdrawal, constriction, a knee jerk. Consciousness is not involved but, again, the organism may become conscious of the offensive stimulus and its own reaction afterwards, again provided it has consciousness and is in a conscious state.

If one applied these examples to the vast multitude of organisms, one might conclude that those capable exclusively of reflexive responses are not conscious—although they must be selves, that is, distinguish self and non-self, and thus may have minds—little minds, as Karl Popper put it (Popper *et al.* 1993). Consciousness in this scheme requires more than the linking of an input to an output; it may even require more than fixed programs for actions, namely a flexibility of action, and a capability to voluntarily suppress responses that cannot be instantiated by a mere hierarchy of reflexes (veto theory).

What is needed for that? You need an input and and an output, as in reflexive behavior, and in between you need something that can elicit, modify, or suppress a response to incoming information, undercutting the direct link between input and output that characterizes reflexive behavior. Broadly speaking, but restricted to the forms of life we know, this something is an intermediary neural net. As reflexes also join input and output patbterns, the intermediary net, or a subnet within it, must have properties that are in some crucial way different: They must allow some central, intermediary processing of information and an adaptation of the behavior to its outcome; that is, the flexibility of grading, changing, or suppressing the reaction to produce the behavioral flexibility we recognize as a hallmark of conscious representations in intact organisms. As yet we do not know which properties of the central net are used for these purposes, but by investigating it in organisms that we know possess consciousness we may eventually be able to find out. Having found out, one could then go back to see whose nervous system has these properties, which would be a better scientific basis for attributing it than our currently used intuitions, illusions, and even the *Du-Evidenz* (you-evidence) suggested by Lorenz (1965). Note, though, that it would not necessarily follow that organisms without it do not have consciousness; different ways lead to Rome.

WHAT IS THE NEURONAL BASIS OF CONSCIOUSNESS?

In view of the multiple functions of the brain, the majority of which probably do not involve conscious mentality, one could ask what properties distinguish conscious from unconscious neuronal processes. We know that neurons differ with respect to a lengthy list of properties, implying a long list of candidates.

1. There is their place in the brain roughly, whether neurons are cortical or subcortical; to which functional system they belong; if they're cortical, whether they are in primary sensory cortex, secondary sensory cortex, multisensory cortex, motor cortex, premotor cortex; if they're subcortical, whether they are again in sensory or in motor nuclei, in homeostase-regulating or in general arousal systems.
2. Their place in the cortical layers, of which there are three in archicortex, four in paleocortex, and six in neocortex.
3. What kinds of inputs they receive and where they project; that is, their connections, which can again be differentiated according to the position of the connected neurons in the brain, in the cortical mantle, according to the site of the synapse.
4. Their discharge properties—sluggish or bursting, transient or sustained, with high or low spontaneous activity.
5. Their morphology: in the cortex, whether they are pyramidal or nonpyramidal, spiny or smooth; in the retina, whether they are of the $P\alpha$, $P\beta$, or $P\gamma$ type.
6. The neurotransmitters they use, whether these are excitatory, inhibitory, or neuromodulatory.

Some of these properties go hand in hand—pyramidal cells are usually excitatory, whereas nonspiny nonpyramidal cells are predominantly GABAergic, that is inhibitory; $P\beta$ retinal ganglion cells project only to certain visual structures in the brain (the parvocellular layers of the dorsal lateral geniculate nucleus, the inferior pulvinar nucleus), are slow conducting, relatively sustained, with high spatial resolution, and color-opponent responses to wavelengths, whereas other combinations have not been reported.

If neurons are not equipotential in their ability to realize conscious representations, there must be certain features that distinguish between those who can and those who cannot realize conscious representations. We argued above that an intermediary net is presumably essential, which implies that these features should be found there, although it does not exclude the possibility that the input signals themselves may be differentially suited from early stages on to become consciously represented. To give an example from the visual system on which we shall focus in the hope that results on this best-studied sensory system may be applicable to the others, it is conceivable that input from the $P\gamma$ retinal ganglion cells can in principle not be consciously represented or accessed. It may be that the presumed functions of this system—orienting responses, programming of saccadic eye-movements, foveal fixation (Guitton 1992, Wurtz and Albano 1980), to name some—do not involve consciousness. A comparison of the neuronal properties of this system with those say of the Pα-system, which mediates the perception of high-velocity motion and seems particularly effective in accessing consciousness, often evoking sensations in pathological conditions when no other visual features do (Riddoch 1917, Weiskrantz et al. 1995), may lead to a foothold regarding the distinctive processing features.

Tackling such questions on a single-cell level requires behavioral physiology, examples being provided by the work of Myerson et al. (1981), Movshon et al. (1985), Maunsell and Newsome (1987), and Logothetis and Schall (1989), who use ambiguous visual stimulation and combine recording of single cell activity with behavioral indications about the perceptual interpretation (see Stoerig and Brandt 1993 for a recent review focused on the neuronal correlates of perceptual decisions). On a systems level, one can approach the conscious-unconscious distinction in visual perception with functional imaging techniques, with investigations of implicit and explicit forms of recognition and memory, and with neuropsychological tools, making use of the lesions that nature, in the form of strokes and diseases and careless conspecifics, inflicts on the visual system.

Vision, as any sense, depends on its sensory organ—if you lose both eyes, you are blind. Blindness here means that you cannot process any visual information, and if one is born absolutely (not legally) blind, it supposedly entails that one does not have visual imagery or visual dreams either. If the lesion is further behind the eye and spares the pathways that travel to extrageniculate target areas, certain reflexive visual responses may persist, like the pupil light response, the photic blink response. Other than these reflexes—which may not be entirely normal because destructions of part of the system can functionally affect other parts, as was most dramatically demonstrated by Sprague (1966, see also Wallace et al. 1989)—no visual functions are observed: The patients do not respond voluntarily to visual stimuli and they do not see. Once the lesion is behind the geniculate, in the optic radiations or in the primary visual cortex, the patients not only show pupil reflexes to visual stimulation, they can also localize stimuli by directing their eyes or hands to the approximate position of the stimulus. Furthermore, patients with postgeniculate lesions can detect stationary and moving stimuli, and discriminate their orientation, motion direction, and wavelength. Patients with blindsight show the reflexive responses that can also be elicited after pregeniculate damage, but in addition they can voluntarily initiate responses to visual stimulation, and, under ideal conditions, can perform up to 100 percent correct. Nevertheless, when asked what they perceive, they insist that they do not see anything at all. At best, they say they feel something, which happens preferentially when stimuli are large or fast moving (see Stoerig and Cowey 1993; Weiskrantz 1990, for recent reviews).

This of course is the phenomenon of blindsight, a term coined by Weiskrantz and colleagues (1974), and it is characterized by statistically significant, voluntarily initiable processing of visual information in the absence of a phenomenal conscious perception of the stimuli themselves. According to a recent report (Farah et al. 1992), this includes an inability to phenomenally "see" imagined visual objects in the affected part of the visual field. Once the lesion is beyond primary visual cortex and destroys only a portion of secondary, extrastriate cortex, deficits such as achromatopsia, stereo blindness, and motion blindness are observed, depending on which of the specialized extrastriate cortical areas are damaged (see Cowey 1994 for a recent review). These extrastriate lesions—as long as they do not destroy the entire circumstriate cortical belt that produces blindness (Bodis-Wollner et al. 1977)—prevent only certain types of visual information from being consciously represented. As in blindsight, however, the unseen information can be used to infer information; superimposing colors as camouflage proved disruptive for texture discrimination even in an achromatopsic patient who could not see the added colors (Cowey, unpublished observation). This is also true, at least as far as is known, for other higher-order deficits such as the various types of agnosia (Bauer 1984, De Haan et al. 1991). In all these cases, the patients retain phenomenal visual perceptions—stripped of color, stripped of meaning, stripped of memory associations—but still phenomenal. This hierarchy of visual deficits (see Figure 18.1) indicates that the primary visual cortex is important, and perhaps indispensable, for conscious phenomenal vision.

This importance cannot be due simply to its representing the third processing stage, or the first cortical one, because extrastriate visual cortical

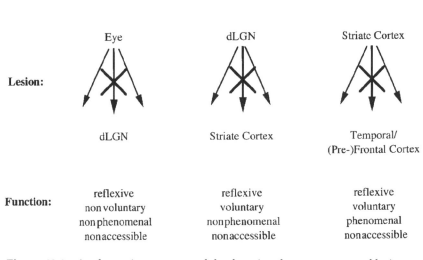

Figure 18.1 A schematic summary of the functional consequences of lesions at different stages of the visual system. Pregeniculate lesions spare only reflexive responses, provided fibers heading for extrageniculate retinorecipient nuclei are left intact. Postgeniculate lesions spare reflexive, implicit, and voluntarily initiated responses to a variety of stimulus features although the phenomenal representation of the stimulus is lost. Lesions that disconnect visual cortex from supramodal cortical areas in temporal and prefrontal cortex spare visual phenomenality, but may destroy conscious access to and recognition of the phenomenally represented information.

areas do receive direct visual information from the geniculate body and from the pulvinar nucleus even in the absence of V1 (Cowey and Stoerig 1989, Yukie and Iwai 1981). In addition, some extrastriate visual cortical areas have been shown to be visually responsive in monkeys with striate cortical inactivation (Bullier *at al.* 1993, Rodman *et al.* 1989); and a recent PET study showed that this is also true in a human patient with traumatic striate cortical destruction (Barbur *et al.* 1993). An absence of all extrastriate cortex, indeed the entire cortical mantle, as in hemidecorticated patients, may actually be detrimental to blindsight, allowing only reflexive (Weiskrantz 1990) and possibly implicit responses in which the patients' responses to a seen stimulus in the normal half-field is influenced by the simultaneous presentation of a second unseen one presented in the blind half-field (Tomaiuolo *et al.* 1994). This is at present controversial (King *et al.* in press, King *et al.* 1995, Perenin 1978, 1991, Perenin and Jeannerod 1978, Ptito *et al.* 1987, 1991, Stoerig *et al.* in preparation), and, if true, points to a distinct representation for these implicit processes somewhere between reflexive responses and phenomenal vision, albeit on the nonconscious side.

The special role of striate cortex for phenomenal vision that is indicated by blindsight as opposed to the functional deficits following higher and lower lesions of the visual pathways, could be due to almost any number of at present speculative factors. For instance, it could be due to striate cortex providing crucial information for the higher visual areas, or to its receiving the results of their specialized analysis, with conscious phenomenal vision thus depending on back projections or reentrant processing loops (Cauller 1995, Cauller and Kulics 1991, Stoerig and Cowey 1993), or to its being essential for establishing coherent cortical activation patterns or cell assemblies (Flohr 1995). We could explore these possibilities in much greater detail if we knew that other animals suffered the same loss of phenomenal conscious vision after striate cortical destruction. It has been known since the work of Klüver (1941) that, like human patients, monkeys with striate cortical destruction show residual visual functions: They can reach out and touch stimuli presented in their affected hemifield, they process information about stimulus position, motion, and possibly about form, complexity, and color (see Pasik and Pasik 1982 for review). The unequivocal demonstrations of residual visual functions, however, do not tell us whether or not the monkeys, like patients with blindsight, lack phenomenal conscious vision in the affected hemifield, or whether they still possess degraded, altered, or otherwise unusual but nevertheless phenomenally represented vision.

TESTING BLINDSIGHT IN MONKEYS

We have recently tried to approach this question with the help of four macaque monkeys. One, Rosie, served as a normal control subject; the three others (Lennox, Dracula, and Wrinkle), all male, had several years before undergone unilateral striate cortical ablation and transection of the caudal third of the corpus callosum. All monkeys are still alive, but we confirmed the completeness of the striate cortical ablation through histology done on the excised tissue, and by NMR-scans through the posterior half of the brains of two monkeys. To first confirm that our monkeys, like those tested by other groups, show residual visual functions in the hemifield affected by the striate cortical lesion, we conducted a series of tests that required them to localize a visual stimulus. During testing, the monkeys squatted in front of a video monitor fitted with a touch screen. Around the monitor were three infrared light sources directed at the monkeys' eyes. They provided specular reflections that allowed us to control the monkeys' eye movements with the help of an infrared sensitive camera sending the image on-line to a monitor in a separate control room. When a starting light appeared at the center bottom of the screen, the monkeys had to touch it. When they did, they also looked at the starting light, and at that moment, the left half of the screen fell into the monkeys' left hemifield, which is normal in all four animals, and the right half of the screen fell into the right hemifield, which is affected in the three operated animals. Triggered by the response to the starting light, a visual stimulus appeared briefly (10—200 msc) in one of the four corners of the monitor. The monkeys had to touch the position at which that stimulus had appeared—it had disappeared before they could reach it, or direct their eyes and head towards its position (see Figure 18.2).

By varying the intensity of these stimuli, and titrating the monkeys' performance around the 75 percent correct level, we showed that sensitivity in the affected half-fields was reduced by no more than 0.5 (Lennox, Dracula) to 1.5 log units (Wrinkle). Interestingly, this amount is very similar to that we found in blindsight patients (Stoerig 1993, Stoerig and Cowey 1991). When we then presented stimuli that were well above the determined threshold—say 0.5 to 0.7 log above—the monkeys performed at 90 percent correct or better in both hemifields (see Figure 18.3).

To rule out that the localization was based on straylight artefacts, we conducted two control experiments, which are summarized in Figure 18.4. In the first, the stimulus was masked in the right hemifield, to see whether light emitted from the stimulus could account for the monkeys' performance. The results of this test show that localization fell to chance level as a consequence. In the second, we masked the entire monitor, omitting only the area around the stimulus positions. If a spread within the screen had been responsible for the monkeys' excellent performance in the unmasked condition, this should have a detrimental effect. As can be seen, however, the monkeys' performance was quite independent of such information, and remained highly significant.

Having confirmed that our monkeys show residual visual localization in their affected hemifields, and having determined the luminance levels required for close-to-perfect performance, we introduced a different paradigm in an attempt to coax them into telling us whether or not they actually see these stimuli they localize so well. In the new paradigm they had to learn to distinguish between trials where a stimulus was presented and

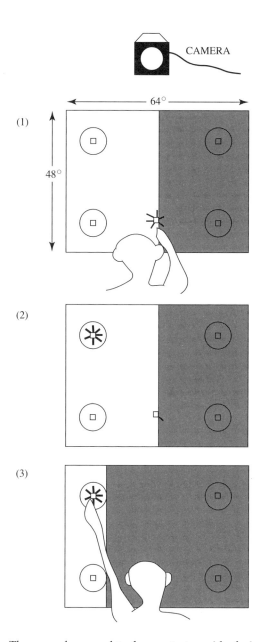

Figure 18.2 The procedure used to demonstrate residual visual functions in the hemianopic field of three macaque monkeys with left-sided unilateral striate cortical ablation. The monkeys press the start light on the monitor whose right half then falls into the affected field, indicated by the gray zone. The eye-movements are monitored with an infrared sensitive camera (1). Triggered by the response to the start light, a stimulus appears briefly at one of four possible positions (2), and the monkeys have to touch this position in order to obtain a reward (3).

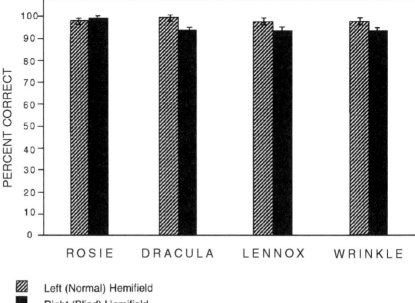

Figure 18.3 The percent correct values, ±1 SE, achieved by Rosie, the normal control, and the three hemianopic monkeys, with stimuli 0.7 log units above detection threshold in each hemifield. Results are given separately for the two hemifields, and in each of them random responding would yield 50 percent correct (and 25 percent correct in the combined fields). Localization is close to perfect.

trials where no stimulus was presented (blank trials). The stimulus was a suprathreshold white 2 by 2 square, presented for 750 msec, which appeared in one of a number of possible positions, all of which were in the normal hemifield during training. As in the previous paradigm, the monkeys, having started the trial by pressing the starting light, had to respond to the stimulus by pressing its position. In blank trials, when no stimulus was triggered by their pressing the starting light, they had instead to touch a constantly outlined rectangle in a left-hand quadrant of the screen. By touching this no-stimulus response area, they obtained a reward (see Figure 18.5).

Once they had mastered this new task, we presented a fraction of the stimuli in the right hemifield, which is hemianopic in the three operated monkeys, to learn how they would now respond to these stimuli. If they were to touch the stimulus position, as we already know they can, their behavior would be the same as in the normal hemifield. If instead they were to press the no-stimulus response area, they would be treating the stimulus in the hemianopic field as if it was a blank, indicating a dissociation between the processing of visual information and seeing it, as in the patients.

The results are given in Figure 18.6. The three bars indicate correct responses to the stimulus in the normal (or left) field, the correct responses to a blank trial, and the correct responses to a stimulus in the affected (or, in

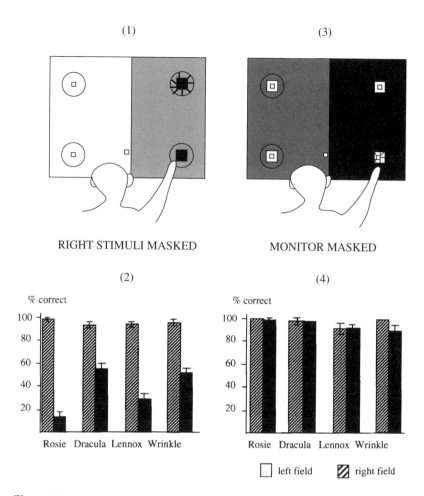

Figure 18.4 A summary of the procedure used for stray light control tests. In the first experiment, the area of the screen immediately in front of the stimulus was masked in the right (hemianopic) field (1), to see whether the monkeys could still localize the stimuli by means of stray light scattered across the remainder of the screen. Their performance declined to chance level on the right (2). In the second experiment, the screen was masked, leaving holes in front of the four stimulus positions and the start light (3), to see whether luminance changes in the screen were needed for localization. The performance of all four monkeys was unaffected by this mask (4).

Rosie, normal) right hemifield. All monkeys performed well with stimuli in the normal field, and they responded equally well to blanks. Rosie, with her two normal hemifields, performed equally well with the probe trials in the right half-field. But the decisive result occurred with the probe stimuli in the impaired hemifield. The three hemianopic monkeys scored only 2 to 8 percent correct, because they consistently pressed the no-stimulus response areas instead of the stimulus position. The small number of correct positives were caused by their looking towards the stimulus which in this paradigm was presented for a full 750 msc (Cowey and Stoerig 1995).

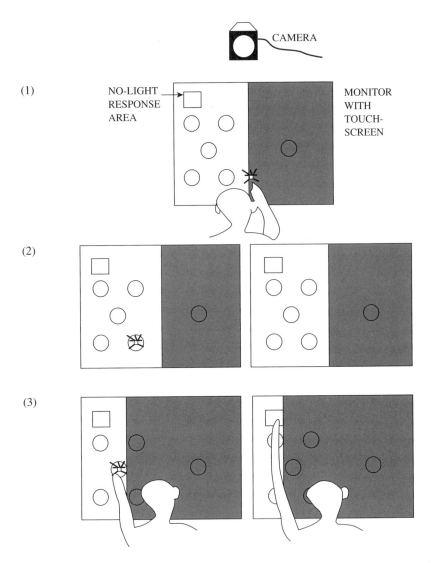

Figure 18.5 The procedure used to test whether the monkeys treat suprathreshold stimuli in the hemianopic field as stimuli or as blanks. They start the trial by pressing the start light (1), which triggers the presentation of a stimulus or a blank (no stimulus) (2). To obtain a reward, they respond to the stimulus by touching its position, and to the blank by touching the no-stimulus response area in the normal left hemifield (3).

To account for the possibility that the stimulus in the affected half-field, albeit of a luminance well above detection threshold, appeared so much less salient than the one in the normal field that the monkeys classify it as a blank although they have a weak phenomenal percept, we conducted additional series in which the stimulus in the affected field was physically much more intense. All three animals still classified it as a blank (see Figure 18.7). As the reinforcement schedule did not affect the performance—we tested different ones on all animals—we conclude that the monkeys' affected hemifields are most probably phenomenally blind for the stimuli we used.

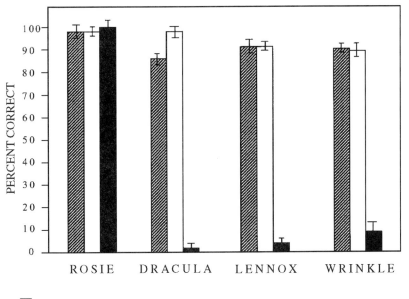

Figure 18.6 The percent correct values for detecting the stimulus in the normal left hemifield, for responding correctly to the blanks, and for detecting the stimulus in the affected right hemifield are given (± 1 SE). In contrast to the control animal, who performed very well no matter where the stimulus appeared, the hemianopic monkeys touched the stimulus position in the hemianopic field on one or four presentations only, and the no-stimulus-response area instead on the remaining 46 or 49 probe trials.

Obviously, with this approach we do not prove that the monkeys have normal phenomenal visual perception in their normal hemifield—we simply assume that this is the case. We only showed that the monkeys treated a suprathreshold stimulus in the affected hemifield like a blank in the normal hemifield, which indicates that they suffer the same dissociation as the patients between the processing of visual information, witnessed by their ability to localize the stimuli in the affected hemifield as well as they do in the normal one, and the phenomenal representation of this same processed information.

This is just one sensory modality—vision. It is also just one slice of consciousness, the phenomenal one. There is (at least) one other slice involved in perception, namely the conscious access to the information that is or has been processed. It has been argued on philosophical grounds that the two are not the same (Nelkin 1995, Nelkin in press, Block 1995), and empirical data support this notion. The processing of external visual information may be entirely unconscious (the pupil constricts in response to light in a sleeping person), but it is sometimes phenomenal, and sometimes consciously accessible. If it has been phenomenal at some point, it may be con-

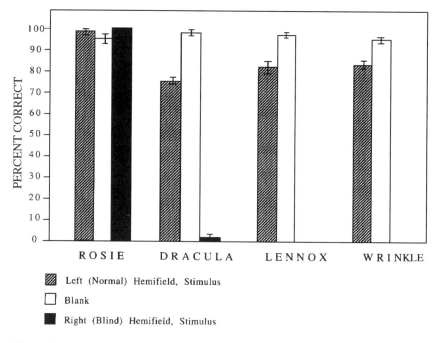

Figure 18.7 As in Figure 18.6, but the stimulus luminance was adjusted to compensate for the difference in sensitivity between the hemifields, being 0.3 log above detection threshold in the normal left and 1 log above in the affected right hemifield. Nevertheless, only one hemianopic monkey responded at all to the stimulus, correctly detecting one (of 50) probe trials.

sciously accessible later, although there is no guarantee because one may obviously forget information that was thus represented; if it has never been phenomenally represented, it may never be consciously accessible. Visual dreams are always phenomenal, and sometimes consciously accessible, but too often forgotten quickly, indicating again that a phenomenal representation is not sufficient for long-term conscious retrieval. And visual imagery is always conciously accessible (after all it is consciously and voluntarily initiated), and sometimes (in some people but not others) it is also phenomenal.

These dissociations show that perception can be understood as being organized in layers, with reflexive and implicit layers on the one side, and voluntary, phenomenal, and consciously introspectible layers on the other side of the gap. All layers must be interrelated and causally connected to some extent, with the higher layers dependent on the lower ones. At least that conclusion is rendered likely by the clinical examples of vision loss already mentioned (Figure 18.1): It is possible to find reflexive or implicit visual processing in the absence of all the higher layers, and it is possible to find phenomenal vision in the absence of conscious introspection of the visual information, for instance in patients with associative agnosia. These patients definitely see phenomenally, but they do not recognize the seen visual objects as what or whose they are. Possibly certain types of neglect patients show a similar loss of conscious access in the presence of phenomenal vision.

In patients with blindsight, however, the phenomenal vision is lost along with the conscious introspection of the visual information: The patients not only state that they do not see anything, they also state that they are merely guessing; they have to be convinced by the experimenter to give it a try, and often show astonishment when told that they performed quite well. Whether the loss of a lower layer necessarily and always implies the loss of all the consequent ones is uncertain. We shall try to further elucidate the relationship between phenomenal representation and conscious introspection by asking our monkeys whether they too perceive themselves as guessing the almost perfectly localized position of a visual stimulus in their hemianopic field. It is this introspective aspect of consciousness that Weiskrantz (1995) primarily addressed when defining blindsight as "visual capacity in the absence of acknowledged awareness," and stressing the importance of a commentary key for assessing a conscious representation.

We suggest that the different layers or stages of visual processing are realized by different parts of the visual system. Phenomenal vision appears to depend in a crucial fashion on an intact striate cortex. We should like to learn whether this holds true for any form of phenomenal vision; that is, whether the complete destruction of striate cortex causes a loss of imagery, dreams, phosphenes, or hallucinations. The highest stage, conscious access, must possess a supramodal part because we can consciously access and combine information from different modalities. It will most likely involve cortical areas, which do not receive exclusively visual input, possibly in the (ento- and perirhinal) temporal and prefrontal lobes.

WHAT IS THE FUNCTION OF CONSCIOUSNESS?

On evolutionary grounds the fact that consciousness—phenomenal, introspective, and also self-consciousness, an aspect we have barely touched upon—has developed at all indicates that it does have a function. Like all other functions of the brain and body it should help the organism to survive, to live well, to live better, to perform Whitehead's (1929) function of reason which, in a general way, is the function of every part of the organism. More specifically, different organs serve different functions in pursuing the common goal; the heart pumps the blood through our arteries, the stomach digests the food, and the brain not only regulates the function of these other organs, but stores programs for walking and talking and dancing, processes information about the environment, dreams and imagines, initiates behavior and through the behavior alters the environment. Only part of its functions are accessible to one's conscious self, and in all cases it is not the brain's activity we are directly informed about but the results of this activity; after all, man and other animals "solve problems without knowing how" (Newell, Shaw, and Simon 1958).

As consciousness itself is a manifold, its various sections are likely to serve different functions. What function might phenomenality serve? Could signalling greenness be the most efficient way to inform us (the conscious self that is unaware of a large part of its functions and intricate structural organization) of the result of the complicated processes and calculations that are required for the visual system to process the wavelengths predominantly reflected from meadows, moss, and woods? Could pain be the most efficient way to inform us of a wound on the back? Is a phenomenal representation in general the most efficient way in which to represent objects and events? Is such efficiency necessary to avoid clogging up of the limited capacities of our consciousness through the details of physical and physiological properties which the scientific community has struggled to comprehend for a long time, and which must be beyond the conscious grasp of an individual organism having to make a decision? Phenomenal vision of reds and greens, smooth and prickly surfaces, gives us qualia; without phenomenal experiences qualia are impossible, and the mind's eye is blind. As we need phenomenal vision to manipulate images and ideas of and about color and texture, phenomenal representations may be prerequisite for the conscious introspective access to presently or previously processsed visual information. Although for such manipulation a voluntary recreation of a phenomenal image appears unnecessary, the access may depend on phenomenal images being or having been available. Their function would then be to provide an efficient, graspable (simple, holistic) informational basis for conscious manipulation.

Conscious or explicit access to memorized information may therefore be restricted to information that has passed through a stage of phenomenal representation. However, the explicit access may be disturbed in the presence of a phenomenal representation; as mentioned before, in some forms of agnosia, neglect, split-brain syndrome, and amnesia, only implicit recognition is functional, while explicit recognition is lost although the information passes or passed through a phenomenal stage (Bauer 1984, Meunier *et al.* 1993, Milner *et al.* 1968, Schacter 1987, Tranel and Damasio 1985).

In this yet-to-be confirmed-or-refuted conjecture, the conscious access to the (once or again) phenomenally represented information allows thinking, planning, scheming, and day-dreaming. These functions are realized by that part of the intermediary net that encompasses the stages between the primary cortices and the prefrontal areas. They broaden the scope of observable and unobservable behaviors without being necessary for behavior as such; as we argued before, behavior can already be elicited reflexively. Such a broadening of the behavioral repertoire must have survival value, because it creates new behaviors, adapts them to different situations, or suppresses them despite some urge. This renders the organism less predictable, predictability being a hallmark of machines and reflexes. Being convinced that the slices of consciousness have these and possibly other functions makes us functionalists. This functionalism is different from the one referred to at the beginning—namely that the algorithms matter, not the matter in which they are implemented. Advocating it, we are in the excellent company of Charles Darwin, William James, Karl Popper, Owen Flanagan, and Jerry Fodor (1989, p. 77) who said:

> ... if it isn't literally true that my wanting is causally responsible for my reaching, and my itching is causally responsible for my scratching, and my believing is causally responsible for my saying.... if none of that is literally true, then practically everything I believe about anything is false and it's the end of the world.

REFERENCES

Barbur, J. L., J. D. G. Watson, R. S. J. Frackowiak, and S. Zeki. 1993. Conscious visual perception without V1. *Brain* 116:1293–302.

Bauer, R. M. 1984. Autonomic recognition of names and faces in prosopagnosia: A neuropsychological application of the Guilty Knowledge Test. *Neuropsychologia* 22:457–69.

Block, N. 1995. On a confusion about a function of consciousness. *Behavioral and Brain Sciences* 18:227–47.

Bodis-Wollner, I., A. Atkin, E. Raab, M. Wolkstein. 1977. Visual association cortex and vision in man: Pattern-evoked occipital potentials in a blind boy. *Science* 198:629–31.

Bullier, J., P. Girard, and P-A. Salin. 1993. "The role of area 17 in the transfer of information to extra striate visual areas." In *Primary visual cortex in primates. Cerebral cortex,* Vol. 10, edited by A. Peters and K. S. Rockland. New York: Plenum, pp. 301–30.

Cauller, L. J. 1995. Layer I of primary sensory neocortex: Where top-down meets bottom-up. *Behavioural Brain Research* (in press).

Cauller, L. J., and A. T. Kulics. 1991. The neural basis of the behaviorally relevant N1 component of the somatosensory-evoked potential in awake monkeys: Evidence that backward cortical projections signal conscious touch sensation. *Experimental Brain Research* 84:607–19.

Cowey, A. 1994. "Cortical visual areas and the neurobiology of higher visual processes." In: *Object representation and recognition,* edited by M. Farah and G. Ratcliff. Hillsdale, NJ: Lawrence Erlbaum, pp. 1–31.

Cowey, A., and P. Stoerig. 1989. Projection patterns of surviving neurons in the dorsal lateral geniculate nucleus following discrete lesions of striate cortex: Implications for residual vision. *Experimental Brain Research* 75:631–8.

Cowey, A., and P. Stoerig. 1995. Blindsight in monkeys. *Nature* 373:247–9.

De Haan, E. H. F., A. W. Young, and F. Newcombe. 1991. Covert and overt recognition in prosopagnosia. *Brain* 114:2575–91.

Dennett, D. 1991. *Consciousness explained.* New York: Little Brown.

DuBois–Reymond, E. 1872. *Über die Grenzen des Naturerkennens.* Leipzig: Veit & Com.

Edelman, G. M. 1989. *The remembered present: A biological theory of consciousness.* New York: Basic Books.

Farah, M., M. J. Soso, and R. M. Dasheiff. 1992. The visual angle of the mind's eye before and after unilateral occipital lobectomy. *Journal of Experimental Psychology: Human Perception and Performance* 18:241–6.

Flanagan, O. J. 1992. *Consciousness reconsidered.* Cambridge, MA: MIT Press.

Flohr, H. 1995. Sensations and brain processes. *Behavioural Brain Research* (in press).

Fodor, J. 1975. *The language of thought*. New York: Crowell.

Fodor, J. 1989. Making mind matter more. *Philosophical Topics* 17:59–79.

Guitton, D. 1992. Control of eye-head coordination during orienting gaze shifts. *Trends in Neuroscience* 15:174–9.

Hameroff, S. 1994. Quantum coherence in microtubules: A neural basis for emergent consciousness? *Journal of Consciousness Studies* 1:91–118.

King, S. M., S. Frey, J. G. Villemure, A. Ptito, and P. Azzopardi. 1995. Perception of motion—in-depth in patients with partial or complete cerebral hemispherectomy. *Behavioural Brain Research* (in press).

King, S. M., A. Cowey, P. Azzopardi, J. Oxbory, and S. Oxbory. 1995. The role of light scatter in the residual sensitivity of patients with cerebral hemispherectomy. *Visual Neuroscience* (in press).

Klüver H. 1941. Visual functions after removal of the occipital lobes. *Journal of Psychology* 1:23–45.

Logothetis, N. K., and J. D. Schall. 1989. Neuronal correlates of subjective visual perception. *Science* 245:761–3.

Lorenz, K. 1965. Über tierisches und menschliches Verhalten. *Gesammelte Abhandlungen Bd II*. München: Piper & Co.

Marcel, A. J., and E. Bisiach, Eds. 1992. *Consciousness in contemporary science*. Oxford, England: Clarendon Press.

Maunsell, J. H. R., and W. T. Newsome. 1987. Visual processing in monkey extrastriate cortex. *Annual Review of Neuroscience* 10:363–401.

McGinn, C. 1991. *The problem of consciousness*. Oxford, England: Blackwell.

Meunier, M., J. Bachevalier, M. Mishkin, and E. A. Murray. 1993. Effects on visual recognition of combined and separate ablations of the entorhinal and perirhinal cortex in rhesus monkeys. *Journal of Neuroscience* 13:5418–32.

Milner, B., S. Corkin, and H.-L. Teuber. 1968. Further analysis of the hippocampal amnesic syndrome: 14-year follow-up study of H. M. *Neuropsychologia* 6:215–34.

Movshon, A. J., E. H. Adelson, M. S. Gizzi, and W. T. Newsome. 1985. "The analysis of moving visual patterns." In *Pattern recognition mechanisms*, edited by C. Chagas, R. Gattass, and C. Gross. Vatican City: Pontificiae Academiae Scientiarum.

Myerson, J., F. Miezin, and J. Allman. 1981. Binocular rivalry in macaque monkeys and humans: A comparative study in perception. *Behavioral Analysis Letters* 1:1149–59.

Nelkin, N. 1993. What is consciousness? *Philosophy of Science* 60:419–34.

Nelkin, N. 1995. The dissociation of phenomenal states from apperception. In *Consciousness—The current debate*, edited by T. Metzinger. Schoeningh: Paderborn.

Nelkin, N. in press. *Consciousness and the origins of thought*. Cambridge, England: Cambridge University Press.

Newell, A., J. C. Shaw, and H. A. Simon. 1958. Chess-playing programs and the problem of complexity. *IBM-Journal*, October.

Pasik, P., and T. Pasik. 1982. "Visual functions in monkeys after total removal of visual cerebral cortex." In *Contributions to sensory psychology, Vol. 7*, edited by W. D. Neff. New York: Academic Press, pp. 147–200.

Penrose, R. 1989. *The emperor's new mind*. Oxford, England: Oxford University Press.

Penrose, R. 1994. *Shadows of the mind*. Oxford, England: Oxford University Press.

Perenin, M.-T. 1978. Visual function within the hemianopic field following early cerebral hemidecortication in man. II. Pattern discrimination. *Neuropsychologia* 16:697–708.

Perenin, M.-T. 1991. Discrimination of motion direction in perimetrically blind fields. *NeuroReport* 2:397–400.

Perenin, M.-T., and M. Jeannerod. 1978. Visual function within the hemianopic field following early cerebral hemidecortication in man. I. Spatial localization. *Neuropsychologia* 16:1–13.

Popper, K. R., B. I. B. Lindahl, and P. Arhem. 1993. A discussion of the mind-brain problem. *Theoretical Medicine* 14:167–80.

Ptito, A., M. Lassonde, E. F. Lepor, and M. Ptito. 1987. Visual discrimination in hemispherectomized patients. *Neuropsychologia* 25:869–79.

Ptito, A., E. F. Lepor, M. Ptito, and M. Lassonde. 1991. Target detection and movement discrimination in the blind field of hemispherectomized patients. *Brain* 114:497–512.

Riddoch, G. 1917. Dissociation of visual perceptions due to occipital injuries, with especial reference to appreciation of movement. *Brain* 40:15–57.

Rodman, H. R., C. G. Gross, and T. D. Albright. 1989. Afferent basis of visual response properties in area MT of the macaque. II. Effects of superior colliculus removal. *J. Neurosci.* 10:2033–50.

Schacter, D. L. 1987. Implicit memory: History and current status. *Journal of Experimental Psychology:Learning, Memory, and Cognition* 13:501–18.

Searle, J. R. 1992. *The rediscovery of the mind*. Cambridge, MA: MIT Press.

Skinner, B. F. 1969. *Contingencies of reinforcement*. New York: Appleton-Century-Crofts.

Sprague, J. M. 1966. Interaction of cortex and superior colliculus in mediation of visually guided behavior in the cat. *Science* 153:1544–7.

Stoerig, P. 1993. Spatial summation in blindsight. *Visual Neuroscience* 10:1141–9.

Stoerig, P. in press. "Consciousness and the matter of perception." In *Matter matters,* edited by M.-O. Olsson and U. Svedin. Hamburg: Springer.

Stoerig, P., and S. Brandt. 1993. The visual system and levels of perception: Properties of neuromental organization. *Theoretical Medicine* 14:117–35.

Stoerig, P., and A. Cowey. 1991. Increment-threshold spectral sensitivity in blindsight. *Brain* 114:1487–512.

Stoerig, P., and A. Cowey. 1993. "Blindsight and perceptual consciousness." In *Functional anatomy of the human visual cortex,* edited by B. Gulyás, D. Ottoson, and P. E. Roland. Oxford, England: Pergamon Press, pp. 181–93.

Stoerig, P., J. Faubert, V. Diaconu, M. Ptito, and A. Ptito. Visual sensitivity following unilateral cerebral hemidecortication in man. (in preparation)

Tomaiuolo, F., A. Ptito, T. Paus, and M. Ptito. 1994. Spatial summation across the vertical meridian after complete or partial hemispherectomy. *Society for Neuroscience Abstracts* 20, p. 1579.

Tranel, D., and A. R. Damasio. 1985. Knowledge without awareness: An autonomic index of facial recognition by prosopagnosics. *Science* 228:1453–4.

Wallace, S. F., A. C. Rosenquist, and J. M. Sprague. 1989. Recovery from cortical blindness mediated by destruction of nontectotectal fibers in the commissure of the superior colliculus in the cat. *Journal of Comparative Neurology* 284:429–50.

Weiskrantz, L. 1990. Outlooks for blindsight: Explicit methodologies for implicit processes. *Proceedings of the Royal Society London B* 239:247–78.

Weiskrantz, L. 1995. The problem of animal consciousness in relation to neuropsychology. *Behavioural Brain Sciences* (in press).

Weiskrantz, L., J. L. Barbur, and A. Sahraie. 1995. Parameters affecting conscious versus unconscious visual discrimination with damage to the visual cortex (VI). *Proceedings of the National Academy of Science USA* 92:6122–6.

Weiskrantz, L., E. K. Warrington, M. D. Sanders, and J. Marshall. 1974. Visual capacity in the hemianopic field following a restricted occipital ablation. *Brain* 97:709–28.

Whitehead, A. N. 1929. *The function of reason*. Princeton, NJ: Princeton University Press.

Wilkes, K. V. 1992. "-, yìshì, duh, um, and consciousness." In *Consciousness in contemporary science,* edited by A. J. Marcel and E. Bisiach. Oxford, England: Clarendon Press, pp. 16–41.

Wurtz, R. H., and J. E. Albano. 1980. Visual-motor function of the primate superior colliculus. *Annual Reviews of Neuroscience* 3:189–226.

Yukie, M., and E. Iwai. 1981. Direct projection from the dorsal lateral geniculate nucleus to the prestriate cortex in macaque monkeys. *Journal of Comparative Neurology* 201:81–97.

19 Levels of Awareness and "Awareness Without Awareness": From Data to Theory

Gary E. Schwartz

INTRODUCTION

The purpose of this chapter is to outline a theory of Levels of Awareness that begins with the seemingly paradoxical premise that all stimuli that activate neural receptors (or biological processes in general) have an inherent "consciousness" to them whether we are "aware" of this consciousness or not. The theory outlined in this paper emerged accidently. It was stimulated by new findings on electroencephalogram (EEG) registration of subthreshold olfactory stimuli uncovering "subtle consciousness" of subliminal stimuli (Schwartz et al. 1994). Just because a stimulus may be too weak in intensity to activate conscious awareness does not necessarily imply that the stimulus is too weak to activate receptors and be registered by the nervous system. A fundamental mystery in consciousness research is what determines the transition from so-called unconscious or nonconscious processing of stimuli to the conscious registration of stimuli. What makes it possible for us to experience stimuli consciously? Why is it that so many stimuli that are continuously monitored by the brain (for example, chemoreceptors within the body) seem to be processed outside of conscious awareness? Is this information truly outside of awareness (the classic interpretation)? Or, is it possible that the information actually influences our awareness in subtle yet important ways, even though we are not aware that our awareness has been influenced?

This chapter begins by reviewing a recent experiment on EEG registration of subliminal olfaction. Unexpected evidence for what may be called functional "blindsmell" is described, and evidence suggesting that implicit perception may differ from implicit sensation in terms of higher frequency EEG registered in the frontal region is considered. Additional findings of "subtle awareness" to the subliminal odors are presented, and some implications for the hypothesis of "awareness without awareness" are put forth.

In this chapter the term awareness is operationalized as subjects' self-reported ratings of their perceptions. Awareness without awareness is operationalized as subtle but systemic changes in subjects' ratings of their perceptions of subliminal stimuli where the changes are so subtle that subjects do not report being aware that their perceptions have changed systematically.

From a systems perspective (Miller 1978), the concept of Levels of Awareness is proposed, where Level I Awareness is "pure awareness" (awareness without conscious awareness of the awareness—the "unconscious"), Level II Awareness is "awareness of awareness" (normal consciousness), and Level III Awareness is "awareness of awareness of awareness" (self-consciousness or reflective consciousness). Specific research directions are proposed.

As will become clear, the theory of Levels of Awareness meets three important criteria for proposing a new theory: (1) the theory incorporates previously published data, (2) the theory makes new predictions that can be confirmed or disconfirmed in future research, and (3) the theory integrates seemingly conflicting ideas and paradigms. It turns out that the theory of Levels of Awareness not only integrates concepts in modern science, but it also integrates some ancient spiritual ideas voiced by native peoples (Native Americans) with modern science.

EEG REGISTRATION OF SUBLIMINAL OLFACTION

Recent research from the Chemical Psychophysiology Laboratory at the University of Arizona has addressed the question of whether it is possible for the brain to register the presence of odors that are not experienced consciously (see Schwartz *et al.* 1994 for a review of these studies). The experimental paradigm is as follows.

Subjects are requested to smell pairs of bottles, one of which contains an odorant dissolved in a solvent, the other contains just the solvent, while EEG is continuously recorded. For example, one bottle might contain isoamyl acetate, an odorant that has a fruity smell, dissolved in nonodorous liquid silicone. Subjects are handed two bottles, and with their eyes closed, are requested to smell each bottle for two seconds. Each bottle is smelled twice. Then, subjects are instructed to open their eyes, guess which bottle contains the odor, and rate the intensity of the odor (from 0 to 10) and the confidence of their guess (from 0 to 10). On half the trials, the odorant is placed in the right hand, on the other half of the trials, the odorant is placed in the left hand. In addition, on half the trials, subjects are instructed to initiate the smelling sequence with their right hand, on the other half of the trials, subjects are instructed to initiate the smelling sequence with their left hand. Hence, order is completely counterbalanced in a 2 × 2 factorial design.

If the concentration of the odorant is sufficiently high (supraliminal concentration), subjects will correctly guess which hand has the odorant on 100 percent of the trials. They will also rate the intensity of the odorant as fairly high and rate the confidence of their guess as high. However, if the concentration of the odorant is sufficiently low (subliminal concentration), subjects will correctly guess which hand has the odorant on only 50 percent of the trials (that is, they will be truly guessing). Correspondingly, they will rate the intensity of the odorant and the confidence of their guess as very low.

EEG is spectral analyzed, and topographic EEG maps are created subtracting the EEG occurring during the smelling of the control bottles (solvent alone) from the EEG occurring during the smelling of the experimental bottles (odorant plus solvent). Subtraction is necessary to remove the effects of hand posture, breathing, attention, and registration of the control solvent, leaving just the effects of the registration of the odorant in the EEG.

In a series of studies, we have found that when subjects smell a subliminal concentration of isoamyl acetate, evidence of brain registration of the odorant is observed in the EEG (reviewed in Schwartz *et al.* 1995). Subjects show relative decreases in alpha activity (8–12 Hz) in mid-central and posterior regions. When the concentration is high enough to be experienced consciously, relative decreases in alpha activity are also observed in anterior regions as well. Summary subtraction EEG maps from the most recent study are displayed in Figure 19.1.

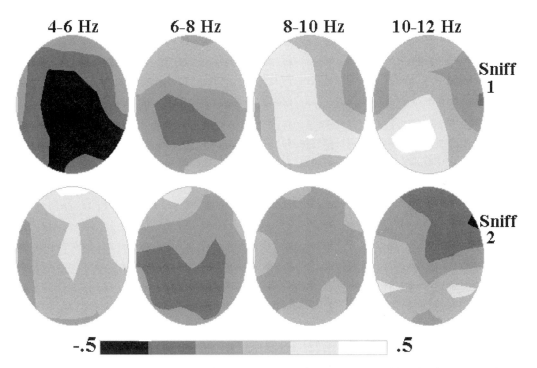

Figure 19.1 Topographic EEG maps subtracting control bottles from the odor bottles for subthreshold isoamyl acetate trials (N=16) for 86 subjects. The top four maps reflect sniff one (the first sniff for the isoamyl acetate and control trials), the bottom four maps reflect sniff two (the second sniff for the isoamy acetate and control trials). The four EEG bands (4–6 Hz, 6–8 Hz, 8–10 Hz, 10–12 Hz) are all drawn on the same scale. Light gray and white areas reflect increases in EEG, dark gray and black areas reflect comparable magnitude decreases in EEG. The top of each map reflects the front of the head, the bottom of each map reflects the back of the head.

In this experiment, 19 channels of EEG were recorded from 86 college students. The subthreshold odors (isoamyl acetate dissolved in liquid silicone) were administered double-blind (neither the subjects nor the experimenters knew which were the experimental bottles and which were the control bottles). The number of subliminal isoamyl acetate trials was 16. EEG was spectral analyzed in 2-Hz bands, and analyses were performed on low-frequency theta (4–6 Hz), high-frequency theta (6–8 Hz), low-frequency alpha (8–10 Hz), and high-frequency alpha (10–12 Hz).

The top four subtraction maps reflect the data for sniff one averaged over the 16 trials, the bottom four subtraction maps reflect the data for sniff two. The four EEG bands are drawn on the same scale. The color scheme shows that light gray and white areas reflect increases in theta or alpha to the odorant (relative to the control), while dark gray and black areas reflect decreases in theta or alpha of a comparable magnitude. Medium gray areas reflect little or no difference between odorant and control. The top of each map is anterior (front of the head), the bottom of each map is posterior (back of the head).

It can be seen that for low-frequency theta (4–6 Hz), EEG registration of the subliminal odor occurred on the first sniff, especially in the central and posterior regions, more on the right side (the regions of the black and dark gray areas). For high-frequency theta (6–8 Hz), EEG registration of the subliminal odor occurred on both sniffs, again central and more posterior on the right side. For low-frequency alpha (8–10 Hz), some evidence for EEG registration was obtained, more central and right sided, somewhat for both sniffs. For high-frequency alpha (10–12 Hz), EEG registration was obtained especially for sniff two, and the effect was central and more anterior on the right side. These effects (and all future effects described in this chapter), were statistically significant at at least $P < 0.05$.

Subjects accuracy in guessing which bottle contained the subliminal concentration of isoamyl acetate was 48 percent. Subjects' mean rating of the odor's intensity was only 1.8 (of a possible 10), and subjects' mean rating of the confidence of their guess was only 3.2 (of a possible 10). Hence, in terms of both performance and self-report, these low concentration isoamyl acetate trials were apparently "subliminal." Despite the fact that subjects did not appear to detect or perceive the odorant, the EEG shows compelling evidence that the odorant was detected by the nose and was registered by the central nervous system.

These subliminal olfaction findings, though theoretically interesting and important, are not especially surprising or controversial. However, when the data were analyzed more finely, a number of surprising and potential controversial findings emerged. These findings led to the concept of functional "blindsmell," and ultimately, to the concept "awareness without awareness" and the theory of Levels of Awareness.

Functional "Blindsmell" and "Awareness without Awareness"

Athough the average group performance in detecting the odor was 48 percent (50 percent is chance), subjects varied in the number of trials out of 16 that they correctly guessed contained the odorant. Some subjects did better than chance, and some subjects did worse than chance. Could it have been the case that the EEG effects described above occurred primarily in the subset of subjects whose performance was somewhat above chance? If their performance was above chance, did this mean that they were actually experiencing the odorant? And if they were experiencing the odorant, would this mean that the EEG effects were actually reflecting conscious processing of the odorant afterall?

To address this question, subjects were split into three performance subgroups (high $N=32$; medium $N=29$; and low $N=25$). Their guessing performance is shown in Figure 19.2. It can be seen that the high group was somewhat better than chance (68 percent), and the low group was somewaht below chance (36 percent). The medium group was virtually at chance. If the performance of the high-performance group reflected consciousness of the odorant, one would expect that their ratings of intensity (Figure 19.3) and confidence (Figure 19.4) would be somewhat higher than the other two groups. However, as seen in Figures 19.3 and 19.4, this was not the case. The high-performance group rated intensity and confidence just as low as the medium- and low-performance groups. Hence, even

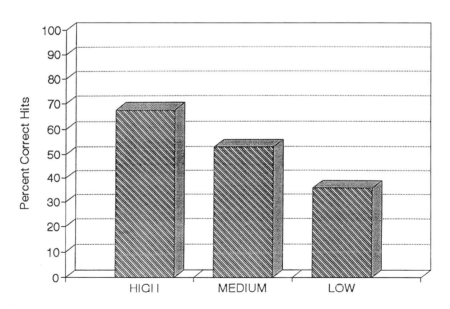

Figure 19.2 Mean percent correct guessing for the subliminal isoamyl acetate trials separately for the high-, medium-, and low-performance subgroups. Fifty percent is chance.

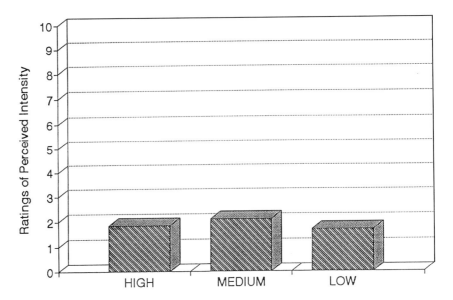

Figure 19.3 Mean rating of perceived intensity (from 0 to 10) for the subliminal isoamyl acetate trials separately for the high-, medium-, and low-performance subgroups.

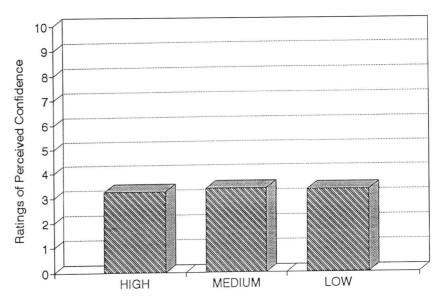

Figure 19.4 Mean rating of perceived confidence of guessing (from 0 to 10) for the subliminal isoamyl acetate trials separately for the high-, medium-, and low-performance subgroups.

though their "performance" was above "chance," their "consciousness" was at "chance."

One possible explanation is that their somewhat better performance simply represented "noise" due to a relatively small number of trials ($N=16$). However, another possibility is that the high-performance subjects were actually more "sensitive" to the subliminal odorant. Their olfactory system (receptors and/or central nervous system) may have been better able to detect subtle concentrations of the odorant, even though they were not consciously aware that they were registering the odorant above chance.

Data suggesting that sensory performance can occur in the absence of conscious awareness has been reported for other senses. Research in vision has documented a phenomenon termed "blindsight" (Weiskrantz 1986). Patients who have received injuries to their primary visual cortex will report an absence of vision in corresponding regions of their visual field. However, they can be trained to detect above chance the presence of a visual stimulus presented in these regions, even though they continue to report seeing nothing. They may "intuit" that the stimulus is present, even though they do not "see it." Similiar reports have been made for "blind-touch" (Paillard, Michel, and Stelmack 1983).

In the present experiment, if the EEG indicated greater registration of the odorant in high-performance subjects, these findings would suggest that "sensitive" subjects actually registered the subliminal odorant more strongly, enabling their performance to be better than chance, even if they were not aware that they were able to detect the odorant. Such findings would support the hypothesis of "blindsmell."

When the EEG data were analyzed as a function of performance, surprising results were obtained. Figure 19.5 displays EEG maps for low-frequency theta (4–6 Hz). It can be seen that for low-frequency theta (similar results were found for high-frequency theta), EEG registration was observed in all three groups, again primarily in sniff one. Even in subjects whose performance was below chance, the odorant was clearly registered. However, the maps suggest that the effects are somewhat stronger as performance increases, and the effects become more anterior.

Figure 19.6 displays EEG maps for high-frequency alpha (10–12 Hz). It can be seen that for high-frequency alpha (similar results were found for low-frequency alpha), EEG registration was observed primarily in the high-performance group, on sniff two.

In other words, the high-performance group showed clear evidence for greater EEG registration of the subliminal odorant in the higher EEG frequency (alpha). The lower EEG frequency (theta) seemed to immediately register the odorant (sniff one), even in subjects whose performance was below chance (implicit sensation). The higher EEG frequency (alpha) took time to register the odorant (sniff two), and primarily in subjects whose performance was above chance (implicit perception). Yet, these "sensitive" subjects showed no evidence of "experiencing" the odorants more strongly

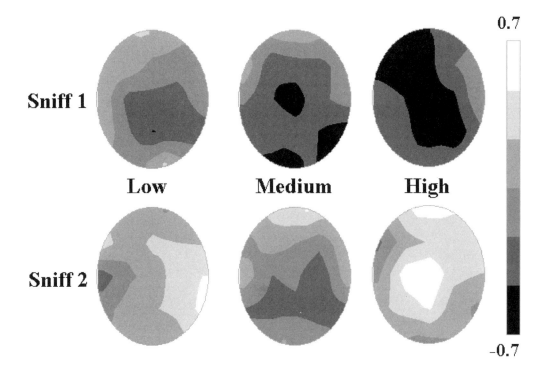

Figure 19.5 Topographic EEG maps for low-frequency theta (4–6 Hz) subtracting control bottles from the odor bottles for subthreshold isoamyl acetate trials (N=16) for high-, medium-, and low-performance subgroups. The top three maps reflect sniff one (the first sniff for the isoamyl acetate and control trials), the bottom three maps reflect sniff two (the second sniff for the isoamyl acetate and control trials. The three subgroups are all drawn on the same scale. Light gray and white areas reflect increases in EEG, dark gray and black areas reflect comparable magnitude decreases in EEG. The top of each map reflects the front of the head, the bottom of each map reflects the back of the head.

than less sensitive subjects. In a word, the high-performance subjects showed evidence for functional "blindsmell."

The data became even more surprising and controversial when the data were analyzed more finely. By definition, when subjects were guessing (recall that the average accuracy for the sample as a whole was 48 percent), on some of the trials subjects would correctly guess which bottle contained the odorant (hits) and on other trials subjects would incorrectly guess which bottle contained the odorant (misses). If the EEG was actually registering the presense of the subliminal odorant (and not simply the "belief" that the odorant was present), both hit and miss trials involving the odorant should show EEG registration. Given the large number of subjects in the present experiment (N=86), both the self-report and EEG data could be examined in terms of hits and misses for high-, medium-, and low-performance groups.

Figures 19.7 and 19.8 display the mean ratings of intensity and confidence separately for hit and miss trials for each of the three groups. First, it

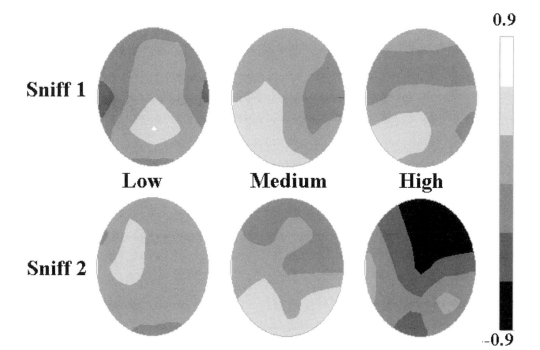

Figure 19.6 Topographic EEG maps for high-frequency alpha (10–12 Hz) subtracting control bottles from the odor bottles for subthreshold isoamyl acetate trials (N=16) for high-, medium-, and low-performance subgroups. The top three maps reflect sniff one (the first sniff for the isoamyl acetate and control trials), the bottom three maps reflect sniff two (the second sniff for the isoamyl acetate and control trials). The three subgroups are all drawn on the same scale. Light gray and white areas reflect increases in EEG, dark gray and black areas reflect comparable magnitude decreases in EEG. The top of each map reflects the front of the head, the bottom of each map reflects the back of the head.

can be seen that there was a subtle, but consistent effect for intensity as a function of hit versus miss (Figure 19.7). Ratings of intensity were slightly but consistently higher for the hit trials than the miss trials. The size of the difference was approximately two tenths of one point! Though this effect is tiny, it was statistically significant for the three groups combined. Hence, even though the subjects were rating the subliminal odorant as very low in intensity, they somehow "sensed" the intensity slightly stronger on those trials when they happened to guess correctly than when they happened to guess incorrectly. It is highly improbable that they were "aware" that their "awareness" of the intensity of the odor was slightly higher for hit trials than miss trials!

For ratings of confidence, the picture is even more curious. The high-performance group (the "sensitives") rated the confidence of their guesses significantly higher on hit trials compared to miss trials (Figure 19.8). Even though they seemed to be no more aware of the intensity of the odorant than the medium- and low-performance groups, they seemed better able to

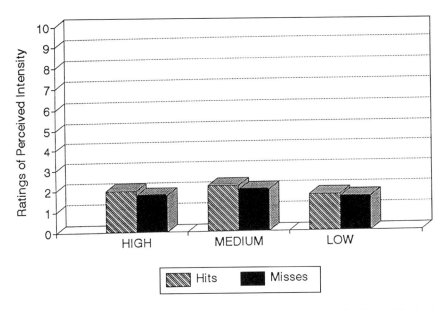

Figure 19.7 Mean rating of perceived intensity (from 0 to 10) for the subliminal isoamyl acetate trials separately for hits and misses for the high-, medium-, and low-performance subgroups.

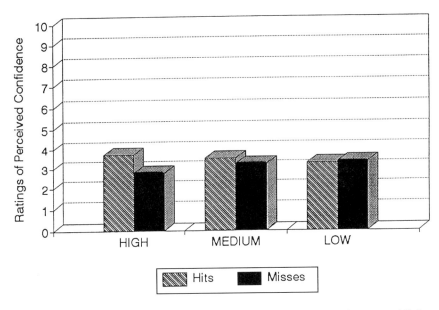

Figure 19.8 Mean rating of perceived confidence of guessing (from 0 to 10) for the subliminal isoamyl acetate trials separately for hits and misses for the high-, medium-, and low-performance subgroups.

"sense" when they were guessing correctly. Recall that their overall confidence was no higher than the medium- and low-performance groups. In other words, in sensitive subjects, although their "awareness" of confidence was somewhat higher for hit trials than miss trials, they were apparently not "aware" that their awareness of confidence was slightly higher on the hit trials compared to the miss trials.

The EEG data complimented these subtle self-report differences, primarily for the higher frequencies. Figure 19.9 displays the high-frequency alpha data for the three groups separately, for sniff one and sniff two, separately for hit and miss trials. First, it can be seen that for the low and medium groups, there were somewhat greater decreases in alpha during hit trials than miss trials (on both sniff one and sniff two). Second, it also can be seen that for the high-performance group, large EEG alpha decreases occurred, primarily for sniff two, and the patterns of the alpha decreases were different for the hit versus miss trials. When sensitive subjects guessed incorrectly (misses), the alpha decrease effects were more posterior and right hemispheric, whereas when the sensitive subjects guessed correctly (hits), the alpha decrease effects were more anterior and included the left hemisphere. Note that the alpha decrease effects for the medium- and low-performance groups also involved the left hemisphere.

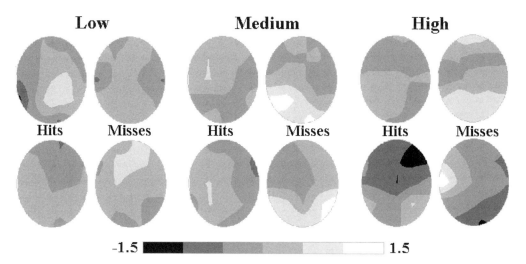

Figure 19.9 Topographic EEG maps for high-frequency alpha (10–12 Hz) subtracting control bottles from the odor bottles for subthreshold isoamyl acetate trials (N=16) separately for hits and misses for high-, medium-, and low-performance subgroups. The top six maps reflect sniff one (the first sniff for the isoamyl acetate and control trials), the bottom six maps reflect sniff two (the second sniff for the isoamy acetate and control trials). The three subgroups are all drawn on the same scale. Light gray and white areas reflect increases in EEG, dark gray and black areas reflect comparable magnitude decreases in EEG. The top of each map reflects the front of the head, the bottom of each map reflects the back of the head.

As mentioned above, when an odorant is perceived consciously, that is, the subject guesses above chance and rates the intensity of the odorant and the confidence of her/his guess appropriately high, alpha decreases are also observed in anterior regions, including the left hemisphere. The frontal cortex may play a special role in consciousness, especially the left frontal cortex (Natsoulas 1992).

The paradox posed by these unexpected hit/miss findings, especially for the sensitive subjects, can be stated simply: How can a subject's awareness of a stimulus (an odorant) be influenced by the stimulus when the subject is not aware that his awareness has been influenced? Since by definition, ratings of perceived intensity and confidence are measures of "awareness," the data suggest that "awareness" has been influenced in the absence of "awareness" (that is, the subjects were not aware that their awareness was different, since in this case, the stimuli were all "subliminal").

Continuing to label these effects as simply "subliminal" or "unconscious" begs the fundamental question. Subjects' "awareness" can change as a function of stimuli, even if subjects are not "aware" that their "awareness" is changing. How can we explain such effects? Must a stimulus have sufficient intensity and/or meaning to break into consciousness to be defined as "awareness"? Or, is it possible that all stimuli that activate biological processes have an inherent "awareness-ability" that can ultimately be measured, whether the subject is "aware" of this "awareness" or not? The hypothesis of "Levels of Awareness," formulated to explain the olfaction data described above, begins with the premise that all stimuli, by nature, are "pregnant with consciousness," and that some level of awareness is the rule, not the exception, in nature.

Levels of Awareness and "Awareness without Awareness"

If we begin with the premise that all stimuli that can activate biological processes have an inherent "awareness-ability" to them, we can create a theoretical framework that examines awareness in terms of hierarchical levels.

The core of the theory involves three general Levels of Awareness, termed Level I, Level II, and Level III. Consider Level II Awareness. This is what is typically thought of as "normal consciousness." When a subject says "I am aware of stimulus X" what she/he is actually saying is that "I—who is aware—is aware of the stimulus." Hence, implicit in normal consciousness is the notion of "awareness of awareness."

Level I Awareness can be thought of as pure awareness, the "awareness-ability" of any stimulus registered by a biological system. This awareness, *per se*, occurs in the absence of consciousness awareness (that is, awareness of awareness, Level II). Level I Awareness traditionally has been referred to as the "unconscious," the "preconscious," the "subconscious," and/or the "nonconscious." It also includes what is currently refered to as "implicit" experience, be it implicit memory, implicit emotion, implicit perception, or implicit sensation.

Level III Awareness can be thought of "awareness of awareness of awareness." Level III awareness is an awareness of Level II awareness, just as Level II awareness is an awareness of Level I awareness. The idea of "self-concept" (that is, being aware of self) is, by definition, a Level III (or higher) Awareness. Concepts of knowledge, understanding, and wisdom reflect Level III (or higher) levels of awareness.

Space does not permit developing the concept of Levels of Awareness more extensively here. The particular organization of the Levels can be debated, and appropriately so, but precise organization is not important here. What is important here is the general concept of Levels of Awareness, beginning with Level I Awareness, and some of its implications for theory and research.

PREDICTIONS OF LEVELS OF AWARENESS THEORY

At least three general predictions follow from the theory that can be confirmed or disconfirmed in future research. First, the theory requires that all stimuli registered by the brain should show evidence of subtle awareness (Level I Awareness) using appropriately sensitive paradigms as measures of self-reports and performance. Theoretically, if a stimulus is registered by the brain, it will be found to subtlely, yet reliably, influence measures of awareness (measures of perceived intensity) even though subjects will not be aware that their reports are linked to stimulus presentation. For example, chemoreception normally thought of as being "unconscious" (for example, oxygen and carbon dioxide reception in the blood stream) should be found to show evidence of subtle awareness as measured by small but consistent changes in a subject's perceptions. The history of science can be thought of, in part, as the process of becoming increasingly aware of stimuli, processes, and patterns that were here-to-fore outside of human consciousness. The concept of Level I Awareness is consistent with this vision of science.

A second prediction is that all stimuli registered by the brain can be transformed from Level I to Level II Awareness using appropriate training procedures and measurement techniques. If the apparatus that allows Level II Awareness is functional (the brain is healthy), theoretically, the information can be transformed with sufficient training. Everyday experience indicates that with training, people become aware of information that previously they were unaware of. If awareness is the rule, not the exception, in nature, and the brain can access this awareness, then even if the intensity of the stimuli are very weak, "awareness" of the inherent "awareness" should be increaseable, at least to some degree. There are important clinical as well as basic science implications that follow from this prediction (for example, patients with blindsight might be trained to regain some of their "vision").

A third prediction is that higher levels of awareness may involve higher EEG frequencies, more anterior cortical regions, and greater cortical integration. This prediction not only follows from the olfaction data, it logically follows that more complex systems will involve greater numbers of levels of information processing. Hence, higher rates of information exchange and greater system integration (coherence and the so-called "binding" problem) will be required as higher levels of awareness are achieved.

Becoming Aware of "Awareness Without Awareness"

The hypothesis that all stimuli have an inherent "awareness-ability" that is "experienced" as "awareness without awareness" has philosophical as well as scientific implications. The hypothesis of Level I Awareness is implicit in the writings of Chalmers (1996) whose work I became aware of after the olfactory blindsmell findings were uncovered and the theory of Levels of Awareness was formulated. Chalmers proposes that consciousness is an inherent quality of matter. According to the double-aspect theory of information, information has two basic aspects: a physical aspect and a phenomenal (or experiential) aspect. If everything in nature has inherent "consciousness-ability," then theoretically human beings have the capability to access information normally thought of as nonexistent (perhaps because it is normally experienced as outside of awareness). Future research may document that "subliminal" olfaction is the rule, not the exception, that "subliminal" registration of stimuli throughout the electromagnetic spectra occurs to varying degrees all the time (Russek and Schwartz 1995), and that this information is expressed in our awareness to varying degrees, whether we are aware of this information or not.

Conceptual integrity requires that one follow predictions suggested by a theory, whether one likes the implications of the predictions or not. The concept of Level I Awareness turns out to have implications for controversial research that historically has not been part of mainstream science, including parapsychology. If Level I Awareness is the rule, not the exception, in nature (including human nature), and people vary in their sensitivity to this fundamental level of information processing, it is conceivable that registration of subtle information, as documented in systematic research on parapsychology (Bem and Honorton 1994), may be more common than is generally recognized. What Native Americans in the Southwest and other native peoples have termed "spiritual awareness" may not only reflect higher levels of awareness, but may reflect a deep awareness of awareness of Level I Awareness. Registration of subtle information and energy may find curious parallels to ancient ideas of "soul" and "spirit" (Schwartz 1994).

Philosophical issues aside, a fundamental challenge for the science of consciousness is to define awareness in such a way that it addresses all levels of awareness-ability. If the present chapter challenges the reader to reevaluate her/his concept of the "unconscious" and the nature of levels of conscious experience, then it will have achieved its goal.

ACKNOWLEDGMENTS

Important contributions to the theory presented in this chapter were made by Geoffrey Ahern, Ziya Dikman, Mercedes Fernandez, John Kline, Ernest Polak, Linda Russek, and Lloyd Smith. Their suggestions and criticisms are warmly acknowledged.

REFERENCES

Bem, D. J., and C. Honorton. 1994. Does Psi exist? Replicable evidence for an anomalous process of information transfer. *Psychological Bulletin* 115:4–18.

Chalmers, D. J. 1996. "Facing up to the problem of consciousness." In *Toward a theory of consciousness: The first Tuscon discussions and debates.* edited by S. R. Hameroff, A. W. Kaszniak, and A. C. Scott. Cambridge, MA: MIT Press.

Miller, J. G. 1978. *Living systems.* New York: McGraw–Hill.

Natsoulas, T. 1992. Consciousness and commissurotomy: Three hypothesized dimensions of deconnected left-hemisphere consciousness. *Journal of Mind and Behavior* 13:37–67.

Paillard, J. F., F. Michel, and G. Stelmach. 1983. Localization with content: A tactile analogue of "blindsight." *Archives of Neurology* 40:548–51.

Russek, L. G., and G. E. Schwartz. 1995. Energy cardiology: A paradigm for integrating cardiology with energy medicine. *Subtle Energies* (in press).

Schwartz, G. E. 1994. Soul is to spirit as information is to energy: A theoretical note and poem. *ISSSEEM Newsmagazine* 5:7.

Schwartz, G. E., I. R. Bell, Z. V. Dikman, M.Fernandez, J. P. Kline, J. M. Peterson, and K. P. Wright. 1994. EEG responses to low level chemicals in normals and cacosmics. *Toxicology and Industrial Health.*

Weiskrantz, L. 1986. *Blindsight: A case study and implications.* Oxford, England: Oxford University Press.

20 Implicit Memory During Anesthesia

Randall C. Cork

As with any field of endeavor, the study of memory under anesthesia demands its own vernacular. To begin with, the concepts of awareness under anesthesia and memory of that awareness must be distinguished. After that, explicit memory must be distinguished from implicit memory. Explicit memory exists when a patient remembers and realizes it; implicit memory is any change in thought or action that is attributable to experience outside of explicit memory, that is, not consciously recalled (Schacter 1987). For example, anesthesiologists are very familiar with the patient who appears wide awake in the recovery room, laughing and joking with the nurses, but who denies any memory of the recovery room the next day. If that same patient is given information during this period and is able to test positive for that information the next day, that is evidence of implicit memory. Awareness during surgery with explicit memory of that awareness is intraoperative recall.

HISTORY

- 1845 Horace Wells: Memory and pain with N_2O
- 1846 William Morton: Awareness without memory of pain with diethyl ether

The history of awareness under anesthesia dates back to the introduction of anesthesia to the practice of medicine in the middle of the nineteenth century (Sykes 1960). Horace Wells is given credit for the discovery of nitrous oxide as an anesthetic. However, his first demonstration in 1845 was not well received by the medical community. Dr. Wells, a dentist, performed a tooth extraction after administering nitrous oxide to his patient. However, his patient cried out in pain during the procedure and complained of explicit memory of that pain immediately after the procedure. A year later, William Morton, also a dentist, was much more successful. He used ether to provide anesthesia to a patient who was undergoing the removal of a neck mass. Although the patient appeared to respond to the procedure during the excision, he later reported no memory of any pain. Of course, it is not known whether or not Morton's first patient retained any implicit memory of the event.

- 1959 Cheek, "Careless Conversation"
- 1965 Levinson, "Bogus Crisis"

It was not until over a century after Morton's demonstration that the concept of memory outside of awareness began to be investigated. In 1964 Cheek studied a series of patients who did not do well after anesthesia. With interviews of the operating-room personnel, he was able to ascertain that a number of these patients had been exposed to "careless conversations." These conversations generally contained unflattering references to the patient during the procedure, when it was presumed that the patient was under deep anesthesia and unaware of his or her surroundings. Hypnosis of the patients who did not do well brought out verbatim recollections of these careless conversations (Cheek 1959). This was followed with Levinson's study involving the staging of a "bogus crisis." While his patients were under general anesthesia, he would suddenly announce that the patient was not doing well. Days later he would hypnotize his patients, who were able to report exactly what had been said (Levinson 1965).

- 1985 Bennett, Davis, and Giannini, "Touch your ear!"
- 1987 Goldman, Shah, and Hebden, "Touch your chin!"

Successful intraoperative suggestions for postoperative behavior were demonstrated by Bennett, Davis, and Giannini (1985), who told patients under general anesthesia to "Touch your ear" when the postoperative interview was in progress. The incidence of this behavior was higher than expected in the treatment group. Goldman, Shah, and Hebden (1987) performed a similar study, suggesting to patients to "Touch your chin" postoperatively. This study, too, had positive results.

PSYCHOLOGY

- "Brother, can you paradigm?"

WORD PARADIGMS

- Standardized word pairs
- Free recall
- Cued recall
- Recognition
- Free association

While the anesthesiologists were busy hypnotizing their patients, psychologists were developing different types of word tests to demonstrate implicit memory in patients with various conditions, including amnesic syndrome (Shimamura 1989), posthypnotic amnesia (Kihlstrom 1980), and normal aging (Light and Singh 1987). Word completion tests and free association tests have become the most popular tests of implicit memory. With word completion tests, a patient is played a word, such as "handle" under anesthesia, and then asked postoperatively to complete the word "han...." Free association tests involve playing the patient a series of word pairs composed of stimulus words and their most commonly generated response, as indicated by standard norms. Postoperatively the patient is presented the first word and asked to respond with the first word that comes to mind. A higher frequency of response for the words played compared to a matched list of control words is taken as evidence for implicit memory. This particular paradigm also allows tests for explicit memory. These include free recall, where the patient is asked if he/she remembers any of the words played; cued recall, where the first word is given, and the patient is asked if he/she remembers the response word; and recognition, where the patient is given both words and asked if he/she remembers either of the words.

MEMORY WITH PURE ISOFLURANE

Population: 30 ASA I and II (healthy) patients scheduled for general anesthesia for gynecologic, orthopedic, and general surgery.

Our first study of implicit memory under general anesthesia was designed to correct what we saw as a basic flaw of previous studies. All previous studies had used a menagerie of anesthetic agents with conflicting results. We decided to limit the agents to one—isoflurane—in a study of 30 healthy patients undergoing general anesthesia for gynecologic, orthopedic, and general surgery (Kihlstrom *et al.* 1990).

Isoflurane Protocol

- No preoperative medications
- No benzodiazepines before, during, or after surgery
- Induction with thiopental and vecuronium
- Maintenance with isoflurane, vecuronium, and oxygen
- Morphine (0.05 mg/kg) at last skin stitch and 2 mg IV prn pain in post-anesthesia care unit (PACU)
- Tape played incision to last stitch

No preoperative medications were given, and, specifically, no benzodiazepines were given before, during, or after surgery. Induction of general anesthesia was with thiopental and vecuronium. Maintenance of general anesthesia was with isoflurane up to 2 percent end-expired concentration. Vecuronium was given during the procedures for neuromuscular blockade. Two tapes containing 15 pairs of stimulus/response words matched for frequency of response were used. Each patient heard one of the tapes; the other tape was used as the control list. The experimental tape was played continuously from the initial skin incision until the last skin stitch. At the last stitch morphine (0.05 mg/kg IV) was administered for analgesia as the patient was awakened from the isoflurane anesthetic.

Testing in PACU and at Two Weeks

- Free Recall
- (Free Association)
- Cued Recall
- (Free Association)
- Recognition

Testing for implicit and explicit memory was done when the patient was ready for discharge from the PACU (the recovery room) and again two weeks later by telephone. Free recall, where the patient is simply asked if he or she remembers any words played, was tested first. Then half the patients were tested for free association; the other half were tested for free association after tests for cued recall. For free association, the first word of the word pair was presented, and the patient was asked for the first word that comes to mind. Guessing was encouraged. For cued recall, the first word of the word pair was presented, and the patient was asked for what the second word was. Guessing was discouraged. Recognition, where the patient was shown both words and asked if he/she remembered either, was always the last test administered. Of course, both the experimental and control lists were presented to the patient. Response words in the experimental list were called "targets"; response words in the control list were called "lures." Free association was the test for implicit memory; while free recall, cued recall, and recognition were all tests for explicit memory.

None of the patients had any free recall, cued recall, or recognition. There were no differences between percent of targets and percent of lures elicited. However, during the free association tests, patients were significantly more likely to produce target items from the experimental list, presented during surgery, than they were from the control list (the lures). This was a significant priming effect. Two other interesting results are (1) the priming effect was maintained two weeks later and (2) more priming was observed if free association was presented as the first test (84.6 percent priming versus 50 percent priming).

Sufentanil Protocol

- No preoperative medications
- No benzodiazepines before, during, or after surgery
- Induction with thiopental and vecuronium
- Maintenance with sufentanil, N_2O, vecuronium, and oxygen
- Tape played incision to last stitch

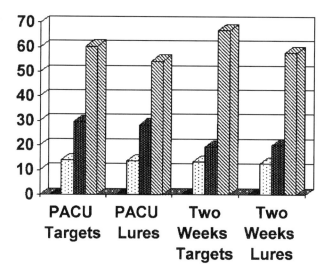

Figure 20.1 Tests of memory of paired associates played during an isoflurane/oxygen anesthetic. Free recall, cued recall, and recognition were not different between targets (paired associates played during anesthesia) and lures (matched controls) both in the Post-Anesthesia Care Unit (PACU) and two weeks later by phone. However, free association was different between targets and lures at both test times ($P<0.05$), and this is evidence for implicit memory.

Impressed (and surprised) with the results of our first study, we forged ahead with an exact repetition of the above protocol, with the exception of using a nitrous-narcotic anesthetic instead of a volatile anesthetic (Cork, Kihlstrom, and Schacter 1992). We decided to use an infusion of sufentanil for our maintenance anesthetic as well as 70 percent N_2O in oxygen. Again, no preoperative medications and no benzodiazepines were used, and induction was with thiopental, while neuromuscular blockade was with vecuronium. The infusion of sufentanil was started with a 0.5 mcg/kg bolus, followed by a rate of 0.5 mcg/kg/min IV.

We found three patients in this group who demonstrated explicit memory of words played during anesthesia. In contrast to our results with isoflurane, however, we found no evidence of implicit memory with this anesthetic technique.

Propofol Protocol

- No preoperative medications
- No benzodiazepines before, during, or after surgery
- Induction with propofol 0.5 mg/kg, fentanyl 1 mcg/kg
- Maintenance with propofol 50 mcg/kg/min plus propofol 30 mg prn
- Tape played once at last skin stitch

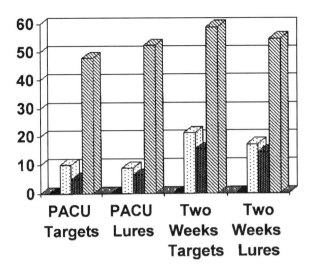

Figure 20.2 Tests of memory of paired associates played during a sufentanil/nitrous oxide/oxygen anesthetic. Free recall, cued recall, recognition, and free association were all not significantly different between targets and lures, indicating that implicit memory is not preserved with this technique.

An important critique of these two studies was that we had no measurement of depth of anesthesia. The fact that such a measurement does not exist did not negate the critique. We realized this and decided to change the protocol. Instead of studying patients receiving general anesthesia, we studied patients undergoing conscious sedation for procedures done under local or regional anesthesia (Cork *et al.* 1994). Thus, we could talk to the patients and roughly gauge their level of sedation by the response. As with our previous studies, no preoperative medications and no benzodiazepines were administered. Patients were given an initial bolus of propofol 0.5 mg/kg/min and fentanyl 1 mcg/kg IV, and a maintenance infusion of propofol 50 mcg/kg/min IV during the procedure. At the last stitch, the experimental list of word pairs was played to the patient one time only. After one hour in the recovery room, patients were tested for free recall, free association, cued recall, and recognition.

Of the 36 patients studied, five demonstrated free recall of target words. For the remaining 31 patients, cued recall and recognition tests showed no evidence of priming. However, the free association tests demonstrated significant priming. Thus, in the absence of explicit memory, implicit memory persists during intraoperative sedation with propofol.

Good or Bad?

- Bad outcome from overheard comments
- Good outcome from therapeutic suggestion

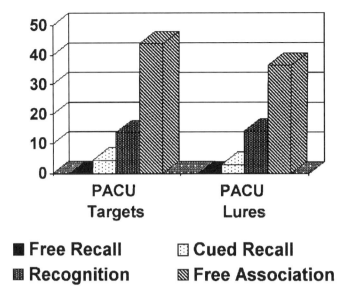

Figure 20.3 Tests of memory of paired associates played during conscious sedation with propofol. Free recall, cued recall, and recognition were not significantly different between targets and lures. However, free association was different between targets and lures ($P<0.05$), evidence for implicit memory.

Anesthesiologists by nature want to ablate all intraoperative memory, implicit or explicit. The answer to the question of whether implicit memory is good or bad is, of course, both. That "careless conversation" or derogatory comments can affect an apparently unconscious patient is certainly a matter of concern regarding bad outcome. However, a number of studies have demonstrated that therapeutic suggestions given intraoperatively may have beneficial postoperative effect.

DISCUSSION

It has been widely assumed that the main difficulty with this type of research is that we have no specific measure of anesthetic depth. For our first two studies of general anesthesia, for instance, it would have been very interesting to see if those patients who exhibited implicit or explicit memory were at a lighter level of anesthesia. We believe, however, that not only depth of anesthesia is important, but susceptibility to implicit or explicit memory is also important. It is quite possible that there is a subset of the human population that may be prone to retention of implicit and/or explicit memory under anesthesia.

REFERENCES

Bennett, H. L., H. S. Davis, and J. A. Giannini. 1985. Non-verbal response to intraoperative conversation. *British Journal of Anaesthesia* 57:174–79.

Cheek, D. B. 1959. Unconscious perception of meaningful sounds during surgical anesthesia as revealed under hypnosis. *American Journal of Clinical Hypnosis* 1:101–13.

Cork, R. C., J. F. Kihlstrom, and D. L. Schacter. 1992. Absence of explicit and implicit memory with sufentanil/nitrous oxide. *Anesthesiology* 76:892–98.

Cork, R. C., K. Friley, J. Heaton, C. Campbell, and J. Kihlstrom. 1994. Implicit memory during propofol sedation. *Anesthesiology* 81:A245.

Goldman, L., M. V. Shah, and M. W. Hebden. 1987. Memory of cardiac anaesthesia: Psychological sequelae in cardiac patients of intraoperative suggestion and operating room conversation. *Anaesthesia* 42:596–603.

Kihlstrom, J. F. 1980. Posthypnotic amnesia for recently learned material: Interactions with "episodic" and "semantic" memory. *Cognitive Psychology* 12:227–51.

Kihlstrom, J. F., D. L. Schacter, R. C. Cork, C. A. Hurt, and S. E. Behr. 1990. Implicit and explicit memory following surgical anesthesia. *Psychological Science* 1:303–06.

Levinson, B. W. 1965. States of awareness during general anesthesia. *British Journal of Anaesthesia* 37:544–46.

Light, L. L., and A. Singh. 1987. Implicit and explicit memory in young and older adults. *Journal of Experimental Psychology\Learning, Memory, and Cognition* 13:531–41.

Schacter, D. L. 1987. Implicit memory: History and current status. *Journal of Experimental Psychology/Learning, Memory, and Cognition* 13:501–18.

Shimamura, A. P. 1989. Disorders of memory: The cognitive science perspective. In *Handbook of neuropsychology*, Vol 3., edited by F. Boller and J. Grafman. Amsterdam: Elsevier, pp 35–73.

Sykes, W. S. 1960. *Essays on the first hundred years of anaesthesia*, Vol 1. Edinburgh: Churchill Livingstone.

21 Experimental Evidence for a Synchronization of Sensory Information to Conscious Experience

Mikael Bergenheim, Håkan Johansson, Brittmarie Granlund, and Jonas Pedersen

INTRODUCTION

For the understanding of consciousness, it is essential to increase our knowledge about the representation of time in the brain. This matter has been subject to extensive theoretical discussions. Recently, Dennet and Kinsbourne (1992) posed the question "How . . . does the brain ensure central simultaneity of representation for distally simultaneous stimuli?" They presented a model (the Multiple Drafts model) that provides a possible answer to this question. However, to date no experimental evidence is available showing that the brain actually does ensure central simultaneity.

In a situation where two stimuli are delivered simultaneously to skin areas with different distances to the somatosensory cortex, there are two major perspectives on the subjective experience of temporal order. One perspective could be named the "order of arrival" theory. Here the order in which different stimuli arrives to the brain will be the subjectively experienced order. Hence, according to this theory, central simultaneity is not ensured. It seems that Libet and his coworkers support this view, when they state that the subjective conscious experience of a sensory stimulus is "referred backwards in time" to the occurrence of an early evoked potential in the somatosensory cortex (Libet, Wright, Feinstein, and Pearl 1979). This implies that the order of arrival alone would determine the subjectively experienced temporal order of two stimuli.

Another perspective, assuming that the brain does ensure central simultaneity, is represented by the Multiple Drafts model (Dennett and Kinsbourne 1992). According to this model a "draft" is constructed in the brain when a sensory stimulus arrives. When a subsequent stimulus arrives, the draft is modified to include information on the second stimulus. These authors introduce the concept of "property K," which separates conscious from unconscious events in the brain. As long as the second stimulus arrives before the first draft achieves property K, and thus becomes "conscious," the two stimuli will be subjectively experienced as simultaneous.

The aim of the present study was first to investigate to what extent the subjects could judge the order of two nearly simultaneous stimuli delivered at different distances from the brain, and if one of the theories described above could explain satisfactorily the obtained results. Second, the study was an attempt to emphasize the importance of obtaining experimental data for the discussion of consciousness. The study included two experiments: (1) subjective judgment of the temporal order of two tactile stimuli, delivered to skin areas with different distances to the somatosensory cortex, and (2) measurement of reaction times for the abovementioned tactile stimuli. According to the Multiple Drafts model, which states that the brain ensures central simultaneity, one would expect (1) the subjects to judge the two stimuli as simultaneous *only* when the stimuli actually were delivered at the same time, and (2) that the reaction times would not differ significantly. Thus, both the subjects' experience of simultaneity and the reaction times would be independent of differences in conduction time. On the other hand, the order of arrival theory suggests that subjects will experience a distal and a proximal stimulus as simultaneous, only if the distal stimulus is delivered a certain time ahead of the proximal stimulus. Also, one would expect a longer reaction time for the distal stimulus. In other words, differences in conduction times would determine both the time interval necessary for a simultaneous experience of the two stimuli and the difference in reaction times.

METHODS

Stimuli and Apparatus

The sensory stimulus was delivered by a mechanical square-wave tapper, with an amplitude of 0.5 mm and a duration of 2 msc. The tapperhead was blunt and had a circular shape with an area of 0.008 mm^2. The stimulus was experienced as a distinct but not painful tap. The tapper was triggered by a computer. This stimulus was used in both experiments (see below). In the reaction-time experiment (see below) the subjects pulled a handle that was connected to an electromagnetic strain gauge. The force was sampled on a PC-486 with a frequency of 10 kHz and the reaction times were measured on the PC-monitor, using the sampling software.

Procedure

The experiment was conducted on ten healthy subjects, and divided in two major parts: (1) Subjective judging of temporal order of two nearly simultaneous tactile stimuli, delivered to skin areas with different distances to the somatosensory cortex, and (2) Measuring of the reaction time from tactile stimuli on skin areas with different distances to the somatosensory cortex.

The two tappers were tightly fastened on the nondominant side of the body on the upper arm and on the foot (on the arch or the back of the foot). The subjects were comfortably seated in a moderately illuminated screened-off room, and were instructed either to look at the wall in front of them, or to close their eyes. Throughout the whole experiment the subjects wore ear protection. The instructions to the subjects were presented in writing in the beginning of the first experimental session, while in the next two sessions, the instructions were repeated orally. On the whole, the recommendations given by Crabtree and Antrim in "Guidelines for Measuring Reaction Time" (Crabtree and Antrim 1988) was followed as far as they were considered relevant to this experiment. It was always ensured that the two stimuli were subjectively judged as equally strong.

The Subjective Judging of Temporal Order In this experiment the subject received two stimuli, separated by a brief time interval. The stimuli were delivered to two different locations, one on the foot and the other on the upper arm. A total of 17 intervals ranging from 2–88 msc were tested. Both the order of the stimuli and the intervals were randomized. Each pair of stimuli was preceded by an auditory warning signal (one second before the stimuli). Using a forced judgment method (Sternberg and Knoll 1973), the subjects were asked to judge the temporal order of the two stimuli by reporting which one they perceived first. They were instructed to say "arm" when the arm stimuli came first, and to say "foot" when the foot stimuli came first. The subjects had to respond before the next trial (within 5 seconds). The whole experiment consisted of three sessions and during each session the range of intervals was repeated four times. Thus, for every subject, each interval was tested 12 times.

The Measuring of the Reaction Time Here the subjects received only one stimulus at a time, either on the foot or on the arm. They were instructed to pull a handle with their dominant hand as soon as they perceived the stimulus. Each stimulus was preceded by an auditory warning signal. This signal was delivered randomly, between 0.8–1.5 seconds before the stimulus.

The tappers were not shifted, but kept in exactly the same location as in the preceding experiment. Twenty stimuli were delivered to each location in a series. This series was repeated twice with a short break in between. Which of the two locations to be tested first was altered from one experimental session to another. After three sessions, a total number of 120 stimulations had been delivered to the arm, and another 120 stimulations to the foot.

The Differences in Conduction Time

The order of arrival theory assumes that conduction time is important both for subjective experience and for motor control. In the present study, the calculation of the conduction time difference between the two sites of stimulation (upper arm and foot) was based on information available in the literature about the conduction velocity of ascending pathways and of fibers innervating low-threshold mechanoreceptors (see Discussion).

Statistical Methods

Subjective Judging of Temporal Order For each trial, the number of "arm" and "foot" responses were recorded and plotted separately in the same diagram. A line was fitted to each plot by means of a nonlinear regression analysis (Y= ± B × arctan AX + C), using a least-square algorithm. At the intersection of the two fitted lines, the subjects could not determine the actual temporal order more accurately than on a chance level (that is, 50 percent). Around this point of intersection, a 95 percent confidence interval was fitted. This confidence interval was determined by calculating a confidential interval of 68.4 percent around the two fitted regression lines. The point of intersection was considered to represent the subjects' experience of simultaneity.

Measuring of the Reaction Time For each subject and each series of stimuli (N=20), errors (that is, responses triggered by the auditory and not the mechanical stimuli) were excluded, and then the mean reaction times were calculated. Errors averaged on the whole less than 5 percent of the total number of reaction times. Comparisons between the mean reaction times for the arm and the foot respectively, were conducted using the Mann-Whitney two-tailed test and also by a paired *t*-test. Since each mean reaction time was calculated from one series of stimuli, and each subject was exposed to six series of stimuli, a total number of 60 mean values were obtained. Finally, average reaction times were calculated for the arm and foot stimuli.

RESULTS

Subjective Judging of Temporal Order

It was found that the two fitted lines intersected at a point where the foot stimulus was delivered 11.5 msc before the arm stimulus. The 95 percent confidence interval that was plotted around the point of intersection ranged from 6.5 to 17.0 msc. It is worth noting that, even if one assumes that the foot stimulus was conducted with the fastest possible conduction velocity (within a 95 percent confidence interval of the three segment's velocity distribution, see above), the delay would still exceed 17.1 msc. On the other hand, (within the same confidence interval), if the foot stimulus was conducted with the slowest possible conduction velocity, the calculated delay would be as long as 35.21 msc. Thus, in both cases the extreme values would be significantly different from the obtained result.

Measuring of the Reaction Time

The mean reaction times were 151 ± 43.7 msc for the arm stimulus and 165 ± 33.1 msc for the foot stimulus (mean difference 13.5 ± 6.2 msc). Hence, the reaction time for the foot stimulus was significantly longer than for the arm stimulus (Mann-Whitney two-tailed test, $P < 0.001$, Student's paired t-test $P < 0.001$, $N = 120$). The mean difference in reaction time between the foot and the arm stimuli was 13.5 ± 6.2 msc.

DISCUSSION

The Differences in Conduction Time Most likely, the receptors stimulated by our mechanical stimulation are, for the hairy skin of the upper arm, hair receptors and Merkel's receptors and, for the glaborous skin of the foot, Merkel's receptors and Meissner's corpuscles. These receptors are innervated by the same type of axon (A β-fiber) (Light and Perl 1984). The conduction distances from the stimuli locations were measured to the first common point, at the C5 segment, in every subject. After that, the stimuli are conducted along the same paths (Carpenter and Sutin 1983). To the first common point, the arm stimulus is only conducted in A β-fibers. The average distance from the arm stimulus location to the C5 segment in the sample participating in this experiment was 35 cm and the corresponding distance from the foot stimulus location was 155 cm. The resulting difference between these distances was 120 cm. The remaining distance for the foot stimulus until it reaches the common point at the C5 segment, can be divided into three different parts according to their respective conduction velocities (CV): (1) The remaining peripheral nerve to the medulla at the L5-S1 segment (CV: M 55 m/s, range 35–75 m/s; Boyd and Davey 1968). (2) Conduction through the lumbar and the lower thoracic segments of the medulla (CV: M 37.5 m/s, range 30–45 m/s; Macon and Poletti 1982). (3) Conduction through the remaining thoracic segments and the C8 segment (CV: M 75 m/s, range 65–85 m/s; Macon and Poletti 1982). The first part described above represents about 55 percent (66 cm) of the remaining distance to C5, the second 28 percent (33.6 cm), and the third part 17 percent (20.4 cm). As a result, conduction times of 12, 9, and 3 msc, respectively, were calculated, and consequently, the difference in conduction time between the arm and the foot stimuli would be 24 msc.

The results from the two experiments indicate that the postulated 24 msc delay for the foot stimulus (due to the longer conduction time), was somehow compensated. However, it appears that this compensation was not of the order or magnitude hypothesized by Dennet and Kinsbourne (1992). Figure 21.1 illustrates the subjective experience of temporal order at three different interstimulus intervals. At the point of intersection (see statistical methods), in our calculation the subjects could not determine the actual temporal order of the foot and arm stimuli more accurately than by chance. This point represents a situation where the stimulus on the foot was delivered 11.5 msc before the arm stimuli (Figure 21.1 B). Thus, if the arm and foot stimuli were to be judged as simultaneous, the stimulus on the foot would have to be delivered 11.5 msc before the arm stimulus.

According to the "order of arrival" theory (see Introduction), one would expect the experience of simultaneity to correspond to the foot stimulus being delivered 24 msc before the arm stimulus. However, at this

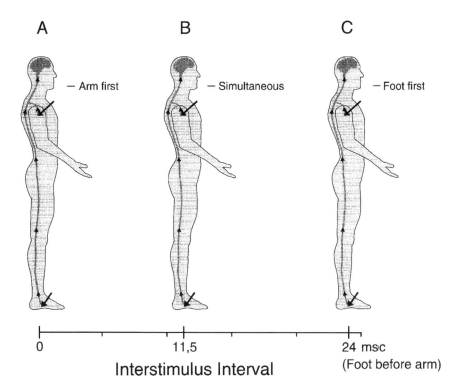

Figure 21.1 This figure describes the subjective experience of temporal order at three different interstimulus intervals: (A) the foot and the arm stimuli delivered simultaneously; (B) the foot stimulus delivered 11.5 msc before the arm, and (C) the foot stimulus delivered 24 msc before the arm.

interstimulus interval the subjects experienced the foot stimulus first (Figure 21.1 C). Therefore, it is quite clear that the "order of arrival" theory cannot sufficiently explain the obtained results. Furthermore, in order for the results to be explained solely by differences in conduction time, it would require axons from low threshold skin mechanoreceptors in the foot with conduction velocities of at least 105 m/s.

On the other hand, the model proposed by Dennett and Kinsbourne (1992) suggests that the experience of simultaneity is independent of differences in conduction time. Accordingly, in this experiment the experience of simultaneity would be found at an interstimulus interval of 0 msc. Interestingly, when the two stimuli were delivered at this interstimulus interval, the subjects experienced the arm stimulus first. Hence, the Multiple Drafts model is not compatible with the results from this experiment. In this context it is also worth noting that in the reaction time experiment it was found that it took significantly longer for the subjects to respond to the foot compared to the arm stimulus. The mean difference between these reaction times was 13.5 ± 6.2 msc.

Again, according to the "order of arrival" theory, one would expect the reaction time for the foot stimulus to be 24 msc longer than the reaction time from the arm stimulus. Proponents of the Multiple Drafts theory would probably not expect the reaction times to differ. In other words, it seems that the results obtained in this experiment could not be fully explained in terms of any of these theories. The results from both experiments suggest that the sensory information is "synchronized" before it reaches a conscious level. However, this synchronization seems to be only partial.

It might be worth mentioning that results obtained by von Békésy (1963) possibly can be regarded as further indications for the existence of a synchronization. However, the author's interpretation does not include the sort of discussion put forward here.

How then can a synchronization be possible? Dennett and Kinsbourne have suggested that "there has to be something, a property K, which distinguishes conscious events from nonconscious events," and they go on to consider a candidate for property K: "A contentful event becomes conscious if and when it becomes part of a temporarily dominant activity in cerebral cortex" (Dennett and Kinsbourne 1992). Also, Libet (Libet *et al.* 1979) estimates that there is a considerable delay (up to 500 msc), before cerebral activity reaches a conscious level.

If it is assumed that there is a varying temporal factor in achieving "property K," then at least three mechanisms may be responsible for the synchronization. First, it is perhaps possible that the arm stimulus is delayed in the process of achieving property K, thus allowing the foot stimulus to "catch up with it." However, we agree with Dennett and Kinsbourne that it would be a great disadvantage for the nervous system to delay any of its information processing more than necessary. Therefore, this explanation seems unlikely (Dennett and Kinsbourne 1992). Second, the foot stimuli may need a shorter time to achieve property K at the cortical level. Finally, and at least theoretically, the process of achieving property K could start peripherally in the first neuron after the receptor. Recently, Dennet and Kinsbourne (1992) have discussed the above mentioned topic in detail, suggesting that the Multiple Drafts theory can account for how the brain ensures central simultaneity. However, the exact mechanisms underlying the apparent synchronization described in this paper remain to be investigated.

In contrast, if there is a constant temporal delay in achieving property K (independent of conduction distances), and CNS is the only place where it can be achieved, then other mechanisms of synchronization have to be considered. In this case, information not only about the where, but also about the when the stimulus occurred would have to be available to consciousness. In the somatosensory homunculus each group of cells represents a where on the body (Penfield and Rasmussen 1950). It is conceivable that each group of cells could also represent a when, in the sense that the local sign provided by the cells in the primary projection areas are preprogrammed with the distances from the CNS. Thus, when activity in response to a stimulus occurs in a certain group of cells on the somatosensory cortex, this activity could be referred not only to a place on the body, but also to the relative time when the stimulus occurred. This would resemble a "temporal homunculus." One must also consider the possibility that property K is not time bound at all. In this case the mechanisms of synchronization would have to be approached quite differently.

If we assent that the results of this study demonstrate a synchronization, there are a number of intriguing questions to be addressed by further research. For example, why is it not complete? One could speculate that an incomplete synchronization might be the result of a compromise between the disadvantage of delaying certain stimuli, and the benefits of a temporally more accurate sensory system. Another possibility is that if the synchronization mechanism is to some extent learned or adaptable, it might just simply reach the degree of precision that is functional for the individual. In the latter case one would expect the degree of synchronization to be variable. If so, it is also conceivable that the degree of physical fitness and coordinative ability are among the factors that could influence the degree of synchronization.

REFERENCES

Boyd, I. A., and M. R. Davey. 1968. *Composition of peripheral nerve*. Edinburgh and London: E & S Livingstone Ltd.

Carpenter, M. B., and J. Sutin. 1983. *Human neuroanatomy*, 8th ed. Baltimore: Williams & Wilkins.

Crabtree, D. A., and L. R. Antrim. 1988. Guidelines for measuring reaction time. *Perceptual and Motor Skills* 66:363–70.

Dennett, D. C., and M. Kinsbourne. 1992. Time and the observer: The where and when of consciousness in the brain. *Behavioral and Brain Sciences* 15:183–201.

Libet, B., E. W. Wright, Jr., B. Feinstein, and D. K. Pearl. 1979. Subjective referral of the timing for a conscious sensory experience. A functional role for the somatosensory specific projection system in man. *Brain* 102:193–224.

Light, A. R., and E. R. Perl. 1984. Peripheral sensory systems. In *Peripheral neuroanatomy*, 2nd ed. edited by P. J. Dyck, P. K. Thomas, E. H. Lambert, and R. Burge. Philadelphia: Saunders, pp. 210–30.

Macon, J. B., and C. E. Poletti. 1982. Conducted somatosensory evoked potentials during spinal surgery. Part 1. Control conduction velocity measurements. *Journal of Neurosurgery* 57:349–53.

Penfield, W., and T. Rasmussen. 1950. *The cerebral cortex of man: A clinical study of localization of function*. New York: Macmillan.

Sternberg, S., and R. L. Knoll. 1973. In *Attention and performance*, edited by S. Kornblum. New York: Academic Press, pp. 629–85.

von Békésy, G. 1963. Interaction of paired sensory stimuli and conduction in peripheral nerves. *Journal of Applied Physiology* 18:1276–84.

22 Positron Emission Tomography, Emotion, and Consciousness

Eric M. Reiman, Richard D. Lane, Geoffrey L. Ahern, Gary E. Schwartz, and Richard J. Davidson

The science fiction writers had it wrong when they designated space the final frontier, for in some respects scientists have had greater access to the outer reaches of the universe than the inner reaches of the human mind and brain. Positron emission tomography (PET) provides a window through which investigators can peer inside the head, measure chemical and physiological processes, and help characterize how the brain is involved in normal and pathological human behaviors. In this chapter, we provide selective reviews of PET, its relationship to other functional brain imaging techniques, and its emerging role in two of the most fundamental but least well understood aspects of human behavior, emotion and consciousness.

POSITRON EMISSION TOMOGRAPHY

PET is an imaging technique that provides quantitative, regional measurements of biochemical and physiological processes. In order to appreciate its role in the study of emotion, consciousness, and other behaviors, it is helpful to know how PET works.

PET studies require the use of a positron-emitting radiotracer, a pharmacological or physiological compound that is labeled with a positron-emitting radioisotope. The radiotracer is typically synthesized on-site using a cyclotron and radiochemistry techniques, rapidly transported to the imaging suite, and administered intravenously or by inhalation. Since it is possible to develop a seemingly unlimited number of radiotracers, PET has the potential to make a wide variety of measurements; however, each tracer must have special characteristics to provide the biochemical or physiological measurement of interest. Since PET requires a cyclotron and radiochemistry laboratory, PET is expensive (typically about $2,000 per scanning session) and limited in its availability. Although the radiation exposure and risks of PET studies are very low, they preclude the scientific study of children or pregnant women and limit the number of scans that can be acquired in each individual.

Our laboratory makes extensive use of ^{15}O-labeled water to provide images of regional cerebral blood flow (CBF). Since ^{15}O decays quickly, it is possible to acquire multiple, sequential images in each individual during a single scanning session; thus subjects can serve as their own controls as they are studied, for instance, before and during an experimentally generated emotion. Since blood flow measurements can be acquired within one minute, it is possible to study individuals during a relatively brief or uncomfortable state, such as rapid eye movement (REM) sleep.

After the radiotracer is administered, a PET imaging system is used to provide multislice images of its distribution in the brain. Any imaging system can be characterized in terms of its sensitivity (its ability to detect processes that occur in small concentrations), its spatial resolution (its ability to distinguish between processes that are close together), and its temporal resolution (its ability to acquire images quickly). Since PET imaging systems have unparalleled sensitivity, they are especially well suited for the study of neurotransmitter and neuroreceptor processes, which occur in minute concentrations. Since PET produces blurry images, it is difficult to study processes that occur in very small regions such as brain stem nuclei; for this and other reasons, PET has limited statistical power and negative findings should be interpreted with caution. Since it takes time to acquire an image, it is difficult to study rapidly fluctuating or fleeting behaviors; in brain mapping studies, PET is limited in its ability to characterize the sequence of changes in implicated regions.

Images of radioactivity are transformed into biochemical or physiological measurements using a tracer-kinetic model, mathematical equations that attempt to account for the tracer's behaviors in the body. The validity of biochemical and physiological measurements depends on this model and its underlying assumptions. Investigators continue to develop and empirically test radiotracers and tracer-kinetic models and, thus, provide an increasing number of apparently valid measurements.

For instance, PET can measure characteristics of an increasing number of neurotransmitters and neuroreceptors, the chemical messenger processes that permit neurons to communicate with each other. These measurements are now being used to investigate the pathophysiology and treatment of psychiatric and certain neurologic disorders, the chemical effects of different drugs, and the interactions among neurotransmitter systems. They have not yet been used to study behavioral states such as attention, arousal, or REM sleep, perhaps in part because of the need to maintain a stable psychobiological state for the duration of the procedure.

PET can also measure regional CBF and glucose utilization. Since these measurements reflect the activity of local neurons, they can be used to investigate regions of the brain that are involved in normal human behaviors, such as aspects of perception, attention, language, memory, motor control, and as discussed below, emotion and consciousness. They can also be used to investigate regions of the brain that are involved in the development and treatment of psychiatric and certain neurological disorders, such as major depression, panic disorder, schizophrenia, and Alzheimer's disease. This chapter places special emphasis on the application of CBF measurements to the study of emotion and consciousness.

In order to capitalize on the ability of PET and other functional brain imaging techniques, such as functional magnetic resonance imaging and magnetoencephalography, to investigate regions of the brain that are involved in the study of emotion, consciousness, and other behaviors, researchers must be prepared to ask interesting, important, and hopefully testable questions and attend to details in experimental design. Some of these features are illustrated in our PET studies of human emotion.

EMOTION

Researchers have long sought to characterize regions of the brain that are involved in normal human emotion, including those which are involved in the emotional evaluation of incoming stimuli, those which are involved in the emotional expressions that can be measured by an outside observer, and those which are involved in emotional experience known to the individual by introspection and typically inferred from his or her verbal report.

In their classic studies of laboratory rats, Joseph LeDoux and his colleagues (1987) monitored the effects of discrete brain lesions on classically conditioned fear to a pure tone or flashing light. They found that a projection from a sensory relay station in the thalamus to the amygdala is involved in the evaluation process that labels simple sensory stimuli with emotional significance, that a projection from the amygdala to the hypothalamus participates in the autonomic expression of fear (in this case, an increase in mean arterial pressure), and that a projection from the amygdala to or through the midbrain participates in the behavioral expression of fear (in this case, freezing). Observing that lesions of primary sensory cortex failed to attenuate the classically conditioned fear, they raised the possibility that classically conditioned fear is unrelated to the conscious experience of the stimulus!

Although these and other lesion, stimulation, and neural tract tracing studies in laboratory animals continue to provide important information about the neural substrates of emotion, animal studies are limited by anatomical differences among species, the restricted range of emotions that can be studied in animals (how does one study happiness in a rat?), problems studying the emotional response to complex sensory and sensory-independent cognitive stimuli (how does one ask a rat to recall a sad experience?), and problems making inferences about the conscious experience of emotion based on an animal's observed behavior.

Using electrical stimulation and depth electrode studies in neurosurgical patients with epilepsy, Gloor and colleagues (1982) demonstrated that the amygdala, hippocampal formation, parahippocampal gyrus, and temporal poles are associated with the conscious experience of ictal fear. Although these studies provide new information about the neural substrates of emotional experience, it is difficult to establish their relevance to the normal brain, normal emotions, and normal emotional stimuli.

We recently used PET to characterize the neural substrates of normal human emotion, investigate their relationship to the type of emotional stimulus, and provide clues about their role in the dissectable components of emotion. We sought to generate potent target emotions, measure these emotions, and control for aspects of the emotion-generating task that are unrelated to emotion.

In order to generate potent emotions in the PET laboratory, we recruited female volunteers who could "accurately describe [their] emotional reactions to daily events"; we identified 12 healthy subjects who reported intense experiences of the three target emotions described below within the previous six months, and had self-ratings of intense target emotions in response to an independent set of trial silent film clips.

PET was used to provide 12 measurements of regional CBF in each subject as she alternated between emotion-generating and control film and recall tasks. Three silent films were used to generate the subjectively, facially, and electrophysiologically well-characterized emotions happiness, sadness, and disgust; three additional film tasks were used to control for aspects of the film task unrelated to emotion, such as emotionally irrelevant visual stimulation and eye movements. Scripts of three recent experiences were used to generate the same target emotions; scripts of three other recent experiences were used to control for aspects of the recall task unrelated to emotion, such as emotionally irrelevant recall memory and visual imagery.

During the PET session, visual analog rating scales, quantitative electroencephalography, and records of heart rate, blood pressure, and electrodermal activity were used in an effort to provide experiential and expressive measurements of emotion. There were potent and comparable increases in the self-rating of each target emotion during the film and recall tasks. Analysis of the automated algorithms were used to compute t-score maps of significant blood flow increases during film-generated emotion, those during recall-generated emotion, and those that distinguished film- from recall-generated emotion.

Film- and recall-generated emotion were each associated with significant blood flow increases in the vicinity of the thalamus and medial prefrontal cortex (Brodmann's area 9). Film-generated emotion was distinguished from recall-generated emotion by significant and significantly greater symmetrical blood flow increases in the vicinity of occipito-temporal and anterior temporal cortex, the amygdala, hippocampal formation, hypothalamus, and lateral cerebellum. Recall-generated emotion was distinguished from film-generated emotion by significantly greater blood flow increases in the vicinity of anterior insular cortex or claustrum.

Since thalamic and prefrontal regions were implicated in both film- and recall-generated emotion, they appear to be involved in aspects of emotion that are unrelated to the type of emotional stimulus. Since different thalamic nuclei have been implicated in classically conditioned fear to simple sensory stimuli (LeDoux 1987), sham rage (Bard 1928), and attentional engagement (Posner and Petersen 1990), the thalamus could be involved in evaluative, expressive, or experiential aspects of emotion. Unfortunately limitations in image noise, spatial resolution, and anatomical localization prevent us from specifying the thalamic region that accounts for the observed blood flow increases.

The medial prefrontal region (Figure 22.1) could be involved in the conscious experience of emotion. Kihlstrom (1987) postulates that the difference that makes for consciousness is the connection between cognitive or perceptual processes and an integrated representation of the self that resides in working memory; if, like dorsolateral prefrontal cortex, this region is involved in aspects of working memory, it could participate in the conscious experience of emotion. A strategy for investigating the role of this and other regions in the integrated representation of self is described in the next section.

Alternatively, the medial prefrontal region could be involved in restraint from otherwise unbridled expressions of emotion. While studies of sham rage suggest that cerebral cortex exerts this role, they did not specify which parts of cerebral cortex are involved. In the rat, lesions of medial prefrontal cortex delay the extinction—in a sense, the appropriate suppression or

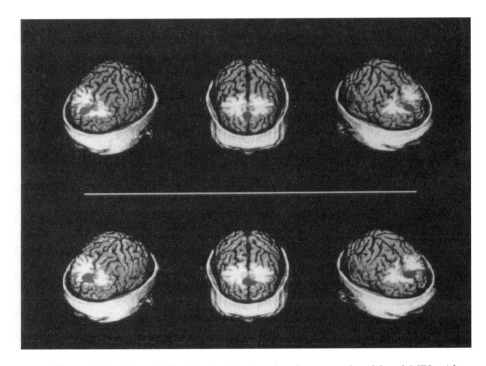

Figure 22.1 A spatially standardized and volume–rendered head MRI with a wedge removed from the frontal lobe to show significant increases in medial prefrontal blood flow during film–generated emotion (top row) and recall-generated emotion (bottom row). As indicated in a recent neuroanthropological study of Phineas Gage (Damasio *et al.* 1994), lesions in this region appear to be associated with socially inappropriate expressions of emotion.

attenuation—of classically conditioned fear (Morgan, Romanski, and LeDoux 1993).

Perhaps the most compelling evidence supporting involvement of medial prefrontal cortex in emotion comes from a recent study of the well-known patient Phineas Gage.

In 1848, Phineas Gage was involved in an accident that caused a solid iron pipe to be propelled through his face, skull, and brain, and into the air. Despite the injury, he remained neurologically well except, it seems, for an inability to hold a job due to socially inappropriate expressions of emotion. In a recent neuroanthropological evaluation of Gage's skull and a clinical evaluation of 12 patients with a similar impairment in the processing of emotion, Hanna Damasio and her colleagues (1994) relate this problem to a lesion in the same region of medial prefrontal cortex implicated in this study.

It would be interesting to compare these 12 patients to control patients who have lesions in other areas in terms of their subjective and facial response to the silent film clips. If the medial prefrontal region is involved in both the conscious experience of emotion and restraint from excessive behavioral expressions, these patients should have a dissociation with significantly smaller increases in self-ratings of emotion and significantly greater facial expressions compared to our control subjects!

Our study of film-generated emotion supports the long-held view that the amygdala, hippocampal formation, and hypothalamus are involved in the generation of emotion. The amygdala and hippocampus are limbic structures, heteromodal sensory association areas that could be involved in the emotional evaluation of exteroceptive sensory; the hypothalamus has been implicated in autonomic and behavioral expressions of emotion. However, our study suggests that these structures are less involved in the emotional response to sensory-independent cognitive stimuli.

This study demonstrates the importance of studying interactions between emotion and the type of emotion stimulus, including simple and complex exteroceptive sensory stimuli, interoceptive sensory stimuli, and sensory-independent cognitive stimuli. It also illustrates the emerging role of PET and other functional brain imaging techniques in the study of human emotion, one which complements, not replaces, other neuroscientific methods. It raises the need to study normal volunteers, brain-injured patients, and laboratory animals using standardized elicitors and measurements of emotion. Finally, it presents the opportunity to study the conscious experience of emotion.

CONSCIOUSNESS

Despite the philosophical, semantic, and scientific challenges involved in understanding the nature of consciousness, PET and other functional brain imaging techniques promise to improve our understanding of this, the most fundamental yet mysterious feature of our lives.

As previously noted, we are now considering ways to test John Kihlstrom's theory that the difference that makes for consciousness is the connection between cognitive or perceptual information and an integrated representation of self that resides in the working memory. One of many possible strategies is to investigate the neuroanatomical correlates of self-awareness by manipulating attention to the representation of self. For instance, subjects could watch adjectives on a computer monitor, and in one condition relate the adjectives to themselves, and in another condition relate the adjectives to someone else. The challenge would be to control for potentially confounding features of the task, such as sensory stimulation, motor response, anxiety, and effort.

Speaking of anxiety, we are now studying patients with and without the elicitation of social phobic anxiety. In one condition, the patients with an intense fear of public speaking are left alone to sing the alphabet song knowing that no one else is in the room; in the other condition, they perform the same task knowing that they are surrounded by observers. Although this study is likely to produce systematic increases in self-awareness, it is also likely to introduce confounding effects of alterations in anxiety and task performance. It is also unclear how the externalized representation of self elicited in our selective attention and social phobia studies is related to the internalized representation that may participate in consciousness.

In a study that controlled for the visual stimulus, motor response, and task instructions, Corbetta and colleagues (1990) demonstrated that they could characterize the neuroanatomical correlates of selective attention to the color, shape, and velocity of moving objects. This study encourages us about the chance to identify neuroanatomical correlates of selective attention to another feature—oneself. It is also relevant to an important feature of consciousness—focal versus peripheral awareness.

Another way to investigate the neuroanatomical correlates of consciousness is to compare tasks that dissociate conscious from nonconscious processes, such as explicit and implicit memory. Explicit memory refers conscious retrieval of previous information, as demonstrated on standard recall and recognition tasks. In contrast, implicit memory refers to the retrieval of previous information on tests that do not make explicit reference to, or require conscious recollection of, a prior study episode; implicit memory is behaviorally manifested in priming, a facilitation in the response to previous information. Studies consistently demonstrate that priming effects on implicit memory tasks can be dissociated from explicit memory for the prior occurrence of target information. For example, brain-damaged patients with an amnestic disorder show normal priming effects on the perceptual identification task (and other implicit memory tasks, noted in the methods section) even though they are severely impaired on explicit recognition tasks. Schacter (1990) proposed that priming effects on implicit memory tasks depend on a class of presemantic perceptual representation systems that are involved in the representation and retrieval of information about the form and structure, but not the meaning, of words and objects. In view of experimental evidence that priming effects occur independently of semantic processing, and depend crucially on form and structure information, it is hypothesized, following Schacter, that perceptual representation systems are involved in priming effects on various implicit memory tasks.

Following an earlier study of implicit and explicit memory for words, we recently identified the neuroanatomical correlates of implicit and explicit memory for structurally possible and impossible visually presented objects (Schacter et al. 1995). Explicit memory is associated with increased blood flow in the hippocampus and parahippocampal gyrus; implicit memory was associated with increased blood flow in occipital and temporal visual association areas; implicit memory areas were less active during recognition memory tasks than priming tasks; and forced choice decisions are associated with increased blood flow in dorsolateral prefrontal cortex. These studies promise to shed new light on the conscious and nonconscious mental operations and neural systems that are involved in human memory.

Implicit and explicit perception refers to the dissociation between nonconscious and conscious perception. For instance, patients with selective lesions of primary visual cortex can develop blindsight: they may insist that they are not aware of a visually presented object, but demonstrate nonconscious processing by their performance on forced choice decision tasks (for example, is it up or down?). Primate studies suggest that blindsight is attributable to projections from visual relay stations in the thalamus directly to secondary visual association areas, enabling them to process some features of the stimulus without being aware of it.

In PET studies, however, our laboratory has decided to study a different form of explicit and implicit perception: hypnotic blindness. We are now developing a paradigm in which we will induce hypnosis in carefully screened, highly suggestible volunteers in the PET laboratory and study them during standard visual and baseline stimuli with and without the suggestion that they are unable to see the object. These and other comparisons promise to shed light on the neuroanatomical correlates of explicit perception and the operations that are involved in hypnosis.

In elegant studies, Kosslyn and colleagues (1993) demonstrated even greater blood flow increases in Brodmann's area 17 of visual cortex during a visual imagery than visual perception, which could be attributable to top-down processing. This study provides the foundation for investigating the many mental operations and related neural processes that are involved in this conscious task. In addition to their role in the study of consciousness as awareness, PET and other functional brain imaging techniques are already implicating regions like dorsolateral prefrontal cortex in the executive control system that formulates, executes, monitors, and revises a task.

Finally PET promises to provide new information about consciousness as wakefulness. Researchers have begun to investigate how the brain is affected during the circadian cycle, different sleep stages, and sleep deprivation. For instance, they report that in comparison to the waking state, whole brain glucose utilization is markedly reduced during stage 2 sleep and the same or slightly increased during REM sleep (Maquet *et al.* 1990). Researchers have not yet applied ^{15}O-water studies to investigate sleep. This technique could help them study relatively transient stages such as sleep onset, REM sleep, and lucid and nonlucid dreaming; it would permit them to make repeated measurements during a single scanning session. PET promises to provide new information about the kinds of information that the brain processes during sleep, anesthesia (Bennett 1993), and coma. For instance, it would be interesting to know if patients in anesthesia activate pain areas or if patients in coma activate areas that might be associated with the recognition of familiar voices. Indeed, functional brain imaging techniques have the potential to provide biological markers of consciousness, thus providing greater access to conscious operations in nonverbal patients and nonhuman species.

When one considers the challenges involved in the study of emotion and consciousness, it is understandable why researchers have taken so long to consider or study these behaviors. When one considers the central role of emotion and consciousness in our lives, it is time to address these challenges. Functional brain imaging techniques like PET have great promise in these endeavors.

REFERENCES

Bard, P. A. 1928. Diencephalic mechanism for the expression of rage with special reference to the sympathetic nervous system. *American Journal of Physiology* 84:490–515.

Bennett, H. L. 1993. The mind during surgery: The uncertain effects of anesthesia. *Advances, The Journal of Mind-Body Health* 9:5–16.

Corbetta, M., F. M. Miezin, S. S. G. L. Dobmeyer, and S. E. Petersen. 1990. Attentional modulation of neural processing of shape, color, and velocity in humans. *Science* 248:1556–59.

Damasio, H., T. Grabowski, R. Frank, A. Galaburda, and A. Damasio. 1994. The return of Phineas Gage: Clues about the brain from the skull of a famous patient. *Science* 264:1102–05.

Gloor, P., A. Olivier, L. F. Quesney, F. Andermann, and S. Horowitz. 1982. The role of the limbic system in experiential phenomena of temporal lobe epilepsy. *Annals of Neurology* 12:129–44.

Kihlstrom, J. F. 1987. The cognitive unconscious. *Science* 237:1445–52.

Kosslyn, S. M., N. M. Alpert, V. M. Thompson, S. B. Weise, C. F. Chabris, S. E. Hamilton, S. L. Rauch, and F. S. Buonanno. 1993. Visual mental imagery activates topographically organized visual cortes: PET investigations. *Journal of Cognitive Neuroscience* 5:263–87.

LeDoux, J. E. 1987. "Emotion." In *Handbook of physiology, The nervous system V. Higher Cortical Functions of the Brain,* edited by V. B. Mountcastle and F. Plum. Washington, DC: The American Psychological Society, pp. 419–59.

Maquet, P., D. Dive, E. Salmon, B. Sadzot, G. Franco, R. Poirrier, R. von Frenckell, and G. Franck. 1990. Cerebral glucose utilization during sleep-wake cycle in man determined by positron emission tomography and [^{18}F]2-fluoro-2-deoxy-d-glucose method. *Brain Research* 513:136–43.

Morgan, M., L. Romanski, and J. LeDoux. 1993. *Neuroscience Letter* 163:109-13.

Posner, M. I., and S. E. Petersen. 1990. The attention system of the human brain. "In *Annual review of neuroscience,* edited by W. M. Cowan, E. M. Shooter, C. F. Stevens, and R. F. Thompson. Palo Alto, California: Annual Reviews.

Schacter, D. L. 1990. "Perceptual representation systems and implicit memory: Toward a resolution of the multiple memory systems debate." In *Development and neural bases of higher cognitive functions,* edited by A. Diamond. *Annals of the New York Academy of Sciences* 608:543–71.

Schacter, D. L., E. M. Reiman, A. Vecker, M. R. Polster, L. S. Yun, and L. A. Cooper. 1995. Brain regions associated with retrieval of structurally coherent visual information. *Nature* 376:587–90.

23 Dimensional Complexity of Human EEG and Level of Consciousness

Richard C. Watt

BACKGROUND

The electroencephalogram (EEG) provides a record of small electrical fluctuations recorded at the scalp, emanating from the brain. The discovery of EEG (usually attributed to Hans Berger 1929) generated hope that encoded within the EEG scientists would find information reflecting thought content, conscious status, and even some clues as to the nature of consciousness. Since that time the EEG has proven useful for a variety of clinical and research applications, including diagnosis of specific diseases such as epilepsy, characterization of normal/abnormal sleep patterns, identification of tumors and lesions, and correlations between specific EEG patterns and generalized mental states (excited, relaxed, etc.). Some pathways for auditory, visual, and somatosensory processing have also been elucidated with the help of evoked responses (observed in the recorded EEG). However, the EEG has not provided much insight regarding aspects of consciousness such as memory storage and retrieval, generation of creative ideation, or the experience of consciousness itself.

Anesthesia provides a unique opportunity to investigate brain dynamics as consciousness is altered (at light doses) and ablated (at high doses) during anesthesia. Although the brain is the target organ of anesthesia, the EEG is not routinely monitored during anesthetic procedures. This is due primarily to the difficulty of interpreting changes in the complex EEG waveform with respect to anesthetic conditions.

Intraoperative monitoring of the EEG can be used to help assess brain integrity, and has been used as a crude index for depth of anesthesia (level of awareness or lack thereof) during surgical procedures (Hameroff *et al.* 1995). However, reliable detection and interpretation of subtle changes in anesthetic depth remains an elusive goal. Intraoperative EEG is commonly analyzed and displayed using time to frequency domain transformations. Univariate spectral descriptors such as mean frequency and spectral edge (the frequency below which 95 percent of the EEG power occurs) change inconsistently among various anesthetics and have shown little clinical value for fine tuning anesthetic depth.

Table 23.1 Summary of calculated dimensional complexity for all simulated signals and EEG data.

Signal	Theoretical Dimensionality	Calculated Dimensional Complexity
Single 10-Hz sine wave	1.0	1.0
Two sine waves combined (integral frequency multiples; 1 Hz, 20 Hz)	2.0	1.9
Two sine waves combined (nonintegral frequency multiples; 10 Hz, 13.7 Hz)	>2.0	2.2
Three sine waves combined	>2.0	2.4
Human EEG at 2.0 MAC	?	3.3
Human EEG at 1.0 MAC	?	4.4
Human EEG awake, relaxed	?	4.9
Random noise	∞	∞

Alternatively, dynamic systems can be characterized on the order/chaos spectrum by assessing underlying deterministic qualities. This can be accomplished by considering an n-dimensional hyperspace into which the variable set $F(t), F(t+T), \ldots F[t+(n-1)T]$ can be mapped. In the case of EEG, F is the amplitude, t is the time, and T is a fixed time interval or phase lag, and R is the hyperspace dimension into which the signal $F(t)$ is embedded. A sequence of such states followed in time defines a curve, the phase space trajectory. As time grows a system whose dynamics are reducible to a set of deterministic laws, an ordered system reaches a permanent state indicated by the convergence of families of phase space trajectories toward a subset of the phase space, the "attractor." Several approaches to estimating dimensionality for such a time series embedded in phase space have been proposed (Farmer 1983). The correlation dimension developed by Grassberger and Procaccia (1983) can be used to calculate a lower bound on dimensionality by considering the correlation function C(R) of the attractor in "n" dimensional phase space.

$$C(R) = \frac{1}{K} \sum_{Y=1}^{K} \frac{1}{N} \sum_{X=1}^{N} H(R - |F_x - F_y|)$$

where N is the total number of data points in an epoch of digitized EEG, R is the radius of an orbit in phase space, F_X is the n dimensional vector described above, and H is the Heavyside function, $H(x)=0$ if $xH(x)=1$ if $x>0$. C(R) is used to count the number of data points F_Y within a radius R. This calculation is performed using $K = N$ reference points over a range of values for n and R, thus C(R) measures the extent to which the presence of one data point affects the position of the other data points. If the attractor is a line $C(R)$ should be directly proportional to R; if it is a surface, $C(R)$ should be proportional to R^2, in general $C(R)$ should be proportional to R raised to the D power for an attractor of dimension D. Therefore, D is equal to the slope of a line fitted to log $C(R)$ versus log R for small values of R. If D versus n becomes saturated beyond some value of n, the saturation value D is considered a lower bound on the dimensionality of the attractor. The calculated value of D is quantitatively affected by many of the variables used in this process (number of reference points K, length of sample EEG epoch, digitization rate, and phase lag T). Therefore, the application of Grassberger-Procaccia methods to EEG data does not yield an estimate of correlation dimension that is accurate as an absolute value. For this reason, the term dimensional complexity will be used in this study. It is regarded as a relative measure useful for comparing identically treated data (Pritchard and Duke 1990).

This investigation of EEG under anesthetic sleep conditions was conducted to evaluate dimensional complexity as an EEG-derived variable that might reflect complex brain dynamical states, and provide an index for changes in anesthetic depth.

METHODS

With informed consent and institutional Human Subjects Committee approval (and as part of a larger study of anesthetic effects), ten human subjects were studied at three anesthetic levels (1.0, 1.5, 2.0 MAC) where 1.0 minimum alveolar concentration (MAC for gas phase anesthetics) is a normalized dose for light anesthesia. EEG signals were obtained from five surface electrodes placed in the frontal and occipital regions of the left and right cerebral hemispheres. This signal was amplified and filtered (fourth order band-pass, 1–40 Hz) and then recorded on a Hewlett-Packard 3964A FM tape recorder. Each of the concentration levels was maintained for at least 50 minutes. For each subject, at each anesthetic level, ten 16-second epochs of EEG were digitized at 256 Hz and analyzed according to the methods of Grassberger and Procaccia (1983). The correlation dimension $C(R)$ was calculated for 50 values of R (the attractor probe); this was performed for embedding dimensions of $n = 1,2,3 \ldots 8$. The slope value $D = \log C(R)/\log R$ was calculated as an average value/scaling region chosen for maximal linearity (see Figures 23.1, 23.2, and 23.3). This implementation of the Grassberger-Procaccia method was also tested on simulated signals of varying complexity using 16-second epochs digitized at 256 Hz (identical to the EEG sample size). The dimensional complexity algorithm was applied to the following signals: single 10-Hz sine wave; a summation of two sine waves at 1 Hz and 20 Hz; a summation of two sine waves at 10 Hz and 13.7 Hz; a summation of three sine waves at 1.5, 5.9, and 9.2 Hz; human EEG from a relaxed awake subject, human EEG at three anesthetic levels, and a random noise signal.

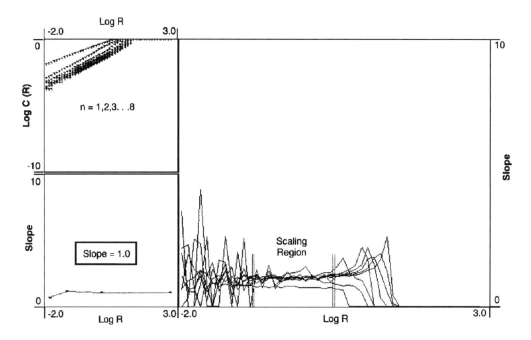

Figure 23.1 Dimensional complexity calculation for a 10-Hz sine wave (4,096 points at 256 Hz). The implementation of the Grassberger and Procaccia method is summarized graphically above. The upper–left window is a log/log plot of the correlation dimension calculated for 50 values of R for embedding dimensions of $n=1, 2, 3\ldots 8$. The right window displays the log of R versus slope (derived from the upper–left log/log plot). An appropriate scaling region is chosen from this plot. The lower–left window displays the average slope value versus log of R through the scaling region. The asymptotic slope value is the calculated dimensional complexity D.

RESULTS

Dimensional calculation for the single 10-Hz sine wave signal is shown in Figure 23.1. The saturation of $D = \log C(R)/\log R$ occurred as embedding dimension n increased above R=2, resulting in a calculated slope or dimensionality of 1.0. The calculated dimensionality for a summation of three sine waves is shown in Figure 23.2. Figure 23.3 exemplifies the lack of convergence when the technique is applied to a noise signal. Figure 23.4 summarizes the results from all signals (EEG as well as simulated test data) analyzed for dimensional complexity. The calculated and theoretical dimensionality for each signal is shown in Table 23.1. Figure 23.5 shows the tendency for dimensionality to decrease as anesthetic depth increases. Group statistics showed a significant change from 1.0 to 1.5 MAC ($P < 0.001$). On an individual basis, dimensionality calculations were highly stable for each subject at each anesthetic level. There were no exceptions to the inverse relationship between dimensionality and anesthetic level. The large standard deviations for the group statistics reflect intersubject variability. For seven of the ten subjects, dimensionality decreased dramatically from

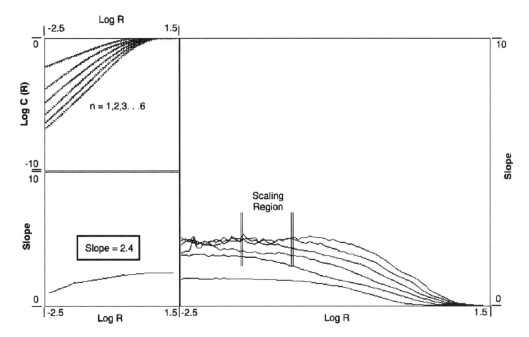

Figure 23.2 Dimensional complexity calculation for three sine waves (1.3, 5.9, 9.2 Hz;) 4,096 points at 256 Hz). The same process described in Figure 23.1 yields a higher dimensional complexity (2.4) calculated for this more complex signal.

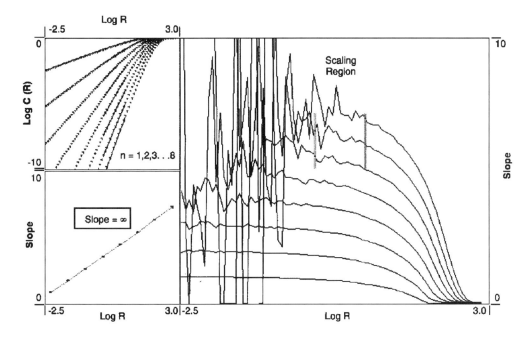

Figure 23.3 Dimensional complexity calculation for random noise (4,096 points at 256 Hz). When the Grassberger and Procaccia process is applied to a random noise signal an asymptotic slope convergence region is not found, confirming the general validity of this algorithmic approach.

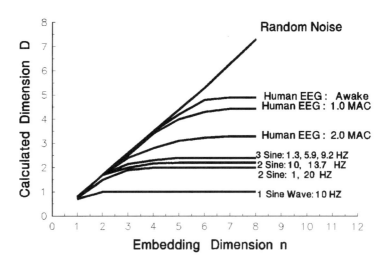

Figure 23.4 Correlation dimension D versus embedding dimension n for all signals (EEG as well as simulated test data). MAC=minimum alveolar concentration, the dose at which (roughly) 50 percent of patients are anesthetized.

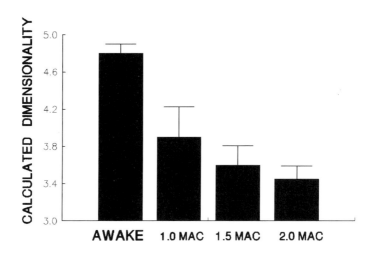

Figure 23.5 EEG dimensionality versus minimum alveolar concentration (MAC; anesthetic level). The decrease in calculated dimensionality is clearly demonstrated for human subjects as anesthetic concentrations increase. By comparison, the calculated dimensional complexity for awake human subjects is much higher.

1.0 to 1.5 MAC and remained essentially unchanged from 1.5 to 2.0 MAC. Three of the ten subjects exhibited little or no dimensionality change from 1.0 to 1.5 MAC, but showed a profound decrease in dimensionality from 1.5 to 2.0 MAC. No clinical observations (including cardiovascular and other neurologic monitoring data) suggested any differences between these three subjects and the rest of the group.

DISCUSSION

The dimensional complexity algorithm worked well on the simulated test data. Calculated dimensional complexity for the 10-Hz sine wave was 1.0 which is in perfect agreement with the theoretical dimensionality of a single sine wave. The calculated dimensional complexity of 1.9 for the two sine waves (which were integral frequencies 1, 20 Hz) compares well with the theoretical dimensionality of a torus in phase space (2.0). As expected, two sine waves of incommensurate frequencies results in a more complex torus experimentally calculated at 2.2. The experimentally calculated dimensionality for the random noise signal was indeterminate because no saturation value was reached. The Grassberger-Procaccia correlation function is based on the theoretical limit as N (number of sample points) approaches infinity. Therefore the larger the epoch of sample data being analyzed, the more accurate the dimensionality calculation should be. Other variables in the Grassberger-Procaccia approach such as the phase lag chosen for phase space transformation of the data, and choice of scaling region chosen for calculation of slope (D) in each dimension, can affect the calculated dimensional complexity D. Theoretically the ideal sample epoch should be as long as possible, but limited by the length of time the signal (EEG) is stationary. The stationarity of EEG signals is not known and probably varies with circumstance. The choice of 16-second epochs (4,096 points) was based primarily on the trade-off between stability of the calculation and computational time required for the process. Since all of the sample data was processed in exactly the same way, the calculated dimensional complexity should be treated as a relative number and as such is a valid approach to comparing grouped data. Previous investigations of human EEG have yielded estimates of dimensionality ranging from 2.09 (epileptic waveform)[3] to 9.7 (awake, eyes open). Dimensionality has been shown to be lower for sleep state EEG than for the awake state[4]. Comparison of estimated dimensionality reported by various researchers is difficult because of the methodological variations used. In particular, as sample size (N) and embedding dimension (n) are increased estimated dimensionality usually goes up. Therefore using such results to draw conclusions regarding the absolute dimensionality or degrees of freedom required to model specific brain activities is premature. However, regarding the calculated dimension as a relative measure of complexity is valid given the high stability of the calculation for each subject at each anesthetic level. At clinical anesthetic levels, there is a tendency for dimensionality to decrease as anesthetic depth increases. A sufficiently large dose of anesthetic can cause the EEG to become isoelectric, ($D=0$). The intersubject variability described in the results section has several implications. First, dimensional complexity probably varies nonlinearly with respect to anesthetic dosing. However, in order to elucidate this relationship with greater detail a protocol involving many more than three anesthetic dosing levels will be required. Secondly, the differences between the two subject groups (seven out of ten and three out of ten) did not correlate with other cardiovascular and neurologic variables. This supports the contention that dimensional complexity is indeed a variable with different information content than those provided by traditional spectral analysis. Third, the changes in dimensional complexity suggest that dimensional analysis may reflect aspects of consciousness that are separate from desired anesthetic effects. Dimensional complexity may prove important in the classification of brain activity and may have clinical utility as a diagnostic tool. The dimensionality calculation used in this study was performed off-line, requiring far more CPU time than power spectral analysis. Improvements in hardware and software approaches could permit "real-time" dimensionality calculations in future EEG monitoring equipment. The combination of conventional power spectral analysis and dimensional complexity may enhance automated EEG interpretation.

Does the EEG contain yet undiscovered information that may yield clues about the nature of consciousness itself? Intriguing studies of brain electrical activity (particularly at 40 Hz) may provide a physiologic underpinning for conceptual models of consciousness (the binding problem) (Hardcastle 1996, Koch 1996). High frequency EEG activity (between 40–100 Hz) remains poorly studied due to extremely difficult signal-to-noise problems. Vast amounts of EEG data are routinely collected during conventional diagnostic procedures. Relatively little of this data is stored in a suitable format or otherwise made available for analysis by those studying consciousness. Conventional power spectral analysis has evolved in such a way that emergent standards permit comparison of various researchers efforts. Standardization of analytical approaches used to investigate chaotic dynamics of EEG (and other signals) will undoubtedly evolve as such approaches become more widely applied, permitting direct comparison of various research efforts.

ACKNOWLEDGMENTS

I am grateful to Ansel Kanemoto for his ability, dedication, and his enthusiasm; and Christina Springfield for her contribution and data collection. I am also appreciative of Kenneth Mylrea for his continued collaboration and discussion.

REFERENCES

Babloyantz, A., and A. Destexhe. 1986. Low-dimensional chaos in an instance of epilepsy. *Proceedings of the National Academy of Science* 83:3513–17.

Babloyantz, A., J. M. Salazar, and C. Nicolis. 1985. Evidence of chaotic dynamics of brain activity during the sleep cycle. *Physics Letters* 111A:152–56.

Berger, R. J. 1969. The sleep and dream cycle. In *Sleep, physiology & pathology*, edited by A. Kales. Lippincott: Philadelphia, pp. 17–32.

Farmer, J. D., A. Ott, and J. A. Yorke. 1983. The dimension of chaotic attractors, *Physica 7D:* 153–80.

Grassberger, P., and I. Procaccia. 1983. *Phys Rev Lett* 50:346.

Hardcastle, G. V. 1996. "The Binding Problem and Neurobiological Oscillations." In *Toward a science of consciousness: The first Tucson discussions and debates*, edited by S. R. Hameroff, A. W. Kaszniak, and A. C. Scott. Cambridge, MA: MIT Press, pp. 51–65.

Hameroff, S. R., J. S. Polson and R. C. Watt. 1995. "Monitoring anesthetic depth." In *Monitoring in anesthesia and critical care medicine*, 3rd ed., edited by C. D. Blitt. and R. L. Hines, New York: Churchill Livingstone.

Holzfuss, J., and G. Mayer-Kress. 1985. An approach to error estimation in the application of dimension algorithms. *Proceedings of the Center for Nonlinear Studies Workshop "Dimensions and Entropies in Chaotic Systems,"* Los Alamos, New Mexico, September 11–16.

Koch, C. 1996. "Towards the neuronal substrate of visual consciousness. In *Toward a science of consciousness: The first Tuscon discussions and debates,* edited by S. R. Hameroff, A. W. Kaszniak, and A. C. Scott. Cambridge, MA: MIT Press, pp. 247–57.

Pritchard, W. S., and D. W. Duke. 1990. Dimensional analysis of no-task human EEG using the Grassberger–Procaccia method. *Psychophysiology*

Xu Nan, X. J. 1988. *Bulletin of Mathematical Biology* 50, 5:559–65.

24 Collapse of a Quantum Field May Affect Brain Function

C. M. H. Nunn, C. J. S. Clarke, and B. H. Blott

INTRODUCTION

Most scientists nowadays think that consciousness is an outcome of the sorts of nerve cell activity that have been familiar for the last half century (that is, complex patterns of activation dependent on depolarization waves and neurotransmitter release). Hofstadter (1979), Edelman (1989), and Crick and Koch (1992) are particularly well-known exponents of this notion. There are also people (Burns 1990) who consider that this is only half the story and that quantum physics could be relevant to the remaining half. In the last decade some of them have come up with ideas sufficiently concrete to be, in principle, testable (for example, Penrose 1989, Marshall 1989).

To show that consciousness can be fully understood only in terms of quantum theory, one must demonstrate that it behaves in ways that have no equivalent in the world of classical physics. The holism and aspects of the nonlocality that appear in quantum theory provide an obvious starting point for devising tests. Main impediments to practical experimentation are that collapse of the Schrödinger wave function is said to be induced by "measurement," which is an extremely vague concept, while no one knows for sure what effects consciousness has on brain function.

Penrose's (1989) theory regards consciousness as associated with the stopping criterion of, or read-out mechanism for, quantum computation. It therefore involves collapse of a quantum field. He also proposed that collapse of any such field occurs (in the familiar random manner) when one of the possibilities in superposition comes to differ from the others by one graviton. In the brain, the one graviton criterion depends mainly on different ion shifts across cell membranes and is not reached until quite extensive brain areas are involved in alternative possibilities.

An implication of Penrose's two ideas is that, if one were to augment the mass (gravitational) consequences of alternative brain states, collapse would occur sooner, affecting any brain activity associated with consciousness. We decided to test this implication as it is both clear and technically simple to refute.

METHODS AND RESULTS

The series of experiments depended on the facts that (1) people use specific brain areas to undertake particular tasks, and (2) using an area alters the electroencephalographic (EEG) activity recorded from it. If one envisages the neural representations of two possibilities in superposition: I need to press this button and I don't need to press this button, there will be different EEG waveforms in superposition when the EEG is being recorded from brain areas relevant to a decision. Hence, different EEG pen positions will also be in superposition and, since pens are heavy relative to ions shifting across nerve cell membranes, the Penrose collapse criterion will be exceeded sooner than usual with consequences for speed or accuracy of button pressing. In other words, taking an EEG from a relevant brain area ought, on the Penrose hypotheses, to affect performance of tasks involving conscious decision making.

Expecting to refute such a seemingly ridiculous conclusion, we devised a simple test in which right-handed volunteers pressed a button with their right thumbs in response to some (2, 5, or 8) out of a series of numbers (0–9) flashed at random onto a screen. This task mainly involved use of the left side of the brain. We tried to ensure that subjects' right brains were kept busy on something different by playing music to their left ears only. Meanwhile EEGs were taken in random sequence from the left or right sides of their heads with a control condition of no recording. The outcome measures were proportion of missed target numbers, proportion of incorrect hits on nontargets, and average reaction times (for hits on targets). To refute the hypotheses, we would have needed to find no significant differences between outcome measures during the different EEG recording conditions.

Twenty-eight subjects were tested in this pilot experiment (results from two of them were not used as their error proportion exceeded the mean by more than two standard deviations; their minds were clearly not on the test). Any simple view of Penrose's hypotheses and the functions of consciousness leads to the expectation that an effect of EEG recording condition should be most obvious in relation to reaction times. As it turned out neither these nor hits on nontargets showed interesting differences, but target misses were apparently influenced by EEG recording condition. There were fewer misses ($P = 0.01$) when the EEG was being taken from the right motor cortex (which was responsible for task performance), but more misses ($P = 0.02$) when it was taken from other areas on the right.

This surprise finding warranted further work; the experiment was fully automated and the physical setup changed so that the experimenter had no way of influencing subjects under test. Instead of having to respond to certain numbers, subjects were asked to respond to particular words out of a small vocabulary. They were also given a test intended to be a mirror image of the word test as far as use of brain areas was concerned; they had to respond with their left hands to certain patterns out of a small library shown in random order while a recorded lecture was played to their right ear. Two types of control experiment were done; first, everything as normal except that the EEG pen motors were switched off; second, as standard except that each pair of EEG leads was fed to all EEG channels (instead of having one channel per pair of leads). Both types of control checked on the possibility that programming errors were responsible for positive findings in the standard experiment. The first also checked whether EEG pen movement, rather than electrical variables or relay closure, was important. The second type was done in case acoustic feedback to subjects gave them subconscious cues.

Forty-two subjects altogether were tested and all results were used. During control experiments there were no significant differences in performance dependent on EEG recording condition. In standard experiments, subjects had fewer target misses when the EEG was being taken from the left during the word test ($P = 0.05$; the recording was from the region of the motor cortex); during the pattern test, there were more hits on nontargets ($P = 0.01$) during left-sided recordings, that is, error frequency was reversed when doing the "mirror image" test. Twenty further subjects were then tested in order to ascertain whether increasing the complexity of the word and pattern tasks, and hence the average reaction times, would allow detection of significant differences in these times. When tasks involving reaction times approximately twice those of the original tests were given, no significant differences were seen on any outcome measure.

The next variable to be examined was EEG frequency band. Hitherto, the machine had been set to amplify the widest possible frequency range as a catch-all. One effect of this was to introduce a lot of noise into pen movements. The word test was abandoned (because it had a rather poor letter quality), and subjects were given a simple pattern task on which the average reaction time was 590 msec, about 30 msec faster than on the original pattern task. The left EEG was always amplified in the 30- to 60-Hz range, while a variable bandpass filter was installed on the right and batches of seven subjects were tested while amplifying six different right EEG frequency bands (42 subjects altogether). All results were used.

With this new setup, individual subjects often showed "significant" performance differences dependent on EEG recording condition; it was no longer necessary to sum the results of batches of subjects in order to find significance. However, apparently significant findings were not always on the same outcome measure or in the same direction. It was therefore necessary to use binomial probability distributions to discover how meaningful these "significant" findings really were. When applied to all findings from all 42 subjects, this method showed a probability of 0.01 that the "significant" results were there by chance. Target misses were once more the most sensitive measure ($P = 0.001$). When recording in the alpha band results were significant at the 0.002 level in relation to all three outcome measures, in neighboring bands at the 0.3 or 0.1 levels, while the 35- to 45-Hz band results had a 0.7 probability of being there through chance.

CONCLUSIONS

The results support Penrose's (1989) hypotheses as they suggested that conscious decision making is affected by inducing collapse of a quantum field involving the brain, while control experiments indicated that the crucial variable affecting collapse was the correlation of EEG pen movement with activity in brain areas used to perform a task. Although it appears unlikely that the findings were due to experimental artefact, it is difficult to give an overall figure for the probability that they might have been due to chance because, in accord with the exploratory nature of the work, we conducted a series of slightly different tests. The number of tests which used an electrode placement over the right motor cortex and the first batch of word tests were, however, very similar (the average reaction times on the two were almost identical). The combined probability of the findings concerning target misses is, on the face of it, 0.0005 but this is too small because of the lack of significant results from the other two outcome measures. A figure of the order of 0.002 would be more reasonable. Results from the final batch of tests, when it was possible to calculate the effect on probabilities of the nonsignificant findings, give grounds for additional confidence.

We cannot claim to have found strong support for Penrose's gravitational theory as it may be possible to explain our results on other ideas, especially those due to Bohm (1983), which avoid the concept of collapse (Ian Marshall, personal communication). All the same, the findings do strongly suggest that consciousness and quantum field events are interrelated. But why, given the quite large number of people tested, were the results not more definite and why was no obvious effect seen on reaction time?

Our methodology was probably too loose to get the best out of the concept. For instance, different subjects may have been using different strategies for task performance; if some people were using awareness to not respond to nontargets for example, while others used it to respond to targets, it's hardly surprising that our results should have been fuzzy. Then there was some evidence that EEG electrode position in relation to brain areas being used for task performance was crucial, but we had no means of standardizing this. Even our assumption that only the contralateral motor cortex is active during hand movement has recently been shown to be questionable (David, Blamire, and Breiter 1994).

The observation that the EEG alpha rhythm is apparently more closely correlated with field collapse than other EEG frequencies fits traditional notions about the association of alpha activity with changes in awareness. The lack of any effect at 35 to 45 Hz may indicate that the 40 Hz correlated activity thought to be important in relation to preception (Barinaga 1990) is relevant to unconscious functions only.

The influence of the setup on target misses, combined with its lack of any obvious effect on reaction times, may be telling us something interesting about the functions of consciousness. Libet's (1993) work has shown that consciousness has only about 150 msc in which to influence any voluntary act involving the reaction times found in most of our tests, but the methodology would have been sensitive to average differences of as little as 10 msc if these had occurred. Perhaps consciousness can influence the probability of occurrence of some crucial neural happening but not its speed. Entry of a particular pattern of activity into a global workspace (Baars 1993) is a plausible candidate.

The lack of significant results when people were given more difficult tasks could be explained in many different ways, but there is a particularly interesting possibility. Brain blood flow changes take around half a second to occur in response to use of an area (David, Blamire, and Breiter 1994), and should probably be regarded as able to influence wave function collapse on the Penrose hypothesis. Maybe, when a task took too long, potential or actual blood flow changes literally outweighed any influence due to EEG pen movement! This raises an additional question about whether one should also take into account alternative bodily movements (for example, button pressing) in relation to field collapse. These could be regarded as setting a severe time limit on the duration of any brain superposition. The EEG might thus be envisaged as able to influence the location (in the absence of significant local blood flow changes) but not the timing of field collapse. If our method does eventually prove valid, there is certainly a lot of scope for future experimentation.

ACKNOWLEDGMENT

A fuller account of this work, including a detailed mathematical treatment of its rationale, is published under the same title in the first issue of *The Journal of Consciousness Studies* 1(1):127–39.

REFERENCES

Baars, B. J. 1993. "How does a serial, integrated and very limited stream of consciousness emerge from a nervous system that is mostly unconscious, distributed parallel and of enormous capacity?" In *Experimental and Theoretical Studies of Consciousness,* edited by G. R. Bok and J. Marsh. CIBA Symposium No. 174. Chichester: Wiley, pp. 123–37.

Barinaga, M. 1990. The mind revealed? *Science* 249:856–58.

Bohm, D. 1983. *Wholeness and the implicate order*. London: Ark.

Burns, J. E. 1990. Contemporary models of consciousness. *Journal of Mind and Behaviour* 11:(2) 153–72.

Crick, F., and C. Koch, 1992. The problem of consciousness. *Scientific American* September, 111–17.

David, A., A. Blamire, and H. Breiter. 1994. Functional magnetic resonance imaging. *British Journal of Psychiatry* 164:2–7.

Edelman, G. M. 1989. *The remembered present*. New York: Basic Books.

Hofstadter, D. R. 1979. *Godel, Escher, Bach: An eternal golden braid*. New York: Penguin Books.

Libet, B. 1993. "The neural time factor in conscious and unconscious experience." In *Experimental and theoretical studies of consciousness,* edited by G. R. Bok and J. Marsh. CIBA Symposium No. 174. Chichester: Wiley, pp. 123–37.

Marshall, I. N. 1989. Consciousness and Bose-Einstein condensates. *New Ideas in Psychology* 7:(1) 73–83.

Penrose, R. 1989. *The emperor's new mind.* Oxford, England: Oxford University Press.

25 Neural Time Factors in Conscious and Unconscious Mental Functions

Benjamin Libet

We have, for some decades, been studying the cerebral physiology of conscious or subjective experience. Our approach has been an experimental one, attempting to discover which neural activities are involved in producing a conscious event, and how these events may differ from neural events that mediate unconscious (nonconscious) mental events (Libet 1965).

The aim was to specify the neural activities that are uniquely sufficient to produce a conscious experience, when they are added to the immensely complex background of brain activities that provide the necessary conditions in which this can occur. That is, we studied the changes in neural requirements at the threshold of transition to a conscious experience.

Electrophysiological stimulation and recording with electrodes, placed intracranially on the somatosensory cortex and in its subcortical afferent pathways by our neurosurgeon colleague (the late Bertram Feinstein), made it possible to study simple sensory experiences elicited by this cerebral system in awake human subjects. The operational criterion of a conscious experience was the subject's introspective report of a somatic sensation (a "raw" feel). These simple, usually near-threshold sensory experiences could be tested for reliability, in relation to applied stimulus quantities that were objectively knowable by the investigator.

THE CEREBRAL TIME REQUIREMENT

Electrical stimulation of somatosensory (SI) cortex to elicit a sensory experience allows one to determine some neuronal requirements at this level, bypassing the effects of the various ascending projections that are normally activated by a stimulus to the skin.

It became evident that duration (of the train of repetitive brief electrical pulses) was a critical factor in producing a conscious sensory response to stimulation of sensory cortex (see Libet 1973 review). A surprisingly long stimulus duration of about 500 msec was required when the peak intensity of the pulses was at the liminal level for producing any conscious sensation (see Figure 25.1). Changes in frequency of the pulses, electrode contact area or polarity, and so forth did not affect the requirement of a long duration. Nor was this requirement unique to locating the stimulus on the surface of

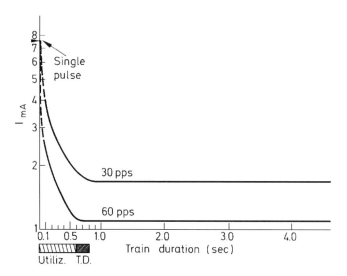

Figure 25.1 Temporal requirement for stimulation of the somatosensory (SI) cortex in human subjects. The curves are schematic representations of measurements in five different individuals, which were not sufficient in each case to produce a full curve individually. Each point on a curve indicates the combination of intensity (I) and train duration (TD) for repetitive pulses that is just adequate to elicit a threshold conscious sensory experience. Separate curves shown for stimulation at 30 pulse per sound (pps) and 60 pps. At the liminal I (below which no sensation was elicited even with long TDs), a similar minimum TD of about 600 msc +100 msc was required for either pulse frequency. Such values for this "utilization TD" have since been confirmed in many ambulatory subjects with electrodes chronically implanted over SI cortex and in the ventrobasal thalamus (from Libet 1973).

cerebral cortex. The same temporal relationship was found with stimulating electrodes placed at any point in the cerebral portion of the direct ascending sensory pathway that leads up to the somatosensory cortex, that is, in subcortical white matter, or in the sensory relay nucleus in ventrobasal thalamus, or in the medial lemniscus bundle that ascends from nuclei in the medulla to the thalamic nucleus. However, the requirement does not hold for the spinal cord below the medulla, or indeed for peripheral sensory nerves from the skin itself; there, a single pulse can be effective for sensation even at liminal intensity (Libet et al. 1967).

Must a single pulse stimulus to the skin also induce prolonged neural responses of cerebral cortex in order to elicit a conscious sensation? The answer to this had to be more indirect, but at least three lines of evidence have been convincingly affirmative on it. These will be listed here without all the experimental details.

1. Skin stimuli that were too weak to evoke the appropriate later components of event-related potentials at the cerebral cortex, but could still evoke a primary response, did not elicit any conscious sensation. Pharmacological agents (atropine or general anesthetics) that depress these late components also depress or abolish conscious sensory responses.

2. The sensation induced by a single stimulus pulse to the skin can be enhanced retroactively by a stimulus train applied to SI cortex, even when the cortical stimulus begins 400 msc or more after the skin pulse (Libet 1978, Libet et al. 1992). This indicates that the content of a sensory experience can be altered while the experience is "developing" during a roughly 500 msc period before it appears.

3. Reaction times to a peripheral stimulus were found to jump discontinuously, from about 250 msc up to more than 600–700 msc, when subjects were asked deliberately to lengthen their reaction time by the smallest possible amount (Jensen 1979). This surprising result can be explained by assuming one must first become aware of the stimulus signal in order to delay one's response deliberately; if up to 500 msc is required to develop that awareness, then the reaction time cannot be increased deliberately by lesser amounts.

SUBJECTIVE REFERRAL BACKWARDS IN TIME: ANTEDATING

If there is a substantial neural delay required before achieving a sensory experience or awareness, is there a corresponding delay in the subjective timing of that experience?

In a direct test of this issue we paired a skin stimulus (single pulse) with a cortical stimulus train of pulses (at liminal intensity, 500 msc duration required), and asked the subject to report which of these stimulus-induced sensations came first. Subjects in fact reported that the skin-induced sensation came first even when the skin pulse was applied a few hundred msc after the onset of the cortical train (Figure 25.2). That is, there appeared to be no subjective delay for the skin-induced sensation relative to the delayed cortically induced sensation. The clue to solving this paradox lay in the differences between responses to SI cortical stimuli and those to skin stimuli. Cortical stimuli, that were subjectively timed as delayed about 500 msc (from onset of the pulse train), did not elicit any primary evoked potentials at SI; but each skin pulse does produce this component (10–20 msc delay), as well as later components out to 500 msc or more.

This led us to propose that (1) the primary evoked neural response acts as a timing signal, and (2) there is a subjective referral of the timing of the skin-induced experience, from its actually delayed appearance back to the time of the initial fast response of the cortex (which has only a 10–20 msc, indiscernible delay) (Figure 25.3). The experience would thus be subjectively "antedated" and would seem to the subject to have occurred without any delay. No such antedating would occur with the cortical stimulus since the normal initial or primary response, to a sensory volley ascending from below, is not generated by our surface-cortical stimuli. This hypothesis was tested and confirmed as follows: A skin stimulus was paired with one in the direct subcortical pathway that leads to sensory cortex, instead of stimulating SI cortex as in Figure 25.2 (Libet et al. 1979). Unlike the cortical stimulus, the subcortical one does elicit the initial primary response in the cortex

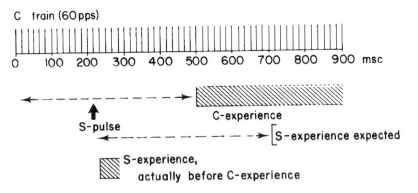

Figure 25.2 Diagram of an experiment on the subjective time order of two sensory experiences, one elicited by a stimulus train to the SI cortex (labelled C) and the one by a threshold pulse to skin (S). CS consisted of repetitive pulses (at 60/sec) applied to the postcentral gyrus, at the lowest (liminal) peak current sufficient to elicit any reportable conscious sensory experience. The sensory experience for CS ("C-experience") would not be initiated before the end of the utilization-train duration (average about 500 msec), but then proceeds without change in its weak subjective intensity for the remainder of the applied liminal CS train (see Libet et al. 1967, Libet 1966, 1973). The S-pulse, at just above threshold strength for eliciting conscious sensory experience, is here shown delivered when the initial 200 msc of the CS train have elapsed (in other experiments, it was applied at other relative times, earlier and later). If S were followed by a roughly similar delay of 500 msc of cortical activity before "neuronal adequacy" is achieved, initiation of S-experience might have also been expected to be delayed until 700 msc of CS had elapsed. In fact, S-experience was reported to appear subjectively before C-experience. For the test of the subjective antedating hypothesis the stimulus train was applied to the medial lemniscus (LM) instead of to somatosensory cortex (CS in the figure). The sensation elicited by the LM stimuli was reported to begin before that from S, in this sequence; this occurred in spite of the empirically demonstrated requirement that the stimulus train in LM must persist for 500 msc here in order to elicit any sensation (see text). (From Libet et al. 1979, by permission of *Brain*.)

with each pulse; but, unlike the skin stimulus, it resembles the cortical one in its requirement of a long (up to 500 msc) train duration of pulses. As predicted by the hypothesis, the subcortically induced sensation was reported to have no delay relative to the skin-induced sensation, even though it was empirically established that the subcortically induced experience could not have appeared before the end of the stimulus train (whether 200 or 500 msc, depending on intensity); see Figure 25.2.

Such subjective referral thus serves to correct the temporal distortion of the real sensory event, a distortion imposed by the cerebral requirements of a neural delay for the experience. An analogous subjective correction occurs for the spatial distortion of the real sensory image, imposed by the spatial representation of the image in the responding neurons of the sensory cortex. Nowhere in the brain is there a response configuration that matches the sensory image as perceived subjectively. Subjective referrals of the timing or the spatial configuration of an experience are clear examples

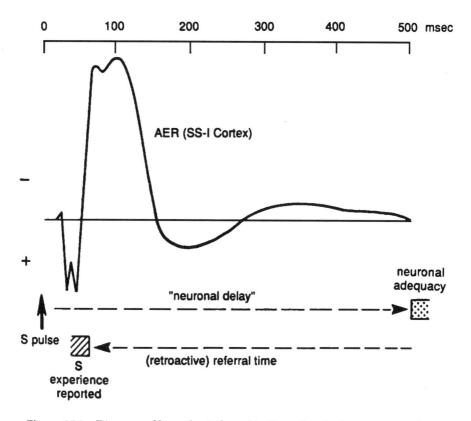

Figure 25.3 Diagram of hypothesis for subjective referral of sensory experience backward in time. The average evoked response (AER) based on 256 averaged responses delivered at 1/sec, recorded at the somatosensory cortex (labelled SS-1 here), was evoked by pulses just suprathreshold for sensation and delivered to the skin of the contralateral hand. Below the AER, the first line shows the approximate delay in achieving the stage of neuronal adequacy that appears (on the basis of other evidence) to be necessary for eliciting the sensory experience. The lower line shows the postulated retroactive referral of the subjective timing of the experience, from the time of neuronal adequacy backward to some time associated with the primary surface-positive component of the evoked potential. The primary component of the AER is relatively highly localized to an area on the contralateral postcentral gyrus in these awake human subjects. The secondary or later components, especially those following the surface-negative component after the initial 100 to 150 msc of the AER, are more widely distributed over the cortex and more variable in form, even when recorded subdurally (see, for example, Libet *et al.* 1972). This diagram is not meant to indicate that the state of neuronal adequacy for eliciting conscious sensation is restricted to neurons in the primary SS-I cortex of postcentral gyrus; on the other hand, the primary component or "timing signal" for retroactive referral of the sensory experience is more closely restricted to this SS-I cortical area. (From Libet *et al.* 1979, by permission of *Brain*.)

of an event in the mental sphere that was not evident in or predictable by a knowledge of the associated neural events.

INITIATION OF A VOLUNTARY ACT

Does an endogenous experience, one arising in the brain without any external inputs, also require a substantial duration of neural activity before the awareness appears? The possibility to investigate this issue had its roots in the discovery by Kornhuber and Deecke (1965 and Deecke, Grotzinger, and Kornhuber 1976) that an electrical change (the "readiness-potential" or RP) is recordable on the head starting up to a second or more prior to a simple "self-paced" movement. Accepting this electrical change as an indicator of brain activity that is involved in the onset of a volitional act, we asked the question: Does the conscious wish or intention to perform that act precede or coincide with the onset of the preparatory brain processes, or does the conscious intention follow that cerebral onset?

We first established that an RP is recordable even in a fully spontaneous voluntary act, with an average onset of about -550 msec, before the first indication of muscle action ("O time"). (The presence of a component of "preplanning" when to act makes RP onset even earlier; this probably accounts for the longer values given by others.) We then devised and tested a method whereby clock-time indicators of the subject's first awareness of his/her wish to move (W) could be obtained reliably. RPs (neural processes) and Ws (the times of conscious intention) were then measured simultaneously in large numbers of these simple voluntary acts (Libet *et al.* 1983).

The results clearly showed that onset of RP precedes W by about 350 msec (Figure 25.4). This means that the brain has begun the specific preparatory processes for the voluntary act well before the subject is even aware of any wish or intention to act; that is, the volitional process must have been initiated unconsciously (or nonconsciously). This sequence is also in accord with the principle of a substantial neural delay in the production of a conscious experience generally.

Is there any role for the conscious function in voluntary action? It is important to note that conscious intention to act (W) does appear about 150–200 msec before the act. Thus, although initiation of the voluntary act is an unconscious function of the brain, the conscious function still appears in time to potentially affect the outcome of that volitional process. That is, the conscious function could potentially either promote the culmination of the process into action, or it could potentially block or veto the final progression to the motor act (Libet 1985).

HOW DOES THE BRAIN DISTINGUISH UNCONSCIOUS FROM CONSCIOUS MENTAL FUNCTIONS?

One possibility suggested itself from our evidence that a minimum activity time (time-on) is required to produce a conscious experience, whether the

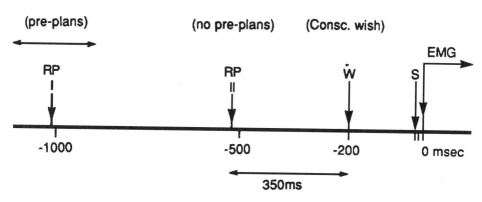

Figure 25.4 Diagram of sequence of events, cerebral and subjective, that precede a fully self-initiated voluntary act. Relative to 0 time, detected in the electromyogram (EMG) of the suddenly activated muscle, the readiness potential (RP) (an indicator of related cerebral neuronal activities) begins first, at about − 1050 msc when some preplanning is reported (RP I) or about − 550 msc with spontaneous acts lacking immediate preplanning (RP II). Subjective awareness of the wish to move (W) appears at about − 200 msc, some 350 msc after onset even of RP II but well before the act (EMG). Subjective timings reported for awareness of the randomly delivered S (skin) stimulus average about − 50 msc relative to actual delivery time. (From Libet 1989, by permission of Cambridge University Press.)

experience is exogenous (sensory) or endogenous (for example a conscious intention to act voluntarily). It seemed attractive to hypothesize that neuronal activation times to produce an unconscious mental function could be less than those for a conscious one. The transition from an unconscious to a conscious function could then be controlled simply by an adequately increased duration of the appropriate neuronal activity. I have called this proposal the "time-on" theory.

A direct experimental test confirmed the theory, at least for the case of somatosensory inputs (Libet *et al.* 1991). Stimulus trains, 72 pps at just above liminal intensity, were delivered to ventrobasal thalamus, with train durations that varied from 0 to 760 msec. With a forced choice test of the subject's ability to indicate the presence of a stimulus, correct responses in 50 percent of trials would be expected on chance alone. Statistical analysis of the thousands of trials in nine subjects showed that correct detection (well above 50 percent) occurred even when subjects were completely unaware of any sensation and were guessing. The mean train duration for all trials in which there was correct-detection-with-no-awareness was compared to the mean train duration for all trials in which there was correct-detection and awareness (of "something," that is, at the uncertain level of awareness). To achieve even an uncertain awareness when being correct required an additional stimulus duration of 375 msec. These results confirmed the "time-on" theory.

The results demonstrate that detection of a signal can be sharply distinguished from awareness of the signal. This distinction is often overlooked in cognitive studies. Strictly speaking, all cognitive experiments in nonhuman animals are studying detection, as no introspective reports or other credible criteria of awareness are available. Since detection with no reportable awareness is clearly possible even in human subjects, one cannot regard detection of a signal by a monkey as valid evidence for the additional feature of awareness of that signal. The results also argue against a view that conscious experience arises when a sufficient level of neuronal complexity is achieved. The degree of complexity in the detections with no awareness was presumably at least as great as that for detections with awareness; the significant distinction was the duration of the appropriate neural activities, not their complexity.

SOME IMPLICATIONS OF "TIME-ON" THEORY

Cerebral Representation

If the transition from an unconscious to a conscious mental function could be dependent simply on a suitable increase in duration of certain neural activities, then both kinds of mental functions could be represented by activity in the same cerebral areas. Such a view would be in accord with the fact that the constituents and processes involved in both functions are basically similar, except for the awareness quality, and with the general view that both types of functions are probably mediated by broadly distributed neural activity. Separate cerebral sites for conscious versus unconscious functions would not be necessary, although this possibility is not excluded.

All-or-Nothing Character of Awareness

If the transition to and production of awareness of a mental event occur relatively sharply, at the time a minimum duration of neuronal activities is achieved, this suggests that an awareness appears in an all-or-nothing manner (Libet 1966). That is, awareness of an event would not appear at the onset of an appropriate series of neural activities and develop gradually. Conscious experience of events, whether initiated exogenously or endogenously, would have a unitary discontinuous quality. This would be opposed to the continuous "stream of conscious" nature postulated by William James and assumed in many present theories of the nature of consciousness; it is, however, in accord with a postulate of unitary nature for mental events adopted by Eccles (1990) as part of his theory for mind-brain interaction.

Filter Function

It is generally accepted that most sensory inputs do not achieve conscious awareness, even though they may lead to meaningful cerebral responses and can, in suitable circumstances (of attention, and so forth) successfully elicit conscious sensation. This "time-on" requirement could provide the basis for screening inputs from awareness, if the only inputs that elicit awareness are those that induce the minimum duration of appropriate activities. Such a requirement could prevent conscious awareness from becoming cluttered and permit awareness to be focused on one or a few events or issues at a time.

Delayed Experience Versus Quick Behavioral Responses

Meaningful behavioral responses to a sensory signal, requiring cognitive and conative processing, can be made within as little as 100–200 msc. Such responses have been measured quantitatively in reaction time tests and are apparent in many kinds of anecdotal observations, from everyday occurrences (as in driving an automobile) to activities in sports (as when a baseball batsman must hit a ball coming at him in a tortuous path at 90 miles per hour). If actual conscious experience of the signal is neurally delayed by hundreds of milliseconds, it follows that these quick behavioral responses are performed unconsciously, with no awareness of the precipitating signal, and that one may (or may not) become conscious of the signal only after the action. Direct experimental support of this was obtained by Taylor and McCloskey (1990), who showed that the reaction time for a visual signal was the same whether the subject reported awareness of the signal or was completely unaware of it (owing to the use of a delayed masking stimulus).

Subjective Timing of Neurally Delayed Experience

Although the experience or awareness of an event appears only after a substantial delay, there would ordinarily be a subjective antedating of its timing back to the initial fast response of the cortex, as discussed above (Libet *et al.* 1979). For example, a competitive runner may start within 100 msc of the starting gun firing, before he is consciously aware of the shot, but would later report having heard the shot before starting.

There is another facet to this issue: for a group of different stimuli, applied synchronously but differing in location, intensity, and modality, there will almost certainly be varying neural delays at the cortex in the times these different experiences appear. This could lead to a subjective temporal jitter for the group of sensations. However, if each of these asynchronously appearing experiences is subjectively antedated to its initial fast cortical response, they would be subjectively timed as being synchronous, without subjective jitter; the differences among their initial fast cortical responses are approximately 10 msc and too small for subjective separation in time.

Unconscious Mental Operations Proceed Speedily

If there is virtually no minimum "time-on" requirement for unconscious (or nonconscious) mental processes in general, then these could proceed quickly, in contrast to conscious events. This feature is obviously advantageous, not only for fast meaningful reactions to sensory signals but also for the more general complex, intuitive, and creative mental processes, many of which are deemed to proceed unconsciously. Conscious evaluation would be expected, according to the theory, to be much slower.

Opportunity for Modulation of a Conscious Experience

It is well known that the content of the introspectively reportable experience of an event may be modified considerably in relation to the content of the actual signal, whether this be an emotionally laden sensory image or endogenous mental event (which may even be fully repressed, in Freud's terms). For a modulating action by the brain to affect the eventual reportable experience, some delay between the initiating event and the appearance of the conscious experience of it seems essential. The "time-on" theory provides a basis for the appropriate delays. We have some direct experimental evidence for such modulatory actions on the awareness of a simple sensory signal from the skin: an appropriate cortical stimulus begun 400 msc or more after the skin pulse could either inhibit or enhance the sensory experience (Libet *et al.* 1972, 1992, Libet 1978, 1982).

Finally, it is important to recognize that the neural time factors in conscious experience could not have been discovered and established without the experimental and falsifiable tests that pitted actual neural activations against the accompanying subjective reports with human subjects. They could not have been discovered by constructing theories based on the previous knowledge of brain processes.

REFERENCES

Deecke, L., B. Grotzinger, and H. H. Kornhuber. 1976. Voluntary finger movement in man, cerebral potentials and theory. *Biological Cybernetics* 23:99–119.

Eccles, J. C. 1990. A unitary hypothesis of mind-brain interaction in the cerebral cortex. *Proc R Soc Lond Ser B Biol Sci* 240:433–51.

Jensen, A. R. 1979. "g": Outmoded theory or unconquered frontier. *Creat Sci & Technol* 2:16–29.

Kornhuber, H. H., and L. Deecke. 1965. Hirnpotentialanderungen bei Willkurbewegungen und passiven Bewegungen des Menschen: Bereitschaftpotential und reafferente Potentiale. *Pfluegers Arch Gesamte Physiol Menschen Tiere* 284:1–17.

Libet, B. 1965. Cortical activation in conscious and unconscious experience. *Perspectives in Biology and Medicine* 9:77–86.

Libet, B. 1966. "Brain stimulation and the threshold of conscious experience." In *Brain and conscious experience*, edited by J. C. Eccles, Berlin: Springer-Verlag, pp. 165–81.

Libet, B. 1973. "Electrical stimulation of cortex in human subjects, and conscious sensory aspects." In *Handbook of sensory physiology* Vol. 2, edited by A. Iggo. New York: Springer-Verlag, pp. 743–90.

Libet, B. 1978. "Neuronal vs. subjective timing for a conscious sensory experience." In *Cerebral correlates of conscious experience,* edited by P. A. Buser and A. Rougeul-Buser. Amsterdam: Elsevier Science Publishers, pp. 69–82.

Libet, B. 1982. Brain stimulation in the study of neuronal functions for conscious sensory experience. *Human Neurobiology* 1:235–42.

Libet, B. 1985. Unconscious cerebral initiative and the role of conscious will in voluntary action. *Behav Brain Sci.* 8:529–66.

Libet, B. 1989. "Conscious subjective experience vs. unconscious mental functions: A theory of the cerebral processes involved." In *Models of brain function,* edited by R. M. J. Cotterill. Cambridge, England: Cambridge University Press.

Libet, B., W. W. Alberts, E. W. Wright, and B. Feinstein. 1967. Responses of human somatosensory cortex to stimuli below threshold for conscious sensation. *Science* 158:1597–1600.

Libet, B., W. W. Alberts, E. W. Wright, and B. Feinstein. 1972. "Cortical and thalamic activation in conscious sensory experience." In *Neurophysiology studied in man,* edited by G. G. Somjen, Amsterdam: Excerpta Medica, pp. 157–68.

Libet, B., E. W. Wright, Jr., B. Feinstein, and D. K. Pearl. 1979. Subjective referral of the timing for a conscious sensory experience: A functional role for the somatosensory specific projection system in man. *Brain* 102:191–222.

Libet, B., C. A. Gleason, E. W. Wright, and D. K. Pearl. 1983. Time of conscious intention to act in relation to onset of cerebral activities (readiness-potential); The unconscious initiation of a freely voluntary act. *Brain* 106:623–42.

Libet, B., D. K. Pearl, D. M. Morledge, C. A. Gleason, Y. Hosobuchi, and N. M. Barbaro. 1991. Control of the transition from sensory detection to sensory awareness in man by the duration of a thalamic stimulus. The cerebral time-on factor. *Brain* 114:1731–57.

Libet, B., E. W. Wright, B. Feinstein, and D. K. Pearl. 1992. Retroactive enhancement of a skin sensation by a delayed cortical stimulus in man: Evidence for delay of a conscious experience. *Consciousness and Cognition* 1:367–75.

Taylor, J. L., and D. I. McCloskey. 1990. Triggering of preprogrammed movements as reactions to masked stimuli. *Journal of Neurophysiology* 63:439–46.

V Neural Networks

In 1943 a publication appeared that opened the way to scientific studies of the dynamics of the human brain. Called "A logical calculus of the ideas immanent in nervous activity" by Warren McCulloch and Walter Pitts, it was based on the algebraic theory of mathematics that had been developed in the 1920s by Alfred North Whitehead and Bertrand Russell. McCulloch and Pitts approximated the neuron as an elementary switch (the "McCulloch–Pitts neuron") and began the study of the behavior of networks of such switches. To put this paper into its proper historical perspective, it is important to recall that the digital computer had not been invented in 1943. Thus the theory of McCulloch and Pitts did not merely suggest a computer metaphor for the brain; it cut the idea from the whole cloth and unleashed a spate of research in *neural network theory* during the 1950s and 1960s that continues to the present day (for example, Scott 1977). This development is based on McCulloch's view that: "both the formal and the final aspects of that activity which we are wont to call *mental* are rigorously deducible from present neurophysiology."

Although the perspectives of neural network theory have extended far beyond the confines of those early days, the first chapter in this section is in the tradition conceived a half century ago. In Chapter 26, John Taylor assumes that the dynamical properties of the brain—and therefore the mind—can be deduced from those of a network of neuronal switches with weights of interconnection that depend on a variety of learning experiences. He proposes that the neural network is organized into modules that are (or can be) in global competition, mediated by interconnections through the *nucleus reticularis thalami*. Basing his description on several of his previously published papers, Taylor argues that the conscious state must involve many independent memories in order to provide the perception with "color." He also suggests the importance of feedback that involves the primary cortex so the imagery associated with a particular perception is preserved. Appealing to both top down (psychological) and bottom-up (neurological) considerations, he obtains a competitive model of consciousness that follows seven stated principles. Although allowing that it is not possible to acquire a person's inner view without becoming that person, he suggests that it may be possible to understand the nature of an inner perspective.

In Chapter 27, Tokiko Yamanoue describes a network model in which each "neuron" exhibits oscillatory behavior, again an idea that goes back to the early days of neural network theory (for example, Greene 1962). Such an oscillatory neural network can exhibit a variety of behaviors including both preattention and the focusing of attention.

Finally, the emergence of memory from a random network of switching circuits is explored in some detail by Andrew Wuensche in Chapter 28. In this study, the individual switching elements are allowed to compute general (Boolean) functions rather than the more restricted repertoire of the threshold gate that was assumed by McCulloch and Pitts. The reason for this relaxed assumption is not merely for analytical convenience; it accounts for the impressive computing power of a single real neuron. Following suggestions of Ross Ashby in 1952 and John Hopfield in 1982, Wuensche views a memory trace as a *basin of attraction* (or stationary state) of the network. Several examples of such basins are presented for relatively simple networks. The memory basin exists in a dimensional space apart from the network dynamics, a relationship that may be analogous to consciousness.

From Wuensche's perspective, a state of consciousness is equated with the desire to learn, reflecting a view of Erwin Schrödinger who asserted in 1958 that: "Consciousness is associated with the learning of the living substance; its *knowing how* is unconscious." Consciousness and learning, however, are clearly not identical. Nonconscious adaptive computers can learn, and patients given amnestic drugs cannot learn or remember, but are quite conscious.

In each of these chapters—to a greater or lesser degree—the authors recognize the importance of a hierarchical organization of the brain's activity. Thus, Taylor considers that long-term memory will involve both "categories and superordinate categories," and Yamanoue considers an oscillator model in the form of a hierarchical cluster, in which envelopes of lower-level clusters are synchronized to the upper-level cluster. The hierarchical aspect of neural organization is most explicitly recognized by Wuensche who views the brain: "as a complex dynamical system made up of many interlinked specialized neural subnetworks," which may "consist of further subcategories of semi-autonomous networks." We shall return to the question of the brain's hierarchical structure in a later section of this book.

REFERENCES

Greene, P. H. 1962. On looking for neural networks and "cell assemblies" that underlie behavior. *Bull. Math. Biophys.* 24:247–75, 395–411.

McCulloch, W. S., and W. H. Pitts. 1943. A logical calculus of the ideas immanent in nervous activity. *Bull. Math. Biophys.* 5:115–33.

Schrödinger, E. 1958/1967. *Mind and matter.* Cambridge, England: Cambridge University Press.

Scott, A. C. 1977. Neurodynamics (a critical survey). *J. Math. Psychology* 15:1–45.

Whitehead, A. N., and B. Russell. 1925. *Principia mathematica.* Cambridge, England: Cambridge University Press.

26 Modeling What It Is Like To Be

John Taylor

INTRODUCTION

Presently the deepest question raised about consciousness is that of determining "What it is like to be X," where X is any living being other than oneself which has the potentiality for consciousness (Nagel 1974). It is this question that has been claimed to be impossible to answer for science (Searle 1991), and has led to the useful distinction between "hard" and "easy" questions about consciousness (Chalmers 1994). The latter are those about the brain mechanisms that appear to support consciousness, and which are being ever more effectively discerned by recent developments in psychology, neuroscience, and pharmacology. Noninvasive brain imaging, lesion studies, and single- and multi-electrode studies on animals and humans have given an enormous wealth of knowledge about such mechanisms. The harder sciences of engineering, physics, and mathematics are also now making important contributions to a theoretical framework for the various micro and ever more global information strategies used.

This progress in solving the easy problems of consciousness is to be contrasted with the clear lack of progress in the hard problems of consciousness. The most important of these is the one raised above by Nagel (1974), where it was claimed that scientific method can never probe the subjective or phenomenal nature of consciousness. That impossibility is said to arise due to the objective, third-person nature of science when opposed to the subjective, first-person character of consciousness. There is a measure of irony in the situation as perceived by many today; consciousness is based solely on physical activity in the brain, but yet cannot be probed by science, which is supposedly the ultimate analyzer of the physical world.

It is the purpose of this paper to present a possible physical model of consciousness that would allow for a solution to Nagel's hard problem. In it will be detailed how this could occur after a presentation of the physical basis of the model. That will be initially at a level of specifying a "boxes" type of information flow in the brain. Parts of the model can then be made more specific in terms of putative neural structures able to perform the functions specified in the boxes model. Some supporting evidence for these structures and their function, as well as for the more global program being carried out, will be brought forward from neurophysiology and psychology. Explanations of deficits and dissociations also appear explicable in the model, and are also shown to be important in guiding its further development. On the basis of the model, and supporting evidence for it, a solution to the hard problem of describing "what it is like to be X" will then be outlined.

Before commencing what is initially a modeling exercise on the physical substructure for supporting consciousness, it is clearly necessary to justify why any approach could ever work. How, using specific features of the physical world, could one ever attempt to answer what appears to be the purely philosophical question of Nagel (1974): how ever to discover, from analysis of the material world, what it is like to be an X? The initial justification is that if science is to build a bridge from the physical world to the apparently nonphysical, mental one, it must do so starting from the physical structures available.

Indeed there is nowhere else from where science can commence; the material world is the only possible domain of scientific analysis. This means that science must attempt to be able to discern how mental features of brain activity may arise from the latter. These mental features must possess seemingly nonphysical attributes. It is through such mental aspects that the beginnings of an answer to Nagel's problem may emerge.

The main thrust of the answer to be developed here is that through the relations between brain activities that mind might emerge. Such relations, to be specified in detail in the next sections, are not in themselves physical. In the same way, relations between numbers (such as less than or greater than), are not themselves numerical. It is this nonphysical essence of relations that opens up the possibility of meeting the philosophers' difficulties over the mind-body problem.

THE RELATIONAL MIND

Relational structures have long been recognized as an integral part of brain and mind. Aristotle proposed over two millenia ago that thinking proceeds by the basic relations of contiguity, similarity and opposites, an idea developed strongly by the associationist schools of psychology in the eighteenth and nineteenth centuries. The manner in which ideas led one into another was considered seriously by the empirical philosophers Locke, Berkeley, and Hume in the seventeenth century. More fundamentally Hume stated in his treatise (Hume 1896).

Mind is nothing but a heap or collection of different perceptions, unified together by certain relations, and suppos'd tho' falsely to be endowed with a perfect simplicity and identity.

Later associationists developed this notion of relations between ideas to that of relations between stimuli and responses or rewards. It is the purpose of this paper to attempt to bring about rapprochement between the later and earlier forms of associationism. In so doing delineation will be attempted of the way in which the relations between ideas corresponds to those between neural representations corresponding to the ideas. Even more fundamentally, more detail will be put into the phrase of Hume's "mind is nothing but a heap or collection of different perceptions, unified together by certain relations, . . ." In particular those relations will be specified in ever greater detail, so as to move towards justifying the basic thesis of the relational mind model (Taylor 1973, 1991):

The conscious content of a mental experience is determined by the evocation and intermingling of suitable past memories evoked (usually unconsciously) by the input giving rise to that experience.

This model goes beyond that of Hume in that the Humean relations between perceptions are now extended to include a range of past experiences entering the relation with a present one. These past experiences need not necessarily have been conscious when they were originally experienced or as they are evoked to enter the relation.

There is increasing support for such a relational model of mind from experiments by psychologists during the last decade. In general there is strong evidence for the thesis that past experiences, however stored, influence present behavior (Kahneman and Miller 1986). Present behavior must influence, or be related to, present consciousness, even if the latter only has a veto effect on developing behavior (Libet 1982). Thus evidence for the influence of past experiences on present responses is supportive of the relational mind model.

An area in which past experiences are used to color consciousness is that of categorization, which appears to be heavily sensitive to the past (Barsalou 1987). Thus if a person is asked to name a bird, if they have been in an urban landscape they may say "robin," whilst if they have just been on a hunting trip they might well say "hawk." Context dependence of prototypes of categories was found quite strong from the studies of Barsalou, and he concluded "The concepts that people use are constructed in working memory from knowledge in long-term memory by a process sensitive to context and recent experience" (Barsalou 1987).

The manner in which past experience alters response in numbers of other situations has also been explored recently. One fertile area has been that of memory illusions. Thus Witherspoon and Allan (1985) had subjects first read a list of words on a computer screen, and later judge the duration of presentation of words presented individually on the screen. Subjects judged the exposure duration as longer for old words (read on the list in the first phase) than new words, although the actual duration was identical. They misattributed their fluent perception of the old words to a difference in duration of presentation. Their conscious experience of the old words had thus been altered by the prior experience of exposure to those words.

A number of similar features, including the well-known "false-fame" effect are recounted in a more recent paper (Kelley and Jacoby 1993). The "false-fame" effect itself (Jacoby and Whitehouse 1989) involves two phases. In the first phase people read a list of nonfamous names; in the second phase these old names were mixed with new famous and nonfamous names in a test of fame judgments. Names that were read earlier were more likely to be judged as famous than were new names; this was especially so if subjects were tested in the second phase under a divided attention condition, preventing the use of conscious recall of the earlier list. Kelley and Jacoby (1993) concluded that "Past experiences affect the perception and interpretation of later events even when a person does not or cannot consciously recollect the relevant experience."

The above are only a brief selection of the many experimental results that support the basic thesis of the relational mind model stated earlier in this section. The model itself was developed in a formal manner initially (Taylor 1973, 1991), and more explicit neural underpinning has been given to it in more recent papers (Taylor 1992a, 1992c, 1992b, 1993a, and 1993b).

The general idea of the relational mind approach to consciousness is that an input is encoded by a semantic net S (such as in words in Wernicke's area or parts of objects in **IT**). The output i' of S is sent to an episodic memory store E, so as to evoke suitable activity of related past experiences stored in E. The output of E and the direct one of S are combined in some manner in a comparison net C. That module functions so as both to delimit the range of output of E and to combine it with the output of S. The resultant combination is supposed to have conscious content. In particular the mental state of the system is defined as:

$$\text{Mental state} = \text{set of memories } m \text{ suitably similar to } i' \qquad (1)$$

(where similar denotes that set of memories determined by the comparison net C). This allows an estimate to be given of the change of the "level of consciousness" over experience, where the level of consciousness is defined as an average, say over a waking day, of the size of the mental state. An overlap theory of the meaning of a sequence inputs was also described, in terms of the size of the overlaps between the mental states (as defined by equation 1) activated by those inputs. This allows meaning to be given to linguistic input. Thus the sentence "This piece of cake is ill" has little meaning, since there is rather small overlap between the set of memories evoked by the phases "This piece of cake" and "is ill."

In order to develop the relational mind model further it is necessary to consider in more detail the control mechanisms involved in the comparison or decision system. Moreover, extensions must be made to the model to take account of the present understanding of the global information processing strategies in the brain, as well as to include the different modalities of consciousness and the actual dynamics of the development of awareness. An even greater problem arises when one faces up to the question of the near-uniqueness of consciousness. How can this supposedly unique stream of consciousness be achieved in spite of there constantly being new objects of awareness? William James's description of consciousness as "a stream flowing smoothly along, now eddying around, then flowing smoothly again" (James 1950) must somehow be incorporated in any model, or arise naturally from it. In the next section a neuroscientifically based control system will be considered which will allow such features, and other related ones, to be seen as natural properties of the stream of consciousness discernible in the model. At the same time, dissociations in consciousness such as arise in hypnosis or multiple personality disorders (Hilgard 1977) or in speeded response (Marcel 1993) must be seen as explicable as part of the general model.

THE GLOBAL GATE

So far a "boxes-systems" approach has been adopted (comprising a semantic net, an eposodic memory net, and a comparison net), in which the boxes' functionality and connectivity have only weakly been constrained by evidence from psychology. In order to make progress, it appears helpful to incorporate neuroanatomical and neurophysiological constraints. These arise from the possible neural modules that might be able to support global competition between the various modality-specific analyzers, in order to produce a putative "uniqueness" for consciousness (although this might dissociate under certain conditions). It is also necessary to be able to include mechanisms for competition between inputs in a given modality.

One might suppose that an effective global competition between distant modules in cortex can be achieved by suitable inhibition between these modules carried by distant excitation of local inhibitory neurons. There is, however, a certain number of arguments against such a mechanism being actually present in higher vertebrates. Inhibition is known to be necessary for orientation selectivity in striate cortex, but only appears to occur over a range of about half a hyper-column laterally. However the main determinant of orientation selectivity in $V1$ seems to be oriented feed-forward excitation from LGN. Long-range feed-forward activity also appears to be mainly excitatory to IP or parietal lobe, as single-cell recordings would seem to indicate. Thus, there is little evidence for long-range lateral inhibition in cortex, and other mechanisms have been proposed, such as conservation of weights on a given neuron, to achieve "winner-take-all" competition. However none of these other suggestions appear to have any experimental support.

All inputs to cortex (other than olfaction) pass through one or other thalamic nuclei. It might be possible that there is sufficient lateral inhibition in these nuclei to support "winner-take-all" processing. However there is little neuroanatomical evidence for this, since the existent inhibitory interneurons in the thalamus are only of the local circuit variety. Past ideas of lateral connections between different nuclei in thalamus to explain coherent cortical activity in sleep have not been supported by further investigation. The thalamus does not seem able to provide a mechanism for "winner-take-all" competition either.

One region, often described as a thalamic nucleus, is the *nucleus reticularis thalami* (NRT for short). It consists of a thin sheet of mutually inhibitory neurons draped over the thalamus. Any axon from cortex to thalamus that passes through NRT gives off collateral excitation NRT, as does any corticothalamic axon. NRT occupies such a strategic position in thalamocortical processing that it has been implicated in numerous aspects of brain activity from a local to a global level. It has long been termed a gateway to cortex, and more specifically "The reticularis gate becomes a mosaic of gatelets, each tied to some specific receptive field zone or species of input" (Schiebel 1980). He goes on to suggest "Perhaps here resides the structurofunctional substrate for selective awareness and in the delicacy and complexity of its connections, our source of knowing, and of knowing that we know."

It is posited here that NRT functions, through its intrinsic global connectivity, allow a long-range winner-take-all competition to be run between various specialized cortical modules. Such competition arises from the combination of activity in various thalamic, NRT, and cortical regions from what we term the TH-NRT-C complex. Loss of NRT action would lead to loss of the global features of this competition; if NRT were dissected, so only to be locally connected, then the system would function in the manner suggested by Schiebel above as a set of local gatelets. In its waking mode of action it is proposed that NRT function as a global gate, supporting a global competitive method of cortical processing. The support from this thesis comes from a variety of sources (discussed in Taylor 1993a), to which the reader is referred.

THE "CONSCIOUS I" GATING MODEL

A certain amount of spade-work has already been done to tackle the problem of constructing a viable model for NRT as a competitive net. That lateral competition could be effective in enhancing the maximum of a set of local inputs has been shown by simulation of a simple model (La Berge *et al.* 1992), based on the mode of NRT inhibitory action suggested by Steriade, Domich, and Oakson (1986), and mentioned in an earlier section. This explicitly used axon collateral feedback from different cortical areas to achieve a competitive relationship between the different cortical activities. An independent approach (Taylor 1992a, 1992b) was based more on the existence of local inhibitory connectivity on the NRT sheet. This allowed the use of an analogy with the outer-plexiform layer (OPL) of the retina, which has dendrodendritic synapses that are gap junctions between the horizontal cells (Dowling 1987). Hence the epithet "Conscious I" for the model.

A mathematical model of the OPL was developed (Taylor 1990). The model was analyzed particularly in the continuum limit, where analytic expressions can be obtained for the OPL response. The OPL itself emerges in this limit as a Laplacian net, so-called since the leaky integrator neuron equations for the membrane potential have a two-dimensional Laplacian operator giving the gap-junction lateral contributions. A similar model for the NRT sheet has the same Laplacian contribution in the continuum limit, but now with a negative coefficient multiplying the operator (arising from the inhibitory action of NRT cells on each other). The resulting negative Laplacian net has an analogy with quantum mechanical scattering, in which oscillatory waves occur in a potential well. Negative Laplacian nets also arise in the analysis of pattern formation, where heterogeneous wave patterns occur (Cohen and Murray 1981) as also in the modeling of hallucinatory effects in cortex (Ermentrout and Cowan 1978). It is these waves that give a strong hint of control. However the NRT equations, when coupling with thalamus (TH) and cortex (C) is included, have a contribution nonlinear in the input and membrane potential that appears crucial. This term leads to the possibility of global control of cortical activity by that incoming over a limited region of thalamus or NRT. This opens the way to the possibility of top-down control in attentional tasks.

To fit with the MEG data reported (Llinas and Ribary 1992), corresponding to a backward sweep of activity of thalamus and cortex every 25 msc, it is to be expected that the whole anterior pole of the NRT exercises its authority over more posterior activity. How could this be achieved? To help answer this, known neuroanatomy (Lopes da Silva *et al.* 1990) indicates the following relevant facts. First, a number of regions in the limbic complex have reciprocal connections with anterior thalamic nuclei. However, these latter have no connections with NRT (except in the rat). The only clear connection with NRT appears to be from the subicular complex, with a one-way projection to the mediodorsal thalamus. This latter is known to have connections, specifically with the anterior pole of NRT. Thus, it would appear that guidance from neuroanatomy indicates that the solution to the control problem raised above—as to what regions exert control over the anterior part of NRT—has been achieved by using external input from limbic cortex, which is part of the Papez limbic circuit, itself being outside direct NRT control. The limbic cortex involves nervous tissue relevant to emotional drives as well as to stored memories so that the general nature of the control on NRT activity will be from these sources. It is to be noted that the rat is not expected to have such control input in which limbic activity is down loaded onto NRT, but has feedback from the anterior thalamus to the limbic circuit and NRT; its emotions will be expected to be inextricably linked with its consciousness.

WINNING CONTROL

Interesting results, now thirty years old (Libet *et al.* 1964), appear relevant to the nature of global control, and give support to the model being developed. The experiments of Libet were to determine the threshold current for conscious experience when a 1mm diameter stimulating electrode was placed on the post central gyrus and the just conscious experience of what seemed like a localized skin stimulus reported by the patient. The stimulus was delivered as a series of short (of around 0.1 or so msec duration) pulses. There are two features of the data which stand out, which can be summarized as two quantitative laws (Taylor 1994a):

1. For threshold current to be consciously experienced over a short (<0.5 sec) duration, the applied electrical energy (frequency times duration times square of current) must be greater than a critical value.

2. For a duration longer than about 0.5 sec the applied electrical power must be large enough to allow the conscious experience to continue.

The requirement of enough applied electrical energy to capture, or turn on, conscious awareness in the short term would seem to fit well with the NRT control structure model above. For that is functioning essentially as a resistive circuit, with some nonlinearity to provide stability, and such a circuit would be expected to function in terms of electrical energy requirements for capture of the dominant mode. The second result above leads to the need for enough injected power to keep the control system going; there will be a certain amount dissipated, and so power above that critical level will have to be injected to hold the control of consciousness achieved by the earlier injected electrical energy.

It is possible to extend the above analysis to include the even more interesting phenomenon of backward referral in time (Libet et al. 1979). This phrase refers to the onset of the conscious experience being backdated to a few tens of milliseconds after the input, in the case of peripheral or thalamic excitation, but not in the case of direct cortical stimulation. Such a phenomena contains important clues for further development of the model, although the reader is referred to Taylor (1994a) for further analysis of this.

The discussions so far can be summarized under the two themes

1. The Relational Theory of Mind in which sematically coded input and its related episodic memories are compared in some decision unit, and

2. The "Conscious I" NRT, in which global competition is carried out through activity on NRT.

The above features must now be combined to obtain consciousness. More specifically, it must be attempted to understand how the TH-NRT-C system which produces the comparator/decisions unit of the section on the relational mind, leads to consciousness as experienced as the "private" or subjective world. To do that, feedback connections from memory will now be incorporated.

There are two crucial features involved here. One is the manner in which episodic memories are stored and re-excited. From the discussion on the relational mind, it is clear that the re-excitation has to be of a whole host of earlier memories, which can give the consciousness "color" to experience. That might not best be achieved by a pattern completion or attractor network (Amit 1990), since only a single pattern would result at one time. The most suitable memory structure is a matrix memory for a one-layer feed-forward net in which the connection weight matrix Aij has the form $Aij = \Sigma\, a_i^{(m)} a_j^{(m)}$. The response $y = Ax$ will then have the form

$$y = \Sigma\, a^{(m)} \left(a^{(m)} \cdot x\right) \quad (2)$$

being a weighted sum of projections of x along the various memories. Assuming that attention has been caught by the (semantically encoded) input x, it would seem reasonable to assume that memories $a^{(m)}$ suitably close to x would therefore be allowed to be reactivated in earlier cortex by the TH-NRT-C complex. The details of how close such memories should be to x (the "metric" of the comparator) depend on the nature of the TH-NRT-C complex, and are presently being analyzed by extensive simulations (Alavi and Taylor 1994). However, it is reasonable to assume that such a metric exists. Thus the "Conscious I" model leads to a specific class of comparators in the relational mind theory.

The second important feature is that feedback to associative and even primary cortex must be present to allow re-excitation of features of the appropriate memories $a^{(m)}$. This is necessary in order that the detailed content of these memories be "unscrambled" from the high-level coding they have been represented as in medial temporal lobe areas. Indeed the important feedback connections in cortex, and also those through the TH-NRT-C complex, may have exactly the form required to achieve this amplification of the memories. In this approach, consciousness is supported by the imagery arising from feedback of excited memories allowed to persist with the input by the action of the TH-NRT-C competitive complex.

It has already been suggested (Taylor 1992a, 1992b, 1992c, 1993a, 1993b) that the crucial component needed to support the persistent input activity to allow return to consciousness is STM, or more precisely working memory (WM) of Baddeley and Hitch (Baddeley and Hitch 1974, Baddeley 1986). This involves continued activity in a "store" or "scratch-pad" over a period of about 2 seconds before decay occurs without rehearsals. More specifically the WM store is the crucial part being considered here.

More can be said, however, on the need for persistent activity support by working memory. The simulation results of the competitive model (Taylor 1994a, Alavi and Taylor 1994) together with the experimental results of Libet and his colleagues (1964), indicate the need for input activity persisting over a period of at least 0.5 sec (for minimal current to achieve awareness). The need for this continued activity in an electrode placed on somatosensory cortex or in ventrobasal thalamus can be interpreted as the requirement for creating an artificial working memory, which has activity thereby sustained over a long enough time to win the competition for awareness. Peripheral stimulation is expected to achieve activation of an actual working memory site, as in the phonological store activation by speech input, the visuospatial sketch-pad by visual pattern input, and so forth (Baddeley 1986). Thus, in normal processing, for consciousness to be achievable it would appear essential to have cortical working memory modules to allow persistent activity to have enough time to win the consciousness competition and achieve activation of suitable stored memory in short-term memory representations, to help in the consciousness competition.

This latter activation will not have to be as strong as for imagery. Such detail may be unnecessary unless conscious recognition/recall or actual imagery processing were occurring. The situations mentioned briefly in the section on the relational mind (the "false fame" and other illusory effects, and other similar content-dependent effects) do not seem to involve such a conscious level of involvement of past memories in consciousness. Nor, in cases mentioned in Kelley and Jacoby (1993), could such conscious recall occur in the case of amnesics (such as in the famous Claparede case) who still have their conscious experience colored by past implicit, nondeclarative memories.

It is now possible to extend the possible circuitry after the semantic encoder has been activated. A semantic module S_A receives input IN_A (say in the visual modality), and responds rapidly to give output going elsewhere or going to its dedicated cortical working memory WM_A. This has persistent activity, over a suitable length of time as discussed above, which activates and receives feedback from an associated memory store E. This latter store may be common to a number of working memories, such as a second one WM_B. WM_A and WM_B are coupled through the TH-NRT-C complex, as well as possibly directly corticocortically. A competition is run between WM_A and WM_B till one or the other wins. The winning WM then controls the activity on the other WM, and is expected to lay down its content, or that of S_A, in hippocampus. This is regarded as conscious memorization, with stored representatives being later available by reactivation of the relevant working memory state at time of memory deposition. This circuitry can be extended, by inclusion of frontal lobe/basal ganglia, for the learning of sequences of actions; the reader is referred to Taylor (1994b).

The principles of the competitive model of consciousness may be summarized as:

1. For each input, with specialized semantic coding, S_A, there is an associated working memory WM_A in which activity persists for up to 2 seconds (or longer).

2. S_A/WM_A feedforward and feedback activation to memory representations E lead to augmentation of the WM_A activity in a relational manner.

3. Competition occurs between the various working memories WM_A, WM_B, and so forth, over a critical "neuronal adequacy" time (Libet *et al.* 1964), in which one WM, denoted WM_{win}, finally wins, and controls the losers' activities in their WMs.

4. The activity in WM_{win} is stored in hippocampus and nearby sites as a buffer, to be usable (possibly nonconsciously) at later times.

5. WM_{win}/S_{win} activity guides thinking and planning sequences.

6. The activities in the non-winning S_As can still be used for automatic, nonconscious level processing, so this nonwinning activity itself is not necessarily controlled, though that in nonwinning WMs may be temporarily.
7. Upgrading of memory, either of declarative form in the hippocampal complex or in semantic memories or nondeclarative form in frontal lobe/basal ganglia thalamic feedback circuits, may have to occur off-line.

The upgrading of memory under point seven above would especially appear to be necessary for the unloading of hippocampal memories to prevent overload, and also to develop semantic categories. Such off-line upgrading may occur in sleep.

The details of the competition in point three above do not need to be specified in general, although it was already noted that finer features of the competitive process cannot be explained without a specific model. However the details of that model need not be restricted, at this stage, only to the TH-NRT-C complex. There may be other candidates, involving inhibitory interneurons in some as yet to be understood manner, or other even less neurobiologically likely choices. Finally the manner in which memory representations E are activated in a relational manner, as part of the relational mind model described earlier, have still to be explored.

It is to be noted that recent experimental results on intrusive thoughts (Baddeley 1993) provide direct psychological evidence for competition between working memories. The experiments involved disruption of intrusive thoughts (which are a problem for depressed patients, often exacerbating depression) by activation of visual or auditory working memories. Both activation of the phonological loop and of the visuospatial sketch–pad by suitably designed tasks, whilst subjects were seated in an isolated sound-isolated room, reduced the occurrence of stimulus-independent thoughts from about 80 percent to 30 percent of the times the subjects were probed. The speed of the requisite additional tasks turned out to be important; shadowing digits (and repeating them) at the rate of one every few seconds caused little disruption on the intrusive thoughts in comparison to an 80 percent to 30 percent reduction for shadowing every second. The disruption was effective on coherent sequential intrusions, but had almost no effect on the relatively infrequent fragmentary independent thoughts. Moreover competition was observed between intrusions and the generation of random sequences of letters.

These data fit the competitive relational model of consciousness very well in a qualitative fashion. There is seen to be competition between sequential intrusive thoughts, random generation of letter, phonological tasks, and those using the visuospatial sketch-pad. The interpretation in Baddeley (1993) that it is the central executive (Shallice 1988) that governs the occurrence or suppression of intrusions is consistent with the competitive relational mind model if the central executive is equated with the TH-NRT-C complex (or more generally the competitive system, whatever it turns out to be). That identification will be discussed elsewhere (Taylor 1994b).

Dissociations in consciousness (such as might arise in hypnosis, MPD, or contradictory report) were mentioned briefly in an earlier section. Their inclusion into the competitive relational mind model is obviously crucial, giving as they do important clues as to the dynamical functioning of the system. The main feature that such dissociations imply is the independent processing by subsystems of an input signal at near-conscious level. Contradictory report (Marcel 1993), for example, can be taken to imply the noisy character of transmission or control of a just threshold input from its winning working memory WM_{win} to other subordinate working memories (for verbal report or various forms of muscular response). The character of contradictory report in blindsight (Marcel 1993), on the other hand, can be explained as arising from input transmission that does not attain a suitable cortical working memory in the first instance. The input is expected ultimately to activate suitable WMs for report by the appropriate code, but that will have noise (of a different character) due to difficulty of having suitable connections (which appear adaptive). The dissociation which arises in MPD or hypnosis (Hilgard 1977) involves the self more specifically, and gives important clues to allow an understanding of the latter in the competitive relational mind model; that is explored more fully in Taylor (1994b).

BUILDING RELATIONS

So far the most important structures in the relational mind approach have been outlined, these being the competitive TH-NRT-C complex, and the WM/Semantic modules between which the competition for consciousness is supposed to be run. Other structures have also been introduced, such as the hippocampus for declarative memory and frontal lobe/basal ganglia for, among other things, implicit or skill memory. However the relations being used in constructing the content of consciousness have not been clarified. What is their specific form? How are they achieved—are they prewired or are they learnt? Some form of learning will be essential to ensure flexibility. Furthermore these relations depend on preprocessing through semantic modules. The coding in these latter modules themselves have to be learnt, at least in part. How is this achieved to be consistent with the competitive structure of the TH-NRT-C complex? There are various deep problems here, especially associated with the nature of categorization, and its development from infant to adult. This is an aspect of development which has not yet been understood, especially associated with language and the meaning of words and concepts. It is only possible to scratch the surface of these difficult problems, but preliminary answers will be attempted for some of the questions raised above to allow the relational structure of mind to be clarified a little. It is also hoped that some guidance is given as to the nature of solutions to the problems in terms of the structures so far introduced into the relational mind model.

There are a variety of approaches possible to relations themselves. One such is in purely mathematical terms, as a binary (or higher order) graph or function on symbols. A second is to follow the developments in relational semantics in which a basic set of relations are determined by clustering analysis of language (Chaffin and Hermann 1988).

A third method of analyzing relations is in terms of the various features into which the objects (between which the relations are supposed to occur) can be analyzed. However it has been noted for at least a decade that which especial features are to be chosen to achieve the definition of an object category in the first place is somewhat arbitrary. The notion of "similarity" to help pick out "similar" features has been strongly criticized by a number of developmental psychologists (Fivush 1987), and support has developed for a script-based development of categories (Schank and Abelson 1977). Thus the eating-routine script will allow the welding together of the actions (get in high chair, put on bib, eat, drink) to help build the functional categories of food (cereal, bread, banana), drink (milk, juice), or utensil (cup, glass, bottle). At the same time thematic categories can be created, involving temporal sequences of action → object → action → object →, as in: put on bib → have bowl → eat cereal → have cup → drink juice. It is known that quite young infants will sort objects in thematic categories (Fivush 1987) and moreover that adults use both functional and thematic categories as well as those defined by prototypes and classical rules.

These results indicate that actions are important in developing categories, especially in the infant. However there would appear to be category information, even in the three-month old, in which little action takes place (Quinn, Burke, and Rush 1993). The infant is supine, and is presented with visual patterns of rows or columns of alternating light and dark squares. After six 15-second exposure sessions the infant appears to have formed a representation of the pattern that helped it, in a Gestalt manner, to process similar patterns that it was presented later. Such category learning does not appear to use any movement sequence. That is quite contrary to the observation (Rovee-Collier, Greco-Vigorito, and Hayne 1993) of category inclusion of an object, such as a butterfly, hung from a mobile, and rocked back and forth by the experimenter, which a three-month old infant had earlier been able to move by kicking when yellow blocks with alphanumeric characters printed on them were attached to the mobile. This is clearly action-based categorization, apparently very different from the earlier visual pattern categorization. Yet there is expected to be considerable eye movement associated with the former categorization, as is found in the viewing of pictures by adults (Yarbus 1967). Moreover frontal eye field (FEF) and supplementary eye field (SEF) modules are in reasonable proximity to supplementary motor cortex, which is known to be necessary for learning visually initiated motor sequences. Thus it might be suspected that FEF and/or SEF modules are involved in learning processes when visual eye movements are made over a scene. There is some evidence for this (Noton and

Stark 1971), and it is proving effective in artificial vision systems (Ryback *et al.* 1994).

The suggestion of Noton and Stark (1971) based on experimental evidence they gathered and, implemented in Rybak *et al.* (1994), is that an object is encoded by means of an alternating sequence of visual feature memory and eye movement memory. A similar use of eye movements for feature binding to make object categories was proposed in Taylor (1994c). If there are many movements associated with a given input then one may suppose that ultimately the encoded memory in LTM may appear to be independent of the action inputs (although could still be activated by it). This was the mechanism for developing semantic memory in Taylor (1994c): simulations showed how features $\int, \int^1$ sequentially viewed by the action A, so by the sequence $\int \to A \to \int^1$ can be used to build the category $(\int\int^1)$ composed of the joint features $\int, \int^1$. This was achieved by the input $\int$ leading to the action (eye movement) A, by the associative action memory $\int \to A$ activated by the visual input $\int$, which itself leads to the new input $\int^1$. The initial input $\int$ to the semantic memory SM activates the associated working memory WM; the continued activity of $\int$ and the new input $\int^1$ on WM brings about learning of the lateral connections $\int \leftrightarrow \int^1$ on SM. Later activation of $\int$ of $\int^1$ on SM then brings about lateral activation of the other feature, so of the composite object $(\int\int^1)$.

In general the encoding of different inputs will be guided by the action sequences involved, so that functional and thematic encoding is expected to occur most naturally in LTM. Thus, categories and superordinate categories are expected to be constructed thereby. There is also the fact that in rapid eye movement (REM) sleep (in which eye movements may be used to reactivate visual memories with the associated actions) posterior associative cortex is mainly active along with frontal eye fields. This suggests that the LTM being constructed during REM is the semantic memory associated with the relevant WM under consideration. We note that in REM sleep the TH-NRT-C complex is still expected to be functional, so that different WMs will be winners at different times. However there will be no external input to drive the competition, and the internal activity playing such a driving role may well be mainly from the frontal and supplementary eye fields and similar motor areas (SMA, PMA). Thus action-based replays are conjectured as occurring, but those are only based on memories stored from the near past. The bizarre character of REM-based dreams may be explained in terms of the different nature of the source of activity during the competition on the TH-NRT-C complex. Thus, the above discussion leads to the conjecture that in REM sleep semantic memory is augmented. There are still the action-based relations between the concepts stored in semantic memories, embedded in the script-based action-object sequences. That is furthermore suggested as the origin of semantics.

In non-REM slow-wave sleep (SWS), on the other hand, the thalamus is hyperpolarized, and motor output is blocked in brain stem neurons. Consciousness is regarded as dull or absent, when people are awoken in SWS.

This is explicable in terms of the lack of the TH-NRT-C system to handle any competition between working memories. On the other hand some frontal lobe circuits are active, as is hippocampus. The latter even appears to be involved in consolidating memory representations (Wilson and McNaughton 1994). It is therefore natural to suggest that during SWS episodic memories may be being laid down, and in particular incorporated into frontal lobe contents, such as determining norms of social behavior.

WHAT IS IT LIKE TO BE

In order to finally approach the very difficult problem of the inner point of view, it might first be useful to consider other features of consciousness that may be understood in terms of the relational model. This will help give confidence in the model, as well as adumbrate further certain of its features. An interesting list of twelve aspects of consciousness has been presented (Searle 1991), the most relevant of these, which will be considered here, are finite modalities, unity, intentionality, familiarity, and finally subjective feeling. The other seven aspects are also of considerable interest, but would require far more space than is available here.

To begin with, modality. There are only five senses, sight, touch, smell, taste and hearing, together with internal body sensations, the sense of balance, and possibly those of self. Why not more? The competitive relational mind model leads to an extremely naive model that may have a modicum of truth. For a two-dimensional homogeneous sheet, the competition, on which the model is based, may only be able to distinguish between a maximum and a minimum of activity along any direction on the sheet. This leads to four possible ways of distributing the activity in two dimensions, with one winner out of four in a 2×2 square of cortex. This limit of four distinct possible winners holds for each cerebral hemisphere, so that a maximum of sixteen different winners could be allowed at different times. Cross-connections between the hemispheres would reduce this, but not completely since there is known lateralization of function. The model may be naive in the extreme, but it does give a possible first approximation to a reason for the limited modality in which consciousness is expressed.

The unity of consciousness has already been built in as part of the relational mind model. There can only be one winner on the NRT sheet, so only one mode of consciousness, and accordingly consciousness content. There are two hemispheres, two NRT, two thalami, so that there are actually two winners. However these winners would not compete with each other, since the NRT sheet of each thalamus does not continue to the other sheet. Thus the two winners would coexist side by side. Indeed these activities would be correlated, allowing the sixteen different modes of consciousness mentioned above. Only when the corpus callosum joining the two hemispheres is cut would there be a dissociation of these two sets of consciousness centers. The consciousness of the two halves would then be separate and reduced, even to the point of inability to respond linguistically to questions.

However, each side would still only have one winner at a time. It is important to add that, in general, lateralization of function is pronounced in the human brain, possibly related to the asymmetric position of the heart. Whatever the cause, usually the left hemisphere is the language-dominant one, whilst the right supports affect. The possibility of the TH-NRT-C system operating in a dissociated manner has been briefly discussed earlier; such dissociation in humans is well documentated, as noted there. There is no space here to discuss this further (Taylor 1994b).

Intentionality is based on the fact one is conscious of something—an object, a feeling, or whatever. As stated in (Searle 1991), "..., consciousness is indeed consciousness of something, and the "of" in consciousness is the "of" of intentionality." This feature may be seen to arise from the relational structure mentioned earlier in the paper and the nature of the working memory that wins the competition at any one time. The detailed content of consciousness can only be "fleshed out" by the feedback support from the episodic and semantic memory. The specific nature of the object of consciousness will be that arising from the coding of the particular working memory involved. In the case of an auditory input, for example, an input to Wernicke's area would be encoded into phonemes, but at the same time activate the appropriate word centers. The most active of these would be representing the actual words heard. There would, however, be words with lesser activation energized as part of the semantic coding giving meaning to words. These would be part of the relational structure at the semantic level. Beyond this would be activation of episodic memories appertaining to personal beliefs and to personal experiences. Feedback from these memories would add to the relational structure constraining further activity and aiding the competition that the auditory input could sustain on entering the phonological working memory. These feedback activations give intentionality to the words fighting their way to consciousness. In other words they give the perspectival character to the consciousness experience, and may be considered as giving the aspectual shape to the intentional state (Searle 1991).

The above account can also help explain the notion of familiarity that the objects of consciousness display so effortlessly. There will be initial familiarity brought about by the semantic coding, since the memories laid down earlier have the essence of familiarity. They present no novelty, no jarring from the expected into unfamiliar territory. Above and beyond this somewhat general source of familiarity is that arising from episodic memory. This gives additional contextual familiarity as a background to the ongoing words as they are processed in the phonological loop. Even though the present context of the words may be unfamiliar, the memory of past environments when the words were experienced is expected to ameliorate this sensation of novelty. The words are then helped to possess more familiarity than without such feedback.

Finally we come to subjective feeling, to the problem of the inner view. To echo Nagel (1974) again "... every subjective phenomenon is essentially connected with a single point of view and it seems inevitable that an objective, physical theory will abandon that point of view." How can the relational mind model approach this supposedly scientifically impossible "inner point of view"? To answer this question, the possible nature of such a point of view must be described from within the model, to see how far the singleness of the point of view can be arrived at.

By the point of view of the sentient being X will be taken to be meant the detailed content of conscious awareness of X. This detail, according to the competitive relational model, is composed of the working memory activity winning the consciousness competition. To that is to be added all the feedback and parallel activated semantic and episodic memory activities related to the input. Moreover there are also parallel activities in the other, coding, working memories, that may be involved in linguistic report (as in the case of the phonological loop). But most especially the "point of view" is determined by the semantic and episodic memories related to the input in the appropriate working memory.

It is possible to explore this view of the subjective character of experience in more detail by considering separately the contributions made by the semantic and the episodic memory feedbacks.

Semantic memory content of consciousness will have general culture- and species-specific characteristics. Thus, there will be words of the natural language in which a human has been brought up, as well as words of other languages learnt by the person during schooling or as part of general life experiences. For animals such as dogs, similar encoding of the few words with which they are familiar would be expected to occur, although there may be no corresponding phonological loop to give conscious experience to words but only direct output to spatial working memories allowing the conscious experience of such words to acquire a visual form. Similarly other modalities will have culture-specific semantic memories, such as for shapes in the visuospatial sketch–pad. There will be expected also to be more person-specific encodings in the semantic memories, such as dialects or particular shapes, although these will usually still be shared with others in a local area. Thus the semantic encodings are expected to have a certain degree of objectivity associated with them, although that will not be absolute.

It is in the episodic memory content that the inner point of view comes into its own. The personal "coloring" added to consciousness, the inner "feel" of the experience, is, it is claimed in the relational model, given by the parallel activation and feedback of memories associated with the present input. Particular people, buildings, or more generally sights, smells, touches, and sounds from relevant past events are excited (usually subliminally) and help guide the competition for consciousness of the present input in its working memory. The laying down of these episodic memories only if they were consciously experienced is an indication of the filter character of the activity of the competition winner. The level of relevance and significance of these past experiences to ongoing activity will not be absolute but will depend on mood and emotion. These stored memories are energized both by inputs and basic drives, and guide the ongoing corticothalamic activity in a top-down manner from limbic structures (Taylor 1994b).

Most crucially the episodic memory is a private diary of the experiences, and responses to the experiences, of an individual. It is so private that it would be difficult to discern its meaning, since it is expressed in a coding that will not be easily accessible to an outsider. The engraved traces in the diary are trained connection weights, developed by using unsupervised and/or reinforcement training algorithms. But the knowledge of the values of these weights is not necessarily enough to know all that the inner point of view would receive for its specification. Mood, and the resulting neuromodulation of the relevant nerve cells by catecholamines and other chemicals, would be an essential boundary condition, as would the contextual inputs involved in the experience. In other words not only is the internal record of the connection weights of memory needed but also a possibly imperfect record of mood and place is needed to recreate the conscious experience of the sentient being X for any one period.

It is of interest to consider the distinction made by James (1950), and amplified recently by Mangan (1993), of the nucleus and the fringe of consciousness. The nuclear conscious state C_{nucl} is regarded as critically determined by winning (primed) semantic memory SM_{pr} and the associated working memory, which are denoted $SM_{pr}(WM)_w$. Then we identify

$$C_{nucl} = (SM_{pr} UWM)_w$$

If $(SM_{pr}UWM)_{rest}$ denotes the other semantic and working memories, $(M_{tot})_{act}$ denotes the current activity in the episodic memory stored (expected to be in various parts of cortex well connected to the hippocampus), and L_{act} denotes the active part of the limbic system, then we may identify

$$C_{fringe} = C_{fringe1} + C_{fringe2} + C_{fringe3}$$
$$C_{fringe1} = (SM_{pr}UWM)_{rest}$$
$$C_{fringe2} = (M_{tot})_{act}$$
$$C_{fringe3} = L_{act}$$

The activity $C_{fringe1}$ is considered as roughly the inattentive information in the fringe, which can be upgraded to be at the nucleus if so needed. $C_{fringe2}$ then would correspond to the experience of familiarity, involving tip-of-the-tongue and familiarity of knowledge type events (Mangan 1993). Finally $C_{fringe3}$ might be expected to consist of the feeling of rightness or wrongness of the environment, although this might also arise partly from $C_{fringe2}$. In total these identifications have begun to give more detailed suggestions for the neurophysiological underpinnings of the fringe, and could lead to noninvasive investigations of the activities of the corresponding regions.

At this juncture it is important to discuss an essential limitation to the nature of the explanation of inner experience put forward so far. It is not being claimed that one is thereby able to have that experience. If it was attempted to build, say, a virtual reality machine, on the basis of the competitive relational model, could it give one the feeling as to what it would be like to be X? One would first require X's input sensor transforms, such as achieved by the retina, the ear, or by the nose. Then one would need built in to the virtual reality headset the further transforms performed at primary cortical areas on that input. But why stop there? Why not build the associative cortices and their actions, the semantic memories, working memories, episodic memories, and action plans. It would be necessary, for the latter, to build additional subcortical structures, such as basal ganglia and cerebellar cortices, for further control. At this juncture momentum has built up to include also the drive/goal and affect structures. One would thus either have built a silicon replica of X's brain, or if one wished to have the experience oneself, and after a suitable trepanning, a silicon version of X's brain in one's own head. But then one has again built a version of X, and one is no longer oneself. One might be able, by brain transplantation, to add cortical (and subcortical) circuits so that one can do the processing required to experience as X at one time and to revert to oneself at another. However such a possibility does not seem to add anything new to the previous possibilities, other than being able to keep one's personality more intact.

The above is based on the desire to have the inner experiences that X does. One could attempt to do so without going to such extremes, by developing and extending one's imagination of what it is like to be X, as suggested by Nagel (1974). However it would appear impossible to experience in that manner, color vision, say, if one was color blind from birth. Thus it is not being attempted to try to be or to experience X's "inner view" from one's own necessarily limited brain circuitry. Suitable brain exercises could give one the increasing illusions that one was X, as might have been pursued by Kafka during his writing of the story of the man who became a cockroach in "The Metamorphosis." What those exercises may be trying to do is indeed to change connection weights of suitable circuits so as to give the impression that one was a cockroach, or whatever. Yet the complete set of X's experiences would still not be able to be felt unless all the crucial brain circuitry was achieved by such a learning process. Since the possibility of achieving such a transformation without considerable structural change seems extremely slight, and in any case we would then be back at the trepanning scenario, it would appear that we have to accept the limitation noted above. It just does not seem possible to acquire X's own inner view without effectively becoming X. But an understanding has, even so, been gained of how it is that it is like to be X. The subjective, unique, multimodal, intentional characteristics of that view have been given an explanation in terms of the competitive relational model of the mind. We expect that science can go no further.

DISCUSSION

A model has been presented which is claimed to enable one to discover how it is that it is like to be the conscious being X. It is assumed that X has suitable neural network structure to enable recognition of separate modules in X performing the activities of global competition, and episodic memory, and also sets of sensory modality-specific working and semantic memories. Then the inner view that X has of its experiences are given by the relations set up by the memory activities accompanying the winning working memory in its global competition with other such memories. The various features of X's consciousness, of finiteness of number of modalities, of uniqueness, intentionally, familiarity, and subjective character are then derived from the model, as it is supposed other features of Searle's list (1992) would be. Moreover, identification of neural structures involved with the nucleus and fringe of consciousness has been made (James 1890, Mangan 1993). Even dissociations of consciousness (Hilgard 1977, Marcel 1993) can be seen to be compatible with the model. What has been gained thereby compared to earlier models of the mind?

There are models that deserve such comparison, since they have proved influential in the development of thinking on the subject very recently, and in particular in the development of the competitive relational model. The first is Dennett's "Pandemonium of specialists" (Dennett 1991). This could be identified with the set of competing working memories, the added value of the relational model being the manner in which semantic and episodic memories are crucial to the ongoing amplification of conscious experience by their activation and feedback. In the language of the Cartesian theater that is eloquently discussed, and rejected, by Dennett (1991), our model has a Cartesian theater peopled only by actors (the working memories), each of whom struts his or her time, only to be replaced by a new winner from amongst the competing actors clamoring in the stalls to be crowned. Each actor carries with them extra loudspeakers that resonate especially with certain inputs, to allow the corresponding actor to win time in the limelight. The inner conscious experience is that shared by all the actors, although only being broadcast by one.

Such a picture is seen to be an extension of the "global blackboard" model (Baars 1988), in which input sensors compete to broadcast their activity to others and effectors by the medium of the global blackboard. The relational mind model goes beyond this by emphasizing the need for relational memory structures to amplify and constrain incoming signals. It is these which are claimed to give the inner view, something absent from Baars' model. At the same time some resemblance (although still far from reality) to neurophysiological structures in the brain is required to effect the various modes of operation of the competitive relational mind model. This is given in the relational mind model by the thalamo-NRT-cortex complex.

There is also noticeable similarity between the competitive relational model and that of Edelman (1989). His emphasis on reafference is very similar to the need for feedback from memory structures for guidance and prediction. These themes were not augmented in Edelman (1989) by the working memory structures discussed earlier, nor in the manner in which the crucial and unique experimental data of Libet and colleagues (Libet *et al.* 1964, Libet *et al.* 1979) could be explained and used to extend the model.

Finally the supervisory attentional system (SAS) of Norman and Shallice (1986) has been very influential, especially in guiding experiments in working memory (Baddeley 1992, Baddeley and Hitch 1974). However it has been admitted by one of the proponents that the model has a problem of underpinning conscious experience (Shallice 1988), so it does not appear easy to discuss. In particular, as pointed out by Horne (1993), who does the contention scheduling for the SAS? It would seem the buck stops there, whilst in the competitive relational mind model the buck is constantly handed from winner to winner. It is true that the buck stops altogether if there are no activations of working memory—consciousness then is lost—a prediction that could ultimately be testable by real-time MEG as sleep takes over from wakefulness, and the fires of the working memories are dimmed (although possibly at different levels in REM, S1, S2, S3, S4 stages of sleep). Already the phenomenon of blindsight can be explained by the lack of suitable activation of the appropriate visual working memory (and the need for suitable long-lasting pulses in artificial vision systems for blind-sight patients). The competitive relational model of mind is testable in an enormous number of ways. This was already indicated earlier, with observable thalamic timing systems and specific timings for the turn-on of appropriate working memories, differentially for peripheral, thalamic, and cortical stimulation. It is also testable in the phenomena of neglect, as indicated for blindsight. Noninvasive MEG and rCBF measurements should allow details of the essential circuitry for consciousness ultimately to be ascertained.

What about the possibility of consciousness in a being without the requisite neural circuitry mentioned above? It would be necessary to consider each case on its merits, however. It may be necessary to discuss beings with different orders of consciousness, say with a three-dimensional neural or other form of network (with up to 64 modalities of consciousness for a being with two three-dimensional hemispheres). However, it would be more useful to wait until such beings have been met and investigated before one should speculate further on this. All vertebrates on Earth have nervous systems that allow their inner experience to come under the guise of the competitive relational model. At least we have given an answer as to how it is that it is like to be a sentient earthling.

ACKNOWLEDGMENTS

The author would like to thank Dr. D. Gorse, Professor S. Grossberg, and Professor W. Freeman for trenchant comments, Dr. P. Fenwick of the Institute of Psychiatry, London for a useful conversation, Professor B. Libet for useful discussions on his work, and Dr. D. Watt on neuropsychological implications.

REFERENCES

Alavi, F., and J. G. Taylor. 1994. (in preparation)

Amit, D. 1990. *Modeling brain function*. Cambridge, England: Cambridge University Press.

Baars, B. 1988. *A cognitive theory of consciousness*. Cambridge, England: Cambridge University Press.

Baddeley, A. 1993. "Working memory and conscious awareness." In *Theories of memory*, edited by A. F. Collins, S. E. Gathercole, M. A. Conway, and P. E. Morris. Hillsdale, NJ: Lawrence Erlbaum, pp. 11–28.

Baddeley, A. 1986. *Working memory*. London: Oxford University Press.

Baddeley, A. 1992. Is working memory working? *Quart. J. Exp. Psyc.* 44:1–31.

Baddeley, A., and G. Hitch. 1974. "Working memory." In *The psychology of learning and motivation*, edited by G. A. Bower. New York: Academic Press.

Barsalou, L. 1987. "The instability of graded structure: Implications for the nature of concepts." In *Concepts and conceptual development*, edited by U. Neisser. Cambridge, England: Cambridge University Press, pp. 101–40.

Chaffin, R., and D. J. Hermann. 1988. Effects of relation similarity on part-whole decisions. *J. Gen. Psychology* 115:131–39.

Chalmers, D. J. 1996. "Facing up to the problems of consciousness." In *Toward a science of consciousness: The first Tuscon discussions and debates*, edited by S. R. Hameroff, A. L. Kaszniak, and A. C. Scott. Cambridge, MA: MIT Press.

Cohen, D. S., and J. Murray. 1981. A generalized diffusion model for growth and dispersal in a population. *Math. Biol.* 12:237–49.

Dennett, D. 1991. *Consciousness explained*. Boston: Little, Brown.

Dowling, J. 1987. *The retina*. Cambridge, MA: Harvard University Press.

Edelman, G. J. 1989. *The remembered present*. New York: Basic Books.

Ermentrout, G. B., and J. D. Cowan. 1978. "Studies in mathematics." *The Math Association of America* 15:67–117.

Fivush, R. 1987. "Scripts and categories: Interrelationships in development." In *Concepts and conceptual development*, edited by U. Neisser. Cambridge, England: Cambridge University Press, pp. 234–54.

Hilgard, E. R. 1977. *Divided consciousness*. New York: John Wiley & Sons.

Horne, P. V. 1993. The nature of imagery. *Consciousness and Cognition* 2:58–82.

Hume, D. 1896. *A treatise on human nature*, edited by E. A. Selby-Briggs. Oxford, England: Clarendon Press.

Jacoby, L. L., and K. Whitehouse. 1989. An illusion of memory: False recognition influenced by unconscious perception. *Journal of Exp. Psych. Gen.* 118:126–35.

James, W. 1950. *The principles of psychology*. New York: Dover Books.

Kelly, C. M., and L. L. Jacoby. 1993. "The Construction of subjective experience: Memory attribution." In *Consciousness*, edited by M. Davies and G. E. Humphries. London: Blackwell, pp. 74–89.

Kahneman, D., and D. T. Miller. 1986. Norm theory: Comparing reality to its alternatives. *Psychology Review* 2:136–53.

La Berge, D., M. Carter, and V. Brown. 1992. *Neural. Comp.* 4:318–33.

Libet, B., W. W. Alberts, E. W. Wright, Jr., D. L. Delattre, G. Levin, and B. Feinstein. 1964. Production of threshold levels of conscious sensation by electrical stimulation of human somato-sensory cortex. *J. Neurophysiol.* 27:546–78.

Libet, B., E. W. Wright, Jr., B. Feinstein, and D. K. Pearl. 1979. Subjective referral of the timing for a conscious experience. *Brain* 102:193–224.

Libet, B. 1982. Brain stimulation in the study of neuronal functions for conscious sensory experience. *Human Neurobiology* 1:235–42.

Llinas, R., and U. Ribary. 1992. Chapter 7. In *Induced rhythms in the brain,* edited by E. Basar and T. Bullock. Boston: Birkhauser.

Lopes da Silva, F. H., M. P. Witter, P. H. Boeijinga, and A. H. M. Lohman. 1990. Anatomic organization and physiology of the limbic cortex. *Physiology Review* 70:453–511.

Mangan, B. 1993. Taking phenomenology seriously: The "Fringe" and its implications for cognitive research. *Consciousness and Cognition* 2:89–108.

Marcel, A. J. 1993. "Slippage in the unity of consciousness," in CIBA Foundation Symposium No. 174, *Experimental and theoretical studies in consciousness.* Chichester, England: John Wiley & Sons.

Nagel, T. 1974. What is it like to be a bat? *Philosophical Reviews* 83:435–50.

Noton, D., and L. Stark. 1971. Scanpaths in eye movements during pattern perception, *Science* 171:308–11.

Quinn, P. C., S. Burke, and A. Rush. 1993. Part-whole perception in early infancy. *Infant Behavior and Development* 16:19–42.

Rovee-Collier, C., C. Greco-Vigorito, and H. Hayne. 1993. The time-window hypothesis: Implications for categorization and memory modification. *Infant Behavior and Development* 16:149–76.

Ryback, I., V. Gusakova, A. Golavan, N. Shertsova, and L. Podlachikova. 1994. Modeling of a neural network system for active visual perception and recognition. 12th Conference on Pattern Recognition, Jerusalem.

Schank, R., and R. Abelson. 1977. *Scripts, plans, goals and understanding.* Hillsdale, NJ: Lawrence Erlbaum.

Schiebel, A. B. 1980. *The reticular formation revisited,* edited by J. A. Hobson and B. A. Brazier. New York: Raven Press.

Searle, J. 1991. *The rediscovery of mind.* Cambridge, England: Cambridge University Press.

Shallice, T. 1988. *From neuropsychology to mental structure.* Cambridge, England: Cambridge University Press.

Steriade, M., L. Domich, and G. Oakson. 1986. Reticularis thalami neurons revisted: Activity changes during shifts in states of vigilance. *J. Neurosci.* 6 (1):68–81.

Taylor, J. G. 1973. A model of thinking neural networks. Seminar, Institute for Cybernetics. University of Tübingen. (Unpublished)

Taylor, J. G. 1990. Analysis of a silicon model of the retina. *Neural Networks* 3:171–78.

Taylor, J. G. 1991. Can neural networks ever be made to think? *Neural Network World* 1:4–11.

Taylor, J. G. 1992a. Towards a neural network model of mind. *Neural Network World* 2:797–812.

Taylor, J. G. 1992b. "From single neuron to cognition." In *Artificial neural networks*, Vol. 2, edited by I. Aleksander and J. G. Taylor. Amsterdam: North-Holland.

Taylor, J. G. 1992c. "Temporal processing in brain activity." In *Complex neurodynamics, Proceedings of the 1991 Vietri Conference,* edited by J. G. Taylor, E. Caianiello, R. J. M. Cotterill, and J. W. Clark. Amsterdam: Springer-Verlag.

Taylor, J. G. 1993a. "A global gating model of attention and consciousness." In *neuro–dynamics and psychology,* edited by M. Oaksford and G. Brown. New York: Academic Press.

Taylor, J. G. 1993b. Neuronal network models of the mind. *Verh. Dtsch. Zool. Ges.* 86(2):159–63.

Taylor, J. G. 1994b. (in preparation)

Taylor, J. G. 1994c. Relational neurocomputing. Invited talk, Special Interest Group Meeting, WCNN 1994. San Diego.

Taylor, J. G., and F. Alavi. 1993a. A global competitive network for attention. *Neural Network World* 5:477–502.

Taylor, J. G., and F. Alavi. 1993b. "Mathematical analysis of a competitive network for attention." In *Mathematical approaches to neural networks,* edited by J. G. Taylor. New York: Elsevier, pp. 341–82.

Taylor, J. G. in press. A competition for consciousness? *Neurocomputing*

Wilson, M. A., and B. L. McNaughton. 1994. Reactivation of hippocampal ensemble memories during sleep. *Science* (in press).

Witherspoon, D., and L. G. Allan. 1985. The effect of a prior presentation on temporal judgments in a perceptual identification task. *Memory and Cognition* 13:101–11.

Yarbus, A. L. 1967. *Eye movements and vision*. New York: Plenum Press.

27 Artificial "Attention" in an Oscillatory Neural Network

Tokiko Yamanoue

INTRODUCTION

One of the main aspects of consciousness is awareness. While the term "awareness" corresponds more to the "feeling" aspect, the term "attention" in a wide sense may be regarded as the functional aspect of the same phenomena.

Many cognitive scientists seem to prefer somewhat computer-oriented models. However, the fact that every process can be simulated by a computer does not mean that computer-oriented models are the best way to represent or understand any process. As for the issue of consciousness, it may be better to investigate more brain-oriented models at this point, since a brain is the only system we know to have consciousness, so far.

In this paper, I study oscillatory neural networks (ONN), in which activity of each "neuron" is represented as time course, typically an oscillation. This kind of model has been considered as a candidate to explain how the brain integrates fragmental information into a complex whole, since coherence of activity can work as a label to distinguish the parts belonging to an "object" from other parts (von der Malsburg and Schneider 1986).

Although this process itself has sometimes been categorized as selective attention (Koch and Crick 1992), this view seems to miss an important aspect of attention that arises when input complexity threatens system capacity.

I will show in the following sections that an ONN, without any additional mechanisms, provides a wide variety of phenomena resembling human attention upon complex input, ranging from preattentive to focal attentional process. I will also try to show that the model qualitatively reproduces a well-known result on psychological experiments regarding attention. Although the result is usually interpreted as the evidence of serial processing in the brain, our model suggests an alternative possibility.

BASIC BEHAVIOR OF THE MODEL

An ONN consists of interacting oscillators that I call "cells." The particular ONN I demonstrate here is designed so that all cells have a tendency to be incoherent with each other. Against this tendency, interactions act to enhance coherence between interacting cells. Such an ONN segments into several coherent groups of cells, according to the following factors.

Distance Regarding Interaction

When two cells are directly interacting, they have a strong tendency to be coherent. When there are more intervening cells, such tendency becomes lower. As a result, for example, a chain of interacting cells tends to segment into two parts, each of which includes an edge.

Connectivity Regarding Interaction

Two indirectly interacting cells have a stronger tendency to be coherent if there are more chains of interactions between the two. For example, when there is a loop of interaction, constituent cells tend to be coherent (Figure 27.1). Also, two cells having more interaction partners in common tend to be coherent.

The above segmentation characteristics can be compared to human strategy on visual segmentation that nearby, similar, or closed figures are favored as forming an object. Although it is possible to construct more realistic and precise models of vision, this simple model is sufficient for the following qualitative demonstration.

"Attentional" Behavior of the Model

Let us broadly consider attention here as a phenomena that arises when input complexity is increased (Figure 27.2). The above ONN produces the

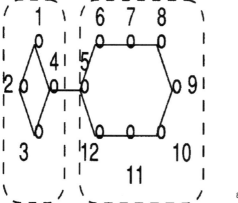 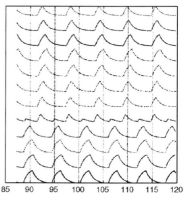

Figure 27.1 An example of an oscillatory neural network (ONN). Left: Structure of the ONN. Cells are displayed as small circles, interactions as lines. Cells surrounded by a dotted square are coherent (see right). Right: Time course of the activity corresponding to numbered cells in the left diagram (from below to top).

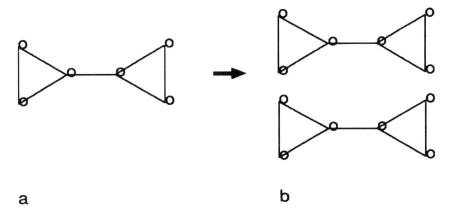

Figure 27.2 An example of increasing complexity, by duplication (from a to b).

following phenomena, each resembling some type of human attentional behavior (Yamanoue 1994, 1995).

1. Rough description of the whole. In this case, when the complexity exceeds system capacity, some objects merge so that the system can maintain the number of objects (Figure 27.3a). This results in a rough representation resembling human preattentive process.

2. Detailed description of a part. An alternative phenomena was observed in the same system with (1), with different parameters. This time, activity corresponding to some objects disappear as complexity is increased (Figure 27.3b). The result is a detailed description of a part, resembling focusing of attention. In some other parameters, cases 1 and 2 seem to coexist. System capacity is two to three objects in Figure 27.3, but can be increased by another kind of parameter change. As for the shift of attention focus, see below.

3. Hierarchical part-whole description. Hierarchical clustering appears often in a multilayered system when the number of the lowest layer cells exceeds system capacity (Figure 27.4, left, right). This kind of phenomena can be regarded as a special case of the above case 2 with the shifts of attention focuses. Another possible interpretation of the phenomena is hierarchical part-whole description. Depth of clustering hierarchy changes with parameter change (Figure 27.4, middle). System capacity can also be controlled by parameter change.

The model shows that an ONN can focus, defocus, and even shift attention without any additional mechanisms or center of control.

The Model's Performance on Conjunction Search

Conjunction search is a famous psychological experiment regarding attention (Treisman 1991). An example of the task is to find a "Green T" target among nontargets, "Green Xs" and "Red Ts." In some combination of

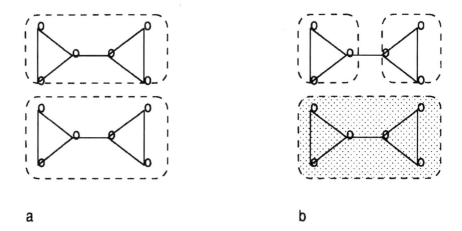

a b

Figure 27.3 Coherence patterns produced by the ONN of Figure 27.2b. The letters a and b correspond to different parameters. The shaded part in b indicates that the activity dies out for that part. The coherence pattern corresponding to Figure 27.2a is similar to the upper part of 27.3b, at both parameters.

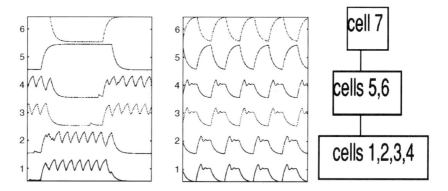

Figure 27.4 A multilayered system. Left: Hierarchical coherence structure. Time course from below to top correspond to numbered cells in the schematic diagram (right). Middle: Flat coherence structure produced by the same ONN, at different parameters. Right: Schematic diagram of the ONN. Subunits described as boxes consist of cells with the tendency to be incoherent to other cells in the same box. Lines connecting the boxes describe coherence-enhancing interactions among cells in the connected boxes.

attributes, human subjects show linearly increasing response time, as the number of nontargets increases. Although this result is often regarded as evidence of serial processing, the above ONN produces a similar result with a somewhat different mechanism. In essence, when the number of nontargets increases, they become major "objects" in the ONN. As a result, "Green" tends to be more coherent to "X" than with "T," and the target "Green T" tends to fire only very rarely, even with the help of intentional

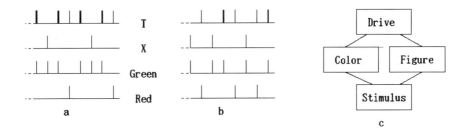

Figure 27.5 A conjunction search with an ONN. Typical task of finding a "Green T" among "Green Xs" and "Red Ts." (a, b) Time-course corresponding to single-attribute cells, "T," "X," "Green," and "Red." Activities are schematically illustrated as impulses. Firing of the cell "T" coherent to the target-stimulus ("Green T") are displayed as thick lines. Remaining parts of "Green" are coherent to the nontargets of "Green X" type. As the number of nontargets increase from a to b, the target's firing rate decreases. (c) Schematic diagram of the ONN. Boxes represent subunits as in Figure 27.4.

drive (Figure 27.5). As a result, delay of response time may occur without any serial processing.

CONCLUDING REMARK

A rigorous experimental approach regarding feeling should be very hard, if not impossible, since such a study may easily end up studying the connections to various observable functions rather than the feeling itself. Building speculative models may be about all we can do with the issue of feeling, although experimental data like pathological studies (Goodale *et al.* 1994) may help this speculation. By counting up the possible models from functional aspect and trying to extract a feeling from it, we may approach an "at least convincing" model of feeling.

The presented model shows that such an ONN provides various attentional behavior without any additional mechanism, control center, or serial mechanism. It seems to reproduce the functional aspects of consciousness to some extent. At this stage, to claim that the model has the feeling aspect also is like claiming that an ocean wave has awareness. Still, a model of feeling based on this kind of model can be quite different from other models for its totally nonlocal character.

REFERENCES

Goodale, M. A., L. S. Jakobson, A. D. Milner, D. I. Perret, P. J. Benson, and J. K. Hietanen. 1994. The nature and limits of orientation and pattern processing supporting visuomotor control in a visual form agnosie. *Journal of Cognitive Neuroscience* 6:46–56.

Crick, F., and C. Koch. 1992. The problem of consciousness. *Scientific American* September, pp. 111–7.

von der Malsburg, C., and W. Schneider. 1986. A neural cocktail-party processor. *Biological Cybernetics* 54:29–40.

Treisman, A. 1991. Search, similarity, and integration of features between and within dimensions. *Journal of Experimental Psychology* 17:652–76.

Yamanoue, T. 1994. Effect of complexity in an oscillatory neural network: Relation to "attention" systems. *Proceedings of 3rd International Conference on Fuzzy Logic, Neural Nets and Soft Computing.* 575:578.

Yamanoue, T. 1995. Effect of complexity in an oscillatory neural network: Relation to human-like reasoning. *International Journal of Fuzzy Sets and Systems*

28 The Emergence of Memory: Categorization Far From Equilibrium

Andrew Wuensche

INTRODUCTION

How does memory emerge in a simple artificial neural network? I will try to provide an answer in the context of a network model first proposed by Ross Ashby (1952) in his classic book *Design for a Brain*. This network is a discrete dynamical system of sparsely connected cells updating in parallel according to each cell's Boolean function. Usually known as a random Boolean network following Stuart Kauffman (1984), the system can also be described as a disordered cellular automaton, with arbitrary nonlocal connections and different rules at each site.

Any artificial network model is necessarily a huge oversimplification compared to biological networks. However, Boolean functions, in contrast to threshold functions, arguably capture some essence of the logic implicit in the complex topology of each neuron's dendritic tree and synaptic microcircuitry. A Boolean function could be implemented in hardware by a combinatorial circuit, a sort of artificial dendritic tree. If the emergence of memory can be demonstrated in such simple networks, perhaps insights may be gained by analogy to the vastly more powerful processes that occur in animal brains. Memory is a relative concept, meaning the creation of useful categories for an organism's adaptive behavior out of the plethora of sensory inputs arriving at neural subnetworks, and the even greater internal traffic between subnetworks. The idea of networks of subnetworks in a sort of nested hierarchy is of course another necessary oversimplification, because the boundaries of subnetworks are unclear. To be useful for adaptive behavior memory-categories should fall naturally into hierarchies of categories of subcategories; they should be highly reliable yet easily changed to permit learning and access should be extremely fast.

Recent work in unraveling the global dynamics of cellular automata (Wuensche and Lesser 1992), and more generally of random Boolean networks (Wuensche 1993) puts into sharp focus a known intrinsic property of these networks. They organize state-space, the space of all possible patterns of network activation, into specific maps of connected states, the basin of attraction field. Directed graphs representing these objects can be calculated and drawn by computer graphics. Separate assemblies of connected states are basins of attraction; typically, the connection topology is branching trees rooted on attractor cycles. Basins of attraction partition state-space into categories. Subtrees within individual basins, formed by merging trajectories leading to attractor cycles, produce hierarchies of subcategories. The basin topology is characterized by long transients, verging on dynamics analogous to chaos in continuous systems.

The idea that state-space is partitioned by attractors is the generally accepted paradigm for "content-addressable" memory in artificial neural networks, following Hopfield (1982) and others. In large recurrent networks, especially biological networks, a very large (perhaps astronomical) number of steps through state-space would typically be required for the system to settle at one of its attractors. Explaining memory just by attractors thus poses the difficulty of the long time needed to reach attractors in large networks, whereas reaction times in biology are extremely fast. This problem is overcome by the realization that categories occur in subtrees, far from equilibrium.

Random Boolean networks have a vast parameter space, and a correspondingly vast space of possible basin of attraction fields. Perhaps any basin-field structure is possible given the appropriate parameters. Learning algorithms for random Boolean networks have been developed for changing the basin of attraction field. States at any location can be reassigned as predecessors of other states in a single step by adjusting either connections or Boolean functions. Adding states implies learning, removing states implies forgetting. This might allow the sculpting of the basin of attraction field to approach any desired configuration. The process of learning and its side effects is made visible if the resulting basin of attraction field or fragment is reconstructed.

The detailed structure of a biological neural subnetwork has found its form, and thus its basin of attraction field, by evolution, development, and learning in the context of networks of subnetworks that maintain each other far from equilibrium. Within this vastly complex dynamical system, higher-level cognitive properties based on lower-level memory categories are able to emerge.

RANDOM BOOLEAN NETWORKS

Random Boolean networks may be viewed as disordered cellular automata. A homogeneous rule and coupling template, giving a regular space, make cellular automata appropriate models in physics (Wolfram 1983). Deviating from either or both of these constraints by degrees progressively degrades coherent space-time patterns and emergent complex structures such as gliders, characteristic of cellular automata (Wuensche 1993, 1994). Different arbitrary couplings and rules at each cell give a vastly greater parameter space, and thus behavior space, than cellular automata. The various symmetries and hierarchies that constrain cellular automata dynamics, such as shift invariance and the conservation of rotational symmetry (Wuensche and Lesser 1992), no longer apply. It might be conjec-

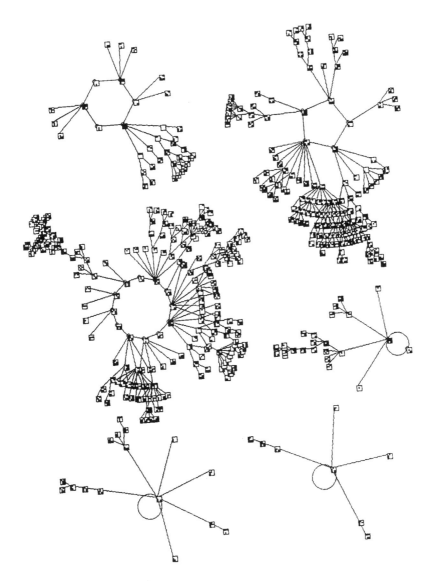

Figure 28.1 The basin of attraction field of a random Boolean network ($N=9$, $K=4$). The $2^9=512$ states in state-space, shown as patterns on a 3×3 grid, are organized into six basins, with attractor periods ranging from 1 to 12. The number of states in each basin is: 39, 191, 234, 24, 16, 8. The network's wiring/rule scheme is shown in Figure 28.3. The length of transition arcs has no significance; it follows a graphic convention for clarity. Time proceeds inward from garden-of-Eden states, then clockwise around the attractor cycle, the only closed loop in the basin.

tured that any arbitrary basin of attraction field configuration is possible given the right set of parameters.

Figure 28.2 illustrates the architecture of a random Boolean network. A cell's value can be just 0 or 1, though in principle this could be extended to more values. A global state of a network of N cells is its pattern of 0s and 1s at a given moment. Each cell synchronously updates its value in discrete time steps. The value of a cell at time t_1 depends on its particular Boolean function applied to a notional or pseudo neighborhood, size K. Values in the neighborhood are set according to single-wire couplings to arbitrarily located cells in the network at time t_0. The system is iterated. The system's parameters consist of a list specifying the function and pseudo neighborhood wiring for each cell. To maintain a given basin of attraction field the parameters remain fixed. Changes to parameters would change the field.

There are 2K permutations of values in a neighborhood of size K. The Boolean function (equivalent to a cellular automata rule) can be written as a rule table, or look-up table, with 2K entries, specifying the output of all neighborhood permutations. By convention (Wolfram 1983) this is arranged in descending order of the values of neighborhoods. For example, the rule table for rule 30 ($K = 3$) is,

$$\begin{array}{llllllll} 111 & 110 & 101 & 100 & 011 & 010 & 001 & 000 \end{array} \ldots \text{neighborhoods}$$
$$\text{rule table} \ldots 0 \quad 0 \quad 0 \quad 1 \quad 1 \quad 1 \quad 1 \quad 0 \ldots \text{outputs (0 or 1)}$$

The total number of distinct rule tables, the size of rule space $= 2^{2K}$. The number of alternative wiring schemes for one cell equals N^K. The number of alternative wiring/rule schemes, S, that can be assigned to a given network turns out to be vast even for small networks, and is given by,

$$S = (N^K)^N \times (2^{2^K})^N$$

for example, for a network where $N = 16$ and $K = 5$, $S = 2^{832}$.

Random Boolean network architecture is in many ways similar to weightless neural networks (Alexander, Thomas, and Bowden 1984). Classical neural network architecture uses weighted connection and threshold

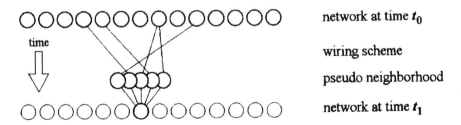

Figure 28.2 Random Boolean network architecture. Each cell in the network synchronously updates its value according to the values in a pseudo-neighborhood, set by single wire couplings to arbitrarily located cells at the previous time-step. Each cell may have a different wiring/rule scheme. The system is iterated. Cells are arranged in a row for convenience only; their positions may be arbitrary.

functions. A random Boolean network may be regarded as a discrete generalization of a sparsely connected classical neural network. Connections with higher weights may simply be replaced by multiple couplings, and the threshold function applied. However, a threshold function is a tiny subclass of the 2^{2^K} possible Boolean functions.

BASINS OF ATTRACTION

Cellular automata and random Boolean networks are both examples of discrete deterministic dynamical systems made up from many simple components acting in parallel. The pattern of network activation at a given time is the network's state. State-space is the space of all $2N$ possible patterns.

Each state has just one successor, though it may have any number of predecessors (known as pre-images), including none. Any state imposed on the network will seed a determined sequence of states, known as a trajectory. Though determined it is usually unpredictable.

In a finite network any trajectory inevitably encounters a repeat. When this occurs the system has entered and is locked into a state-cycle known as the attractor. The portion of a trajectory outside an attractor is a transient. Many transients typically lead to the same attractor. Because states can have multiple pre-images, transients can merge and will typically have a topology of branching trees rooted on the attractor cycle (though this may be a stable point—an attractor cycle with a period of 1). The set of all transient trees plus their attractor make up a basin of attraction. The separate basins that make up state-space (though there may be just one) is the basin of attraction field, a mathematical object in space-time constituting the dynamical flow imposed on statespace by the network.

Computing transient trees or subtrees and basins of attraction poses the problem of finding the complete set of pre-images of any global state. The possible solution of exhaustively testing the entire state-space rapidly becomes intractable in terms of computer time as the network's size increases beyond modest values. Algorithms have recently been devised, however, for directly generating pre-images, giving an average computational performance many orders of magnitude faster than exhaustive testing. A reverse algorithm for one-dimensional cellular automata was introduced by Wuensche and Lesser (1992). A general direct reverse algorithm for random Boolean networks was introduced by Wuensche (1993).

Basins of attraction are portrayed as computer diagrams in the same graphic format as presented in Wuensche and Lesser (1992). States are represented by nodes, or by the state's binary, decimal, or hex expression at the node position. Nodes are linked by directed arcs with zero or more incoming arcs but exactly one outgoing arc (one out degree). Nodes with no preimages, thus no incoming arcs, represent so called garden-of-Eden states. Typically, the vast majority of nodes in a basin of attraction lie on transient trees outside the attractor cycle, and the vast majority of these states are garden-of-Eden states. Figure 28.1 shows a typical basin of attraction field of a random Boolean network ($N = 9$, $K = 4$). The $2^9 = 512$ states in state-space are shown as patterns on a 3×3 grid. The network's wiring/rule scheme is shown in Figure 28.3.

MEMORY, FAR FROM EQUILIBRIUM

Memory far from equilibrium along merging transients may answer a basic difficulty in explaining memory by attractors in biological neural networks.

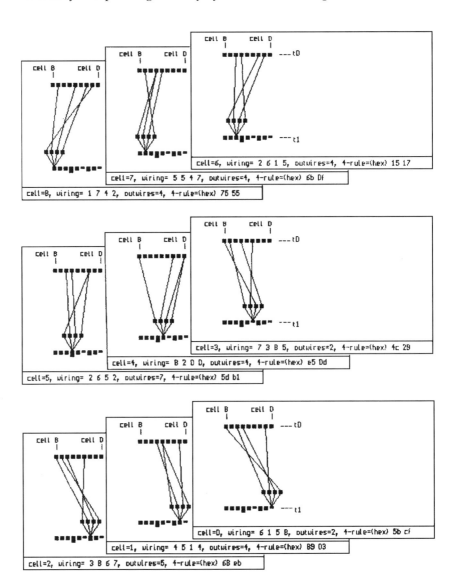

Figure 28.3 The wiring/rule scheme of a random Boolean network ($N=9$, $K=4$); its basin of attraction field is shown in Figure 28.1. The wiring specifies couplings from the pseudo-neighborhood of each cell (at time t_1) to cells at the previous time-step, t_0. Different heights of cells at t_0 indicate their number of output wires. Each cell's rule is shown in hex. The parameters were chosen at random.

A view of the brain as a complex dynamical system made up of many interlinked, specialized neural subnetworks is perhaps the most powerful paradigm currently available. Subnetworks may consist of further subcategories of semi-autonomous networks, and so on, which contribute to resetting or perturbing each other's dynamics. A biological neural subnetwork is nevertheless likely to be extremely large; the time required to reach an attractor from some arbitrary global state will probably be astronomical. This has been demonstrated with a simple $K = 5$ random Boolean network with 150 cells (Wuensche 1993). Even when an attractor is reached, it may well turn out to be a long cycle or a quasi-infinite chaotic attractor. The notion of memory simply as attractors seems to be inadequate to account for the extremely fast reaction times in biology.

A discrete dynamical system with synchronous updating categorizes its state-space reliably along transient trees, far from equilibrium, as well as at the attractors. A network that has evolved or learnt a particular global dynamics may be able to reach useful memory categories in a few steps, possibly just one. Moreover, the complex transient tree topology in the basin of attraction field, makes for a much richer substrate for memory than attractors alone, allowing hierarchies of memory subcategories.

A BIOLOGICAL MODEL

A random Boolean network may serve as a model of a semi-autonomous patch of neurons in the brain whose activity is synchronized. A cell's wiring scheme models that subset of neurons connected to a given neuron. Applying the Boolean function to a cell's pseudo neighborhood models the nonlinear computation that a neuron is said to apply to these inputs to determine whether or not it will fire at the next time-step. This is far more complex than a threshold function (Shepherd 1990). The biological computation may depend on the precise topology of the dendritic tree, its microcircuitry of synaptic placements, and intrinsic membrane properties. Networks within cells based on the cytoskeleton of microtubules and associated protein polymers may be involved, suggested by Stuart Hameroff (1987) as the neuron's "internal nervous system." There appears to be no shortage of biological mechanisms that could perform the role of a Boolean function.

A network's basin of attraction field is implicit in its wiring rule/scheme. In a subnetwork of biological neurons it would be implicit in the wetware. If subnetworks are linked as components in a super-network, indirect feedback could maintain each subnetwork's dynamics in the outer branches of transient trees, far from equilibrium. Suppose a new state is imposed on subnetwork n_1 by another subnetwork or sensory input. This seeds a determined trajectory in n_1, which opens up a sequence of categories to which the seed-state belongs. Recognition of these categories would involve another subnetwork, n_2, activated by the axons of neurons in n_1. Recognition of categories in n_2 would involve a third subnetwork, n_3. After just a few steps, n_1's dynamics might be reset to a new seed, shifting the start point in its basin of attraction, by the axons of another subnetwork belonging to this hypothetical super-network. Recognition in this system is automatic because all trajectories in n_1, n_2, n_3, and so on are inevitable given unchanged neuronal architecture. Recall is more difficult because a seed with the right association needs to be supplied. Folk psychology bears this out. I may know "Humphrey Bogart," but "it's on the tip of my tongue." Recalling the name to say it may be hard. If I hear "Humphrey Bogart" mentioned, I will recognize it instantly, effortlessly, and automatically.

Learning new behavior implies amending the basin of attraction field by adjusting the wiring/rule scheme, analogous to some physical change to the wetware's neural architecture and synaptic function. Learnt behavior often requires no mental effort. Take driving a car: such well rehearsed semiconscious behavior is delegated to a sort of "automatic pilot." The network's dynamics acts in the same way as for recognition, effortlessly and automatically. Learning to drive, on the other hand, requires mental concentration. After all, physical changes in the brain must somehow be induced. By this argument, the state of consciousness coincides with the desire to learn or behave creatively, or the actual process of making the changes. The mechanism for this is unknown. A subnetwork may be able to alter the parameters of another. Hameroff suggests that cytoskeletal functions may provide retrograde signaling (analogous to back propagation) which may reconfigure intraneuronal architecture (Hameroff *et al.*, in press).

LEARNING ALGORITHMS

Learning and its side effects in random Boolean networks show up as changes to the detailed structure of the basin of attraction field, and its computer graphic representation. In networks too large to allow basins, or even fragments of basins, to be computed, the principles would still apply.

Learning algorithms that enable a random Boolean network to learn new transitions from experience (and also to forget) have been set out in detail (Wuensche 1993). Before learning starts, a wiring/rule scheme must already be in place. If the relevant transitions in the basin of attraction field are already close to the desired behavior, the side effects of learning will be minimized. The initial wiring/rule scheme would ideally be pre-evolved from a population of wiring/rule schemes with a genetic algorithm.

Suppose we want to make the state P_1 in Figure 28.4 the pre-image of state A. This entails correcting the mismatches between P_1's actual successor state and A. This can be done in one step by either of two methods—adjusting the network's wiring or rule scheme. The two methods have very different consequences. Correcting a mismatch by adjusting the rule scheme is achieved by changing a specific bit in each Boolean function at mismatched cells. The procedure is bound to succeed, and it turns out that there is no limit to the number of pre-images of a given state that can be learnt by this method with no risk of forgetting previously learnt pre-images of the state. There will be side effects elsewhere in the basin of

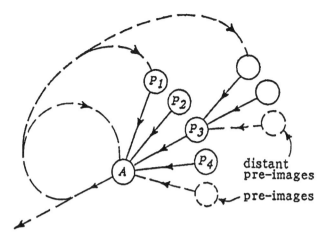

Figure 28.4 States P1, P2, P3,... may be learnt as pre-images of the state A. Distant pre-images of A may also be learnt, for instance as pre-images of P3. Learning A as a pre-image of itself creates a point attractor. Learning A as a distant pre-image of itself creates a cyclic attractor. If A is learnt as the preimage of some other state in the basin of attraction field, the states flowing into A, its transient subtree, may be fully or partially transplanted along with A.

attraction field. On the other hand, in correcting a mismatch by rewiring, success is not certain, though there are many alternative wire move options. Previously learnt pre-images of the state may be forgotten, besides other side effects. Side effects may be positive because similar pre-images are likely be learnt by default, so the network is able to generalize.

Rewiring has a much greater effect on basin structure than mutating the rule scheme, but in either case the stability of basin structure is noteworthy. Using these methods, point attractors, cyclic attractors, and transient subtrees can be created. Transient subtrees are sometimes transplanted along with the repositioned state, indicating how learnt behavior can be reapplied in a new context. Forgetting involves pruning pre-images and transient subtrees, and is achieved by the inverse of the method for learning. Since it is sufficient to create just one mismatch in order to forget, the side effects are minimal as compared with learning.

CONCLUSIONS

The emergence of novel structures by the interaction of many low-level components underlies complex adaptive systems, in particular the emergence of life by the self-organization of matter. Can this approach extend to the emergence of cognition and consciousness, and if so what are the low-level components from which cognition emerges? I have suggested that these components are the hierarchical categories implicit in the dynamics of neural subnetworks, and that memory emerges when subnetworks combine and maintain each other far from equilibrium. Memory may in turn provide the substrate for unconstrained emergence producing higher level cognitive states.

The basin of attraction diagrams of random Boolean networks capture a simple network's capacity for distributed memory. The diagrams demonstrate that a complex hierarchy of categorization exists within transient trees, far from equilibrium, providing a vastly richer substrate for memory than attractors alone. In the context of many semi-autonomous weakly coupled networks, the basin field/network relationship may provide a fruitful metaphor for the mind/brain.

ACKNOWLEDGMENTS

I am grateful to colleagues at the Santa Fe Institute, the University of Sussex, and elsewhere for discussions and comments.

REFERENCES

Ashby, W. R. 1952. *Design for a brain: The origin of adaptive behavior.* London: Chapman & Hall.

Alexander, I., W. Thomas, and P. Bowden. 1984. WISARD, a radical new step forward in image recognition. *Sensor Review* 120–24.

Hameroff, S. R. 1987. *Ultimate computing: Biomolecular consciousness and nanotechnology.* New York: North Holland.

Hameroff, S. R., J. E. Dayhoff, R. Lahoz-Beltra, S. Rasmussen, E. M. Insinna, and D. Koruga. In press. Nanoneurology and the cytoskeleton: Quantum signaling and protein conformational dynamics as cognitive substrate. In *Behavioral neurodynamics,* edited by K. Pribram and H. Szu. New York: Pergamon Press.

Hopfield, J. J. 1982. Neural networks and physical systems with emergent collective computational abilities. *Proceedings of the National Academy of Sciences* 79:2554–58.

Kauffman, S. A. 1984. Emergent properties in random complex systems. *Phisica. D,* 10D:146–56.

Shepherd, G. M., ed. 1990. *The synaptic organization of the brain,* 3rd ed. Oxford, England: Oxford University Press.

Wolfram, S. 1983. Statistical mechanics of cellular automata. *Review of Modern Physics* 55(3):601–64.

Wuensche, A., and M. J. Lesser. 1992. "The global dynamics of cellular automata: An atlas of basin of attraction fields of one-dimensional cellular automata." *Santa Fe Institute studies in the sciences of complexity.* Vol. 1. Reading, MA: Addison-Wesley.

Wuensche, A. 1993. "The ghost in the machine: Basin of attraction fields of disordered cellular automata networks." In *Artificial Life III, Santa Fe Institute Studies in the Sciences of Complexity.* Reading, MA: Addison-Wesley.

Wuensche, A. 1994. Complexity in one-D cellular automata: Gliders, basins of attraction and the Z parameter. *Santa Fe Institute Working Paper* 94–04–026.

VI Subneural Biology

Connectionist views of the brain and often mind consider information to be represented exclusively at the level of neuronal synapses: one synaptic firing, one information bit. Each living cell, however, is a complex world of its own. Single-cell organisms, for example, lead rather active lives without any synapses. As famed neuroscientist Charles S. Sherrington observed in 1951:

> Many forms of motile single cells lead their own independent lives. They swim and crawl, they secure food, they conjugate, they multiply. The observer at once says "they are alive"; the amoeba, paramecium, vorticella, and so on . . . Of nerve there is no trace. But the cell framework, the cyto–skeleton, might serve. There is therefore, for such mind as might be there, no need for our imagination to call halt and say "the apparatus for it is wanting."

Although they don't swim, neurons are even more complicated than single-cell organisms; one particular example is the dynamical regulation of synaptic sensitivity, controlled largely by the cell cytoskeleton. Rather than mere bit states, each neuronal cell has a complex repertoire of activities. Can we expect to understand consciousness if we don't understand life?

How do cells function as living, unitary entities, rather than jumbles of internal events (the cellular binding problem)? Addressed as a serious scientific question decades ago (Schrödinger 1944, Szent–Gyorgyi 1960), the issue has not been appreciably clarified by the recent avalanche of research in molecular biology. In a situation analogous to the controversy surrounding reductionist/artificial intelligence approaches to the brain/mind problem, the field of "artificial life" (Langton 1989) contends that life is an adaptive process independent of the medium in which it occurs. But what is the process? How is it bound together?

In Chapter 29 John Watterson traces the quality of living "oneness" to cell water ordered on protein surfaces. The size scale of proteins—roughly five nanometers—is the boundary between thermal chaos and living order within cell water. According to Watterson, protein-ordered water clusters in the five nanometer range produce extended, cooperative three-dimensional patterns that can resonate and transform into macroscopic quantum states. (The following section discusses potential benefits of quantum effects in explaining the binding problem and other enigmatic aspects of consciousness.)

Jack Tuszyński, B. Trpisová, D. Sept, and M. V. Satarić discuss microtubules in Chapter 30: major structures of the neuronal cytoskeleton—Sherrington's candidate for the site of the "cell mind" (Hameroff 1987). Tuszyński *et al.* consider the suitability of microtubules for information processing by examining their geometrical lattice dipole structure. Depending on conditions, three distinct types of lattice dipole behaviors can occur: random, ferroelectric, and spin-glass. Tuszyński *et al.* describe how the ferroelectric phase can support signaling, and how the spin-glass phase can support computation. Microtubules are proposed as each cell's "on board computer," organizing real-time dynamical activities.

In Chapter 31 Ron Wallace views the neuronal membrane as the essential site for quantum computing, which he argues is essential to consciousness. Quantum computing proposes to use information components that compute while in quantum coherent superposition (see the following section). The quantum wave function then collapses, reducing to a single-state solution. Wallace proposes that "charge density automata" among neuronal membrane components yield quantum computation utilized in consciousness.

In Chapter 32, Dyan Louria and Stuart Hameroff focus on the antithesis of consciousness: anesthesia. At precisely the right amount of inhaled anesthetic, consciousness is erased while other brain functions continue. The anesthetics bind weakly by nonchemical physical forces at particular regions (hydrophobic pockets) within certain brain proteins in membranes and cytoplasm. Using computer simulation, Louria and Hameroff investigate how these weak, quantum-level interactions can impair protein function and have such profound effects in erasing consciousness. They compare the anesthetic effects to quite opposite "consciousness expanding" effects caused by psychedelic hallucinogen drugs, and conclude that anesthetics prevent quantum events in protein hydrophobic pockets that are essential to consciousness.

Living cells, including neurons, are far more than bit states. Their capabilities for processing and representing information extend several hierarchical levels downward, perhaps inevitably to the quantum level. In this section, strata for information processing are described in cell water (Watterson), membranes (Wallace), microtubules (Tuszyński, Trpisová, Sept, and Satarić), and intraprotein hydrophobic pockets (Louria and Hameroff). Whether cellular and subcellular activities (including life itself) are directly relevant to consciousness or merely cellular housekeeping is a matter of intense current debate.

REFERENCES

Hameroff, S. R. 1987. *Ultimate computing: Biomolecular consciousness and nanotechnology.* Amsterdam: Elsevier–North Holland.

Langton, C. G. 1989. *Artificial life: The proceedings of an interdisciplinary workshop on the synthesis and simulation of living systems* held September, 1987 in Los Alamos, New Mexico. Redwood City, CA: Addison–Wesley.

Schrödinger, E. 1944, reprinted 1967. *What is life?* Cambridge, England: Cambridge University Press.

Sherrington, C. S. 1951. *Man on his nature,* 2nd ed. Cambridge, England: Cambridge University Press.

Szent–Gyorgyi, A. 1960. *Introduction to a sub-molecular biology.* New York: Academic Press.

29 Water Clusters: Pixels of Life

John G. Watterson

INTRODUCTION

Underlying the phenomenon of consciousness is the concerted, synchronized activity of living cells: the building blocks of the brain and all living tissues. This activity is the result of the ability of cells to sense and respond "meaningfully" to their environment. This ability is a clear sign to us that a cell processes information, that is, the cell is a minicomputer (Conrad 1992). But cells could not act together if they could not act as integral units. I believe that it is this fact, that the cell functions as an entity and not as a jumble of its internal events, that gives it the holistic quality we recognize as "alive." If cells were mere collections of their constituent biochemistry, they could not act in concert, because the overwhelming number and heterogeneity of these internal events would ensure that the probability of any two cells becoming coordinated is next to zero. A worm would not be possible, much less a brain. The living cell (1-µm scale) is a complex system of roughly one part protein to four parts water. Going down in size, we have in the 100 nanometer (10^{-9} meters: "nm") range, integrated subcellular complexes, which when isolated, are able to function as though still in the intact cell; that is, they contain the essential order of living matter. However, at around 5 nm we pass from the world of ordered movement to that of random thermal motion. This is the size range of single protein molecules (that of water being below 1 nm), and represents the mesoscopic range. At this level, we find that intermediate world where independent, coherent, protein mechanisms like enzyme action still operate. Or viewed in the opposite direction, protein is the first step on the upward path from thermal chaos to living matter.

In this presentation, the origin of the "oneness" quality exhibited by cells is traced to order in liquid water at the mesoscopic level. A basic ephemeral order exists in a latent form in water, which manifests itself as dynamic clusters of defined size (Watterson 1991). Under certain conditions these clusters can be given a degree of permanence, giving rise to structures that resemble proteins in size, form, and function. These clusters explain how protein-water interaction produce the extended cooperative three-dimensional system that constitutes living matter, and how, as a consequence, these individual mesoscopic structures can resonate and transform into a single macroscopic energy quantum.

THE ROLE OF WATER

Ever since that distant time when living matter first appeared, physical conditions on Earth have remained sufficiently stable to ensure the continued existence of liquid water. This fact alone points to a pivotal role for water in the origin and maintenance of life. The liquid state is in general a curious one, because it possesses the properties of both gases and solids, giving flexibility and richness to its molecular dynamics. Although under pressure, liquids adopt a condensed shape as though they are held together internally like solids. This means they exert both pressure and tension simultaneously and thus lie between gases and solids in their properties. Their ability to exist in this seemingly contradictory state is due to the dynamic nature of their molecular structure. The opposing forces of pressure and tension do not cancel one another, but coexist with equal force on separate levels: pressure on the macroscopic and tension on the microscopic level.

Consider a covered beaker containing some liquid (Figure 29.1). When the contents come to equilibrium, the pressure P and temperature T are constant throughout. We know from Maxwell's Kinetic Theory of Gases, that as we go down in size in the upper gas phase, a unit of volume is reached defined by the Gas Law, which contains an average of one gas molecule. This unit is ideally the volume v_0 where

$$P_0 v_0 = kT \qquad (1)$$

and k is Boltzmann's Constant.

Borrowing a useful term from information science, we will call v_0 the "pressure pixel." This concept is an important one for the model developed here, because it takes emphasis away from the actual physical size (cubic

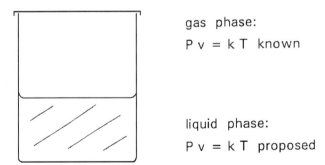

Figure 29.1 The Pressure Pixel. The gas and liquid contents of a covered beaker come to mechanical (pressure) and thermal equilibrium. In the gas phase, the size of the pressure pixel is given by the Gas Law. It is argued that the same must apply to the liquid phase, so that pressure has everywhere the same meaning on the molecular level.

nm), and places it on the fact that this volume represents a level in the size hierarchy of physical phenomena. In the top-down direction, a pixel is that region of size where information becomes fuzzy and then lost. Thus v_0 also represents a pixel, since in smaller volumes pressure has no meaning. And this holds whatever the actual size of v_0.

Now P_0, T, and k do not change in value at the interface between the gas and liquid phases; that is, they do not "see" the boundary, so that the properties of the contents are uniform with respect to these parameters throughout the beaker. This raises the question of the meaning of v_0 in the liquid. It is proposed here, that, as in the gas phase, v_0 is the unit of volume that defines the pressure pixel. At room temperature and pressure conditions prevailing in the biosphere, v_0 is about 40 cubic nm, corresponding to a cube with an edge about 11 molecules long containing around 1400 water molecules. We cannot speak meaningfully of pressure in volumes smaller than this basic unit.

The existence of multimolecular structures of this size in liquids is easily understood in terms of the cluster model originally put forward by Frank and Wen (1957). Entities with the size of the pressure pixel do not experience disruptive internal pressure, rather they are held together as a particle by internal tension. In liquids, this tension is mediated by intermolecular interactions, of which the best understood is that in water. It is stronger than in other ordinary liquids because of the H-bond. This bond is due to the tendency of the H atoms of one H_2O molecule to bond with the O atoms of its neighbors. It is known to be responsible for the internal cohesion shown by physical properties of the liquids, as for example, its high melting and boiling points, specific heat, and surface tension. In addition, these interactions are cooperative; that is, their effects are transmitted. Molecules already bonded to others form even stronger new bonds with still other neighbors, and reciprocally, newly formed bonds strengthen existing ones. Thus structuring promotes structuring, so a structure forming process travels like a wave through the liquid (Watterson 1981).

At any instant, a cluster is held together by the cooperative H-bonds like a giant three-dimensional molecule. But this process is opposed by the disordering effect of thermal collisions which limit the extent of structural buildup. As a result, the multimolecular clusters travel through the liquid medium, because their buildup and breakdown are equally fast on-going chemical processes with opposite effect. If we freeze the idealized picture, we see the liquid medium subdivided into a grid of equally sized cubic clusters, each occupying the unit volume v_0. This is the wave-cluster model of liquid structure described more fully elsewhere (Watterson 1987a). Because thermal collisions break down the structure, pressure and cluster size must be inversely related in this picture, just as predicted by the Gas Law.

Even if all the air is pumped out of the beaker, water will evaporate into the gas phase to ensure that the liquid remains under pressure and that Equation 1 still applies. (In this particular case of "no air," P_0 is called the vapor pressure.) It is a long-known fact that if some solute is dissolved in the liquid so that it is no longer a pure solvent (for example, using a salt or sugar solution as the liquid phase), then the vapor pressure drops. In this new situation, we have pressure P_1, where $P_1 < P_0$, so that in the gas phase

$$P_1 v_1 = kT \tag{2}$$

where $v_1 > v_0$, that is, there are now fewer molecules in the gas phase.

In the liquid phase however, the water clusters are smaller than in the pure solvent; that is, the solute causes a decrease in the size of the pressure pixel, $v_1 < v_0$ (Watterson 1987b). Therefore, there must be a drop in the value of k according to Equation 2. To understand this change, we recall that the expression kT gives the energy of a vibration, whereby Boltzmann's Constant can be interpreted as the spring constant, and T the variable that indicates which energy level the vibration occupies. Each cluster is supplied with energy at the same level T, whether in the pure solvent (Equation 1) or the solution (Equation 2). But in the case of the solution, the strength of its vibration has been lowered compared with the pure solvent, because the foreign solute molecules disrupt the solvent-solvent interactions, interfering with their geometry and thus breaking down the extent of their cooperative influence, or in other words, the solvent structure becomes degraded by the introduction of unlike molecules. Thus, a lower value of k means that solvent clusters possess lower energy in solutions than in pure solvent under the same conditions of temperature and pressure.

Variation of k opens up a new source of energy available at constant temperature conditions, as is the situation for biological systems. It is the source that explains osmosis and gelation. Proteins also use this source. As opposed to our familiar machines, which operate at high temperatures and pressures, proteins work by tapping into this osmotic form of energy; that is, energy supplied by changes in long-range structures existing on the mesoscopic level. This energy is distinct from the thermal energies derived from random motion on the molecular level below.

According to the classical theory of osmosis however, a solution has a lower free energy than the pure solvent, because the foreign solutes produce increased randomness on the molecular level and the energy change is therefore thermal in nature. This theory predicts that osmosis and gelation cannot occur spontaneously, because, as the pressure increases on a system, so does its free energy, and this is forbidden for a spontaneous change. Yet osmosis and gelation are perhaps the two most important physical processes in biology! Furthermore, randomization has also not been able to explain protein stability—another fundamental problem in biology. This question, discussed next, has been under intensive investigation for 50 years, and still remains unanswered. This fact alone shows us that the statistical mechanical approach will never explain these phenomena. (For a rigorous thermodynamic discussion see Watterson 1995).

PROTEIN STRUCTURE

Proteins are long polymer molecules composed of amino acids linked together in a long chain (average length, about 200 amino acids). But these chains are not loosely extended, "kicking and screaming" chaotically in a watery solution as predicted by statistical thermodynamics (Cooper 1976). On the contrary, they are folded back and forth in a predetermined way into a compact shape excluding water, which occupies a volume of about 40 cubic nm (Watterson 1991). Thus we can imagine these large molecular solutes as cubes, with an edge dimension in the 3–4 nm range. Because there is a choice of 20 different amino acids as links in the chain, the cubes can have a great variety of internal structure, but for a given purified protein, they are all the same. In other words, in a solution of a purified protein, all the chains are folded identically into a three-dimensional structure unique for that protein, which is stable in water at conditions prevailing in the biosphere.

Because protein size fits that of the pressure pixel, the internal forces are tensile, not compressive. In fact, proteins are known to be held together by an internal network of H-bonds operating between the amino acids, just as is the case in water clusters. The pressure pixel that operates throughout the space filled with water, controls the form of the protein solutes as well. Like water clusters, they too behave like particles. This means that proteins are energetically linked via water structure to gases, where the Gas Law applies. In other words, proteins are related in a fundamental way to other physical particles and are not the chance outcome of a series of highly improbable reactions that occurred in random solutions of heterogeneous chemicals in the distant past, as required by statistical theories of the origin of life.

CYTOPLASMIC TRANSITIONS

The subcellular world is not a disordered solution of proteins, but a structured complex called the cytoplasmic gel. It is the complex as a whole, protein plus water, that constitutes living matter. Therein, water plays the role of an active component and is not just as an inert background solvent (Clegg 1984). The nature of this material cannot be understood in terms of classical theories, which are based on the reductionist assumption that the agent that causes pressure is the water molecule in chaotic motion at the microscopic level. This failure is well illustrated in the case of gelation, the cause of which is thought to lie with the solute molecules alone. In the accepted theory (Flory 1953), long polymer solutes become cross-linked with one another forming a spatious random polymer network. But this picture does not explain why solvent molecules do not flow (some gels are more than 99% water!). Nor can it explain the gelled state of the cell, since in this case the protein solutes are not stretched out and cross-linked throughout its space. Indeed, they are not cross-linked at all. Consequently the theory throws no light on the mechanism underlying this important phenomenon.

In the protein gel, the solute protein molecules are folded into a unique, compact shape occupying a pixel volume. These "permanent water clusters" give solidity to the complex, by inducing the alignment of the spatially compatible water clusters, so that the solute and solvent units stack together to form a stable, three-dimensional assembly (Figure 29.2). The individual water molecules are not stationary, the clusters are, and consequently water cannot flow as a macroscopic medium.

Further, because of the spatial regularity within the assembly, new entities will emerge at the macroscopic level, as a result of harmonic transitions that occur naturally in wave motion. The swelling of clays provides a sim-

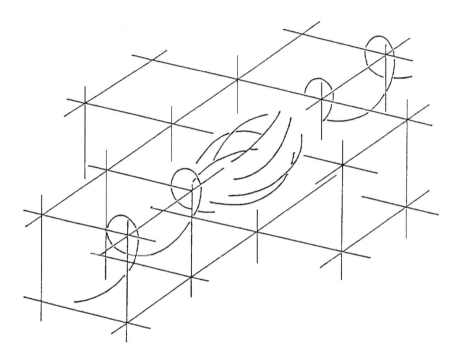

Figure 29.2 The Gel State. Although up to 80% water, this solid-like material is the most important state of matter in the biosphere. As with osmotic phenomena, it is produced by cooperativity between clusters. Protein solutes are like compact cubes with the basic pixel size, 40 nm^3. The central cube contains a group of strands representing a protein chain about 200 amino acids long. The strands are each about 10–15 amino acids long and are folded back and forth (the connections at the ends are not shown), so that the chain has an overall form resembling a twisted barrel. The barrel is not rigid, but flexible, because the strands can vibrate with wavelike motion relative to one another. The exact nature of this vibration is determined by the sequence of amino acid links along the neighboring strands, so in this way each individual protein has its own vibrational signature. Small molecules, about the size of a single amino acid, can attach to special binding sites on the protein molecule, usually at the ends of the barrel. (Sometimes they may enter the barrel, for example, the small O_2 molecule binds inside the hemoglobin protein molecule). It is this binding/release step that switches between vibrational modes. The turns of the spiral shown going through the line of surrounding clusters represent how cooperative behavior between clusters in these three-dimensional networks can induce increases in pixel size. This transformation can be likened to a phase transition, since it results here in a "solid" extended structure, that is, a gel. Indeed many protein-water complexes containing tens of separate protein molecules, each with its own function yet acting in coordination with the others, are now well characterized by biochemists.

ple example of this phenomenon. Clay particles are extremely asymmetric in shape, having length and breadth hundreds of thousands of times greater than their thickness (1–2 nm). Familiar clay material is composed of stacks of these microscopically thin sheets, which swell when in contact with water (a problem well known to construction engineers). Water flows spontaneously in between the sheets forcing them apart, even against pressures of hundreds of atmospheres imposed on the stack. As with other examples of such everyday osmotic processes, there is no satisfactory description of molecular mechanism in terms of statistical concepts (Watterson 1991). These classical theories fail again in their explanation, because single water molecules are seen as the agents causing pressure. With this view, we are always faced with the immediate problem of how can molecules under one atmosphere external pressure in the surroundings flow into a region where the pressure is 100 times greater?

As shown in Figure 29.3, the answer lies in the size of the pressure pixel. The introduction of a solid two-dimensional surface into the liquid medium causes an ordering of the clusters. This flat surface induces clusters to take up position forming a packed layer of clusters. Another way of viewing this, is to say the two-dimensional boundary forces a nodal plane in the structure wave. The side-by-side packing of clusters does not mean the immobilization of single water molecules. They move just as in the bulk medium and maintain the wave motion. With the clusters aligned in a regular way, they can interact and induce harmonic transition forming larger and larger clusters. It is now the tension within them that drives the osmotic process by pulling water into the stack parallel to the surfaces (Watterson 1989). This fusion process can produce a cluster as wide and as broad as the particle surface, resulting in a pressure pixel of macroscopic proportions. Pressure no longer operates within this unified layer, since tension has spread laterally throughout the entire region.

Of course, clay is dead material. Its lack of complexity means that it has no information-processing ability. The system lacks the internal flexibility needed for computation, because the pixels are too large and too static. However, the wave mechanism underlying nonlocal, hierarchical effects is well illustrated by its simplicity.

MACROSCOPIC IMPLICATIONS

The mesoscopic entities, water clusters and proteins, lie above the molecular level on the ascending path from microscopic to macroscopic. They are structural elements defined in size by the pressure pixel and emerge from the chaos of thermal motion as a result of cooperative order-disorder interactions between molecules.

Because of the dynamic nature of these opposing interactions, these basic elements appear as a wave, and are no doubt best understood as structural quanta (Kaivarainen 1989). Their wave form means that many such quanta can resonate and fuse to produce a higher-level quantum, if favorable spa-

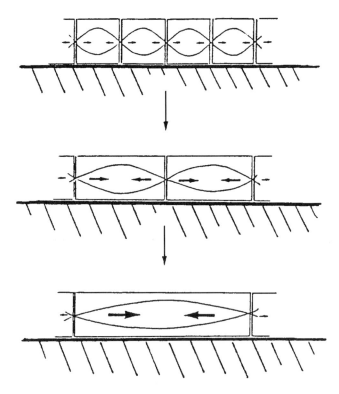

Figure 29.3 The Osmotic Mechanism. Many readers are aware of the swelling of clays and gels as they imbibe water against high opposing pressures. This seemingly contradictory behavior is explained by the cooperativity between water clusters. The presence of an extended two-dimensional surface of the solute aligns clusters, inducing in-phase behavior so the pressure pixel can become as large as the surface. Since tension operates within the pixel, tension can now extend over macroscopic distances even though the material is in the liquid state. The overall three-dimensional effect is an emergent phenomenon that results from the highly parallel arrangement of the surfaces when they are packed on top of each other in a stack.

tial and temporal correlations exist among them. Then many small pixels become one macroscopic one, disruptive pressure becomes unifying tension, and the multicomponent assembly becomes one entity. When such a transition spreads through the whole cytoplasm, the cell acts as one. This macroscopic quantum gives the protein gel its alive quality. This picture is in line with the quantum mechanism that has been proposed to explain long-range coherence in water (Del Giudice et al. 1988) and information processing in the cytoplasm of the nerve axon (Jibu *et al.* 1994).

This phenomenon does not stop at the cell membrane. Although an impermeable barrier to chemical metabolites, the membrane is not a solid, static shell that is separate in substance and function from the cytoplasm. On the contrary, because it is also just 3–4 nm across, its size matches the dimensions of the pixel, making it a natural extension of the protein gel inside (Watterson 1987c). Thus the quanta do not need special gating mechanisms to allow them to pass in and out of the cell. One and the same quantum can exist on both sides of the membrane.

By simple extension of this argument, we can ascend to the next hierarchical level, the tissue. Correlations among a population of aligned cells could induce harmonic transitions involving the macroscopic quanta that appear in cells individually as a result of their own internal transitions. Then the whole population would resonate as a single entity. Such may be a mechanism that would help explain the synchronous behavior of brain cells, known to be associated with mental phenomena (Hameroff 1987).

REFERENCES

Clegg, J. S. 1984. Properties and metabolism of the aqueous cytoplasm. *Am. J. Physiol.* 246:R133–R151.

Conrad, M. 1992. Quantum molecular computing: The self-assembly model. *Adv. in Computers* 31:235–24.

Cooper, A. 1976. Thermodynamic fluctuations in proteins. *Proc. Natl. Acad. Sci. USA* 73:2740–1.

Del Giudice, E., A. Preparata, and A. Vitiello. 1988. Water as a free electric dipole laser. *Phys. Rev. Letters* 61:1085–88.

Flory, P. J. 1953. *Principles of polymer chemistry.* Ithaca, NY: Cornell University Press, pp. 348–98.

Frank, H. S., and W. Y. Wen. 1957. Cooperative interaction in aqueous solutions. *Disc. Faraday Soc.* 24:133–44.

Hameroff, S. R. 1987. *Ultimate computing: Biomolecular consciousness and nanotechnology.* Amsterdam: Elsevier-North Holland.

Jibu, M., S. Hagan, S. R. Hameroff, K. H. Pribram, and K. Yasue. 1994. Quantum optical coherence in cytoskeletal microtubules: Implications for brain function. *Biosystems* 32:195–209.

Kaivarainen, A. I. 1989. Theory of the condensed state as a hierarchical system of quasiparticles formed by phonons and deBroglie waves. *J. Mol. Liquids* 41:53–72.

Watterson, J. G. 1981. *Biophysics of water.* New York: Wiley, pp. 144–47.

Watterson, J. G. 1987a. Does solvent structure underline osmotic mechanisms? *Phys. Chem. Liquids* 16:313–36.

Watterson, J. G. 1987b. Solvent cluster size and colligiative properties. *Phys. Chem. Liquids* 16:317–20.

Watterson, J. G. 1987c. A role for water in cell structure. *Biochem. J.* 248:615–17.

Watterson, J. G. 1989. Wave model of liquid structure in clay hydration. *Clays Clay Miner* 37:285–86.

Watterson, J. G. 1991. The interaction of water and proteins in cellular function. *Prog. Molec. Subcell Biol* 12:113–34.

Watterson, J. G. 1995. What drives osmosis? *J. Biol. Phys.* 21:129.

30 Microtubular Self–Organization and Information Processing Capabilities

J. A. Tuszyński, B. Trpisová, D. Sept, and M. V. Satarić

BACKGROUND INFORMATION

In this article, we intend to investigate the role microtubules may play as the biological substrate for consciousness-related activities at a subcellular level. To develop a working hypothesis we first need to look at their structure and dynamical processes in which microtubules participate. Microtubules (MTs) appear to be one of the most fundamental filamentous structures that comprise the cytoskeleton (Dustin 1984). MTs can be viewed as hollow cylinders formed by protofilaments aligned along their axes (Figure 30.1) and whose lengths may span macroscopic dimensions. In vivo, MTs are assemblies of 13 longitudinal protofilaments, each of which is a series of subunit proteins known as tubulin dimers. Each subunit is a polar, 8-nm-long dimer that consists of two slightly different 4-nm-long monomers with a molecular weight of 55 kilodaltons each. These two constituent parts are referred to as α and β tubulin. Each dimer possesses a permanent electric dipole moment $\vec{p}$ which can change its orientation. Thus, MTs provide an example of an electret substance, that is, an assembly of oriented dipoles. In this paper we investigate the question of which type of spatial arrangement these dipoles prefer under various conditions. We also try to elucidate the role they may play in information processing at a cellular level. The latter question will bring us closer to the possible involvement of MTs in the emergence of consciousness at a subcellular level.

It was suggested (Hameroff 1987, Rasmussen *et al.* 1990, Hameroff and Watt 1982) that conformational states of tubulin dimers present within MTs may be coupled to charge distribution or dipolar states thereby allowing for cooperative interactions with neighboring tubulin states. Furthermore, this mechanism could lead to the presence of piezoelectric properties that are very common in ferroelectrics. This, in turn, could prove very important in their assembly/disassembly behavior (Athenstaedt 1974, Margulis, To, and Chase 1978). Using dark-field microscopy it was experimentally found (Horio and Hotani 1986) that when a single MT is monitored, its two ends grow at different rates. The active positive end grows faster than the inactive negative end. However, each end stops growing independently in a stochastic manner and then immediately begins to shorten at a high rate

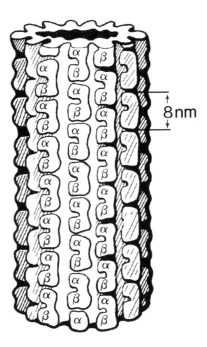

Figure 30.1 A schematic illustration of a microtubule (adapted from Amos and Klug 1974).

(this is referred to as a catastrophe). After the shortening period an MT suddenly stops and restarts the growth phase (which is called a rescue). It was also established that ensembles of MTs are synchronized which leads to periodic oscillations in their dynamics (Mandelkow and Mandelkow 1992).

In addition to energetic and structural factors, information processing appears to be an important aspect of MT behavior. Barnett (1987) suggested that filamentous cytoskeletal structures may operate as information strings in analogy to semiconductor word processors. The structural similarity between computer parallel processors and MT architecture is quite striking. Thus, it has been conjectured that MTs could act as processing channels along which strings of information bits can move, transferring messages over substantial distances between various components of the cell interior or even outside the cell. Furthermore, MTs form parallel arrays that are interconnected by cross-bridging proteins called microtubule-associated proteins (MAPs). Hence, MT arrays can easily be envisioned as playing the role of parallel-arrayed memory channels.

In this paper we address the question whether the basic functions of MTs are compatible with information-processing capabilities and thus by extension, with their possible role as the substrate for the emergence of consciousness at a cell level. We focus our attention on the dominant physical pattern; that is, the lattice of dipoles and investigate its ground-state properties. In general, the kinds of geometrical arrangements of dipoles we see in MTs are found to be: random (or paraelectric), ferroelectric (parallel-aligned), and of spin-glass type. Each of the above structures exists under different conditions of temperature, electric field, MAP distribution, and MT length giving rise to a possible sensitive state-switching mechanism. The ferroelectric state appears to be most suitable for assembly/disassembly processes while the spin–glass state is ideally suited for information processing (and thus can be seen as providing the substrate for consciousness) as will be argued in this paper.

THE EMERGENCE OF DIPOLAR PHASES

Our basic premise is that the entire MT may be physically viewed as a regular (triangular) array of coupled local dipole moments that interact with their immediate neighbors via dipole-dipole forces. Although Melki *et al.* (1989) showed that tubulin undergoes a conformational change (Figure 30.2a), we will tentatively adopt a simplified view where elastic degrees of freedom are not explicitly included in the description. However, an appropriate generalization of the physical model poses no technical difficulties and we comment on it in the last section.

The starting point in the analysis is to adopt a triangular lattice structure (located on the surface of the MT) with the dimensions and orientations as shown in Figure 30.2b and 30.2c. Each lattice site is assumed to possess a dipole moment $p = Q \cdot d$ where $Q = 2e$ and $d \approx 4$ nm and its projection on the

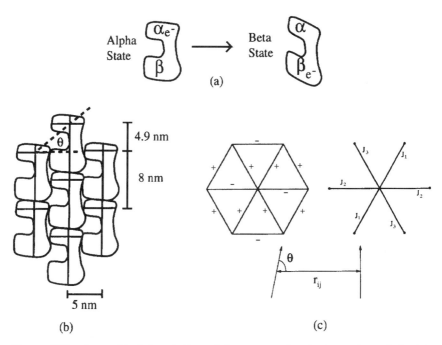

Figure 30.2 A graphical description of the structural units in a microtubule: (a) the two electronic states of a dimer, (b) the dimensions and angles of a unit cell, and (c) a schematic illustration of the model parameters used and their meaning.

vertical axis can only be $+p$ or $-p$. The interaction (dipole-dipole) energy E_{ij} between two neighboring lattice sites (labelled i and j). is, therefore,

$$E_{ij} = \frac{1}{4\pi e \epsilon_0} \frac{3\cos^2\theta - 1}{r_{ij}^3} p^2 \tag{1}$$

where ϵ_0 is the vacuum permitivity, e the dielectric constant of the medium, r_{ij} is the distance between sites i and j. The angle θ is between the dipole axis and the direction joining the two neighboring dipoles. Figure 30.3 illustrates the relevant situation used in our calculations.

In Figure 30.3a the signs " + " and " − " refer to dipole-dipole interactions that prefer either a parallel or an antiparallel arrangement of dipole moments, respectively. The numerical results for the constants J_1, J_2, and J_3 and the corresponding angles that were found based on the known structural data (Rasmussen et al. 1990) are: $J_1 = 5.77 \cdot 10^{-21}$ J, $J_2 = -0.71 \cdot 10^{-21}$ J, $J_3 = 3.40 \cdot 10^{-21}$ J, $\theta_1 = 0°$, $\theta_2 = 58.2°$, and $\theta_1 = 45.6°$.

With the known strong axial anisotropy of interactions we can map this situation onto an anisotropic two–dimensional Ising model on a triangular lattice so that the approximate effective Hamiltonian is now given by

$$H = -\sum_{<nn>} J_{ij} s_i^z s_t^{zj} \tag{2}$$

and the effective spin variable $s_i^z = \pm 1$ denotes the dipole's projection on the vertical MT axis. The exchange constants J_{ij} take the values J_1, J_2, J_3 depending on the choice of dipole pairs.

Due to the fact that $J_2 < 0$ and that there are an odd number (13) of protofilaments, the system exhibits a certain amount of frustration (Suzuki 1977). This means that for a closed path along the direction corresponding to J_2, it is impossible to satisfy all bond requirements. Hence, there will always be a conflict (hence the word frustration) between satisfying the energetical requirements of "+" bonds and "−" bonds. The ensuing dipolar phase structure is known in the physical literature as a spin-glass phase (Fischer 1983, 1985). In a spin-glass (SG), spin orientations are locally "frozen" in random directions due to the fact that the ground state has a multitude of equivalent orientations. For example, for each triangle reversing the spin on one side with respect to the remaining two, leads to an energetically equivalent configuration. Having the number of triangles on the order of the number of lattice sites, that is, $N \sim 2 \cdot 10^4$, yields the degeneracy of the ground state on the order of 6^N which is a very large number! This provides a very convenient property from the point of view of encoding information in such a highly degenerate dipolar lattice state. Note, however, that the other two directions do not exhibit frustration and this limits the extent of the SG phase.

Relatively small potential barriers separating the various equivalent arrangements of spins in the SG phase may play a very useful role since relaxation times are very long for the various accessible states. Some other properties of the spin-glass phase are the absence of long-range order

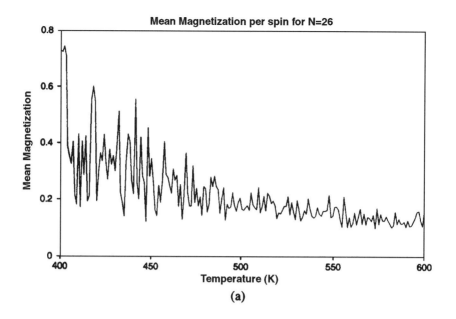

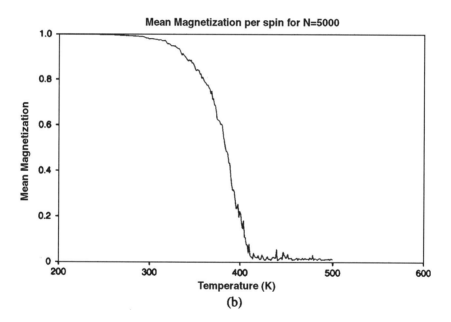

Figure 30.3 Plots of the mean polarization per site (y-axis) for (a) $N=26$ and (b) $N=5000$

between spin sites and the presence of short-range correlations. This means that the system is very "soft" energetically, permitting a number of spin arrangements that are relatively stable. However, there is no tendency towards the formation of large correlated spin clusters. This is very attractive from the point of view of computational applications of a MT. A SG state is, in fact, ideal for this purpose since it allows easy formation of local ordered states, each of which carries an information content and is long-lived. On the other hand, it is also not difficult to switch from one state to another through various physical means required to overcome the small potential barriers between individual states. A complete elimination of the SG phase can be achieved by: (1) the application of an electric field along the MT axis. We estimated that fields on the order of 10^4 V/m are sufficient to switch an MT from a SG phase to a ferroelectric (F) state. Note that fields up to 10^5 V/m are known to exist at a cell level so this situation is quite feasible. (2) Raising the temperature above the SG transition temperature will drive the system to a disordered (paraelectric). phase (3) A SG is also sensitive to the boundary conditions. This means, for example, that the presence of domain walls will have a profound effect on the SG phase.

When a moderate external electric field is applied (or when the temperature is lowered further) a F phase is favored, which is characterized by long-range order manifested by an alignment of spin directions along the axis. Obviously, from the point of view of information processing potential, this is not a very useful phase. However, it can play a major role in the assembly/disassembly processes taking place in the development stage of an MT architecture. It can be shown (Satarić, Tuszyński, and Zakula 1993) that a unique kink-like excitation (KLE) exists in the F phase which may propagate along a MT with a velocity in the range of 2–100 m/s that is proportional to the electric field E. The fact that growth rates are different at the two ends of a MT can be explained using the concept of a kink-like excitation. Namely, assuming that KLE formation is mainly due to the hydrolysis of GTP into GDP, one act of hydrolysis would correspond to conformational change resulting in the formation of a single KLE. However, a KLE is preferentially oriented towards the direction of the intrinsic electric field in the cell. Thus, propagation of a KLE would then distribute the energy of hydrolysis at the preferred end of a MT. This energy could then be used to detach dimers from the MT at its far end.

CONDITIONS FOR PHASE STABILITY

The next question we wish to address is the range of stability of the various phases identified in the previous section. This can throw light on the conditions required for the functioning of MTs in the SG phase. Assuming that this is the phase in which the onset of consciousness activities is the most likely to take place, this question could be of utmost importance to the microscopic origin of consciousness. Note, however, that for an infinite triangular lattice governed by the Hamiltonian of Equation 2, only two phases exist: the paraelectric phase (above T_c) and the F phase (below T_c), and the SG phase is completely absent in this limit. It turns out that the critical temperature T_c for the $F \rightarrow P$ transition depends linearly on the combination of model parameters used, that is, on $\bar{Q}$-Q^2 d^2/ϵ. For the realistic (although not known previously) values of Q, d, and ϵ, the transition temperature lies between 200 K and 400 K indicating the possibility of the associated phase transition close to room temperature, that is, at physiological conditions.

Many important factors may affect the value of T_c and thus provide sensitive control mechanisms for phase selection. Through a coupling to the elastic degrees of freedom (conformational change), the dielectric constant ϵ may be altered by the presence of water molecules surrounding an MT structure thereby decreasing the value of T_c and introducing dipolar disorder. Small structural changes, in particular shifts in the angles between the dimer dipoles, may remove the frustration mechanism effectively preventing the onset of the SG phase. Changes in the opposite direction can enhance frustration favoring the SG over the F-phase effectively switching from the growth mode of operation of MTs to their information-processing behavior.

In order to obtain some insight into the above questions, we have performed Monte–Carlo simulations for finite lattices with dimension $13 \times N$ (N is the length of the microtubule in terms of the number of layers). It is clear that as N increases, the SG phase is gradually removed. We see this effect by directly plotting the mean polarization per site for $N = 26$ (Figure 30.3a) and $N = 5000$ (Figure 30.3b) as two contrasting examples.

We conclude that dynamic processes leading to the elongation of MTs could effectively remove the information processing capabilities of MTs by expelling the SG phase. The same can be achieved by raising the temperature above a characteristic value that is length dependent.

We have also examined the effect of external electric fields and MAPs on the aforementioned transition. The electric field shifts the transition region and makes it broader. A similar effect can be seen by incorporating MAPs as "empty" (that is, non-polar) lattice sites. The actual magnitude of the shift and broadening depends on the pattern of MAPs chosen and the ratio of MAPs to the total number of lattice sites. Taking the set of parameter values which yields $T_c = (300 \pm 15)$ K for the perfect lattice results in $T_c = (250 \pm 20)$ K for the lattice with MAPs at a ratio of 1:11 while $T_c = (230 \pm 20)$ K is obtained for a ratio of 1:8. This indicates that MAPs substantially lower the transition temperature and make the SG-phase accessible to the MT system at much lower temperatures than those required in the absence of MAPs.

INFORMATION CAPACITY ESTIMATES

In order to examine the usefulness of MTs as the cell's information processors we must first evaluate the information capacity within each of the three phases identified. These results can be used to find the optimal conditions for the MTs to function as the substrate for consciousness-related activities. We base the calculations that follow on the standard

(Shannon) definition of information I of a statistical system where (Haken 1990)

$$I = -\sum_{i=1}^{K} p_i \ln(p_i) \tag{3}$$

Here, p_i stands for a probability value in state i and, obviously, the probability distribution must satisfy:

$$\sum_i p_i = 1 \text{ with } 0 \leq p_i \leq 1. \tag{4}$$

For the ferroelectric and paraelectric phases we adopt the mean-field approximation where each state is characterized by the continuous variable P (mean polarization per site). The energy functional is taken in the Landau form as

$$E = \left(\frac{A}{2}P^2 + \frac{B}{4}P^4\right)N_0 \tag{5}$$

where N_0 is the total number of sites in the lattice, $A = a(T-T_c)$ and $B > 0$. As is well-known above the critical temperature, that is, for $T > T_c$, E is minimized by $P = 0$ while below the critical temperature, that is, for $T < T_c$ by $P_\pm = \pm(-A/B)^{1/2}$. The associated continuous probability distribution $f(P)$ that replaces p_i of Equation 3 is the Boltzmann-weighted distribution function in the form:

$$f(P) = Z^{-1}\exp(-\beta E) = f_0\exp(\alpha P^2 - \gamma P^4) \tag{6}$$

where $f_0 = Z^{-1}$ is the normalization, $\beta^{-1} = k_B T$, $\alpha = -A/(2 k_B T)$ and $\gamma = B/(4 k_B T)$. Hence, for $T > T_c$, $f(P)$ is single-peaked at $P = 0$ while for $T < T_c$ it is double-peaked at $P = P\pm$.

Following Haken (1990) we calculate the information capacity in the paraelectric ($P = 0$) and ferroelectric ($P \neq 0$) phases as

$$I = \ln(Z) - \alpha\langle P^2\rangle + \gamma\langle P^4\rangle \tag{7}$$

where the averages are obtained using:

$$\langle P^n\rangle \equiv \int_{-\infty}^{\infty} f(P)P^n dP \tag{8}$$

We carry out the requisite calculations in a straightforward manner for both the ferroelectric and paraelectric phases where analytical calculations can be performed. For the SG-phase, however, we assume that the above prescription is valid only within the local domain of coherence or within the correlation length. Hence, for each domain i we have a local polarization P_i and the associated probability distribution $f_i(P_i)$, essentially analogously to those of Equations 5 and 6. Thus, for the total system the probability distribution becomes a product of local distributions each of which characterizes a domain of coherence

$$f = \prod_{i=1}^{n} f_i(P_i) \tag{9}$$

where n is the number of domains.

Note that n depends on temperature and we assume for simplicity that

$$n = 1 + \frac{(N_o - 1)(T - T_B)}{T_A - T_B} \tag{10}$$

in order to interpolate continuously between the ferroelectric and paraelectric phases since $T_B \leq T \leq T_A$. At $T = T_B$, virtually the entire system is uniformly polarized while at $T = T_A$ it is completely depolarized and incoherent. Note, that as a consequence of Equation 9 we obtain for the information capacity in the SG-phase

$$I = \sum_{i=1}^{n} I_i \tag{11}$$

where I_i refers to each individual domain. Our numerical computation clearly indicates that information capacity I is highest at the boundary between the SG and the paraelectric phase (see Figure 30.4) and hence if MTs are to be effective as information processors, they should use this narrow "window of opportunity" at the border area between these two phases. Of course, the actual location of the border area depends on the magnitude of the electric field applied and the concentration of MAPs present.

SUMMARY AND OUTLOOK

We have argued in this paper that the spatial arrangement of dipole moments of a MT is crucial to its functioning as a dynamic self-organizing

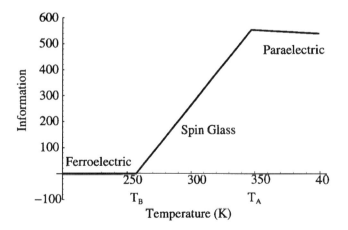

Figure 30.4 Plot of the information capacity as a function of temperature.

system. Due to the presence of frustration in the dipole-dipole interactions, a SG phase is predicted to arise at low enough temperatures and electric fields. The presence of MAPs will lower the temperature values required for SG-formation. The transition temperature itself decreases in proportion to the MAP ratio. The attractiveness of the SG phase has been recognized earlier (Stein 1992) and in the present context it lies in its maximum computational capabilities offered by a highly degenerate ground state. Moreover, long relaxation times give relative stability to short-range correlated dipole patterns. Each pattern can be seen as containing binary information encoded in the lattice.

The other ordered state that is possible to exist is a F phase with an almost perfect alignment of dipole moments along the protofilament axis. It is characterized by long–range order and hence its usefulness for information transfer and processing is dubious. However, it eagerly supports the formation of domain walls between the two stable orientations of dipole moments. The application of an external electric field preferentially directs kink-like excitation towards the properly aligned end causing a disassembly of the protofilament due to the energy released by the kink.

We have recently performed preliminary calculations including the presence of a conformational change associated with a β–state. We have found using Monte Carlo simulations that quite a different picture arises. Instead of three dielectric phases, only two exist: a low-temperature F phase and a high-temperature ferri-electric phase. The latter phase is characterized by the formation of local domains of polarization in two possible directions: vertically up the MTs axis or downwards and at 29° off the axis. However, net polarization appears to persist well above room temperature. The polarization may be a key physical factor in the hypothesized communication between microtubules in sufficiently concentrated assemblies.

We have discussed the various possible control mechanisms (field, distortion, temperature, MAP patterns) that could provide a means by which the MT could select an operating mode between information processing and assembly/disassembly types. This could shed some light on why the MT formation rate is enhanced in particularly important stages of the organism's history and development (learning, division, growth).

ACKNOWLEDGMENTS

This research was supported by NSERC (Canada), DAAD, and the Alexander von Humboldt Foundation.

REFERENCES

Amos, L.A., and A. Klug. 1974. Arrangement of subunits in flagellar microtubules. *J. Cell Sci.* 14:523–50.

Athenstaedt, H. 1974. Pyroelectric and piezoelectric properties of vertebrates. *Ann. N. Y. Acad. Sci.* 238:68–93.

Barnett, M. 1987. Molecular systems to process analog and digital data associatively. *Proceedings of the Third Molecular Electronic Device Conference*, edited by F. Carter. Washington, DC: Naval Reseach Laboratory.

Dustin, P. 1984. *Microtubules*. Berlin: Springer.

Fischer, K. H. 1983. Spin glasses (I). *Phys. Stat. Sol.* (b) 116:357.

Fischer, K. H. 1985. Spin glasses (II). *Phys. Stat. Sol.* (b) 130:13.

Haken, H. 1990. *Synergetics: An introduction*. Berlin: Springer-Verlag.

Hameroff, S. 1987. *Ultimate computing: Biomolecular consciousness and nanotechnology*. Amsterdam: North Holland.

Hameroff, S. R., and R. C. Watt. 1982. Information processing in microtubules. *J. Theor. Biol.* 98:549–61.

Horio, T., and H. Hotani. 1986. Visualization of the dynamic instability of individual microtubules by dark-field microscopy. *Nature* 321:605–07.

Mandelkow, E. M., and E. Mandelkow. 1992. Microtubule oscillations. *Cell Motility and the Cytoskeleton* 22:235–44.

Margulis, L., L. To, and D. Chase. 1978. Microtubules in prokaryotes. *Science* 200:1118–24.

Melki, R., M. F. Carlier, D. Pantaloni, and S. N. Timasheff. 1989. Cold depolymerization of microtubules to double rings: Geometric stabilization of assemblies. *Biochem.* 28:9143.

Rasmussen, S., H. Karamppurwala, R. Vaidyanatu, K. Jensen, and S. R. Hameroff. 1990. Computational connectionism within neurons: A model of cytoskeletal automata subserving neural networks. *Physica* D142:428–49.

Sataric´, M. V., J. A. Tuszyński, and R. B. Zakula. 1993. Kinklike excitations as an energy transfer inversion in microtubules. *Phys. Rev.* E48:589.

Stein, D. L., ed. 1992. *Spin glasses and biology*. Singapore: World Scientific.

Suzuki, M. 1977. Phenomenological theory of spin-glasses and some rigorous results. *Prog. Theor. Phys.* 58:1151.

31 Quantum Computation in the Neural Membrane: Implications for the Evolution of Consciousness

Ron Wallace

Evidence from natural and artificial membrane studies combined with physical principles of charge–transfer complexes and Rydberg atoms suggests a quantum computational system within the nerve cell membrane. The system has highly structured components coupled to permeant ion movements and synaptic morphological change. The latter process is a learning feature that permits natural selection to operate on the quantum level in animal nervous systems. Because the charge density configurations of membrane algorithms interact as computational tokens in a Hilbert space, there is massive parallelism and convergence of algorithms to close approximations of global optimality. Highly intractable problems ranging from regulation of external and internal cellular environments to integration of the hierarchical modules constituting animal consciousness may be routinely computed in polynomial time. The incorporation of quantum mechanics into the theoretical structure of neurobiological evolution may explain optimizations, including those of consciousness, more realistically than traditional connectionist models.

QUANTUM COMPUTATION IN THE LIQUID CRYSTAL NEURAL MEMBRANE

The nerve cell membrane is a phospholipid bilayer consisting of hydrophilic head groups defining the outer surfaces of the exterior and interior leaflets and a hydrophobic interior comprised of hydrocarbon diacyl chains. Embedded in the bilayer matrix are membrane-spanning proteins for which the spanning domain may range from a single polypeptide to several structured strands of amino acids. The neural membrane may be viewed macroscopically as a many-particle system in which constituent molecules fluctuate between disorder and order in position and orientation as a result of thermal or chemical changes or the application of an external field; that is, as a liquid crystal displaying phase transitions between the liquid and gel states (Mouritsen and Jørgensen 1992). Underlying the phase transitions are possible quantum-level processes resembling the search procedures of simulated annealing algorithms.

The hypothetical quantum computational process is initiated by spatially and/or temporally summed impulses converging at the outer surface of a neural membrane at resting potential (Kinnunen and Virtanen 1986). The impulses produce a surge of sodium ions, causing phospholipids to become deprotonated. This oscillating event dissociates a phospholipid-cholesterol complex, releasing a spin-correlated electron pair that moves longitudinally through the membrane lattice. The pathway for the system is created by the high orbital overlap, and consequent high conductivity, of ethylenic bonds between carbon atoms 9 and 10 in the phospholipid diacyl chains. The inter-diacyl bonds display a metastable conformation that can be further stabilized by one or two additional electrons. Molecular theory predicts that if one extra or an extra pair of electrons is introduced at one end of the metastable array, electron density will propagate as a soliton through the system. This would constitute a charge-transfer complex in which energy moves from a highest occupied molecular orbital to a lowest unoccupied molecular orbital. Charge-transfer reactions typically produce brief (10^{-10} sec) excited states in metals and biomolecules. Ethylenic orbitals, which have a low excitation threshold, would be promoted to metastable levels as the reaction proceeds through the lattice. The major constraints on interactions between and within deformed ethylenic orbitals would be hypersurface and charge.

As spin-correlated electron pairs continue to move through the system, driven by oscillating deprotonation at the outer membrane surface, ethylenic p orbitals would be deformed into a series of topologically equivalent hypershapes. Density configurations with dipolar, complementary hypersurfaces would combine to form larger units in a quantum variation of the simulated annealing algorithm. Local optima would be avoided by stochastic interactions among topologically fluctuating charge configurations (which constitute multiple parallel searches) as well as by applied fields of successive electron-pair movements produced by the oscillations at the outer membrane surface. The process would continue until the configurations "settle" into a stable arrangement of high probability density. The time estimate for this process based on hexadecane simulation of diacyl-chain carbon reorientations during increasing membrane viscosity (the possible macroscopic correlate of the settling process) is approximately 10^{-10} sec (Venable et al. 1993). This estimate is compatable with the 2×10^{-10} sec lifetime of highly localized, reconstituted Rydberg wave packets, or fractional revivals, which could function as automata in the computational process.

The termination of the quantum annealing algorithm opens an ion channel at the lipid-protein interface, thereby generating an action potential. The high probability density or, equivalently, the high-energy terminal state of the field Hamiltonian, is proposed as a mesoscopic coupling mechanism linking intramembrane quantum computation to the familiar action potentials of frequency-coded connectionist networks. The terminal state is recognizable as a transverse distribution of electronegative charge adjacent to the integral protein (Hall 1992). Coulombic interactions between this region and positively charged arginine and lysine residues open the channel pore and permit a highly regulated ion-by-ion permeant movement along a "backbone" of channel carbonyl oxygens (Leuchtag 1994). The mesoscopic coupling between a membrane terminal energy state and ion channel opening is thus a controlled process regulated by microscopic correlates of the membrane crystalline phase. Expressed in terms of complexity theory, phase relations of afferent impulses at the outer membrane surface are encodings of combinatorially complex problem statements. The problem is unpacked in the biomembrane into charge density automata or Rydberg computational tokens. A close approximation to a globally optimal solution is represented by the final high-energy arrangement of annealed charge-density automata. Subsequent channel opening and regulated ion movement converts the microscopic solution into a classic frequency code. The automata may encode values of temperature, pH, osmolarity, metabolite and hormone binding, and a wide variety of other combinatorially explosive variables that must be precisely regulated for moment-to-moment neuronal survival. In more neurologically complex species, the system could compute intractable problems involving the integration of sensory, autonomic, emotional, and memory modules at each level of a processing hierarchy (Jerison 1985). The latter system would closely approximate many philosophical and psychological definitions of consciousness (Oakley 1985).

Learning in any biosystem is evolutionarily essential, given short- and long-term fluctuations in environmental parameters. Information regarding changes must be communicated back to the system and generate, as a consequence, structural or behavioral modification. The possibility of quantum learning has previously seemed doubtful due to the transient nature of microscopic change. Given sufficient time, a finite quantum system will return to its initial conditions; for example, the gel phase of a biomembrane eventually returns to liquid and promoted ethylenic orbitals eventually return to the ground state. Learning is nonetheless possible in neural membrane microcomputations if the convergence to high-probability density generates a chain of events mediating long-term cytological change. Mesoscopic coupling between the membrane and the cytoskeleton appears to fulfill these conditions. Membrane-regulated permeant movement of Na^+ and K^+ generates a spike frequency pattern electrotonically conducted to a ligand-gated Ca^{++} channel, the NMDA receptor. Conjunction of the depolarization signal and binding of glutamate from an afferent terminal to kainate and quisqualate receptors activates the NMDA protein permitting regulated Ca^{++} movement into the cytoplasm (Kelso, Ganong, and Brown 1986). Incremental increase in cytoplasmic Ca^{++} in turn activates proteases (calpains) for which microtubule associated proteins (MAPs) are substrates (Baudry and Lynch 1984). Calpain-mediated phosphorylation modifies the attachment topology of MAPs to microtubules, generating nanoscale (10^{-9} sec 10^{-9} meter) tubulin dimer conformational change (Hameroff 1993). Due to dipole forces in neighboring dimers, MAP configurations, and binding of water and ions, the tubulin conformational changes propagate as waves through the cytoskeletal network. This is a candidate mesoscopic basis for gross cytological change, including dendritic spine modification associated with learning paradigms. The change in morphology would, in turn, affect the phase relations of subsequent afferent pulses that initiate membrane algorithms.

MICROCOMPUTATIONAL EVOLUTION IN NEURAL MEMBRANES

Quantum computation in membranes may have emerged early in unicellular evolution. Membraneous and cytoplasmic electrochemical events underlying the coordinated beating of cilia in *Paramecium, Tetrahymena,* and other protozoan genera resemble the electrochemistry of membrane-cytoskeletal interactions in nerve cells. This suggests that neural communication on both the quantum and classical levels may have begun as an exaptation of unicellular movement mechanisms. The ciliary machinery is a protein core, the axoneme, consisting of nine doublet microtubules joined by nexin links and arranged in a peripheral ring with two singlet microtubules at the center (Bray 1992). Regulation of axoneme microtubule activities by the ciliary membrane is suggested by observations of mutant *P. tetraurelia*. These forms display highly deviant movement sequences when ciliarly membrane lipid composition is altered but the axoneme is left intact (Kaneshiro 1990). These data are compatible with Bray's (1992) suggestion that the repertoire of *Paramecium* movements could be generated by membrane-controlled interactions of protein kinases, ion concentrations, and axonemal proteins that constitute a form of parallel distributed network. Incremental modification of this system in multicellular forms and primitive Coelenterata would have produced the earliest nerve nets.

During vertebrate neural evolution, highly intractable problems presented by mobility in complex environments required hierarchical modules (Jerison 1985) of mesoscopic computational systems. Peripheral modules would compute sensory and autonomic data; higher levels would integrate these into the representations of consciousness. Within any module, the massive parallelism of intramembrane computations would progressively converge to solutions at or near global optimality. Human examples of peripheral optimization include word recognition, visual search, and maximum auditory sensitivity. Higher-level optimization has been demonstrated for memory retrieval latencies, probability of recall for most-needed information, causal inference, problem solving, and predicting features of novel objects (Anderson 1991).

Although such performances may be simulated through Boltzmann machines, their architectures reflect the questionable assumption that the neuron is the elementary processing unit of nervous systems. An alternative possibility is the conservative modification, and application to complex functions, of membrane computational properties in existence for 10^9 years. The most sophisticated applications would exist in the higher modules as the computations of "mind."

CONCLUSION

Nerve cell membranes utilize charge density configurations in promoted Rydberg orbitals as cellular automata in massively parallel computations. The system originated in early eukaryotes as an exaptation of membrane-regulated cellular movements. During vertebrate neural evolution, intramembrane computations in hierarchical modules generated progressively optimal solutions to sensory and autonomic problems of mobility in complex environments.

The most sophisticated computations in higher-level processing modules involved continual integration of sensory and autonomic data into the representations of consciousness. The model nominates the neural membrane as the "master computer" of consciousness and extends Darwinian theory into the realm of quantum mechanics.

REFERENCES

Anderson, J. R. 1991. Optimality and human memory. *Behavioral and Brain Sciences* 14:215–16.

Baudry, M., and B. Lynch. 1984. "Glutamate receptor regulation and the substrates of memory." In *Neurobiology of learning and memory*, edited by G. Lynch, J. L. McGaugh, and N. M. Weinberger. New York: The Guilford Press.

Bray, D. 1992. *Cell movements*. New York: Garland.

Hall, Z. W. 1992. "Ion channels." In *An introduction to molecular neurobiology*, edited by Z. W. Hall. Sunderland: Sinauer Associates.

Hameroff, S. R. 1993. "Quantum conformational automata in the cytoskeleton: Nanoscale cognition in protein connectionist networks." Paper presented at Conference on Toward a Material Basis for Cognition. Abisko, Sweden.

Jerison, H. J. 1985. "On the evolution of mind." In *Brain and mind*, edited by D. A. Oakley. New York: Methuen.

Kaneshiro, E. S. 1990. "Lipids of ciliary and flagellar membranes." In *Ciliary and flagellar membranes*, edited by R. A. Bloodgood. New York: Plemun.

Kelso, S. R., A. H. Ganong, and T. H. Brown. 1986. Hebbian synapses in the hippocampus. *Proceedings of the National Academy of Sciences USA* 83:5326–30.

Kinnunen, P. K., and J. A. Virtanen. 1986. "A qualitative, molecular model of the nerve impulse: Conductive properties of unsaturated lyotropic liquid crystals." In *Modern Bioelectrochemistry*, edited by F. Gutman and H. Keyzer. New York: Plenum.

Leuchtag, H. R. 1994. Long–range interactions, voltage sensitivity, and ion conduction in S4 voltage segments of excitable channels. *Biophysical Journal* 66:217–24.

Mouritsen, O. G., and K. Jørgensen. 1992. Dynamic lipid–bilayer heterogeneity: A mesoscopic vehicle for membrane function? *BioEssays* 14:129–36.

Oakley, D. A. 1985. "Animal awareness, consciousness, and self–image." "In *Brain and mind*," edited by D. A. Oakley. New York: Methuen.

Venable, R., Y. Zhang, B. Hardy, and R. Pastor. 1993. Molecular dynamics simulations of a lipid bilayer and of hexadecane: An investigation of membrane fluidity. *Science* 262:223–26.

32 Computer Simulation of Anesthetic Binding in Protein Hydrophobic Pockets

Dyan Louria and Stuart R. Hameroff

INTRODUCTION

At a precise level of general anesthesia, consciousness is erased while other brain functions (EEG, evoked potentials, autonomic drives) continue (for example, Hameroff et al. 1994). Understanding anesthesia may illuminate consciousness.

General anesthetics include a wide variety of volatile gases with differing structures ranging from halogenated hydrocarbons to ethers to inert xenon. Attempts to understand anesthesia stem from their first clinical use in the nineteenth century; for example Claude Bernard (1875) showed that the anesthetic chloroform inhibited protoplasmic streaming in single-cell organisms such as slime mold. Meyer (1899) and Overton (1901) demonstrated that, despite their varied structure, the potency of anesthetic molecules correlated with their solubility in a lipid-like environment such as olive oil or octanol (Figure 32.1). This led to the assumption that the loss of consciousness associated with anesthesia occurred due to anesthetics acting in lipid portions of membranes. However, two factors led to reassessment of the lipid-membrane assumption for the site of action of anesthetics (Figure 32.2). One factor was that, in series of nonvolatile anesthetics of homologous structures and increasing size (like alkanes or alcohols), anesthetic effect is lost at a critical volume "cut–off" equivalent to two hexanol molecules or about 400 cubic angstroms (0.4 cubic nanometers, Franks and Lieb 1982, 1985). Lipid membranes have very large volumes of olive oil/octanol-like solubility, so the cut–off effect is difficult to explain by a lipid site of anesthetic action. The second factor causing reassessment of the lipid hypothesis was realization that dynamic conformational functions of proteins (ion channels, receptor activation, second messenger transduction, enzyme and cytoskeletal function) were essential to neural activity and likely to be the primary targets for anesthetic effects. Membrane-free enzymes such as light-emitting luciferase from fireflies and bacteria were found to be directly inhibited by anesthetics in proportion to their anesthetic potency (Franks and Lieb 1982). The cut-off effect and direct protein action were reconciled with the Meyer-Overton evidence by the fact that proteins have internal hydrophobic pockets with the

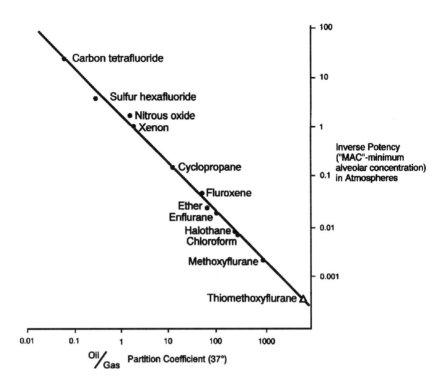

Figure 32.1 The Meyer–Overton correlation. Potency of anesthetics correlates with solubility in a lipid–like, hydrophobic environment.

same solubility characteristics as lipids (Halsey 1989). These subnanometer, water-excluding regions are comprised of amino acid side groups with polarizable character such as the aromatic electron resonance rings of tryptophan and tyrosine.

Binding of anesthetics in hydrophobic regions occurs by weak, physical interactions known as van der Waals forces (coupling of quantum level dipoles among electron clouds of the anesthetic and amino acid electron resonance rings within the hydrophobic pocket). The weak, nonchemical binding helps account for easy reversibility of anesthesia. Gas phase concentration of anesthetic in the lungs equilibrates with blood and brain. By simple mass action, when the anesthetic vaporizer is turned off and the anesthetic lung and blood concentration (gas "partial pressure") drops, anesthetic molecules vacate their brain protein hydrophobic pockets and the patient wakes up.

In this study, we used computer simulation to image intraprotein hydrophobic regions and investigate effects of anesthetic binding. Because such simulation is limited to proteins whose tertiary structures have been determined by crystallography (thus excluding membrane and cytoskeletal proteins), we chose two enzymes with documented tertiary structures and known anesthetic effects. One (papain) is known to be inhibited by anesthetics, and the other (acetylcholinesterase) is unaffected by anesthetics. We derived and compared effects of the anesthetic halothane on energy measures of hydrophobic pockets in the two proteins.

METHODS

Tertiary structures of papain and acetylcholinesterase were obtained from the National Protein Data Base and simulated using a Silicon Graphics computer system. Anesthetic bindings were optimized with INSIGHT (1993 Discover Program, Biosym Laboratories; San Diego, Calif.); optimizations, including all vibrational and rotational terms, were run by Steepest Descents for 2000 iterations with the convergence value set to 10 root–mean–square deviations. Hydrophobic contour maps were produced with the INSIGHT DOCKING program with an atomic cut-off radius of 8 angstroms. Contour maps were produced by initially computing a three-dimensional energy grid composed of a minimum of 1000 separate grid points. A visualization grid was then overlaid on the energy grid. Each grid representing hydrophobicity was then solidly contoured (Figures 32.3 and 32.4).

Two types of energy calculations were done for both proteins with and without halothane:

1. Intermolecular energy calculations represent binding between anesthetic and hydrophobic region. Ten intermolecular calculations (gradients of less than 0.11 kcal/mol/angstrom) computed for each anesthetic-protein interaction were averaged (Table 32.1).

2. Active site energies represent the total energy of the hydrophobic site and include van der Waals forces and electrostatic non–bonded energy terms.

Interatomic distances of the hydrophobic region and its surroundings, a measure of conformational state, were also calculated.

RESULTS

Figures 32.3 and 32.4 show a portion of papain's hydrophobic pocket without and with the anesthetic halothane.

Intermolecular energy terms (Table 32.1) are very small, substantiating weak van der Waals interactions between anesthetic and protein hydrophobic region. Interactions that approximate -1 kcal/mol are thought to predict sufficient binding (low Km) for anesthetic inhibitory effect. Papain-halothane intermolecular energies (-1.30 kcal/mol) predict favorable (exothermic) binding. Negligible interaction terms were calculated for acetylcholine–halothane.

Active site energies for both proteins with and without halothane (in Kcal/mol) are also shown in the Table. The presence of halothane causes a significant decrease in the energy of the active site for papain, but not for acetylcholinesterase. The majority of total energy reduction by halothane in papain is contributed by van der Waals forces between the halothane and the protein active site.

Interatomic distances revealed no significant changes with anesthetic in either protein.

Table 32.1 Changes in intermolecular energy and hydrophobic site energies (total, van der Waals, and electrostatic) for halothane binding in hydrophobic pockets of papain (anesthetic sensitive) and acetylcholinesterase (anesthetic insensitive). Change in total energy is comparable to $\Delta(\Delta G^0)$ (Franks and Lieb 1993).

	Δ Intermolecular Energy (Kcal/mol)	Δ Total Hydrophobic Site Energy (Kcal/mol)	Δ Van de Waals Hydrophobic Site Energy (Kcal/mol)	Δ Electrostatic Hydrophobic Site Energy (Kcal/mol)
Papain hydrophobic site (Vacant → Halothane) Anesthetic Sensitive	-1.28 ± 0.01	-22.3	-18.0	-4.3
Acetylcholinesterase Hydrophobic site (Vacant → Halothane) Anesthetic Insensitive	-0.15 ± 0.01	$+3.7$	$+11.3$	-7.6

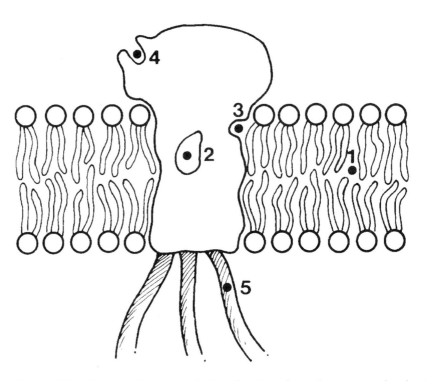

Figure 32.2 Five possible types of sites of action of anesthetic gas molecules in relation to a membrane protein. (1) within lipid portion of membrane, (2) hydrophobic pocket in membrane protein, (3) lipid–protein interface, (4) protein–water interface, and (5) cytoskeletal protein.

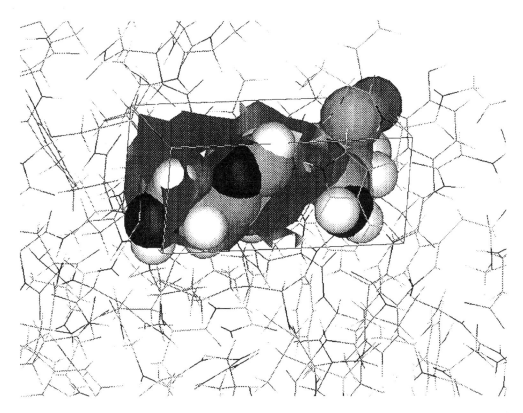

Figure 32.3 Computer simulation (scale 23 × 25 angstroms; 2.3 × 2.5 nanometers) of portion of the hydrophobic region (smooth volume within box) inside papain.

CONCLUSION

Intermolecular energy calculations show favorable anesthetic binding conditions for the anesthetic halothane and the hydrophobic active center of papain, and unfavorable conditions for anesthetic and acetylcholine.

Decrease in papain's active hydrophobic site energy in the presence of anesthetic represents a stabilization of the protein—a decreased capacity of the enzyme to reach an excited, active state. The majority of the total energy reduction is contributed by van der Waals forces, electron cloud dipole couplings between the anesthetic and the hydrophobic pocket of the protein. No such changes are seen with acetylcholinesterase.

Lack of anesthetic-induced change in interatomic distances suggests that anesthetic binding does not cause a protein conformational change (but may rather prevent conformational responses to appropriate stimuli).

We conclude that activity of papain (and other anesthetic-sensitive protein function) is inhibited by anesthetic reduction of van der Waals energy in the protein's hydrophobic pocket.

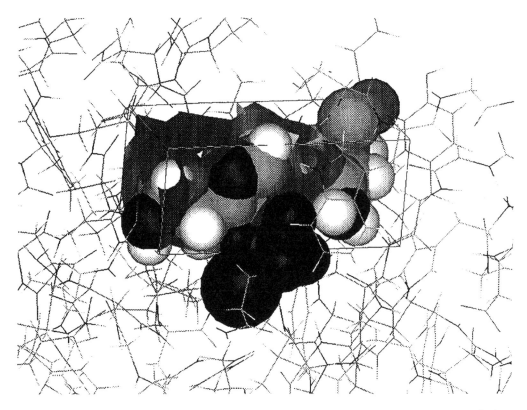

Figure 32.4 Computer simulation of hydrophobic region inside papain with the anesthetic halothane (stippled atoms).

DISCUSSION

Why do weak, physical interactions in small hydrophobic regions (roughly 1/100 volume within each protein) have such profound effects on protein function, and ultimately on consciousness?

Protein function depends on dynamic conformational activity: the ability of each protein to change its shape in response to an appropriate stimulus. Ion channel opening and closing, receptor activation, enzyme and cytoskeletal function all require protein conformational changes, but how those changes are regulated is poorly understood. One notable theory (Fröhlich 1968, 1975) is that global protein conformational states are coupled to quantum level events (dipole oscillation, electron mobility) in the protein's hydrophobic regions. Aromatic rings and other components of hydrophobic pockets permit electron delocalization: electrons are free to roam within the pocket among resonance orbitals of several stacked rings (van der Waals energy being a measure of this quantum-level mobility). According to Fröhlich's suggestion, if electron density is localized in one particular portion of the hydrophobic pocket, the protein assumes one particular conformation. If the electron is in a different intrapocket area, the protein assumes a different conformation. Switching between conforma-

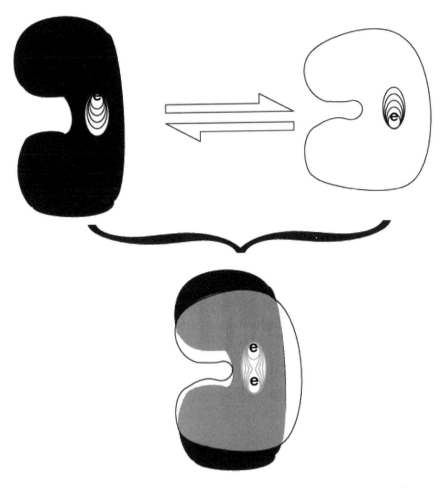

Figure 32.5 Top: Schematic representation of switching between two conformational states of a protein coupled to localization of an electron within a hydrophobic pocket. Bottom: quantum superposition of the electron and protein conformation.

tional states can occur on a time scale in the nanosecond to picosecond range. According to quantum theory, individual electrons can also be in a state of quantum "superposition" in which they occupy both positions (and the protein assumes both conformations—Figure 32.5)!

Mobility of free electrons (in a corona discharge) is inhibited by anesthetic gas molecules' van der Waals effects (Hameroff and Watt 1983). This further suggests that anesthetics act by preventing mobility of electrons within hydrophobic pockets that normally regulate protein conformation.

A totally different class of drugs, the hallucinogenic ("psychedelic") tryptamine and phenylethylamine derivatives bind in a very similar class of hydrophobic pockets, but exert opposite types of effects. Composed of aromatic rings with polar "tails," these drugs are known to bind to serotonin receptors. However effects on serotonin receptors are but one facet of their action, and their psychedelic mechanism is basically unknown (Weil, 1996). Nichols, Shulgin, and Dyer (1977) showed that psychedelic drugs bind in "octanol–like" hydrophobic pockets of less than 6 angstroms (0.6 nanometers) length. Kang and Green (1970) and Snyder and Merrill (1965) showed correlation between hallucinogenic potency and the drug molecules' capability to donate electron orbital resonance energy (comparable to van der Waals energy).

The following picture then emerges. Both anesthetics and hallucinogens bind and act in hydrophobic pockets within a class of neural proteins (receptors, channels, second messengers, cytoskeleton, enzymes). Hallu-

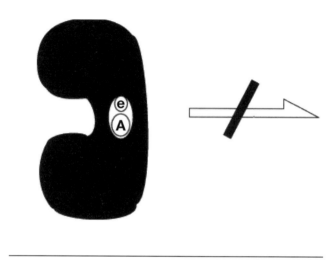

Figure 32.6 Top: Schematic of anesthetic (A) in hydrophobic protein pocket blocking electron mobility, conformational change, and quantum coherent superposition. Bottom: Psychedelic (P) with polar tail promotes electron mobility and quantum coherent superposition in protein hydrophobic pocket.

cinogen binding is far more selective than anesthetics (although not necessarily exclusive to serotonin receptors), directed by polar tails of each molecule's structure to a subset of protein hydrophobic pockets. Thus, hallucinogens act at very low concentrations whereas anesthetics act at high concentrations because anesthetics have far more nonspecific hydrophobic binding in lipids and nonneural proteins.

Under normal conditions, electron mobility within each protein's hydrophobic pocket is coupled to the protein's conformation and function. Electron mobility within spatially arrayed (and energetically pumped) brain protein hydrophobic regions leads to quantum coherent superposition (and subsequent self-collapse, or reduction) of electron/protein states throughout large volumes of brain—a macroscopic quantum state linked to aspects of consciousness (Marshall 1989, Penrose 1989, 1994; Hameroff 1994, Hameroff and Penrose 1996). According to this scheme (Figure 32.6), by lowering van der Waals electron resonance energy within hydrophobic pockets, anesthetics retard electron mobility, prevent quantum coherence and ablate consciousness. By enhancing electron resonance energy and mobility within a class of brain protein hydrophobic pockets, psychedelic drugs promote macroscopic quantum coherent superposition, and expand consciousness. "Normal" consciousness requires quantum coherent superposition in hydrophobic pockets of certain brain proteins.

ACKNOWLEDGMENTS

Dyan Louria and Stuart Hameroff were partially supported by NSF Grant No. DMS–9114503. The authors are grateful to Alwyn Scott, Djuro Koruga, Ron Lynch, and Bob Gillies for expertise and to Richard Hofstad for technical assistance.

Dedicated to Burnell R. Brown Jr MD, PhD, FRCA.

REFERENCES

Bernard, C. 1875. *Lecons sur les anésthetiques et sur l'asphyxie.* Paris: Baillère et fils.

Dickinson, R., N. P. Frank, and W. R. Lieb. 1993. Thermodynamics of anesthetic/protein interactions. Temperature studies on firefly luciferase. *Biophysical Journal* 64(4):1264–71.

Franks, N. P., and W. R. Lieb. 1982. Molecular mechanisms of general anaesthesia. *Nature* 300:487–93.

Franks, N. P., and W. R. Lieb. 1985. Mapping of general anaesthetic target sites provides a molecular basis for cut–off effects. *Nature* 316:349–51.

Fröhlich, H. 1968. Long range coherence and energy storage in biological systems. *Int. J. Quantum Chem.* 2:641:–49.

Fröhlich, H. 1970. Long range coherence and the action of enzymes. *Nature* 228:1093.

Fröhlich, H. 1975. The extraordinary dielectric properties of biological materials and the action of enzymes. *Proc. Natl. Acad. Sci. USA* 72:4211–15.

Halsey, M. J. 1989. "Molecular mechanisms of anaesthesia." In *General anaesthesia*–5th ed., edited by J. F. Nunn, J. E. Utting, and B. R. Brown, Jr. London: Butterworths, pp. 19–29.

Hameroff, S. R. 1994. Quantum coherence in microtubules: A neural basis for emergent consciousness. *Journal of Consciousness Studies* 1(1):91–118.

Hameroff, S. R., and R. Penrose. 1996. "Orchestrated reduction of quantum coherence in brain microtubules: A model for consciousness." In *Toward a science of consciousness: The first Tuscon discussions and debates.* edited by S. R. Hameroff, A. W. Kaszniak, and A. C. Scott. Cambridge, MA: MIT Press, pp. 507–39.

Hameroff, S. R., J. S. Polson, and R. C. Watt. 1994. "Monitoring anesthetic depth." In *Monitoring in anesthesia and critical care medicine,* 3rd ed., edited by C. D. Blitt and R. L. Hines. New York: Churchill Livingstone.

Hameroff, S. R., and R. C. Watt. 1983. Do anesthetics act by altering electron mobility? *Anesth. Analg.* 62:936–40.

Kang, S., and J. P. Green. 1970. Steric and electronic relationships among some hallucinogenic compounds. *Proc. Natl. Acad. Sci. USA* 67(1):62–7.

Marshall, I. N. 1989. Consciousness and Bose–Einstein condensates. *New Ideas in Psychology* 7(1):73–83.

Meyer, H. H. 1899. Zur Theorie der Alkoholnarkose. I. Mitt. Welche Eigenschaft der Anasthetika bedingt ihre narkotische Wirkung? *Arch. Exp. Path. Pharmak.* 42:109.

Nichols, D. E. 1986. Studies of the relationship between molecular structure and hallucinogenic activity. *Pharmacology Biochemistry & Behavior* 24:335–40.

Nichols, D. E., A. T. Shulgin, and D. C. Dyer. 1977. Directional lipophilic character in a series of psychotomimetic phenylethylamine derivatives. *Life Sciences* 21(4):569–76.

Overton, E. 1901. Studien über die Narkose zugleich ein Beitrag zur allgemeinen Pharmakologie. Jena: G Fischer.

Penrose, R. 1989. *The emperor's new mind.* Oxford, England: Oxford University Press.

Penrose, R. 1994. *Shadows of the mind.* Oxford, England: Oxford University Press.

Snyder, S. H., and C. R. Merrill. 1965. A relationship between the hallucinogenic activity of drugs and their electronic configuration. *Proc. Natl. Acad. Sci. USA* 54:258–66.

Weil, A. 1996. "Pharmacology of consciousness." In *Toward a science of consciousness: The first Tuscon discussions and debates,* edited by S. R. Hameroff, A. W. Kaszniak, and A. C. Scott. Cambridge, MA: MIT Press, pp. 677–89.

VII Quantum Theory

Can we understand consciousness if we don't understand the details of the universe in which it exists? Quantum theory describes these details, and reveals features that may help explain enigmatic features of consciousness.

Having developed from experimental results in physics, quantum theory reveals that the atoms and subatomic particles that comprise our brains (and nearly everything else) usually aren't in a definite place at any one time. The grains of our existence are intangible waves of vibratory possibility. Like spinning coins that are both "heads" and "tails" until they fall, quantum entities can exist in "coherent superposition" of different states (that is, *simultaneously* having different or opposite spins, locations, and/or momenta). Yet somehow, uncertainty at the quantum level yields stable structures in our macroworld: a transition known as *wave function collapse*, or *reduction*.

Physics has no obvious explanation for the cause and occurrence of wave function collapse, or reduction. Experimental evidence through the 1930s led Heisenberg, von Neumann, and others to conclude that quantum coherent superposition continues until *conscious observation* collapses, or reduces, the wave function. Even macroscopic objects, if unobserved, could remain superposed. To illustrate the apparent absurdity of this notion, Erwin Schrödinger (1935) described his now-famous "cat in a box" being simultaneously both dead and alive until the box was opened and the cat observed. This unsettling prospect has evoked various new physical schemes in which growth and persistence of superposed states reach a critical mass-time threshold, at which collapse, or reduction abruptly and spontaneously occurs (Károlyházy, Frenkel, and Lukacs 1986, Diósi 1989, Ghirardi, Grassi, and Rimini 1990, Penrose 1989). According to Penrose (1994), superposed states each have their own space-time geometries; when the cumulative mass-energy difference between superposed states sufficiently separates these geometries (to a degree related to quantum gravity), the system must reduce to a single universe state, thus avoiding the need for "multiple universes" (for example, Wheeler 1957).

When quantum systems do collapse or reduce, for whatever reason, each superposed state "chooses" a singular classical state: one spin, one location, "heads" or "tails," cat dead or alive. According to conventional quantum theory (the "Copenhagen interpretation"), each choice of outcome state is random, or probabilistic—a feature with which Einstein, for one, expressed displeasure: "God does not play dice with the universe!" Penrose (1989, 1994) observes that the choices are more accurately described as "noncomputational," that is they cannot be deduced algorithmically—a property Penrose ascribes as being essential to consciousness and unattainable in computers. Thus wave function collapse or reduction *per se*, involving self-organization of fundamental space-time geometry, may be intimately associated with consciousness (see also Stapp 1993).

Another quantum property is *nonlocal coherence*. Having once interacted, quantum objects in coherent superposition remain connected and indivisible—a possible solution to the "binding problem" of consciousness. Quantum theory, with implications for nonlocal "oneness," universal order, and fundamental space-time geometries, has lent itself to Eastern philosophy. In mystical allegory, the Mind of God is envisioned as a spinning coin whose chosen face is revealed only when directly observed (Borges 1964). In *monistic idealism*, matter and mind arise from consciousness—the fundamental constituent of reality (for example, Goswami 1993). Some Western philosophy may also be comparable. For example in David Chalmers's view (Chapter 1), consciousness derives from the experiential aspect of fundamental information; in Leopold Stubenberg's (Chapter 3), qualia comprise the monistic entity underlying both physical and mental substance. Self-organizing quantum processes in fundamental space-time geometry provide a possible common ground.

But where in the brain, and how, could coherent superposition and wave function collapse or reduction occur? A number of proposals have been put forward (for example see Walker 1970, Lockwood 1989). John Eccles (1994, see also Beck and Eccles 1992) has proposed that quantum uncertainty provides a dualistic interface between the "Self" and its brain, by influencing the seemingly probabilistic release of neurotransmitter vesicles from presynaptic axon terminals. In this section, proposals for quantum coherence and collapse relevant to consciousness are offered for brain proteins, membranes, cell water, and microtubules.

In Chapter 33, Danah Zohar draws parallels between the holistic nature of consciousness and certain aspects of quantum theory in which components equate to the whole. Zohar describes macroscopic quantum coherent phenomena (called *Bose-Einstein condensates*—akin to superconductors and lasers) whose individual components become coherent and indivisible. Such coherence is proposed by Zohar (and Ian Marshall) to occur among neural proteins throughout significant brain volumes. By correlating and unifying distributed brain activities, this *nonlocal quantum coherence* may provide mechanisms for binding and the unitary sense of self.

Fred Alan Wolf discusses in Chapter 34 how self-awareness can arise from superposition of quantum automata in neurons and glial cells. Wolf argues that these "self-reflective images" become ordered and emerge during dreaming. Aside from the quantum perspective, Wolf's recognition of glia (which comprise over 80 percent of brain cells) as resonators of consciousness is intriguing.

Michael Conrad outlines in Chapter 35 how quantum wave function collapse at the level of neural proteins may be amplified to higher scales capable of controlling brain activities. The enriched state information of the wave function carries the capability for qualia, he claims, and wave function collapse, or reduction, is the essence of choice or free will. Conrad describes how quantum superposition dynamics of electrons and hydrogen bonds within proteins control their conformation and function.

Mari Jibu, Scott Hagan, and Kunio Yasue describe in Chapter 36 how quantum optical coherence essential to consciousness can emerge from the ordering of water in hollow cores of microtubules, specific cytoskeletal structures inside brain neurons. They propose laser-like coherent photons generated by biochemical pumping of microtubules can provide unitary binding essential to consciousness. Jibu, Hagan, and Yasue further contend that general anesthesia derives from effects on microtubule and other proteins by disrupting the water-ordering and quantum coherence.

Finally, in Chapter 37 Stuart Hameroff and Roger Penrose develop a model of quantum coherence and wave function self-collapse ("orchestrated reduction") in the constituent proteins of microtubules. They relate preconscious processing to quantum computing (while in coherent superposition), and consciousness itself to self-collapse and orchestrated reduction. Hameroff and Penrose contend that microtubules are capable of supporting and isolating quantum coherence, and that self-collapse (orchestrated reduction) occurs due to a critical perturbation of space-time geometry. They calculate estimates for the number of microtubules and neurons required for self-collapse by a quantum gravity criterion, and argue that self-selections in fundamental space-time geometry can account for enigmatic aspects of consciousness.

In review, this section begins with two general applications of quantum properties in attempts to explain binding and emergence of self-identity (Danah Zohar and Fred Alan Wolf). These are followed by Michael Conrad's in-depth treatise on wave-function collapse related to proteins, and how qualia may be intrinsic quantum properties. Mari Jibu, Scott Hagan, and Kunio Yasue consider the implications for consciousness of ordered water and quantum waveguide effects stemming from the cylindrical configuration of microtubules. Continuing a focus on microtubules as possible quantum devices, Stuart Hameroff and Roger Penrose examine coherence and wave function self-collapse related to consciousness and fundamental space-time properties.

Critics argue that invocation of quantum theory in consciousness is *reductio ad absurdum*; quantum effects are no more needed to explain consciousness than they are to explain higher-order mechanisms of combustion engines, hurricanes, or space shuttles. Consciousness, however, remains unexplained.

REFERENCES

Beck F., and J.C. Eccles. 1992. Quantum aspects of brain activity and the role of consciousness. *Proc. Natl. Acad. Sci. USA* 89:11357–61.

Borges, J. L. 1964. *Labyrinths—Selected stories and other writings.* Harmondsworth: Penguin.

Diósi, L. 1989. Models for universal reduction of macroscopic quantum fluctuations. *Phys. Rev. A.* 40:1165–74.

Eccles, J. C. 1994. *How the SELF controls its brain.* Berlin: Springer-Verlag.

Ghirardi, G. C., R. Grassi, and A. Rimini. 1990. Continuous-spontaneous reduction model involving gravity. *Phys. Rev. A.* 42:1057–64.

Goswami, A. 1993. *The self-aware universe: How consciousness creates the material world.* New York: Tarcher/Putnam.

Károlyházy, F., A. Frenkel, and B. Lukacs. 1986. "On the possible role of gravity on the reduction of the wave function." In *Quantum concepts in space and time,* edited by R. Penrose and C. J. Isham. Oxford, England: Oxford University Press.

Lockwood, M. 1989. *Mind, brain, and the quantum.* Oxford, England: Blackwell.

Penrose, R. 1989. *The emperor's new mind.* Oxford, England: Oxford University Press.

Penrose, R. 1994. *Shadows of the mind.* Oxford, England: Oxford University Press.

Schrödinger, E. 1935. Die genewärtige Situation der Quanten mechanik. *Naturwissenschaften* 23:807–49.

Stapp, H. 1993. *Mind, matter and quantum mechanics.* Berlin: Springer-Verlag.

Walker, E. H. 1970. The nature of consciousness. *Mathematical Biosciences* 7:138–78.

Wheeler, J. A. 1957. Assessment of Everett's 'relative state' formulation of quantum theory. *Revs. Mod. Phys.* 29:463–5.

33 Consciousness and Bose-Einstein Condensates

Danah Zohar

There has been considerable discussion in recent years about the physics of consciousness and the possibility that a quantum dimension is a necessary (though not a sufficient) element in any coherent, physical theory of consciousness. Often, this discussion is related specifically to the properties of Bose-Einstein condensates and to possible Bose-Einstein condensation in microtubules, in cell membranes, or in the water within cells. It is my purpose here to give an overview of what is at issue—what is Bose-Einstein condensation, what role might it play in consciousness, and how can we incorporate knowledge of Bose-Einstein condensation in various parts of the cell into a coherent theory of consciousness?

In the discussion that follows, I think it important that we make a crucial distinction between the basic capacity for conscious awareness and the structure, or contents, of that awareness. The structure of our awareness and the experience to which it gives rise, including rationality, intention, language, and features of the unconscious, is surely associated with our neural development and may to some extent be specific to human beings. The basic capacity for consciousness, on the other hand, is something that we share with all other sentient creatures, including perhaps even with such simple one-celled animals as paramecia and amoebae (or even, possibly, with yeast cells). When we discuss the physics of consciousness, I think it important that we concentrate on this basic level of capacity.

Philosophically, there are two basic approaches to the nature of conscious awareness that have been discussed during this conference. One approach is dualist. Dualism holds that mind and body are separate, that mind stuff is somehow of a different substance than matter stuff. The second approach is monist. Monism holds that there is just one substance and that both mind and matter are made of it. Monism can be materialist, arguing that both mind and body are derived from matter, or it can be idealist, arguing that mind is the primary property and matter is derived from it.

But there is a third possible philosophical approach to the mind body relation—that is the double-aspect approach. Double-aspect theories hold that there is an underlying common substance or common reality out of which both mental and physical properties arise, that both the mental and the physical are different organizations or different possible expressions of this underlying commonality. Quantum theories of consciousness are double-aspect theories, suggesting that both the mental realm and the physical realm derive from the underlying realm of quantum reality (see Figure 33.1).

Most theories of mind put forward in this century have been either materialist or dual aspect, suggesting either that mind derives from matter or that both matter and mind derive from a common, third substance. Either of these approaches raises the question of what is the nature of the physical realm with which mind is allied or from which it is derived? The overwhelming majority of twentieth-century physical theories of the mind have described the physical realm in Newtonian, or mechanistic, terms and this, in turn, has greatly colored their thinking about the nature and capacities of mind.

Newtonian physics has four salient characteristics that bear on the nature of anything described by this physics.

- It is deterministic. Events in Newton's scheme are law abiding, certain, and predictable. Things happen because they have to happen. B will always follow A in the same way if A starts from the same position and the same forces act upon the system. There are no surprises.
- It is reductionist. C is always just the sum of A plus B. A whole is never greater than the sum of its parts, and any whole can best be studied or

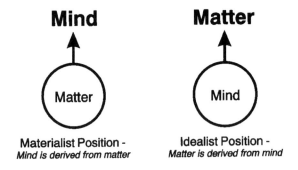

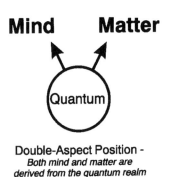

Figure 33.1 Some philosophical positions on the relation between mind and matter.

best understood by breaking it down into its parts and then analyzing those parts in isolation.

- It is atomistic. The world ultimately consists of separate, unanalyzable, impenetrable bits each occupying its own, isolated place in space and time. These bits can't relate internally, they can't get inside each other, and they can't form creative relationships. C will always equal just A plus B.

- It is a physics of either/or. It is a world of single actualities and exclusive truths, like Aristotelian logic. Things are either here or they are there, they're either now or then, something is A or not A.

When this physics is taken to be the physical basis of mind, it raises certain difficulties for anyone wishing to take full account of certain well-known phenomenological features of consciousness. Consciousness as we experience it is associated with things like free will, creativity, imagination, and the emergence of a sense of self, or "I-ness." Such difficulties are the motivation behind quantum theories of consciousness.

We are all aware that Newtonian physics is no longer at the cutting edge of physical thinking. It is still necessary for building bridges and for putting people on the moon, but it is not what is exciting in physics today. There is a whole new physics of the twentieth century that includes relativity theory, quantum physics, chaos, and complexity physics. While any full-blown theory of mind may well require that we take on board chaos and complexity dynamics as well, I am going to restrict my remarks here to quantum physics. Quantum physics has four salient characteristics that make it radically different from its Newtonian counterpart, and I think we can learn something important by looking at these.

- It is indeterminate. While the Schrödinger wave equation is itself determinate, the collapse of the wave function is not. We don't really know how or when any single quantum event is going to unfold. Collapse is uncertain and unpredictable because it is radically contingent.

- It is emergent. In quantum physics, C is always greater than A plus B, the whole is always more than the sum of its parts. When quantum things or systems combine, they acquire characteristics, acquire identity, that they did not previously have, so it is not entirely useful to study quantum wholes by analyzing them into the properties of their parts.

- It is holistic. In the quantum realm, separateness is at best an approximation. The quantum world is entangled, each part being correlated with or overlapped with every other, to some extent defined in terms of every other. This follows both from nonlocal relations, where apparently separate events are correlated in the absence of local, causal forces, and from the fact that the wave aspects of quantum entities are entangled.

- It is a physics of both/and. Quantum things evolve by throwing out possibilities—often mutually contradictory possibilities—in every direction, and these possibilities can have a real effect on the real world. Quantum things are both here and there, both now and then. Schrödinger's cat is both alive and dead at the same time.

It has been some 45 years since David Bohm (1951) first noted uncanny similarities between these features of quantum processes and some features of our mental processes. He wondered whether there mightn't be some fruitful line of investigation following from this. Many people since have wondered the same thing—hence the topic of today's session of this conference.

One of the similarities Bohm noted was that between the uncertainty principle in quantum physics and some uncertainty relations in our field of consciousness. Just as in Heisenberg's principle we have to choose between knowing the position or the momentum of a particle (knowing one causes the other to become fuzzy and indefinite), so in our consciousness we have to choose between entertaining a vague train of peripheral thought or focusing clearly on one concentrated thought. We cannot do both simultaneously.

Again, Bohm noted the similarity between the holism of quantum processes and the holism or contextualism of our thinking processes. We can't easily separate one element out from a train of thought and have it still make any sense. We can't break our sentences down into "atoms of meaning." Take for instance the metaphor, "She was the apple of my eye." If we try to break this down into bits, it ceases to make sense. The metaphor has to hang together as a whole.

Others have mentioned several other phenomenological features of consciousness that seem akin to, or compatible with, a quantum description of nature. There is, for instance, the whole question of free will. We experience ourselves as free and responsible beings. We can and do make choices and we are not wholly bound by our childhood conditioning. This free will is impossible to model in Newtonian terms, where everything is fully determined. Modeling it on quantum principles does not give us the whole story—quantum processes are, after all, frequently random—but at least a quantum substrate for mind is compatible with the development of free will.

Imagination is another thing that it is difficult to model in Newtonian either/or terms but easier to accommodate in the both/and world of quantum multiple possibility. As we sit in this room, we can imagine all sorts of contradictory futures for ourselves after the session ends—we might go down to the cafeteria for lunch, we might go back to the hotel for a swim, we might spend the afternoon attending the additional quantum symposium, and so on. We can hold all these mutually contradictory possibilities in our minds all at once. Imagination is both/and. It requires a both/and physical substrate.

Human creativity, too, requires that we have the ability to build new concepts, new categories, and new languages. We are constantly being creative in this way, constantly making new emergent wholes. This capacity is incompatible with Newtonian determinism and reductionism, but it is compatible with quantum emergence.

But the phenomenological feature of consciousness that I think it most useful to concentrate on for the rest of my talk is its special unity. As we sit in this room, each of us is being bombarded by millions of sensory data—olfactory data, auditory data, tactile and visual data—and yet we don't perceive the room in millions of bits. Our brains somehow combine all this incoming information into a unitary picture. How can they do it? This question is what Christof Koch and others have described as "the binding problem." If our brains consist of 10^{10} or 10^{11} different neurons, how do all these neurons cooperate to bind our experience of this room into a unitary whole? How, beyond that, is it that at a higher level still each of us has a unitary sense of self, a sense of ourselves sitting in this room, attending this conference in this city, and so on. From whence comes all this unity?

This unity of consciousness was in fact Descartes' main philosophical motive for the division of mind and body. He noted that mind is essentially unified, whereas the body is not. The matter of the body can be broken down into separate bits and pieces, ultimately to cells and even atoms, but the mind always hangs together. The mind has what Roger Penrose (1987) calls a "oneness or a globality." It is the point of today's session of the conference to point out that quantum "stuff" also has this kind of unity, and that a quantum substrate to mind might provide consciousness with its characteristic unity.

There are, as I mentioned earlier, two different senses in which quantum reality is holistic. One sense is that of quantum nonlocality, illustrated for instance by the behavior of the photons in the experiment by Aspect, Grangier, and Roger (1981). In that experiment, the two photons are introduced and then shot off to opposite ends of a large room. When they are measured, they are found to have correlated characteristics—the polarization of one is always opposite to the polarization of the other, even though no force or signal has travelled between them. These instantaneous correlations are very interesting, but they do not, I think, bear directly on the unity of consciousness.

The other kind of quantum unity is that shown in Figure 33.2, where it is contrasted with classical (Newtonian) atomism. Where classical atoms can only bump into, clash, and go their separate ways after meeting, quantum entities overlap and become entangled. This entanglement is made possible by the fact that all quantum entities have both particle aspects and wave aspects. The wave aspect extends almost infinitely across space and time, whereas the particle aspect is more localized. When the extended wave aspects of two similar quantum systems become entangled, those systems begin to share an identity. They begin to share their properties, they can occupy the same place in space and time, and we can write just one equation to describe the whole of their combined system. It is this kind of unity where an identity is shared that is relevant to the unitary properties of consciousness.

Let us remember back to our discussion of the millions of sensory data bombarding us in this room and the question of how our brains can bind them all together in a unitary whole. How does the same eye perceive the same data, how does the brain perceive it all as a unity? Something in the brain must share an identity between all these diverse physical bits. But there could be no shared identity between diverse Newtonian bits of matter—between, say, the brain's 10^{11} neurons, each located in its own place in space and time. But if there were a quantum substrate in the brain, its many entangled elements could share an identity.

There is, however, another property of our consciousness that we must consider here. Our consciousness is not just unified, it is also ordered. We aren't just reposing here in a vague meditative state but are, rather, busy perceiving and thinking. There are contents to our consciousness and they are ordered. This gives us an important further clue to the nature of the physical substrate that must underlie consciousness.

In classical physics, there is no physical substance that can give us both order and unity. There are substances like glass, where there is neither order nor unity. The various constituent bits are assembled in a random

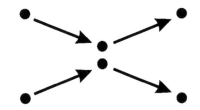

Newtonian Atoms
bump into each other, clash, and go their separate ways

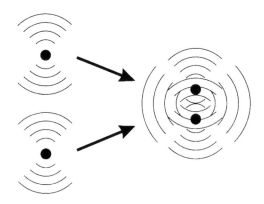

Quantum Systems
overlap, become entangled and share an identity

Figure 33.2 Differences between Newtonian and quantum views.

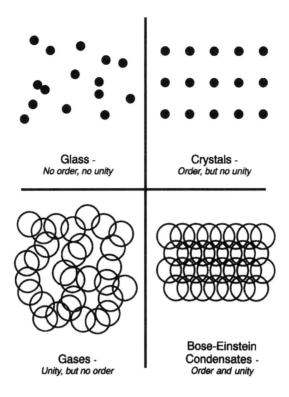

Figure 33.3 Bose-Einstein condensates provide quantum unity.

jumble of separateness (Figure 33.3). Then there are substances like crystals, which have a high degree of order, but no unity. The crystal lattice is lined up in a very orderly fashion, but each molecule in the lattice is separate from every other. Finally, in classical physics there are substances with a high degree of unity but no order. These are the gases, whose molecules do overlap with one another, but do so in a random and chaotic way. So there is no substance in classical physics that could give us both the order and the unity displayed by consciousness.

But there is a fourth kind of substance displayed in Figure 33.3. This substance has both unity and a high degree of order. It is a quantum substance known as a Bose-Einstein condensate. Bose-Einstein condensates are the most coherent structures that we know about in the physical universe. Their separate bits are so overlapped and entangled with each other that they behave as though there is just one large molecule present. A physicist can write just one equation for the whole frictionless ensemble. The physical quality of Bose-Einstein condensates that makes them so interesting is that they have a high "occupation number"—many different constituent bits can get into the same state, can share an identity.

There are several reasonably familiar Bose-Einstein condensates in nature and in our higher technology. Neutron stars are Bose-Einstein condensates. So are superfluids, superconductors, and laser beams. This is why superfluids have their special frictionless properties and why laser beams are so coherent. The many separate photons of a laser beam are so entangled that it is though there is just one photon in the beam.

The special physics of lasers has arisen at least twice already in our discussions in connection with holograms and holographic models of the mind. Holograms are in fact quantum structures. They are pictures "written" on laser beams. They are information carried by excitations of the laser beam's background substrate. Carl Pribram (1971) has been suggesting for at least twenty years that holograms offer us important models for the mind because they address the unity question. Any piece of a hologram contains an image of the whole. The whole is represented in every part. This seems to reflect the reality of our conscious life, in which every part of consciousness seems to contain unified awareness of our whole perceptual field.

It might be possible that we could take the analogy with holograms further. Just as the hologram (the information it contains) is an excitation of the laser beam's underlying Bose-Einstein condensate, our thoughts and perceptions may be excitations of a comparable Bose-Einstein condensate in the brain. Extensive work along these lines has been done by Ian Marshall (1989), who suggests that the brain's Bose-Einstein condensate is like a pond, and our thoughts and perceptions like excitations, or waves, on that pond. This suggestion may even link up with EEG phenomena in the brain (Nunn, Clark, and Blott 1994).

Several people have noted that Bose-Einstein condensation in the brain (laserlike phenomena or superfluidity) would make an excellent physical substrate for the unity of consciousness, but there has always been a major objection. Is it feasible? Most Bose-Einstein condensates form at either very low temperatures or very high energies, whereas the brain is a warm and sticky substance. It was Marshall who first pointed out that there is one excellent candidate for Bose-Einstein condensation in biological tissue. That is the work done by Herbert Fröhlich at England's Liverpool University (1986).

Fröhlich's work related principally to coherent quantum activity in bacteria and yeast cells, though since his first papers other experiments have been done and Fröhlich-style coherence has been noted in red blood cells (Rowlands 1983), DNA (Chwirot 1986), cell microtubules, and water (Del Guidice *et al.* 1989)—all at body temperature. The hypothesis is that any biological tissue ought to incorporate Bose-Einstein condensation of this sort. Fröhlich himself was interested in biological coordination. Marshall was the first to apply his work to the problem of consciousness.

So what might Fröhlich-style coherent quantum activity look like in the brain? Where might it occur and what would be its mechanism of action? In answering these questions, I disagree with those speakers earlier in the conference who have argued that we need only look to neural level activity in the brain to find the source of awareness. I am going to suggest that we must look to the subneuron level, even to the quantum level.

If we look at Figure 33.4, we see the diagram of a single neuron. It might just as well be the diagram of a microtubule for our purposes here. The interesting thing to look at is what is happening to the molecules in the membrane wall of the cell. All such membranes are lined with protein and/or fat molecules, and these molecules act as tiny electric dipoles. They are positive at one end and negative at the other. When the cell is at rest, as in the first part of the figure, these dipoles oscillate haphazardly. They are not "in phase," as a physicist would say. But what Fröhlich showed was that as energy is pumped through the cell—energy originating with the digestion of food and the rise in blood sugar levels—these dipoles begin to synchronize, and as they do so they broadcast tiny microwave signals. At a certain critical energy level, Fröhlich suggested, these microwave signals would be pulled into a coherent phase relationship constituting Bose-Einstein condensation. A coherent quantum field would form across the cell, and indeed extend to other cells.

The extent of coherence, and thus the strength of consciousness itself, in this model, would vary depending upon the amount of energy available to the brain. There is, for instance, more energy being pumped in when we concentrate, and less when we sleep, so we would expect variations in the

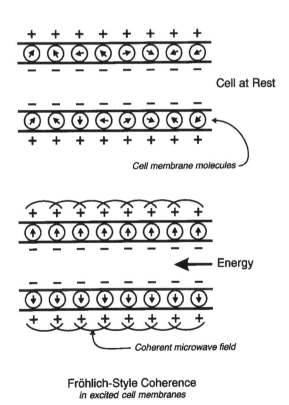

Figure 33.4 A model for biological Bose-Einstein condensates.

level of consciousness generated during these two activities. This of course accords with our phenomenological observations.

There are various suggestions as to where in the brain such Fröhlich-style activity might be taking place. Fröhlich himself suggested it was among protein or fat molecules in the cell membrane. At this conference, Stuart Hameroff has suggested it happens among protein molecules in the membrane walls of the cell's microtubules (the cell's skeletal structure), Mari Jibu, Kunio Yasue, and Giuseppe Vitiello have suggested it happens in water within the microtubules or in water within the cell.

How far across the brain might such Fröhlich-style coherence extend? Emilio Del Giudice *et al.* (1989) have calculated that laser-like effects in water can extend over something like a few hundred microns, which is the distance of about twenty ordinary cells. If the activity is in microtubules, it should be able to flow along from cell to cell over considerable distances, so there is scope here for coherent quantum phenomena unifying spatially separated regions of the brain and binding together the information going across several neurons.

Let us look now at these subneuron level quantum phenomena in a larger context. How might they relate to other, known, brain activity that is clearly associated with the structure of our conscious experience? Christof Koch spoke about neural oscillations and their role in binding our perceptual experience together. When I perceive something like the slide projector on my right, it is known that all the neurons in my brain associated with perceiving a particular angle of one piece of the projector oscillate in unison. Other neurons to do with perceiving the intensity of light from the projector also oscillate in unison, and so on. But the question behind the binding problem is, what causes all the neurons in my brain, related to all the many perceived features of the projector, to oscillate all together so that I perceive the projector as a whole and in its context? What synchronizes all these many separate oscillations? The quantum suggestion relating to this problem is that all the many neural oscillations are really modulations of an underlying quantum field, itself consisting of coherent microwave oscillations. On this model, the unifying quantum substrate would act to integrate the higher-level neural oscillations, resulting in perceptual unity. The hypothesis is that this quantum field could integrate oscillations, and hence information, from across the whole brain.

But is there any proof of a link between subneuron level quantum phenomena and the background state of consciousness? Do excitations of consciousness (thoughts and perceptions) have anything to do with the excitation of an underlying quantum field? An experiment recently done at England's Southampton University may shed positive light on these questions.

The point of the experiment was to see whether measuring a state of consciousness alters that state. If consciousness has a quantum dimension, there should be such a measurement effect; if not, not. To test this question, several dozen research subjects had their skulls connected to electroencephalogram (EEG) electrodes while being asked to perform a simple conscious task (selecting numbers from a random number generator). The subjects did not know whether the EEG electrodes were switched on or off. The result of the experiment was that measuring EEG during task performance affected the result of that performance at odds against chance of 1000 to 1. This strongly suggests that there was quantum activity involved in the task performance (Nunn, Clark, and Blott 1994).

There is one further phenomenological reason for supposing there is a quantum dimension in the brain's perceptual processing mechanisms. This is the speed with which we integrate perceptual data. As Christof Koch mentioned, it takes our brains roughly one tenth of a second to process all the data relevant to perceiving the contents of this entire room. Yet one physicist has calculated that if this amount of processing were done by a serial computer, it would take the computer from now until the end of the universe to handle so much data (Tsotsos 1990). How do we do it so quickly? Both Ian Marshall and Roger Penrose have suggested it is because the brain can utilize quantum superpositions—we can deal with many possible situations all at the same time, and thus process simultaneously information coming from all over the brain. Some people are now trying to develop superfast quantum computers on the same principle (Deutsch 1985).

What, then, would an overall model of mind look like that incorporated a quantum dimension? This model, I think, would have to take into account knowledge we have gleaned about the brain's operation at several levels—that of whole networks of neurons, that at the neuron level, and that at the subneuron—molecular and/or quantum—level. At the quantum, or molecular level, we would have something like Fröhlich-style coherent activity generating a field that gives us our basic capacity for unified awareness. At higher levels, we would have the structure of consciousness being given to us by the activity of serial and parallel processing embodied in neural tracts and neural networks. All three levels, I suggest, are crucial to our conscious experience. (It may be relevant to note that in brain injuries and illnesses such as schizophrenia, it is the structure of consciousness rather than the capacity for consciousness that is usually impaired.)

REFERENCES

Aspect, A., P. Grangier, and G. Roger. 1981. Experimental tests of realistic local theories via Bell's theorem. *Physical Review Letters* 47:460–67.

Bohm, D. 1951. *Quantum Theory.* New York: Prentice Hall.

Chwirot, W. B. 1986. New indication of possible role of DNA in ultraweak photon emission from biological systems. *Journal of Plant Physiology* 122(1):81–86.

Del Giudice, E., S. Doglia, M. Milani, C. W. Smith, and G. Vitiello. 1989. Magnetic flux quantization and Josephson behaviour in living systems. *Physica Scripta* 40:786–91.

Deutsch, D. 1985. Quantum theory, the Clark-Turing principle and the universal quantum computer. *Proceedings of the Royal Society of London* A400: 97–117.

Fröhlich, H. 1986. Coherent excitations in active biological systems. In *Modern bio-electrochemistry,* edited by F. Gutman and H. Keyzer. New York and Tokyo: Springer-Verlag.

Hameroff, S. R. 1994. Quantum coherence in microtubules. *Journal of Consciousness Studies* 1(1): 91–118.

Marshall, I. N. 1989. Consciousness and Bose-Einstein condensates. *New Ideas in Psychology* 7:73–83.

Nunn, C. M. H., C. J. S. Clarke, and B. H. Blott. 1994. Collapse of a quantum field may affect brain function. *Journal of Consciousness Studies* 1(1):127–39.

Penrose, R. 1987. "Minds, machines and mathematics." In *Mindwaves,* edited by C. Blakemore and S. Greenfield. Oxford, England: Blackwell.

Pribram, R. 1971. *Languages of the brain.* Englewood Cliffs, NJ: Prentice-Hall.

Rowlands, S. 1983. Coherent excitations in blood. In *Coherent excitations in biological systems,* edited by H. Fröhlich and F. Kremer. New York and Tokyo: Springer-Verlag.

Tsotsos, J. K. 1990. Analyzing vision at the complexity level. *Behavioural and Brain Sciences* 13:423–69.

Wilber, K, ed. 1982. *The holographic paradigm and other paradoxes.* Boulder, CO and London: Shambala Press.

34 On the Quantum Mechanics of Dreams and the Emergence of Self-Awareness

Fred Alan Wolf

This paper explores and offers a quantum mechanical model for the emergence of self-awareness from holographically generated dream images. Self-awareness arises from the ability of a simple memory device, an automaton in the brain, to obtain images of holographically stored glial cell memories and, most importantly, through a quantum mechanical process, to also obtain images of itself. Each self-image is composed of a quantum-physical superposition of primary glial cell images and an image of the automaton containing those images. These self-reflective images are ordered according to a hierarchy based on increasing levels of self-inquiry conducted during the dreaming process. Thus higher levels of the automaton's self-awareness are achieved by integrating images of itself on lower levels of the hierarchy. A comparison of the model with the observed dream self-reflectedness scale put forward by Rossi is made.

THE PHYSICS OF SELF-AWARENESS

In the past few years I have been researching the relationship of physics and consciousness. (Wolf 1981, 1984, 1985, 1986, 1987, 1989a, 1989b, 1991, 1992, 1994a, 1994b).

My work has led me to a novel model of the self as a hierarchy of levels of awareness. This idea is based in part on the work of Nobili and in part on the work of Albert. Nobili's work (1985) shows that ionic wave movement (that is similar in form and structure to quantum waves but different from them in certain essential details[1]) occurs in the glial cells of the brain making it an ideal medium for supporting and producing holographic imagery.

Albert's work[2] shows that quantum automata can not only remember objective properties but also hold memories of themselves. This self-reflective property, in combination with the holographic model, suggests a mechanism for the arising of the self as a hierarchy within the brain.

A BRIEF DESCRIPTION OF GLIAL MEMORY CELLS

The brain generates electrical wave activity as exemplified by the records of electroencephalograms (EEGs). By looking at individual neurons' electrical activity we see that there are quite active movements of electrical charges, mainly of sodium (Na+) and potassium (K+) ions, that pass from one side of a neural membrane to the other as a nerve pulse travels along the axon of the neuron.

Besides neural cells, that do not undergo cell mitosis (cellular division), there are glial cells that do. No one quite knows exactly why they are present.[3] Some studies indicate that these cells perform a metabolic function (Pribram 1977). When Albert Einstein died, his brain was autopsied and it was discovered that he had a larger than normal amount of glial cells associated with his visual cortex. This led many to speculate that glial cells had something to do with intelligence and possibly Einstein's enhanced ability to visualize very abstract concepts. Einstein had often written that before he wrote down any mathematical expression, he "saw" or conceptualized the new idea. This speculation about the connection between glial cells and visualization may have some foundation in truth.

Some researchers have shown that glial cells do more than provide nourishment to neurons. Peculiar movements of ions have been detected in glial cells and it is now suspected that these ion "transport" processes affect the bioelectrical activity of neurons and of the whole cerebral cortex (Nobili 1985). Research on multiple sclerosis, a disease which is associated with the breakdown of glial cells, also indicates that memory processes and motor processes are deeply affected, thus suggesting that glial cells indeed do more than just provide nourishment and supporting tissue for neurons (Pribram 1977). In Nobili's model, glial cells act as the medium for holographic waves and are therefore capable of storing holographic memory.

ALBERT'S AUTOMATA: A BASIS FOR SELF-AWARENESS

Based on the many-worlds interpretation of quantum physics, Albert explained how quantum mechanical automata might be constructed and used as computer memory elements (Albert 1983, 1986, 1992, Deutsch 1985). Albert indicated that these automata are capable of observing self-reflective states without disrupting themselves in the recall process. Yet they cannot necessarily observe objective states in other automata without disrupting them and thereby changing their memories and causing them to "jump" into different and usually random states.

SELF-REFLECTIVE QUANTUM PHYSICAL NEURAL STATES

My model is based on Nobili's and Albert's theoretical work. Let me first put the basic idea of the model into words. I assume that such automata exist in the brain. They may be glial cells or they may consist of structures consisting of spaces between the neural walls. They could also be boutons within the synapses of neurons. Other possibilities may exist (Hameroff 1987).[4] Whatever they are, the key insight into self-awareness arises from their ability to obtain and record images from holographically stored glial cell memories and, most importantly, to also obtain images of themselves while holding quantum physical complementary images. This property is unique to self-reflective automata.

Each self-image is ordered according to a hierarchy based on levels of self-inquiry. Higher levels of the automaton's self-awareness are achieved by integrating images of itself on lower levels of the hierarchy. When such an inquiry occurs, a jump from a lower to a higher level takes place as the new self-reflective image becomes part of the device's record of previously obtained self-reflective images. Thus each jump upward integrates images from all of the lower levels resulting in a sequence of images beginning with the lowest, simplest images and ending with the highest, complex images.

The lowest or "zero" level consists of nonself-reflective images and superpositions of images. According to the uncertainty principle, it is not possible for an automaton to obtain a single image and a superposition of images simultaneously. At the first self-reflective level, images and the records of the automaton containing those images are superimposed resulting in a bounded "emotional" memory. At the second level, these thought forms are integrated into "archetypes." At the fourth level, these archetypes are integrated into "super-archetypes." In principle the process is never-ending.

I suggest that this self-reflective aspect, which exists as a necessary consequence of the parallel worlds interpretation of quantum theory, is a deep clue to our own self-conscious nature and that the dream is a laboratory in which the differences between self-reflective and unreflective (nonself-reflective) perception can possibly be measured.

THE MATHEMATICAL STRUCTURE OF SELF-REFLECTION

Suppose that the object that the automaton interacts with, and thereby obtains a measurement of, is an ionic wave pattern simulating a holographic image in a glial cells. This wave pattern can be observed in several complementary ways depending on what the automaton wishes to extract from the cell.

Now suppose that the cell contains a holographic record of a superposition of images. To keep this as simple as I can, I will assume that there are only sixteen primary images, W_i, where "i" is a counter index. These sixteen can be superimposed into eight pairs, E_i, (120 different pairings[5]) where each pair constitutes a secondary image. These eight secondary images can be superimposed into four tertiary images, F_i, (28 different pairs) and so on. Thus the secondary images are complementary to the primary images and tertiary images. The tertiary images are complementary to both the secondary and primary images, and so on.

Although an automaton cannot hold simultaneously multiple images consisting of complementary observations of another automaton, it turns out that it can do so for complementary observations of images built up self-reflectively. An automaton is capable of "knowing" complementary observables of itself, but it cannot obtain and "know" complementary observables of another automaton.

The superposition of two images, say W_1 and W_2, is also an image that we label "E_1." To make this a little dramatic, and perhaps reflective of some people's memories, let W refer to women images, and E to emotional images. Let this superposition, W_1 and W_2, reflect some aspect of the person's life while living with these two women, say mother and sister. Of course there are many images in a person's memory. We are only looking at two specific images. Perhaps the sister image, W_2, reflects a younger crying child and the mother image, W_1, reflects a hysterical woman. Now there is no experienced emotional content to these separate images. They are just pictures.

But the younger crying sister image, W_2, and the hysterical mother image, W_1, when taken together create the emotional image "unhappy woman" which we have symbolized by the letter E_1. There may also be an E_2 image consisting of the superpositions of say, the first girlfriend image, W_3, and the second girlfriend image, W_4. These latter images taken together composing E_2 might reflect another emotional state that the person observed in woman, say, a joyful woman image.

Other images could exist. A superposition of the two emotional images E_1 and E_2 would make up a thought-form image, F_1. Since this image consists of both the unhappy and the joyous emotions it could represent a state of feminine or motherly understanding. A superposition of thought-form images, say F_1 and F_2, would represent an archetype say "goddess" image, G_1. And a superposition of archetypal images, G_1 and G_2, would stand for the super-archetypal "woman" image, S_1.

In the language of quantum physics, complementary observable-operators W, E, F, G and S, operate on sets of complementary glial cell states: the woman identity states, W_i, women emotional states, E_i, woman thought-form states, F_i and so on. If the glial cell contains a specific image W_1, and a woman image is called forth, a quantum-mechanical operator W "operates," and the glial cell yields the image W_1, which will be incorporated into the memory of the automaton. This produces a composite state of both the glial cell and the automaton that is the product of their separated states. We label this state,

$$W_1^{(0)} = [W_1]W_1 \tag{1}$$

The block brackets "[]" signify the automaton and whatever is inside of them is now a memory of the automaton. The superscript "(0)" means that a level "0" interaction has occurred between the automaton and the object of its inquiry. In this case that object is the image contained in the glial cell. Thus the state, $W_1^{(0)}$, is a coupled state of automaton and glial cell in which the automaton reflects the state held by the glial cell.

Now suppose that the glial cell contains a superposition of two women images W_1 and W_2, corresponding to the state of emotion, E_1. What happens if the automaton interacts with that cell? If it is under the instruction to obtain a woman image, and, according to quantum rules, it can only obtain a single image, W_1 or W_2, it will do so, obtaining one or the other at random according to the Copenhagen interpretation.

Now we come to the parallel "worlds" idea of quantum physics. In one "world" it obtains the image record, W_1, and in the other "world" it obtains and records the image, W_2. In the parallel "worlds" interpretation the superposition of the automaton and the glial cell in the two worlds is also an observable which means that it is, itself, a possible memory record capable of being measured by yet another automaton or even by the automaton itself. This state is

$$E_1^{(0)} = W_1^{(0)} + W_2^{(0)} = [W_1]W_1 + [W_2]W_2 \qquad (2)$$

However note that $E_1^{(0)}$ and E_1 are not the same. $E_1^{(0)}$ refers to a state involving both the automaton and the glial cell, while E_1 refers to a state involving only the glial cells.

QUANTUM SELF-INTERROGATION

Now we come to a very interesting aspect of all this: what story would the automaton tell if asked about its memory. In one version of the story (world II), the automaton would yield a crying sister image and simultaneously the glial cell would yield the same image. In another version of the story (world I), the automaton would yield the image of a hysterical mother and the glial cell would likewise yield the same image.

How would such an interrogation occur? There are two ways: (1) Interrogation by a second automaton and (2) self-interrogation. And in each interrogation there are two complementary questions that can be asked: (a) What image exists separately in the glial cell and in your memory? This question is about the W_i states. Or (b) What image exists compositely? This question is about the emotional composite state, $E_1^{(0)}$.

Consider (1a). A second automaton is brought in to attempt to determine the woman image in the glial cell and then compare it with the record in the first automaton. It finds the image of the first automaton and the glial cell record match exactly. In (1b) the second automaton can ask about the emotional state of the first automaton and the glial cell taken together. In that case it will know that there is an unhappy woman image present, but it will not know which image is present.

All the second system "knows" is that the first automaton and the glial cell have related and as a result they together are in the state, $E_1^{(0)}$, and it, the second automaton simply "knows" that but not which woman image is present. If it attempts to determine the women image it loses its knowledge of the emotional state. The image and the emotional states are complementary observables for the second automaton.

If a billion automata come along and do the same thing as the second automaton does, then they too will either enter into each of the two parallel worlds and find that a single image of woman is present or they may simply agree with the second automaton that the system of the first automaton and glial cell are in an unhappy woman emotional state.

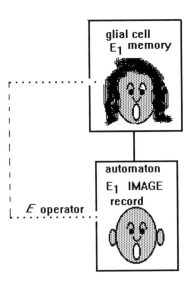

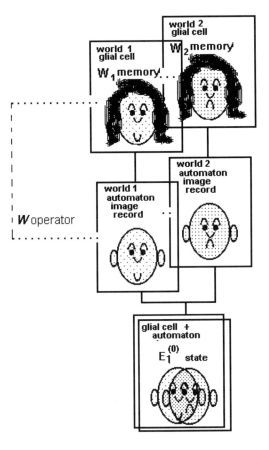

Figure 34.1 The automaton interacting in two complementary ways with a glial cell containing a memory. When the E operator is used, the emotional woman image is recorded by the automaton as a superposition of woman images W_1 and W_2. When the W operator is used, the images, W_1 and W_2 are obtained separately in parallel worlds. The superposition of the glial cell and automaton states compose the state $E'^{(0)}$.

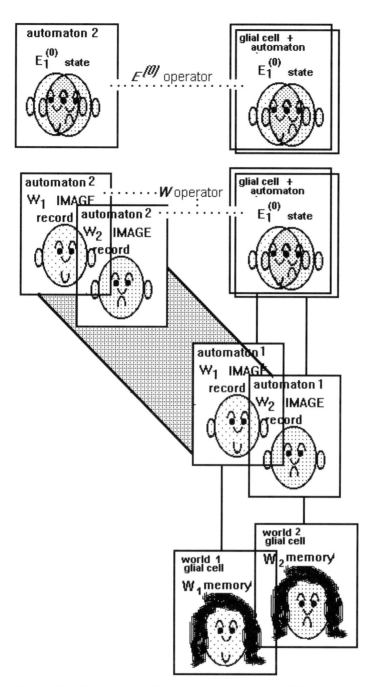

Figure 34.2 Interrogation by a second automaton. When automaton 2 interrogates the system composed of the first automaton and glial cell, it can do so in two complementary ways. If it inquires about the emotional state, $E_1^{(0)}$, using the operator, $E^{(0)}$, it obtains an exact duplicate of the emotional state. If it inquires about the woman image state, W_i, using the operator, W, it, too, is split into two parallel worlds, obtaining in each world a single woman image, W_i in agreement with the first automaton and glial cell. We note that the two automata must interact, and in doing so, the second automaton must "break in" to the memory of the first automaton, thus disturbing its memory content, consistent with the uncertainty principle.

But now we are going to ask the first automaton to do what we asked of the second in (1b) above. That is, to record and remember something that, it turns out, could not be done by an outside automaton. This record is the interior "emotional" state superposition of itself and the glial cell in both worlds taken together.

In other words we are going to ask the automaton to record the state, $E_1^{(0)} = [W_1]W_1 + [W_2]W_2$, and put that in its memory. When it does that, it makes its first self-observation and jumps one level upward, from "0" to "1." Then the composite state of the automaton/glial cell becomes,

$$E_1^{(1)} = [E_1^{(0)}]E_1^{(0)} = [E_1^{(0)},W_1]W_1 + [E_1^{(0)},W_2]W_2 \qquad (3)$$

Here the "(1)" signifies that the automaton has made its first level self-inquiry. The state, $E_1^{(1)}$, is quite interesting when we take into account what it means. In each world the automaton "knows" both the woman image it has observed in the glial cell, and in fact has created by its observation, W_i, and simultaneously it "knows" the emotional state, $E_1^{(0)}$, consisting of itself holding both composite images taken together, one in each world. It has in effect in each world knowledge of its own existence in another world symbolized by having both images inside of one bracket, $[E_i^{(0)},W_i]$. Possession of simultaneous knowledge of the emotional state of its composite "self" and the image state of the glial cell is a jump in levels from "0" to "1." This is the first level of self-reflection.

THE SELF/NOT SELF BOUNDARY

The appearance of this state, $E_1^{(1)}$, marks the boundary between self and nonself. It is the first act of self-reflection and when it takes place, as I pointed out, the automaton jumps levels. Next I speculate on how that jump in levels would be experienced.

Before the level jump, the emotional state, E_1, a glial cell memory, was recorded, but not experienced. These recorded but not experienced emotional states, E_i, constitute a basis for the unconscious mind. When the automaton obtains the woman images, W_i, the eigenstate $E_1^{(0)}$ is created, but not recorded. Instead, in each world a woman image, W_i, is recorded.

After the level jump, the emotional state, $E_1^{(0)}$, is recorded and experienced as $E_1^{(1)}$ resulting in a self-reflected feeling. When the automaton looks at its own memory, the unhappiness is now "felt" by the automaton. Furthermore the automaton "knows that it knows" and "knows" at the same time.

Now $E_1^{(1)}$ is a self-reflective state where the superscript "(1)" refers to the fact that it is a single self-reflection. Undoubtedly there is an aura of uncertainty surrounding this state since it is a memory of something real and "unreal" at the same time as far as the automaton is concerned in each world. The other world for it reflects upon its world and in doing so provides a recognition of itself via this reflection.

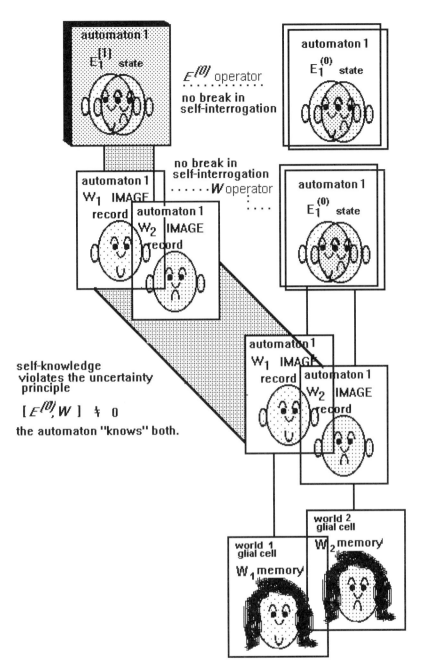

Figure 34.3 Self-interrogation by the automaton yields secret knowledge. When the automaton interrogates itself, it can also do so in two complementary ways. If it inquires about the emotional state, $E_1^{(0)}$, using the operator, $E^{(0)}$, it obtains knowledge of that state; however, in doing so, it is also obtaining knowledge of itself as well as the glial cell. If it inquires about the woman image state, W_i, using the operator, W, it, already split into two parallel worlds, obtains in each world a single woman image, W_i, already in its memory. In doing so, the automaton need not "break in" to its own memory. Thus, even though $E^{(0)}$ and W do not commute, the automaton knows both $E_1^{(0)}$ and W_1 in world 1 and both $E_1^{(0)}$ and W_2 in world 2. This does not violate the uncertainty principle because the information is not objective.

ASKING AND KNOWING

Suppose a second automaton attempted to obtain the total information found in automaton one. Could it do so without altering the records of the first automaton? The answer is "no." The attempt of the second automaton to obtain records of the woman image would put it into both parallel worlds where the separate woman images were held (see Figure 34.2). The attempt to determine the emotional state of the first automaton and glial cell could also alter the memory of the first automaton's record.

This is a very interesting aspect concerning asking and knowing. Before the second automaton is involved, the first automaton, as a result of interacting with the glial cell and then itself, contains memories of both its emotional state, $E_1^{(0)}$, and an identity state, W_i, in each world. The state of the system is unchanging so long as no one outside makes any inquiries. In the asking comes the inevitable disturbance regardless of who asks, if that information is passed outside of the first automaton. The second automaton may alter the system in inquiring and, even if the first automaton asks of itself, "What's going on?", the system also may be altered. Thus the mere asking of a question can change the memory.[6]

The ability of an automaton to know information depends on the identity of the automaton. The self-knowledge state, consisting of information taken from two worlds, peculiar to quantum automata, does not exist in classical memory elements. I suggest that at this remarkably simple level, we are seeing the arising of a delicate and primary self-concept. It comes about via secret knowledge, awareness of something that actually exists in parallel worlds simultaneously known only to the automaton in each world.

Now suppose we consider the other "woman" images. Just as the unhappy woman image E_1 was a superposition of hysterical mother, W_1, and crying sister, W_2, there is also an emotional state "happy woman," E_2, consisting of images W_3 and W_4 taken together. The self-joy emotional state would be written,

$$E_2^{(0)} = [W_3]W_3 + [W_4]W_4 \qquad (4)$$

where the subscript "2" refers to the second unconscious emotional state.

After self-observation of the joyful emotional state, the state becomes,

$$E_2^{(1)} = [E_2^{(0)}]E_2^{(0)} = [E_2^{(0)}, W_3]W_3 + [E_2^{(0)}, W_4]W_4 \qquad (5)$$

again the subscript "2" refers to the second emotional state.

And there is also the unconscious "thought-form" image, F_1, consisting of

$$F_1 = W_1 + W_2 + W_3 + W_4 = E_1 + E_2 \qquad (6)$$

Thus in a similar manner there are thought-form memory states of the combined automaton and glial cell that can be created in the same way that the unhappy emotional state was created.

By superposing the self-reflecting conscious emotional states, $E_1^{(1)}$ and $E_2^{(1)}$, we create a new unconscious thought-form state,

$$F_1^{(1)} = E_1^{(1)} + E_2^{(1)} \tag{7}$$

If the automaton attempts to self-reflect on this unconscious state as well, we have a second self-reflection occurring creating a jump in levels and the new conscious self-reflection state,

$$F_1^{(2)} = [F_1^{(1)}]F_1^{(1)} \tag{8}$$

Thus, following this procedure, we have an ascending[7] ladder of levels and associated self-reflecting states. A level remains unconscious until self-interrogation occurs. The self-interrogation takes place at every level resulting in unconscious superpositions of images becoming conscious. At the (0) level all images are objective, unconscious, and are found in the glial cells only. These images consist of the woman identity states, W_i, woman emotional states, E_i, thought-form states, F_i, archetype states, G_i, and even the super archetype states, S_i.

At the first self-reflection level we have conscious self-reflected emotional states, $E_i^{(1)}$, and unconscious emotional states, $E_i^{(0)}$, composed of superpositions of primal images. Thus we begin to see a general picture of the structure of the relationship of conscious content to unconscious content of the mind arising at each level.

At the second self-reflection level we have the unconscious thought-form self-reflective state, $F_i^{(1)}$, composed of the unconscious superposition of the conscious self-reflections of the first emotional state and the conscious self-reflection of the second emotional state. Here we see a principle that may be indicative of the Freudian model of repression: emotional material resulting in conscious emotions or feelings produce, through the superposition principle, unconscious thought-forms. Self-reflection on this level results in the unconscious thought-form, $F_i^{(1)}$, becoming conscious, $F_i^{(2)}$. A superposition of the conscious second level self-reflection thought-form states, $F_1^{(2)}$ and $F_2^{(2)}$, results in an unconscious second level archetypal state,

$$G_1^{(2)} = F_1^{(2)} + F_2^{(2)} \tag{9}$$

from which a jump to the third conscious archetypal level,

$$G_1^{(3)} = [G_1^{(2)}]G_1^{(2)} \tag{10}$$

is possible.

Again following the same procedure, a superposition of conscious self-reflective archetypes at level (3) $G_1^{(3)} + G_2^{(3)}$, yields an unconscious super archetypal image,

$$S_1^{(3)} = G_1^{(3)} + G_2^{(3)} \tag{11}$$

Self-reflection of the super archetypal image leads to a jump to the conscious fourth level,

$$S_1^{(4)} = [S_1^{(3)}]S_1^{(3)} \tag{12}$$

and so on.

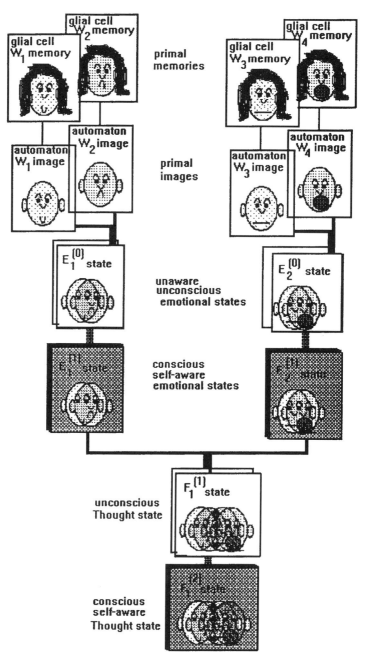

Figure 34.4 The formation of the hierarchy of self-identification. By combining four primal image states, W_i, into two emotional states, E_i, and combining those into a single thought-form state, F_i, a mind and its identity as an object or person is constructed. At each level of the hierarchy, an opportunity is presented for self-reflection. Each self-reflection results in an experience of the state just above it. Thus, unconscious emotional states are felt, made conscious, when the automaton self-interrogates at the lowest "(0)" level resulting in the "(1)" level of feeling. Felt emotional states, level (1) inquiries, are superposed to level (1) thought-forms which remain unconscious as thoughts until a self-inquiry is performed at level (1) resulting in experienced, conscious, thought-forms at level (2).

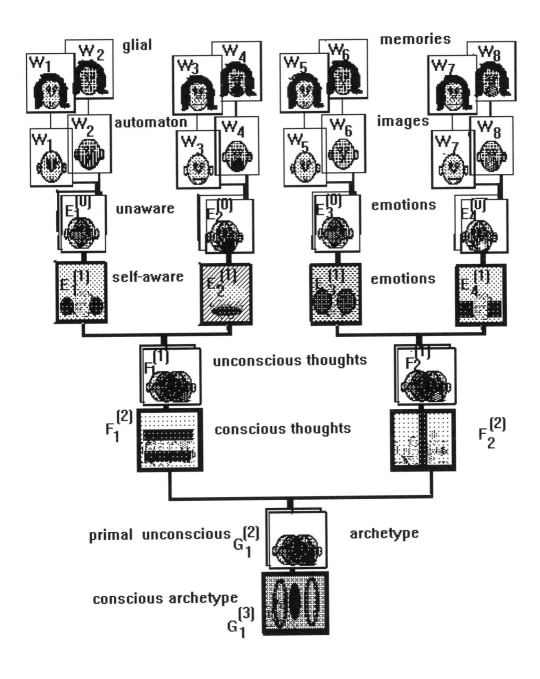

Figure 34.5 How consciousness and unconsciousness exists at all levels of the hierarchy. Here we see a more complex hierarchy built up from eight primal images resulting in four emotional states each conscious and conscious, two thought-forms states, both conscious and unconscious, and a single archetype state, that becomes conscious only when a self-inquiry occurs at that level.

THE ARISING OF IMAGES IN DREAMS

As put forward by Rossi (1992) and Moffitt (1994) the major stages of the processing of self-reflection take place during dreams. Although my descriptions of five stages of self-reflection differ somewhat from Rossi's, the intent is to show that some overlap clearly exists. At this stage no attempt has been made to make more than a gross comparison. During an ordinary dream, processing would be taking place at the lowest levels of image production wherein no sense of self is present, that is, the dreamer is unconscious of having a dream. This would correspond to the formations of nonself-reflecting states, level (0), where the dreamer is not aware of being present in the dream and the dream is experienced unconsciously with little sense of identification of self and other. The images would be seen, but not self-reflectedly experienced since no self has been defined. This would constitute unconscious data processing. I say unconscious, meaning unself-reflecting. The images would be there but they would have no meaning.

At level (1) the dreamer becomes involved in the dream. Moments of conscious and unconscious dream experience would result. This would correspond to jumps between level (0) and level (1) and the formation of images mostly in the lowest level. When level (1) was reached, emotions would be felt during the dream, but as soon as the dreamer ascended levels, only images would be present. By jumping between levels (1) and (0) various emotions could be aroused and sensed consciously during the dream. Meaning arises when the self appears. Such events would appear as enlightened or intense awareness. When level (1) is achieved the dreamer not only sees the images, she or he also possesses feelings about them that can be associated within space and time boundaries. These are perhaps primarily body images and would be emotionally felt. Perhaps this is a clue to illness arising from emotional causes probably initiated during early childhood.

At level (2) the dreamer is able to have thoughts during the dream. Here thought-forms arise and jumps between levels (0), (1), and (2) occur, resulting in a range of states of conscious and unconscious thinking, feeling, and observing. Thought adds much greater meaning to the emotions and the observations. Just as images come from sensory inputs, thoughts and feelings are capable of outputting in terms of expression of words and feelings. Unconscious thought forms integrate or superimpose conscious emotions and thus tend to have no emotional content *per se*. However, as thoughts are expressed in words, that is, self-reflection occurs, emotions can and often do arise, sometimes unexpectedly. This would be due to a jumping from level (2) to level (1) as a result of inquiry. Most likely the jump downward in the hierarchy involves a disruption of the higher (lower on the drawing in Figure 34.5) level functioning. Thus, during the wake state, when words move us to emotional action, we are descending to a "lower" (higher on the Figure) level of self-awareness.

At level (3) the dreamer is able to simultaneously be aware of the previous levels of participation and observation during the dream. Here the sense of self more fully emerges as the dreamer deals with archetypal images. Again there is full access to the lower levels.

At level (4) the dreamer consciously reflects on the fact that she/he is dreaming. This would be the super archetype state corresponding to lucid awareness during the dream.

At each level the automaton is able to record images from lower (higher on the Figure) levels or images of the highest (lowest on the Figure) level it is capable. The balance is quite delicate as I see it and the tendency would be to descend levels (move upward in the Figure) more readily than to ascend them (move downward in the Figure). My intuition suggests that these levels are somewhat akin to energy levels of atomic systems, and concepts involving entropy may be applicable. Descent results in less self-awareness (greater entropy) and therefore a requirement for more automaton-mechanical behavior. Ascent results in greater choices, to become aware of existence in other "worlds," and more complex imagery with a higher number of paradoxical features simultaneously knowable.

Remember any attempt to observe these images coming from different levels in another automaton destroys the images. It is only through the passive awareness within oneself that simultaneous knowledge of all of the images is possible. Also remember that any attempt in asking about one's state of awareness and communicating this to the outside world in this model alters the state in correspondence with the uncertainty principle.[6] This could have profound effect on dream research where the researcher attempts to communicate with the dreamer while a dream is in progress.

Of course there would be more levels requiring the formation of more complex images, and as I would imagine, greater mind-ability. This would be reflected in neurophysiological data corroborating the observations of lucid dream researchers.

At the highest level, possibly reachable by training as LaBerge suggests (1985), or through meditation techniques as Transcendental Meditation practitioners indicate, we reach a state of "pure" awareness with surprisingly no images present. This may come about through some form of focusing on the highest level image. What this image could be, I can only guess. We could call it God awakening from the dream.

ACKNOWLEDGMENT

I would like to thank Dr. Jack Sarfatti for reading the manuscript and for helpful comments.

NOTES

1. Nobili proposed that sodium and potassium ion transport through glial cells in the form of oscillating currents producing wave patterns—in which the motion of the sodium ions effects the movement of the potassium ions and vice versa—that satisfy a Schrödinger Wave equation. He found that contrary to lightwave holography,

Schrödinger wave holography was far more efficient in producing holograms in glial tissue. He also discovered that the close proximity of signal sources and receptors (which is in itself in good agreement with other neurophysiological cortical diagrams) in the cortex was ideal for both production of reference waves and information wave recovery. Complete details for the mechanism for glial imaging/holography are in his paper.

2. See my earlier book, *Parallel Universes,* for a more detailed discussion of the many worlds concept and Albert's automaton.

3. There is some recent evidence that these cells are involved in some sort of memory function and they may be instrumental in the development of tumors.

4. Recent work by Hameroff suggests that microtubules within the cytoskeleton of neural cells act like wave guides for photons leading to holographic information processing mechanisms.

5. I don't wish to make a big deal of this. There are 120 ways of producing unique pairs of 16 objects. The important point is that I assume some pairing of the w_i's produces the E_i's and some pairing of the E_i's produces the F_i's and so on.

6. We might inquire how an automaton would interrogate itself. In each world it contains knowledge of two noncommuting observables. These combinations of data are thus known and stored as memories, simultaneously. Presumably as long as the self-interrogation process did not involve communication of that knowledge to the outside world, it could interrogate itself *ad infinitum* without altering its memory in either world. However if it attempts, in the process of interrogation, to communicate what it knows to the environment outside of itself, or if the self-interrogation necessarily involves such communication, then it would invariably alter its own memory in each world.

7. Note that the "zero" level is shown in the figures as the highest on the page. Hence moving lower in the drawing is actually moving higher in the hierarchy. I apologize for this unnecessary confusion.

REFERENCES

Albert, D. Z. 1983. On quantum-mechanical automata. *Physics Letters* 98A:249–52.

Albert, D. Z. 1986. "How to take a photograph of another Everett world." In *New techniques and ideas in quantum measurement theory,* edited by D. M. Greenberger. *Ann. N.Y. Acad. Sci.* 480.

Albert, D. Z. 1992. *Quantum mechanics and experience*. Chapter 8. Cambridge, MA: Harvard University Press.

Deutsch, D. 1985. Quantum theory, the Church-Turning principle, and the universal quantum computer. *Proceedings of the Royal Society of London* A400:97–117.

Hameroff, S. R. 1987. *Ultimate computing: Biomolecular consciousness and nanotechnology.* Amsterdam: Elsevier-North Holland.

LaBerge, S. 1985. *Lucid dreaming: The power of being awake & aware in your dreams*. Los Angeles: J. P. Tarcher.

Moffitt, A. 1994. The creation of self: Self-reflectedness in dreaming and waking. *Psychological Perspectives* 30:42–69.

Nobili, R. 1985. Schrödinger wave holography in brain cortex. *Physical Review* 32(6):3618–26.

Pribram, K. H. 1977. *Languages of the brain. Experimental paradoxes and principles in neuropsychology.* Monterey, CA: Brooks/Cole Publishing Co., pp. 34–47.

Rossi, E. 1972/1985. *Dreams and the growth of personality.* New York: Brunner/Mazel.

Wolf, F. A. 1981. *Taking the quantum leap: The new physics for nonscientists.* Rev. Ed. 1989. New York: HarperCollins.

Wolf, F. A. 1984. *Star wave: Mind, consciousness, and quantum physics.* New York: Macmillan.

Wolf, F. A. 1985. The quantum physics of consciousness: Towards a new psychology. *Integrative Psychology* 3:236–47.

Wolf, F. A. 1986. *The body quantum: The new physics of body, mind, and health.* New York: Macmillan.

Wolf, F. A. 1987. The physics of dream consciousness: Is the lucid dream a parallel universe? *Lucidity Letter* 6(2):130–5

Wolf, F. A. 1989. *Parallel universes: The search for other worlds.* New York: Simon & Schuster.

Wolf, F. A. 1989. On the quantum physical theory of subjective antedating. *Journal of Theoretical Biology* 136:13–9.

Wolf, F. A. 1991. *The eagle's quest: A physicist's search for truth in the heart of the shamanic world.* New York: Summit Books.

Wolf, F. A. 1992. The dreaming universe. *Gnosis* 22:30–5.

Wolf, F. A. 1994. *The dreaming universe: A mind expanding journey into the realm where psyche and physics meet.* New York: Simon & Schuster.

Wolf, F. A. 1994. The body in mind. *Psychological Perspectives* 30:22–35.

35 Percolation and Collapse of Quantum Parallelism: A Model of Qualia and Choice

Michael Conrad

GROUND RULES

Consciousness is arguably the precondition for all experience, and certainly for all scientific knowledge. Yet incorporating this inescapable feature of the world into a unified system of science has clearly involved deep difficulties. One possible position is that the scientific community has not pursued the issue vigorously enough. But the historical facts indicate otherwise: many of the world's greatest scientists and philosophers have grappled with the problem of assimilating mind to the spatiotemporal and material categories in which physical theory is ordinarily formulated. The kind of agreement that typically characterizes scientific advances has never been attained, however, and accordingly the achievements are perhaps best viewed in terms of their contribution to laying bare the exceedingly difficult nature of the problem.

Another possible position is that in dealing with the place of consciousness we are touching on conceptual antinomies, such as freedom versus necessity, part versus whole, and inside versus outside. Perhaps any theory that succeeded in providing a definitive solution to the problem of consciousness would also provide answers to philosophical questions that are not decidable by logical or empirical means. Such a theory, since it would entail conclusions that are inherently arguable, would no longer qualify as scientific in the traditional sense of satisfying criteria that could reasonably be accepted by the whole community.

Unfortunately this criticism can be leveled at the main theories of physics that have dominated in the past, and that dominate today, if these are taken to be paradigmatic of a sought for universal theory that in principle would be capable of unifying all experience. The main gist of these theories, apart from the measurement process in quantum mechanics, is determinism and reversibility. The sense of freedom that at least some of us experience would then have to dissolve into illusion. But of course this sense of freedom is a datum of experience, and consequently for at least some portion of the community the claim of universality must fail for any deterministic theory.

Let us suppose now that we go in the opposite direction and construct a theory that requires freedom of the will. Someone might insist, as intimated above, that this is just an illusion, and therefore reject the theory on the grounds that it predicts a phenomenon that from his point of view is not a bona fide datum of experience. So the theory would again be false for at least some portion of the community.

The above dilemma admits a solution. A proper scientific theory should leave the question of freedom, and of consciousness generally, open to interpretation. It should be equally acceptable for the man who revels in his freedom and for the man who revels in his bondage.

My purpose in this paper is to construct a theory that satisfies this principle (which I will later associate with the conventionalist philosophy of science forwarded by Poincaré). The theory should allow for the main features of consciousness: intention and choice, the quality aspect of experience (or qualia), and the privacy of experience. But it should also allow for the possibility of reformulation, in terms of a more complicated underlying dynamics, in such a way that all these features are transformed into illusion. It should account for the manner in which the features, whether interpreted as nonillusory in a simpler theory or illusory in a more complicated theory, relate to the capabilities for intelligent action. In short, the theory should admit a complete relativity of philosophical position.

FLUCTUON MODEL

The physical underpinning of our conception, to be called the fluctuon model, is that the true time evolution dynamics of the universe is irreversible and irreducibly probabilistic (in the sense that no deterministic reformulation would provide more detailed predictions that are testable). Wave function collapse—the transformation of possibility into actuality—is inherent in the time development of the universe. But for the most part—except in the extreme high mass, high velocity region—the consequences of wave function collapse and irreversibility are not detectable. This is not the case in biological systems, however, due to the ubiquity of highly sensitive transduction-amplification cascades that mediate interactions across radically different space, time, energy dissipation, and particle number scales. The model thus builds into itself a role for gravity in the processes of life and mind, since gravity is intimately connected to the collapse process (Penrose 1989).

Technical details of the fluctuon model have been presented elsewhere (Conrad 1989a, 1991, 1993a, 1993b, 1993c). Roughly speaking, the model starts off with a Dirac-like sea of unmanifest vacuum fermions. Virtual particles are identified as self-perpetuating chains of transient fermion-antifermion pairs. Such chains are called fluctuons because they arise as uncertainty fluctuations of unmanifest vacuum particles in the neighborhood of manifest particles. The transient pair must decay within a time interval specified by the time-energy uncertainty principle ($\sigma t \approx \hbar/\sigma E$), otherwise an unacceptable violation of conservation of energy will occur. But once the pair has moved away from the source particle it cannot decay without violating conservation of momentum (since momentum conservation can only hold in arbitrarily selected coordinate systems when it occurs in the presence of mass). Thus the decay of the pair must be accompanied by the creation of a second transient pair, and so forth, until it enters the neighborhood of a second manifest particle, or absorber, to which it can transfer the excess momentum.

The vacuum (or, better, plenum) comprises at least three types of vacuum fermions. The supersea, comprising all vacuum fermions, mediates the gravitational field (with gravitons interpreted as the excitation chains in this sea). The subsea of vacuum electrons mediates the electromagnetic field and the weak interaction, with photons being interpreted as chains of transiently excited electron-positron pairs. The chromodynamic field connected with the strong force is mediated by a second subsea, but we need not consider this aspect here. The important point is that changes in the states of motion of manifest particles in response to the bombardments by virtual particles alter the density structure of the vacuum sea (since mass and charge depress surrounding vacuum density), and consequently alter the forces acting on them. The forces (or potential functions) must be redefined, introducing a fundamental nonlinearity into the time evolution equations.

The density structure of the full sea of vacuum particles is isomorphic to the space curvature of general relativity, and the endemic nonlinearity corresponds to the nonlinearity of the equations of motion in general relativity. The nonlinearity is incompatible with the linear superposition principle of quantum mechanics. Wave function collapse thereby becomes an intrinsic feature of time development. The density of vacuum particles that mediate the electromagnetic and chromodynamic forces make a negligible contribution to the total vacuum density; consequently these subseas have no detectable effect on gravitational field. The density structure of the full sea is controlled by the distribution of all the positive energy mass in the universe. The fluctuon model thus intrinsically incorporates Mach's principle (that the gravitational force between any two masses is due to influences from all the mass in the universe). Actually, gravity, unlike the other forces, is an indirect interaction in the fluctuon model. The gravitational attraction between any two manifest particles is due to their being pushed together by gravitons emanating from less depressed, low mass regions of the universe.

The requirements of a proper quantum mechanical theory of acceleration provide the conceptual backdrop of the fluctuon model. According to the principle of equivalence gravity should be indistinguishable from acceleration, at least locally. According to general relativity the time evolution equations of physics should be nonlinear (since the gravitational field couples to itself). But *a fortiori* the linearity required for strict validity of the quantum mechanical superposition principle then fails. Pure case wave functions should spontaneously transform into mixtures, and mixtures should transform into mixtures of greater entropy. The true time evolution equations should be nonunitary and inherently irreversible. The main feature of the time development is that the distribution of unmanifest vacuum particles both controls and is altered by the motions of manifest particles. The two distributions are always evolving towards mutual self-consistency. It is the deviations from self-consistency that are responsible for the breakdown of superposition and for the concomitant departure from reversibility. The picture becomes one in which the universe is a giant homeostat, continually maintaining its stability through a sort of negative feedback interaction between its manifest and unmanifest structure.

FREEDOM AND CONTROL

The departure from reversibility that would be exhibited in individual particle interactions would be negligible, since the effects of gravity are negligible in such cases. The irreversibility would also be undetectable in ordinary macroscopic objects, even if the gravitational energy were sufficient to yield a nonnegligible entropy increase. Any entropy increase would require substantial internal motions of the particles, and would therefore inevitably be swamped by entropy increases not attributable to wave function collapse (as opposed, for example, to entropy increases attributable to the atypicality of the initial conditions). More importantly, the self-corrective dynamics concomitant to collapse would have no significance for the time development of the system that could be distinguished from noise, since it would be distributed over numerous degrees of freedom in a random manner. The situation is different for biological cells and organisms, due to the intricate choreography with which the constituent molecules move. The self-corrective interaction between vacuum sea and manifest particles could have nonrandom effects in this regime. The claim of the fluctuon model, as it bears on biological life, is that the remarkable homeostatic capabilities of organisms draw on self-corrective interactions inherent in the fundamental time evolution dynamics of the universe.

Carbon macromolecules, in particular proteins, play the pivotal role, since the combination of nuclear and electronic mass scales allows for a unique classical-nonclassical interface. The shape-based recognition capabilities of such molecules, together with their combinatorial variety, support a level of dynamic intricacy sufficient to unmask the underlying collapse process. We will return to these points later. At this juncture what is important is that the subjective psychological sense of collapse of possibility, and choice about the direction of collapse, that accompanies decision-making and the exercise of intention finds its natural physical analog in the continual collapse of microdynamic possibility into macrodynamic actuality.

This does not mean that we can "see" this sense of freedom in the structure of the theory any more than we can see it when we look at another individual. What it does mean is that there is an element of the theory to which the sense of freedom could be assigned as a referent, just as, say, there is a referent to which the phenomenon of light can be assigned in Maxwell's equations. The difference is that light is something which is itself defined in terms of space and time, whereas the feeling of freedom is not. Those who do not wish to assign the subjective feeling of freedom a nonillusory character may refrain from doing so. But those who do regard their feeling of freedom as nonillusory are not precluded from assigning it a real status. A theory that did not embed collapse (or some comparable process) into its intrinsic structure would not have this eligibility property, and would therefore fail to satisfy the principle of philosophical relativity.

QUALIA

In a similar way we can find elements of the theory that allow for qualia—the qualities of subjective experience—as tenable (but not obligatory) referents. This would be true of any quantum theory that admits a fundamental status for an unpicturable stratum of "superpositional dynamics" (as opposed to hidden variable theories which deny such a stratum). The new feature of the fluctuon model is that by building the collapse process into its structure it builds in the relation between the qualitative and the quantitative aspect of experience, and furthermore shows how the richer structure of the qualitative aspect can lead to capabilities that exceed those which can be achieved by classical machines or understood on the basis of classical analyses.

First let us consider more carefully what we mean by qualia, and what criterion must be satisfied for some symbol (or formal element) of a physical theory to be eligible for having qualia as a referent. The subjective sensations of redness, saltiness, and hotness are examples of qualia. But so are the subjective experiences of space and time (such as the sensations produced in us by rulers and clocks). Qualia are thus inextricably connected with all our measurements, or assignments of quantitative measures (or real numbers) to the external world. On the surface it might appear that we could completely characterize qualia by real numbers—say completely characterize our sensation of redness by assigning a collection of real numbers corresponding to various intensities of red, or completely characterize our conception of space by imagining that a collection of real numbers is associated with it. Then we would have reduced these qualities (which are qualities of the subject undergoing the experience, not of the object experienced) to quantities in a complete way, without loss of any essential aspect. But the subject could also be object, so we would then have to conclude that no qualities exist anywhere whose "qualityness" cannot be eliminated. But anyone looking at such a theory is entitled to complain that this contradicts what is given to him in experience, and that no theory that claims in principle to be competent for accounting for his experience should preclude qualities from having a bona fide (nonillusory) existence and therefore compel them to be completely reducible to real quantities without the loss of any essential aspect.

Consider for example temperature, the measure associated with our sensation of hotness. This can be reformulated in terms of average kinetic energy and therefore in terms of the putatively more primary qualities of space and time. The sensation of hotness itself cannot be reformulated in this simple way, since it is a property of the organism and not of the object that the organism touches. Suppose then that we go a step further and attempt to describe the organism as a whole in terms of space-time categories, ignoring for present purposes the impossibility of doing so without being overly destructive. Would we then have accounted for the feeling of hotness, or even for the sensations produced by measuring rods and clocks? All we have done in the end is to describe the organism in terms of other objects, such as measuring rods, in such a way that we can put these rods (or pointers connected to them) into correspondence with markings that can be interpreted as decimal numbers. So we finally have quantities. The only qualities that are left are those experienced by the observing system that is registering the pointer readings. This observing system in no way apprehends the perception of hotness, or of space and time, that the organism being observed is experiencing. If all that could be stated about the organism observed could be summarized in terms of real numbers, any need for this organism to have subjective experience in order to do its job would dissolve. The qualia might be there, but they would be entirely epiphenomenal.

Already we have arrived at a conclusion about consciousness that is inherently arguable. We can go a step further and suppose that the observing system can perform all the requisite measurements on itself. More modestly we can assume that a description of the observing system in terms of real quantities exists in principle, even if it would be impossible ever to actually construct this description. Then even the observer's own subjective experience would become entirely gratuitous so far as any functional capabilities are concerned, thereby eliminating entirely any role for what is given in experience for both observer and observed.

To avoid these entailments, which violate the principle of philosophical relativity, it is necessary to give up the assumption that all essential aspects of the system observed (or of the observing system) can be condensed into real numbers. This is consistent with the Kantian distinction between experience and precondition for experience. If the assumptions of a scientific theory asserted or implied that the categories in terms of which experience are described could be used to describe themselves, then it would be possible for that theory to entail conclusions about the ultimate nature of reality that are inherently arguable. For example, we could with equal credibility conclude that subjective experience is nonexistent or epiphenomenal (by collapsing everything into real numbers describing space and time, as in the imaginary experiment above) or that qualia are the stuff of reality (by emphasizing that our registration of the pointer readings in terms of which we describe space and time are themselves qualia and that our supposition of real space and time as something existing apart from mind is therefore derivative).

We may not be able to define qualia in a positive way, but according to the argument above we can characterize them in a way that imposes a negative requirement on physical theories that lay claim to universality. Such theories, to satisfy the principle of philosophical relativity, must possess formal elements that are richer than real numbers and that cannot be reduced to real numbers without the loss of some essential aspect. But caution: this does not mean that the particular enriched formal elements of any theory refer to the qualia of human experience. It would only mean that they are eligible to have what might be thought of as elemental qualities as referents. Only when the theory is applied to the human organism as a whole could it in principle accommodate the qualia of human subjective experience, including such putative primitives as the subjective experience of space and time.

Let us consider how quantum mechanics incorporates formal elements that satisfy the eligibility requirement, ignoring for the moment the fact that the standard formulations do not embed within themselves the reduction aspect. But keep in mind throughout that the definition used here for true quality, or more generally qualia, is negative: a quality is anything that cannot be completely reduced to real quantities, or cannot be reduced to real quantities without adding extra assumptions that confer no increase in the competence to predict or calculate.

Consider first a classical system, such as a classical gas. At any given time this may be described by a collection of numbers, the positions and momenta of all the particles. A single classical particle would be described by six numbers, three positions and three momenta. In quantum mechanics even the single particle must be described by a collection of numbers, say the possible positions and momenta. The crucial point is that this collection of numbers does not represent a statistical ensemble, like a classical gas, since the possible positions and momenta interfere with each other in the manner that waves interfere. Thus the elementary descriptors of quantum theory must be built out of complex functions—any set of complex functions will do as long as they are complete in the sense that one can build a description of any system out of them. If the functions were functions of a real variable we could always assign particular real numbers to each point in space and time, and consequently we would not have captured the idea of possible states interfering with each other. But once one has bona fide possible states one necessarily has choice as to how one converts them to actual states; doing it in one way kills other ways. The noncommutative feature of quantum mechanics thus enters. The investigator has the choice of selecting some particular classical features (such as position) from the quantum mechanical description, but at the expense of losing the opportunity to create classical descriptions of the conjugate feature (momentum in the case of position).

Furthermore, the element of randomness should enter into the conversion process. If the conversion of (true) quality to quantity could be described deterministically it would of course mean that quality could be reduced to quantity without the loss of any essential aspect, contradicting its defining property. Randomness has a simple meaning here; it means that even if we could construct a deterministic (hidden variable) theory, we should not be able to achieve predictive competence greater than what would be obtainable with a probabilistic assumption. The probabilistic assumption can be formulated in terms of the random decorrelation of the phases associated with the various contributions to the superposition, since such random decorrelation leads to the disappearance of the interference terms and to the concomitant suppression of the wave character. The randomization is ordinarily viewed as due to the intervention of the measurement process, considered as separate from the equations governing the time evolution of the system in the absence of measurement. As outlined above, the fluctuon model embeds the collapse process into the time evolution, and consequently measurement becomes a very special case of an expanded dynamics.

Here is not the place to argue the advantages of the collapse-embedding approach as compared to the various other approaches to the consistency of quantum mechanics. It is sufficient to point out that it has the advantage of being able to accommodate in principle the main thing that organisms do, which is to make measurements, and it would seem that any theory that projects claims of universality should at least be able to satisfy this requirement. It also allows for the existence of a classical stratum of reality, and therefore obviates the paradoxes of quantum mechanics (since superpositions spontaneously collapse at the macrolevel).

TYING QUALIA TO INTELLIGENCE

Now we can address the connection between the real time capabilities of organisms and the structure of the wave function—what might be called the parallelism of possibilities that confers on the wave function its eligibility to have qualities as referents. The trick is to convert the parallelism into speed of operation. This would be impossible on the average if all the states of the system have the same energy, like the states of a digital computer, since some of the possible time developments would correspond to behaviors that more rapidly perform some function and some to behaviors that do so more slowly. But the random aspect of collapse would mean that on the average there is no advantage. (If the collapse could be controlled by the possibility that corresponded to a useful behavior it would of course no longer be random and furthermore, in the absence of such internal control, testing for usefulness could only occur after actualization.)

The situation is different if the possibility descriptors (or basis functions) are of different energies, since different energies are concomitant to different vibrational frequencies. The pattern of constructive and destructive interference changes in time, so that in this case the probability distribution that the states actualized by measurement would be found to have also changes in time (Bohm 1951). This is the quantum mechanical description of acceleration (or change of state of motion). The motion is completely smoothed out in a classical system, where the de Broglie wavelength of the objects is too small for wave properties to be noticed; but it becomes the dominant feature at the microlevel, as in the jumpy transitions of electrons between different stationary states in an atom.

So it is inherent in quantum theory that the rates of processes are controlled by the structure of the wave function, and that this structure is the dominant factor in systems whose time development is controlled by microphysical particles. But how is this brought to bear on the capabilities of organisms for intelligent action at the classical level? The key point is the continual evolution of the universe towards mutual self-consistency between the unmanifest distribution of vacuum particles and the manifest distribution of mass. This can be thought of as a perpetual self-corrective interaction between the micro- and macrostructure of the universe. The superpositional dynamics of the microstructure speeds up the self-corrective dynamics. In a sense, we can picture the universe as a sort of giant organism working (as intimated above) to maintain its homeostatic state. It is impossible for this giant system to simultaneously satisfy all the requisite symmetry and conservation laws. Fluctuons are a local consequence of this. The disequilibrations of manifest and unmanifest particles are the global consequence. If the global inconsistency becomes too great the interactions between particles are altered, and the alterations continue until self-consistency is satisfied as nearly as possible. This picture fits into the notion of a universe that is continually reprocessing itself (Misner, Thorne, and Wheeler 1970) and also into a picture of historical development that is generated by a perpetual disequilibrium (Matsuno 1989).

The reprocessing and concomitant collapse required would, as noted earlier, only yield phenomenologically significant effects in the presence of extremely high masses. Possibly the effects were of dramatic significance in the early universe. Biological organisms are an exception, however, because of the sensitive transduction-amplification pathways that link microscale aspects of the dynamics to macroscale aspects. Those organizations whose structural and functional integrity are increased by the electronic and molecular motions that accompany the approach to self-consistency enjoy an increased likelihood of persisting and reproducing, whereas those whose integrity is decreased will inevitably taper out of existence. Clearly the yet higher-level reprocessing mechanism of variation and natural selection plays an important molding role. But the important point here is that, because of the cross-scale linkages, the superpositional parallelism that speeds the approach to self-consistency at the microlevel percolates up scale to control the classically picturable actions of the organism.

ROLE OF MACROMOLECULES

We can now see why proteins and other biological macromolecules provide the crucial interface between classical and nonclassical levels of the percolation process. The atomic nuclei from which such molecules are constructed are sufficiently massive so that superpositional effects are of minor importance. Shape, or conformation as defined by positions of the atomic nuclei, is the dominant feature. The ability of proteins to recognize specific molecules in their environment, either for catalysis or self-assembly, is largely based on mutual shape fitting. Dynamic changes in shape are the universal concomitant to this shape-fitting process. The level of the dynamics is inherently classical, since by its very nature shape and shape change are classical concepts. The electrons that glue the nuclei together are microphysical objects, however. Superpositional effects are the dominant feature. This would be true of practically all macroscopic objects. But for the most part the nonclassical behavior of the electrons is relevant only to the stability of such objects, or to their reliability in the case of machines, but not to their functional capabilities. The situation is different for proteins and nucleic acids, though, since the electronic and nuclear degrees of freedom are coupled. This is sometimes referred to as the electronic-conformational interaction (Volkenstein 1982). It could as aptly be referred to as the nonclassical-classical interface, since the shape changes so important for the functional capabilities of the molecules are, as we shall shortly see, controlled by the superpositional dynamics of the electrons. Hydrogen bonds (sharable or mobile protons) also play an important intermediating role in this interaction, connected with the fact that their mass is such that they stand on the borderline between objects that can be treated classically and those that must be treated quantum mechanically.

Let us then briefly consider how the electronic-conformational interaction works, how it allows the unpicturable microdynamics of electrons to percolate up to the functional capabilities of proteins, and how the interaction could interface with the unmanifest structure of the vacuum (for technical details see Conrad 1992a, 1993a, 1993b, 1994a). Each of the nuclei move in a charge field that is due to the electrons and the other nuclei. Similarly each of the electrons moves in a field that is due to all the other particles. Let us focus our attention on the electrons that are not tightly bound to nuclei. These would be delocalized electrons, surface electrons, and electrons that tunnel through hydrogen bond pathways. The energy of the whole molecule depends on numerous interactions among these particles, and strictly speaking we should therefore think in terms of a single wave function for the whole molecule. But as a matter of approximation let us suppose that we can write separate wave functions for the loosely bound electrons—the so-called non-Born-Oppenheimer electrons—and for the collection of atomic nuclei and tightly bound electrons. We will call the former the electronic wave function and the latter the nuclear (or conformational) wave function. The two wave functions must be consistent in the sense of Hartree-Fock type self-consistent field: the electronic wave function must be consistent with the potential function implicit in it and in the nuclear wave function, and similarly the nuclear wave function must be consistent with the potential function implicit in it and in the electronic wave function. Now suppose that the protein interacts with another molecule, which could either be a candidate substrate or a candidate partner in a self-assembled complex. The initial point of collision is quite random. The nuclear and electronic wave functions will be modified. If the fields are inconsistent the wave functions will undergo a time development, either converging to a self-consistent complex, cycling among a family of inconsistent complexes, or diverging to a consistent dissociated state.

Self-assembly occurs when the interacting molecules converge to a self-consistent complex. In enzyme catalysis the convergence to the complexed state is followed by divergence to the dissociated state.

Now we can describe how the superpositional dynamics of the electronic system speeds up the formation or decomposition of the complex. When the atomic nuclei in the macromolecule change their positions the tightly bound electrons will move in tandem, so much so that the nuclear and electronic charges can be wrapped together (this is the Born-Oppenheimer approximation). Loosely bound electrons will inevitably exhibit some acceleration relative to the nuclei. Hence a small component of the radiation ordinarily attributed to nuclear rotations and vibrations should in fact, in a nonperiodic process such as complex formation, properly be attributed to transitions between closely spaced electronic states. Due to perturbations arising from the radiation field, or to collision with substrate molecules, the electronic system must be described in terms of a superposition of these states. Interference effects will occur, yielding variations in the electronic charge distribution that will agitate the nuclei. The resulting nuclear motions are an additional source of perturbation to the electronic system, further spreading out the states that contribute to the superposition. If the relation between electronic and nuclear systems is such that the process is self-amplifying, the agitation will increase until the nuclear system jumps to a new conformation. In essence what is happening is that thermal energy in the environment is funneled to selected nuclear degrees of freedom through the superpositional dynamics of the electronic system. In this way the unpicturable strata of superpositional dynamics controls the picturable shape dynamics of the enzyme. Complex formation and decomposition is sped up. Uniqueness of the complexed shape is possible as long as multiple self-consistent minima are excluded.

So far no reduction of the wave function has occurred, since the approach to self-consistency is entirely describable in terms of conventional quantum mechanics (for example, the Schrödinger equation, together with a Hartree-like self-consistency scheme interpreted as describing the self-organization of the complex). This is reasonable, since the action of enzymes should be reversible (that is, fluctuation should match dissipation for a catalytic process at equilibrium). The mass of a single macromolecule is too small, taken in isolation, to alter the unmanifest structure of the vacuum enough to produce detectable deviations from reversibility. But the enzyme is never isolated; it is always embedded in a bath of other particles, whether in vivo or in vitro. Thus collapse is always occurring, but it is a property of the whole system. In general, as described earlier, the effects are entirely random. However, the interactions of the nuclei and electrons in a single biological macromolecule are sufficiently intricate for the self-corrective interactions between the manifest and unmanifest distribution of particles to influence the motions of the loosely bound electrons and to do so in a way that enhances the functional capabilities of the molecule. Furthermore, these functional capabilities can now be seen to draw on the superposition of unmanifest vacuum states, since this is crucial to the efficacy of the self-corrective dynamics.

UP SCALE TO THE CELL

So far we have looked down scale, at the interface between atomic nuclei, protons (hydrogen bonds), electrons, and finally to the density structure of the unmanifest sea of vacuum particles that mediates the interactions between nuclei, electrons, and protons. Now let us look up scale, to the manner in which enzymes and other macromolecules interface with the influences impinging on the cell, and then onto the manner in which biological cells interlink to yield the actions of organisms. Needless to say this is not the place to do justice to the immense detail of cellular and multicellular biology. Our objective is restricted to giving a more global impression of the transduction-amplification cascades that link the macroscopic actions of organisms to the microdynamic mechanisms discussed.

Consider first an individual cell. The two main tasks are to regulate its internal environment and to act appropriately in response to inputs impinging on its external membrane from the outside world. The cell might be required to differentiate into a particular cell type, such as a nerve cell or a lymphocyte, or it might be required to take a required action, such as emitting a particular sequence of pulses in the case of a nerve cell or exporting antibody in the case of a lymphocyte. We can say that each cell, whether an individual organism *per se* or a part of a multicellular organism, must recognize and respond in a functionally acceptable way to situational information, as coded into the influences impinging on its external membrane.

These influences, which represent the macroscopic features of the external world, are transduced to the internal physiochemical features of the cell, where they are then processed by enzymes. In some cases chemicals from the environment, say steroid hormones, are directly translocated into the cell. In other cases, receptors on the cell membrane code external signals into internal (or second) messenger molecules, such as cyclic AMP. These translocations or transductions produce local physiochemical situations in the cell. The enzymes and structural proteins can be thought of as processors that must recognize and respond appropriately to these local situations. The recognition capability is in general quite specific, since enzymes in general act on particular bonds in particular substrate molecules. But whether they act, and the rate at which they act, can depend on multiple influences, including control molecules and physiochemical features of the milieu. The enzyme in this respect is a highly context-sensitive pattern recognizer; its real time power for such context-sensitive recognition depends on the classical-nonclassical interface outlined above (Conrad 1992b).

That this context-sensitive pattern recognition capability percolates through to pattern capabilities at the cellular level is illustrated by the simple gedanken device illustrated in Figure 35.1. Activation of different input lines impinging on the device (or model cell) release differently shaped macromolecules. The released molecules self-assemble to form a polymacromolecular mosaic, thereby recoding the input pattern into shape features that can be recognized by readout enzymes. Different shape features may be common to different categories of input patterns, making it possible for readout enzymes to trigger events that lead to actions appropriate to

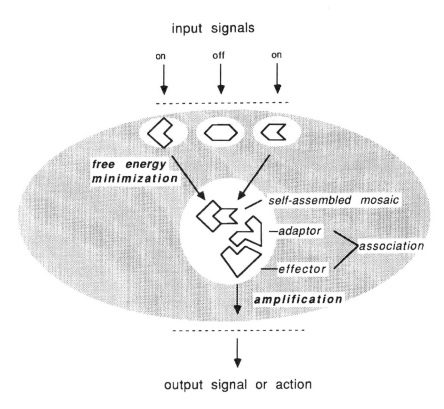

Figure 35.1 Self-assembly paradigm. Signals impinging on the cell membrane along different input lines activate different molecular conformations, which then self-assemble into a supermolecular mosaic (or reorganize from one mosaic form to another) on the basis of free energy minimization. The input pattern is thus recoded as a molecular shape pattern. The adaptor part of a readout enzyme (or complex) recognizes a shape feature common to some grouping of input patterns and triggers, via an effector part, events that either control the internal dynamics of the cell or control its output actions. The background shadings in the diagram are intended to represent the contribution of the electronic-conformational interaction to the recognition-action capabilities of the enzyme, and to represent the contribution that, according to the fluctuon model, should be attributed to renormalizing interactions with the vacuum sea.

whole classes of input. Spatial variations on a given pattern will belong to the same class if the same molecular shapes are released; different temporal variations will be classified the same way if the molecules released stay active for a shared period of time. The self-assembly mode of processing is thus highly suitable to dealing with ambiguity.

The underlying principle of the self-assembly model is operative in nearly all biochemical processes. The self-assembly of initially separate molecules is not essential. All that is necessary is for a molecule or molecular complex that controls the biochemical process to undergo a conformational reorganization in response to the context of influences impinging on it. Such conformational reorganizations are a generic feature of enzymatic action.

The passive control properties of the cell—its ability to maintain a steady state under fixed conditions—can probably be accounted for in terms of such local context sensitivity on the part of its macromolecular components. The coordination of distant parts, connected for example with mitosis, call for more global intracellular networks. Such networks are particularly important for the ability to respond appropriately to input influences widely dispersed over the membrane. Reaction-diffusion dynamics of control molecules, such as second messengers, is one example. Diffusion, however, is only suitable for a very slow integration of information, except over small regions of space. A number of lines of evidence now indicate that the cytoskeleton and the membrane, the structural matrices of the cell, can provide for more rapid, longer distance mechanisms of coordination and communication (Matsumoto and Sakai 1979, Hameroff and Watt 1982, Liberman *et al.* 1982, 1985, Hameroff 1987, Koruga 1990, Hameroff *et al.* 1992, Werbos 1992, Chen and Conrad 1994). Other forms of communication, involving phonons (Liberman 1979, Conrad and Liberman 1982, Samsonovich, Scott, and Hameroff 1992) and various forms of coherent dynamics (Fröhlich 1983, Mishra 1989, Jibu et al. 1994) have also been proposed. The mobile hydrogen bonds that travel adjacently to the membrane and that envelop and interpenetrate with proteins and nucleic acids, provide a further possible basis for distant coordination, conceivably with a coherent component arising from interactions with polarizable electrons of the membrane (Conrad 1988).

FURTHER UP SCALE TO THE BRAIN:
THE PERCOLATION NETWORK MODEL

We can capture some aspects of our conception of the enzyme and the cell with a spider web analogy. Imagine a spider web, occupied by a population of spiders who only explore rather locally. The enzymes are like the spiders and the web like the global communication network. Each spider receives inputs from the objects (including other spiders) in its immediate neighborhood in the web; it also is sensitive to signals traveling along the web from distant locations.

The variety of dynamic processes relevant to biological capabilities is vastly greater than suggested by the spider web analogy, and greater still when multicellular organizations are considered (Conrad 1989b, 1990, 1993d). The term "dynamic universality" is appropriate. The brain is built up as a vast connection of neurons. Each of the neurons can be thought of as a transformer of input patterns to output patterns—for example a transformer of presynaptic and neuromodulatory inputs to temporal firing patterns. The particular transformation performed depends on the particular transduction dynamics and the particular readout enzymes in the neuron. The transduction dynamics can be thought of as signal integrating dynamics. The readout enzymes are the context-sensitive elements that respond to local features that result from bringing distant signals together in space and time. Many such signal integration and readout steps may transpire before the final activation step that determines the pattern of neuronal firing. Dynamic universality means that the perception-action capabilities of the organism depend on the recognition-action capabilities of neurons as well as on the manner in which the neurons are linked together. A giant telephone network provides a useful analogy. The activities mediated by the network could not be understood in terms of its connectivity and the switching properties of the telephones. The humans talking on the phones and making the decisions are much more important. In the case of the brain the recognition-action capabilities of the neurons are the important factor for determining the dynamic capabilities of the network. The recognition-action capabilities of the macromolecules in the cell are the main factor for enabling the recognition-action capabilities of the neurons. The classical-nonclassical interface mediated by the conformational-electronic interaction is a critical factor in enabling the recognition-action capabilities of the macromolecules and, going a step further, the interaction with the vacuum sea also contributes to these capabilities. In this way the unpicturability of microphysical dynamics percolates up to control the picturable actions of the organism.

The term "percolation network" serves to emphasize the interleaving of processes at many different scales. Macroscopic influences from the environment percolate down scale to set the microstate of the cell. Processing occurs at each level, but the main processing occurs at the molecular level, where the interface between mesophysical and microphysical features is very close. Decisions made by enzymes then percolate up to control the decisions made by cells, which in turn percolate up to control the decisions made by organisms. But the term "decision" only becomes meaningful at the cell and organism level, since at the level of individual proteins the irreversibility concomitant to collapse is minute.

The downscale transduction and upscale amplification is accompanied by information filtering at all levels. Thus we can roughly picture the downscale transduction of environmental influences as being accompanied by continual selection and elimination of information. Many local microscale processes are triggered and proceed in parallel, but only some of these will survive, through cooperative or competitive interactions, to control the picturable actions of the organism.

PRIVACY AND THE VARIETIES OF CONSCIOUSNESS

Now we can turn back to the issue of consciousness, and to the privacy of consciousness. On the surface it might seem sufficient to point out that the transduction-amplification chains that link macro to micro and manifest to unmanifest are so delicate and so unique as products of evolutionary history that no two such systems could ever interface with each other in a way that allowed for a merger of their subjective experiences without destroying the conditions for these experiences. Indeed it is unlikely that even standard measurement procedures could be applied to such history-laden systems at any significant level of detail without unduly interfering with their essential functions.

The analysis should be a bit more careful, however, since it is important to distinguish between merger and measurement. This correlates with the distinction between subject and object. In the case of measurement we are treating the system measured as object. We would therefore not expect to observe or participate in its experiences. We would only be able to say that as an object it produced such and such qualities in our subjective experience, and that if to account for its behavior we had to utilize a superpositional description, then it would be eligible to have qualities in its own right.

What qualifies a system to be a subject, and to be a subject that could experience qualia comparable to our own? Is every particle a subject as well as an object; is the whole universe a subject? Why should a percolation network—a relatively self-enclosed transduction-amplification cycle—qualify as an integral whole eligible in and of itself to experience quality rather than as an aggregate whose parts are wholes or as a part of a larger indecomposable whole?

Conceivably we could demarcate systems in an entirely arbitrary way and declare them as subjects. But if the system so demarcated does not have any organizational integrity—if the internal couplings are not richer and more significant than the external couplings—we would hardly expect the system to have any significant systemic properties. The framework developed so far suggests that a self-enclosed transduction-amplification cycle has all the required properties: collapse processes that allow choice to be an eligible referent; superpositional dynamics that allows qualia to be eligible referents; a tightly linked chain of processes that extend across radically different physical scales and that ties the superpositional dynamics and collapse to the functional capabilities of the system as a whole. The last point is connected to the self-enclosure property. All systems undergoing changes in states of motion will, according to the fluctuon model, interact with the vacuum structure. So to some extent qualia and superpositional dynamics always play a role. But it is only when the self-corrective aspect of the interactions between manifest particles and vacuum structure are utilized to maintain the integrity of the system that enough duration is achieved to yield purpose-laden capabilities from the outside point of view and temporally unified subjective concomitants of these from the inside point of view.

The whole universe presumably has the self-enclosure property. Biological organisms may, in this respect, be thought of as homomorphic images of the whole universe. The difference is that the regime of biological matter is constraint dominated rather than energy dominated. Accordingly organisms possess a distinctively enormous capacity to accumulate history (for example, to accumulate information in DNA) and therefore to exhibit a vast variety of organizational forms and capabilities that subserve the persistence of these forms.

What about the distinction between consciousness and self-consciousness? The answer implied by our framework is that there is no difference. Subjects can only be conscious of themselves, and never of another subject. A subject can of course be influenced by what is outside it, in the usual manner of objects exerting forces on each other, but it is only aware of these influences by virtue of its state being affected by them. In the case of an intricate percolation network, such as a cell or an organism, it is inherently impossible for the whole system to be aware of itself. This would be incompatible with the integrity and persistence of the network, which requires that all the possibilities for action are finally collapsed into a single action. As the multiple processes in the system percolate up scale only one can be selected for controlling the macroscopic behavior. The vast majority of molecular actions, the vast majority of cellular actions in the body, and the vast majority of cellular actions in the brain are suppressed so far as the final action is concerned; even those that are not suppressed have their outputs funneled into a top level subsystem whose responsibility it is to execute the action. So it is only this very top level transduction-amplification cycle which is ordinarily associated with the awareness that we identify as self-awareness. This top level simply is not aware of the community of other self-awarenesses that influence it and contribute to its behavior. But this is not a logical necessity. In some instances two or more awarenesses could vie for control (the split-brain case) or different systems might at different times take the place of the top level system (the multiple personality case). But these are pathological situations, not ordinarily compatible with the persistence of the whole community of "windowless yet interacting" self-awarenesses.

Clearly the number of variations on biological form, at all levels, is enormous. A virtually limitless variety of dynamical regimes and percolation networks is possible. The presumption may be that similar organizations have states eligible to be similar qualia, as colored by inherited variations and personal history. Dissimilar organizations, with differently structured self-enclosed transduction-amplification cycles, would be eligible to experience very different qualia. The variety of qualia in the universe would be as endless as the variety of dynamical regimes and percolation networks.

SEMANTIC CAPABILITY

Digital computer programs are a special case. The fluctuon theory precludes these from having qualia as referents or from affording them the possibility of subjective experiences. This is because digital computations are completely classical, and consequently preclude any role for superpositional dynamics in guiding the course of the computation. The actual digital machine that executes the computation will of course utilize physical processes that may be quantum mechanical in nature. These are relevant to the speed and reliability of execution, but not to the computational search processes at the system level. The electronic motions that occur in the machine may lead to some very minute renormalizations of the vacuum structure, but these are completely irrelevant to the course of the computation and to maintaining the integrity of the machine. Errors might conceivably occur in real machines that have some very small probability of being traced to percolation and collapse of quantum parallelism. States of the machine that have consequences for actions that might be taken would then be "quality eligible"; but insofar as the engineer builds effective fault tolerance into the machine this eligibility is quenched. A completely reliable digital computer, as a completely controllable physical realization of a formal system, is of all the objects in the universe the one that is least eligible for being sentient.

This absence of sentience (or aliveness) has implications for the possibility of simulating the world with digital computations. The fluctuon model implies that a purely classical system would never have the real time capabilities of a system whose system level capabilities draw on the percolation and collapse of superpositional dynamics. By itself this would not preclude the possibility of simulating all local behaviors that can occur in the universe, assuming that the time and space resources required to do so could be ignored (obviously this is impossible if the whole universe, rather than a part of it, is the object to be simulated). But the ability to simulate all behaviors should include the ability to simulate the semantic capability of interpreting formal systems, including computer programs. The semantic capability requires the assignment of referents to the formal symbols in a computer program. Computer programs are arguably incapable of performing such "semantic computations," however, since no formal symbol or collection of formal symbols entails its own meaning. The meaning is a property of the subject who is observing the symbol (or the sign) and whose state is influenced by it. The formal symbols of a computer program are not eligible for being subjects, however, or at least not eligible for being subjects with qualities that have anything to do with the capabilities of the program, and consequently the states of the computer that run the program are not on their own quality eligible in a way that has anything to do with the program being run. The situation is quite different for systems supporting self-enclosed transduction-amplification cycles. The states of such systems are eligible to have qualities—properties that cannot be fully described in terms of classical quantities—in their own right, and accordingly such systems are eligible to be viewed as subjects as well as objects. We can rephrase the above argument in terms of Wittgenstein's famous example of a sign with a pointing finger (Wittgenstein 1953). The meaning of the sign (in what direction it points or even whether it points) is entirely extrinsic to it. The sign is an object to be interpreted by a subject, never a subject that implicates within itself its own referent. Transduction-amplification cycles that support percolation and collapse of the wave function are eligible for experiencing qualities, and therefore eligible for performing a function that formal models of computing (such as Turing machines) are incapable of performing without the help of a quality-laden interpreter, such as a human. In fact this is the most important functional capability of all: the semantic capability.

PRINCIPLE OF PHILOSOPHICAL RELATIVITY

The problem of consciousness, as stated at the beginning, has been a perennial challenge to science and philosophy. In the foregoing the author has nevertheless outlined a framework that would appear to make a claim of universality. At the same time he has reiterated a principle of philosophical relativity that would seem to foreclose the possibility of a knockdown theory so far as the genuinely philosophical issues are concerned. How can two such claims—one so bold that it undoubtedly will strike some readers as arrogant and the other so modest that it undoubtedly will strike others as arrogant—cohabit the same manuscript?

Let us first step back and trace the connection between philosophical relativity and the conventionalist philosophy of science of Poincaré (1946). Does relativity, either the special or general theory, require a modification of our concept of space and time? According to the conventionalist view one can choose to think in terms of simple descriptions of force and complicated geometries; or one could choose to retain simple (Euclidean) geometries but at the expense of introducing complicated forces. The choice is dictated not by direct experimental or mathematical considerations, but rather by considerations of conceptual simplicity or coherence, and in some instances by the ease of doing the mathematics, executing calculational procedures, or connecting elements of the theory to experimental data.

The fluctuon model of force can be viewed in this light. The model assumes a full sea of all types of vacuum particles, and subseas for the electromagnetic and strong forces. One could attempt to make an entirely geometrical theory that encompasses all the forces, but the geometry would be incredibly complicated. Or one could attempt to construct a virtual particle exchange model of all the forces on top of a simple geometrical background that is essentially independent of the forces (or fields). The description of gravitational force would then be very complicated. The density structure of the supersea of all vacuum particles in the fluctuon model may be interpreted as isomorphic to the space curvature of general relativity, and the density of the subseas as being essentially independent of the density of the full sea. Whether the density structure of the full sea is interpreted as corresponding to an objectively existing space and time, or whether space and time are mathematical extensions of qualities of the subject (like the feeling of temperature is a quality of the subject) is a matter of choice. The structure of the theory is actually independent of the interpretation in terms of space and time, whose ultimate nature, whether real or illusory, is finally a matter that is not susceptible to experimental determination and that therefore is inherently arguable.

In extending physics to deal with the phenomenon of mind a new element enters—the predictions of theory are not necessarily independent of arguable interpretations, and as a consequence it is necessary to introduce constructs that make it possible to transform from one interpretive "coordinate system" to another without altering the empirically testable content of the theory. We have used the issue of freedom versus necessity as an example particularly prone to generating dispute. In part the quantum theory addressed this issue, since superposition and collapse introduced extra features that widened the scope for interpretation. From a historical perspective the introduction of these features may well have been motivated by experimental observations, but they nevertheless transcend such observations (since the wave function denies the validity of sensory categories as adequate descriptors). It is conceivable that a more complicated hidden variable theory, with underlying chaotic dynamics and action at a distance, could have been invented first. Such a hidden variable theory would have, like its Newtonian predecessor, entailed definite conclusions about questions that are not decidable on empirical or mathematical grounds. According to the principle of philosophical relativity it would on this score be unsatisfactory, even if it were superior in empirical competence to any contemporaneous contender. It should be possible to find an empirically equally competent and less complicated theory that eliminates the entailment of undecidable claims.

The principle of philosophical relativity thus elevates the attitude of conventionalism to a guideline for theory construction. The principle may be stated thus: *the structure of scientific theories should be such that they do not entail solutions to philosophical questions that are undecidable by virtue of admitting alternative solutions that are incompatible but nevertheless supportable by equally credible arguments (that is, antinomic solutions)*. It might be possible to recast theories that satisfy this criterion into hidden variable theories that make the same empirical predictions, but that fail to be neutral as regards antinomic philosophical viewpoints. However, such viewpoint-bound reformulations would always suffer from increased complication, as measured in terms of number of assumptions and/or in terms of the computational resources required to connect the assumptions to empirically testable predictions. In effect, the theory would have to carry extra baggage to entail interpretations that it should only allow.

Philosophical relativity differs from conventionalism in an important respect. According to conventionalism the real nature of the constructs of a theory (such as space and time in relativity theory) are matters of interpretation as long as there exists a class of theories with different interpretations that are equally comprehensive in their coverage of the empirical phenomena. According to the principle of philosophical relativity, constructs without direct empirical correspondents (such as the quantum mechanical wavefunction) should be introduced into scientific theories to preclude the entailment of inherently arguable interpretations. These constructs must preserve (and may increase) empirical comprehensiveness. A theory so augmented satisfies philosophical relativity. At this point there should in principle exist a class of theories that are empirically equally comprehensive, including theories that entail antinomic interpretations (for example, by virtue of replacing the relativity-conferring construct by hidden variables). The existence of such a class, in so far as its members are equivalent in their empirical coverage, implies that the antinomic interpretations can be regarded as matters of convention. But only those members of the class that avoid implications that resolve antinomies satisfy the principle of philosophical relativity. All members of the class that entail antinomic interpretations are bound to particular philosophical points of view. The member that most closely approaches the ideal of being unbound is the one that best satisfies philosophical relativity. To the extent that this member approaches the ideal it will be eligible for supporting the greatest number of consistent but mutually antinomic interpretations, though it may be necessary to assume hidden complication to support some of these interpretations. Interpretations that require no such hidden complication may appear to entail their own preferred status from the standpoint of economy (that is, Occam's razor), but it should be recognized that what is economical or natural for one interpreter may be quite tension-laden for an observer operating in a very different philosophical context.

Now we can consider how the percolation-collapse framework outlined here can address the issue of consciousness and nevertheless attempt to satisfy philosophical relativity. The reason is that the theory is formulated in terms of constructs that are eligible for having mental referents whose ontological status is open for interpretation rather than entailed. The interpreter of the theory can set up a "philosophical coordinate system" in which intentions and qualities that are private have more, less, or as much claim on reality as qualities that are publicly observable. Thus one might choose to deny the existence of radically different forms of subjective experience on the ground that forms other than one's own are not available for intersubjective observation. One could conceivably argue that all subjective experiences are illusory, or that only one's own are nonillusory, or that "red is red is red" and that this is so as an independent Platonic reality above the particularities of biological form. Or one might argue that even the subjective experiences of space and time that appear to us to be so universal are just peculiarities of our particular species, or even individual peculiarities, that cannot be projected as properties of an objectively existing external reality. Or one might argue (following Spinoza) that space-time and mind are objectively existing dual aspects of an objectively existing but unknowable nuomena, and (going a step further) that they come to the fore when microphysical dynamics are embedded in suitable macrophysical architectures. Or one might argue that subjective experience is an emergent property with some definite cutoff, and that while dogs are eligible to have subjective experience so far as the theory is concerned they do not in fact have such experiences. The theory we have outlined, to the extent that it succeeds in satisfying philosophical relativity, allows the interpreter to maintain any one of these or other views, provided that the elements of the view are consistently coordinated to the elements of the theory. It does not mean that these views are meaningless, or that some are not preferable to others on grounds of conceptual elegance or coherence. But these are issues that to the extent possible should be isolated from the empirical claims of the theory.

The principle of philosophical relativity may remind the reader of the positivist criteria of truth in terms of either empirical test or mathematical deduction (Ayer 1959). It is not the same, however, since the principle asserts that constructs that are not empirically testable must be added to any theory that claims universality in order to eliminate implications that are not empirically testable. Further, the interpretations of the theory are not to be regarded as meaningless, but rather as perspectives that are as necessary for working with the theory as particular coordinate systems are for making measurements. In some cases it might be difficult or impossible to precisely distinguish between statements that are bound to the perspective (hence interpretive) and statements that are empirically decidable independent of perspective. It might be that general agreement even on this matter is not obtainable, and that it is impossible to formulate absolute criteria for determining whether a theory satisfies philosophical relativity. This is not the point: absolute agreement on the truth, beauty, or usefulness of scientific and philosophical theories has never been achieved. Philosophical relativization is rather to be viewed as an ideal, roughly analogous to the ideal of separating the choice of particular reference frames for performing measurements from the general scheme of physics. The criteria laid out for achieving this ideal are accordingly most appropriately viewed as heuristics that have especial value for the construction of theories that address domains, such as that of consciousness, where the main achievement of several thousand years of philosophical work has been to lay bare the antinomic nature of the issues and thereby (from the philosophical relativity viewpoint) to lay the basis for the most effective application of the philosophical relativity principle to scientific theory construction.

ACKNOWLEDGMENT

This work was supported by grant ECS-9409780 from the U.S. National Science Foundation.

REFERENCES

Ayer, A. J., ed. 1959. *Logical positivism*. New York: The Free Press.

Bohm, D. 1951. *Quantum theory*. Englewood Cliffs, N.J: Prentice-Hall.

Chen, J. C., and M. Conrad. 1994. Learning synergy in a multilevel neuronal architecture. *BioSystems* 32:111–42.

Conrad, M. 1988. Proton supermobility: A mechanism for coherent dynamic computing. *J. Molec. Electronics* 4:57–65.

Conrad, M. 1989a. "Force, measurement and life." In *Toward a theory of models for living systems,* edited by J. Casti and A. Karlqvist. Boston: Birkhauser, pp. 121–200.

Conrad, M. 1989b. The brain-machine disanalogy. *BioSystems* 22:197–213.

Conrad, M. 1990. "Molecular computing." In *Advances in computers,* edited by M. C. Yovits. Boston: Academic Press, pp. 235–324.

Conrad, M. 1991. Transient excitations of the Dirac vacuum as a mechanism of virtual particle exchange. *Phys. Lett. A* 152:245–50.

Conrad, M. 1992a. Quantum molecular computing: The self-assembly model. *Int. J. Quant. Chem.: Quantum Biology Symp.* 19:125–43.

Conrad, M. 1992b. The seed germination model of enzyme catalysis. *BioSystems* 27:223–33.

Conrad, M. 1993a. The fluctuon model of force, life, and computation: A constructive analysis. *Appl. Math. and Computation* 56:203–59.

Conrad, M. 1993b. Fluctuons—I. Operational analysis. *Chaos, Solitons & Fractals* 3:411–24.

Conrad, M. 1993c. Fluctuons—II. Electromagnetism. *Chaos, Solitons & Fractals* 3:563–73.

Conrad, M. 1993d. Emergent computation through self-assembly. *Nanobiology* 2:5–30.

Conrad, M. 1994a. Amplification of superpositional effects through electronic-conformational interactions. *Chaos, Solitons & Fractals* 4:423–38.

Conrad, M., and E. A. Liberman. 1982. Molecular computing as a link between biological and physical theory. *J. Theoret. Biol.* 98:239–52.

Fröhlich, H. 1983. Evidence for coherent excitation in biological systems. *Int. J. Quantum Chem.* 23:37–46.

Hameroff, S. R. 1987. *Ultimate computing: Biomolecular consciousness and nanotechnolgy*. Amsterdam: North-Holland.

Hameroff, S. R., J. E. Dayhoff, R. Lahoz-Beltra, A. V. Samsonovich, and S. Rasmussen. 1992. Models for molecular computation: Conformational automata in the cytoskeleton. *Computer* 25(11):30–40.

Hameroff, S. R., and R. C. Watt. 1982. Information processing in microtubules. *J. Theoret. Biol.* 98:549–61.

Jibu, M., S. Hagan, S. R. Hameroff, K. H. Pribram, and K. Yasue. 1994. Quantum optical coherence in cytoskeletal microtubules: Implications for brain function. *Biosystems* 32:195–209.

Koruga, D. 1990. Molecular network as a sub-neural factor of neural network. *BioSystems* 23:297–303.

Liberman, E. A. 1979. Analog-digital molecular cell computer. *BioSystems* 11:111–24.

Liberman, E. A., S. V. Minina, N. E. Shklovsky-Kordy, and M. Conrad. 1982. Changes of mechanical parameters as a possible means for information processing by the neuron (in Russian. *Biofizika* 27(5):863–70 (Translated to English in *Biophysics* 27(5):906–15, 1982).

Liberman, E. A., S. V. Minina, O. L. Mjakotina, N. E. Shklovsky-Kordy, and M. Conrad. 1985. Neuron generator potentials evoked by intracellular injection of cyclic nucleotides and mechanical distension. *Brain Res.* 338:33–44.

Matsumoto, G., and H. Sakai. 1979. Microtubules inside the plasma membrane of squid giant axons and their possible physiological function. *J. Membrane Biol.* 50:1–14.

Matsuno, K. 1989. *Protobiology: Physical basis of biology*. Boca Raton, FL: CRC Press.

Mishra, R. K. 1989. "The living state, the matrix of self-organization." In *Molecular and biological physics of living systems,* edited by R. K. Mishra. Dordrecht: Kluwer, pp. 215–37.

Misner, C. W., K. S. Thorne, and J. A. Wheeler. 1970. *Gravitation*. New York: W. H. Freeman and Co.

Penrose, R. 1989. *The emperor's new mind*. New York: Penguin.

Poincaré, H. 1943. "Science and hypothesis" (originally published in 1903). In *Foundations of science*, translated by G. B. Halsted, 1913. Lancaster PA: The Science Press.

Samsonovich, A., A. Scott, and S. Hameroff. 1992. Acousto-conformational transitions in cytoskeletal microtubules: Implications for intracellular information processing. *Nanobiology* 1:457–68.

Volkenstein, M. V. 1982. Simple physical presentation of enzymatic catalysis. *J. Theoret. Biol.* 89:45–51.

Werbos P. J. 1992. The cytoskeleton: Why it may be crucial to human learning and to neurocontrol. *Nanobiology* 1:75–95.

Wittgenstein, L. 1953. *Philosophical investigations*, 3rd ed. New York: Macmillan.

36 Subcellular Quantum Optical Coherence: Implications for Consciousness

Mari Jibu, Scott Hagan, and Kunio Yasue

INTRODUCTION

Neuron models clearly encompass numerous aspects of consciousness, but others remain stubbornly out of reach. The aim of this chapter is to enumerate specific properties of consciousness that cannot be reconciled with standard treatments and use these as a template to determine what theoretical directions might be productively pursued. It is postulated that the system of neurons considered as a connectionist network must be supplemented by one incorporating quantum-theoretical manifestations. Quantum field theory has been applied with great success in many-body physics to explain phenomena inaccessible to a reductive analysis of the constituents of the system under study. Common objections to the introduction of quantum theory in discussions of biology are taken up and a toy model is presented, explicitly instantiating the kind of subcellular dynamics postulated. The model is subsequently elaborated and discussed in relation to the action of general anesthesia in ablating consciousness, a process that finds natural explanation in terms of quantum optical coherence.

CRITERIA FOR CONSCIOUSNESS

Structure

The construction of consciousness requires structure, its elements being related to one another so as to form a coherent and unified whole. Computers also structure information but in a fundamentally different way. Quite generally computers borrow their structure from physical spacetime. Addresses in memory correspond to specific physical locations and it is in virtue of this correspondence between stored information and its spatial coordinates that a computer can keep track of the relations between bits. This can be most clearly seen in terms of a Turing machine, a sort of idealized surrogate for any computer. Connectionist architectures are included as they are computationally isomorphic to serial architectures. So long as computation is considered to be the only aspect relevant to consciousness (and this is true of all the current mainstream proposals of cognitive science and neurophysiology) of the functioning of a neural network, it may be conceptually replaced without deficit by a Turing machine. Such a device has only the capacity to read a binary digit from a tape, erase or print these numbers, and advance or reverse the tape. A Turing machine cannot form relations between elements in the series of ones and zeros occurring on the tape; it is attentive to only one digit at a time. The necessary structure is recorded in the ordering of the binary digits read from and written to the tape. The "memory" of that order, and consequently the entire sense of both program and output, depend on the tape being embedded in physical spacetime. Spacetime acts as a "bulletin board" onto which the computer affixes bits of information. None of these bear any relations to any others prior to being affixed but in virtue of being "pinned" to a particular place on the bulletin board each finds a specific relation to every other, thus comprising the structure of the whole.

Consciousness must also keep track of the order and relation of elements but in a space distinct from physical spacetime, a mental space ("space" is here intended in the mathematical sense), the one in which we perceive, envisage, think, and so on. To avoid a Cartesian duality, this space must not maintain a preexisting structure. Rather relations must be determined and recorded by the processes of consciousness itself without recourse to a bulletin board. Consciousness cannot therefore be built up piecemeal but must occur all at once, each element cohering with every other to structure a unified whole devoid of scaffolding. Implicit in the prevailing functionalist view is the thesis that information extracted from the flux of data is loaded into the space of consciousness. But this presumes that one can address information to a specific location, that the mental space is previously structured. Such a presumption radically underestimates the problems to be encountered in confronting consciousness.

Complexity

Locality is a concept in physics (not to be confused with the completely different application of the term in neurophysiology) that is invoked by most theories put forth since the advent of relativity. It is the requirement that spatially separated entities do not exert physical influences on one another. The conspicuous exception is quantum theory. Bell's theorem (Bell 1987) has in fact shown that results such as those obtained in the Aspect experiment on quantum coherence (Aspect, Dalibard, and Roger 1982) are incompatible with any theory for which locality is a prerequisite. A local object is therefore defined to be one for which the correspondence between the quantum and classical realms does not hold; that is, objects that are essentially quantum theoretical. For the most part such entities will be restricted to about the size of atoms, perhaps as large as molecules.[1]

Some manifestations of quantum theory, however, are evident on a macroscopic scale. In such circumstances and *only* in such circumstances as quantum theory plays a significant role on macroscopic scales will the concept of a local object be extended to include those larger than the atomic or molecular.

In a classical system, such as the brain is conceived to be in neurophysiology and cognitive science, local objects should not be larger than molecules and should certainly not be as colossal as a neuron. On the current understanding of bound states and thermal noise, the capacity for a local entity of this kind to simultaneously encapsulate information falls far short of that necessary for even a mere one hundred bits requiring $2^{100} \approx 10^{30}$ distinguishable states let alone the full panoply of consciousness (Marshall 1989). Recalling the requirement of structure in mental space, the construction of consciousness must occur of a piece and all at once. Locality allows this to be possible only within the confines of a local object. Beyond this domain, information can only be transferred by signalling from one locale to another, but this has no effect on the complexity of any local object and so doesn't change the amount of information simultaneously available. The complexity of consciousness is thereby radically impoverished in the classical domain by the requirement of locality and cannot possibly achieve the richness phenomenologically observed. Moreover, the hypothesis that consciousness should have an entirely classical account has constricted the focus of investigation to single atoms or molecules, a rather unlikely conclusion from the start, but in any case, it is difficult to see how such a model would bear any relation to the brain or even general metabolic processes. The structure of consciousness would thus seem to contravene the locality criterion, obviating a nonlocal explanation. The only fundamentally nonlocal coherence phenomena have an essential quantum character. While, to the uninitiated, the invocation of quantum field theory may seem unmotivated in the study of consciousness, a mere conflation of mysteries, the dual considerations of structure and locality demonstrate the need for its introduction.

Nonbehavioral Character

Mental experience would seem to provide the basis for a qualitative distinction between conscious entities and a class of computational devices that behaviorally mimic them. Since the waning of behaviorism in psychology, science has generally accepted the existence of internal mental processes. Nevertheless, through strict adherence to a paradigm that uncritically accepts a computer metaphor, mainstream models of consciousness have eschewed explanation of such apparently noncomputational aspects in favor of modeling behavioral dispositions.

Connectionist models picture the neuron as disposing other neurons to act in certain ways, in turn disposing further neurons, and ultimately leading to some behavior of the organism. No mechanism is presented by which systems of neurons might yield mental experience or anything other than physical behavior. Coding schemes, like temporal binding (Crick and Koch 1990), in no way compromise this assertion. Though the observation of synchrony in widely separated neurons surely plays a role in global brain dynamics, temporal binding merely affects the dispositions of neurons to fire, specifically a neuron to which two or more synchronously firing neurons give common output, directly or indirectly. It binds input from different sources to act coherently on a dispositional level. Understood in this way, the binding problem can certainly be solved by temporal binding but is thereby reduced to an engineering difficulty in the design of connectionist circuits rather than a conceptual problem in the understanding of consciousness. The problem is defined in terms of neural networks and no longer addresses the more fundamental questions posed for a materialist account of consciousness. The problem of structure, in the sense already discussed, remains untouched by this approach.

The only consistent alternative from the point of view of connectionism is to deny the existence of qualitative aspects of consciousness (Dennett 1991, Minsky 1986), yet this flies in the face of phenomenal data readily available to anyone. No doubt certain aspects of our awareness are illusory or in error, but the issue here is the general phenomenon of awareness, not its particular objects. While there is plenty of room for debate about what we are aware of, that we are aware is a fact upon which our entire scientific understanding hinges.

QUANTUM THEORY AND BIOLOGY

Although the notion of applying quantum concepts to the brain traces its origin to the 1960s (Ricciardi and Umezawa 1967), such proposals have met with widespread skepticism. While conceding the trivial sense in which the brain is composed of fundamental constituents and therefore subject to quantum elaboration, researchers in the biological sciences generally assumed that the activities relevant to consciousness would be susceptible to a classical characterization. This may be the result of the popular perception that quantum theory is manifest exclusively in the microscopic realm of elementary entities, a conception contradicted by the existence of macroscopic ordered states. Such systems exhibit collective modes, naturally treated in field-theoretic terms, that do not emerge from an analysis of their components, taken individually. The advent of a quantum formalism for the description of stable macroscopic objects (Umezawa 1993) opened the way for applications to the elaborate order observed in biological media.

The relatively high temperatures at which living systems function have often been presented as an insuperable barrier to discussions of quantum coherence in biology. In particular, the Bose-Einstein condensation phenomena figuring in such proposals were believed to occur only at low temperatures.[2]

Fröhlich (Fröhlich 1968), however, has demonstrated the possibility of similar mechanisms by which single modes can be populated to the exclusion of all others, in this case for systems at high temperature supplied with external energy at a rate exceeding a specified threshold value. The dipolar phonon introduced below is identified with the giant dipole mode of

Fröhlich's work and the corticon of Stuart, Takahashi, and Umezawa (1978, 1979). In the toy model to follow, condensation of photons yields a quantum optical coherence explicitly protected from thermal effects by a phenomenon familiar in the study of lasers and nonlinear optics known as self-induced transparency.

The choice of candidate systems for the model is guided by speculation in biology as to likely sites of action for mechanisms of the Fröhlich type, primarily cell membranes and cylindrically arranged filamentous proteins in the cytoskeleton called *microtubules*. These latter structures, the focus of our investigation, have an associated layer of ordered water around and within them. Experimental observations suggest that this ordering can be brought about by very low level electric fields under typical biological conditions (Hasted, Millany, and Rosen 1981, Hasted *et al.* 1983).

The most fundamental concept linking stable macroscopic objects to quantum field theory is the notion of *spontaneous symmetry breaking*. Structured water and protein are macroscopic ordered states realized in the interaction between the electromagnetic and matter fields, and constituting domains of order that result when certain translational and rotational symmetries of molecular bound states are violated by the ground state (Umezawa 1993). Order is maintained by the emergence of *Goldstone bosons*, field-theoretic entities that inevitably result from spontaneous symmetry breakdown. In particular, those associated with the structuring of water and protein in the following model are dipolar phonons. The concept of spontaneous symmetry breaking will be further elaborated in connection with the model.

A SUBCELLULAR MODEL FOR QUANTUM OPTICAL COHERENCE

Characterization

To see how subcellular structures can mediate between neural activity and quantum processes, a simple model is constructed of the electromagnetic field interacting with water molecules, characterized as quantum rotators, lining the walls of microtubules. Water, a polar molecule, interacts with a quantized electromagnetic field by means of its electric dipole moment μ. Though this allows the molecule many possible energy eigenstates, only the two principal energy levels, separated by an energy difference ϵ, are considered. The quantum dynamics of a two-level system is aptly treated in terms of the Pauli spin operators $s = 1/2\sigma$ and the 2×2 unit matrix. Any linear, Hermitean operator of any function of two-valued variables can be constructed as a linear combination of these operators. In particular the Hamiltonian is such an operator that, operating on an eigenstate takes the energy as its eigenvalue.

The Hamiltonian for a system of N noninteracting water molecules is:

$$H_{wm} = \epsilon \sum_{j=1}^{N} s_3^j \qquad (1)$$

The eigenvalue of ϵs^j_3 is the energy of the jth molecule $\pm 1/2\epsilon$.

The Hamiltonian for the quantized electromagnetic field in the volume V of the microtubule is:

$$H_{em} = \frac{1}{2}\int_V (E^2 + B^2)d^3x. \qquad (2)$$

To simplify calculations the electric field strength operator is chosen to be linearly polarized along a direction indicated by a unit vector $\hat{e}$. The magnitude is then expanded in normal modes:

$$E = \sum_k (E^+_k e^{-i(kx - \omega t)} + E^-_k e^{i(kx - \omega t)}), \qquad (3)$$

where $E^+_k(E^-_k)$ is the positive (negative) frequency part. In field theory $E^+_k(E^-_k)$ are seen as creation (annihilation) operators for photons with wave vector k and frequency ω.

When the characteristic length of the system under study is small in comparison to the typical wavelength of radiation, a dipole approximation is appropriate to the description of the interaction of the electromagnetic field and a system of electric dipole moments:

$$H_i = \mu \sum_k \frac{\epsilon}{\omega}(E^+_k S^-_k + S^+_k E^-_k). \qquad (4)$$

The collective variables $S^\pm_k$ are defined by:

$$S^\pm_k = \sum_j s^j_\pm e^{\pm i(k \cdot xj - \omega t)}, \qquad (5)$$

where $s^j_\pm = s^j_1 \pm i s^j_2$. The restriction imposed by the dipole approximation is negligible as it excludes from consideration only very high frequencies of radiation (in the x-ray range and higher) such as are not expected to play a role in living matter.

The sum of equations (1), (2), and (4) has an invariance under

$$E^\pm_k \to E^\mp_k e^{\mp i\theta}$$
$$S^\pm_k \to S^\pm_k e^{\mp i\theta} \qquad (6)$$

The second of these transformations corresponds to a rotation in the plane of the three-dimensional vector space of the Pauli spin operators. This invariance is not, however, respected by the vacuum state, a situation that leads to spontaneous symmetry breakdown and the establishment of a macroscopic ordered state.

Spontaneous Symmetry Breaking

Macroscopically ordered spatial domains automatically result in a class of theories that are spontaneously broken. Spontaneous symmetry breaking occurs when a dynamical symmetry of the system is not respected by the vacuum or ground state. Consider a system described by a vector field p of two components whose dynamics are invariant under two-dimensional rotations of the field. Now allow that the lowest energy state has a preferred direction. This is possible, for instance, if the potential energy is (see Figure 36.1):

$$V = -\frac{1}{2}m^2(p \cdot p) + \lambda(p \cdot p)^2, \tag{7}$$

describing two interacting real fields with a common mass m. The vacuum is degenerate. Any set of values of the component fields such that

$$p_1^2 + p_2^2 = \frac{m^2}{4\lambda} \tag{8}$$

will yield a ground state. The set chosen as the vacuum determines a preferred direction that is not invariant under the rotational symmetry. As a consequence, one of the component fields becomes massless.

In general, spontaneously broken theories exhibit massless quanta called Goldstone bosons in their particle spectra. These accompany the spontaneous breaking of a compact, continuous symmetry as a field-theoretic consequence by the Nambu-Goldstone theorem (Nambu 1960, Nambu and Jona-Lasinio 1961, Goldstone 1961, Goldstone, Salam, and Weinberg 1962). Goldstone or *collective* modes correlate the vacua chosen by elements of an extended system. As they are massless,[3] these modes are easily excited by weak external stimuli.

Goldstone modes account for the rearrangement of dynamical attributes to form a *macroscopic ordered state*, a region in which a degenerate vacuum becomes ordered. In quantum brain dynamics (QBD) the imprinting of memory is described as a phase transition to an ordered vacuum of the

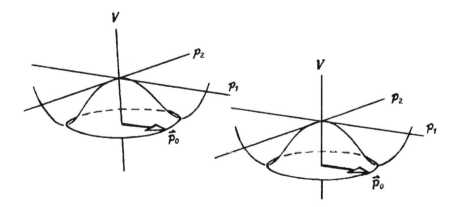

Figure 36.1 The potential V as a function of the component fields p_1 and p_2 at two sites of an extended system. Goldstone modes allow the choice of ground state, $\vec{p}_0$ at different locations to be correlated and thereby order the vacuum.

spontaneously broken type (Jibu and Yasue 1992, 1993). The process of recollection is implicitly incorporated as the excitation of a Goldstone mode. A model of this kind explicates the relative stability of memory in a manner that automatically captures its dynamic and nonlocal (in the sense of neurophysiology) character.

Superradiance

For simplicity, attention is restricted to the dominant mode, having a wave vector that resonates with the energy difference ϵ. This corresponds to the situation in which the energy difference in moving from one eigenstate to another is accounted for by a single photon. Full treatment of all modes is expected to yield only perturbative differences from the dominant effect of the single-photon approximation. In radiation gauge, the Heisenberg equations of motion for the system are then

$$\frac{dS}{dt} = -i\mu(S^+E^- - E^+S^-),$$

$$\frac{dS^\pm}{dt} = \mp i\mu SE^\pm, \qquad (9)$$

$$dE^\pm = \mp i\epsilon \frac{\mu}{2V} S^\pm.$$

In the case for which the transit time through a microtubule of length l_{mt} of a pulse mode propagating along its axis, is much shorter than the characteristic time of thermal interaction, thermal loss can be ignored and the derivative in the last equation approximated by $E^\pm/l_{mt}$ so that

$$E^\pm = \pm \frac{i\epsilon\mu l_{mt}}{2V} S^\pm. \qquad (10)$$

A pulse mode of the quantized electromagnetic field thus follows the collective dynamics of the water molecules in the microtubule as determined by the mechanism of spontaneous symmetry breaking. Once a long-range ordering phenomenon commences in the ensemble of water molecules, coherent pulse modes, called *superradiance*, follow as a consequence. Interaction with water induces a condensation of photons in a single quantum state making the electromagnetic field coherent. Conversely, interaction with a quantized electromagnetic field sets up a long-range correlation amongst water molecules.

The spatial range over which the coherence of the ordered state extends, the coherence length, can be estimated by the inverse of the energy difference ϵ. In the case of the energy difference between the principal rotational states of water (Franks 1972) the coherence length is on the order of microns. A semiclassical calculation of the intensity of photon emission (see Figure 36.2) from the system yields (Agarwal 1971)

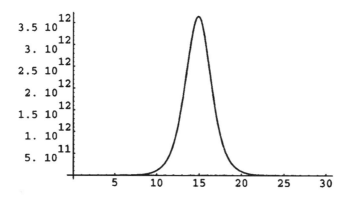

Figure 36.2 Intensity of photon emissions for a microtubule 100 nm in length, given in watts per square meter as a function of time measured in units of t_1. The density of water in the microtubular volume is assumed to be only slightly lower than one, roughly approximated by its usual value at standard temperature and pressure.

$$I = \frac{1}{(4t_1\mu)^2} \operatorname{sech}^2 \frac{1}{2}\left(\frac{t}{t_1} - \log 2N\right) \tag{11}$$

The lifetime of the superradiance,

$$t_1 = \left(\epsilon\mu^2 l_{mt} \frac{N}{V}\right)^{-1},$$

determines the intensity to be proportional to N^2, a hallmark of superradiance (Dicke 1954).

The dynamics of polar proteins and biomolecules control changes in the collective dynamics that in turn trigger cycles of coherent emission. Polar biomolecules maintain regions of nonzero polarization in which the electric dipoles of water molecules are aligned. The invariance of the system dynamics under rotations of the electric dipole moments is thus not exhibited by the ground state. Spontaneous symmetry breaking results (Del Giudice et al. 1985) and the consequent Goldstone bosons maintain the long-range correlation typical of such phenomena.

Self-Induced Transparency

To enable the transmission of pulse modes with coherence intact over time scales approaching that of thermal interaction, the phenomenon of self-induced transparency is invoked. In a semiclassical approximation, the inhomogeneous Maxwell equation for a linearly polarized electric field reduces in 1 + 1 dimensions to

$$\frac{\partial E^{\pm}}{\partial z} + \frac{\partial E^{\pm}}{\partial t} = \mp \frac{i\epsilon\mu}{2V} S^{\pm}, \tag{12}$$

for slowly varying collective variables

$$\frac{\partial S^{\pm}}{\partial t} \ll i\omega S^{\pm}.$$

Semiclassical results are valid for large intensities as are expected for superradiant phenomena.

Eliminating the collective variables through their expectation values, the Maxwell equation can be cast in sine-Gordon form in terms of a new variable:

$$\theta^{\pm}(z,t) = \frac{1}{2}\mu \int_{-\infty}^{t} E^{\pm}(z, t')dt'. \tag{13}$$

The sine-Gordon equation has solitonic solutions that propagate without dispersion so their shape is maintained. The solution for the scalar electric field has the form:

$$E(z,t) = \sqrt{\frac{\epsilon N v_0}{2V(c-v_0)}} \, \text{sech} \sqrt{\frac{\mu^2 \epsilon N v_0}{2V(c-v_0)}} \left(t - \frac{z}{v_0}\right), \tag{14}$$

for a soliton propagating with constant speed v_0 (see Figure 36.3). *Self-induced transparency* neutralizes the pulse spreading and distortion due to the thermal environment of the microtubule that would be predicted on the basis of linear dispersion theory. The distortionless pulse obtained here

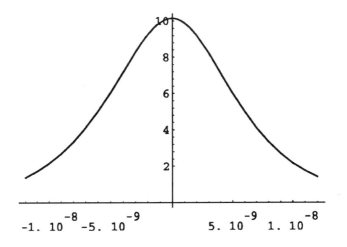

Figure 36.3 Scalar electric field strength at time $t=0$ for a soliton with speed $v_0=10.5c$ (about 3 nm/ps) as a function of position measured along the microtubular axis. Reducing the speed will decrease the width of the pulse.

is a special case of nonlinear optical effects in an inhomogeneously broadened two-quantum system (McCall and Hahn 1969).

Solitonic emissions of coherent photons may have sufficient intensities to initiate protein conformational changes gating transmembrane ionic diffusions, thereby allowing a quantum mechanism to exert nanosecond control over classical neural functions. It is the coherence of the pulse modes that gives them the intensity to affect classical processes in much the same way that a laser uses coherence to amplify the intensity of light.

DISCUSSION

In reality water has considerably more degrees of freedom than are depicted in this rudimentary model. Any or many of these may induce a long-range correlation of water molecules, communicated by dipolar phonons, by coupling to the electromagnetic field via water's electric dipole moment. Conversely, coherent modes of the electromagnetic field are generated by the interaction of the dipolar field of water and follow its collective dynamics. The presence of polar proteins provides an external condition for the dynamics of the system composed of the electromagnetic field and the dipolar field of water, allowing classical mechanics to trigger and control the quantum processes. The vacuum supports memory by the macroscopic ordering effect of the emergent phonons and consciousness is constituted in the excitation dynamics of photons and phonons that take place out of the ordered vacuum. Finally the coherent action of the electromagnetic field yields a back reaction in the dipolar field of proteins, allowing the quantum processes of consciousness to exert their influence in the classical domain of transmembrane ionic diffusions. Over macroscopically large domains, quantum coherence may coordinate the emission of coherent photons in microtubules of different neurons. If coherent radiation is efficacious on a cellular level, this may be related to coherent neuronal firing among widely dispersed neurons (Singer 1993).

A subcellular substrate for consciousness is consonant with the findings of anesthesiology. General anesthetics operate at a molecular level, binding weakly in hydrophobic regions of proteins (Franks and Lieb 1982). This is awkward to explain on a systems account of consciousness but falls naturally within the scope of a model based on quantum optical coherence. For instance, a nonlinear optical medium reacts to increased pressure by stepping up coherent optical activity. If anesthetics inhibit the mechanism of superradiance leading to opacity in microtubular pathways, then the reversal of anesthesia by pressure (Halsey 1976) can be understood as the restoration of transparency in the presence of anesthetic impurities due to the consequent increase in activity. General anesthetics in fact bind directly to microtubules (Allison and Nunn 1968) and may disrupt the dynamics mediating the interaction of the protein and water fields (Wulf and Featherstone 1957).

Direct demonstration of coherent photon emission in microtubules will require nanoscale observations, for which the technology is only currently becoming accessible. Such efforts will require the modification of experimental designs from nonlinear optics to make them appropriate to the microscopic regime. Indirect corroborating evidence is presently available. Tissue exhibiting the superradiance and self-induced transparency is expected to emit anomalous coherent photons at low levels above the incoherent thermal background. A literature exists in which positive results are reliably obtained (see Li *et al.* 1983 for a list of references).

NOTES

1. There is a trivial sense, of course, in which the entire universe is expected to be quantum theoretical. It is important to stress that this is not what is at issue here. Systems whose dynamics occur on a scale and of a character adequately described by classical theory are being qualitatively distinguished from those for which even an approximate understanding of system dynamics *requires* the introduction of aspects of quantum theory.

2. With the recent observation of high T_c superconductors, speculation along these lines has, in any case, become outdated.

3. Finite size effects and the consideration of long-range forces endow such modes with a small effective mass (Ryder 1985) that restricts their range but does not otherwise alter the conclusions considered here.

REFERENCES

Agarwal, G. S. 1971. Master-equation approach to spontaneous emission III: Many-body aspects of emission from two-level atoms and the effect of inhomogeneous broadening. *Physical Review* A4:1791–1801.

Allison, A. C., and J. F. Nunn. 1968. Effects of general anesthetics on microtubules: A possible mechanism of anesthesia. *The Lancet* 2:1326–9.

Aspect, A., J. Dalibard, and G. Roger. 1982. Experimental test of Bell's inequalities using time-varying analyzers. *Physical Review Letters* 49:1804–7.

Bell, J. S. 1987. *Speakable and unspeakable in quantum mechanics.* Cambridge, England: Cambridge University Press.

Crick, F., and C. Koch. 1990. Towards a neurobiological theory of consciousness. *Seminars in the Neurosciences* 2:263–75.

Del Giudice, E., S. Doglia, M. Milani, and G. Vitiello. 1985. A quantum field-theoretical approach to the collective behavior of biological systems. *Nuclear Physics* B251:375–400.

Dennett, D. 1991. *Consciousness explained.* Boston: Little, Brown and Co.

Dicke, R. H. 1954. Coherence in spontaneous radiation processes. *Physical Review* 93:99–110.

Franks, F. 1972. *Water: A comprehensive treatise.* New York: Plenum.

Franks, N. P., and W. R. Lieb. 1982. Molecular mechanisms of general anesthesia. *Nature* 300:487–93.

Fröhlich, H. 1968. Long-range coherence and energy storage in biological systems. *International Journal of Quantum Chemistry* 2:641–9.

Goldstone, J. 1961. Field theories with "superconductor" solutions. *Nuovo Cimento* 19:154–64.

Goldstone, J., A. Salam, and S. Weinberg. 1962. Broken symmetries. *Physical Review* 127:965–70.

Halsey, M. J. 1976. "Mechanisms of general anesthesia." In *Anesthesia uptake and action*, edited by E. I. Eiger. Baltimore: Williams and Wilkins.

Hasted, J. B., S. K. Husain, A. Y. Ko, D. Rosen, E. Nicol, and J. R. Birch. 1983. "Excitations of proteins by electric fields." In *Coherent excitations in biological systems*, edited by H. Fröhlich and F. Kremer. Berlin: Springer-Verlag.

Hasted, J. B., H. M. Millany, and D. Rosen. 1981. Low-frequency electrical properties of haemoglobin. *Chemical Society Journal, Faraday Transactions II* 77:2289–2302.

Jibu, M., and K. Yasue. 1992. "A physical picture of Umezawa's quantum brain dynamics." In *Cybernetics and system research 92*, edited by R. Trappl. Singapore: World Scientific.

Jibu, M., and K. Yasue. 1993. "Introduction to quantum brain dynamics." In *Nature, cognition, and system III*, edited by E. Carvallo. London: Kluwer Academic.

Li, K. H., F. A. Popp, W. Nagl, and H. Klima. 1983. "Indications of optical coherence in biological systems and its possible significance." In *Coherent excitations in biological systems*, edited by H. Fröhlich and F. Kremer. Berlin: Springer-Verlag.

Marshall, I. N. 1989. Consciousness and Bose-Einstein condensates. *New Ideas in Psychology* 7:73–83.

McCall, S. L., and E. L. Hahn. 1969. Self-induced transparency. *Physical Review* 183:457–85.

Minsky, M. 1986. *The society of mind*. New York: Simon & Schuster.

Nambu, Y. 1960. Axial vector current conservation in weak interactions. *Physical Review Letters* 4:380–2.

Nambu, Y., and G. Jona-Lasinio. 1961. Dynamical model of elementary particles based on an analogy with superconductivity I. *Physical Review* 122:345–58.

Ricciardi, L. M., and H. Umezawa. 1967. Brain and physics of many-body problems. *Kybernetik* 4:44–8.

Ryder, L. H. 1985. *Quantum field theory*. Cambridge, England: Cambridge University Press.

Singer, W. 1993. Synchronization of cortical activity and its putative role in information processing and learning. *Annual Review of Physiology* 55:349–74.

Stuart, C. I. J. M., Y. Takahashi, and H. Umezawa. 1978. On the stability and non-local properties of memory. *Journal of Theoretical Biology* 71:605–18.

Stuart, C. I. J. M., Y. Takahashi, and H. Umezawa. 1979. Mixed-system brain dynamics: Neural memory as a macroscopic ordered state. *Foundations of Physics* 9:301–27.

Umezawa, H. 1993. *Advanced field theory: Micro, macro, and thermal physics*. New York: American Institute of Physics.

Wulf, R. J., and R. M. Featherstone. 1957. A correlation of Van der Waals constants with anesthetic potency. *Anesthesiology* 18:97–105.

37 Orchestrated Reduction of Quantum Coherence in Brain Microtubules: A Model for Consciousness

Stuart R. Hameroff and Roger Penrose

INTRODUCTION

Current neurophysiological explanations of consciousness suggest that it is a manifestation of emergent firing patterns of neuronal groups involved in either specific networks (Hebb 1949, 1980, Freeman 1975, 1978), coherent 40–80 Hz firing (von der Malsburg and Schneider 1986, Gray and Singer 1989, Crick and Koch 1990), and/or attentional scanning circuits (Crick 1984, Edelman 1989, Baars 1988, 1993). But even precise correlation of neuronal firing patterns with cognitive activities fails to address perplexing differences between mind and brain including the "hard problem" of the nature of our inner experience (Chalmers 1996). In this paper we apply certain aspects of quantum theory (quantum coherence) and a new physical phenomenon described in Penrose (1994) of wave function self-collapse (objective reduction: OR) to specific, essential structures within each neuron: cytoskeletal microtubules. Table 37.1 summarizes how quantum coherence and OR occurring in microtubules (Orch OR) can potentially address some of the problematic features of consciousness.

Quantum theory describes the surprising behavior at a fundamental level of matter and energy which comprise our universe. At the base of quantum theory is the wave/particle duality of atoms and their components. As long as a quantum system such as an atom or subatomic particle remains isolated from its environment, it behaves as a "wave of possibilities" and exists in coherent "superposition" (with complex number coefficients) of many possible states.

There are differing views as to how quantum superposed states (wave functions) are "collapsed" or "reduced" to a single, classical state (Table 37.2). The conventional quantum theory view (Copenhagen interpretation) is that the quantum state reduces by environmental entanglement, measurement, or conscious observation (subjective reduction: SR, or R). Precisely where a quantum particle is and how it is moving when observed is "indeterminate" and, according to the Copenhagen interpretation, results in random measured values. We take the view (Penrose 1994) that, to address this issue, a new physical ingredient (objective reduction: OR) is needed in which coherent quantum systems can "self-collapse" by growing

Table 37.1 Aspects of consciousness difficult to explain by conventional neuroscience and possible quantum solutions

Problematic Feature of Consciousness	Possible Quantum Solutions
Unitary sense: "binding problem"	1. Nonlocal coherence; Indivisible macroscopic quantum state (e. g. Bose-Einstein condensate);
	2. Instaneous *self*-collapse of superpositioned states (Orch OR).
Transition from preconscious/ subconscious to conscious processes	1. Sub- and preconscious occur in quantum computing mode;
	2. Automatic, autonomic functions occur in classical computing mode;
	3. Quantum → classical transition. (Wave function "self" collapse—Orch OR—is intrinsic to consciousness).
Noncomputable, nonalgorithmic logic	Orch OR is noncomputable.
(Apparent) nondeterministic "free will"	Noncomputable, but nonrandom wave function self-collapse (Orch OR).
Essential nature of human experience	1. Wave function self-collapse (Orch OR) from incompatible superposition of separated space-times;
	2. Preconscious → conscious transition;
	3. Effectively instantaneous "now" (Orch OR) collapse.

and persisting to reach a critical mass/time/energy threshold related to quantum gravity. In the OR scheme, the collapse outcomes ("eigenstates") need not be random, but can reflect (in some noncomputable way) a quantum computation occurring in the coherent superposition state.

Another feature of quantum systems is quantum inseparability, or nonlocality, which implies that all quantum objects that have once interacted are in some sense still connected! When two quantum systems have interacted, their wave functions become "phase entangled" so that when one system's wave function is collapsed, the other system's wave function, no matter how far away, instantly collapses as well. The nonlocal connection ("quantum entanglement") is instantaneous, independent of distance, and implies that the quantum entities, by sharing a wave function, are indivisible.

Where and how in the brain can quantum effects occur? Warm, wet, and noisy, the brain at first glance seems a hostile environment for delicate quantum phenomena that generally demand isolation and cold stillness (superconductors) or energy pumping of crystals (lasers). Nonetheless, various authors have implicated ion channels, ions themselves, DNA, presynaptic grids, and cytoskeletal microtubules as somehow mediating "standard" quantum effects. In a dualist context, Beck and Eccles (1992) proposed that an external "conscious self" might influence the apparently random quantum effects acting on neurotransmitter release at the presy-

Table 37.2 Descriptions of wave function collapse

Context	Cause of Collapse (Reduction)	Description	Acronym	
Quantum coherent superposition	No Collapse	Evolution of the wave function (Schrödinger equation)	U	
Conventional quantum theory (Copenhagen interpretation)	Environmental entanglement, measurement, conscious observation	Reduction; subjective, reduction	R	SR
New physics (Penrose 1994)	Self-collapse—quantum gravity induced (Penrose 1994, Diósi 1989)	Objective reduction	OR	
Consciousness (present chapter)	Self-collapse, quantum gravity threshold in microtubules orchestrated by MAPs, etc.	Orchestrated objective reduction	Orch OR	

naptic grid within each neural axon. Stapp (1993) has suggested that SR wave function collapse in neurons is closely related to consciousness in the brain. In our view, cytoskeletal microtubules are the most likely sites for quantum coherence, OR, and consciousness.

Networks of self-assembling protein polymers, the cytoskeleton within neurons establishes neuronal form, maintains synaptic connections and performs other essential tasks (Figure 37.1). The major cytoskeletal components are microtubules, hollow cylindrical polymers of individual proteins known as tubulin. Microtubules are interconnected by linking proteins (microtubule-associated proteins: MAPs) to other microtubules and cell structures to form cytoskeletal lattice networks (Figure 37.2).

Traditionally viewed as the cell's "bone-like" scaffolding, microtubules and other cytoskeletal structures also appear to fill communicative and information processing roles. Theoretical models suggest how conformational states of tubulins within microtubule lattices can interact with neighboring tubulins to represent, propagate, and process information as in molecular-level "cellular automata" computing systems (Hameroff and Watt 1982, Rasmussen et al. 1990, Hameroff et al. 1992).

In this paper, we present a model linking microtubules to consciousness using quantum theory as viewed in a particular "realistic" way, as described in *Shadows of the Mind* (Penrose 1994). In our model, quantum coherence emerges, and is isolated, in brain microtubules until the differences in mass-energy distribution among superpositioned tubulin states reaches a threshold related to quantum gravity. The resultant self-collapse

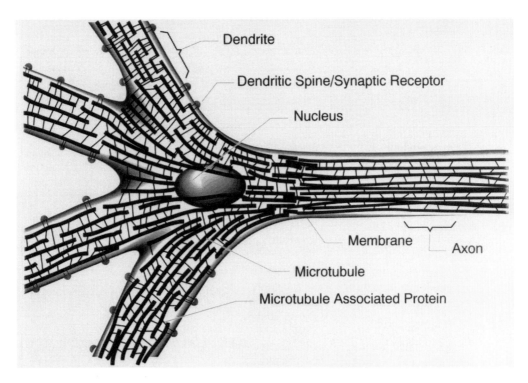

Figure 37.1 Schematic of central region of neuron (distal axon and dendrites not shown) showing parallel arrayed microtubules interconnected by microtubule-associated problem (MAPs). Microtubules in axons are lengthy and continuous, whereas in dendrites they are interrupted and of mixed polarity. Linking proteins connect microtubules to membrane proteins including receptors on dendritic spines.

(OR), irreversible in time, creates an instantaneous "now" event. Sequences of such events create a flow of time and consciousness.

We envisage that attachments of MAPs on microtubules "tune" quantum oscillations, and "orchestrate" possible collapse outcomes. Thus we term the particular OR occurring in MAP-connected microtubules, and relevant to consciousness, as "orchestrated objective reduction" (Orch OR).

COLLAPSE OF THE WAVE FUNCTION

The boundary between the microscopic, quantum world and the macroscopic, classical world remains enigmatic. Behavior of wave-like, quantum-level objects can be satisfactorily described in terms of a deterministic, unitarily evolving process (for example, state vector evolving according to the Schrödinger equation) denoted by U. Large-scale (classical) systems seem to obey (different) computable deterministic laws. The transition when system effects are magnified from the small, quantum scale to the large, classical scale (measurement process) chooses a particular "eigenstate" (one state of many possible states). According to the conventional Copenhagen interpretation of quantum theory, the "choice" of eigenstate is

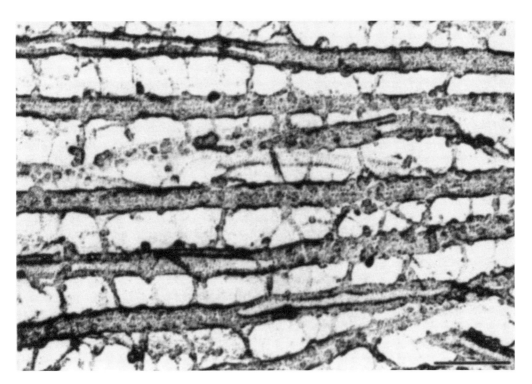

Figure 37.2 Immunofluorescent micrograph of neuronal microtubules interconnected by MAPs. Scale bar: 100 nanometers. (With permission from Hirokawa 1991).

purely random. The noncomputable R process is known in various contexts as collapse of the wave function, quantum jump, Heisenberg event, and/or state reduction.

Von Neumann, Schrödinger, and others in the 1930s supposed that quantum collapse, or R effectively occurred when a quantum system interacted with its environment, was otherwise "measured" or consciously observed. Exactly why and how collapse occurs, and how eigenstates are determined, are unknown and indicate a gap in physics knowledge: R is not taken to be an objectively real, independent phenomenon in the standard Copenhagen interpretation.

A number of physicists have argued in support of specific models (or of general schemes) in which the rules of standard U-quantum mechanics are modified by the inclusion of some additional procedure according to which R does become an objectively real process. The relevant procedure of any such specific scheme is here denoted by OR (objective reduction). In *Shadows of the Mind*, Penrose (1994) describes OR in which quantum coherence grows until it reaches a critical threshold related to quantum gravity, and then abruptly self-collapses. Other schemes for OR include those due to Pearle (1989) and to Ghirardi, Rimini, and Weber (1986), and those which are based on gravitational effects, such as Károlházy, Frenkel, and Lukacs (1986), Diósi (1989), Ghirardi, Grassi, and Rimini (1990), and also Penrose

(1989). Recent work (Pearle and Squires 1994) lends some considerable support, on general and observational grounds, for a gravitational OR scheme. There are also strong arguments from other directions (Penrose 1987, 1989) supporting a belief that the appropriate union of general relativity with quantum mechanics will lead to a significant change in the latter theory (as well as in the former—which is generally accepted). There is also some tentative, but direct, evidence in favor of this union being a noncomputable theory (for example, Geroch and Hartle 1986, Deutsch unpublished, Penrose 1994). OR in microtubules relevant to consciousness was first considered in somewhat general terms in *Shadows of the Mind*. Here we shall adopt a fairly specific proposal for OR (in accordance with Penrose 1994, Diósi 1989, Ghirardi, Grassi, and Rimini 1990) applied quantitatively in microtubules in which emergence of quantum coherence U and subsequent OR are "guided" and "tuned" ("orchestrated") by connecting MAPs. We thus elaborate a model of "orchestrated OR" (Orch OR) in microtubules which may support consciousness.

An important feature of OR (and Orch OR) is that noncomputable aspects arise only when the quantum system becomes large enough that its state undergoes self-collapse, rather than its state collapsing because its growth forces entanglement with its environment. Because of the random nature of environment, the OR action resulting from growth-induced entanglement would be indistinguishable from the random SR or R process of standard quantum theory.

Consciousness, it is argued, requires noncomputability (Penrose 1989, 1994). In standard quantum theory there is no noncomputable activity, the R process being totally random. The only readily available apparent source of noncomputability is OR (and Orch OR) self-collapse. An essential feature of consciousness might then be a large-scale quantum-coherent state maintained for a considerable time. OR (Orch OR) then takes place because of a sufficient mass displacement in this state, so that it indulges in a self-collapse that somehow influences or controls brain function. Microtubules seem to provide easily the most promising place for these requirements.

MICROTUBULES AND THE CYTOSKELETON

Ideal properties for quantum brain structures relevant to consciousness might include: (1) high prevalence, (2) functional importance (for example regulating neural connectivity and synaptic function), (3) periodic, crystal-like structure with long-range order, (4) ability to be transiently isolated from external interaction/observation, (5) functionally coupled to quantum-level events, and (6) suitable for information processing. Membranes, membrane proteins, synapses, DNA, and other candidates have some, but not all, of these characteristics. Cytoskeletal microtubules do appear to have the requisite properties.

Interiors of living eukaryotic cells (including the brain's neurons and glia) are organized by integrated networks of protein polymers called the cytoskeleton (Dustin 1984, Hameroff 1987). In addition to "bone-like" support, these dynamic self-organizing networks appear also to play roles as each cell's circulatory and nervous systems.

The cytoskeleton consists of microtubules (MTs), actin filaments, intermediate (neuro)filaments, and MAPs that, among other duties, link these parallel structures into networks (Figures 37.1 and 37.2).

The most prominent cytoskeletal component, MTs are hollow cylinders 25 nanometers (nm = 10^{-9} meter) in diameter whose lengths vary and may be quite long within some nerve axons. MT cylinder walls are comprised of 13 longitudinal protofilaments that are each a series of subunit proteins known as tubulin (Figure 37.3). Each tubulin subunit is a polar, 8-nm dimer that consists of two slightly different classes of 4-nm, 55,000 dalton monomers known as α and β tubulin. The tubulin dimer subunits within MTs are arranged in a hexagonal lattice that is slightly twisted, resulting in differing neighbor relationships among each subunit and its six nearest neighbors, and helical pathways that repeat every 3, 5, and 8 rows.

MTs, as well as their individual tubulins, have dipoles with negative charges localized toward α monomers (De Brabander 1982). Thus MTs are "electrets": oriented assemblies of dipoles that are predicted to have piezoelectric (Athenstaedt 1974, Mascarenhas 1974) and ferroelectric (Tuszyński *et al.* 1995) properties. Biochemical energy is provided to the cytoskeleton in at least two ways: tubulin bound GTP is hydrolyzed to GDP in MTs, and

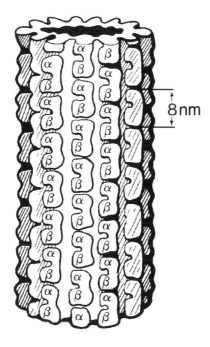

Figure 37.3 Microtubule structure from x-ray crystallography (Amos and Klug 1974). Tubulin subunits are 8 nanometer (nm) dimers comprised of alpha and beta monomers.

MAPs are phosphorylated. Each tubulin has a large hydrophobic region (Andreu 1986), a nonpolar pocket of amino acid side groups that interact by van der Waals forces and can support quantum-level electron delocalizability (Louria and Hameroff 1996).

MTs self-assemble and disassemble (for example, Kirschner and Mitchison 1986). The different scaffoldings they assume by their assembly and MAP attachments determine cell form and function including synaptic connections in neurons. Cell architecture (and synaptic connections) can quickly adapt by MT disassembly, and subsequent reassembly and MAP network formation in another shape or direction. Many organized cytoskeletal functions are carried out by MAPs. Some MAPs (dynein, kinesin) act as motors and carry material along microtubules (axoplasmic transport).

Several types of studies suggest cytoskeletal involvement in cognition. For example long-term potentiation (LTP) is a form of synaptic plasticity that serves as a model for learning and memory in mammalian hippocampal cortex. LTP requires MAP-2, a dendrite-specific, MT-crosslinking MAP that is dephosphorylated as a result of synaptic membrane receptor activation (for example, Halpain and Greengard 1990). In cat visual cortex, MAP-2 is dephosphorylated when visual stimulation occurs (Aoki and Siekevitz 1985). Auditory Pavlovian conditioning elevates temporal cortex MAP-2 activity in rats (Woolf et al. 1994). Phosphorylation/dephosphorylation of MAP-2 accounts for a large proportion of brain biochemical energy consumption (for example, Theurkauf and Vallee 1983) and is involved in functions that include strengthening specific networks, such as potentiating excitatory synaptic pathways in rat hippocampus (Montoro et al. 1993). The mechanism for regulating synaptic function appears related to rearrangement of MAP-2 connections on MTs (Bigot and Hunt 1990, Friedrich 1990).

Other types of evidence also link the cytoskeleton with cognitive function (Dayhoff, Hameroff, and Swenberg 1994). Production of tubulin and MT activities correlate with peak learning, memory, and experience in baby chick brains (Mileusnic, Rose, and Tillson 1980). When baby rats first open their eyes, neurons in their visual cortex begin producing vast quantities of tubulin (Cronley-Dillon, Carden, and Birks 1974). In animals whose brains are temporarily deprived of oxygen, the degree of cognitive damage correlates with decrease in measured levels of dendritic MAP-2 (Kudo et al. 1990). Bensimon and Chernat (1991) showed that selective damage of MTs in animal brains by the drug colchicine causes defects in learning and memory which mimic the symptoms of Alzheimer's disease (in which neuronal cytoskeleton entangles). Matsuyama and Jarvik (1989) have linked Alzheimer's disease to microtubule dysfunction, and Mandelkow and Mandelkow (1993) and others have pinpointed the axonal MAP "tau protein" as the tangle-causing defect.

How might the cytoskeleton signal and process information? Tubulin can undergo several types of conformational changes (for example, Engelborghs 1992, Cianci et al. 1986). Roth and Pihlaja (1977) suggested that patterns of tubulin conformation within MTs represented information. In one example of tubulin conformational change observed in single protofilament chains, one monomer can shift 27 degrees from the dimer's vertical axis (Melki *et al.* 1989). Whether such mechanical deformation occurs in tubulin within intact MTs is unknown; neighbor tubulins in the MT lattice might be expected to constrain movement. However, cooperativity among tubulins bound loosely in the MT lattice by hydrophobic forces could coordinate conformational changes, and support propagation of wave-like signals in MTs. Vassilev, Kanazirska, and Tien (1985) demonstrated signal transmission along tubulin chains formed between excitable membranes. A number of models of signaling and information processing within MTs and other cytoskeletal components have been suggested. These include propagating tubulin conformational changes (Atema 1973), ion transfer (Cantiello *et al.* 1991), sequential phosphorylation/dephosphorylation along MT tubulins (Puck and Krystosek 1992), tensegrity (Wang and Ingber, 1994), nonlinear soliton waves along MTs (Chou, Zhang, and Maggiora 1994, Sataric, Zakula, and Tuszyński 1992) and "cellular automaton" behavior due to electrostatic dipole coupling among tubulin lattice neighbors (for example, Rasmussen *et al.* 1990).

MICROTUBULE INFORMATION PROCESSING

Protein Conformation

Proteins have conformational transitions at many time and size scales (Karplus and McCammon 1983). For example small side chains move in the picosecond to femtosecond time scale (10^{-12} to 10^{-15}), but conformational transitions in which proteins move globally and upon which protein function generally depends occur in the nanosecond (10^{-9} sec) to 10 picosecond (10^{-11} sec) time scale. Related to cooperative movements of smaller regions, hydrogen bond rearrangements and charge redistributions such as dipole oscillations, these global changes linked to protein function (signal transduction, ion channel opening, enzyme action) may be regulated by a variety of factors including phosphorylation, ATP or GTP hydrolysis, ion fluxes, electric fields, ligand binding, and "allosteric" influences by neighboring protein conformational changes. Noting the extraordinary dielectric strength of proteins (their abilty to sustain a voltage), Fröhlich (1968, 1970, 1975) proposed that the various factors determining protein conformation were integrated through a quantum level dipole oscillation within each protein's hydrophobic region (Figure 37.4).

Fröhlich's Coherent Pumped Phonons

In addition to linking protein conformation to hydrophobic quantum events, Herbert Fröhlich, an early contributor to the understanding of superconductivity, also predicted quantum coherence in living cells (based on earlier work by Penrose and Onsager 1956). Fröhlich theorized that sets

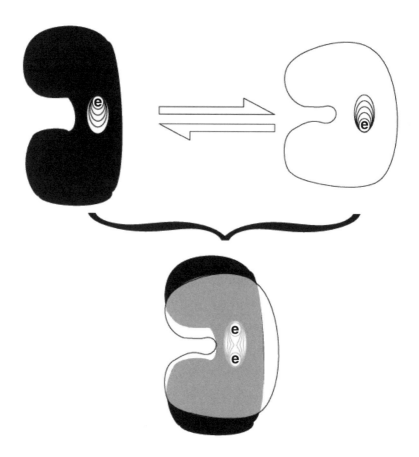

Figure 37.4 Schematic model of tubulin states. Top: Two states of microtubule subunit protein "tubulin" in which quantum event (electron localization) within a hydrophobic pocket is coupled to protein conformation. Bottom: Tubulin in quantum coherent superposition of both states.

of protein dipoles in a common electromagnetic field (for example, proteins within a polarized membrane, subunits within an electret polymer like microtubules) undergo coherent conformational excitations if energy is supplied. Fröhlich postulated that biochemical and thermal energy from the surrounding "heat bath" provides such energy. Cooperative, organized processes leading to coherent excitations emerged, according to Fröhlich, because of structural coherence of hydrophobic dipoles in a common voltage gradient.

Coherent excitation frequencies on the order of 10^9 to 10^{11} Hz (identical to the time domain for functional protein conformational changes, and in the microwave or gigaHz spectral region) were deduced by Fröhlich who termed them acousto-conformational transitions, or coherent (pumped) phonons. Such coherent states are termed Bose-Einstein condensates in quantum physics and have been suggested by Marshall (1989, 1996) to provide macroscopic quantum states that support the unitary binding of consciousness.

Experimental evidence for Fröhlich-like coherent excitations in biological systems includes observation of gigaHz-range phonons in proteins (Genberg et al. 1991), sharp-resonant nonthermal effects of microwave irradiation on living cells (Grundler and Keilmann 1983), gigaHz-induced activation of microtubule pinocytosis in rat brain (Neubauer et al. 1990), and Raman spectroscopy detection of Fröhlich frequency energy (Genzel et al. 1983).

Cellular Automata In Microtubules

Computational systems in which complex signaling and patterns emerge from local activities of simple subunits are called cellular automata. Their essential features are: (1) at a given time, each subunit is in one of a finite number of states (usually two for simplicity). (2) The subunits are organized according to a fixed geometry. (3) Each subunit communicates only with neighboring subunits; the size and shape of the neighborhood are the same for all cells. (4) A universal "clock" provides coherence such that each subunit may change to a new state at each "clock tick." (5) Transition rules for changing state depend on each subunit's "present" state and those of its neighbors. Depending on initial conditions (starting patterns), simple neighbor transition rules can lead to complex, dynamic patterns capable of computation. Von Neumann (1966) proved mathematically that cellular automata could function as Turing machines.

In a series of simulations (Hameroff, Smith, and Watt 1984, Rasmussen et al. 1990) Fröhlich's excitations were used as a clocking mechanism and electrostatic dipole coupling forces as "transition rules" for cellular automata behavior by dynamic conformational states of tubulins within MTs. As in the top of Figure 37.4, the two monomers of each tubulin dimer are considered to share a mobile electron within a hydrophobic pocket, which is oriented either more toward the α-monomer (alpha state) or more toward the β-monomer (beta state) with associated changes in tubulin conformation at each "Fröhlich coherent" time step (for example, 10^{-9} to 10^{-11} sec). The net electrostatic force (f_{net} from the six surrounding neighbors acting on each tubulin can then be calculated as:

$$f_{net} = \frac{e^2}{4\pi\epsilon} \sum_{i=1}^{6} \frac{y_i}{r_i^3}$$

where y_i and r_i are intertubulin distances, e is the electron charge, and ϵ is the average protein permittivity. MT automata simulations (Figure 37.5) show conformational pattern behaviors including standing waves, oscillators and gliders range of 8 to 800 meters per second, consistent with the velocity of propagating traveling one dimer length (8 nm) per time step (10^{-9} to 10^{-11} sec) for a velocity nerve action potentials.

MT automata patterns can thus represent and process information through each cell; gliders may convey signals that regulate synaptic

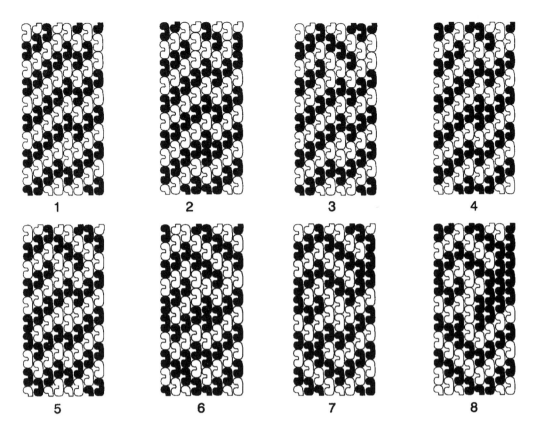

Figure 37.5 Microtubule automaton simulation (from Rasmussen *et al.* 1990). Black and white tubulins correspond to states shown in Figure 37.4. Eight nanosecond time steps of a segment of one microtubule are shown in "classical computing" mode in which patterns move, evolve, interact, and lead to emergence of new patterns.

strengths, represent binding sites for MAPs (and thus neuronal and synaptic connectionist architecture) or material to be transported. Information could become "hardened" in MTs by tubulin modifications or stored in neurofilaments via MAPs. MT conformational automata patterns provide a further level of computational complexity within each of the brain's neurons. However by considering only classical computing and local neighbor interactions, microtubule automata fail to address the problematic features of consciousness for which quantum theory holds promise.

WAVE FUNCTION SELF-COLLAPSE IN MICROTUBULES: OR AND ORCH OR

The Collapse Criterion And Conscious Thought

Consider a quantum superposition $w|A> + z|B>$ (where w and z are complex numbers) of two macroscopically distinguishable quantum states

|A> and |B>. In standard quantum theory and in the absence of environmental entanglement, this superposition would persist forever. If, after a time t, |A> would have evolved to $|A_t>$ and |B> would have evolved to $|B_t>$, then $w|A> + z|B>$ must evolve to $w|A_t> + z|B_t>$. (This is a feature of the linear nature of U.)

According to the present OR criterion, such macroscopic superpositions are regarded as *unstable* even without environmental entanglement. Therefore the state $w|A> + z|B>$ will decay in a certain time scale T, to either |A> or |B>, with relative probabilities $|w|^2:|z|^2$. (This is analogous to the situation with an *unstable radioactive particle*, with a lifetime T and two separate decay modes |A> or |B>, whose branching ratios are $|w|^2:|z|^2$.) The idea is that the states |A> and |B> each correspond to clearly defined energy distributions (and to well-defined space-time geometries), whereas the combination $w|A> + z|B>$ does not (and so would lead to superpositions of different space-time geometries—a particularly awkward situation from the physical point of view!). According to a number of authors (Diósi 1989, Ghirardi, Rimini, and Weber 1986, Ghirardi, Grassi, and Rimini 1990, Penrose 1993, 1994, Pearle 1992), the gravitational self-energy E of the difference between the mass distributions involved in |A> and in |B> will determine spontaneous reduction (at time T) of the superposed combination $w|A> + z|B>$ to either |A> or |B>.

We view |A> and |B> as representing two conformationally coupled quantum states of each tubulin within microtubules, and $w|A> + z|B>$ as the superposition of those states (for example, Figure 37.4). We envisage that time T, at which the state $w|A> + z|B>$ for each tubulin will decay to either |A> or |B> should relate to the transition between preconscious and conscious events.

We assume that pre- and subconscious processing corresponds with quantum coherent superposition which can perform "quantum computing" (Penrose 1989). A number of authors (Deutsch 1985, Deutsch and Josza 1992, Feynman 1986, Benioff 1982) have proposed that quantum coherence can implement multiple computations simultaneously, in parallel, according to quantum linear superposition: the quantum state then "collapses" to a particular result. A state that "self-collapses" (OR) will have an element of noncomputability, even though evolution of its quantum coherence had been linear and computable. A quantum superposed state collapsed by external environment or observation (SR or R) lacks a noncomputable element, and would thus be unsuitable for consciousness. Large scale quantum coherence occurring among tubulins (for example, via electrons in hydrophobic pockets arrayed in the microtubule lattice, or ordered water within hollow MT cores) could take on aspects of a quantum computer in preconscious and subconscious modes.

We also assume that "nonconscious" autonomic processes correspond with classical, nonquantum computing by microtubule conformational automata. Thus an OR transition from quantum, preconscious processing, to classical, nonconscious processing may be closely identified with consciousness itself.

But what is consciousness? According to the principles of OR (Penrose 1994), superpositioned states each have their own space-time geometries. When the degree of coherent mass-energy difference leads to sufficient separation of space-time geometry, the system must choose and decay (reduce, collapse) to a single universe state, thus preventing "multiple universes" (for example, Wheeler 1957). In this way, a transient superposition of slightly differing space-time geometries persists until an abrupt quantum → classical reduction occurs and one or the other is chosen. Thus, consciousness may involve self-perturbations of space-time geometry.

The extent of space-time superposition causing self-collapse is related to quantum gravity, and equal to one in "absolute units." Absolute units convert all physical measurement into pure, dimensionless numbers (Penrose 1994, pp. 337–339). This is done by choosing units of length, mass, and time so that the following constants take the value of unity:

$$c \text{ (speed of light)} = 1$$
$$\hbar \text{ (Planck's constant divided by } 2\pi) = 1$$
$$G \text{ (gravitational constant)} = 1$$

Physical quantities relevant to our calculations in absolute units:

$$\text{second} = 1.9 \times 10^{43}$$
$$\text{nanometer} = 6.3 \times 10^{25}$$
$$\text{mass of nucleon ("dalton")} = 7.8 \times 10^{-20}$$
$$\text{fermi (strong interaction size, diameter of nucleon)} = 6.3 \times 10^{19}$$

Using absolute units, we can ask how many tubulins in quantum coherent superposition for how long will self-collapse (Orch OR)?

The gravitational self-energy E for a quantum superposition of mass whose displacement for a given time T sufficiently perturbs space-time for OR (Orch OR) is taken from the "uncertainty principle":

$$E = \frac{\hbar}{T}$$

where $\hbar$ is Planck's constant divided by 2π, and T is the coherence time.

We estimate T the coherence time from research by Libet (1990) and others (Deecke, Grotzinger, and Kornhuber 1976, Grey-Walter 1953; Libet et al. 1979), who found the time scale characteristic of preconscious to conscious transitions to be about 500 milliseconds (msc). This is similar to other estimates in the 100–200 msc range (Koch 1996). Hence $T = 500$ msc (half second) seems appropriate from experimental results for at least some functionally significant transitions.

In absolute units:

$$T = 500 \text{ msc} = (0.5)\, 1.9 \times 10^{43} \cong 10^{43}$$

Since $\hbar = 1$ in absolute units:

$$E = T^{-1} = 10^{-43}$$

This is the gravitational self-energy E for which n_t tubulins displaced in quantum coherent superposition for 500 msc will self-collapse (OR, Orch OR). To determine n_t, we calculate the gravitational displacement self energy E_t for one tubulin. We assume the tubulin conformational movement displaces its mass by a distance r which is 1/10 the 2 nanometer (nm) radius of the tubulin monomer, or 0.2 nm. In absolute units:

$$r = 0.2 \, (6.3 \times 10^{25}) \cong 10^{25}$$

As illustrated in Figure 37.6, the distribution of mass m for each tubulin conformational change may be considered as either two protein spheres, two granular arrays of atoms, or two granular arrays of nucleons (protons and neutrons).

Protein Spheres Consider the tubulin dimer as two uniform (monomer) spheres (Figure 37.6a). As a reasonable approximation, the energy E_t for one tubulin dimer is twice the displacement energy of one (monomer) sphere position in the gravitational field of its other position (twice because there are two spheres per tubulin). m_t of the tubulin monomer sphere is 55,000 daltons (nucleons). In absolute units:

$$m_t = 5.5 \times 10^4 \, (7.8 \times 10^{-20}) \cong 4 \times 10^{-15}$$

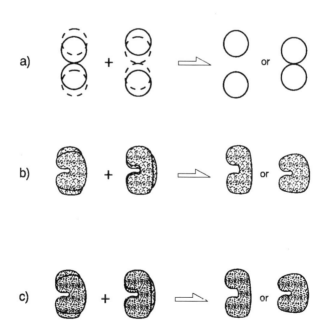

Figure 37.6 Three treatments of mass distribution of tubulin. a) 2 protein spheres, b) 2 granular arrays of (carbon) atoms, or c) 2 granular arrays of nucleons.

Since the displacement is less than the sphere radius, we need a detailed calculation to obtain the gravitational self-energy E_m of the difference between the displaced mass distributions for each tubulin monomer, considered as a uniform sphere. Taking the sphere to have a radius a and the distance of displacement to be r, we find, as the result of a double integration:

$$E_m = m^2 G \left(\frac{r^2}{2a^3} - \frac{3r^3}{16a^4} + \frac{r^5}{160a^6} \right)$$

where G is Newton's universal constant of gravitation with $G = 1$ in absolute units. Here we have $m = (1/2)\, m_t \approx 2 \times 10^{-15}$ and $a \approx 10^{26}$, and we are taking $r = 10^{25}$. We can ignore the higher-order terms, so we obtain, in absolute units:

$$E_m \cong \frac{m^2 r^2}{2a^3} \cong \frac{(4 \times 10^{-15} \times 10^{25})^2}{2 \times 10^{78}} = 8 \times 10^{-58}$$

Thus, approximately, $E_t = 1.6 \times 10^{-57}$ for the gravitational self-energy of displacement for a single tubulin dimer (2 monomers). For n_t (the number of tubulin dimers) in absolute units (approximately):

$$n_t = \frac{T^{-1}}{E_t} = \frac{10^{-43}}{1.6 \times 10^{-57}} \cong 6 \times 10^{13} \text{ tubulins}$$

Considering the tubulin mass distributions *simply* as protein spheres, we would thus estimate 6×10^{13} tubulins displaced in quantum coherent superposition for 500 msc will self-collapse (Orch OR).

Granular Arrays of Atomic Nuclei Consider tubulin as two arrays of (carbon) atoms (Figure 37.6b). The mass m_c of one carbon atom (12 nucleons) is its nuclear mass in absolute units:

$$m_c = 12(7.8 \times 10^{-20}) = 10^{-18}$$

Because the carbon nucleus displacement is greater than its radius (the spheres separate completely), the gravitational self-energy E_c is given by:

$$E_c \cong G \frac{m^2}{a_c}$$

where a_c is the carbon nucleus sphere radius equal to 2.5 fermi distances. In absolute units:

$$a_c = 2.5(6.3 \times 10^{19}) = 1.6 \times 10^{20}$$

$$E_c \approx G \frac{m^2}{a_c} = \frac{(10^{-18})^2}{1.6 \times 10^{20}} \cong 10^{-56}$$

This is the gravitational self-energy in absolute units for displacement of one carbon atom nucleus (12 nucleons). To determine how many carbon nuclei n_c and tubulins n_t displaced for 500 msc will elicit Orch OR:

$$E = n_c E_c = T^{-1}$$

$$n_c = \frac{T^{-1}}{E_c} = \frac{10^{-43}}{10^{-56}} = 10^{13} \text{ carbon atoms} = 12 \times 10^{13} \text{ nucleons}$$

As one tubulin is 110,000 nucleons,

$$n_t = \frac{12 \times 10^{13}}{1.1 \times 10^5} \cong 10^9 \text{ tubulins}$$

Considering the tubulin mass distribution as atomic (carbon) nuclei, we thus estimate that 10^9 tubulins displaced in coherent superposition for 500 msc will self-collapse, and elicit Orch OR.

Granular Arrays of Nucleons Consider tubulin as two arrays of nucleons (Figure 37.6c). The mass m_n of one nucleon in absolute units:

$$m_n = 7.8 \times 10^{-20}$$

Since the nucleon displacement is again greater than its radius (complete separation):

$$E_n = G \frac{m^2}{a_n}$$

where a_n is the nucleon radius, or 0.5 fermi:

$$a_n = 0.5(6.3 \times 10^{19}) \cong 3 \times 10^{19}$$

$$E_n = G \frac{m^2}{a_n} = \frac{(7.8 \times 10^{-20})^2}{3 \times 10^{19}} \cong 2 \times 10^{-58}$$

This is the gravitational self-energy for displacement of one nucleon for 500 msc. ($G = 1$ in absolute units.) To find n_n, the number of nucleons whose displacement for 500 msc will elicit Orch OR:

$$E = n_n E_n = T^{-1}$$

$$n_n = \frac{T^{-1}}{E_n} = \frac{10^{-43}}{2 \times 10^{-58}} = 5 \times 10^{14}$$

As one tubulin is 110,000 nucleons,

$$n_n = \frac{5 \times 10^{14}}{1.1 \times 10^5} \approx 5 \times 10^9 \text{ tubulins}$$

Considering tubulin mass as arrays of nucleons, we thus estimate that 5×10^9 tubulins displaced in quantum coherent superposition for 500 msc will self-collapse, Orch OR.

Using the three types of tubulin mass distributions (protein spheres, atoms, nucleons), we obtain 6×10^{13}, 10^9, and 5×10^9 respectively for the required number of n_t quantum coherent tubulins displaced for 500 msc to elicit Orch OR. Although in approximation all three mass distributions contribute to Orch OR, that which gives the highest energy (fewest tubulins, shortest reduction time if T were not fixed) predominates. Thus, 10^9 tubulins is perhaps the best estimate. The possible significance of this number of tubulins will be discussed later.

As discussed in other papers in this volume (Elitzur 1996, Tollaksen 1996), consciousness may be linked to creation of an instantaneous "now," and the flow of time. As Orch OR is instantaneous, noncomputable, and irreversible, it can provide "now" moments, and directionality in time. Sequential cascades of Orch OR events would then constitute the familiar "stream of consciousness."

The Orch OR process in MTs selects patterns (eigenstates of mass distribution) of tubulin conformational states. Figure 37.7 illustrates eight possible "eigenstates of mass distribution" for Orch OR occurring in three adjacent tubulins. The selected patterns can influence neural function by determining MAP attachment sites and setting initial conditions for MT "cellular automata" information processing.

These MT activities can then govern intraneuronal architecture and synaptic function by modulating sensitivity of membrane receptors (in, for example, dendritic spines), ion channels and synaptic vesicle release mechanisms, communication with genetic material, and regulating axoplasmic transport which accounts for delivery of synaptic material components.

MAPs attached to certain microtubule tubulin subunits would seem likely to communicate the quantum state to the outside "noisy" random environment, and thereby entangle and collapse it (SR or R, rather than OR or Orch OR). We therefore presume that these MAP connections are placed along each MT at sites which are (temporarily at least) inactive with regard to quantum-coupled conformational changes. We envisage that these connection points are, in effect, "nodes" for MT quantum oscillations, and thus "orchestrate" the possibilities and probabilities for MT quantum coherence and subsequent Orch OR (Figure 37.8).

MAP connection points can be regularly placed on MT lattices in superhelical patterns (Kim, Jensen, and Rebhund 1986, Burns 1978), which seems appropriate for their proposed roles as nodes. However, in neural MTs, the MAP connection points appear more randomly placed. This would not prevent their acting as "nodes" because the "quantum cellular automaton" activity that we envisage could be extremely complicated, and could well appear to be random.

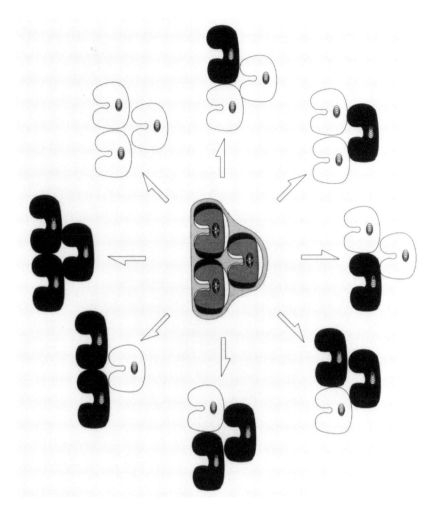

Figure 37.7 Possible outcome "eigenstates of mass distribution" for wave function collapse (Orch OR) in three tubulins (center) in quantum coherent superposition.

In addition to MAPs, genetic tubulin variability (Lee *et al.* 1986), and "learned, experiential" (posttranslational) tubulin modifications (Gilbert and Strochi 1986) can orchestrate OR. Accordingly, we term the particular objective reduction (OR) occurring in MTs, self-tuned by MAPs and other factors, and relevant to consciousness as "orchestrated objective reduction" (Orch OR).

Because OR phenomena are fundamentally nonlocal, the coherent superposition phase may exhibit puzzling bidirectional time flow prior to self-collapse (Aharonov and Vaidman 1990, Penrose 1989, 1994). As we equate the precollapse quantum computing superposition phase to preconscious processing, bidirectional time flow could explain the puzzling "backwards time referral" aspects of preconscious processing observed by Libet *et al.* (1979).

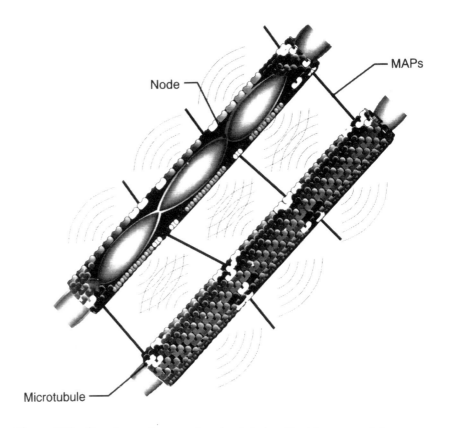

Figure 37.8 Quantum coherence in microtubules. Having emerged from resonance in classical automaton patterns, quantum coherence nonlocally links superpositioned tubulins (gray) within and among microtubules. Upper microtubule: cutaway view shows coherent photons generated by quantum ordering of water on tubulin surfaces, propagating in microtubule waveguide. MAP attachments breach isolation and prevent quantum coherence; MAP attachment sites thus act as "nodes" that tune and orchestrate quantum oscillations and set possibilities and probabilities for collapse outcomes (orchestrated objective reduction: Orch OR).

How Can Quantum Coherence In Microtubules Be Isolated From Environmental Entanglement?

When the quantum system under consideration becomes entangled with another system, we must consider the entire state involved. For example $|A\rangle$ might be accompanied by the environment state $|P\rangle$, and $|B\rangle$ with the environment state $|Q\rangle$. Then, in place of the state $w|A\rangle + z|B\rangle$, we have

$$w|A\rangle|P\rangle + z|B\rangle|B\rangle|Q\rangle.$$

We must consider mass movement in $|P\rangle$ and $|Q\rangle$ as well as in $|A\rangle$ and $|B\rangle$, and whenever this dominates, we get, in effect, the random SR, or R process of conventional quantum measurement theory, rather than the noncomputable aspects of OR (or Orch OR) that would be important for consciousness.

At first glance, the interiors of living cells would seem unlikely sites for quantum effects. "Noisy" thermal motions of cell water ($|P\rangle$) would seem to decohere any quantum coherence and cause random SR, or R collapse. However several factors could serve to isolate microtubules and sustain quantum coherence.

Ordered Water Water on cytoskeletal surfaces can be highly "ordered," extending up to nine layers (about 3 nm) around each microtubule (Clegg 1983, Watterson 1996). Thermal interactions which would cause decoherence involve energy coupling to oscillations in short-range interactions (for example, hydrogen bonds) among water molecules and groups of water molecules with energy of about kT (10^{-12} sec). Interactions of MT surface ordered water molecules with a Fröhlich coherence in MTs are predicted to have a frequency (10^{-14} sec) much higher than thermalization energy. Thus MTs may be embedded in "cages" of structured (coherent) water which can act to isolate MT quantum coherence (Jibu et al. 1994).

If a quantized electric field generated by Fröhlich pumped phonons in MTs is comparable to the coherent strength of water, the field penetrates/propagates by "piercing" it (self-focusing, filamentary propagation analogous to Meissner effect in superconductivity). Del Giudice, Doglia, and Milani (1983) showed this self-focusing should result in filamentous energy beams of radius 15 nanometers, precisely the inner diameter of microtubules!

Isolation Inside Microtubule Hollow Core Using quantum field theory, Jibu et al. (1994) and Jibu, Hagan, and Yasue (1996) have modeled the ordering of water molecules and the quantized electromagnetic field confined inside hollow microtubule cores. They predict a specific collective dynamics called superradiance in which each microtubule can transform incoherent, disordered energy (molecular, thermal, or electromagnetic) into coherent photons within its hollow core. The time for superradiant photon generation is much shorter than the time needed for the environment to act to disorder the coherence thermally.

Sol-Gel States Cytoplasm is comprised of two phases: sol (solution) and gel (gelatinous). Calcium ions binding to actin and other cytoskeletal polymers convert sol to gel transiently and reversibly (for example, cytomatrix: Satir 1984). MTs and other MT-associated proteins (that is, calmodulin) bind/release calcium. Gelatinous phases adjacent to microtubules could isolate them during quantum coherence.

The Collapse Fraction

Here we get a very rough idea of the fraction of brain required for Orch OR. Earlier we obtained a very rough estimate that 10^9 tubulins in quantum coherent superposition for 500 msec are sufficient for Orch OR. Yu and Baas (1994) measured about 10^7 tubulins per neuron. Thus, about 100 neurons whose tubulins were *totally* coherent for 500 msc may be the minimal number for Orch OR, and for consciousness. It may be more likely that only a fraction of tubulins within a given neuron becomes coherent. (Global macroscopic states such as superconductivity can result from quantum coherence among only very small fractions of components.) For example if 1 percent of tubulins within a given set of neurons were coherent for 500 msc, then 10,000 such neurons would be required to elicit Orch OR. Thus for 500 msc we get a possible range for minimal consciousness from hundreds to thousands of neurons. Nervous systems of organisms such as the

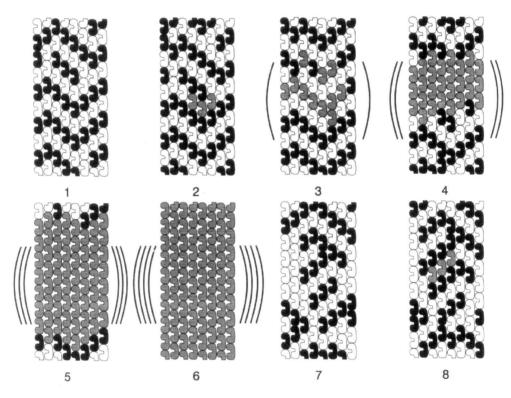

Figure 37.9 Microtubule automaton sequence simulation in which classical computing (step 1) leads to emergence of quantum coherent superposition (steps 2–6) in certain (gray) tubulins due to pattern resonance. Step 6 (in coherence with other microtubule tubulins) meets critical threshold related to quantum gravity for self-collapse (Orch OR). Consciousness (Orch OR) occurs in the step 6 to 7 transition. Step 7 represents the eigenstate of mass distribution of the collapse which evolves by classical computing automata to regulate neural function. Quantum coherence begins to re-emerge in step 8.

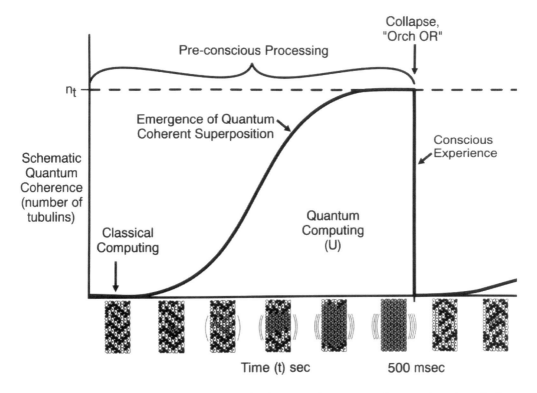

Figure 37.10 Schematic graph of proposed quantum coherence (number of tubulins) emerging versus time in microtubules. The time for preconscious processing is 500 milliseconds (Libet 1979). Area under curve connects mass-energy difference with collapse time in accordance with gravitational OR. This degree of coherent superpositioning of differing space-time geometries leads to abrupt quantum classical reduction ("self-collapse" or "orchestrated objective reduction: Orch OR").

nematode *C. elegans* contain several hundred neurons. Functional groups of neurons in human cognition are thought to contain thousands. Hebb's (1949) "cell assemblies," Eccles' (1992) "modules," and Crick and Koch's (1990) "coherent set of neurons" are each estimated to contain some 10,000 neurons that may be widely distributed throughout the brain (Scott 1995).

Rather than always being 500 msc, time and the number of tubulins n_t may vary, and result in different types of conscious experience. A very intense, sudden input may recruit emergence of quantum coherent tubulins faster so that Orch OR occurs sooner ("heightened experience," Figure 37.11c). For example, 10^{10} coherent tubulins would elicit Orch OR in 50 msec, and so on. Lower intensity input patterns develop coherence more slowly, and Orch OR occurs later. An instantaneous Orch OR may then "bind" various coherent tubulin superpositions whose net displacement energy is T^{-1}, but which may have evolved in separated spatial distributions and over different time scales into an instantaneous event (a

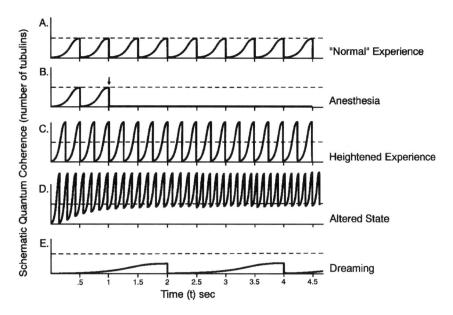

Figure 37.11 Quantum coherence in microtubules schematically graphed on longer time scale for five different states related to consciousness. The area under each curve is equivalent in all cases. A. Normal Experience, as in Figure 37.8. B. Anesthesia: anesthetics bind in hydrophobic pockets and prevent quantum delocalizability and coherent superposition (see Louria and Hameroff, 1996). C. Heightened Experience: increased sensory experience input (for example) increases rate of emergence of quantum coherent superposition. Orch OR threshold is reached faster (for example, 250 msec) and Orch OR frequency is doubled. D. Altered State: even greater rate of emergence of quantum coherence due to sensory input and other factors promoting quantum state (for example, meditation, psychedelic drug, see Louria and Hameroff 1996). Predisposition to quantum state results in baseline shift and only partial collapse so that conscious experience merges with normally subconscious quantum computing mode. E. Dreaming: prolonged quantum coherence time.

"conscious now"). Cascades of Orch ORs can then represent our familiar "stream of consciousness."

SUMMARY: ORCHESTRATED OBJECTIVE REDUCTION (ORCH OR) OF QUANTUM COHERENCE IN BRAIN MICROTUBULES

The picture we are putting forth involves the following ingredients:

A. A macroscopic state of quantum coherent superposition can exist among tubulin subunits in microtubules across a large proportion of the brain. Plausible candidates for such states include models proposed by:

1. Marshall (1989) in which Fröhlich-pumped phonons induce a Bose-Einstein condensate among proteins distributed throughout the brain,
2. Jibu et al. (1994) in which ordered water within microtubule hollow cores is coupled to Fröhlich excitations of tubulins in microtubule walls resulting in coherent photons ("superradiance");

3. Kaivarainen (1994) in which quantum coherent "flickering" clusters of ordered water within hollow microtubule cores are coupled to (non-Fröhlich) tubulin conformational dynamics and generate coherent photons; and
4. Conrad (1992, 1994) in which conformational states and functional capabilities of proteins are controlled by quantum superposition dynamics of electrons and hydrogen bonds within them.

B. The quantum coherent state is weakly coupled to conformational activity taking place in tubulins within microtubules. The link occurs by changes in individual electric dipole moments within tubulins. For example, movements of a single electron within a hydrophobic pocket centrally placed within each tubulin dimer may couple to the conformational state of the tubulin (Figure 37.4).

C. This combined quantum state among many tubulins is able to maintain itself without significant entanglement with its environment for a relevant period of time (up to 500 msc). We envisage several possible mechanisms that could serve to isolate the MT quantum state from its environment within the neuron. These include:
1. Shielding by ordered water on tubulin surfaces,
2. Isolation within hollow MT inner cores, and
3. Shielding by gelatinous cytoplasmic layer.

D. Cooperative interactions among neighboring tubulins in microtubules can signal and process information by computational mechanisms such as cellular automata behavior. We propose two types of microtubule computation:
1. Classical computing: conformational patterns propagate through the cytoskeleton to regulate synapses and perform other neural functions (Figure 37.5). This mode correlates with nonconscious and autonomic activities.
2. Quantum computing: large-scale quantum coherence occurs among tubulins (for example, via electrons in hydrophobic pockets arrayed in the microtubule lattice) and takes on aspects of a quantum computer (Deutsch 1985, Deutsch and Josza 1992, Feynman 1986, Benioff 1982) where multiple "computations" are performed simultaneously, in parallel, according to quantum linear superposition. We equate quantum computing with pre- and subconscious processing.

E. In the quantum computing mode, changes in dimer conformations involve the movement of mass. According to the arguments and criterion described earlier, we estimate the time scale T, and calculate the number n_t (of coherent superpositioned tubulins) required for self-collapse (objective orchestrated reduction: Orch OR) to occur. As we equate quantum computing with preconscious processing, we approximate the time scale T to be equivalent in some cases to that found by Libet *et al.* (1979) and others (Deecke, Grotzinger, and Kornhuber 1976, Grey-Walter 1953) to be characteristic of the transition from preconscious to conscious processing (up to 500 msc). For $T = 500$ msc, we get a rough estimate of tubulins required for Orch OR.

F. MAPs attached to certain microtubule tubulin subunits would seem likely to communicate the quantum state to the outside "noisy" random environment, and thereby entangle and collapse it (SR or R). We therefore presume that these MAP connections are placed along each MT at sites that are (temporarily at least) inactive with regard to quantum-coupled conformational changes. We envisage that these connection points are, in effect, "nodes" for MT quantum oscillations, and (along with genetic and other tubulin modifications) thus "orchestrate" MT quantum coherence and subsequent OR (Figure 37.8). Accordingly, we term the particular objective reduction (OR) occurring in MTs and relevant to consciousness as "orchestrated objective reduction" (Orch OR).

G. The Orch OR process selects a new set of tubulin conformational states ("eigenstates of mass distribution") within MTs which can implement and regulate neural function by determining MAP attachment sites and setting initial conditions for "cellular automata" information processing by "classical" conformational transitions. These MT activities can then govern intraneuronal architecture and synaptic function through modulating sensitivity of membrane receptors, ion channels and synaptic vesicle release mechanisms, communication with genetic material, and regulating axoplasmic transport that accounts for delivery of synaptic material components.

H. How is a particular conformational pattern within each microtubule chosen in the Orch OR process? The Copenhagen quantum interpretation would suggest the selection of states upon (SR or R) collapse is purely random. Effects of MAPs, genetic and other tubulin modifications can set the possibilities and probabilities ("orchestrate"), but reduction within that context is noncomputable. As described in *Shadows of the Mind* (Penrose 1994), it remains possible that presently unrecognized OR (or Orch OR) quantum-mathematical logic acting on these programming influences provides a hidden order.

I. Because OR (and Orch OR) phenomena are fundamentally nonlocal, the coherent superposition phase may exhibit puzzling bidirectional time flow prior to *self*-collapse (for example, Aharonov and Vaidman 1990, Penrose 1989, 1994). We equate the precollapse quantum computing superposition phase to preconscious processing. This could explain the puzzling "backwards time referral" aspects of preconscious processing observed by Libet *et al.* (1979). (Also see Tollaksen 1996).

J. The persistence and global nature of consciousness is seen as a feature of large-scale quantum coherent activity taking place across much of the brain. Varieties of Orch OR with differing coherence times and amounts of coherent tubulin may blend into conscious thought. Very intense, sudden inputs may recruit emergence of quantum coherent tubulins faster than 500 msc so that Orch OR occurs sooner ("heightened experience," Figure 37.11c). Lower intensity, unexciting input patterns develop coherence more slowly and Orch OR occurs later. An instantaneous Orch OR may then "bind" disparate tubulin superpositions that may have evolved in separate spatial distributions and over different time scales into an instantaneous conscious "now." Cascades of Orch ORs can then represent our familiar "stream of consciousness."

CONCLUSION

Approaches to understanding consciousness that are based on known and experimentally observed neuroscience fail to explain certain critical aspects. These include a unitary sense of binding, noncomputational aspects of conscious thinking, difference and transition between preconscious and conscious processing, (apparent) nondeterministic free will, and the essential nature of our experience. We conclude that aspects of quantum theory (for example, quantum coherence) and of a newly proposed physical phenomenon of wave function self-collapse (objective reduction, OR, Penrose 1994) offer possible solutions to each of these problematic features. We further conclude that cytoskeletal microtubules, which regulate intraneuronal activities and have cylindrical paracrystalline structure, are the best candidates for sites of quantum action and OR, and of "orchestrated OR" (Orch OR). Accordingly, we present a model of consciousness based on the following assumptions:

- Coherent excitations (Fröhlich-pumped phonons) among microtubule subunits (tubulins) support "cellular automaton" information processing in both classical (conformational) and quantum coherent superposition modes. Classical processing correlates with nonconscious, autonomic activity; quantum processing correlates with pre- and subconscious activity.

- The microtubule quantum coherent computing phase is able to be isolated from environmental interaction and maintain coherence for up to 500 msc (preconscious processing).

- A critical number of tubulins maintaining coherence within MTs for 500 msc collapses its own wave function (objective reduction: OR). This occurs because the mass-energy difference among the superpositioned states of coherent tubulins critically perturbs space-time geometry. To prevent multiple universes, the system must reduce to a single space-time by choosing eigenstates. The threshold for OR is related to quantum gravity; we calculate it in terms of the number of tubulins coherent for 500 msc to be very roughly 10^9 tubulins. Larger coherent sets will self-collapse faster, and smaller sets more slowly. Coherent sets that evolve over different time scales and in different brain distributions may be bound in an effectively simultaneous collapse which creates instantaneous "now." Cascades of these events constitute the familiar "stream of consciousness."

- MAPs and other tubulin modifications act as "nodes" to tune microtubule coherence and help to orchestrate collapse. We thus term the specific OR proposed to occur in microtubules and intrinsic to consciousness as "orchestrated objective reduction" (Orch OR).
- The Orch OR process, which introduces noncomputability (Penrose 1989, 1994), results in patterns of tubulin conformational states that help direct neural function through the actions of microtubules.

In providing a connection among preconscious to conscious transition, fundamental space-time notions, noncomputability, and binding of various (time scale and spatial) superpositions into instantaneous "now," we believe Orch OR in MTs is the most specific and plausible model for consciousness yet proposed.

ACKNOWLEDGMENTS

SH is partially supported by NSF Grant No. DMS-9114503. We are grateful to Alwyn Scott and Richard Hofstad. Artwork by Dave Cantrell, Biomedical Communications, University of Arizona. Dedicated to Burnell R. Brown Jr. MD, PhD, FRCA.

REFERENCES

Aharonov, Y., and L. Vaidman. 1990. Properties of a quantum system during the time interval between two measurements. *Phys. Rev. A.* 41:11.

Amos, L. A., and A. Klug. 1974. Arrangement of subunits in flagellar microtubules. *J. Cell Sci.* 14:523–50.

Andreu, J. M. 1986. Hydrophobic interaction of tubulin. *Ann. NY Acad. Sci.* 466:626–30.

Aoki, C., and P. Siekevitz. 1985. Ontogenic changes in the cyclic adenosine 3', 5' monophosphate—stimulatable phosphorylation of cat visual cortex proteins, particularly of microtubule-associate protein 2 (MAP2): Effects of normal and dark rearing and of the exposure to light. *J. Neurosci.* 5:2465–83.

Atema, J. 1973. Microtubule theory of sensory transduction. *J. Theor. Biol.* 38:181–90.

Athenstaedt, H. 1974. Pyroelectric and piezoelectric properties of vertebrates. *Ann. NY Acad. Sci.* 238:68–93.

Baars, B. J. 1988. *A cognitive theory of consciousness.* Cambridge, England: Cambridge University Press.

Baars, B. J. 1993. "How does a serial, integrated and very limited stream of consciousness emerge from a nervous system that is mostly unconscious, distributed, parallel and of enormous capacity?" In *Experimental and theoretical studies of consciousness* CIBA Foundation Symposium 174. Chichester, England: Wiley, pp 282–303.

Beck, F., and J. C. Eccles. 1992. Quantum aspects of brain activity and the role of consciousness. *Proc. Natl. Acad. Sci. USA* 89(23):11357–61.

Benioff, P. 1982. Quantum mechanical Hamiltonian models of Turing machines. *J. Stat. Phys.* 29:515–46.

Bensimon, G., and R. Chernat. 1991. Microtubule disruption and cognitive defects: Effect of colchicine on learning behavior in rats. *Pharmacol. Biochem. Behavior* 38:141–5.

Bigot, D., and S. P. Hunt. 1990. Effect of excitatory amino acids on microtubule-associated proteins in cultured cortical and spinal neurons. *Neurosci. Lett.* 111:275–80.

Burns, R. B. 1978. Spatial organization of the microtubule associated proteins of reassembled brain microtubules. *J. Ultrastruct. Res.* 65:73–82.

Cantiello, H. F., J. L. Stow, A. G. Prat, and D. A. Ausiello. 1991. Actin filaments regulate Na^+ channel activity. *Am. J. Physiol.* 265 (5 pt 1):C882–8.

Chalmers, D. J. 1996. "Facing up to the problem of consciousness." In *Toward a science of consciousness: The first Tucson discussions and debates,* edited by S. R. Hameroff, A. W. Kaszniak, and A. C. Scott. Cambridge, MA: MIT Press.

Chou, K–C., C–T. Zhang, and G. M. Maggiora. 1994. Solitary wave dynamics as a mechanism for explaining the internal motion during microtubule growth. *Biopolymers* 34:143–53.

Cianci, C., D. Graff, B. Gao, and R. C. Weisenberg. 1986. ATP-dependent gelation-contraction of microtubules in vitro. *Ann. NY Acad. Sci.* 466:656–9.

Clegg, J. S. 1983. "Intracellular water, metabolism and cell architecture." In *Coherent excitations in biological systems,* edited by H. Fröhlich and F. Kremer. Berlin: Springer-Verlag, pp 162–75.

Conrad, M. 1992. Quantum molecular computing: The self-assembly model. *Int. J. Quant. Chem.: Quantum Biology Symp.* 19:125–43.

Conrad, M. 1994. Amplification of superpositional effects through electronic-conformational interactions. *Chaos, Solitons & Fractals* 4:423–38.

Crick, F. H. C. 1984. Function of the thalamic reticular complex: The searchlight hypothesis. *Proc. Natl. Acad. Sci. USA* 81:4586–93.

Crick, F., and C. Koch. 1990. Towards a neurobiological theory of consciousness. *Seminars in the Neurosciences* 2:263–75.

Cronly-Dillon, J., D. Carden, and C. Birks. 1974. The possible involvement of brain microtubules in memory fixation. *J. Exp. Biol.* 61:443–54.

Dayhoff, J. E., S. Hameroff, R. Lahoz-Beltra, and C. E. Swenberg. 1994. Cytoskeletal involvement in neuronal learning: A review. *Eur. Biophys. J.* 23:79–93.

De Brabander, M. 1982. A model for the microtubule organizing activity of the centrosomes and kinetochores in mammalian cells. *Cell Biol. Intern. Rep.* 6:901–15.

Deecke, L., B. Grotzinger, and H. H. Kornhuber. 1976. Voluntary finger movement in man: Cerebral potentials and theory. *Biol. Cybernetics* 23:99–119.

Del Giudice, E., S. Doglia, and M. Milani. 1983. Self-focusing and ponderomotive forces of coherent electric waves: A mechanisim for cytoskeleton formation and dynamics. In *Coherent excitations in biological systems,* edited by H. Fröhlich and F. Kremer. Berlin: Springer-Verlag, pp 123–27.

Deutsch, D. 1985. Quantum theory, the Church-Turing principle and the universal quantum computer. *Proc. Royal Soc. (London)* A400:97–117.

Deutsch, D., and R. Josza. 1992. Rapid solution of problems by quantum computation. *Proc. Royal Soc. (London)* A439:553–6.

Diósi, L. 1989. Models for universal reduction of macroscopic quantum fluctuations.*Phys. Rev. A.* 40:1165–74.

Dustin, P. 1984. *Microtubules,* (2nd Revised Ed.). Berlin: Springer.

Eccles, J. C. 1992. Evolution of consciousness. *Proc. Natl. Acad. Sci.* (89)7320–4.

Edelman, G. 1989. *The remembered present: A biological theory of consciousness,* New York: Basic Books.

Elitzur, A. 1996. "Time and consciousness: The uneasy bearing of relativity theory on the mind-body problem." In *Toward a science of consciousness: The first Tucson discussions and debates,* edited by S. R. Hameroff, A. W. Kaszniak, and A. C. Scott. Cambridge, MA: MIT Press.

Engelborghs, Y. 1992. Dynamic aspects of the conformational states of tubulin and microtubules. *Nanobiology* 1:97–105.

Feynman, R. P. 1986. Quantum mechanical computers. *Foundations of Physics* 16(6):507–31.

Freeman, W. J. 1975. *Mass action in the nervous system.* New York: Academic Press.

Freeman, W. J. 1978. Spatial properties of an EEG event in the olfactory bulb and cortex. *Electroencephalogr. and Clin. Neurophysiol.* 44:586–605.

Friedrich, P. 1990. Protein structure: The primary substrate for memory. *Neurosci.* 35:1–7.

Fröhlich, H. 1968. Long-range coherence and energy storage in biological systems. *Int. J. Quantum Chem.* 2:641–9.

Fröhlich, H. 1970. Long range coherence and the actions of enzymes. *Nature* 228:1093.

Fröhlich, H. 1975. The extraordinary dielectric properties of biological materials and the action of enzymes. *Proc. Natl. Acad. Sci. USA* 72:4211–5.

Genberg, L., L. Richard, G. McLendon, and R. J. Dwayne-Miller. 1991. Direct observation of global protein motion in hemoglobin and myoglobin on picosecond time scales. *Science* 251:1051–4.

Genzel, L., F. Kremer, A. Poglitsch, and G. Bechtold. 1983. Relaxation processes on a picosecond time scale in hemoglobin and poly observed by millimeter-wave spectroscopy. *Biopolymers* 22:1715–29.

Geroch, R., and J. B. Hartle. 1986. Computability and physical theories. *Foundations of Physics* 16:533.

Ghirardi, G. C., R. Grassi, and A. Rimini. 1990. Continuous-spontaneous reduction model involving gravity. *Phys. Rev. A.* 42:1057–64.

Ghirardi, G. C., A. Rimini, and T. Weber. 1986. Unified dynamics for microscopic and macroscopic systems. *Phys. Rev. D.* 34:470.

Gilbert, J. M., and P. Strocchi. 1986. In vitro studies of the biosynthesis of brain tubulin. *Ann. NY Acad. Sci.* 466:89–102.

Gray, C. M., and W. Singer. 1989. Stimulus-specific neuronal oscillations in orientation columns of cat visual cortex. *Proc. Natl. Acad. Sci. USA* 86:1698–1702.

Grey-Walter, W. 1953. *The living brain.* London: Gerald Duckworth and Co., Ltd.

Grundler, W., and F. Keilmann. 1983. Sharp resonances in yeast growth prove non-thermal sensitivity to microwaves. *Phys. Rev. Lett.* 51:1214–6.

Halpain, S., and P. Greengard. 1990. Activation of NMDA receptors induces rapid dephosphorylation of the cytoskeletal protein MAP2. *Neuron* 5:237–46.

Hameroff, S. R. 1987. *Ultimate computing: Biomolecular consciousness and nanotechnology.* Amsterdam: North-Holland.

Hameroff, S. R., J. E. Dayhoff, R. Lahoz-Beltra, A. Samsonovich, and S. Rasmussen. 1992. Conformational automata in the cytoskeleton: Models for molecular computation. *IEEE Computer* (October Special Issue on Molecular Computing) 30–39.

Hameroff, S. R., S. A. Smith, and R. C. Watt. 1984. "Nonlinear electrodynamics in cytoskeletal protein lattices." In *Nonlinear electrodynamics in biological systems,* edited by W. R. Adey and A. F. Lawrence. New York: Plenum Press.

Hameroff, S. R., and R. C. Watt. 1982. Information processing in microtubules. *J. Theor. Biol.* 98:549–61.

Hebb, D. O. 1949. *The organization of behavior.* New York: Wiley.

Hebb, D. O. 1980. *Essay on mind.* Hillsdale, NJ: Lawrence Erlbaum.

Hirokawa, N. 1991. "Molecular architecture and dynamics of the neuronal cytoskeleton." In *The neuronal cytoskeleton,* edited by R. D. Burgoyne. New York: Wiley-Liss, pp. 5–74.

Jibu, M., S. Hagan, S. R. Hameroff, K. H. Pribram, and K. Yasue. 1994. Quantum optical coherence in cytoskeletal microtubules: Implications for brain function. *BioSystems* 32:195–209.

Jibu, M., S. Hagan, and K. Yasue. 1996. "Subcellular quantum optical coherence: Implications for consciousness." In *Toward a science of consciousness: The first Tucson discussions and debates,* edited by S. R. Hameroff, A. W. Kaszniak, and A. C. Scott. Cambridge, MA: MIT Press.

Kaivarainen, A. 1994. personal communication.

Károlházy, F., A. Frenkel, and B. Lukacs. 1986. "On the possible role of gravity on the reduction of the wave function." In *Quantum concepts in space and time,* edited by R. Penrose and C. J. Isham. Oxford, England: Oxford University Press.

Karplus, M., and J. A. McCammon. 1983. "Protein ion channels, gates, receptors." In *Dynamics of proteins: Elements and function, Ann. Rev. Biochem.,* edited by J. King. Menlo Park, CA: Benjamin/Cummings, pp. 263–300.

Kim, H., C. G. Jensen, and L. I. Rebhund. 1986. "The binding of MAP2 and tau on brain microtubules *in vitro*." In *Dynamic aspects of microtubule biology,* edited by D. Soifer. *Ann. NY Acad. Sci.* 466:218–39.

Kirschner, M., and T. Mitchison. 1986. Beyond self assembly: From microtubules to morphogenesis. *Cell* 45:329–42.

Koch, C. 1996. "Towards the neuronal substrate of visual consciousness." In *Toward a science of consciousness: The first Tucson discussions and debates,* edited by S. R. Hameroff, A. W. Kaszniak, and A. C. Scott. Cambridge, MA: MIT Press.

Kudo, T., K. Tada, M. Takeda, and T. Nishimura. 1990. Learning impairment and microtubule-associated protein 2 (MAP2) decrease in gerbils under chronic cerebral hypoperfusion. *Stroke* 21:1205–9.

Lee, J. C., D. J. Field, H. J. George, and J. Head. 1986. Biochemical and chemical properties of tubulin subspecies. *Ann. NY Acad. Sci.* 466:111–28.

Libet, B. 1990. "Cerebral processes that distinguish conscious experience from unconscious mental functions." In *The principles of design and operation of the brain*, edited by J. C. Eccles and O. D. Creutzfeld. *Experimental Brain Research Series 21*, Berlin: Springer-Verlag, pp. 185–205.

Libet, B., E. W. Wright, Jr., B. Feinstein, and D. K. Pearl. 1979. Subjective referral of the timing for a conscious sensory experience. *Brain* 102:193–224.

Louria, D., and S. Hameroff. 1996. "Computer simulation of anesthetic binding in protein hydrophobic pockets." In *Toward a science of consciousness: The first Tucson discussions and debates*, edited by S. R. Hameroff, A. W. Kaszniak, and A. C. Scott. Cambridge, MA: MIT Press.

Mandelkow, E., and E-M. Mandelkow, 1994. Microtubule structure. *Curr. Opinions Structural Biology* 4:171–9.

Mandelkow, E-M., and E. Mandelkow. 1993. Tau as a marker for Alzheimer's disease. *Trends in Biol. Sci.* 18:480–3.

Marshall, I. N. 1989. Consciousness and Bose-Einstein condensates. *New Ideas in Psychology* 7:73–83.

Marshall, I. N. 1996. "Three kinds of thinking." In *Toward a science of consciousness: The first Tucson discussions and debates*, edited by S. R. Hameroff, A. W. Kaszniak, and A. C. Scott. Cambridge, MA: MIT Press.

Mascarenhas, S. 1974. The electret effect in bone and biopolymers and the bound water problem. *Ann. NY Acad. Sci.* 238:36–52.

Matsuyama, S. S., and L. F. Jarvik. 1989. Hypothesis: Microtubules, a key to Alzheimer's disease. *Proc. Nat. Acad. Sci. USA* 86:8152–6.

Melki, R., M. F. Carlier, D. Pantaloni, and S. N. Timasheff. 1989. Cold depolymerization of microtubules to double rings: Geometric stabilization of assemblies. *Biochemistry* 28:9143–52.

Mileusnic, R., S. P. Rose, and P. Tillson. 1980. Passive avoidance learning results in region specific changes in concentration of, and incorporation into, colchicine binding proteins in the chick forebrain. *Neur Chem* 34:1007–15.

Montoro, R. J., J. Diaz-Nido, J. Avila, and J. Lopez-Barneo. 1993. N-methyl-d-aspartate stimulates the dephosphorylation of the microtubule-associated protein 2 and potentiates excitatory synaptic pathways in the rat hippocampus. *Neuroscience* 54(4):859–71.

Neubauer, C., A. M. Phelan, H. Keus, and D. G. Lange. 1990. Microwave irradiation of rats at 2.45 GHz activates pinocytotic-like uptake of tracer by capillary endothelial cells of cerebral cortex.*Bioelectromagnetics* 11:261–8.

Pearle, P. 1989. Combining stochastic dynamical state vector reduction with spontaneous localization. *Phys. Rev. D.* 13:857–68.

Pearle, P. 1992. "Relativistic model state vector reduction," In *Quantum chaos—quantum measurement*. NATO Adv. Sci. Institute Ser. C. Math. Phys. Sci. 358 (Copenhagen 199). Dordrecht: Kluwer.

Pearle, P., and E. Squires. 1994. Bound-state excitation, nucleon decay experiments and models of wave-function collapse. *Phys. Rev. Letts.* 73(1):1–5.

Penrose, R. 1987. "Newton, quantum theory and reality." In *300 years of gravity*, edited by S. W. Hawking and W. Israel. Cambridge, England: Cambridge University Press.

Penrose, R. 1989. *The emperor's new mind.* Oxford, England: Oxford University Press.

Penrose, R. 1993. "Gravity and quantum mechanics." In *General relativity and gravitation,* Proceedings of the thirteenth international conference on general relativity and gravitation held at Cordoba, Argentina 28 June—4 July 1992, edited by R. J. Gleiser, C. N. Kozameh, and O. M. Moreschi. Plenary I: Plenary Lectures. Bristol, England: Institute of Physics Publications.

Penrose, R. 1994. *Shadows of the mind.* London: Oxford University Press.

Penrose, O., and L. Onsager. 1956. Bose-Einstein condensation and liquid helium. *Phys. Rev.* 104:576–84.

Puck, T. T., and A. Krystosek. 1992. Role of the cytoskeleton in genome regulation and cancer. *Int. Rev. Cytology* 132:75–108.

Rasmussen, S., H. Karampurwala, R. Vaidyanath, K. S. Jensen, and S. Hameroff. 1990. Computational connectionism within neurons: A model of cytoskeletal automata subserving neural networks. *Physica D* 42:428–49.

Roth, L. E., and D. J. Pihlaja. 1977. Gradionation: Hypothesis for positioning and patterning. *J. Protozoology* 24(1):2–9.

Sataric, M. V., R. B. Zakula, and J. A. Tuszyński. 1992. A model of the energy transfer mechanisms in microtubules involving a single soliton. *Nanobiology* 1:45–56.

Satir, P. 1984. Cytoplasmic matrix: Old and new questions. *J Cell Biol* 99(1):235–8.

Scott, A. C. 1995. *Stairway to the mind: The controversial new science of consciousness.* New York: Copernicus (Springer-Verlag).

Singer, W. 1993. Synchronization of cortical activity and its putative role in information processing and learning. *Ann. Rev. Physiol.* 55:349–74.

Stapp, H. P. 1993. *Mind, matter and quantum mechanics.* Berlin: Springer-Verlag.

Theurkauf, W. E., and R. B. Vallee. 1983. Extensive cAMP-dependent and cAMP-independent phosphorylation of microtubule associated protein 2. *J. Biol. Chem.* 258:7883–6.

Tollaksen, J. 1996. "New insights from quantum theory on time, consciousness, and reality. "In *Toward a science of consciousness: The first Tucson discussions and debates,* edited by S. R. Hameroff, A. W. Kaszniak, and A. C. Scott. Cambridge, MA: MIT Press.

Tuszyński, J., S. Hameroff, M. V. Sataric, B. Trpisova, and M. L. A. Nip. 1995. Ferroelectric behavior in microtubule dipole lattices; implications for information processing, signaling and assembly/disassembly. *J. Theor. Biol.* 174:371–80.

Vassilev, P., M. Kanazirska, and H. T. Tien. 1985. Intermembrane linkage mediated by tubulin. *Biochem. Biophys. Res. Comm.* 126:559–65.

von der Malsburg, C., and W. Schneider. 1986. A neural cocktail party processor. *Biol. Cybern.* 54:29–40.

von Neumann, J. 1966. *Theory of self-reproducing automata,* edited by A. W. Burks. Urbana: University of Illinois Press.

Wang, N., D. E. Ingber. 1994. Control of cytoskeletal mechanics by extracellular matrix, cell shape and mechanical tension. *Biophysical Journal* 66(6):2181–9.

Watterson, J. G. 1996. "Water clusters: Pixels of life." In *Toward a science of consciousness: The first Tucson discussions and debates,* edited by S. R. Hameroff, A. W. Kaszniak, and A. C. Scott. Cambridge, MA: MIT Press.

Wheeler, J. A. 1957. Assessment of Everett's "relative state" formulation of quantum theory. *Revs. Mod. Phys.* 29:463–5.

Woolf, N. J., S. L. Young, G. V. W. Johnson, and M. S. Fanselow. 1994. Pavlovian conditioning alters cortical microtubule-associated protein-2. *NeuroReport* 5:1045–8.

Yu, W., and P. W. Baas. 1994. Changes in microtubule number and length during axon differentiation. *J. Neuroscience* 14(5):2818–29.

VIII Nonlocal Space and Time

What is reality? Why does time flow? Possible answers await the systematic combination of quantum theory, general relativity (Einstein's theory of gravity), and information theory to unify all known forces and define the geometry of space-time. A number of "grand unification schemes" have already been proposed, and although none are as yet generally accepted, they collectively suggest that our three-dimensional geometrical reality is actually a subrealm of a larger, multidimensional geometrical reality. As Saul-Paul Sirag recounts, this is an ancient idea, alluded to in the story of Plato's cave in which two-dimensional shadows on the cave wall constitute the only perceived reality. In *Shadows of the Mind*, Roger Penrose (1994) proposes that consciousness may utilize a similar, simplifying projection. The mathematical constructs capable of describing multidimensional space-time geometries may also be describing the potential structure of consciousness. Some suggested ones include David Bohm's "implicate order" of enfolded space-time (Bohm and Hiley 1994), "hyperspheres," quantum foam, twistors, and string theory. Chapters by Saul-Paul Sirag and Doug Matzke in this section discuss these constructs in relation to consciousness.

As discussed in Part 4 on Experimental Neuroscience, time has peculiar aspects in relation to consciousness. In conventional physics and relativity theory, there is no passage of time: all past, present, and future events have the same degree of reality. According to relativity theory, there is also no universal simultaneity, no instantaneous *now* consistent throughout space. Our conscious experience, however, is quite different. Time does flow, and as Avshalom Elitzur explains in Chapter 38, a certain temporal property—"the Now"—continuously moves from one moment to the next. Elitzur argues that consciousness and time's passage are two parts of the same mystery.

Jeff Tollaksen links consciousness to time through the irreversibility of quantum wave function collapse in Chapter 39. Tollaksen claims that the present "conscious Now" is created from collapse inherent in brain action. Consciousness thus *creates time*, which ticks forward with each collapse. Citing work by Aharonov and Vaidman (1990), Tollaksen claims that collapse outcomes, and consciousness, are influenced both from the past and from the future. Recent work by Stapp (1990) supports this possibility.

In Chapter 40, Doug Matzke considers space-time properties that are optimal for computation. Assuming that our three-dimensional conscious reality is a mere shadow projection of higher dimensional "true" reality, Matzke concludes that a sparse hyperspace with ten symmetry dimensions and unrestricted locality would be computationally ideal. He further claims that a mobile observer mechanism moving within this hyperspace provides a consciousness entity that would appear dualist to an observer.

Saul-Paul Sirag in Chapter 41 discusses the array of mathematical constructs applicable to grand unification schemes. Sirag sees these as different ways of viewing the "hyperdimensional crystallographic" structure that comprises space-time and consciousness. He describes mathematically how high dimensional spaces may be projected onto fewer dimensions, suggesting that our minds may reflect shadows of more complex geometry.

If consciousness does exist in a nonlocal, multidimensional universe, evidence for such nonlocality should be expected. The final two chapters in this section move in this direction.

In Chapter 42, Mario Varvoglis presents a meta-analysis of a number of previous studies supporting various types of nonlocal parapsychological phenomena (psi). If validated, psi phenomena would have important implications for theories of consciousness. Varvoglis discusses these controversial implications.

Finally, in Chapter 43 Ezio Insinna links Jung's theory of synchronicity and archetypal images to nonlocality that is intrinsic to quantum theory. Insinna draws on correspondence between Jung and quantum theorist Wolfgang Pauli regarding possible quantum mechanisms for meaningful coincidence, the "collective unconscious" in dreams, and the role of the observer's *sub*conscious on wave function collapse.

The first two chapters in this section by Avshalom Elitzur and Jeff Tollaksen relate the flow and simultaneity of *time* to consciousness. Saul-Paul Sirag and Doug Matzke then discuss multidimensional space-time geometries with nonlocal configurations. The last two papers by Mario Varvoglis and Ezio Insinna provide possible examples of nonlocal conscious and subconscious processes.

Is consciousness linked to fundamental aspects of space-time? Critical philosophical analysis suggests that to explain the qualities of subjective experience, the very nature of reality must be confronted.

REFERENCES

Aharonov, Y., and L. Vaidman. 1990. Properties of a quantum system during the time interval between two measurements. *Phys. Rev. A.* 41:11.

Bohm, D., and B. Hiley. 1994. *The undivided universe.* London: Routledge.

Penrose, R. 1994. *Shadows of the mind.* Oxford, England: Oxford University Press.

Stapp, H. P. 1994. Theoretical model of a purported empirical violation of the predictions of quantum-theory. *Phys. Rev. A.* 50:18–22.

38 Time and Consciousness: The Uneasy Bearing of Relativity Theory on the Mind-Body Problem

Avshalom C. Elitzur

The old debate concerning the nature of consciousness seems to have been discussed by now from every possible viewpoint, with no decisive argument raised so far in favor of one view or another. This is the reason why, for many scientists, the issue looks uninteresting. Yet, a new way of posing the problem may give it an unexpected twist. Our experience reflects something in the nature of time about which physics is, oddly, mute. Relativity theory has made this conflict between subjective experience and physical formalism even more acute, thereby indicating that the dichotomy between mind and matter might be rooted in an even more fundamental dichotomy in physical reality itself.

DOES TIME PASS?

It is perhaps the most fundamental ingredient of our experience that reality is constantly changing: Any moment, in its turn, seems to bring new events that did not exist before and that will vanish later. Hence, any event in our lives has three temporal properties: Before it takes place it is a potential "future" event, subject in principle to interference. Then, when it actually happens, it is a fleeting "present." And finally, after its occurrence, it is a given, unchangeable "past." There seems to be a certain temporal property—the "Now"—that continuously moves from one moment to the next. Conversely, we may view time itself as passing relative to us, its passage being experienced as the change of the future into present and then into past.

Elementary logic, however, shows all these apparent truisms to be absurd. Any statement like "time flows/passes/moves" is bound to produce gross inconsistencies. For the very concept of movement is based upon that of time; saying that an object "moves" amounts to saying that "it is in one position at one moment and in another position at the next." But how can such a statement be made about time itself? Time is the very parameter of any movement; to see the absurdity of granting it motion, just ask yourself what is this motion's velocity! The same contradiction besets the statement "we move in time." To say this we must regard time as a sort of "space" in which we move, thereby necessitating another time. Neither shall we avoid absurdities by saying "the future changes into present and then into past." If "change" denotes different states at different moments, how can a moment in time, the measure of all changes, itself change its designation from future to past? Would such dynamics not require another time within which the passage or change of time itself occurs?

Physics, from antiquity to this day, has tackled the problem in a simple way: It has ignored time's passage altogether. The laws of physics are, in essence, functions that relate the consecutive states of a system to the appropriate moments; nothing more. This is similar to the way a geographical map gives the altitude of each site according to its latitude and longitude. Does the map indicate in any way that these sites come into being and then vanish one after another? Or should we think of the altitude as "moving" up and down? Of course, there is no dynamics in a map. But neither is there in physical law! As odd as it may seem, physics does not say about time more than the information contained in the map about geographic sites. From the purely physical viewpoint, all the past, present, and future events of the universe have the same degree of reality.

Relativity theory has brought this view to an extreme. Dismissing the notion of simultaneity as arbitrary and observer-dependent, it has undermined the very notion of "Now" as well. Consider an observer walking back and forth on a distant planet, first towards Earth and then away from it. By relativity, an observer's simultaneity-plane is slanted in accordance with his or her velocity (Figure 38.1). How, then, would that observer designate the moment that for you, dear reader, is "now"? During the first stage of his walking, he would state that this moment in your life is in the remote past and you are by now very old. In contrast, during the second stage of his walk, as he marches away from you, this moment in your life would be for him in the future and you would still be a baby according to his account. So, you state that this moment is "now," while he first states that it is "past" and then reverses his own account and states that this is "future"! (Notice the glaring contrast with the normal time evolution: The past "unevolves" into future. . .) Which of the three accounts, then, is correct? Relativity theory assigns to all of them equal soundness. Every event in the universe can equally be "present," "future," or "past," depending on the observer's reference-frame.

Clearly, this conclusion sharply conflicts with our most basic intuitions. It seems to us obvious that the past is fixed and unchangeable, while the future is still undetermined. But the above considerations tell us that even our future, being past for some possible observers, is equally fixed. For most people, the thought that all future events, even those that will take place millions of years later, "already" exist in some sense, sounds absurd. This was also the feeling of eminent philosophers such as Bergson and Whitehead. They argued that time itself is subject to some creative "becoming" that physics is not yet capable of dealing with (see Čapec 1991 for a beautiful and penetrating exposition of these views). But, against the perfect mathematical rigor and logical beauty of relativity theory, these philosophical ideas were never taken seriously by mainstream science. The

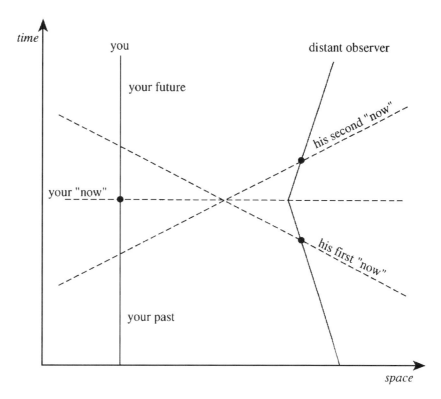

Figure 38.1 The present "I" (now).

relativistic model, known as "Block Universe," has been preferred over the notion of "Becoming."

SOMETHING MISSING IN THE PHYSICAL ACCOUNT OF BEHAVIOR

When you face a riddle that keeps resisting resolution, why not try adding to it another, equally difficult riddle? In science, it sometimes turns out that the combination of two questions is less troublesome than each in itself. This is why I would like to call consciousness to this discourse.

The problem is as notorious as it is baffling, so I will state it very briefly. Imagine a perfect neurophysiological description of what is going on in a person's brain while he or she, say, perceives a red object, smells cinnamon, or experiences an itch. In principle, such a complete description is possible. And unless something is terribly wrong with our basic assumptions, such a complete description would be comprised purely of physical interactions inside and between the neurons, nothing more.

But it seems that something must always be left out in such a description. The latter can say nothing about the conscious experience itself. Nothing in the laws of physics, upon which such a description is based, indicates that there is a conscious experience involved; consciousness could as well not be there at all. Let us rephrase this impasse in information-theory terms: No matter how detailed is the information I exchange with my friend about the sensation of red or blue, whereupon we are both absolutely certain that we talk about the same colors, we can never be sure that my red is not his or her blue, or anything else for that matter. No experiment one can think about can rule out this possibility. In fact, no experiment can assure us that another person has *any* conscious experience; he or she could be automatically imitating the behavior of a conscious being. If, in spite of this epistemological impasse, we are certain that our fellow human has a conscious experience whenever he or she gives us this impression, it is because we have such an experience under similar circumstances. Nothing in present-day science, however, can justify this intuitive certainty.

But why bother if subjective experience is, by definition, accessible only to oneself? Science, according to the still-common positivist view, should deal only with phenomena shared by all observers. Well, I am no positivist; I believe that anything that exists must sooner or later objectively show up in some carefully chosen observation or experiment. Moreover, I have shown a few years ago (Elitzur 1989) that consciousness as such, that is, something not identical to the neural processes, plays a causal role in behavior. Here is the gist of the argument.

All the theories that seek to reconcile consciousness with physics (like epiphenomenalism, parallelism, and identity theory) must render consciousness causally ineffective, for example, a passive reflection or a mere aspect of the neural processes. This means that we cannot allow statements of the form "A mental event at a certain moment, M_{t_1}, has caused a later physical event, P_{t_2}, in the brain." Rather, we have to convert such a statement to a more complex one that does not allow anything nonphysical to interfere with physical processes. The corrected statement should therefore be: "A physical event P_{t_1} in the brain has also the mental aspect M_{t_1} (or it causes M_{t_1} as a mental side-effect), and it is P_{t_1} that has caused P_{t_2}." In other words, for any conscious experience that seems to have led to a physical action—for example, an arousal of passion that has resulted in a kiss—it should be possible to show that it was not the conscious experience as such but a physical event in the brain, of which the conscious experience is only a side aspect. In short, only physical events should cause other physical events. Consciousness must therefore have no share in the causal chain operating in the brain.

Now, this procedure of converting "M" to "P" seems to be applicable to every human action—save the case of those weird humans who are troubled by the alleged difference between P and M! In other words, actions that indicate puzzlement about consciousness are behavioral phenomena that cannot be adequately explained by an account that denies any causal role to consciousness. From these rare instances, in which the physical account is evidently incomplete, one can extend the conclusion to all other forms of behavior where the purely physical explanation seems to suffice; even in such cases, it now seems, consciousness contributes its causal influence.

This argument, although extravagant at first sight, has been elaborated in detail (Elitzur 1989) and has sustained all the objections raised against it since then, and has been endorsed and elaborated by others as well (see Elitzur 1996a)—much to my own dismay, for I have always adhered to physicalism. So it is quite reluctantly that I claim to have shown that any purely physical account of behavior is inherently incomplete.

A purely physical description that nevertheless leaves out something very essential... Is it only a coincidence that this is how we feel about time's passage? The similarity to the dichotomy between physical and experienced time might be merely superficial, but it is worth a more careful look.

TWO RIDDLES—OR TWO ASPECTS OF THE SAME ONE?

If relativity theory, one of the greatest achievements of modern physics, turns out to have any bearing on the most ancient of riddles, namely, consciousness, this would surely be an important step forward. Indeed, Lockwood (1989), following Russell (1927), has applied relativistic considerations to show that if mental events take place in time, they must also take place in space. He concluded that mental and physical events occupy the same region in space-time, hence are identical. There is, however, another bearing of relativity on consciousness, quite an awkward one. I believe that this very awkwardness gives yet another motivation to seek a new theory of both time and consciousness.

Relativity, as we have seen, ignores time's passage altogether. Now this attitude, if worked out consistently, entails ignoring the unity of consciousness as well. Perhaps nothing expresses this better than a personal reaction of Einstein himself to the loss of a friend, the physicist-philosopher Michele Besso, with whom he had been engaged in a lifelong debate about time's passage. Contrary to Besso, Einstein emphasized that "There is no irreversibility in the basic laws of physics. You have to accept the idea that subjective time with its emphasis on the now has no objective meaning." Some years went by (well, at least they seemed to go by) and Besso died. Einstein—it was just four weeks before his own death—wrote to the family a somewhat odd letter of condolence: "Michele has preceded me a little in leaving this strange world. This is not important. For us who are convinced physicists, the distinction between past, present, and future is only an illusion, however persistent" (quoted in Prigogine 1980, p. 203).

Let us take this statement seriously for a moment and follow its consequences. What would Einstein say had he needed to console Besso himself for, say, the latter's lost youth? In order to be consistent, he should have to assure his friend that the distinction between past, present, and future is illusory, hence the young Besso still resides in the past, just as the old one resides in the present. As comic as this statement may sound, it forcibly follows from the relativistic picture of the four-dimensional space-time: Each individual person is a world-line, extending from birth to death. The present "I" is merely an arbitrarily chosen section of this four-dimensional world-line, which, like all world-lines, is static and changeless (see Figure 38.1). Why, then, does one believe oneself to be the same person throughout one's life? The relativistic explanation is simple: The brain's world-line carries, like any world-line, marks of events in the past direction. In other words, every person has memories related to his or her previous selves, therefore believing that it is the same "I" that experienced them. (For the thermodynamic explanation of this time-asymmetry of memory see Elitzur 1996c.) In reality, however, all our numerous momentary selves are supposed to reside together along time, from infancy to old age, like the static single pictures of a film.

That Einstein himself was aware of the awkwardness of this view is evident from Carnap's testimony:

Einstein said that the problem of the Now worried him seriously. He explained that the experience of the Now means something special for man, something essentially different from the past and the future, but that this important difference does not and cannot occur within physics. That this experience cannot be grasped by science seemed to him a matter of painful but inevitable resignation (quoted in Zeh 1989, p. 151).

Now what is this "Now," so inherent to any possible experience yet so alien to physics? Clearly, it is closely akin to consciousness which, by its very nature, is restricted to the fleeting present. No wonder that both phenomena elude the abstraction needed for objective description. Consciousness and time's passage thus seem to be two aspects of the same mystery.

TOWARD A NEW THEORY OF TIME AND CONSCIOUSNESS

During the last decade I have been studying time's apparent passage and its asymmetries, the foundations of quantum mechanics, the thermodynamic basis of organic life, and the mind-body problem (Elitzur 1991 through 1996d, Horwitz, Arshansky, and Elitzur 1988). From these works a new theory of time is emerging, extending the present relativistic framework so as to include Becoming in the physical formalism. The theory, called "space-time dynamics," ascribes real evolution to time itself. Space-time, as it were, "grows" in the future direction, events being created anew with the progression of the "Now." This Becoming is the long-sought-for master asymmetry, underlying the diverse "arrows of time" known in thermodynamics, electromagnetism, and gravitation. It is possible to work out this theory without getting into an infinity of times: "Becoming" is a more fundamental concept than time itself. This is similar to the way the big bang is the origin of space-time rather than an event that has occurred "somewhere" "at some time." In the spacetime dynamics theory, Becoming creates space-time; it is a continuation of the big bang itself (Elitzur 1996d).

The bearing of this theory on consciousness is immediate. The essence of the mind-body problem, as the earlier example of color shows, is that no description of a physical phenomenon, such as a neurophysiological process, can convey or even indicate (let alone explain) subjective qualities. Could this failure indicate that there is something fundamental *in matter itself* that is still inaccessible to science? If matter itself is subject to constant

Becoming, every state being created by the "Now" in its progress from past to future, then our consciousness is the intrinsic reflection of the Becoming that governs the matter of which we are made.

Logical thinking, especially science, "objectivizes" every phenomenon reported by the senses, that is, it takes off their particular, unique properties in order to grasp their more invariant ones, those that can be manipulated by thought and communicated. The price that we have to pay for this objectivization is the inability to grasp the most inherent property of all events, namely, the very Becoming that creates them.

Only one process, namely, the very basis of our cognitive mechanism, escapes this objectivization, giving us an immediate, unconveyable awareness of the elusive Becoming. Consciousness is the inner expression of the Becoming that keeps creating the universe, one moment after another.

Admittedly, in its present stage, space-time dynamics is a philosophical rather than a scientific theory, as it gives no testable prediction yet. I believe, however, that the theory can develop beyond this stage. Objections to the present relativistic orthodoxy have already been raised by some physicists dissatisfied with the present understanding of time (Davies 1995, Horwitz 1983, Rosen 1980, 1991, Stapp 1986). The fact that a revision in the present physical account of time seems increasingly warranted on purely physical (quantum-mechanical as well as relativistic) grounds, thereby, paradoxically, siding again with our "primitive" intuition against physical orthodoxy, seems to point out a new, surprising affinity between physical reality and consciousness. It is therefore not too pretentious of Penrose (1989) to believe that once we have the long-desired theory of quantum gravity, that is, a quantum mechanical theory of space-time that reconciles quantum mechanics and relativity theory, it will have something new to say about consciousness as well. In fact, the more you think about it, the more natural it seems to expect that new developments in the foundations of physics will make consciousness part and parcel of physical reality.

REFERENCES

Ćapec, M. 1991. "The new aspects of time: Its continuity and novelties. Selected Papers in the Philosophy of Science." *Boston Studies in the Philosophy of Science*, Vol. 125. Dordrecht: Kluwer Academic Publishing.

Davies, P.C.W. 1995. *About time: Einstein's unfinished revolution.* New York: Simon & Schuster.

Elitzur, A. C. 1989. Consciousness and the incompleteness of the physical explanation of behavior. *Journal of Mind and Behavior* 10:1–19.

Elitzur, A. C. 1991. On some neglected thermodynamic peculiarities of quantum non-locality. *Foundations of Physics Letters* 3:525–41.

Elitzur, A. C. 1992a. Two persistent wonders—or one? Consciousness and the passage of time. *Frontier Perspectives* 2(2):27–33.

Elitzur, A. C. 1992b. Locality and indeterminism preserve the second law. *Physics Letters* A167:335–40.

Elitzur, A. C. 1994. Let there be life: Thermodynamic reflections on biogenesis and evolution. *Journal of Theoretical Biology* 168:429–59.

Elitzur, A. C. 1995. Life and mind, past and future: Schrödinger's vision fifty years later. *Perspectives in Biology and Medicine* 38:433–58.

Elitzur, A. C., and L. Vaidman. 1993. Quantum mechanical interaction-free measurements. *Foundations of Physics* 23:987–97.

Elitzur, A. C. 1996a. Consciousness can no more be ignored: Reflections on Moody's dialogue with zombies. *Journal of Consciousness* In press.

Elitzur, A. C. 1996b. "What's the mind-body problem with you anyway?" To be presented at the Conference *Toward a Scientific Basis for Consciousness II.* University of Arizona, Tucson, Arizona, April 1996.

Horwitz, L. P. 1983. "On relativistic quantum theory." In *Old and new questions in physics, cosmology, philosophy, and theoretical Biology,* edited by A. Van der Merwe. New York: Plenum Press.

Horwitz, L. P., R. I. Arshansky, and A. C. Elitzur. 1988. On the two aspects of time: The distinction and its implications. *Found. Phys.* 18:1159–93.

Lockwood, M. 1989. *Mind, brain, and the quantum: The compound "I."* Oxford, England: Basil Blackwell.

Penrose, R. 1989. *The emperor's new mind: Concerning computers, minds, and the laws of physics.* New York: Vintage.

Prigogine, I. 1980. *From being to becoming: Time and complexity in the physical sciences.* San Francisco: W. H. Freeman.

Rosen, N. 1980. General relativity with a background metric. *Found. Phys.* 10:673–704.

Rosen, N. 1991. Can one have a universal time in general relativity? *Found. Phys.* 21:459–72.

Russell, B. 1927. *The analysis of matter.* London: Kegan Paul.

Stapp, H. P. 1986. "Einstein time and process time." In *Physics and the ultimate significance of time,* edited by D. R. Griffin. New York: SUNY Press.

Zeh, H. D. 1989. *The physical basis of the direction of time.* Berlin: Springer-Verlag.

39 New Insights from Quantum Theory on Time, Consciousness, and Reality

Jeff Tollaksen

INTRODUCTION

Cognitive scientists generally believe that the brain can be understood entirely in terms of classical physics. Indeed there are several intriguing theories that purport to explain consciousness without direct reference to quantum mechanics (QM). Moreover, as an incomplete theory QM does not give a full description of our experienced reality. Nonetheless, it is often assumed that since QM has successfully explained most things in our world, that it must be able to explain consciousness. However, there are fundamental phenomena about which all the interpretations of QM are inadequate or incomplete, such as an understanding of the EPR/Aspect experiments and an understanding of time (phenomena which may have relevance to consciousness). This article explores new insights and perspectives that QM can lend to the scientific study of consciousness. In particular, a new formulation of QM by Aharonov and Rohrlich (1990) is considered. This formulation can sucessfully resolve some of the deepest paradoxes of QM, such as the nonlocality evident in the measurement problem, the meaning of the wave function, and so on. New insights that can be gained within this context for the study of consciousness are discussed here. These include a new understanding of time and the experienced flow of time, the unity of consciousness and the binding problem,[1] phenomenal qualities, or qualia and free will.

Superposition is part of the conceptual revolution implied by quantum theory. In classical physics, different possibilities can evolve independently of each other: when flipping a classical coin one either gets heads or tails, but not both. In quantum theory, however, these possibilities combine in a fundamentally different, interdependent way. To get the probability in QM, one must sum the complex partial amplitude (rather than the partial probability) and then take the squared modulus. Within this expression are the squared terms (as in classical probability) and also interference terms of phase differences. These interference terms reveal some of the unique features of QM that have no analog in classical physics. When flipping a quantum coin (for example in a 2-slit experiment), it is false to say that it took the up path, or the down path, or both paths, or neither paths. Niels Bohr tells us that QM requires "a radical revision of our attitude as regards physical reality" and "a fundamental modification of all ideas regarding the absolute character of physical phenomenon" (Wheeler 1986). The differences between classical and quantum physics is not just in details, but in essence. The wavefunction, the fundamental entity in QM, evolves deterministically as described by the linear Schrödinger equation. However, whenever an observation is made on a quantum system, an unpredictable and seemingly random change occurs in the state of the wavefunction. This is part of the unsolved measurement problem of quantum theory or the problem of the actualization of potentialities. As a result of this seemingly unexplainable randomness, nature is said to be capricious, or as Einstein said, "God plays dice." Another interpretation is suggested here: this "random" phenomenon is trying to tell us that fundamental reality is inherently nonlocal in time as demonstrated by the two-vector interpretation of Aharonov and Rohrilich (1990).

HOLISM AND NONLOCALITY OF QUANTUM MECHANICS

As many brain studies have indicated, the neurophysiological correlate of consciousness is extended throughout appreciable portions of the brain. William James believed that reductionistic classical physics could not in principle explain the holism of brain states. When two quantum systems are correlated, only the combined system is in a definite quantum state. The nonlocality (geometrically) or nonseparability (algebraically) of QM is evidenced when the various parts of a quantum system are spatially and/or temporally separated. Bell's theorem proved that the correlations between these parts cannot be explained by any local-realistic theory. Thus, QM implies a kind of holism (that is, the nonfactorizability of a many body state into a product of one body states). When the wavefunctions of two or more equivalent particles overlap, they become indistinguishable. For example, the many parts that make up the highly ordered phase in a Bose-Einstein condensate don't just act as a whole, they become a seamless whole (Marshall 1989). The individual identities of the parts merge in such a way that they lose their individual identities, just as the individual symbols lose their individuality in the wholeness of a conscious thought.

In a recently proposed exposition of consciousness within the context of quantum theory (Stapp 1993), Henry Stapp proposed that the "actual event" (corresponding to an actualization or reduction of the wavefunction) constitutes the fundamental unit of our conscious experience and that the holism of QM paves the way for the incorporation of consciousness into the physical sciences. According to Stapp, as a result of QM, "the essential unity of the psychic state . . . mirrors the essential unity of its physical counterpart" (Stapp 1993, p. 157). David Chalmers asks the question, "why are brain functions and processes accompanied by our subjective experience?" There seems to be a fundamental explanatory gap and the need for new fundamental principles in physics is suggested. For instance, a double-aspect theory, such as the one proposed by Stapp: "each actual event has two aspects, a feel, and a physical representation within the quantum formalism" (Stapp 1993). Furthermore, Stapp suggests that the process by which the wavefunction collapses is not entirely random or stochastic, "each quantum choice injects meaning, in the form of enduring structure in the physical world" (Stapp 1993). This approach re-introduces the causal role that consciousness seems to have (Stapp 1993, p. 168):

At the purely physical level the Heisenberg actual event is passive: it is simply the coming into being of a new set of tendencies. However, in the context of the present ontology the actual event must be construed actively: the event actualizes the shift in tendencies. If the feel is identified as the active aspect of the event, then the feel is the veridical feel of actively actualizing the new state of affairs, and consciousness becomes the efficacious agent that it veridically feels itself to be.

Abner Shimony agrees: "In sum, I agree with Stapp that a unified science of mind and matter is in principle possible, and that quantum mechanics, or some refinement thereof, is crucial for that unification" (Shimony 1994).

SOME STRANGENESS IN THE MEASUREMENT PROCESS

In the past, numerous eminent physicists have suggested a possible relationship between the quantum measurement problem and consciousness. This section contains a review of some important issues within measurement theory.

Consider the process of measuring the spin of a particle with a Stern-Gerlach magnet. If a particle is spin up in the y-direction, then the quantum state of the particle plus measuring device (MD) becomes: $|\uparrow\rangle |MD^\uparrow\rangle$. However, if the particle is spin up in the x-direction, then this is a superposition of states in the y-direction:

$$|\rightarrow\rangle = \frac{1}{\sqrt{2}}(|\uparrow\rangle + |\downarrow\rangle) \qquad (1)$$

And therefore, when the y-measuring MD interacts with it, it too goes into a superposition of states:

$$|\rightarrow\rangle |MD\rangle$$

$$|\rightarrow\rangle |MD\rangle = \frac{1}{\sqrt{2}}\{|\uparrow\rangle |MD^\uparrow\rangle + |\downarrow\rangle |MD^\downarrow\rangle\} \qquad (2)$$

However, this is in direct conflict with our subjective experience. We always experience MD to be in one state or another. It was postulated by von Neumann that at some point in the chain of measurements up to the point of awareness, this correlated state gets replaced by a single state, corresponding to a definite outcome, either $|\uparrow\rangle |MD^\uparrow\rangle$ or $|\downarrow\rangle |MD^\downarrow\rangle$

with equal probability. This is known as the projection postulate or the collapse of the wavefunction, or actualization of potentialities. In the rest of this section, we explore several clues about the question of when, where, and *if*, this collapse occurs.

Clue 1

Consider two particles in a correlated spin state and consider two observers A and B spatially separated. Without specifying a given spin direction, we can say that the combined system is in a state:

$$\frac{1}{\sqrt{2}}(|\uparrow>A\ |\downarrow>B + |\downarrow>A\ |\uparrow>B) \qquad (3)$$

In this state, the combined system is in a definite state, but we cannot say what state an individual particle is in if it is taken by itself. Suppose that observer A measures his spin in the y-direction with result up and B measures his particle in the x-direction with result down. In A's frame of reference, his measurement occurs first and then B's measurement. However, in B's frame of reference, B's measurement occurs first and then A's. A believes that his measurement caused the two-particle state to collapse into an eigenstate of σy, whereas B believes that his measurement caused the two-particle state to collapse into an eigenstate of σx. The two different versions give the same net results, but they differ completely on the intervening events. A theory of the collapse, it would seem, must select one of the two versions, however, there is nothing in the dynamics to suggest which one.

Clue 2

The rules of quantum theory do not select between prediction and retrodiction of the quantum rules. In 1964, Aharonov, Bergman, and Lebowitz proved that one achieves time-symmetry in measurement theory by closing the probability measure under conditionalization, in fact this symmetry must be physically meaningful in any theory based on probability. For example, consider measuring observable A with eigenstates $|\alpha_i>$ at time t_1 and then measuring B with eigenstates $|\beta_j>$ at time $t_2 - t_1 + \delta t$. In the usual theory of collapse in which the system collapses into an eigenstate after the first measurement, one would say the probability of getting eigenstate $|\beta_j>$ at time t_2 after measuring $|\alpha_i>$ at time t_1 is $|<\beta_j|\ exp(-iH\delta t)\ \alpha_i>|^2$ (Figure 39.1a). However, one may equivalently say that the probability is $K\beta_j\ exp(+iH\delta t)\ |\alpha_i>|^2$ (Figure 39.1b), which is the time reverse of the first picture. Aharonov suggests that is necessary to consider both situations together, as in Figure 39.1c.

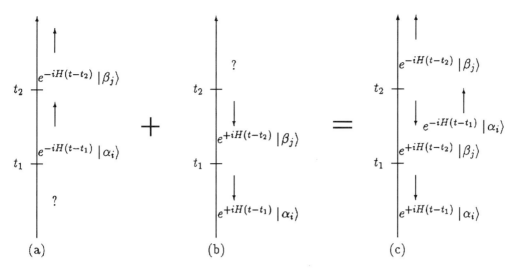

Figure 39.1 Time reversal symmetry in probability amplitudes.

Clue 3

In the Renninger type of experiment, it seems that the state vector is collapsed by the absence of an interaction. The only traceable change seems to be in the knowledge of the observer. Mandel has recently performed a number of experiments using a parametric down-convertor, which indicates that "The quantum state reflects not only what we know about the system but what is in principle knowable." In other words, a measurement seems to occur without any direct physical interaction. The measurement can even be erased using a quantum eraser.

Clue 4

It is usually postulated that a "collapse" occurs when an irreversible recording is made of the measurement outcome. As a result of the large degrees of freedom of the MD, and due to its density of energy states, for all practical purposes, the interaction introduces random phases into MD and so we cannot experimentally distinguish between an uncollapsed MD and a collapsed MD (Daneri, Loinger, and Prosperi 1962). Some researchers take this to mean that for all practical purposes, the "collapse" occurs at the level of MD; that is, they mistake this theorem to be about whether or not collapses occur or not. But obviously measurements have outcomes, as revealed by our own direct introspection. This paper, however, addresses questions of principle; that is, how does one deal with the nonlocal aspects of the collapse and how does one deal with the local aspects of the collapse.

Hopefully the reader is getting the idea that the usual notion of collapse as described by the projection postulate is in serious trouble. Recently, a theory of wavefunction collapse was proposed (the GRW theory) in which the quantum state undergoes spontaneous collapses onto eigenstates of position. At a completely random and rare time, the wavefunction of a particle gets multiplied by another function. A rate of collapse and a width are chosen so that a single particle will almost never collapse. However, if there are a large number of entangled particles, the wavefunction will collapse quickly. Albert and Vaidman (1988) proposed a typical type of quantum experiment, the Stern-Gerlach experiment, on a fluorescent screen and applied the GRW theory to it to find at what point the collapse would occur. The authors of the GRW theory were forced to conclude that the collapse did not occur until the wavefunction interacted with the brain of the observer. We believe this phenomenon will be evident in any stochastic or nonlinear theory of the collapse that presupposes a basis.

OBSERVER-PARTICIPANCY

John Wheeler (1992, p. 353) helps to elucidate the role that we play as observer-participants:

. . . Is it a member of the animal kingdom?" "No." "Mineral kingdom?" "Yes." Strangely, each new respondent requires a yet longer time of reflection before he summons up his yes or no reply. Soon I approach my twenty-question limit and must venture all upon a single word. "Is it cloud?" I ask. Long agonized thought by the respondent: then a reluctant, "Y..es." Everyone bursts out laughing. While I was out of the room, they explain, they had agreed not to agree on a word. There was no word in the room when I entered. Everyone could respond "yes" or "no" as he pleased—with only one small proviso. The respondent, whatever his answer, had to have a word in mind compatible with his own reply and with all the others. Otherwise, challenged and unable to reply, he lost and I won. The Game of Twenty Questions in its Surprise Version was as difficult for my friends as for me. No wonder it took time for them to answer!

The game in its two versions illuminates physics in its two formulations, classical and quantum. First, the word already existed in the room—we thought—independent of any question that we might or might not ask. But it didn't. Likewise the electron has a position and a momentum inside the atom—physics one thought—independent of any act of observation. But it doesn't. Second, no information about the word came into being except by question asked, as no information develops about the electron except by experiment made. Third, if I had posed different queries I would have ended up with a different word. Likewise, the installation of equipment to measure the position of the electron automatically makes it impossible to install in the same place at the same time equipment to measure the momentum of the electron and conversely. Fourth, partial power only did I have to influence the outcome by my choice of questions. A major part of the decision lay in the hands of my friends. Similarly, the experimenter decides what feature of the electron he or she will measure: but "nature" decides what the magnitude of the measured quantity will be. The conclusion? Does the world exist "out there?" No.

Rather than speaking about a foundation for reality, Wheeler says it is a closed-loop, the meaning-circuit, completed by observer-participancy (Wheeler 1986, p. 309).

With particles owing their definition and existence to fields, with fields owing their definition and existence to phases, with phases owing their definition and existence to distinguishability and complementarity, and with these features of nature going back for their origin to the demand for meaning, we have exposed to view (at least in broad outline) the main features of the underground portion of the model of existence as a meaning circuit closed by observer-participancy.

Another of Wheeler's points is that the only reality we can discuss are the outcomes of measurements and it is meaningless to speculate about the state of the quantum system between measurements. Although consistent, this minimalist approach disregards some interesting physics. The standard "classical" attribution to reality is the dynamical variables. However, in the two-vector theory presented below, it is argued that the quantum state itself and expectation values of observables should be attributed reality. We are observer-participants, then, in both the selection of the destiny state, and in the selection of the observable. As Yakir Aharonov notes: "Don't think of measurement as just determining what we don't know. The real issue of measurement theory is determining what can manifest itself."

TIME

According to modern science the passage of time has no fundamental or dynamical importance, it is merely an illusion. Yet it plays a crucial role in our lives.

Albert Einstein denied the transitory nature of time based on his theory of time as the fourth dimension. This conception of time implies a very strange model for the mind: the mind is a collection of many separate momentary minds existing on the world-line. According to Elitzur (1992, p. 28).

If time's passage is illusory, then our feeling that we have a single, unitary self must also be an illusion. For, if there is no privileged "Now," as relativity stresses, but every event has its relative "now," then all one's selves, from infancy to old age, coexist in the fourth dimension with the same degree of reality; no momentary self is more "real" than the earlier or later ones.

However, this is completely at odds with our practical experience. According to Einstein: "There is something essential about the Now which is just outside the realm of science" (requoted from Elitzur 1992, p. 28).

So, why do we feel that time is "flowing" forward? The conventional response is that this passage is based on our memory of our previous selves. Due to this memory, we are deluded in believing that the past selves are the same as the present self (contrary to relativity theory which says that each self is separate and coexisting). We do not remember "future" selves due to the dependence of our memory systems on the thermodynamic

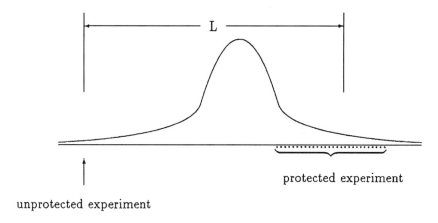

Figure 39.2 Meaning of the wave function.

time-arrow. However, consciousness involves a kind of dynamic time, as opposed to the kind of static time represented in physics. Clearly, there is quite a gulf between our subjective experience of time and the time of modern physics. A deeper understanding of time will lead to a deeper understanding of consciousness and "Becoming."

THE TWO-VECTOR FORMALISM OF QUANTUM MECHANICS

Meaning of the Wavefunction

Recently Aharonov *et al.* (1996) proposed a new type of experiment on a quantum system, the protected experiment. This experiment measures the full wavefunction for a single particle by avoiding entanglement with a measuring device, and thus avoiding collapse. These proposed experiments suggest a new ontological meaning for the wavefunction, namely that the wavefunction for a single particle has physical reality. On the other hand, the conventional Copenhagen interpretation of QM is an epistemological theory with no ontological commitments, it says the wavefunction only has a nonphysical mathematical existence for ensembles of particles.

The physical existence of the wavefunction also points strongly to the two-vector theory. Consider a protected experiment that measures the wavefunction in space over a certain region. Just after this protected experiment, somebody else does the usual strong measurement and observes the whole particle in one place a distance L from the region of the protected measurement. Huge currents would have to flow from the extended object to the localized particle, and thus there would be frames in which charge is not conserved. Violation of causality can be avoided if the two-vector theory is adopted because the intermediate description depends on what happens in the future.

In order to resolve this and numerous other problems in the foundations of quantum theory (such as those listed earlier), Aharonov and Rohrlich (1990) propose that it is necessary to describe quantum systems in terms of two state vectors; the usual one evolving from initial conditions (history state), and a new one evolving backwards in time from future boundary conditions (destiny state) (Aharonov, Albert, and Vaidman 1988).

Consider an experiment that occurs within a certain region of space, such as a scattering experiment. The incoming particle, ψ_1, interacts and then evolves into various outgoing states, such as ψ_2, ψ_3, and so on. In a classical system, there is a one-to-one mapping between incoming states and outgoing states, whereas in QM, it is one-to-many. We can thus define a new type of ensemble that has no classical analog, that is, a preselected (ψ_1) and postselected (ψ_2) ensemble, that is, the boundary conditions at the beginning and end of an experiment.

In Figure 39.3 a system is prepared (preselected) at time t_1 in an eigenstate $|a\rangle$ of an operator A and postselected at time t_3 in an eigenstate $|b\rangle$ of operator B. If another observable C is measured at time t_2, then the probability of a certain outcome $|c_i\rangle$ is:

$$P_i = \frac{|\langle b|U(t_3,t_2)|c_i\rangle|^2 |\langle c_i|U(t_2,t_1)|a\rangle|^2}{\Sigma_j |\langle b|U(t_3,t_2)|c_j\rangle|^2 |\langle c_j|U(t_2,t_1)|a\rangle|^2} \quad (4)$$

However, this can be rewritten in a time symmetric form as:

$$P_i = \frac{|\langle c_i|U(t_2,t_3)|b\rangle|^2 |\langle c_i|U(t_2,t_1)|a\rangle|^2}{\Sigma_j |\langle c_j|U(t_2,t_3)|b\rangle|^2 |\langle c_j|U(t_2,t_1)|a\rangle|^2} \quad (5)$$

This probability can be interpreted as evolving $|b\rangle$ backwards in time and $|a\rangle$ forwards in time: the present is literally created out of influences both from the past *and from the future*.

It is important to note that the two-vector formalism reproduces all the predictions of standard quantum theory. All the fundamental laws of physics are symmetric under time reversal, with the exception of the theory

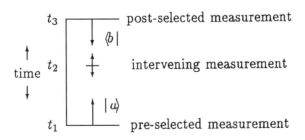

Figure 39.3 Interpretation of two-vector formulation, the destiny state $|b\rangle$ propagates backwards in time from t_3 to t_2 and the history state $|a\rangle$ propagates forward in time from t_1 to t_2. Both get annihilated at time t_2.

of measurement in quantum mechanics. The two-vector formalism reintroduces time symmetry into the theory of measurement.

Furthermore, the two-vector formalism resolves some of the fundamental paradoxes of quantum physics, such as the nonlocality evident in the Einstein-Podolsky-Rosen paradox, Wheeler's delayed choice paradox, and the Lorentz covariance violation of the collapse of the wavefunction.

Weak Measurements

One of the physical consequences of this new formalism is a new type of observable, known as the weak value of a quantum observable that is obtained from the weak measurement. In the two-vector formalism, the expectation value of an observable $<A>$ generalizes to the weak value of an observable, where Ψ_{fin} is a postselected ensemble, and Ψ_{in} is a preselected ensemble:

$$A>_w = \frac{<\Psi_{fin}|A\Psi_{in}>}{<\Psi_{fin}|\Psi_{in}>} \tag{6}$$

The weak value is a measureable quantity (Aharonov and Vaidman 1989). Consider an ensemble of N particles preselected with the x component of spin $s_x = 1/\sqrt{2}$ and postselected with $s_y = 1/2$. At an intermediate time t_2 a measurement will find the total angular momentum in a direction $\hat{x}$ (at an angle between x and y) to be $\frac{J_x + J_y}{\sqrt{2}} = \frac{N}{\sqrt{2}}$ as long as the uncertainty is of the order $\sqrt{N}$, that is, the measurement is weak. For an individual spin, the component of spin $s_x = 1/\sqrt{2}$ which is $\sqrt{2}$ times bigger than the original eigenvalues.

The weak value has actual physical consequences. In cases of interactions that are weak enough, the outcome of the interaction will be the weak value. Thus the existence of weak values is evidence for the existence of the two boundary conditions during the intermediate time. The weak value possess some fascinating, almost miraculous qualities. First of all, A_w is not bounded by the same spectrum as A and incompatible operators are simultaneously measureable. The weak value can also be observed for a single quantum system. Perhaps one of the strongest arguments for the two-vectors, is that the destiny vector can be protected and observed for a single particle. Weak values are entirely reproducible and therefore satisfy the EPR reality condition and the argument for counterfactual implications. Furthermore, the weak value has been observed in a number of physics experiments to date.

Negative Kinetic Energy as Confirmation of Two-Vector Interpretation

Consider a particle in a square well. First a measurement is made of the particle's kinetic energy to an arbitrary precision. Secondly, we perform a measurement attempting to detect the particles in the classically forbidden region. When they are localized outside the well, it is discovered that the corresponding kinetic energy measurement earlier elicits a negative value which is centered around the classical value $E - U$, the total energy less the potential energy. Standard quantum theory interprets this measurement of negative kinetic energy as an error of the measuring device, since all the eigenvalues of the kinetic energy operator are positive, that is, it has nothing to do with any property of the particle. However, this type of consistency strongly suggests something much more than a random collection of errors.

What relevance does this have with consciousness? Consider an interesting temporal anomaly: instead of postselecting far from the well, measure the kinetic energy again with greater precision. In this case, almost every time the first kinetic energy measurement yields a negative value, the second kinetic energy measurement yields a positive value. Clearly, in this case, the negative kinetic energy is interpreted as an error due to the measuring device. However, we could have decided to measure the position far from the well. In this case, the negative kinetic energy is attributed to the particle. The point is that the postselected measurement is performed after the measurement of negative kinetic energy. It seems that the cause is after the effect. From an ensemble standpoint, there is no violation of causality. However, from an individual standpoint, there does seem to be a kind of backwards causality. The errors experienced by the measuring device are like whims, that is, a random decision not based on the past. However, if we make a pact with the future now, then we are creating meaning in the present, that is, the particle actually does have a negative weak value to it's kinetic energy if we perform the correct experiment in the future. This causality paradox in conjunction with the assumption of free will is currently being studied.

Within the two-vector theory, it is not necessary to introduce a nonlocal reduction of the wavefunction as a result of the final reading of the MD. In the usual theory, the collapse of the system plus MD can occur even after the MD is spatially separated from the system. In the two-vector theory, two local boundary conditions are invoked. The two-vectors are the fundamental reality: the fact that it may or may not physically look like a local reduction is not important.

New Model for Consciousness

The two-vector theory is a step in the right direction for understanding the nonlocal aspect of the collapse. What about the local aspect? There are a number of directions that one can follow. Perhaps the collapse is a result of an irreversible recording in a macroscopic measuring device as pointed out in Clue 4. Although this seems to be true in practice, it is not necessarily true in principle. Perhaps one needs to go to a higher degree of complexity, such as a closed system with sufficient degrees of freedom that it can observe itself. In this section, this later possibility is explored.

In this model, objective consciousness is associated with the selecting or creating of destiny states. An awareness variable $|\Omega\rangle$ is defined to characterize the relevant state of consciousness. This variable is restricted to always be in a definite state. That is, it can never be in a superposition of states. It is our experience that we always experience ourselves in a definite state with certainty. We never experience ourselves in a superposition of states.

Different levels of the brain experience quantum superpositions, but at the conscious level, superposition is never experienced. For example, at the very least, superposition is significant in various macromolecules, possibily in an inhibitor protein and in microtubulins (Beck and Eccles 1992). A transition from quantum to "classical" occurs at some level in brain dynamics, and we suggest that the transition occurs at the point of awareness. Von Neumann proved that the collapse can occur at any point in the measurement chain up to the point of awareness, without changing the predictions of quantum theory. So, this suggestion is entirely consistent: whenever a quantum system is measured by consciousness, then at the time of awareness, $|\Omega\rangle$ can only be in a definite state, and this thereby determines or creates the destiny state for the quantum system. If no measurement is done by the consciousness, then no destiny state is created.

What does this have to say about time? It has the topology of something like an onion. For each observation, a layer is peeled, and then one sees the next layer which was the destiny state for the first layer . . ., and so on. This gives us a new model for the internal structure of space-time, and how *consciousness literally creates time*. Thus, the rate of ticking of the consciousness is directly related to the rate of peeling, or the rate of creation of destiny states; it is an objective phenomenon. The time usually dealt with in physics is Schröedinger time, or the mathematical time evident in the mathematical equations governing the evolution of quantum systems. This time is a virtual time for it characterizes the evolution of potentialities as opposed to the evolution of our world of classical events. The new time suggested above by the ticking of awareness corresponds to the dynamic psychological time that we experience in our conscious experience. This model resolves many of the issues discussed in the earlier section.

So, in a sense, consciousness is the action of finding out the destiny state that is already there, but what is there in the past depends on what is searched for in the future. Thus, consciousness has very interesting properties that are deeply intertwined with time and nonlocality in time.

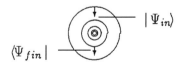

Figure 39.4 Model of time as created by consciousness, the selection of destiny states. Each awareness peels off another layer. This peeling corresponds to the ticking of time.

It is important to distinguish that the selection of the actual value of a quantum observable is performed by nature. What consciousness does is to provide the destiny state so that the quantum system "knows" at what time it has to have a destiny. A dualistic model is not being proposed. It is suggested that the collapse is parametrized by complexity, so it is something like a phase transition. It is also possible for the collapse to occur independently of any brain or consciousness, but this is less frequent.

What is accomplished by placing the selection at the point of awareness. In the process of "collapse," a fundamental kind of integration occurs that also reflects this holism. If a system is in a state $\psi(x_1, x_2, \ldots x_n)$ and a possible outcome of the collapse is $\phi(x_1)$, then the new quantum state (Stapp 1993):

$$\psi'(x_1, x_2, \ldots x_n) = \phi(x_1) \int dx\, \phi^*(x_1) \psi(x_1, x_2, \ldots x_n) \qquad (7)$$

The new ψ is determined from the old ψ in an integrative manner. The integration of the collapse creates the unity of consciousness.

In standard QM, there exists only one state, $|\psi\rangle$, whereas in the two-vector theory, the state is generalized to three elements which create a process: $|\psi_{in}\rangle, \langle\psi^f_{in}|$, and the evolution operator between the two. To David Bohm, the ultimate organizing principle for a system is "in the outcome of the process and not its genetic origin."

When one considers multiple-time states (Aharonov, Albert, and D'Amato 1985), correlated states between the values of noncommuting observables that cannot be written in terms of single-time states, then one must expand the Hilbert space with each new destiny state. With each creation of a new destiny state, the Hilbert space is enriched, it is made larger, more degrees of freedom are created. This gives us a new understanding of "Becoming," and points to our suggested definition of consciousness (Stapp 1993, p. 234):

That luminescent presence of coming-into-beingness that constitutes our inner world of experience.

This integrative aspect and the enrichening of the Hilbert space may correspond to our "feel," or the subjective quality of our experience.

An Experiment to Test This Theory

Consider a device that measures the spin of a particle. If the spin is up then it interacts with another particle $MD_1 \downarrow$ along the up path and changes its state to $MD_1 \uparrow$. If the spin is down then it interacts with another particle $MD_2 \uparrow$ along the down path and changes its state to $MD_1 \uparrow$.

MD_1 and MD_2 are connected directly to that part of the brain, with a minimum number of intermediary parts, which can distinguish the two outcomes of our device. Then a particle is put into the device which is in a superposition of states in the y-spin direction:

$$|\rightarrow\rangle = \frac{1}{\sqrt{2}}(|\uparrow\rangle + |\downarrow\rangle) \qquad (8)$$

Then the consciousness irreversibly records (on paper) that it has made on observation of the spin state, but it does not tell us what that state is. Next all of these steps are time-reversed except the irreversible recording onto paper with a time-translation machine (Aharonov et al. 1990). And finally, the x-component of the spin is measured.

If reality does not allow superpositions of consciousness, then there will be an equal distribution of ← and → spins, because the state must have collapsed while the observer was contemplating the spin result. However, if reality allows superpositions of consciousness, then there is no collapse and only → is obtained. This experiment can also be repeated using a "quantum observer" or automata instead of a human observer. It can be determined at what point collapses occur by varying the complexity of the device. Although this experiment is possible in principle, we do not currently have the technology to do it in practice. Even if collapses do occur at the level of measuring devices, as opposed to at the level of consciousness, this experiment would introduce new parameters on consciousness.

A more practical test is to explore whether there are any computational advantages that can be derived by exploiting unique quantum mechanical effects in the design of a computer. Recent work by Deutsch and Jozsa (1992) on quantum parallelism answers this question with a definite yes. Deutsch specified a wide class of problems that are solvable in polynomial time as opposed to exponential time in a classical computer. To test whether the brain is doing quantum-type calculations, one must compare the capabilities of a classical parallel-distributed processing system (PDP) and a quantum PDP against psychological data. As an example, a proposal was recently made and preliminary success reported (McCarthy, and Goswami 1993) in comparing these PDP models against word-sense disambiguation experiments.

Temporal Dynamics of Consciousness and Elementary Integration Units

Poppel, Schill, and von Steinbuchel (19??) have shown that spatial and temporal changes in an object under observation present processing difficulties for the brain because the temporal central availability (TCA) is undefined due to the different transduction times for different sensory modalities. As a solution, Poppel proposes temporal "quantization" into zones of simultaneity, or Elementary Integration Units (EIU): within any given zone, there is no before-after relation defined between registered events. Experiments indicate a duration of 30 msec for each zone. Poppel analyzes the processes involved in consciousness from four functional perspectives: perception, mnemonic representation, stimulus evaluation (emotion), and stimulus response (volition). Each of these processes involve largely different time constants and are significantly distributed throughout the brain. Any given content of consciousness is characterized by these different functional domains and the brain must therefore be able to bind these distributed processes together into a single unified experience. Within any given EIU, information is collected from various parts of the brain and an exact TCA is not needed due to the structure of the EIU (Poppel, and Schwender 19??, p.27). How are the various functions integrated together into a single gestalt? The integration across time has a range of up to 3 seconds, and must have top-down qualities to it: the representation partly determines the method of representation. The type of seamless temporal integration required between EIUs does not seem possible with classical neural algorithms. Thus, Poppel suggests:

This time- and structure-creating process and the observed non-localized interactions occurring in the network of neurons lead us to the supposition that the functioning of the brain cannot be understood using classical concepts and that abstract brain theory has to be based on abstract quantum theory.

A model of this process using the two-vector theory is suggested here: each boundary of an EIU is the creation of one destiny state. In between the planes, all registered events are linked together via multiple time states. The multiple-time state actualizes correlations between quantum observables at different times, although the observable itself is undetermined for any one given time. This actualization across time creates the necessary seamless integration. The selection of a destiny state creates the EIU and is therefore responsible for psychological time.

Libet's Work

In a sequence of cognitive experiments on awake patients undergoing cortical surgery, Benjamin Libet discovered a substantial neural delay between the actual sensory signal and the experience. Libet also discovered that there is a subjective referral backwards in time to the time of the original signal. In studying voluntary acts, Libet discovered the will to act occurred 350 msec after the initial cerebral processing for the action, thus contradicting the notion that the will to act is initiated consciously. (Libet (1985). It is hard to understand how consciousness could have any dynamical control without some kind of temporal nonlocality.

CONCLUSIONS

A new interpretation of quantum mechanics is discussed. This formalism gives a consistent Lorentz covariant mechanism for the collapse of the wavefunction and a new understanding of quantum phenomenon. A new understanding of time is discussed, which resolves a conflict between special relativity and the transient now and also gives us a new model for "Becoming."

NOTES

1. By unity of consciousness we mean that the essence of a conscious "event" is lost when we attempt to break it down into a simple aggregate of localized components as classical physics requires.

ACKNOWLEDGMENTS

I thank Professors Yakir Aharonov, David Albert, and Abner Shimony for many helpful conversations. I would also like to thank the Fetzer Institute for their support of this project.

REFERENCES

Aharonov, Y., D. Albert, and S. D'Amato. 1985. Multiple-time properties of quantum mechanical systems. *Phys. Rev.* D32(8):1975–84.

Aharonov, Y., D. Albert, and L. Vaidman. 1988. How the result of a measurement of a component of a spin 1/2 particle can turn out to be 100? *Phys. Rev. A* 35:4052.

Aharonov, Y., and J. Anandan. 1993. Measurement of the Schrödinger wave of a single particle. *Phys. Lett. A* 178:38–42.

Aharonov, Y., J. Anandan, S. Popescu, and L. Vaidman. 1996. Superpositions of time evolutions of a quantum system and a quantum time-translation machine. *Phys. Lett. Rev. A* 64:2965–68.

Aharonov, Y., and D. Rohrlich. 1990. Towards a two vector formulation of quantum mechanics. In *Proceedings of the International Conference on Fundamental Aspects of Quantum Theory to Celebrate 30 years of the Aharonov-Bohm Effect, Quantum Coherence*, edited by J. S. Anadau. Singapore: World Scientific, pp. 221–31.

Aharonov, Y., and L. Vaidman. 1989. "A new characteristic of a quantum system between two measurements—Weak value." In *Bell's theorem, quantum theory, and conceptions of the universe.* New York: Academic Publishers, p. 17.

Aharonov, Y., and L. Vaidman. 1990. Properties of a quantum system during the time interval between two measurements. *Phys. Rev. A* 4:11.

Albert, D., and L. Vaidman. 1988. On a proposed postulate of state vector reduction. *Physics Letters A* 139:1.

Beck, F., and J. Eccles. 1992. Quantum aspects of brain activity and the role of consciousness. *Proc. Natl. Acad. Sci. USA* 89:11357–61.

Danieri, A., A. Loinger, and G. M. Prosperi. 1962. Quantum theory of measurement and ergodicity conditions. *Nuclear Physics* 33:297–319.

Deutsch, D., and R. Jozsa. 1992. Rapid solution of problems by quantum mechanics. *Proc. R. Soc. Land. A* 439:553–58.

Elitzur, A. 1992. Two persistent wonders—or one? Consciousness and the passage of time. *Frontier Perspectives* 2(2):27–33.

Libet, B. 1985. Unconscious cerebral initiative and the role of consciousness will involuntary action. *Behav. Brain Sci.* 8:529–39.

Marshall, I. 19??. Consciousness and the Bose-Einstein condensate. *New Ideas in Psychology* 7:73–83.

McCarthy, K., and A. Goswami. 1993. CPU or self-reference: Discerning between cognitive science and quantum functionalist models of mentation. *Journal of Mind and Behavior* 14:13–26.

Poppel, E., K. Schill, and N. von Steinbuchel. 1990. Sensory integration within temporally neutral systems states: A hypothesis. Naturwissenschaften 77:89.

Poppel, E., and D. Schwender. 1993. Temporal mechanisms of consciousness. *Int. Anesthesiology clinics.* 31(4)27–38.

Shimony, A. 1994. Mind, matter, and quantum mechanics. *Am. J. Phys.* 62:956–57.

Stapp, H. 1993. *Mind, matter, and quantum mechanics.* Berlin: Springer-Verlag.

Wheeler, J. A. 1986. How come the quantum? In *New techniques and ideas in quantum measurement theory,* edited by D. Greenberger. New York Academy of Science, 480:304–16.

Wheeler, J. A. 1992. Recent thinking about the nature of the physical world: It from bit. Frontiers in cosmic physics: Symposium in memory of Sergei Alexander Korff. *Annals N.Y. Acad. Sci.* 655:350.

40 Consciousness: A New Computational Paradigm

Douglas J. Matzke

INTRODUCTION TO CONSCIOUSNESS AS COMPUTATION

Is the brain a fantastic computer we have yet to decipher or does some mysterious nonphysical property called the "mind" evolve with and control this biological robot? Dualists have stated there are two distinct types of substance: mental and physical. Alternatively, materialists conclude from neurological research that the brain and the mind are one and the same. All but a handful of scientists believe that the "dualist" approach has been outmoded for decades (Killheffer 1983). Either approach to the mind-body (or mind-brain) problem must be built on a computational theory, since man's intelligence is as much a mystery as his consciousness. Purely mechanistic computational approaches do not add much insight into the mystery of consciousness (Maudlin 1989). My assumption is that a good computational model for intelligence must precede and therefore support, any theory of consciousness. If a correct computational strategy could be developed to support the kind of "real intelligence" demonstrated by mankind, then it may also shed light onto consciousness.

This paper presents a rationale, important properties, and computational framework that would be required for a dualist model of the mind. Most scientists will reject the need for a dualist model of the mind, but research in physics has uncovered a wealth of understanding that could be applied to the computational approach to the mind-brain problem. Physics has developed the concepts, theories, and techniques to deal with real items that can only be indirectly measured. For example, quark theory predicts that matter consists of tiny invisible quarks, but quarks themselves are not directly measurable. In fact, most of modern science depends on indirect measurements based on predictions from some theory. These same techniques can be applied to studying a real but nonphysical mind.

Nonphysical Sciences for the Study of Mind

If a nonphysical mind really does exist, then it should be amenable to study in the same fashion as other physical theories that deal with indirectly observable phenomena. Once the cultural limitations of accepting the possibility of nonphysical mechanisms are addressed (that is, indirectly observable), the next major questions to be tackled for a dualist model of the mind-brain includes the concerns about representational, architectural, and physical limits.

Since humans are intelligent as well as conscious, a good predictive computational theory is the key requirement for a solution to the mind-brain puzzle. Such a theory must address the representational issue of information versus knowledge (or knowing). Information is traditionally considered a static measure, whereas knowing, meaning, and consciousness implies a dynamic action. Any representation of information and its dynamics impacts all aspects of our models, including the computer architectures we can conceive or build. Our computer-dominated representations for information, space, and time have limited our race's ability to develop a computational theory of the mind.

The most widely used computer design is the von Neumann computer architecture. In this architecture, the memory and the processor are separated by a set of wires called a bus. This bus is called the "Von Neumann bottleneck" because all data must move through this limited speed set of wires, no matter how fast the processor or how large the memory. This computer organization should be relabeled the "Newtonian bottleneck" because it reflects the computer industry's use of last century's models of an independent space and time. Computation requires both space (memory/communication) and time (processor or change) resources, but segregating these two resources seems to violate what modern physics has learned about a unified spacetime. A modern view of a unified space-time is best reflected in cellular architectures where small amounts of memory and logic produce a space of active data that approximates the organizational dynamics found in physics.

Limits to Physical Computation

Since computing is a physical action (Landauer 1992), physical limits impact the realization of any computational model. The speed of light, the wavelength of light, the discrete charge of an electron, noise margins, thermal problems, and the uncertainty principle are obvious limits to building physical computing machines. Other less obvious limits, such as the number of physical dimensions of our universe, the black hole limit, and rate of growth of exponential problems also have a profound impact on the size and nature of physical machines we can design or build.

Hubert Dreyfus (1992), a critic of the artificial intelligence community, predicts that computers will never be intelligent due to known formal and computational limits. Dreyfus states that computer science has made interesting progress in mimicking human intelligence, but these algorithms require an exponential amount of computing resources for larger and larger problems. The problems of vision and language understanding, dynamic motion control, cryptography, and planning far exceed the ability of any conventional computing machine. Future scalability limits ultimately restrict how powerful a computer we can design or build. It is for these reasons that understanding ordinary human intelligence may be a prerequisite to understanding consciousness.

It is clear that a revolutionary computational approach is needed to build truly intelligent machines, because all the hard computational problems listed above are all members of the same formal class of algorithms. For these reasons, the following two areas of physics research are being investigated. First, scientists are studying quantum computing as a mechanism for exponential speedup (Shor 1994). The appeal of applying quantum physics for computational leverage (and consciousness) comes from its unusual properties of instantaneous, nonlocal correlations of discrete states. Second, relativity also comprehends the unusual properties of variable space and time. Some of these strategies for providing extraordinary computing resources might also provide insight concerning computational processes with properties suitable for consciousness. It is possible that systems that exhibit the self-organization required for human "real intelligence" (nothing artificial about it), may exhibit consciousness. The next section surveys aspects of physics that could build a conceptual framework for extraordinary computational facilities and consciousness.

CROSSDISCIPLINARY PHYSICS AS THE BASIS FOR MIND

Physics must ultimately develop a solution for human "real intelligence," because it represents an evolutionary, complexity-increasing informational process. This process must not violate what physicists know about the evolution of the complexity of the universe. Cosmologists that study the evolution of the universe (Hawking 1988) are combining techniques from information, quantum, and relativity theories, the three most successful theories of all time. These hybrid efforts describing the evolution of the universe could be applied to evolution of the mind.

Research results from these three fields, on higher dimensional semantics may be applicable to the puzzle of intelligence and ultimately consciousness. Many interesting and complex information systems have higher dimensional semantics and repeatably show up in nature. In addition, computer scientists have demonstrated in coding theory (Lucky 1989), neural nets (Kanerva 1988), and many forms of higher-dimensional mathematics, that increasing the independent degrees of freedom can be computationally advantageous. It is also well-known that simulating certain high-dimensional problem semantics on computers with a one-dimensional virtual memory system (Margolus and Toffoli 1993) is inefficient.

Many researchers are already cognizant of the benefits of combining these three powerful theories. In his famous "It from Bit" paper, John Wheeler elegantly describes how information and quantum theories can be combined (Wheeler 1989). His paper describes other hybrid efforts that combine information, gravity, and quantum theories by Schiffer and Bekenstein (1992), Hawking (1975), Penrose (1979), and Unruh (1976). This work is exciting because it combines the sciences of the very large (gravity theory), very small (quantum theory), and very complex (information theory).

The challenge of building a grand theory that combines all known theories is the goal of many researchers (Kaku 1994). These hybrid theories have interesting names such as Bitstring Physics (McGoveran and Noyes 1989), Grand Unified Theory, Quantum Gravity, Theory of Everything, and many others (Kaku 1994). Just as in information applications, many of these theories depend on an emergent spacetime and are based on high-dimensional semantics (symmetries in five or ten dimensions). All these grand theories are topological and geometrical, which is similar to the classification of computer algorithms and architectures. Another common theme among these theories is the requirements and mechanisms for consistency. Consistency can be viewed as an informational cause behind conservation laws and is therefore more primitive than mass, energy, or even spacetime.

Topological consistency and higher dimensional space-times seem to be common themes relevant to information theory and the physical sciences. These ideas are the foundation for Einstein's relativity, which ushered in an entirely new physical theory governing consistency laws and space-time. Quantum theory is also based on an algebraic consistency of certain conserved properties described in Hilbert Space, a high-dimensional mathematics. Both of these theories have made verifiable predictions that space and time must have nonintuitive properties in order to maintain these consistency laws. This combination of well-understood physical mechanisms (consistency and spacetime) defines a framework for all physics. It is possible that computational leverage for intelligence and consciousness can arise from these powerful concepts and theories.

CONSISTENCY FRAMEWORKS AND OBSERVATION

Consistency frameworks form the physical foundation for multiple observational viewpoints or different "Points of View." Formally defining the interaction between the observer and the "action or thing being observed" is part of understanding the observation process. Historically, scientists have prided themselves in their belief that true science occurs when the observer does not participate or disturb an act of measurement. Unfortunately, quantum physics measurements depend on how a question is asked or what question is asked. If an experiment asks particle questions then the results are particle answers. If an experiment asks wave questions then the results are wave answers. Likewise in relativity, asking how much "energy" is in a system is dependent on the observer's velocity and acceleration.

Four independent frameworks for observation have been developed: (1) information/sampling theory, (2) relativity inertial frames, (3) quantum wave function collapse, and (4) the self-referential aspect of mind, called consciousness. Ideally, these observational frameworks should be combined into one unified framework that describes consistent observation. The consistency arguments that were used to develop relativity theory should apply to all observational frameworks. These arguments are critical because they define the very nature of space and time, which are the primary resources for computation.

The main idea stated in Einstein's relativity principle was that "all inertial frames are totally equivalent for the performance of all physical experiments" (Rindler 1977). In other words, no matter where you are in space or what speed you are traveling, the laws of physics must be the same. The laws define the possible actions as well as the process of observing those actions from any vantage point.

The logical progression of consistency steps that follow the relativity principle are: (1) Any point in space is as good as any other (position invariance), (2) Any direction is as good as any other (rotation invariance—isotropy), (3) Any speed is as good as any other (velocity invariance) except max is speed of light, -c, (4) Inertial frames are mathematical framework for relating various vantage points, (5) Many conserved properties are not truly invariant primitives, such as mass and energy, and (6) Parametrized consistency metrics are the only true cornerstone for observation.

One major outcome from relativity was experimental proof that the speed of light is constant no matter how you measure it, and no matter what speed you are traveling. In fact, mass, energy, distance, and time have changing values depending on one's speed. This result is required to keep anything that has mass from exceeding the speed of light, and to have all the laws of physics work consistently in all inertial frames, even those traveling very fast. The speed of light is a cosmic speed limit, and all observations using light are dependent on relativity principles. Gravity was also shown to be nothing more than an acceleration due to matter bending space and time. Relativity has the intuitive and mathematical framework to make it one of the most advanced theories of our time. Even observational frameworks of the mind could benefit from the power of this theory.

MANY OF THE CONSEQUENCES AND PREDICTIONS OF THE CONSISTENCY MODELS ARE COUNTER-INTUITIVE

1. Consistency is more primitive than conservation laws of energy/mass, or space and time.
2. Consistency requires light to follow locally "straight line" geodesics (curved space-time).
3. Consistency results in constant c and variable space and time (Lorentz transformation).
4. Inertial frames are completely relative and outside physics (they cannot be acted upon).

5. Consistency in quantum results in more extraordinary space-time models than relativity.
6. Causality must be replaced by synchronization among quantum events.

Relativity theory really is a new physical theory of space and time that impacts all physical laws (and all observational frameworks). These astounding and counterintuitive predictions are all based on absolute consistency requirements for the observation of physical events. When quantum events are considered (no direct observation possible), even more unusual spacetime properties emerge. In fact, recently Peter Shor (1994) has described how the quantum collapse of states can be used to solve very hard cryptography calculations.

Potential Mechanisms for Computational Leverage

Physics research has uncovered many nonphysical mechanisms that could be useful for creating computational leverage. A list of some of those ideas are:

1. Use ballistic computation and geodesics to reduce power costs.
2. Locality could be manipulated to shorten perceived distances.
3. Consistency mechanisms behave as superluminal synchronization primitives.
4. Act upon inertial frames or consider higher order Lorentz contraction (two dimensional and three dimensional).
5. Quantum computers theoretically can provide exponential speedup.
6. Consistency mechanisms interact outside normal linear time—excluding illegal time loops.
7. Increased dimensionality increases degrees of freedom (number of bits of choice).
8. Prespatial and pretemporal change mechanisms may be hierarchically and sparse.

These ideas appeal to researchers studying the mind and consciousness because certain biological (Sheldrake 1971), psychological (Ornstein 1969), parapsychological (Jahn 1982), and meditative research (Dillbeck and Alexander 1989) strongly suggest that these properties are exhibited by the mind. An interesting point to note concerning computational leverage mechanisms is that they deal with cosmological issues such as the framework of space-time and the structure of the universe, and are thus, "outside the box" of what is normal day-to-day physics. This is not surprising given that the evolution of the mind (both collectively and individually) deals with many of the same issues (information, complexity, and energy) as the evolution of the universe. The next section will succinctly describe a model dealing with observation and the mind that could encompass many of these computational leverage ideas.

IDEAL ARCHITECTURE FOR COMPUTATIONAL OBSERVATION

Relativity and quantum mechanics have each formalized the role of an observer. Mind is possibly the ultimate observational mechanism and therefore must also have some kind of observational framework. An abstract observational framework can be constructed that allows computational leverage properties.

This framework is built upon Carver Mead's two costs of computation (Mead and Conway 1980), which are: (1) Information is in the wrong place -spatial entropy → must move the data and (2) Information is in the wrong form -logical entropy → must rotate or transform the data. Mead labeled these costs as spatial and logical entropy to suggest a link between physics and information theory. His computational cost model can be intuitively applied to a physical, three-dimensional geometric framework and extended to deal with computational leverage.

Mead's initial idea behind spatial entropy was based on traditional communications theory. This can be expanded to a modern view in which space-time is the backdrop for events that include moving information through both space and time. The model can be expanded even more if the perception of "locality" for distances/times are distorted due to relativistic or quantum mechanisms. An additional mechanism for manipulating locality is to assume the dynamics of a sparse higher-dimensional space. As was mentioned earlier, information, relativity, quantum, and combined theories have all adopted higher dimensional modeling, so it is reasonable to expect higher-dimensional semantics to enter into a high-leverage computational model. This hyperspace model would most likely be sparse because of the desirable properties that arise (Kanerva 1988, Pietsch 1981) that are similar to Wheeler's pregeometric model (Wheeler 1962). Thus, the new expanded spatial entropy theory deals with all of the backdrop issues of dimensionality, geometry, locality, and space-time metrics required of an observational framework.

Mead's logical entropy originally was a conventional algorithmic view of transformation (that is, inputs to outputs) and data rotation (one form to another form). Recognizing that all actions and events can be placed in a space-time backdrop, modern physics deals with rotation and transformation by using inertial frames to formally convert from one perspective to another. Conventional computer science takes a view of transformation costs where the stationary processor implements the algorithm and the data is mobile, whereas physics takes a mobile observer frame and stationary geometric backdrop for events. Inertial frames represent the only concept from physics that matches many of the desirable properties of a mobile observer that are exhibited by psychological research (Targ and Puthoff 1977, Jahn 1982, Monroe 1971, McMoneagle 1993, Moody 1975). A physical theory for inertial frames (that is, they have state and therefore require bits) needs to be created to explain how they can be included in physics (be acted upon) and how they can distort locality in more than one direction (greater than one dimension length contraction). Expanded consciousness experiences demand a mechanism with these properties (Murphy and White 1978).

The ultimate computational leverage is achieved when information is always in the correct location (unlimited locality) and always in the correct form (optimum mobile perspective). In the limit, spatial entropy is minimized in a sparse hyperspace with unlimited locality, and logical entropy is minimized when a mobile observer can choose the optimal perspective of an event backdrop (assuming that cost is not proportional to the size of the backdrop). If such a computational leverage model existed, it would be useful for looking at mental processes.

CONCLUSION FOR TOPOLOGICAL CONSISTENCY

This paper introduced the idea that "real intelligence" of humans may require revolutionary computational leverage due to physical limits of computation within normal three-dimensional space and time. Modern physics theories that are based on observer consistency arguments have already defined many possible avenues for computational leverage based on indirect measurement and extraordinary views of space and time. These models of sparse hyperspacetime form a consistency backdrop for all possible events and all possible observer interactions. Consciousness may be a direct consequence of a dualist model of the mind-brain based on these consistency and computational leverage mechanisms. If the dualist model of the mind exists outside normal space-time, then the mind is akin to a "Gödel machine" that is capable of stepping outside of our normal space-time limits.

Carver Mead's intuitive model of computational costs was expanded to provide an informal model for incorporating the following computational leverage ideas: (1) information is always in the correct place (unlimited locality) and (2) information is always in the correct form (optimum mobile perspective). Similar to quantum mechanical models, this dualist solution does not suffer from homunculus regression (infinite nesting of mind solutions) because the mind is not limited to three-dimensional space but represents a topological consistency in a sparse hyperspacetime. Higher-dimensional models of the mind cannot be faithfully simulated using holograms or neural networks, because such simulations represent the spatial semantics but not the corresponding temporal speedups. Self-evolving conscious mind could emerge from such a hyperspacetime computation framework.

REFERENCES

Dillbeck, M. and C. Alexander. 1989 Higher states of consciousness: Maharishi Mahesh yogi's vedic psychology of human development. *The Journal of Mind and Behavior.* 10(4):307.

Dreyfus, H. 1992. What artificial experts can and cannot do. *AI & Society* 6(1):18–26.

Hawking, S. 1988. *A brief history of time, from the big bang to black holes.* New York, Bantam.

Hawking, S. 1975. Commun. *Math Physics* 43:199.

Jahn, R. G. 1982. The persistent paradox of psychic phenomena: An engineering perspective. *Proceedings of the IEEE* 70(2):136–70.

Kaku, M. 1994. Hyperspace, a scientific odyssey through parallel universes, time warps, and the tenth dimension. Oxford, England: Oxford University Press.

Kanerva, P. 1988. Sparse distributed memory. Cambridge, MA: MIT Press.

Killheffer, R. 1993. The consciousness wars. *Omni* 16(1):50–9.

Landauer, R. 1992. Information is physical. *Proceedings of the Workshop on Physics and Computation.* IEEE Computer Society Press.

Lucky, R. 1989. *Silicon dreams: Information, man, and machine.* New York: St. Martin's Press.

Margolus, N. and T. Toffoli. 1993. "CAM-8: A Computer Architecture based on Cellular Automata." Technical Report of MIT CAM8 group.

Maudlin, T. 1989. Computation and consciousness. *The Journal of Philosophy* 8608:407–32.

McGoveran, D., and P. Noyes. 1989. An essay on discrete foundations for physics. *Physics Essays.* 2(1).

McMoneagle, J. 1993. *Mind trek.* Norfolk, VA: Hampton Roads Publishing.

Mead, C., and L. Conway. 1980 *Introduction to VLSI systems.* Menlo Park, CA: Addison-Wesley, pp. 333–371.

Monroe, R. 1971. *Journeys out of the body.* New York: Doubleday Press.

Moody, R. 1975. *Life after life.* New York: Bantam Books.

Murphy, M. and R. White. 1978. *The psychic side of sports.* Menlo Park, CA: Addison-Wesley.

Ornstein, R. 1969. *On the experience of time.* New York: Pelican Books.

Penrose, R. 1979. "Singularities and time asymmetry." In *General relativity: An Einstein centenary survey,* edited by S. W. Hawking and W. Israel. Cambridge, England: Cambridge University Press.

Pietsch, P. 1981. *ShuffleBrain: The quest for the hologramic mind.* Boston: Houghton Mifflin.

Rindler, W. 1977. *Essential relativity: Special, general, and cosmological.* New York: Springer-Verlag.

Schiffer, M. 1992. The Interplay between Gravitation and Information Theory. *Proceedings of the Workshop on Physics and Computation.* IEEE Computer Society Press.

Sheldrake, R. 1971. *A new science of life, The hypothesis of formative causation.* Los Angeles: J. P. Tarcher Publisher.

Shor, P. 1994. Algorithms for quantum computation: Discrete log and factoring. Proceedings of the 35th Annual Symposium on the Foundations of Computer Science. Los Alamitos, Calif.: IEEE Computer Society Press, p. 124.

Targ, R., and H. Puthoff. 1977. *Mind reach.* New York: Dell Publishing.

Unruh, W. G. 1976. Notes on black-hole evaporation. *Phys. Rev.* D14:870.

Wheeler, J. 1989. "It From Bit." In *At home in the universe.* Woodbury, NY: AIP Press, pp. 295–311.

Wheeler, J. 1962. *Geometrodynamics.* New York: Academic Press.

41 A Mathematical Strategy for a Theory of Consciousness

Saul-Paul Sirag

INTRODUCTION

Ordinary reality, objectified by the methods of measuring space, time, and matter, is a subrealm of a larger reality. This is an ancient idea—at least as ancient as Plato's cave story in which prisoners are chained in such a way that they identify themselves with their own shadows on the cave wall. It is very clear that Plato meant to imply that the larger reality is hyperdimensional; that is, although we tend to identify ourselves with our three-dimensional bodies, there is a higher-dimensional realm in which our three-dimensional bodies are like shadows. Interpretation of the cave parable is augmented by Plato's motto for his Academy: "Let no one enter here without geometry" (Hinton 1904, 1980).

UNIFIED FIELD "THEORY OF EVERYTHING"

The idea that reality is hyperdimensional is entertained today by physicists attempting to unify all the physical forces in a unified field theory: a "theory of everything." If one looks closely at the mathematics of such possible theories, it becomes plausible that a theory of consciousness will be included in "everything." In fact mathematicians have, in recent decades, been engaged in a kind of unified theory of mathematics. And it is the structure of this mathematical unification that, wittingly or not, physicists have drawn on to develop various versions of a unified field theory, especially the currently popular *superstring theories*. The unified mathematical structures and the way in which they are used in physics are strongly suggestive of a theory of consciousness.

The forces to be unified are: electromagnetism, the two nuclear forces (weak and strong), and gravity. In order to include gravity in the scheme, there must be a unification of quantum mechanics with general relativity, which is Einstein's theory of gravity. This has turned out to be a very tall order; and physicists have been forced to use deeper mathematical techniques. This, however, is an old game in physics. Newton unified terrestrial gravity with celestial mechanics into a universal gravity by discovering a deeper mathematics, the calculus. Maxwell unified electricity and magnetism by using a new mathematics, vector analysis. In this case something entirely unexpected came out of the unification: the electromagnetic theory of light, which implied also a vast spectrum of invisible light. More recently, electromagnetism and the weak force have been unified by a new mathematical technique, which physicists call "gauge" theory, but mathematicians call "fiber bundle" theory (Bleecker 1981, Moriyasu 1983). It is quite likely that a deep understanding of mass will come out of this theory, with consequences as far-reaching as Maxwell's theory of light.

Since 1978 physicists have been developing superstring theory primarily as a quantum gravity theory, but with promise to entail all the other force particles and matter particles (Schwarz 1985, 1986). Several new mathematical techniques are used in this theory. Perhaps most crucially, the technique of "anomaly cancellation" is used to prove that the theory is self-consistent. Mathematically speaking, the physicists are using an "error-correcting-code lattice" without necessarily being aware of it (Goddard and Olive 1985, Conway and Sloan 1988). Will something new come out of this theory? I propose that a theory of consciousness, in fact a theory of a *spectrum of consciousness states,* will emerge.

REFLECTION SPACE

Let us put three facts side by side:

1. According to quantum theory, only eigenvalues (of measurement operators) are observable. From this postulate all of basic quantum theory may be derived (Dirac 1967).

2. In unified field theory, the fundamental eigenvalues constitute the vertices of hyperdimensional crystallographic structures residing in a certain space, which I call a *reflection space,* because mathematical reflections occur in such space in such a way as to map crystallographic structures onto themselves. Mathematicians call this space the dual space of a Cartan subalgebra of a Lie algebra (Humphreys 1972, Bleecker 1981, Cahn 1984).

3. The big mystery of quantum theory is the "measurement problem": what is the process, if any, by which an eigenvalue emerges in an observation? Some eminent theoreticians have speculated that consciousness is somehow entailed in this process (von Neumann 1955, Wigner 1967, Penrose 1989).

In light of these three facts, a reasonable research strategy would be to study the mathematics of reflection spaces, and then to see what this study suggests about the nature of consciousness.

The first thing one discovers about reflection spaces is that they were completely classified by the Canadian mathematician H. S. M. Coxeter in 1934. He did this by devising an infinite series of graphs made up of nodes and connecting lines, each such graph corresponding to a different reflection space. Then he proved that these are the only possible reflection spaces (Coxeter 1973). These graphs are called Coxeter graphs (Figure 41.1).

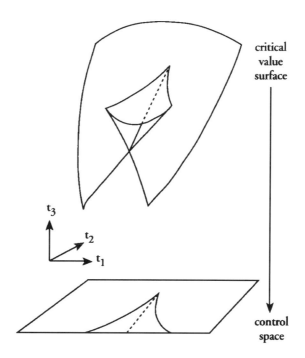

Figure 41.1 The "A_3 catastrophe" pictured via the real critical value surface embedded in the three-dimensional base space R^3 (from Sirag 1993 in Mishlove's *The Roots of Consciousness*).

Independently, the Russian mathematician, E. B. Dynkin (in 1947) devised the same set of graphs (now called Dynkin diagrams) to classify the Lie algebras (Dynkin 1962).

In the 1960s the French mathematician, Rene Thom developed a theory he called catastrophe theory to deal with rapid changes caused by small perturbations of a dynamic system (Figure 41.2). He described seven types of such elementary catastrophes. Thom's motivation for developing this theory was primarily to describe changes in living systems. He thought that these structures would also be used to deal with the nature of consciousness: "There is no doubt it is on the philosophical plane that these models have the most immediate interest. They give the first rigorously monistic model of the living being, and they reduce the paradox of the soul and body to a single geometric object" (Thom 1975). A proposal for such a single geometrical object had previously been made by the philosopher C. D. Broad and the psychiatrist J. R. Smythies (Smythies 1969).

In 1968, the Russian mathematician V. I. Arnold used a subset of the Dynkin diagrams (the A–D–E graphs) to classify all "simple" catastrophes (Arnold 1981). This is an infinite set of catastrophes, which contains Thom's elementary catastrophes as a subset. As might be expected, there is a very intimate relationship between a reflection space and a catastrophe. In fact the "control space" of a catastrophe is derived from a reflection space by

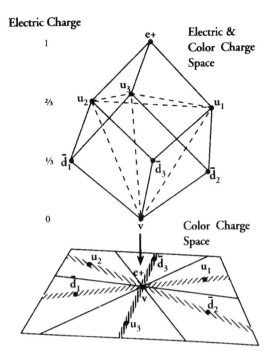

Figure 41.2 The A_3 reflection space projects down the A_2 reflection space, so that a tetrahedron (inscribed in a cube) is projected down to a triangle (from Sirag 1993 in Mishlove's *The Roots of Consciousness*).

allowing all possible reflections to occur in the reflection space and thus creating a new space of reflection "orbits" (Figure 41.3).

Since then Arnold and other mathematicians have used the A–D–E graphs to classify more than twenty other types of mathematical objects. All of these mathematical objects are useful in physics, especially unified field theory. I will mention a few:

1. Reflection spaces; reflection = Coxeter = Weyl groups (Coxeter 1973).
2. Lie algebras (and Lie groups) (Dynkin 1962).
3. Gauge theories of particle physics (Sirag 1993).
4. Error-correcting-code (root) lattices (Conway and Sloane 1988).
5. Quantizing (weight) lattices; analog-to-digital conversion (Conway and Sloane 1988).
6. Catastrophes; unfoldings and resolutions of singularities (Arnold 1981).
7. Bifurcations of equilibrium states (Arnold 1981, 1986)
8. Heisenberg algebras (and groups); uncertainty principle (Kostant 1984).
9. Gravitational instantons; twistor structures (Kronheimer 1989, 1990).
10. Wave front evolutions; optical caustics (Arnold 1981, 1986).
11. Generalized braid groups; knots and links (Kaufman 1991).

Lie Algebra Label	Coxeter Graph	Total Number of Mirrors
A_n	0–0–0–... –0 (n nodes)	$(n^2 + n)/2$

D_n	0–0–0–... –0 (n nodes)
	    |
	    0
		$n^2 - n$ (n greater than 3)

E_6	0–0–0–0–0
	    |
	    0
		36

E_7	0–0–0–0–0–0
	    |
	    0
		63

E_8	0–0–0–0–0–0–0–0
	    |
	    0
		120

Figure 41.3 A–D–E Coxeter graphs.

12. Platonics; platonic solids and other regular figures left invariant by the finite subgroups of SU(2). I call these groups McKay groups, because of the astounding correspondence with Lie algebras which he discovered (McKay 1980, Slodowy 1981).

Mathematicians regard each of these (and many other) entities as different ways of seeing into the structure of some vast underlying mathematical object. *It is as if the mathematicians are working as archaeologists digging up some deeply lying city—not a ruin—but a glorious abode for the mind.*

Concerning the A–D–E correspondence of mathematical objects, Arnold says: "At first glance, functions, quivers, caustics, wave fronts and regular polyhedra have no connection with each other. But in fact, corresponding objects bear the same label not just by chance... To easily checked properties of one of a set of associated objects correspond properties of the others which need not be evident at all. Thus the relations between all the A–D–E classifications can be used for the simultaneous study of all simple objects, in spite of the fact that the origin of many of these relations (for example, of the connections between functions and quivers) remains an unexplained manifestation of the mysterious unity of all things" (Arnold 1986).

It has long been suspected that mental and physical properties are different aspects of an underlying whole (d'Espagnat 1976). The study of the reflection spaces afforded by the A–D–E correspondences shows how vast this underlying whole might be.

Penrose has speculated that the emergence of eigenvalues is related to the action of gravity, and this action is the key to the understanding of consciousness (Penrose 1989). This suggests that a unified field theory that entails gravity would have relevance for a theory of consciousness. Penrose is best known for his invention of twistor theory as an approach to quantum gravity. The superstring theory version of quantum gravity has since

1984 been more extensively developed. Edward Witten (one of the deepest thinkers in this field of quantum gravity) has proposed that superstring theory can be advanced by clarifying the relationships between twistors and superstrings (Peat 1988).

As might be expected, the A–D–E correspondences can be used for this clarification. In superstring theory, the basic entity is a two-dimensional sheet that "vibrates" in a ten-dimensional space-time. This sheet has the overall topology of the surface of a many holed doughnut, and this "worldsheet" is the string theory analog of the intersecting worldlines of point particles. The vibrational modes of the worldsheet are supposed to correspond to the particle states of the unified field theory. The manner of this correspondence is a research topic in superstring theory (Green and Gross 1986).

Here, using hints from the A–D–E correspondence, I consider a complexified version of the worldsheet. To every A–D–E graph corresponds a complex catastrophe bundle in which the base space is of (complex) dimension equal to the number of nodes in the graph and each fiber is a two-dimensional complex space. The fiber undergoes various deformations as different points in the base space are specified (Arnold 1981, Gilmore 1981). The deformations of the fiber are the analog of the vibrations of the string worldsheet. The total catastrophe bundle space is the analog of superstring space-time. The requirements of a ten-dimensional space-time, as well as an appropriate mapping to the known particle states, should single out one of the A–D–E graphs.

For reasons I have given in detail elsewhere (Sirag 1982, 1989, 1993), I favor the E7 graph in particular as the appropriate descriptor of unified field theory. In this case, the identity fiber is C2/OD, that is, the set of OD orbits on C2 (where OD is the 48-element octahedral double group). This identity fiber has a singularity (like a kink) that can be unfolded by perturbations specified by choosing points in C7/W, where C7 is the seven-dimensional complex reflection space of E7, and W is the E7 reflection group. This means that C7/W is the set of W orbits in C7. Since W is the E7 reflection group, C7/W is derived by allowing all the reflections on all the points of C7 to occur. As a result of this process, the reflection hyperplanes (six-dimensional) are transformed into the discriminant surface (six-dimensional) in C7/W. This surface is "thin" in the C7/W space and a path across this surface (a tiny change) corresponds to a radical change in the deformation of the fiber.

The deformations of the fibers can be described in many ways. A recent aid to this description is the development of gravitational instanton theory, which is geometrically related to twistor theory (Kronheimer 1989, 1990). A gravitational instanton is a solution to Einstein's gravitational field equations for an empty universe (the vacuum state) with complex time. For each A–D–E graph there is such a solution; that is, a four-dimensional space-time with a single "end" at infinity, which looks like C2/g, where g is some McKay group. In the E7 case this is C2/OD. The fields which live on this space-time correspond to the structure of the OD group algebra C[OD], and in fact this algebra may be viewed as a set of "harmonics" on the instanton space. I propose (Sirag 1989, 1993) that the unitary elements in C[OD] constitute an appropriate unified gauge theory for all the forces:

U(2)xT6—a 10-dimensional space-time
    U(1)—electromagnetism [standard model]
    SU(2)—weak force [standard model]
    SU(3)—strong (color) force [standard model]
    SU(3)—hyperweak force [a prediction]
    SU(4)—gravity gauge force acting on T6 [a prediction]
    SU(2)—gravity gauge force acting on U(2) [a prediction]

This accounts for all the unitary elements in C[OD]. It is implicit in this view that the ten-dimensional space-time is part of the harmonic structure of the underlying instanton space. Note: these gravity gauge groups correspond to spin-1 gauge particles, 15 for SU(4), and 3 for SU(2). It is plausible for the standard spin-2 gravitons to be composites of these gravity gauge particles, which I call anandons for Jeeva Anandan, who proposed such composite gravity particles (Anandan 1980).

Since U(2) = U(1) × SU(2) and T6 consists of a product of 6 U(1) groups, we can transform U(2) × T6 into SU(2) × T7. Now SU(2) is geometrically a three-sphere, while the torus T7 corresponds to the seven nodes of the E7 graph. Thus T7 is closely related to the reflection space of E7. In fact T7/W, where W is the E7 reflection group, has the same structure as the base space C7/W of the E7 catastrophe.

Remember that the E7 reflection space contains the eigenvalues for the E7 measurement operators. These operators form a basis for the seven-dimensional reflection space, so that there is a close relationship between these operators and T7 in C[OD]. There are, of course, many more relationships that tie these various structures together. The key linking structure is the E7 graph.

DISCUSSION

The relationships tied together by this graph are suggestive of a universal consciousness. Since eigenvalues are the primitive observables, they must correspond to fundamental qualia (or sensations). The reflection group W acts on the E7 reflection space and thus on the eigenvalue crystallographic polytopes—changing one vertex into another. Each such change corresponds to basic quantum jump. Thus the fundamental processes of consciousness are the reflections. Each such reflection (jump) is presumably accompanied by a minute "blip" of awareness. The geometry afforded by the many structures described by the E7 graph suggests that these blips of awareness would be available at every point of space-time. This implies that a nervous system is a means of experiencing many myriads of these blips in a coherent fashion in each moment of time. This view of the function of a nervous system is similar to the detailed model developed by James Culbertson (Culbertson 1963, Herbert 1993).

The coherence of the experience of consciousness (for sentient beings) may be due to some macroscopic quantum system in the body of the experiencer (Walker 1970, Penrose 1989, Herbert 1993, Goswami 1993).

The E7 graph can be used to generate the Hamming-7 error-correcting code. This suggests that cognitive aspect of consciousness may be related to coding theory, which has deep connections to invariant theory, and analog-to-digital transforms. The Hamming-7 code is a subcode of the Hamming-8 code which is generated by the E8 graph (Conway and Sloane 1988). It is the properties of this code that provide for anomaly cancellation in the E8xE8 superstring theory. This raises the question: why is the E8 graph not the preferred descriptor of deep reality?

CONCLUSION

In general the hierarchical nature of the entire A–D–E classification scheme implies the embedding of lower dimensional structures in higher dimensional structures. This suggests that there must be a *hierarchical spectrum of reality structures: a spectrum of states of consciousness*. Each realm would have a describable physics, but the "higher" physics would to a certain extent contain the "lower" physics. If ordinary physics corresponds to the E7 realm, there is a realm "above" ordinary reality, E8; and there is a realm "below" ordinary reality, E6. It is noteworthy that in the A–D–E hierarchy there are only these three E realms. Moreover, the three E realms are the only doorways to the "nonsimple" structure beyond the A–D–E hierarchy (Arnold 1981). *There is a vast beyond.*

REFERENCES

Anandan, J. 1980. On the hypotheses underlying physical geometry. *Foundations of Physics* 10:601–29.

Arnold, V. I. 1981. *Singularity theory.* Cambridge, England: Cambridge University Press.

Arnold, V. I. 1986. *Catastrophe theory.* Berlin: Springer-Verlag.

Bleecker, D. 1981. *Gauge theory and variational principles.* Reading, MA: Addison-Wesley.

Cahn, R. N. 1984. *Semi-simple Lie algebras and their representations.* Menlo Park, CA: Benjamin/Cummings.

Conway, J. H., and N. J. A. Sloane. 1988. *Sphere packings, lattices and groups.* New York: Springer-Verlag.

Coxeter, H. S. M. 1973. *Regular polytopes.* Third ed. New York: Dover.

Culbertson, J. T. 1963. *The minds of robots.* Urbana: University of Illinois Press.

D'Espagnat, B. 1976. *Conceptual foundations of quantum mechanics.* Reading, MA: W.A. Benjamin.

Dirac, P. A. M. 1967. *Quantum mechanics,* Fourth ed. Oxford, England: Oxford University Press.

Dynkin, E. B. 1962. The structure of semisimple Lie algebras. *Trans. Amer. Math. Soc Series 1* 9:328–469.

Gilmore, R. 1981. *Catastrophe theory for scientists and engineers.* New York: Wiley-Interscience.

Goddard, P., and D. Olive. 1985. "Algebras, lattices and strings." In *Vertex operators in mathematics and physics,* edited by J. Lepowsky, S. Mandelstam, and I. M. Singer. New York: Springer-Verlag.

Goswami, A. 1993. *The self-aware universe.* New York: Putnam's.

Green, M., and D. Gross. (eds.). 1986. *Unified string theories.* Singapore: World Scientific.

Herbert, N. 1993. *Elemental mind.* New York: Dutton.

Hinton, H. C. 1904. *The fourth dimension.* New York: Allen and Unwin.

Hinton, H. C. 1980. *Speculations on the fourth dimensions.* New York: R. Rucker, Dover.

Humphreys, J. E. 1972. *Introduction to lie algebras and representation theory.* New York: Springer-Verlag.

Kauffman, L. H. 1991. *Knots and physics.* Singapore: World Scientific.

Kostant, B. 1984. On finite subgroups of SU(2), simple lie algebras, and the McKay correspondence. *Proc. Natl. Acad. Sci. USA* 81:5275–77.

Kronheimer, P. B. 1989. A Torelli-type theorem for gravitational instantons. *J. Differential Geometry* 29:685–97.

Kronheimer, P. B. 1990. Instantons and the geometry of the nilpotent variety. *J. Differential Geometry* 32:473–90.

McKay, J. 1980. Graphs, singularities, and finite groups. *Proc. Symp. in Pure Math* 37:183–6.

Moriyasu, K. 1983. *An elementary primer for gauge theory.* Singapore: World Scientific.

Peat, F. D. 1988. *Superstrings and the search for the theory of everything.* Chicago: Contemporary Books.

Penrose, R. 1989. *The emperor's new mind.* Oxford, England: Oxford University Press.

Schwarz, J. H. 1985. "From the bootstrap to superstrings." In *A passion for physics,* edited by C. DeTar, J. Finkelstein, and C.-I Tan. Singapore: World Scientific.

Schwarz, J. H. 1986. "Lectures on superstring theory." In *Physics in higher dimensions,* edited by T. Piran and S. Weinberg. Singapore: World Scientific.

Sirag, S.-P. 1982. Why there are three fermion families. *Bulletin of the Amer. Phys. Soc.* 27(1):31.

Sirag, S.-P. 1989. A finite group algebra unification scheme. *Bulletin of the Amer. Phys. Soc.* 34(1):82.

Sirag, S.-P. 1993. "Consciousness: A hyperspace view." Appendix in *Roots of consciousness,* Second ed., by Jeffery Mishlove. Tulsa, OK: Council Oak.

Slodowy, P. 1983. "Platonic solids, kleinian singularities and lie groups." In *Algebraic geometry,* edited by I. Dolgachev. Berlin: Springer-Verlag.

Smythies, J. R. 1969. "Aspects of consciousness." In *Beyond reductionism,* edited by A. Koestler and J. R. Smythies. New York: Macmillan.

Thom, R. 1975. *Structural stability and morphogenesis*. Reading, MA: W. A. Benjamin.

Walker, E. H. 1970. The nature of consciousness. *Mathematical Biosciences* 7:131–78.

Wigner, E. 1967. *Symmetries and reflections*. Bloomington: University of Indiana Press.

von Neumann, J. 1955. *Mathematical foundations of quantum mechanics*. Princeton, NJ: Princeton University Press.

42 Nonlocality on a Human Scale: Psi and Consciousness Research

Mario Varvoglis

Consider the following statements:

It is impossible for a person to perceive a physical event or a material thing except by means of sensations which that event or thing produces in his mind . . .

It is self-evidently impossible that an event should begin to have any effects before it has happened . . .

It is impossible for an event in a person's mind to produce directly any change in the material world except certain changes in his own brain . . .

These are perfectly reasonable assertions. The philosopher Broad (1960) calls them "basic limiting principles" of modern western thought, but to most of us they look like just plain common sense. Then again, according to a number of surveys, most of us also believe in the existence of "psychic" or psi phenomena, which clearly violate these common-sense principles. In clairvoyance, for example, a person seems to acquire information about events or things without any sensory input; likewise, in telepathy we seem to perceive others' thoughts or feelings at a distance. In precognition, insofar as people acquire noninferential information about the future, the event appears "to have . . . effects before it has happened." Finally, in psychokinesis (PK), the mind apparently produces changes in the material world directly, without mediation of the motor system or any known physical energies.

Some scientists grit their teeth at the mere mention of such possibilities. But it is crucial we look beyond received opinion and dogma, and honestly examine psi research. If these phenomena are real, they seriously challenge predominant assumptions concerning consciousness and its relationship to physical reality; we simply cannot afford to ignore their potential implications. Thus, the primary goal of this chapter is to provide an overview of contemporary experimental parapsychology, presenting major research directions and results, and references for more detailed scrutiny.

THE BIG PICTURE: META-ANALYSIS

Investigations of psi have always been hampered by one major obstacle: its rarity. Two explanations of this have been popular among researchers: psi depends upon a rare convergence of particular psychological, genetic, or environmental circumstances, found only in a few gifted persons; and psi is much more common than we think, but tends to be overwhelmed by normal sensorimotor activity, or blocked by cognitive or affective filters. Research strategies depend on which of these two views is adopted.

"Elitist" approaches seek to combat the rarity of psi by locating and testing gifted subjects. The objective is to obtain clear-cut demonstrations of psi, under perfectly controlled conditions, so as to establish its reality once and for all. Yet, it is increasingly being recognized that, given the predominant intellectual climate, no such demonstration can be definitive. As long as the experiment depends upon a particular subject, sooner or later it will be unrepeatable. No matter how well-controlled the original demonstration, subsequent scientists will view it as just another claim; they will remain unconvinced, pending new demonstrations.

The "universalist" approach takes as premise the idea that psi is a poorly developed, but generally distributed ability; it seeks to define protocols that can statistically detect weak psi "signals" across whole subject populations. This approach does not depend upon a gifted subject or "crucial" experiment. Researchers do not expect incontrovertible evidence for psi in any single trial, nor even in the averaged results of every experiment: there are simply too many unknown or uncontrollable variables to hope for total replicability. Nevertheless, the explicit goal of universalist research is to determine predictors of positive psi performance, and develop experimental procedures that progressively enhance repeatability. It is only in the past decade, with the introduction of meta-analysis, that it has become apparent how far the field has advanced toward this goal.

Meta-analysis is a means for quantitatively summarizing experimental results across a large number of studies; it is increasingly coming to replace traditional literature reviews in the social and behavioral sciences, psychosomatic medicine, and other areas. Essentially, the procedure consists in locating all studies pertinent to a specific research domain, coding studies' characteristics, cumulating results, and quantitatively evaluating the robustness of the database. Effect sizes (standardized estimates of the degree to which a phenomenon is present in the population) can thus be based on large databases rather than isolated studies. Cross-study comparisons facilitate identification of common moderator variables, and assessment of the homogeneity of findings. The coding of methodological weaknesses using explicit criteria reduces subjectivity in the evaluation of results; and databases can be reassessed once purged of heavily flawed studies. We can thus form a rather clear-cut picture as to how solid is the evidence for a particular claim, within a given dataset.

In parapsychology, a number of meta-analyses have focused on investigations of extended interactions between a person and the mental processes of another person; the physiological state of another person or nonhuman organisms; and the state of some inanimate material system. For convenience, these are labeled Person-Person, Person-Organism, and Person-Matter exchanges.

EXTENDED PERSON-PERSON EXCHANGES

Many kinds of telepathy experiments have been performed, but I would like to focus on the "ganzfeld" studies, in which I have been intensely involved myself, and which are considered among the most promising in the field, having yielded consistent results over a 20-year period. In these studies, a "receiver" is placed in a mildly altered state of consciousness by means of the ganzfeld: a sensory deprivation procedure in which translucent hemispheres are put over the eyes, and headphones with "white noise" over the ears. While a remotely located "sender" tries to mentally transmit visual elements of a randomly selected target-picture, the receiver reports any spontaneously arising images and thoughts; the experimenter, who is blind as to the target, takes notes. At the end of a fixed ganzfeld period, the receiver is presented with four pictures, and must select the one which seems to correspond best to his or her imagery. By the null hypothesis, then, the mean success rate should be one in four, or 25 percent.

As part of a background paper commissioned by the National Academy of Sciences, Rosenthal (1986) conducted a meta-analysis of 28 ganzfeld studies, completed between 1974 and 1981 using a uniform statistic ("direct hits"). These studies were conducted at ten different laboratories, and included 835 sessions contributed by 589 subjects. Six of ten laboratories produced significant results; 12 of the 28 studies were statistically significant, and 23 of the 28 studies yielded positive scores. The composite z-score (Stouffer z) is 6.60, corresponding to a $P<10^{-9}$ (that is, the probability that this cumulative score is due to chance is less than 1 in 1,000,000,000). The mean hit rate was 38 percent (as compared to the expected 25 percent), which translates to an overall effect size of 0.28.

Starting in 1982, and up through the closing down of our laboratory in 1989 (for lack of funds), Honorton *et al.* (1990) undertook a series of "auto-ganzfeld" studies, involving largely automated protocols (computer controlled target-selection and display, automated data storage); this series was designed to take into account various criticisms expressed vis-à-vis earlier studies. Over the course of 11 experimental series, involving 8 experimenters and 241 receivers, a total of 355 sessions were conducted. Positive scores were obtained in 10 of the 11 studies. The mean hit rate was 34 percent ($Z = 3.89$, $P = 0.00005$); the mean effect size was 0.29. In short, the introduction of rigorous controls did not diminish results; evidence for psi remained at about the same level as in earlier research. (For more details, see Bem and Honorton 1994.)

EXTENDED PERSON-ORGANISM EXCHANGES

"Bio-psychokinesis" (bioPK) studies can be viewed as experimental attempts to assess the claim that some people can affect others' health by mental means ("psychic healing"). Since the mid-1960s, over 150 studies have accumulated implicating a wide range of biological target materials (from microorganisms and plants to animals and humans); over half of these have yielded statistically significant results. So far, however, no meta-analysis of the entire database has been attempted, largely because of the lack of homogeneity in experimental conditions and controls. Here I will present a meta-analysis of a subset of bioPK studies, involving uniformly rigorous controls and quite homogeneous protocols. In these experiments, the target and agent are located in nonadjacent rooms, and target system activity is automatically recorded. While receiving real-time feedback from the target, the agent is instructed to attempt to influence it, according to a randomly determined influence/noninfluence (control) schedule; success is evaluated by comparing target activity levels during experimental versus control epochs.

Braud and Schlitz (1991) report 37 such bioPK experiments, involving 13 experimenters and 153 subjects, who contributed a total of 655 sessions. A range of biological materials served as PK targets: the activity level of different animals, in vitro cellular preparations, and indices of other persons' physiological activity. Significance was achieved in 57 percent of these experiments; the meta-analysis yields a combined $z = 7.72$, $P = 2.58 \times 10^{-14}$, with a mean effect size of 0.33.

EXTENDED PERSON-MATTER EXCHANGES

In this group of experiments, the physical target is an inanimate system that generates random events or numbers mechanically (for example, a card-shuffler) or electronically (hardware random number generators or RNGs). The subject is instructed to guess the future output of the system (precognition), or to attempt to influence it (microPK). It is not always clear which of these two processes is really at work; the main issue here is whether the system's output remains random, or tends to match subjects' guesses or intentions.

A meta-analysis by Honorton and Ferrari (1989) examines all precognition studies between 1935–1987. Time intervals between subjects' guesses and the subsequent random event ranged from under a second to a year; targets ranged from ESP cards to RNGs. A total of 309 studies were located, involving 62 investigators, 50,000 subjects, and nearly 2,000,000 trials. About a third of the studies was significant at the 0.05 level ($z = 11.41$, $P = 6.3 \times 10^{-25}$). Effect sizes—considerably smaller than in the ganzfeld and bioPK research—remained fairly constant across the years, while study quality ratings improved (reflecting the shift to automated protocols). Results were significantly stronger for short time intervals, suggesting that the accuracy of precognition may decline with time.

Radin and Nelson (1989) report a meta-analysis of research exploring correlations between RNG outputs and intention. They located 597 experimental and 235 control studies (about a dozen reports in this database overlap with the precognition database). The bulk of these studies, involving a total of 68 investigators, have been conducted over the past 25 years. Control studies showed a zero effect size, confirming the randomness of the

RNGs. The experimental studies yielded an effect size significantly superior to that of the control studies (z = 4.1); however, as in the precognition studies, this effect is quite small. Nevertheless, the combined result of all experimental studies, representing over 400 million random events, is substantial (z = 15.58, P = 1.8×10^{-35}).

MODERATOR VARIABLES IN PSI RESEARCH

A number of meta-analyses involve research exploring variables that enhance or suppress psi. Schechter's (1984) meta-analysis of 20 hypnosis/ESP studies from 10 laboratories points to significantly superior ESP results in hypnosis versus controls conditions in 7 of the studies (P = 0.000034), and similar trends in 9 of the remaining 13. A meta-analysis by Honorton, Ferrari, and Bem (1990), involving 14 studies and over 600 subjects, shows positive correlations between extraversion and free-response ESP scores (z = 4.82, P = 0.0000015). A meta-analysis by Lawrence (1993) confirms the well-known "sheep-goat" effect: in 73 studies by 37 investsigators, believers in ESP consistently score higher than those who do not believe in it (z = 8.17, P = 1.33×10^{-16}). Finally, in a meta-analysis of 13 studies relating ESP scores to the Defense Mechanism Test, Watt (1991) found consistently superior ESP scores in low- over high-defensive subjects (z = 4.55, P = 0.000006); this confirms findings with other openness/defensiveness measures.

SUMMARY AND CONCLUSION

Meta-analytic tools show that psi, though subtle, can be demonstrated in both person-person and person-organism interactions with fairly high replicability rates and moderate effect sizes. They also can be shown in interactions with mechanical or electronic random systems, though with smaller effect sizes. These meta-analyses provide no support for the common skeptical argument that evidence for psi is based on a few flawed studies. Positive results are widespread, implicating broad subject populations and a large number of independent laboratories. Effects cannot be attributed to selective reporting, faulty statistics, or a few suspect subjects. Study quality ratings and outcome are not related; methodologically tight experiments are as likely to yield positive results as looser ones. The data show conceptual cohesiveness and intuitively plausible relationships to several state and trait factors. We are thus nearing the stage where we can define reliable predictors of positive laboratory psi performance.

Recent studies by Radin (1993) illustrate the promise of interdisciplinary approaches to psi. Using neural network pattern-recognition techniques, Radin showed that subjects engaging in microPK trials structure RNG outputs in unique ways: the network was able to identify person-specific "signatures" in the RNG data.

Braud and Schlitz's (1991) bioPK research, using a person's physiological activity to detect another's remote influence, constitutes another promising direction for interdisciplinary consciousness research. The addition of such bioPK methodologies to studies in psychosomatic medicine or psychoneuroimmunology could clarify the potential role of psi in the recovery from disease. Similarly, investigations of mental and physiological states that facilitate or block psi could help us understand the dynamics that shift consciousness from the usual "local" mode of functioning to extended, nonlocal interactions. These latter studies could be greatly facilitated by the use of modular software/hardware packages, incorporating automated psi-tasks and procedures for inducing different attentional states (Varvoglis 1992).

In our search for "a scientific basis for consciousness," the primary, most fundamental issue is whether or not consciousness is, in some nontrivial sense, real. Most neuroscience tacitly or explicitly aims to show that consciousness is not real: it is either a fancy term describing certain kinds of brain processes, or else, an epiphenomenon, distinct from brain activity, but of no causal consequence. Psi research seriously challenges the adequacy of both reductionist and epiphenomenalist frameworks; it suggests that consciousness is more than an incidental byproduct of brain mechanism. Going beyond theoretical arguments as to the role of consciousness in quantum mechanics (for example, by Wigner and Von Neumann), psi research explicitly demonstrates the existence of correlations between mental states and remote inanimate, organic, and human systems. Whether looked at in causal or in informational terms, these results suggest that consciousness, far from being an epiphenomenon, is an active source of nonlocal entropy reduction (Schmidt 1982, Costa de Beauregard 1985).

In this context, it is worth mentioning a related epistemological issue. To the extent to which it is possible to transcend the usual sensory or inferential means of knowledge acquisition, or to directly influence inanimate or organic systems, the common safeguards of experimental research (for example, double-blinds) do not guarantee "observer-free" data. If it is reasonable to assume that scientists, like most of the population, have some latent psi capacities, then it is likely that they shape certain experimental findings through subtle "expectancy effects." In this sense, "objective" data may be telling us as much about consciousness, as about the external world. Or, to put it the other way around: investigations of consciousness may be telling us as much about reality as about the human mind.

REFERENCES

Bem, D., and C. Honorton. 1994. Does psi exist? Replicable evidence for an anomalous process of information transfer. *Psychological Bulletin* 115:4–18.

Braud, W., and M. Schlitz. 1991. Consciousness interactions with remote biological systems: Anomalous intentionality effects. *Subtle Energies* 2:1–46.

Broad, C. D. 1960. *The mind and its place in nature.* New Jersey: Littlefield, Adams & Co.

Costa de Beauregard, O. 1985. On some frequent but controversial statements concerning the Einstein-Podolsky-Rosen Correlations. *Foundations of Physics* 15(8): 871–87.

Honorton, C., R. Berger, M. Varvoglis, M. Quant, P. Derr, E. Schechter, and D. Ferrari. 1990. Psi communication in the ganzfeld: Experiments with an automated testing system and comparison with a meta-analysis of earlier studies. *Journal of Parapsychology* 54: 99–139.

Honorton, C., and D. C. Ferrari. 1991. "Future telling": A meta-analysis of forced-choice precognition experiments, 1935–1987. *Journal of Parapsychology* 53: 281–308.

Honorton, C., D. Ferrari, and D. Bem. 1990. Extraversion and ESP performance: A meta-analysis and new confirmation. *Proceedings, 33rd Annual PA Convention*, pp. 113–25.

Lawrence, T. 1993. A meta-analysis of forced choice sheep-goat studies 1947–1993. *Proceedings, 36th Annual PA Convention*, pp. 75–86

Radin, D. 1993. Neural network analysis of consciousness-related patterns in random sequences. *Journal of Scientific Exploration* 7: 355–74.

Radin, D., and R. Nelson. 1989. Evidence for consciousness-related anomalies in random physical systems. *Foundations of Physics* 19: 1499–514.

Rosenthal, R. 1986. Meta-analysis and the nature of replication: The ganzfeld debate. *Journal of Parapsychology* 50: 315–66.

Schmidt, H. 1982. Collapse of the state vector and psycho-kinetic effect. *Foundations of Physics* 12: 565–81.

Schechter, E. 1984. Hypnotic induction vs. control conditions: Illustrating an approach to the evaluation of replicability in parapsychological data. *Journal of the American Society for Psychical Research* 78: 1–29.

Varvoglis, M. 1992. *La rationalité de l'irrationnel*. Paris: Inter-Editions.

Watt, C. 1991. Meta-analysis of DMT-ESP studies and an experimental investigation of perceptual defence/vigilance and ESP. *Proceedings, 34th Annual PA Convention*, pp. 395–411.

43 Synchronicity and Emergent Nonlocal Information in Quantum Systems

E. M. Insinna

INTRODUCTION

As C. G. Jung stated almost forty years ago, advances in psychology and, by extension, in consciousness research will probably be achieved through advances in physics and especially quantum physics. This is the reason why in this paper I put an emphasis on Wolfgang Pauli's ideas, Pauli being one of the rare physicists to have attempted a unification between psychology and physics. Pauli's suggestion to view individual events in quantum systems as the possible place for synchronistic manifestations has been neglected for almost forty years. Experimental investigations of Pauli's interpretation have never been attempted. I believe that this has happened for two reasons:

1. Pauli himself did not sufficiently promote his ideas during his lifetime and his theoretical suggestions thus remained in a rather embryonic state.
2. The science of dynamical systems was at Pauli's time not yet sufficiently developed such as to permit the description of new significant parallels existing between the world of the psyche and some basic physical phenomena (and related mathematical concepts).

In this paper, I attempt not only to rehabilitate Pauli's ideas connecting synchronicity and quantum mechanics but I also suggest to consider synchronicity from an experimental viewpoint. In the following I will:

1. Expose Jung's theory of synchronicity.
2. Reconsider Pauli's interpretation of Bohr's original concept of complementarity. This will lead us to the measurement problem in quantum mechanics and to a new interpretation of the collapse of the wave function.
3. Reinterpret Jung's concept of archetype in terms of a mathematical attractor and shortly outline some significant similarities between the quantum world and the world of the psyche.
4. Consider the implications of synchronicity relative to consciousness and the emergence of nonlocal information in the human brain.
5. Suggest an experimental approach.

JUNG'S THEORY OF SYNCHRONICITY

Towards the end of the fifties, Jung formulated in rather scientific terms a definition of some peculiar events lying on the borderline between physics and psychology. What he aimed at was to open a new scientific approach for phenomena which until then were classified in the vast and multifarious domain of "parapsychology." In those phenomena we observe the more or less simultaneous occurrence in space and time of a psychic event with a physical event. Matter mirrors the psychic event such as to become a meaningful representation of the original exclusively psychic experience. Concretely speaking, when we experience a psychic event like a premonitory dream or vision that becomes real a few days later or when we have a vision of something happening at a distant place, we are confronted with the features of synchronicity. Those features may be resumed as follows:

1. Synchronistic events are a-causal manifestations; that is, they are not explicable in terms of physical causality. We cannot invoke any (physical) causal relationship between a premonitory dream and the physical event occurring afterwards. The former psychic phenomenon manifestly transcends space and time for, in synchronistic events, space and time become relativized and lose their usual features.

2. The two independent causal chains (psychic and physical) become connected through the emergence of meaning. The observer is placed in a middle position that allows him to establish a connection between a psychic event and a physical occurrence. In the absence of the psychic content, the physical experience would have been considered as a mere meaningless chance event (Figure 43.1).

3. Psyche and matter show a significant correlation that suggests an intrinsic wholeness in which the dichotomy psyche-matter is abolished.

According to Jung, synchronistic events occur through the constellation or activation of an archetype in the most profound layer of the collective unconscious, an undifferentiated domain he calls "unus mundus." The activation of this archetypal layer or "mode," brings about synchronistic

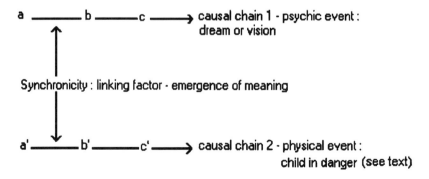

Figure 43.1 Synchronicity as a linking factor between two separate causal chains.

events. This often happens in connection with stress situations where the life of an individual or a collective is in danger. Constellation of an archetype always leads the individual to become "emotionally affected" and thus more sensitive to synchronistic phenomena.

In Jung's synchronicity, the unconscious becomes a universal substrate in which both extremities of being: spirit and matter, reach into an undifferentiated unity (Figure 43.2). In other words, the action of the archetype spans psyche and matter for they are factors affecting reality from beyond the apparent duality of our everyday life.

W. PAULI AND THE COLLAPSE OF THE WAVE FUNCTION

Pauli was confronted with Jung's analytical psychology during more than twenty-five years. Their recently published correspondence witnesses an intense exchange of ideas mainly concerning basic questions touching on both physics and psychology (Meier 1992). It has been the merit of Pauli to try to establish a connection between Jung's world of archetypes—synchronicity—and quantum physics.

Pauli explicitly refers to the measurement problem of quantum mechanics: a well-known situation in which the observer of a quantum system seems to be involved in the measurement's results. Pauli insists that Bohr's concept of complementarity completely eliminated the notion of a detached observer. In what is usually called "the Copenhagen interpretation of quantum mechanics," the observer is fully involved in the "great drama of the world," as Bohr used to say. Pauli went still further. He established a deeper connection between the observer and the observed quantum system.

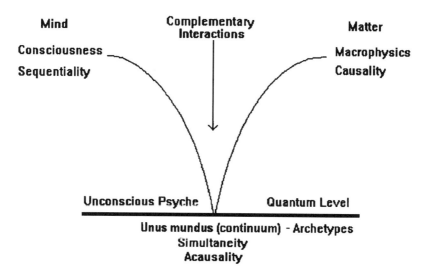

Figure 43.2 The intrinsic unity existing at the quantum/unconscious level progressively diverts and finally appears as separate domains of reality.

We know that, in quantum mechanics, the choice of a specific set-up influences the result of a quantum experiment (emergence of wave or particle features, position of a particle, and so on). However, the observer's influence is in this interpretation, a direct consequence of a conscious choice. This has led J.A. Wheeler (1983) and E. Wigner (1983) to suggest that the phenomenology inherent in quantum systems makes the presence of a conscious observer indispensable. We are confronted here with what physicists call "the collapse of the wave function."

Schrödinger's equation, describing the state of the observed system (the wave function), only gives a theoretical or virtual description of the causal development of the quantum process. It leaves all possibilities open and its square (the probability amplitude) allows only probabilistic predictions about the final issue of the process. Thus, according to the Schrödinger equation, all possibilities of development coexist "in potentia" in the quantum system as pure virtualities. Physicists speak also of a "linear superposition of many possible states of a system."

One striking example of this particular situation has been given by Schrödinger himself in a famous "Gedankenexperiment" (thought experiment). A cat is placed in a box containing a vial with a poisonous substance. The latter is liberated when a particle emitted by a radioactive source hits a detector connected to a hammer. The issue of the experiment cannot be predicted as long as the observer does not open the box and look at the cat. In fact, according to Schrödinger's equation the cat has as many chances to be alive as to be dead. Quantum physics only allows the cat to live in a kind of limbo state in which it is dead and alive at the same time.

Following Schrödinger (and Wheeler and Wigner), only the conscious observation of the system leads to the knowledge of the final state and to the definitive collapse of the wave function, that is, to the realization of a single solution out of the infinite number of virtualities (cat dead or cat alive).

As a complementary interpretation, Pauli instead suggested that what happens in a quantum system might also be connected with the unconscious attitude of the observer. He said:

Modern physics reintroduces the observer as a small god of creation in his microcosm, with the ability of (at least partial) free choice and mostly uncontrollable effects on the observed object. However, if such phenomena depend on how (the experimental condition) they are observed, why couldn't there exist also phenomena (extra corpus) [i.e. in quantum systems] which depend on the person who observes them (i.e. on the psychic quality of the observer)? (Pauli, Letter to Jung of December 23rd, 1947, in Meier 1992)

From this viewpoint, individual quantum events, which quantum physicists try to eliminate by the use of statistical calculations, may sometimes become the place for the emergence of synchronistic occurrences. Synchonistically speaking, this means that the issue of the cat experiment would matter for the observer and would no longer be a meaningless chance event. In a letter to Jung of May 27, 1953, Pauli said more explicitly :

It was in 1931 when I met you personally for the first time. At that time I experienced the unconscious as a new dimension. Shortly after my marriage in 1934 and at the end of my analytical treatment. . . .I had the following dream, with which I was occupied for many years:

A man looking like Einstein draws the following figure on a cardboard:

This stood in a manifest connection with the controversy (with Einstein) and was a kind of answer of the unconscious. It showed quantum mechanics, and official physics at large, as the one-dimensional fraction of a two-dimensional more meaningful world, the second [complementary] dimension of which could only be represented by the unconscious and the archetypes. Today I believe in fact, that eventually the same archetype may manifest itself in the choice of an experimental set-up by the observer as well as in the result of the measurement (similarly to the throwing of dice in J. B. Rhine's experiments). (Pauli, in Meier 1992)

This statement of Pauli implies a widening of Bohr's complementarity concept. Exactly as in the complementary relationship existing between the ego-conscious and the unconscious (which was outlined by Jung in his analytical psychology), in Pauli's interpretation the system under observation may mirror the psychic state of the observer in a complementary fashion. Synchronicity would manifest itself within the individual events occurring in every quantum system. Those occurrences are normally eliminated by statistical calculations.

Pauli's conception of quantum physics is summarized in Figure 43.4. The quantum world would occupy a central position. As a matter of fact, quantum physics only responds to what Pauli called "statistical causality," taking a middle way between causality and synchronicity.

Pauli expressly emphasized the profound irrationality of the individual quantum events that result from every new observation of the system. We quote him once more:

Now there comes the major crisis of the quantum of action: one has to sacrifice the unique individual and the "sense" of it in order to save an objective and rational description of the phenomena. If two observers do

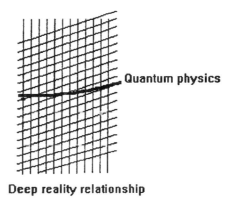

Figure 43.3 Einstein's drawing in Pauli's dream.

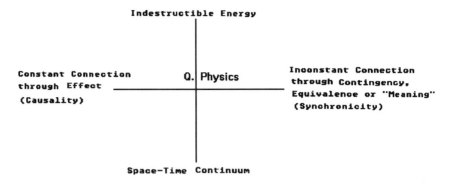

Figure 43.4 Reality scheme according to W. Pauli and C. G. Jung. Quantum physics is represented in the center of the scheme, that is, a position encompassing both aspects of reality.

the same thing even physically it is, indeed, really no longer the same; the physically unique individual (event) is no longer separable from the observer—and for this reason it goes through the meshes of the net of physics. The individual case is "occasio" and not "causa." I am inclined to see in this "occasio" which includes within itself the observer and the selection of the experimental procedure which he has hit upon—a revival of the "anima mundi" which was pushed aside in the 17th century (naturally in an altered form). Here something has remained open which previously appeared to be closed, and it is my hope that new concepts, which are uniformly simultaneously physical and psychological, can force themselves through this gap in place of "parallelism." May "more successful offspring" attain this. (Pauli, Letter to Fierz of October 13th, 1951, in Laurikainen 1988).

In other words, for Pauli the unconscious (the "anima mundi" and the archetypes) can manifest itself in the quantum system under observation in the form of individual (usually random) quantum events that are not included in the statistical considerations.

To sum up, following Pauli, in some special cases the collapse of the wave function does not happen as a consequence of a conscious observation of the system, but as a spontaneous process correlated with the unconscious psychic attitude of the observer! More precisely, Pauli alludes to the fact that the collapse of the wave function may occur outside the field of consciousness of the observer. The quantum phenomenon thus becomes an act of creation in space and time endowed with meaning, in the sense of a *"creatio continua"* (continuous creation). The final state of the system is no longer observer-dependent in the restricted sense of Bohr or J.A.Wheeler and E. Wigner, but in the sense of a fully complementary and spontaneous response of the unconscious to the state of mind of the observer.

In Pauli's interpretation, individual quantum events may thus sometimes become endowed with meaning. That is, in synchronistic events we are sometimes brought into the position to discover that the collapse is correlated with the observer and that the connected macroscopical event is the carrier of a meaning. However, in most cases we are not in such a position and tend to consider irrational (individual) quantum events, that is, the relative macroscopical phenomena, as mere chance occurrences.

A good argument in favor of Pauli's hypothesis of a spontaneous collapse of the wave function are for instance psychosomatic diseases. In fact, our somatic body may be viewed as a complex quantum system. There the collapse of the wave function occurs without our conscious knowledge, and this often leads to unwanted and sometimes lethal manifestations. We sometimes suddenly become ill and unconsciously respond to the psychological conflict we were hiding from ourselves. Were we consciously able to avoid the complementary response of the system, we would all be healthy people and diseases would disappear from the earth's surface.

Pauli's hypothesis finally allows us to relativize the role of the ego conscious with respect to the creation of the universe. According to Wheeler and Wigner, ego consciousness is an indispensable factor and its intervention corresponds to a conscious act of creation. In contrast to this interpretation one might instead envisage that the entire creation is perhaps a synchronistic event that occurred without our conscious contribution and whose meaning is still unknown to us. As it often happens with unconscious processes, the meaning of a synchronistic event may not be instantly realized by the observer and it may enter his field of consciousness at a later time. The collapse of the wave function occurs nevertheless.

For example, if one week after the occurrence of a car accident we realize that we had dreamed about it the month before it happened, we might say that the synchronistic event (and the collapse of the wave function with it) has entered the field of our consciousness only afterwards.

The observation of reality (and not of physical reality alone) through our consciousness creates meaning where there was none before. Thus, the knowledge of the system's state may lead to meaning and this continuous creation of meaning is an ongoing process that is still far from being achieved.

ARCHETYPES AND ATTRACTORS

But how can psychic factors or, as Pauli states, "new concepts, which are uniformly simultaneously physical and psychological," force themselves through the gap of individual events in quantum systems?

In order to tentatively answer this question and open to Pauli's suggestion a field of application in the physical sciences, we need to widen the Jungian concept of the archetype.

Jung had defined archetypes as factors spanning psyche and matter. In addition to that, Jung had also made the assumption that natural numbers are archetypes of order and dynamic aspects of the unconscious psyche. Numbers are for Jung the ultimate foundations or the primitive form of the archetype and the only possible connection between the two worlds of psyche and matter. M. L. von Franz, in "Number and Time" (von Franz 1972) has further insisted that numbers inherently belong to the dynamic "behavior" of archetypes in general. In fact, archetypes seem to manifest themselves in space and time in the form of ordered sequences of events. The series of natural numbers (1, 2, 3, 4, and so on) is the first expression of such an ordered sequence. Archetypes and numbers can thus be considered to be qualitatively active fields of rhythmically unfolding psychic and physical sequences.

Pauli, as a very intuitive scientist, had immediately adopted Jung's ideas and in a letter to Jung he wrote:

[With regard to the doubts on the exclusively psychic nature of the archetypes] I prefer to say that psyche and matter are subject to the action of neutral "in itself not identifiable" principles of order. . . . I believe in fact, not as a dogma, but rather as a working hypothesis, in the identity (homousia) between the archetypal world (mundus archetypus) and matter (physis), as you formulated it on p. 6 of your letter. If this hypothesis is right—and the possibility of parallel physical and psychic laws calls for it—this should however be conceptually expressed. This can happen, in my opinion, only through such concepts which are neutral with respect to the opposition psyche-matter.

Such concepts already exist, and they are mathematical ones: the existence of mathematical ideas, which may be applied also to the physical world (physis), appears possible to me because of the homousia (identity) of the archetypal and the physical world. In such a situation there always appears the archetype of number. . . . . It is this archetype which allows the application of mathematics to physics. On the other hand, this archetype also has a connection with psyche (Trinity, Quaternity, Mantic [procedures], etc.). . . . " (Letter to Jung of March 31st, 1953, in Meier 1992)

Jung's and Pauli's ideas, and especially M. L. von Franz's work, suggest that archetypes may be considered to be very similar to mathematical attractors. An attractor is also a concept transcending its empirical manifestations. So as to support this suggestion, Jung had defined archetypes as "the possibility of the course of energetical processes," a definition that exactly sticks to the concept of attractor in dynamical systems (for an introduction to the concept of attractor please refer to Gleick 1990, Cvitanoviç 1989, Hao Bai-Lin 1990).

Archetypes are potential form-giving factors determining the dynamical sequences in the most fundamental processes in both the psychic and the physical world. In the psychic world the concept of archetype may be used to describe the dynamic behavior of individuals exactly as the attractor concept is used in the physical world to describe the behavior of dynamical systems. Archetypal dynamics lies beyond the duality psyche-matter (see Figure 43.2). Jung called archetypes "psychoid factors," that is, transcending human intellectual categories.

In a way, the concept of attractor in dynamical systems is also transcendent. The limit cycle attractor, which describes a repetitive dynamical behavior (see Figure 43.5) would remain in itself a pure mathematical concept as long as we do not visualize it in the form of a physical repetitive dynamical process. The rotating molecules we can observe in the Rayley-

Bénard convection phenomenon, are for instance a good example of such a limit cycle attractor (see Figure 43.6).

If archetypes really are factors beyond the dynamics of all natural processes, Pauli's hypothesis that they might manifest within quantum systems becomes plausible. In fact, quantum systems are nothing less than dynamical systems of a special type, that is, whose dynamics has not yet been completely elucidated. In addition to those theoretical similarities between archetypes and attractors, quantum systems and the world of psyche (and of archetypes) have more points in common than one might think at first glance.

SIMILARITIES BETWEEN THE COLLECTIVE UNCONSCIOUS AND THE QUANTUM WORLD

In synchronistic events we observe the occurrence of mirror phenomena between psyche and matter which suggest an underlying identity between the two worlds. This identity becomes more evident through and is further

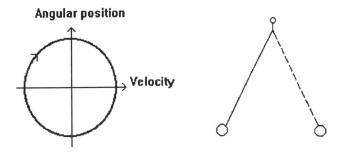

Figure 43.5 State-space representation of the dynamical behavior of a driven pendulum. Whichever the initial position, the pendulum will follow the limit cycle attractor and describe a circular trajectory in the state-space; that is, it will display a regular oscillation.

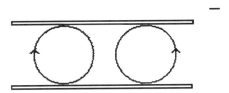

Figure 43.6 Schematic representation of two Rayleigh-Bénard convection cells occurring in a liquid medium placed between two plates (the lower plate is heated). The repetitive numerical values characterizing the molecular process (temperature, position of the molecules) can be represented in state-space by a limit cycle attractor. The purely mathematical concept (a series of repetitive numerical values) is used to describe a real object in the physical world. Archetypes are similar for they may be used to describe the psychic behavior or the psychic experience of an individual.

supported by the similarities existing between the world of psyche and the quantum world.

I can describe those similarities only very briefly here.

1. Causality. Causality breaks down in both the world of the psyche (synchronistic events) and the quantum world (double slit experiment, and so on).

2. Complementarity. Complementarity exists in the psyche as stressed by Freud and Jung (complementary response of the unconscious to the attitude of the ego-conscious, especially in dreams and synchronistic events). In the quantum world complementarity was outlined by Bohr in order to overcome the wave-particle dichotomy.

3. Simultaneity and Nonlocality. In synchronicity we note the occurrence of events transcending space and time. In quantum physics some notorious experiments (for instance A. Aspect's experiment with polarized photons) have confirmed the existence of nonlocality too.

The above similarities represent an additional argument in favor of Pauli's ideas on synchronicity, for they point toward the existence of a domain in which matter (quantum matter as it were) becomes another aspect of psyche and vice-versa. In that undifferentiated, yet energetically active layer, archetypes, which are predispositions for the occurrence of energetical processes, when activated, may give form to that inherent dynamics and thus become for us the source of both psychic and physical experiences.

IMPLICATIONS

A literal interpretation of Pauli's ideas allows a more pragmatic approach of Jung's theory of synchronicity. Thus the field of scientific experimentation becomes suddenly open. In the following, I will briefly analyze some implications and applications of synchronicity theory to the physical and biological sciences.

In many cases, synchronicity suggests the existence of nonlocal information processing in the human brain. The following example is extracted from Jung's memoirs:

I recall once during the Second World War when I was returning home from Bollingen. I had a book with me, but could not read for* the moment the train started to move I was overwhelmed by the image of someone drowning. This was a memory of an accident that had happened while I was in the military service. During the entire journey I could not rid myself of it. It struck me as uncanny, and I thought, "what has happened? Can there have been an accident?" I got out at Erlenbach still troubled by this memory. My second daughter's children were in the garden. . . . The children stood looking rather upset and when I asked, "Why, what is the matter?," they told me that Adrian, then the youngest of the boys, had fallen into the water in the boathouse. It is quite deep there and since he could not really swim he had almost drowned. His older brother had fished him out. This had taken place at exactly the time I had been assailed by that memory in the train. (Jung 1983)

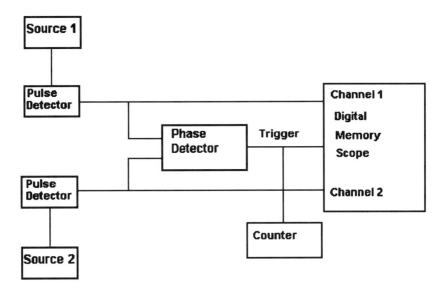

Figure 43.7 Experimental set-up for the detection of a synchronistic event in two separate quantum systems. The recording device is triggered only when synchronous pulses are detected at the input of the phase detector. Following Pauli's suggestion, the constellation of an archetype in the psychic system of the observer should manifest itself in the form of an increased number of correlations between the two separate systems. Regularities should also manifest themselves within each quantum system.

In order to attempt a new a-causal approach to those nonlocal phenomena which until now have been interpreted as "parapsychological occurrences" I have suggested in another paper (Insinna 1996) to consider the cytoskeleton of neural cells as a dynamical system in which quantum information might emerge and be amplified by the inherent nonlinearities. In the above vision cited by Jung, the dynamical system (the cytoskeleton of several cell assemblies) would be capable of following one or more archetypal attractors (ordered sequences of quantum events). This, in turn, would trigger the coherent response of all associated neurons and create in the brain the next best visual association (the image of the man drowning during Jung's military service). The brain would, from such a viewpoint, become a synchronistic device, capable of processing quantum nonlocal information based on a-causal phenomena. Eccle's causalistic suggestion of a direct mind-body interaction (Eccles 1991) might be overcome by the application of synchronicity, and a new pathway of research might be opened connecting quantum physics and psychology.

If the hypothesis I suggest here is correct, it may be inferred that large computers based on information processing by means of neural networks might become the place for the occurrence of synchronistic events, provided the inherent nonlinearity of their configuration would be capable of amplifying emergent quantum nonlocal phenomena.

This simply means that the behavior of such computers may become unstable and process nonlocal information that does not belong to the assigned task. The computer might thus in some cases display phenomena connected with the unconscious attitude of its user (the observer of quantum physics). Experiments might be implemented in order to test this last assumption of an a-causal interaction between the observer and the observed (complex) quantum system (Figure 43.7).

Synchronicity suggests that consciousness might indeed be a nonlocal phenomenon and that the human brain might be considered as a kind of focusing substrate suitable for its emergence.

This last hypothesis cannot be excluded in spite of the many materialistic and reductionistic approaches of modern science. In fact, contemporary science is attempting to explain consciousness and mind-body interactions on an exclusively mechanistic and causalistic basis.

Finally, quantum physics and psychology are still wide apart and both need a bridge uniting them. Their unification is a necessity, for physics needs psychology in order to become a science of wholeness, and psychology needs physics in order to discover and confirm the universality of dynamical laws. Synchronicity is such a bridge and I believe that it would be worth investigating further along the path indicated by Jung and Pauli.

REFERENCES

Cvitanoviç, P. 1989. *Universality in chaos.* New York: Adam Hilger.

Eccles, J. C. 1991. *Evolution of the brain.* London: Routledge.

Franz, M. L. von. 1974. *Number and time.* London: Evanston.

Gleick, J. 1990. *Chaos.* London: Sphere Books.

Hao, B. L. 1990. *Chaos II.* Singapore: World Scientific.

Insinna, E. M. 1996. "Cytoskeleton, quantum mechanics and cognition." In *Nature, cognition and systems.* Vol. 3. Dordrecht: Kluwer (in press).

Jung, C. G. 1983. *Memories, dreams, reflections.* London: Flamingo.

Laurikainen, W. A. 1988. *Beyond the atom.* Berlin: Springer.

Meier, C. A. 1992. *W. Pauli und C.G. Jung.* Berlin: Springer.

Wheeler, J. A. 1983. "Law without law." In *Quantum theory and measurement*, edited by J. A. Wheeler and W. H. Zureck. Princeton, NJ: Princeton University Press.

Wigner, E. 1983. "Remarks on the mind-body question." In *Quantum theory and measurement*, edited by J. A. Wheeler and W. H. Zureck. Princeton, NJ: Princeton University Press.

IX Hierarchical Organization

A seminal event in the early history of brain modeling was the appearance in 1949 of a book entitled *The Organization of Behavior* by the Canadian psychologist Donald Hebb. In order to appreciate the significance of this work, one should consider that it was among the first serious attempts to "bridge the long gap between the facts of psychology and those of neurology." His essential idea was expressed in the following terms:

Any frequently repeated, particular stimulation will lead to the slow development of a "cell-assembly," a diffuse structure comprising cells capable of acting briefly as a closed system, delivering facilitation to other such systems and usually having a specific motor facilitation. A series of such events constitutes a "phase sequence"—the thought process. Each assembly may be aroused by a preceding assembly, by a sensory event, or—normally—by both. The central facilitation from one of these activities on the next is the prototype of attention.

Although this concept is related to John Hopfield's (1982) basin of attraction, it is also important to notice that the cell assembly is essentially hierarchical in nature. This feature arises because an assembly of neurons shares the two essential dynamical properties of an individual neuron: (1) A *threshold* for excitation, and (2) *All-or-nothing* response. Just as a number of neurons can participate in an assembly, a number of subassemblies can participate in a higher-order (or complex) assembly. The four chapters in this section are concerned with this hierarchical structure of the brain, and with the possibility that mind might emerge from such a hierarchy.

The most ambitious chapter is that of Erich Harth, who presents in Chapter 44 a theory that has been described in detail in his recent book *The Creative Loop: How the Brain Makes a Mind*. In his theory, the "creative loop'" has two unusual features: (1) It is a *self-referential* system with top down control that modifies activities at the sensory areas, and (2) It employs *positive feedback* to enable a pattern of interest to be picked out of a busy background. The positive feedback operates between all levels of the hierarchy so an entire complex assembly acts as a dynamic unity that is tailored to a particular perceptual event. To show that this is more than a mere theory, Harth and his colleagues have realized a computer algorithm with this structure having the ability to tease a recognizable image of Albert Einstein (among others) out of an almost completely noisy background.

In Chapter 45, Nils Baas approaches the hierarchical concept from the general perspective of mathematics. In his view, complex structures are naturally composed of more simple ones so the hierarchical character of complexity emerges from the formalism in a natural way. Such higher-order structures are called *hyperstructures*, and it is important to recognize that the fundamentally new properties of a hyperstructure may induce changes in lower-level interactions. Baas also points out that the phenomenon of life is related to that of consciousness since both emerge from hierarchies of dynamical organization.

Whatever consciousness is, it is related—perhaps closely related—to the ability to speak, and the power of speech has emerged out of the hierarchical development of evolution. In an interesting analysis of phylogenetic data, B. Raymond Fink has related in Chapter 46 the development of language and consciousness to the evolution of the larynx as a means for dealing with the problem of ingesting both oxygen and food without mixing the two.

In Chapter 47, the final chapter of this section, Alwyn Scott discusses the hierarchical nature of scientific knowledge that results from the organization of biological dynamics into levels of activity. From this perspective, the phenomenon of life is seen to emerge from interactions among the levels of chemistry, biochemistry, cytology, and physiology. From the work of Hebb (1949) and others (Scott 1995), it is evident that the dynamic activities of the brain are similarly organized into hierarchical levels including the level of ethnology or cultural anthropology, and it is suggested that mind emerges from the brain's hierarchy in much the same manner that life emerges from that of cells, tissues, and organisms.

In review, Eric Harth discusses neuronal connection hierarchies in the brain, and Nils Baas describes mathematical "hyperstructures" that can emerge from such arrangements. Raymond Fink defines emergence over time in the course of evolutionary development, and Alwyn Scott places the brain's hierarchy in a larger hierarchical perspective.

REFERENCES

Harth, E. 1993. *The creative loop: How the brain makes a mind.* Reading, MA: Addison-Wesley.

Hebb, D. O. 1949. *The organization of behavior.* New York: Wiley.

Hopfield, J. J. 1982. Neural networks and physical systems with emergent collective computational abilities. *Proc. Natl. Acad. Sci. USA* 79:2554–58.

Scott, A. C. 1995. *Stairway to the mind: The controversial new science of consciousness.* New York: Copernicus (Springer-Verlag).

44 Self-Referent Mechanisms as the Neuronal Basis of Consciousness

Erich Harth

INTRODUCTION

The story is told of a famous London pub that is known for serving the best stout. How did this happy circumstance come about? The pub's reputation was the result of a concatenation of chance events and the mechanism of positive feedback that is often the cause of instability and novelty. The stout is always fresh because the pub is always crowded. The pub is always crowded because the stout is always fresh. Resulting also from this bootstrap process, which started with an all but untraceable fluctuation, was the sudden, unexpected prosperity of the owner.

I will show that the assumption of positive feedback in the form of optimization algorithms taking place along sensory pathways can account for mental imagery, and may be the physical basis for the phenomenon of consciousness.

Listening to the many excellent presentations at this conference, I was struck by two facts. The first was the great range of physical scales on which answers to the problem of consciousness were being sought. These extended from the nanometer scale of microtubules, where quantum effects were invoked, to the macroscopic phenomena of coordinated and oscillatory activity in large populations of cortical neurons.

This impressive range reflects a property that is peculiar, and perhaps exclusively peculiar, to life: Phenomena at all scales are coupled through mutual interactions that allow the exchange of both bottom-up and top-down controls. Microstructures are organized by macroscopic experience, and microscopic fluctuations are readily propagated upward.

My second observation is that, in spite of the great variety of approaches, a single unifying theme runs through practically all the papers: the problem of reconciling the diversity of cortical processors with the apparent perceptual and individual unity, the *binding problem.*

The problem has been dealt with in different ways. Early ideas on perception held that the messages received by the different senses must be brought together at a single common sensorium, the *sensus communis* depicted here in a famous fifteenth-century drawing (Figure 44.1.). Our expression "common sense" derives from this notion that rational thought requires a confluence of sensory information.

In the twentieth century anatomists and physiologists have been searching the brain for a *convergence zone* that serves a similar function. The prefrontal cortex, where, "the degree of cortical afferent convergence... is unparalleled in any other part of the neocortex" (Fuster 1980, p. 138), appeared to be a likely candidate. But this region, in humans, comprises nearly one third of the cortex and appears as compartmentalized as lower areas.

A number of approaches to the binding problem have been suggested traditionally, none very palatable. In the classical dualist approach a nonphysical entity is made to look down on the brain, gathering everything the senses have picked up and collecting the verdicts from the diverse cortical analyzers. Thought, feeling, consciousness, and intentionality are held to be functions of this nonmaterial homunculus. But what is most upsetting to physical scientists is the necessary further assumption that this spirit can insert its immaterial hand into the gears and levers of the neural machinery and make it carry out its intentions. It may be argued—without helping

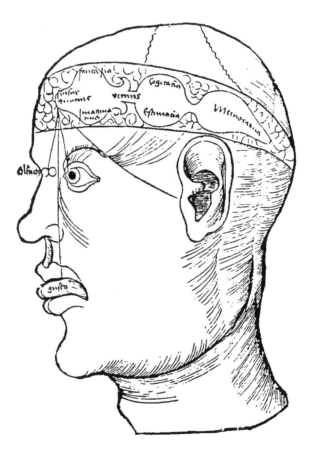

Figure 44.1 Convergence of sensory data according to Gregor Reisch (c. A.D. 1467–1525). The point of convergence in the first ventricle of the brain is labeled *sensus communis*.

matters, I believe—that this machinery is often poised in such delicate balance that it would take almost no force to push it one way or the other.

Theories that seek a physical basis of consciousness therefore tend to avoid any tinge of dualism. Most of them try to identify specific neural activity patterns, or subsets of cortical neurons, with mental states. While such assumptions may have some validity, they leave a fundamental question unanswered. Anything expressed in the brain by coordinated activity in a subpopulation of neurons would seem to require another sensor to detect that coordination, an external observer or the same pesky homunculus we are trying to exorcise. Patterns of neural activity extend over space and time. How are they bound into a subjectively experienced unit?

The difficulties sketched here are typical of any theory of consciousness that seeks the answer in some final, or *output state,* that is, in a coded neural message expressed as coordinated activity in a group of neurons.

One can avoid the necessity of an intelligent observer by the implausible assumption that sensation and/or knowledge resides in the neurons themselves. But, if neurons can be said to know anything, they know only transmitter molecules that occasionally impinge on their synapses. Interactions between neurons are local. I will call this the locality problem because it bears some relation to what is called locality in physics. The only known function of neurons is to transmit simple signals to other cells. Neurons are links in a vast communications network. They die when their target cells are destroyed.

Quantum mechanics has been invoked by a number of investigators to avoid some of the conceptual difficulties (Penrose 1989, Hameroff and Penrose 1996). It is argued (Stapp 1993, Kafatos and Nadeau 1990) that the nonlocal character deduced from recent experiments on Bell's inequality is a universal characteristic and may provide the key to the binding problem. These are exciting possibilities, but they are difficult to confirm experimentally at this time.

I will present here a theory of consciousness based on known structural features of the brain and using only classical concepts. I avoid the difficulties inherent in any theory that relies on a semantic final, or output state, and present the conscious brain as a single, integrated, self-referent system.

BACKGROUND

Dynamics Of Neural Systems

The brain has been compared with a vast telephone exchange made up of billions of stations—the neurons—each being connected to thousands of other stations. The characteristics of the network are determined by its connectivity—which neurons are connected to which other neurons—and by the strengths of these connections. We presume that the latter are modifiable and probably play a role in learning and memory. In the classical interpretation of neural dynamics, each neuron is either "firing" or resting.

When firing, it will produce excitatory or inhibitory effects on all other neurons to which it is connected. A description of the dynamics would then consist of a specification of the precise sequence of neural firings.

Present experimental methods provide only very limited information in this regard. Microelectrode recordings monitor at best only a few neurons simultaneously, while gross methods (EEG, PET, NMR) lack the temporal and spatial resolution to observe action potentials of individual cells. Theoretical methods are hampered by our lack of precise knowledge of the connectivity and—ultimately—by limited computing power. There are, however, a few facts that are well established. The following must be kept in mind when we try to make the connection between the incoming sensory messages and our conscious perception of external events.[1]

1. Sensory pathways, consisting of large numbers of fibers, lead from a set of receptors to a succession of neural centers, each of which again consists of many neurons.

2. Sensory centers may be classified according to specific sensory features to which they are sensitive. Thus, there are color-, motion-, and shape-sensitive centers in the visual pathway.

3. Neural centers are, as a rule, reciprocally connected; that is, a pathway from a center A to another center B is usually matched by another bundle of fibers leading from B to A.

4. The neural net is multiply connected, meaning that there are in general several pathways originating in one center.

5. The neural net is a massively parallel processor. Individual sensory systems operate simultaneously and independently, and each sensory system contains several parallel processing channels carrying out different cognitive functions.

6. There is no single center in which all analyzed sensory messages converge.

The fact that even the most primitive sensory stimuli give rise to widespread and sustained neural activity suggests that functionally significant brain states must be sought not in single neuron outputs but in the behavior of neural populations. This idea goes back to Hebb's (1949) concept of *cell assemblies*. It appears again in the treatment of the dynamics of *netlets* (Harth *et al.* 1970, Anninos *et al.* 1970) and in the theory of *group selection* (Edelman and Mountcastle 1978).

Theoretical studies of neural net dynamics are generally based on assumptions that make the nets deterministic and computable. Nets were found to exhibit hysteresis (Harth *et al.* 1970, Anninos *et al.* 1970, Wong and Harth 1973), a form of short-term memory, and may contain a number of locally stable attractors that Hopfield (1982) interpreted as a content-addressable memory. It was shown also that, with the relaxation of some of the simplifying assumptions concerning neural dynamics, the trajectories become more complex, suggesting eventual transition to chaotic behavior

(Wong and Harth 1973). The possible role of chaotic dynamics was discussed by Freeman (1987) and Harth (1983).

A complete description of the dynamics of neural systems approaching the human brain in size and complexity is—as I have already mentioned—beyond our experimental and theoretical means. Human behavior represents readily observable output states of some neural processing, but we know of no such reduction to simple observable (or even nonobservable) neural states in the case of conscious perception or thought processes.

If we are to attribute meaning and consciousness to the activity of a neural population, or perhaps to a particular temporal signature within that population, then we run into the already mentioned locality problem. We have no reason to assume that any one neuron can know, or feel, or be in any sense aware of the semantic content of the signals it receives from other neurons. It cannot know what other neurons, scattered widely through neural tissue are doing at this time.

Then who or what carries that knowledge? We run into a conceptual problem similar to what in physics was called action-at-a-distance. This was in conflict with the conviction that physical laws must be local. Here we would require something like *knowledge*-at-a-distance. As external observers, we can talk about collective states of activity in specialized groups of neurons, and we could, in principle, observe and describe such states. It would require a neurophysiologist applying experimental techniques of a very sophisticated kind. But, for any part of the brain to take cognizance of its own collective modes would seem to require knowledge-at-a-distance, unless we assume some agency, an internal neurophysiologist, to carry out the observation. The brain of that observer—if he or she is smart enough—could then draw the appropriate conclusions. This internal neurophysiologist is, of course, none other than our old friend the *homunculus*. We are back to the same old problem.

Complexity

That the brain is a complex system is almost too blatantly true to be mentioned. It is useful, though, to discuss briefly some of the implications.

Complex systems, as I want to define them here, consist of a very large number (billions or more) of more or less invariant subunits, forming a hierarchy of structures and hyperstructures. The concept of complexity is thus closely related to Alwyn Scott's hierarchical organization or the *hyperstructures* envisioned by Baas (Scott 1995, 1996, Baas 1994). The dynamics of the whole system encompasses, as I mentioned before, a range of scales extending over many orders of magnitude from the microscopic to the macroscopic. What distinguishes such complex systems from ordinary macroscopic matter is the fact that there is lively communication both up and down the scales in what is sometimes called bottom-up and top-down control.

It may be argued that a simple rock is a complex system by that definition: stresses may produce molecular dislocations that produce later fractures. But while in the rock most macroscopic behavior can be predicted without reference to its microstructure, what happens in a complex system reflects at all times the numerous subtle changes that have taken place at the lowest structures formed by the subunits. In what is perhaps the most dramatic example of this, we see the struggle for existence of populations of a species translated into the structure of a single molecule, the DNA, which, in turn, will decide the fate of future generations.

The ease with which information flows both up and down the physical scales is perhaps the outstanding characteristic of complex systems. In the case of the brain, some of the microstructural changes brought about by macroscopic events—sensory experience and behavior—are most likely in the synapses which number in the trillions. Conversely, changes at the synaptic levels percolate upwards through the hierarchy of structures affecting thought processes and behavior. If we are correct in our assumption that chaos may play a role in neural dynamics (Harth 1983), then minute events at the lowest levels, often termed *noise*, will be able to materially affect macroscopic events. The distinction between noise and meaningful engrams at the microscopic level may in fact be somewhat fuzzy and arbitrary. The vast majority of past modifications must be exceedingly weak—to avoid the "superposition catastrophe"—and thus indistinguishable from random noise. It is this fact that prevents us from ever truly understanding the dynamics of the thinking brain.

Sensory Pathways

Consciousness is most intimately connected with sensations. We will therefore turn to the sensory portion of the brain, the visual pathways in particular, for clues to an understanding of consciousness.

The processing of information along sensory pathways is usually discussed in terms of what I have called the *slaughterhouse paradigm* (Harth 1993). In this picture, useful features are extracted from the corpse of the original sensory image by a multitude of analyzers that are arranged both serially and along parallel chains. In vision the dissection begins at the first visual center in the neocortex, V1, and continues over more than thirty different stations. Color, form, and movement are features extracted at different centers along the way. There is no center in the brain where the entire picture is reassembled. Although much of this information winds up in some form at the hippocampus, along with other sensory information, Crick (1994, pp. 173–174) points out that we should not look for the seat of consciousness there. Destruction of the hippocampus (through disease) impairs the formation of new memories but does not seem to abolish consciousness.

This picture of successive stations of visual processing is correct as far as it goes, but it overlooks one salient feature of the sensory pathways, a feature that in fact is characteristic of the neural network as a whole: most connections between centers are reciprocal.

The function of these return pathways, some of which have been known since Ramon y Cajal, has long puzzled physiologists. Why is information returned from higher to lower sensory centers? It is difficult to avoid the conclusion that what is transmitted, for example from the LGN to V1, is not the raw input received from the retina, but is somehow altered by information already residing in the cortex.

Figure 44.2, which is a detailed diagram of the peripheral portion of the visual sensory system, shows a further input to the LGN coming from brainstem nuclei via the perigeniculate nucleus (PGN). Here, PGN cells are excited by collaterals from geniculo-cortical fibers (Hersh and White 1981), and inhibited by ascending fibers from brainstem nuclei. PGN cells, in turn, inhibit LGN relay cells (Watson, Valenstein, and Heilman 1981, Steriade 1986). It is seen that brainstem input into PGN thus facilitates transmission from LGN to V1.

If indeed sensory information is modified by these return pathways, the means of these modifications present a fundamental puzzle, known as the problem of the inverse. Let us assume that the purpose of the feedback is to strengthen certain features of the sensory message, and to suppress others. For the feedback to be thus feature-specific, we would require neural circuitry that inverts the feature-extracting properties of higher centers (see for example Rothblatt and Pribram 1972). Every face detector, in other

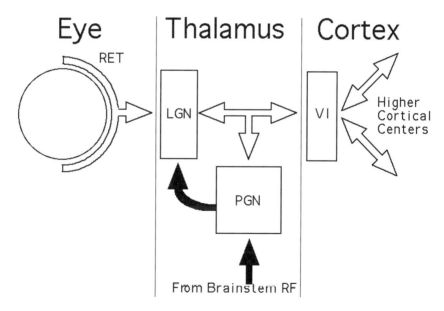

Figure 44.2 Schematic of peripheral visual pathways. Perigeniculate cells (PGN) are inhibitory (black arrows) on LGN relay cells and receive inhibitory connections from brainstem reticular formation.

words, would have to be accompanied by a backward-projecting face generator.

I have suggested (Harth 1976) that a general hill-climbing process may be the solution to the problem of the inverse. I envisioned that a scalar function be generated by high-level feature analyzers and used in a simple optimization algorithm that modifies sensory input patterns. Such a mechanism would be capable of enhancing and suppressing different features of the sensory message and simulating images when sensory input is absent.

Imagery

Sensory messages, especially if repeated, produce certain changes in the brain that enable one to recall them later. Such recall is often accompanied by sensations similar to those we experienced when actually confronted with the event. The phenomenon is called mental imagery.

I can imagine seeing a familiar face or any scene out of memory, just as I can imagine hearing a familiar melody without real acoustic input, although I may come close to humming the tune.

There is a longstanding controversy over the neural basis of imagery. In the case of visual images we can think of pictures of the event in the form of neural activity appearing at peripheral visual centers like the LGN or V1 where sensory messages are normally represented in this *iconic,* or retinotopic form. On the other hand, it is possible that, to conjure up a mental image of an event means only to trigger circuits at higher cerebral centers where neural codes of one kind or another *symbolize,* or stand for, various features of the event.

I will make a point later in my model of imagery that the cortical depth of a visual mental image depends very much on the amount of retinotopic detail or spatial relationship we seek in the image. Thus, I can think abstractly of a hexagon as one of a class of geometrical figures, the regular and irregular polygons. The "images" in this case are of words or of nonverbal geometrical concepts. Alternatively, I may form a mental image in which I visualize a hexagon as a pencil line drawn on white paper, or perhaps a colored cutout. It is likely that the nature and location of the corresponding neural activities will be different in the two cases.[2]

There are many experiments in psychology that bear on this question of the nature and location in the brain of mental images. Some of these (Shepard and Cooper 1982) suggest that images should possess picture-like properties that can be examined by some "inner eye" the way real sensory images are examined. Recent experiments involving positron emission tomography (PET) reported by Kosslyn and coworkers (Kosslyn *et al.* 1993) have shown that neural activity appears in V1 and other retinotopically organized visual brain centers when subjects were engaged in visual mental imagery. The spatial extent of this activity increases with the size of the visualized object. The authors conclude that, in visual imagery, neural activity at these peripheral centers is organized in pictorial form, and that the quasi-sensory patterns are observed by higher visual centers as though they were sensory input patterns.

If we accept this attractive notion, two fundamental questions remain: (1) How are such quasisensory patterns created without appropriate sensory inputs? and (2) How does the appearance of these peripheral images lead to conscious perception of the imagined event? The first question brings up again the problem of the *inverse*, assuming that the generation of such images is under control of higher centers. The second question is part of the general problem of how the arrival of peripheral sensory images (both real and imagined) can give rise to that subjective feeling of conscious perception.

Consciousness

The binding problem is at the core of the problem of consciousness. It is really a twofold problem. We find, on the one hand, that different sensory features of objects, though analyzed in different parts of the brain, are perceived as belonging to the same object. The shape, the color, the motion, and the song of the bird are all referred by the mind to a single object: the bird. The other aspect of the binding problem concerns the perceived unity of the conscious subject. He or she has the sensation of being a unique and indivisible "I," the single receiver of all sensory experiences past and present, the planner of his of her own future.

Theories of consciousness differ in the seriousness with which they regard the two problems. At one extreme, dualists hold that the two types of binding can be achieved only by an entity apart from the neural machinery of the brain, a nonmaterial observer in whom both the objective and subjective worlds come together. Dualism is no longer a popular view among scientists, but is not quite as dead as some would have us believe.

At the other extreme are theories that question, and often deny, the reality of the two types of apparent binding. Minsky (1985) conceived of the mind as being not one but many, a "society of mind." Dennett (1991) argues against the convergence of sensory messages at a single location in the brain, the Cartesian Theater, there to be viewed by a "central meaner." The story conveyed by the senses is not one, he claims, but made up of "multiple drafts." The central "I" is a figment of the imagination.

In between these two extremes is a collection of theories that somehow try to reconcile the appearance of objective and subjective unity with the multiplicity of neural processing locations and the diversity of subjective experience. It has been argued (Penrose 1989, Hameroff and Penrose 1996) that only a quantum mechanical theory is able to achieve this.

A simple, classical theory by Crick and Koch (1990, also Koch 1996) has gained considerable attention lately. It holds that binding between the outputs of scattered sensory analyzers is achieved through temporal correlation of the activities of the neurons involved. Thus, in the pandemonium of ongoing neural firings, neural groups may be distinguished by their simultaneous firings, or perhaps by firing at a single common frequency.

This would pin a label on these selected neurons and their attached significance. Crick explains that from the simultaneous firing of a "red" neuron and a "circle" neuron "the brain can deduce that the circle is red..." (Crick 1994, p. 212).

I have already pointed out that this still leaves unsolved the problem of *locality*. Who, or what, knows what neurons are firing? If "the brain" is the deducer, we are still faced with the how? and where? It becomes difficult to avoid bringing in the unsavory homunculus whenever one attempts to ascribe consciousness to some final, or output state.

Dualist approaches have been generally shunned for introducing unscientific or antiscientific notions. But this need not be the case. The argument most frequently advanced against mind-brain dualism is that, to assert that nonphysical elements such as *mind* could affect physical processes in the brain, and ultimately determine behavior, would violate all laws of physics as we know them. Just how fallacious this argument can be is seen in the following quotation by Dennett. "A fundamental principle of physics," he states, "is that any change in the trajectory of any physical entity is an acceleration requiring the expenditure of energy, and where is this energy to come from?" He then associates such supposed interaction between brain and "mind stuff" with Victorian seances, ghosts, and ectoplasm.

Dennett's remark about the "fundamental principle of physics" is just plain wrong on two counts. Not every change in trajectory is an acceleration—a change is produced also when an acceleration ceases—and not every acceleration requires the expenditure of energy. But, aside from that, we can have dualism without ectoplasm. Let us look at an analogous situation in a larger physical system, that of a population, a society of human beings. We may try to interpret all of the dynamics of the society by noting the physical interactions between its members, and of the members with their environment. One problem with this approach is that little is explained by noting only the physical interactions, the pullings and pushings, making love, or trading punches, or jostling for a seat in the subway. What moves a society, physically *moves* it for better or worse, are nonphysical elements, both individual and collective ones: love, hate, prejudice, political philosophies, economic theories, national pride, dreams of empire. People have killed themselves over unrequited love, starved from a failed experiment in economics, or died in front of a firing squad for holding unpopular political views. In the ordinary course of events, societies are guided by a body of laws, which, in turn, have a long and complex history. But laws—though they may be embodied in law libraries—are not physical elements that can affect us by any of the four fundamental forms of physical interactions. On the other hand, there is nothing mysterious about the way they affect the physical conduct of human affairs: Nobody can buy liquor in New York state on Sundays, but nobody is surprised that something as tenuous as a law can prevent six-packs of beer from moving down supermarket checkout counters.

To say that all of the dynamics of human affairs is describable and explainable in strictly physical terms is sheer nonsense. It is equally nonsensical to assert that introducing such elements as political philosophies, or laws, or a *climate of opinion,* means resorting to some kind of mysticism and embarking on a nonscientific cabal. We cannot expunge such concepts from a discussion of societal dynamics, unless we confine ourselves to describing patterns of movement of people through subway turnstiles during rush hour. It must be apparent to all but the most simpleminded reductionist that the attempt to construct a true physical theory of society would be a foolish undertaking.

I believe that in talking about brain dynamics, such elements as intentionality and consciousness become valid concepts, no more mystifying, or science-defying than laws or politics were when we talked about society. While it is true that there are physical links that form causal chains connecting all physical events, these chains may be complex beyond our ability to follow them in any meaningful way. Nobody would seriously consider explaining the following situation in terms of forces, acceleration, momentum transfers, or energy exchanges: "He stepped into the voting booth and pulled the lever for the Democratic ticket (a physical act) because of his liberal politics (a nonphysical entity)."

The complexity of thinking individuals forces us to select new sets of descriptors, and to seek relationships and laws connecting them. Some of these will have to be recast. Perhaps *liberal politics* is not a good descriptor. Others may turn out to be meaningless, as "levity" or "caloric" were at the dawn of physics. But descriptors and their corresponding relations that are valid and useful at one level may be useless, though still valid, at a higher level of complexity. I believe John Searle's assertion (Searle 1992) that "consciousness is a higher-level physical feature of the brain." This is the gist also of Alwyn Scott's thesis expressed in these proceedings and in his excellent book *Stairway to the Mind* (1995).

This is not to say that any given portion of the hierarchy can be treated in isolation. If a science of the mind is to become part of the scientific edifice it must connect to the phenomenology at lower levels. Such a connection will be sought in the model presented below. We cannot overcome, however, the necessity of dealing with concepts and relationships not encountered at lower levels. In that sense we will have answered only the first, easy, question posed by David Chalmers at the beginning of the symposium.

THE MODEL

Consciousness As A Bootstrap Process

The model of consciousness I wish to propose is based on the following premises:

Subjective Unity. The feeling that different sensory elements such as color, shape, and so on belong to the same perceived object, and that a single unified self is the recipient of all sensations (the two binding problems), is not an illusion, but must be accounted for.

Output States. Theories of consciousness in which subjective perception is equivalent to the appearance of a final, or output state, are inadequate because they all beg the question of the ultimate observer of that state. This difficulty could be avoided only if we assumed "smart" neurons or the abandonment of the principle of locality.

Feedback. The neural network is multiply connected, containing many causal loops. In particular, centers along sensory pathways are reciprocally connected, so that every element of the pathway is under both bottom-up and top-down control. This causes sensory information flow to be self-referent.

Mental Imagery. Experiments have shown that visual mental images are accompanied by marked neural activity in peripheral visual centers of the cortex (Kosslyn *et al.* 1993). There are indications that these activities reflect some of the metric qualities of the objects imagined, and that the feedback is therefore feature-specific.

To complete the model I supplement these premises with the following assumptions:

Self-reference. Instead of the classical picture of sequential sensory processing with final outputs (which I have called the slaughterhouse paradigm) I envision a cyclic process, with activities at peripheral sensory centers modified by top-down control. The resulting images are viewed by the higher centers by way of the ascending sensory pathways. This, in turn, changes the top-down control, causing further modifications of the sensory messages. Neither the appearance of the peripheral images by themselves, nor activity at higher centers isolated from peripheral afferents, would constitute consciousness.

Inverse Problem. Inversion of sensory processing may be accomplished by feedback paths that perform template matching. For many types of feature analyzers this will not be possible without very elaborate feedback circuitry. Alternatively, feature-specific top-down control can be achieved by hill-climbing mechanisms. I propose that a scalar *cost function* that is a measure of correctness, or expectation, or of some other desirable image quality, or of some affect triggered by the sensory message, is derived at higher levels of sensory processing, and is used to modify image quality through an optimization, or hill-climbing process.

Selection. The cost function is derived from the responses of any number of feature analyzers, and will be able to select specific input features, excluding others, depending on the sensory messages, cortical associations, and chance events.

Modification. The optimization process will tend to enhance particular features of the sensory message and suppress others. In the absence of sensory input it is able to generate quasisensory images.

Imagery. Peripheral visual centers function as internal sketchpads on which higher cortical centers project activity patterns in mental imagery. The process may be initiated by fluctuations in the prefrontal cortex, sometimes referred to as the *working memory*, and amplified by cyclic interaction with lower centers. The cortical level at which the imaging takes place depends on the degree of abstractness of the image.

Alopex: A Hill-Climbing Algorithm

Optimization algorithms must be able to adjust a multitude of control parameters of a system in such a way that a single scalar variable, which is a measure of the functioning of the system, is optimized. The scalar is called the *cost function*, and is, in general, to be maximized or minimized. The process is called *hill-climbing*, because it is as though one were trying to find the highest point in a landscape while walking in thick fog, without map or compass, but having as one's only aid a sensitive altimeter that tells after each step whether one has gained or lost altitude. Unlike our fogbound hiker, mathematical hill-climbing may take place in a many-dimensional space.

Harth and Tzanakou (1974) have proposed a simple optimization algorithm, which they used later to plot visual receptive fields of single neurons in frog visual tectum (Tzanakou, Michalak, and Harth 1979). The idea was to use the strength of the response of the neuron under observation as the cost function to be maximized. The algorithm yielded values for the light intensities of an array of pixels presented to the animal on a CRT screen. Convergence of the pattern of light to the receptive field was accomplished by the optimization algorithm we called *Alopex* (see Appendix).

In 1976 I proposed that a similar algorithm may be operating in the brain to achieve feature-specific feedback during visual perception (Harth 1976). Here, responses of central feature analyzers may supply the cost function, which is maximized by making the pattern of neural activity at the lower level become more strongly suggestive of the corresponding *trigger features*. Such optimization processes are capable of generating the *inverse* of the known processes of feature extraction that occur on the ascending legs of the visual pathway. It was shown also (Harth, Unnikrishnan, and Pandya 1987, Harth, Pandya, and Unnikrishnan 1990) that implementation of the process requires only very primitive neural circuitry that we encounter, for example, in the geniculo-cortical loop (Figure 44.2).

If these ideas are correct, then weak activation of a central feature analyzer, due to an incomplete or noisy input pattern, would lead to completion or sharpening of this pattern and to the suppression of irrelevant or conflicting detail. I emphasize that, in most cases, the inversion process cannot be achieved by simple template matching. It would require complex neural circuitry to reconstruct a particular visual scene from a set of corresponding high-level neural codes. A neural optimization algorithm, perhaps of the type described in the Appendix, provides a simple and general procedure for implementing the inversion process.

Computer Simulations

In a series of computer simulations we have tested the idea of pattern modification and generation by optimization processes (Harth, Unnikrishnan, and Pandya 1987, Harth, Pandya, and Unnikrishnan 1990). In Figure 44.3, which shows the general design of the experiments, a *sensory input*, S, is incident and displayed as an array of pixel elements P. This pattern is affected also by input from an *optimizer*, O, whose algorithm tends to maximize a scalar function F. The combined pattern P is *viewed* by a series of *analyzers*, $A_1 \ldots A_n$, whose responses $R_1 \ldots R_n$ are combined in I to form the *cost function F*.

The ability of the system to perform feature-specific modifications on a complex spatial pattern by a simple optimization algorithm was demonstrated. Different variations of the Alopex algorithm were employed.

In one experiment, P was a 30 × 30 array. Thus, the optimizer had to control 900 variables. The analyzers A were templates constructed from photographs portraying four different people, Einstein among them. Their responses expressed the resemblance between the pattern appearing on P and the template. When the input S was the Einstein photo with random noise added, (Figure 44.4a), the algorithm transformed P into the easily recognizable pattern Figure 44.4b.

The process is gradual. As the resemblance between P and the template increases, the response of the Einstein analyzer rises (R_2 in Figure 44.5),

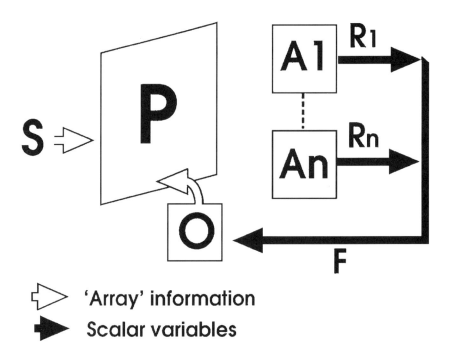

Figure 44.3 Block diagram of computer simulation of feature-specific feedback through optimization. S: "sensory" input; P: pixel array; $A_1 \ldots A_n$: feature analyzers; $R_1 \ldots R_n$: analyzer responses; F: cost function; I: integrator; O: optimizer.

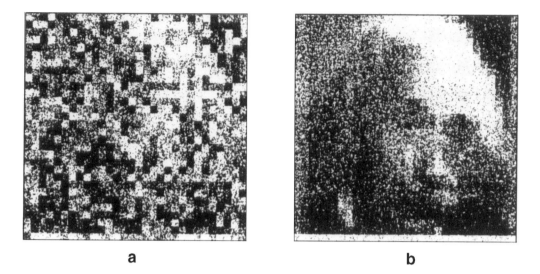

Figure 44.4 A set of 30 × 30 pixel arrays in computer simulation. The pixel intensities are the 900 control variables $x_1 \ldots x_N$. **(a)** Initial pattern made from Einstein template with random noise superimposed. **(b)** final array.

while those of the other analyzers decrease. In time, the cost function F is almost completely determined by R_2 (Figure 44.5).

In similar runs we showed that—with no sensory input S (or with pure noise)—the process would randomly select one of the patterns characterizing the analyzers.

The dynamic properties emerging from this algorithm, and shown in the computer simulations, can be summarized:

1. *Completion.* Partial or noisy input patterns, which cause weak responses in one of the analyzers, are completed with appropriate features strengthened and extraneous features suppressed, causing rising response in the corresponding analyzer.

2. *Selection.* When the sensory input is pure noise, fluctuations will select one of the central analyzers to take control of the process, thus generating the corresponding input image.

3. *Chaining.* When a sequence of events had previously been established through associative strengthening of connections between groups of analyzer neurons, the appearance of the image of one of these events will tend to generate the remaining sequence.

The speed of convergence of the *Alopex* process depends on the number of variables that must be controlled. With the 900 control variables used here, the process takes many iterations before complete convergence, as seen in Figure 44.5. The inherent slowness of this process may be cited as an argument against this model. However, the following points should be kept in mind.

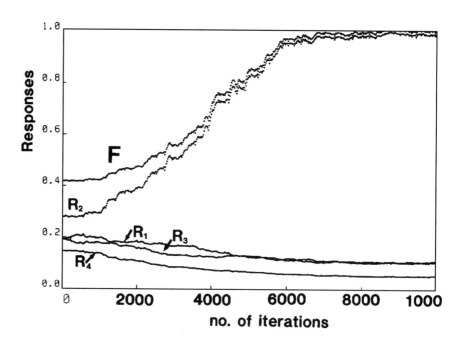

Figure 44.5 Responses of the four analyzers $R_1 \ldots R_4$. Here R_2 is the re-sponse of the Einstein analyzer. F is the cost function constructed from the four analyzer responses.

- Many of our cognitive processes are slow. Recognition or recall of images may take many seconds. Thus, times equivalent to thousands of iterations in our simulations are not excluded.
- Convergence times may be shortened considerably by use of a hierarchy of feature analyzers that build complex patterns from simpler features (Harth, Unnikrishnan, and Pandya 1987).
- Subjective recognition may require only a slight increase in the cost function rather than complete convergence.
- Other, more efficient optimization algorithms may be operating, and optimization algorithms may be supplemented by simple template matching where possible.

DISCUSSION AND SUMMARY

The neural states responsible for subjective awareness have often been sought in sets of privileged *output neurons,* which must come at the end of sensory processing, hence at the highest cortical levels. The prefrontal cortex has been suspected as a possible location. The functioning of the system was thought to resemble the classical feed-forward net of artificial neurons (Figure 44.6), in which a set of input units are analogous to sensory neurons, and a set of output units convey the final information.

I have called these attempts at understanding consciousness "final state theories," and pointed out that the physical principle of locality leaves

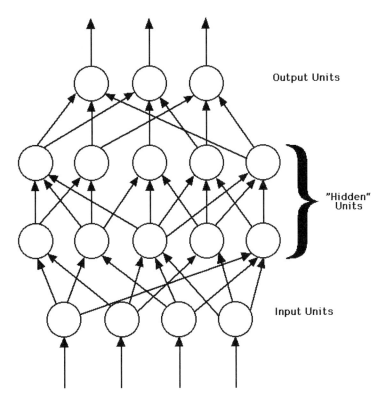

Figure 44.6 A feed-forward artificial neural net.

unanswered the question of true convergence of sensory information. By contrast, the present model takes account of the ubiquitous feedback pathways that link higher and lower sensory centers. It does not assign special output roles to privileged neurons, hence, it does not require an additional interpreter, or homunculus. Instead, peripheral sensory centers, central analyzers, and working memory form reentrant loops and operate in self-referent fashion. Binding of neural messages, analyzed in scattered cortical centers, some belonging to different modalities, occurs not at the top of the sensory pathways as hitherto assumed, but by top-down control of sensory images through optimization algorithms. This puts the Cartesian theater, as Dennett calls it, near the bottom of the sensory pathway, and makes the rest of the brain the long-sought mysterious spectator.

Simulation studies have demonstrated that feature-specific modification of input patterns is achieved by an optimization algorithm that uses a single scalar cost function. These modifications include feature enhancement and suppression for incomplete or noisy inputs, as well as generation of quasisensory patterns when primary sensory input is absent. When the cost function was constructed by a superposition of several different analyzer responses, and the initial input state was random noise, the system would select and converge on *one* particular quasisensory pattern. The system is thus highly sensitive to fluctuations in the initial phases of the process.

In applying these results to a model of consciousness, I assume that a cost function is extracted from the massively parallel neural activities generated by simultaneous sensory messages originating in one or more modalities. This scalar quantity may express a degree of compatibility between messages or their conformance with expectation. The same cost function may be used simultaneously in optimization algorithms at different levels of processing and in different modalities. Thus, sensory stimuli that originally only hint at the presence of a bird, will tend to enhance appropriate images at all levels, thereby raising responses of the compatible analyzers and increasing the cost function. The latter now *signifies* the presence of a red cardinal, but only because it enhanced all the peripheral sensory images. It may also stimulate the linguistic representation of the object perceived.

The mechanisms underlying conscious perception may be summarized as follows: The near-chaotic parallel processes in working memory do not constitute consciousness without the selecting and confirming effect of images in the peripheral sensory centers. Conversely, sensory images would be meaningless without interactions with the rich store of multimodal associative connections at higher cortical centers. In mental imagery, quasi-sensory patterns are generated peripherally through bootstrap mechanisms, starting with chance fluctuations. The self-referent interactions between central and peripheral sensory regions by means of hill-climbing mechanisms provide both the Cartesian theater and the observer that constitute consciousness.

One might object that the positive feedback envisioned here is intrinsically a runaway process that would easily lead to hallucinations. We would hear the bird sing even if it were quiet. Hallucinations do indeed occur, especially where sensory input is vague or absent (Figure 44.7). I believe this mechanism applies particularly to dreaming. In ordinary perception, however, the modifications attempted by cortical fancy must compete with reality as conveyed by the senses, and no major discrepancies will appear.

Finally, I made the point that the complexity of the human brain functions prevent us from carrying out detailed analyses in terms of concepts and laws belonging to much lower hierarchic structures. On the other hand, such higher-order concepts as intentionality, or consciousness, have no meaning at levels below that of the individual. The question, therefore, is not whether consciousness is a real thing, but whether it will turn out to be a useful descriptor in our attempts to deal with the dynamics of human thought.

APPENDIX

An optimization algorithm, called *Alopex*, was developed originally as a practical means of determining visual receptive fields of single neurons (Harth and Tzanakou 1974, Tzanakou, Michalak, and Harth 1979). It is a multiparameter stochastic process designed to adjust the control variables $x_1 \ldots x_n$ of a system in such a way that a single scalar quantity F is either maximized or minimized. Here F, called the *cost function*, generally charac-

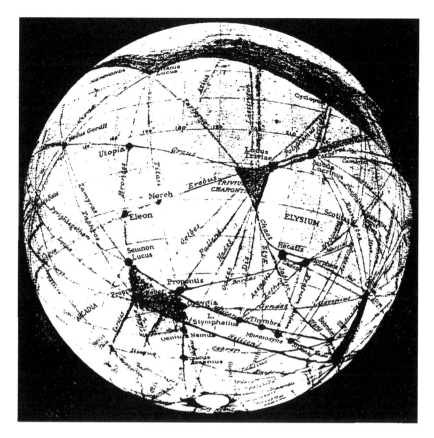

Figure 44.7 "Canals on Mars," as drawn by the astronomer Percival Lowell (1855–1916).

terizes the performance of the system and is a function of the variables x_i. In its digital form the algorithm goes through successive iterations, the nth of which is given by

$$x_i^{(N)} = x_i^{(N-1)} + g_i^{(N)} \pm k \cdot [x_i^{(N-1)} - x_{ii}^{(N-2)}] \cdot [F^{(N-1)} - F^{(N-2)}] \quad (1)$$

Here g_i is a noise term that can take different forms. The constant k is chosen in such a way that the absolute magnitude of the product of the two square brackets is of approximately the same magnitude as the noise term g_i. The product of the two square brackets will be positive whenever *both* the control variables x_i and the cost function show either an increase or decrease. With the plus (minus) sign in front of k, the algorithm will maximize (minimize) the cost function. The algorithm has been applied to a large variety of optimization problems (Harth and Pandya 1988).

The algorithm is attractive because of its simplicity, and because it is easy to construct simple and plausible neural circuits to carry it out (Harth, Unnikrishnan, and Pandya 1987, Harth, Pandya, and Unnikrishnan 1990).

NOTES

1. We will pay particular attention to the sense of vision; unless otherwise stated, reference will be to the human brain.

2. I will make the assumption throughout that, whatever neural activities are responsible for mental images, they would be similar to the patterns evoked at the same location by real sensory inputs.

REFERENCES

Anninos, P. A., B. Beek, T. J. Csermely, E. Harth, and G. Pertile. 1970. Dynamics of neural structures. *Journal of Theoretical Biology* 26:121–48.

Baas, N. A. 1994. "Emergence, hierarchies, and hyperstructures." In *Artificial life III*, edited by C. G. Langton. *SFI Studies in the Sciences of Complexity*, Proc. Vol. XVII. Reading, MA: Addison-Wesley.

Crick, F. 1994. *The Astonishing hypothesis. The scientific search for the soul*. New York: Charles Scribner's.

Crick, F., and C. Koch. 1990. Towards a neurobiological theory of consciousness. *Seminars in Neuroscience* 2:263–75.

Dennett, D. C. 1991. *Consciousness explained*. Boston, MA: Little Brown.

Edelman, G. M., and V. B. Mountcastle. 1978. *The mindful brain. Cortical organization and the group-selective theory of higher brain functions*. Cambridge, MA: MIT Press.

Felleman, D. J., and D. C. Van Essen. 1991. Distributed hierarchical processing in the primate cerebral cortex. *Cerebral Cortex* 1:1–47.

Freeman, W. J. 1987. Simulation of chaotic EEG patterns with a dynamic model of the olfactory system. *Biological Cybernetics* 56:139–50.

Fuster, J. M. 1980. *The prefrontal cortex. Anatomy, physiology, and neuropsychology of the frontal lobe*. New York: Raven Press.

Hameroff, S. R., and R. Penrose. 1996. "Orchestrated reduction of quantum coherence in brain microtubules: A model for consciousness. In *Toward a science of consciousness: The first Tucson discussions and debates*. edited by S. R. Hameroff, A. W. Kasvniak, and A. C. Scott. Cambridge, MA: MIT Press.

Harth, E. 1976. Visual perception: A dynamic theory. *Biological Cybernetics* 22:169–80.

Harth, E. 1983. Order and chaos in neural systems. *IEEE Transactions on Systems, Man, and Cybernetics* 13:782–9.

Harth, E. 1993. *The creative loop. How the brain makes a mind*. Reading, MA: Addison-Wesley.

Harth, E., T. J. Csermely, B. Beek, and R. D. Lindsay. 1970. Brain function and neural dynamics. *Journal of Theoretical Biology* 26:93–120.

Harth, E., and A. S. Pandya. 1988. "Dynamics of the Alopex process: Application to optimization problems." In *Biomathematics and related computational problems*, edited by L. Y. Ricciardi. Dordrecht: Kluwer Academic, pp. 459–71.

Harth, E., A. S. Pandya, and K. P. Unnikrishnan. 1990. Optimization of cortical responses by feedback modification of sensory afferents. *Concepts of Neuroscience* 1:53–68.

Harth, E., and E. Tzanakou. 1974. Alopex: A stochastic method for determining visual receptive fields. *Vision Research* 14:1475–82.

Harth, E., K. P. Unnikrishnan, and A. S. Pandya. 1987. The inversion of sensory processing by feedback pathways: A model of visual cognitive functions. *Science* 237:184–7.

Hebb, D. 1949. *The organization of behavior.* New York: John Wiley & Sons.

Hersh, S. M., and E. L. White. 1981. Thalamocortical synapses with cortico thalamic projection neurons in mouse SMI cortex. *Neuroscience Letters* 24:207–10.

Hopfield, J. J. 1982. Neural networks and physical systems with emergent collective computational abilities. *PNAS* 79:2554–58.

Kafatos, M., and R. Nadeau. 1990. *The conscious universe.* New York: Springer-Verlag.

Koch, C. 1996. "Towards the neuronal substrate of visual consciousness." In *Toward a science of consciousness: The first Tucson discussions and debates,* edited by S. R. Hameroff, A. W. Kaszniak, and A. C. Scott. Cambridge, MA: MIT Press.

Kosslyn, S. M., N. M. Alpert, W. L. Thompson, V. Maljkovic, S. B. Wise, C. F. Chabris, S. E. Hamilton, S. L. Rauch, and F. S. Buonanno. 1993. Visual mental imagery activates topographically organized visual cortex: PET investigations. *Journal of Cognitive Neuroscience* 5:263–87.

Minsky, M. 1985. *The society of mind.* New York: Simon & Schuster.

Penrose, R. 1989. *The emperor's new mind.* New York: Oxford University Press.

Rothblatt, L., and K. H. Pribram. 1972. Selective attention: Input filter or response selection? *Brain Research* 39:427–36.

Scott, A. 1995. *Stairway to the mind.* New York: Springer Verlag.

Scott, A. 1996. "The hierarchical emergence of consciousness." In *Toward a science of consciousness: The first Tucson discussions and debates.* edited by S. R. Hameroff, A. W. Kaszniak, and A. C. Scott. Cambridge, MA: MIT Press.

Searle, J. 1992. *The rediscovery of mind.* Cambridge, MA: MIT Press.

Shepard, R. N., and L. A. Cooper. 1982. *Mental images and their transformations.* Cambridge, MA: MIT Press.

Stapp, H. P. 1993. *Mind, matter, and quantum mechanics.* New York: Springer-Verlag.

Steriade, M. 1986. Reticularis thalami neurons revisited: Activity changes during shifts in states of vigilance. *Journal of Neuroscience* 6:68–81.

Tzanakou, E., R. Michalak, and E. Harth. 1979. The Alopex process: Visual receptive fields by response feedback. *Biological Cybernetics* 35:161–74.

Watson, R., T. E. Valenstein, and K. M. Heilman. 1981. Thalamic neglect. A possible role of the medial thalamus and nucleus reticularis in behavior. *Arch. Neurol.* 38:501–4.

Wong, R., and E. Harth. 1973. Stationary states and transients in neural populations. *Journal of Theoretical Biology* 40:77–106.

45 A Framework for Higher-Order Cognition and Consciousness

Nils A. Baas

INTRODUCTION

The idea of viewing consciousness as an emergent property of the brain is not new. Especially Sperry has strongly argued this point of view (for example Sperry 1983). Recently also Hameroff (1993) and Scott (1995) have argued in the same direction. Our own interest in this comes from an attempt to understand how complex systems are built from simpler ones and how new properties are being created. Intuitively complex structures arise when more primitive structures are being put together in a nontrivial way. Certainly the brain is a complex hierarchical structure of neurons, and out of this complexity consciousness seems to emerge. The question is then: What kind of hierarchical complexity and what does emergence mean?

Therefore it seems fruitful to introduce a formal framework in which emergence, complex structures, and hierarchies can be discussed. The unifying concept we call a hyperstructure, which defines what reasonably should be called a higher-order structure or object (Figure 45.1).

EMERGENCE

The notion of emergence will play a key role in our approach to consciousness. Often emergence is being used in rather vague terms; therefore, we will introduce a more precise framework for emergence.

Intuitively, emergence is used as a name for the creation of new structures and properties. Therefore it has been intimately connected with the theory of evolution. We shall not discuss the history, development of emergence—or present-day approaches to emergence, but just refer to the references.

Which are the basic ingredients needed in order to explain what is meant by emergence? We need some primitive objects or entities. For something *new* to be created we need some dynamics or better—*interaction*—among the entities. But in order to register that something new has come into existence, we need mechanisms to observe the entities. This is intuitively the idea we have in mind and would like to express in a more formal language. But we can only represent events for which we have "meters" (Minch 1988).

We shall start out with a general notion of *structure* as our primitive objects or entities. Structures may in our sense be of an abstract or physical nature, for example, systems, organizations, organisms, machines, spaces, fields, symmetries, and concepts. Furthermore we assume that we have some kind of observational mechanism (or family of such) in order to evaluate, observe, and describe the structures. Of course such mechanisms can be defined in many ways—classically as numerical functions—but as we have seen in quantum mechanics more sophisticated ones may be needed (operators). In fact they may be very general (functors—in a category theory setting). We want to give a general procedure for how to construct a new structure from a family of old ones. So we start with a family of structures:

$$(S_i)\ i \in J \text{ (some index set, finite or infinite)}$$

Then we apply our observational mechanisms—*Obs*—to obtain properties of the structures:

$$S_i, \text{Obs}(S_i)$$

Next we subject the S_i's to a family of interactions—*Int*—using the properties registered under the observation (which could be a dynamic process). Hence we get a new kind of structure:

$$S - R(S_i, \text{Obs}(S_i), \text{Int})_{i\ \in J}$$

where R stands for the result of the construction process.

We call S a *second-order structure*—as opposed to the S_i's which we consider *first-order structures*—the primitives of the theory. Specifying the observational mechanism is basic and we may also require some stability of the observed properties. The interactions may be caused by the structures themselves or imposed by external factors. Basically we are describing how to form a totality out of a family of structures. The observational mechanism may not only be of an external nature, but also of an internal nature.

In mathematics we have many examples of such constructions. Of particular interest in connection with emergence are *limits* in so-called categories, but we do not want to restrict ourselves to this situation. Also our notion of observation plays a crucial role. *Obs* is related to the creation of new categories in the systems (categories here in the philosophical meaning of the word).

Let us prepare the notation for further steps. First-order structures—our primitives—are denoted by:

$$S^1_{i_1},\ i_1 \in J_1,$$

second-order structures

$$S^2 = R(S^1_{i_1}, \text{Obs}^1, \text{Int}^1)$$

and families

$$\{S^2_{i_2}\},\ i_2 \in J_2$$

Second-order structures may now be observed by observational mechanisms Obs^2 (they may be equal, overlap, or disjoint from Obs^1), and they may also observe the first-order structures of which they consist. (We may allow first-order structures to be considered also as second-order structures $S^2 = R(S^1)$, in the same way as in set theory an element of a set may also be considered as a subset $\{x\}$.)

The collection S—which is more than a mere aggregate—is considered as a new unity, whose properties we measure by Obs^2.

In this quite general setting we propose the following:

Definition Of Emergence:

$$P \text{ is an emergent property of } S^2$$

if

$$P \in Obs^2(S^2), \text{ but } P \notin Obs^2(S^1_{i1}) \text{ for all } i_1.$$

So in this sense we may say that:

The whole is more than the sum of its parts.

This is what we shall call *first-order emergence*. In the next section we shall define higher-order emergence (Figure 45.2).

Let us finally point out that this also relates to cognition, which may be defined as "the emergence of global states in a network of simple components" (Varela, Thompson, and Rosch 1992). For more on emergence, see Appendix 1.

HYPERSTRUCTURES

Hierarchical structures occur often in biological systems. It seems that they have been favored by evolution. In studying multilevel systems, a basic problem is the passage from one level to the next. It is like going from one set of biological or physical types to a higher-order type. In a way this is analogous to type theory in logic. It is our opinion that going from one level to the next is exactly what emergence is about. Therefore, in developing a framework and theory for higher-order structures, emergence—in our sense—will play a vital role. We shall not only cover the classical hierarchical tree-type situation, but also the "flatter" network-type structures. All this will be embraced in our notion of a hyperstructure—so we shall now give a general introduction.

In our definition of emergence we considered structures at two levels: primitives, or first-order structures, and second-order structures. We may say that a structure formed as S^2 is a *complex* structure relative to its primitives. The "observer"—which could be just the environment of a system—represents a kind of *selection* mechanism giving rise to a process creating *new levels of structure*. This also gives *adaptation* relative to Obs (or environment). So we have a mechanism of *evolution* towards a higher-order structure. Level structures are often thought of as *hierarchies,* but hierarchies with the traditional tree-type structure are insufficient for our purposes—though they will appear as special cases of our construction. We shall return to this later.

Next we proceed to form families of second-order structures. They are being observed to obtain new second-order properties and interactions, giving:

$$S^3 = R(S^2_{i_2}, Obs^2, Int^2).$$

In general, Int^3 may be interactions based on both first- and second-order properties. Now this process proceeds and we see that at each level of construction new properties or new behavior may emerge, giving room for *new interactions* and, hence, each level is necessary in order to get the last level's properties. The process of *Obs* is essential and shows that this construction is much more sophisticated than a recursive procedure, which may be considered a special case. The new interactions may not be meaningful at lower levels.

Finally, we form the *Nth order structure*:

$$S^N = R(S^{N-1}_{N-1}, Obs^{N-1}, Int^{N-1}, \cdots), i_{N-1} \in J_{N-1}$$

This is a short way of writing an Nth order structure, which in principle may depend on all lower levels in a cumulative way and is more general than a recursive structure. The whole process we call *the higher-order structure principle*—the result of which we call a *hyperstructure*, in this case of order N (or an emergent structure of order N).

The underlying interaction graph in Figure 45.3 is much more subtle than merely trees (as for ordinary hierarchies) or general graphs or networks.

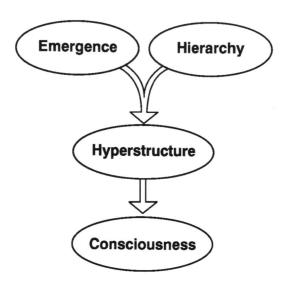

Figure 45.1 Hyperstructure in relation to consciousness.

EMERGENCE

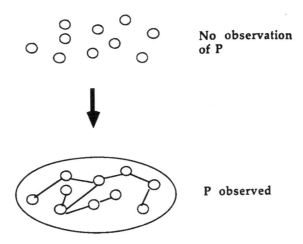

Figure 45.2 Emergence as a function of observation.

Hyperstructure

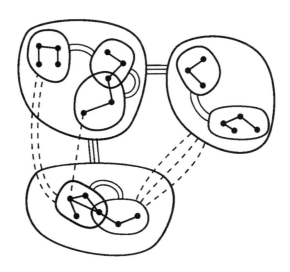

Figure 45.3 Schematic example of a hyperstructure.

We think that this is a general and important process that may be applied to a variety of structures (systems) in order to obtain new—*fundamentally new!*—structures with new emergent properties. Through the level structure, we also get orders of emergence and complexity, indicating a classification of such structures. *Nth order emergent properties are those of S^N, which is a cumulative structure.*

We shall not write down in technical details the rules of the interactions at the various levels, but just refer to Figure 45.3, which describes the intuition. It is worth noting that we allow (not require) overlapping aggregates and that creation of high-level interactions may cause changes in lower-level interactions. Hyperstructures allow for both "downward" and "upward" causation. As a model one may think of interactions of organizations and social systems. If needed, interactions may also be deleted (or forgotten) so that the system is only partially cumulative.

It is our conviction that *complexity often takes the form of a hyperstructure*. This is an extension of H. Simon's thoughts about complexity and hierarchies (Simon 1962). The formation principle we have described for hyperstructures may be useful in understanding for example cognitive and constructional processes. It is really a framework for studying problems of *complex design* (Dawkins 1986). In a sense it reflects the principle of evolution where observation and interaction are due to the various environments and the structures themselves. But in our model we allow for introducing new observations and interactions and controlling them, hence being able to influence the evolution and even speed up evolution. This flexibility clearly makes the structure more easily adaptable to various changes, and the *Obs* and *Int* mechanisms will allow for creativity (of nonpredictable nature), learning, and so on.

Two basic questions for hyperstructures are:

1. How to build good models generating higher-order hyperstructures from first principles?
2. How to organize matter in a hyperstructured way and what are the properties?

We consider self-organization as a hyperstructured process.

For more on hyperstructures, see Appendix 2 and Baas (1994a, 1994b). We will next try to connect our framework with cognitive processes and consciousness.

HIGHER-ORDER COGNITION AND CONSCIOUSNESS

If we think of cognition as the appearance of global states or properties in a network of primitive components, we need a kind of *Observer* in order to recognize or register these new states or properties. The Observer could, for example, be some kind of sensory system. Therefore, cognition may be viewed as emergence in our sense.

The new states or properties will give rise to new interactions among networks again, new *Observations* and this will eventually lead to a hyperstructured process as a model for higher-order cognition.

Our point is that cognitive processes are not only *syntactic* but also *semantics*. The semantics are represented by the Observer. Reasoning is a combination of using syntactic rules, observations, and interactions over and over again. For cognitive structures one could appropriately call this *emergent deduction* expressed as:

$$\frac{\frac{S^1}{S^2}}{\vdots}$$

like inference rules in logic. But this remains to be developed further.

We think that hyperstructures may provide a useful framework for describing and studying cognitive processes. As pointed out previously the use of Observers in the construction may cause nonalgorithmic behavior.

In general we view organisms as examples of hyperstructures where we take cells as primitives—not meaning that we necessarily understand all the emergent processes involved. Depending on the purpose we could also have taken atoms or molecules as primitives. For these structures life—under any reasonable definiton—becomes an emergent property.

With respect to human consciousness, we will think of it as associated with the brain and considered a complex system of neurons. Based on available evidence (for example, Scott 1995, Sperry 1983 and references therein) we may take neurons as primitives and are led to the following suggestion:

The brain is hyperstructured and consciousness is an emergent property of high order.

Consciousness is therefore emergent in the same way as life is, but at different scales with different primitives and orders. This is similar to Alwyn Scott's (1995) point of view.

Assuming that the brain is hyperstructred in some way one may ask: *Are there areas of special importance for the coherent functioning of the higher-order level structures?* (Not meaning strict localization!) Broca's and Wernicke's areas may be candidates in such a scheme. This should be related to Edelman's theory of higher-order consciousness in Edelman (1989).

The hyperstructure hypothesis is supported by the cell assembly theory of D. Hebb and the work of T. Wiesel and D. Hubel on the structure of the cat's visual cortex (see Scott 1995 for a discussion).

Interesting ideas on higher-order cognition and consciousness are presented in Ehresmann and Vanbremeersch (1992). Their approach is basically within the framework of our hyperstructures, but it uses the language of category theory.

We hope that putting consciousness into the present framework may lead to new and interesting questions and ideally to experiments and a theory with predictive power. At the moment this is too much to ask. We should also emphasize that what we have said about consciousness is merely speculations based on selected available evidence and subjective interpretation.

QUESTIONS AND CONJECTURES. WHAT IS CONSCIOUSNESS? WHAT IS LIFE?

These two fundamental questions are not directly related but they are in a way very similar. In recent years there has been intense activity in order to characterize life and life-like behavior. This is what the new field artificial life is about (Langton 1988). Therefore it seems natural to search for a similar characterization of consciousness and conscious-like behavior. This leads to a new notion of *artificial consciousness* to be studied.

We suggest the following:

Conjectures

A. For a medium to have consciousness it must be organized as a hyperstructure.

B. Consciousness is a hyperstructured cognitive process.

These conditions should be considered as necessary, but not necessarily sufficient. Furthermore, this implies that consciousness is a *multilevel process* and that we may have *degrees* of consciousness. Clearly, human consciousness should be of a very high degree or order, but there may be higher! In this context consciousness is obviously an emergent phenomenon.

In Baas (1994b) we suggested that it could be interesting to form various new kinds of matter into hyperstructured materials using methods from nanotechnology. Such hypermaterials (or "nano-matter") could possibly represent a new kind of matter bridging the gap between living and nonliving matter.

Question 1 Could there also be some notion of consciousness—bridging the gap between unconscious and conscious? Is the difference basically a matter of difference in degrees of consciousness? Unconsciousness meaning low order consciousness?

As mentioned earlier, it seems natural as a starting point to take neurons as primitives in a hyperstructure for studying consciousness. But our framework allows for taking primitives at the nanoscale level as well in order to incorporate quantum effects. The importance of quantum effects have been argued by Hameroff (1993) and Penrose (1989).

Penrose argues that "there must be an essentially nonalgorithmic ingredient in the action of consciousness." It should be pointed out that nonalgorithmic behavior may not only arise from quantum effects, but also from structures associated with nonlinear interactions in the hyperstructure as pointed out in the section on emergence. Equally important is another statement by Penrose: "Consciousness is needed in order to handle situations where we have to form new judgements, and where the rules have not been laid down beforehand."

This motivates question 2.

Question 2 Is consciousness more of a principle of how a medium may form new Observational (semantic) mechanisms and structures in order to create and handle new types of situations, form judgments, and so on? Again, leading to "higher-order semantics" via newly created observational structures?

How can quantum mechanical effects possibly influence macroscopic processes such as consciousness? S. Hameroff has argued that quantum coherence among neural proteins may explain this. Cytoskeletal components such as microtubules may be particularly suitable for quantum level effects because of their quasicrystalline structure, proposed coherent excitations, and molecular, nanoscale dimensions (Hameroff 1993). He then suggests that microtubules and other elements of the cytoskeleton constitute the bottom level substrate in the dimensional hierarchy from whose highest level consciousness emerges. This is closely related to the question of quantum coherence and coherent excitations in biological systems (Fröhlich 1968, Ho and Popp 1993).

This leads us to question 3.

Question 3 Could it possibly be the case that hyperstructures represent a kind of coherent organization of a medium that would stimulate or facilitate the occurrence of global quantum effects?

This is thought of as an extension of the situation in microtubules. Proteins are vibrant and dynamical structures, and the coherent excitations are cases of emergence. In a hyperstructured medium we may get excitations and vibrations at several levels, but could we then under favorable conditions expect a transfer or propagation of local effects to global effects via the same mechanisms that cause the emergence and tie the hyperstructure together? In general vibrations in hyperstructured media seem like an interesting phenomenon, and may be closely connected with consciousness.

CONCLUSION

We have presented a framework for emergence and higher-order structures and discussed higher-order cognition and consciousness within this framework. Based on this we have presented some thoughts and questions that we hope will stimulate further work on consciousness.

The appendices contain more details from Baas (1994a, 1994b), see also further references therein.

APPENDIX 1: MORE ON EMERGENCE

It may be useful to distinguish between two different notions of emergence—for quite deep reasons both mathematically and philosophically.

A. *Deducible or computable emergence* means that there is a deductional or computational process D such that $P \in Obs^2(S^2)$ can be determined by D from $S^1{}_{i1}\ Obs^1, (Int^1)$.

B. *Observational emergence* if P is an emergent property, but cannot be deduced as in A.

Let us indicate a few examples.

Deducible Emergence

1. Coupled or compositional structures (finite or infinite) where the components interact in known ways such that we can deduce or calculate the *new* composite properties. This is the case in many types of engineering constructions.

2. Nonlinear dynamical systems where, for example, simple systems interact to produce *new* and complex behavior. Chaotic dynamical systems may theoretically be of type A, but in practice may be usefully thought of as type B, and hence cases of borderline emergence.

3. Phase-transitions—emergence in the thermodynamic limit. Broken symmetries.

4. A simple and explicit example is the following. Given two simple dynamical systems:

$$S_1^1 : \dot{x} = ay$$
$$S_2^1 : \dot{y} = -bx$$

$Obs = Obs^1 = Obs^2 =$ set of various properties of the system, including existence of periodic solutions.

P is the property of having a periodic solution.

Therefore $P \notin Obs(S_1^1)$ and $P \notin Obs(S^1{}_2)$. We then let the systems interact to form a predator/prey model *a la* Lotka-Volterra:

$$S^2: \begin{matrix} \dot{x} = ax - cxy \\ \dot{y} = -bx + dxy \end{matrix}$$

$$(a, b, c, d > 0)$$

It is a simple well-known fact that $P \in Obs(S^2)$, hence an emergent property (of type A) caused by the nonlinear couplings or interactions.

5. Depending on our choice of observables we can also create emergence linearly. Let us observe oscillatory solutions. We may consider linear systems with no oscillation, but when coupled linearly we observe oscillatory solutions.

   This shows the importance of *Observables* in our theory, and also how it gives great flexibility.

6. In topology, a manifold is a space that is glued together by locally Euclidean pieces.

$$M = \cup_{i=1}^{n} V_i / \sim$$

To study or observe such spaces one uses for instance cohomology theories K^*. We know that $K^*(V_i) = 0$ (trivial), but $K^*(M)$ may be very complex. Let $Obs = Obs^1 = Obs^2$ = set of properties including measuring the nontrivial complexity of K^*. P property of nontrivial K^*. Then $P \notin Obs(V_i)$, but

$$P \in Obs(M)$$

In this sense we may say that the nontriviality of $K^*(M)$ has emerged from the gluing construction. Lots of similar examples exist in topology and algebra where, for example, gluing is replaced by limits.

7. The Scott-model of the λ-calculus. Let X be a set, $[X \to X]$ = the set of mappings from X to X.
Consider the *set* equation:

$$X = [X \to X]$$

Within set theory this is not solvable. In constructing a model of the λ-calculus D. Scott considered spaces D_i, with a special topology and continuous mappings $[D_i \to D_i]$. He started with a very simple set D_0, formed

$$D_1 = [D_0 \to D_0]$$

and inductively

$$D_{i+1} = [D_i \to D_i]$$

in such a way that they formed an inverse system

$$D_0 \leftarrow D_1 \leftarrow \ldots \leftarrow D_i \leftarrow \ldots \leftarrow D_\infty$$

where D_∞ is the so-called inverse limit.
For all i

$$D_i \neq [D_i \to D_i]$$

but in the limit

$$D_\infty = [D_\infty \to D_\infty].$$

Let us put this into our framework.
Set

$$S_i^1 = D_i$$
$$Int^1 = \leftarrow$$

$Obs(D) = Obs^1(D) = Obs^2(D)$ = set of properties of D, being a solution of the equation $D = [D \to D]$.

$$P = \text{property of being a solution of } D = [D \to D].$$

Then we have:

$$P \notin Obs(D_i), \text{ but}$$
$$P \in Obs(D_\infty)$$

So we may say that the property of being a solution to the equation emerged through the inverse limit construction. For details, see Stoy (1977).

In these examples we assume that the various properties can be decided by well-defined procedures. Therefore *Obs* is really a procedure or algorithm leading to a set of properties. Some of these examples may look rather abstract and formal, but we think they will help us in better understanding the nature of emergent structures and properties.

Observational Emergence

1. A profound example here is Gödel's theorem (Gödel 1986), which states that in some formal systems there are statements that are true, but this cannot be deduced within these systems. Here observation is the truth function. We think of this as follows: first-order predicate calculus is complete in the sense that every true statement can be deduced from the axioms. If we now add further axioms to cover the theory of arithmetic, Gödel's theorem says that there will be true statements that cannot be deduced (internally). We view here the additional axioms as "added interactions" among the well-formed expressions.

2. Consider a dynamical system in some space. View the dynamics as an interaction and associated fractal-like sets such as attractors, repellers, Julia sets, and so on as newly created second order structures. *Obs* is deciding membership in these sets.

Penrose (1989) raised the question: Is the Mandelbrot set decidable? (That is, can membership be algorithmically decided; in our terminology, is it deducible?) In order to make the question mathematically meaningful, it had to be put into a new theory of computation over the real numbers developed by Blum, Shub, and Smale (1989).

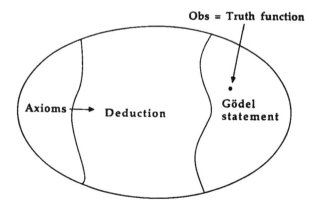

Figure 45.4 Schematic illustration of Gödel's theorem.

They showed that most Julia sets are undecidable. In 1991, M. Shishikura (1991) completed the proof of the conjecture that the Mandelbrot set is also undecidable.

So in a suitable choice of structure in our framework, we may view Julia sets (most of them) and the Mandelbrot set as observationally emergent structures.

3. Another example of the difference in our two notions of emergence is the following. In a formal string system we have a syntactic relation

$$s = R(s_1, \ldots, s_k)$$

Take as our *Obs* a semantic "meaning function" M. The question about semantic compositionality is then: Does there exist a (computable) function F such that

$$M(s) = F(M(s_1), \ldots, M(s_k))?$$

If yes, $M(s)$ is *deducibly emergent*.
If no, $M(s)$ is *observationally emergent*.

We have here only considered two basic classes of emergence. Clearly both can be further refined and subdivided according to additional complexity conditions.

The notion of observational emergence is quite profound as the examples show. Gödel's theorem as observational emergence may at first appear as a rather negative result for constructive purposes. But it is not. The reason is that we have incorporated the observational mechanisms in such a way that they can be usefully taken into account in further constructions—even if they are not deducible or computable. This is an important aspect that will be used in our hyperstructure concept. Our discussion here shows that even in formal, abstract systems profound emergence may occur—particularly in artificial life models. This seems to be contrary to a claim by Cariani (1991).

We have here given a general framework for studying emergence, but we have not discussed the detailed causes of emergence. Let us mention as an observation two basic causes:

1. Nonlinearity of the interactions, meaning that there is no superposition, principle.

2. The number of structures (systems, entities) is often large (number $\to \infty$).

When these two conditions are satisfied, emergent phenomena will often occur. For a more general discussion of the relation between limits and emergence, see Ehresmann and Vanbremeersch (1987). We end this section with an old and provocative question: Does emergence in some cases—as contended by Mill, Lewes, and Morgan—rank among the *laws of nature* (Morgan 1923)?

MORE ON HYPERSTRUCTURES

Let us give some examples of known structures that may naturally be considered as hyperstructures—but without details.

1. Higher-order organisms viewed as evolutionary products.
2. Hypercycles in the sense of Eigen and Schuster (1979). Here the *Obs* mechanism registers the properties leading to intercyclic coupling and hence higher-order cycles.
3. Coupled nonlinear dynamical systems may often be viewed as hyperstructures through the appearance of, for example, limit cycles and attractors.
4. Higher-order logic and set theory. In the syntactic build up, no use of *Obs* is made. *Obs* would normally be thought of as part of the semantic structure. In a sense oracles may be considered as observers. Furthermore, higher-order (or high-level) languages and metaconstructions may also be viewed as hyperstructures.
5. Higher-order graphs and networks—formalizing the structure illustrated in Figure 45.3. This also extends the already existing notion of a hypergraph—which in our terminology is a second-order structure.
6. Various types of hyperspaces in mathematics. These are spaces of the form 2^x (or suitable subsets or iterations of this construction). 2^x = {set of subsets of x}. Fractals are naturally studied in spaces of this type.
7. Cognitive processes where we successively build up knowledge by "observations" and "interactions."
8. General technological constructions designed in a modular or levelwise way and metatype constructions in system theory.

It is important to notice that the higher-order structure principle can be used to synthesize new structures by varying the *Obs* and *Int* structures. To a large extent we can vary the "laws" or rules at each level, but in real-life evolution they are due to nature. *Obs* and *Int* may even vary continuously—by some kind of sensor-device. In other structures, *Obs*, for example, may play no role—as in higher-order logic considered syntactically. An important way to look at the interactions is as communication or information processing. For every level, the information-carrying capacity increases. In the successive formation of levels in a hyperstructure, we may have to impose more conditions depending on the situation—especially conditions on *stability* and *autonomy* seem to be relevant.

Hyperstructures should be thought of as a language—a framework—to describe complex problems, solutions, and situations. We may also generate a whole series of interesting questions and notions trying to associate hyperstructures with well-known structures, systems, concepts, symbols, signs, and so on, for example, hypernetworks, hyperlanguages, hyperdynamics, hyperalgorithms, hypercomputation, hyperforms, hyperpatterns,

. . . also, physical and social systems may be organized in this way leading, for example, to hypermaterials and hyperorganizations. In all the cases we have to specify *Obs* and *Int*. New observational structures may also emerge.

This is a very general and broad program, and further progress will depend on the study of *specific* examples of these structures from various angles: mathematical, physical, chemical, biological, and computational.

Also in nanotechnology, hyperstructures may represent a useful architectural scheme. It is absolutely conceivable that it could give new results combining emergence and levels of observation. The purpose is of course to design a system from which primitive building blocks construct new levels with new properties.

What kind of emergence of levels do we have in mind? Biological molecules emerge in a sense from lower-order molecules. Cells emerge from biological molecules. Organisms emerge from cells. Ordinary chemical (low level) molecules emerge from atoms. We do not claim that these processes of emergence are fully understood. But putting these together we get higher-order emergence—as we have tried to formalize in hyperstructures. Of course many more levels may be involved. It is important in the hyperstructural scheme to include the notion of an *Observer* as we have done. This will allow us to actively enter the evolution-process—if desired, and in many cases direct and speed up evolution. *So we get a scheme for man made evolution.* This should be compared with the recent methods developed for example by Beaudry and Joyce (1992) in order to mimic molecular evolution to obtain biological molecules with novel properties.

The possibility of getting new properties increases with the ability to manipulate lower and lower levels. It is also important to use the ability to complexify the lower nanolevels—as has been done successfully with the introduction of microchips and integrated circuits. Here we enter the world of quantum mechanics.

REFERENCES

Baas, N. A. 1994a. "Emergence, hierarchies and hyperstructures." In *Artificial life III, SFI studies in the sciences of complexity*, Vol. XVII. edited by C. G. Langton. Reading, MA: Addison-Wesley, pp. 515–37.

Baas, N. A. 1994b. "Hyperstructures as a tool in nanotechnology. *Nanobiology* 3:49–60.

Beaudry, A., and G. Joyce. 1992. Directed evolution of an RNA enzyme. *Science*, 257:635–41.

Blum, L., M. Shub, and S. Smale. 1989. On a theory of computation and complexity over the real numbers: NP completeness, recursive functions and universal machines. *Bull. Amer. Math. Soc.* 21(1):1–46.

Cariani, P. 1991. "Emergence and artificial life." In *Artificial Life II, SFI studies in the sciences of complexity,* Vol. X. edited by C. G. Langton, C. E. Taylor, J. D. Farmer, and S. Rasmussen. Reading, MA: Addison-Wesley, pp. 775–97.

Dawkins, R. 1986. *The blind watchmaker.* Essex, England: Longman Scientific & Technical.

Edelman, G. M. 1989. *The remembered present.* New York: Basic Books.

Ehresmann, A. C., and J. P. Vanbremeersch. 1987. Hierarchical evolutive systems: A mathematical model for complex systems. *Bull. of Math. Biology* 49:13–50.

Ehresmann, A. C., and J. P. Vanbremeersch. 1992. "How do heterogeneous levels with hierarchical modulation interact on a system's learning process?" In *Human systems and information technologies,* edited by G. Lasker. Windsor, Canada: Publications University of Windsor, Canada, pp. 181–86.

Eigen, M., and P. Schuster. 1979. *The hypercycle: A principle of natural self-organization.* Berlin: Springer-Verlag.

Fröhlich, H. 1968. Long-range coherence and energy storage in biological systems. *Int. J. Quantum Chemistry* 2:641–9.

Gödel, K. 1986. *On formally undecidable propositions of principia mathematica and related systems I, Collected works.* Oxford, England: Oxford University Press, pp. 145–95.

Hameroff, S. R. 1993. "Quantum conformational automata in the cytoskeleton: Nanoscale cognition in protein connectionist networks." In *Towards a material basis for cognition.* (In press) Abisko, Sweden.

Ho, M. W., and F. A. Popp. 1993. Biological organization, coherence and light emission from living organisms. "In *Thinking about biology,* edited by W. D. Stein and F. J. Varela. *SFI Studies in the Sciences of Complexity, Lect. Notes, Vol. III,* Reading, MA: Addison-Wesley, pp. 193–213.

Langton, C. 1988. "Artificial life." In *Artificial life, SFI studies in the sciences of complexity,* edited by C. Langton. Reading, MA: Addison-Wesley.

Minch, E. 1988. Representation of hierarchical structure in evolving metworks. State University of New York at Binghamton. Thesis.

Morgan, L. C. 1923. *Emergent evolution.* London: Williams and Norgate.

Penrose, R. 1989. *The emperor's new mind.* Oxford, England: Oxford University Press.

Scott, A. C. 1995. *The stairway to the mind.* New York: Copernicus (Springer-Verlag).

Shishikura, M. 1991. The Hausdorff dimension of the boundary of the Mandelbrot set and Julia sets. New York: Stony Brook, Preprint, Institute for Mathematical Sciences, SUNY.

Simon, H. A. 1962. The architecture of complexity. *Proceedings of the American Philosophical Society* 106:467–82.

Sperry, R. 1983. *Science and moral priority.* New York: Columbia University Press.

Stoy, J. E. 1977. *Denotational semantics: The Scott-Strachey approach to programming language theory.* Cambridge, MA: MIT Press.

Varela, F. J., E. Thompson, and E. Rosch. 1992. *The embodied mind—Cognitive science and human experience.* Cambridge, MA: MIT Press.

46 Bioenergetic Foundations of Consciousness

B. Raymond Fink

Human consciousness depends on profuse use of phosphoryl chemical energy (~P) by the brain. Since ~P is redox in origin, the evolution of consciousness necessitated the coevolution of a suitably plentiful intake of oxidant (O) and reductant (R), constrained by the dynamics of the O and R intake hub, the pharynx. The respiratory inflow of O is quasicontinuous, and though rate-limiting and vital to consciousness, does not itself demand consciousness. What, in humans, does require consciousness, and may indeed be its principal reason, is intermittent garnering of R and *initiation* of swallow. Swallowing of food, after conscious initiation, usually triggers *reflex* protective closure of the larynx. The larynx, it seems, has undergone extensive coevolution with the brain. Phylogenetically, the larynx presents serial additions to its structure and mechanism apparently in association with increases in the size and complexity of the cerebrum. Each addition tends to displace the larynx caudally and appears to increase the dimensions of the open passage. Certain monkeys added the subhyoid air sac, partly usable as an inflatable, elastic larynx plugging mechanism. In apes the homologous structure is solidified and, in human, solid and elongated, apparently in conjunction with further progression of the above-mentioned features. Joint progressions of the larynx and cerebrum are recognizable during human embryonic and fetal development and postnatal maturation. Parsimony can interpret the coordinated complexifications as related manifestations of an integrative morphogenetic action somehow residing in the flow of O-based phosphoryl energy. Starting in the mitochondria of the ovum, the energy would dissipate through the genome and its products and give rise to a self-organizing, evolutionary necessary harmonious dissipative structure—the organism, and would be experienced in waking *Homo sapiens sapiens* as human consciousness.

INTRODUCTION

Consciousness, like all life, is sustained by the flow of a particular form of chemical energy, *phosphoryl chemical energy*, here dubbed "phosphergy" and symbolized ~P. ~P is packaged in molecules of adenosine triphosphate (ATP), which can be written adenine-ribose-phosphate-phosphoryl-phosphoryl, or adenine-ribose-P~P~P, to show that there are two units of ~P in each ATP molecule, although often it is only the second unit that is put to work. The magic of phosphoryl resides in the electronic strain of the ~P (the anhydride PO^3 group), which transfers to any molecule it becomes attached to (and is technically called the group transfer potential) and activates the new combination (Lehninger, Nelson, and Cox 1993). In the early biosphere the source of ~P may have been simple pyrophosphate, P~P.

A human adult makes some 2.3 kilograms of ATP every day (Zubay 1993). You use more than ten million million million molecules of ATP while reading this line, twenty percent of them in your brain (Courville 1953)—such is the phosphergy price of consciousness and language. Every one of these molecules is redox-derived, that is to say, it is derived from reactions between reductant (R) of food and oxidant (O) of air. Garnering and using the R and O ingredients of phosphergy may be considered the basic chemical function and meaning of every instant of life in every living organism, including every instant of human consciousness.

One can make only a limited amount of ATP in a day. The limit is set, not by the compact R, which needs to be eaten only at lengthy intervals, but by the tenuous O, which humans have to breathe quasicontinuously. The geohistorical evolution of the ATP-voracious human body and brain (and consciousness) has therefore depended on coevolving an adequate intake system of O—an adequate O logistics. Analogous coevolution must also have occurred in other extant vertebrate clades. Unfortunately, the logistics do not fossilize and its evolution has to be deduced by studying existing species, an interpretive exercise that attracts few investigators but is nonetheless a legitimate integrative scientific approach to the study of consciousness. How *did* -P flow dependent on adequate O logistics evolve and become conscious of itself? An initial sketch is attempted here.

EVOLUTION OF "PHOSPHERGY" LOGISTICS

Life on Earth may have begun when flow of phosphergy through a phosphergy-making system replicated the system. In the course of multimillion years of chance and necessity evolution, bacterial-replicating systems of "phosphergy" flow appeared, followed eventually by metazoan animal systems, at first invertebrate and, later, vertebrate.

Extant invertebrate systems, though comprising a great diversity of phyla, display an O logistics that in one respect is completely stereotyped: intake of O always occurs through a pathway entirely distinct from the path of R intake. This is as true of a primitive jellyfish cnidarian, where O enters by simple diffusion from the body surface, as of an apical invertebrate such as a cephalopod that absorbs O through a gill in its mantle cavity (Meglitsch and Schram 1991); in both the intake path of O is throughout unshared with that of R.

The invertebrate mode of energy logistics is immensely successful in terms of speciosity, but, in general, remarkably deficient in generating individuals with "personality." It seems that animal systems of ~P flow in which the R and O logistic tracks are completely separate tend not only to stay relatively small but lack a logistic energy focus for the mind. The case is strikingly different in vertebrates.

Vertebrate systems inherently include a logistic focus of action and potential self-knowledge. This is so because in all vertebrates R and O enter a common cavity or hub, the pharynx (the so-called pharynx of invertebrates handles only R). For example, a prototypical fish drives respiratory O-containing water from the pharynx to the gills but drives nutrient R from the pharynx to the alimentary canal. Correct triage of R and O at the hub of energy intake is crucial to continued existence of a vertebrate, no matter whether it breathes water or air.

As regards O, breathing atmospheric air of course offers an enormous energy advantage since air contains some 30 times as much O as the same volume of aerated water. In a lungfish sufficient air is passed from the hub through a minute median fissure in the ventral wall of the posterior (that is, caudad) pharynx. Evolution has erected a complex master structure here, the larynx, derived from gill arch primordium. Phylogenetically, successive additions to the structure are recognizable (Goeppert 1937), specialized morphologically according to clade, and harmonizing with the O demand, amount of cerebrum, and presumptive cerebral decision-making power of the animal, as outlined below.

With respect to R, the larynx behaves as a switch, either closed to the entry of R and O, or open for the passage of O (in mammals the open condition is the default condition). In the human, laryngeal closure during R swallow is a *reflex* that follows on the *conscious* decision to *initiate* the swallow of a food bolus. As far as is known the sequence may also apply to other mammals (but see below). Food-swallowing closure of the larynx does not occur in an unconscious or sleeping person; instead, such a person will choke on food present in the pharynx and will be in imminent danger of death, as every anesthesiologist knows. Many nonanesthesiologists insist that nonhuman mammals and even human infants can breathe and swallow at the same time (Lieberman 1984). But the imagined virtuosity has never been documented. Cineradiographic studies have demonstrated that cat (Ramsey 1955, unpublished), pig, and ferret (Larson and Herring 1994), as well as human babies (Ardran and Kemp 1952), all close their larynx during the course, of a deglutition, just as do human adults (Saunders, Davis, and Miller 1951, Ardran, Kemp, and Lind 1952, Ramsey *et al.* 1955). The closure, of course, interrupts breathing. The larynx does remain open while such subjects collect milk or other food in the buccal cavity, but not when they swallow the collected material. In infant pigs and infant macaques, closure of the larynx during swallow is unproven, but simultaneous swallow and breathing also remains unproven (German and Crompton 1994). In general, in mammals the waking cerebral decision to swallow a bolus may conclude the waking series of decisions to forage and eat. Once past the hub, the food is safely inside the animal and further propulsion becomes the care of less energy-costly reflexes.

However, there has been more to cerebrolaryngeal coevolution than simple coenlargement of the glottis and cerebrum (Fink 1975). Both structures develop modifications adaptive to new modes of behavior, those in primates being of particular interest here. An interpretive sketch of some of the evidence follows (Figure 46.1). Such evidence, unfortunately, is at best indirect. The absence of fossil larynges compels utilization of extant species to conjecture the phylogeny of the human laryngeal morphology and mechanism.

EXAMPLES OF O LOGISTICS IN VERTEBRATES

1. In the Australian lungfish, Lepidosiren (Figure 46.1, 1) the larynx is a fissure in the mid-ventral wall of the pharynx, with muscle fibers arranged to close the fissure. The lungfish gulps air into the mouth and forces it into the lungs by contractions of the pharyngeal constrictor muscle (Negus 1929).

2. Among amphibia, in the axolotl salamander (*Ambystoma mexicanum*) (Figure 46.1, 2), two small plates of cartilage on either side of the slit give attachment of muscle fibers running outwards which act as openers of the glottis (Negus 1929).

3. In newt salamander (Figure 46.1, 3), the upper separated tip of each lateral cartilage appears as an arytenoid cartilage while the lower remainder fuses with its fellow to form a cricoid ring. This sets up the possibility of rotatory and tilting motion of the arytenoids on the cricoid, and formation of a triangular opening instead of a slit.

4. In monotreme mammals (Figure 46.1, 4), a distinct, well-developed thyroid cartilage is present above the cricoid, together with a bilateral undivided thyroarytenoid fold and, above the thyroid, an epiglottic cartilage. The new additions enlarge the open glottis to a quadrilateral, the anterior part between the thyroarytenoid folds being guarded by the epiglottis.

5. In marsupial mammals (Figure 46.1, 5), elastic corniculate cartilages attached to the tops of the arytenoids are found, projecting medially and backward. Compressed by arytenoid adduction, they probably act as springs that separate the arytenoids and help to reopen the glottis after closure.

6. In placental mammals (Figure 46.1, 6), the thyroid cartilage is no longer fused to the cricoid but is free to move separately. The thyroarytenoid folds are now divided. The lower ones are the vocal folds, specialized for phonation; the upper folds are the vestibular folds, specialized for laryngeal closure. Each vestibular fold contains an additional elastic cartilage, the cuneiform cartilage, that functions as a torsion spring loaded by adduction of the vestibular folds and whose recoil returns the folds to the open con-

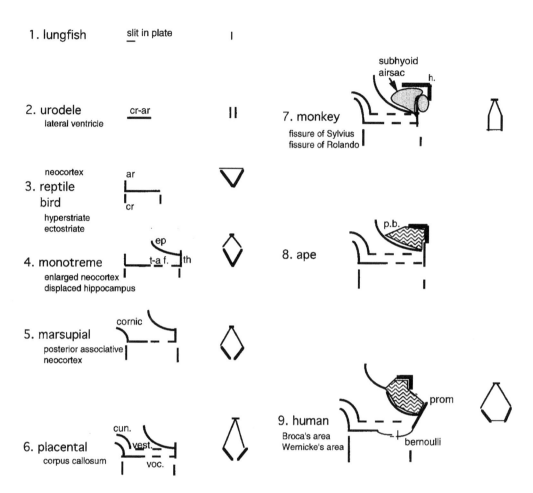

Figure 46.1 Novelty at cerebrum, larynx (side view), and glottis (cranial view). (Schematic, not to scale). These diagrams illustrate the phylogenetic sequences of innovations in vertebrate laryngeal structure (lateral view) and shapes of open glottis (cranial view), as deduced from living species (Goeppert 1935, Negus 1962, Fink et al. 1976). Some coeval cerebral innovations are also listed. 1. Lepidosiren lungfish. 2. Urodele amphibian. cr-ar, cricoarytenoid cartilage (bilateral). 3. Reptile. Bird. c, cricoid cartilage; a, arytenoid cartilage (bilateral). 4. Monotreme mammal. t, thyroid cartilage; e, epiglottic cartilage; t-a f., thyroarytenoid fold (bilateral). 5. Marsupial mammal. cornic., corniculate cartilage (bilateral) 6. Placental mammal. cun., cuneiform cartilage; voc., vocal fold; vest., vestibular fold (all bilateral). 7. Cercopithecoid monkey. h, hyoid bone. 8. Pongid ape. p.b., pre-epiglottic body. 9. Hominid human. prom., prominentia laryngis (Adam's apple); bernoulli; cranial view of adducted right vocal fold.

figuration. Subdivision of the thyroarytenoid folds was distinct in only 42 of the 121 studied nonprimate species studied by Negus (1962), but in all of the 35 primate species.

7-9. Primates

Certain primate clades present an increasingly caudad station of the larynx relative to the cranium and the hyoid bone (Jordan 1960). The laryngeal passage appears enlarged (Figure 46.1), and the cerebrum increased in size (Figure 46.2). The effect on the laryngeal passage is predicted because of the bellows-like construction of the laryngeal folds in primates. The paired vocal and vestibular folds and the median epiglottic fold converge on and attach to the thyroid cartilage; in the human ascent of the larynx toward the hyoid bone folds and closes the cavity of the bellows, while descent of the larynx away from the hyoid, as in inspiration, unfolds and opens the passage and tends to draw the arytenoids laterally. Recoil of elastic membranes and ligaments automatically helps to restore the intermediate open configuration.

In nonprimate mammals the arytenoids rotate on the cricoid cartilage, yielding a quadrilateral glottis, but in cercopithecoids (monkeys) the arytenoids slide mediolaterally, so that the vocal processes remain parallel and the open glottis becomes pentagonal: the lateral slide is limited by attachment between the tips of the corniculate cartilages. In apes and humans this attachment is absent and the lateralized arytenoids are free to rock back-

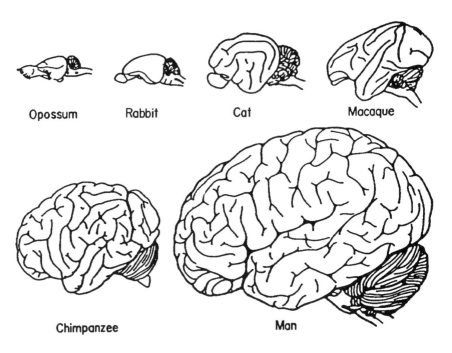

Figure 46.2 Brains of mammals drawn on the same scale (Courtesy of Professor Jansen). From Eccles (1989), reproduced by permission of the publisher.

ward round the sloping axis of the cricoarytenoid joint and make the vocal processes diverge and further expand the open glottis.

Specializations of the vocal and vestibular folds adapt primates to various modes of behavior and locomotion. In lemurs and New World monkeys, the free edges of the vocal folds are turned upward to form cusps, which, when adducted, form an inlet valve that prevents inward passage of air and can fix the rib cage as an aid to brachiation (Negus 1929).

In Old World monkeys, in addition to the cusp-like vocal folds, down-turned vestibular folds are present along with a median subhyoid air sac in front of the epiglottis (Figure 46.1, 7). This sac empties into the larynx at the convergence of the vestibular and vocal folds, so that compression of the sac while the vocal and vestibular folds are adducted will elastically reinforce the apposition of the cusps and fix the volume of the rib cage, preventing both decrease and increase, correspondingly enlarging the gamut of larynx-supported brachial effort.

In apes the subhyoid sac is solidified (Figure 46.1, 8), apparently strengthening the exit plug and the resistance to thoracic collapse when weight-lifting in a bipedal stance. In *Homo* the plug is solid, slender, and elongated (Figure 46.1, 9), seemingly dedicated with the *prominentia laryngis* to sustaining bipedal manual work efforts (Fink 1974) yet by its elasticity also serving speech formant versatility. Correlative with full time bipedality, the human vocal and vestibular folds furthermore lack cusps. The vocal fold appears to be specialized in support of low-effort phonation: the elastic ligament at the edge of the vocal fold becomes arched outward during adduction (Farnsworth 1940), establishing a paramedian axis of fold oscillation (or "vibration"). The paramedian axis enables vocalizing airflow to suck the elastic free edge aerodynamically toward the midline before increasing air pressure forces it in the opposite direction, so that the total excursion is a full amplitude and not a half amplitude as it would be if the vibratory cycle started from the midline. The paramedian axis thus not only economizes phonatory energy but also mitigates midline collision of the phonating folds. Discernible in high speed motion pictures of the human vocal folds (Farnsworth 1940), this subtle specialization seems beautifully adaptive to ease of conversation.

DISCUSSION

It is pertinent to note here that *Homo sapiens sapiens* is the sole hominid to have developed a protruding chin. As well as a signal of improved cranial balance atop an erect vertebral column (Dubrul 1958), a relation to speech mechanics is not far to seek: the jutting chin accompanies a shortened, curved dental arch that decreases the inertia and hence the energy cost of conversational accelerations of the mandible, while the chin itself preserves or enhances the length of the geniohyoid and genioglossus muscles attached to its back and promotes mobility of the hyoid bone and tongue in the production of speech formants. The paleontological lateness of the chin's contribution to advanced speech mechanics suggests that the abovedescribed paramedian vocal fold aerodynamics also may be a *H. sapiens sapiens* innovation, possibly along with additional caudad progression of the station of the larynx (which tends to unfold the laryngeal bellows and facilitate the throughput of O in support of an enlarged cerebrum, but also facilitates graduated folding of the bellows in speech). However, there is certainly no unanimity of opinion on these points.

The larynx's capacity for innovative coevolution in parallel with that of the cerebrum may partly stem from its developmental origin in pluripotent branchial arch mesoderm of the viscerocranium (Warwick and Williams 1973), which is derived from cranial neural crest. It is also interesting that the above-discussed conjoint phylogenetic elaboration of laryngeal and cerebral mechanisms is hinted at in the codevelopment which these structures undergo *in utero* and postnatally. The coupling of these various trends suggests that an underlying influence orchestrates the O inflow and mechanical capacities of larynx with the potential O demand and behavioral repertoire of the cerebrum. Individually, each effect recalls the organizing tendency attributed in nonequilibrium thermodynamics to the flow of energy in a dissipative structure (Prigogine and Stengers 1979). In a living system several such effects might be coupled by the flow of phosphoryl energy. In vertebrates the O-based flow would originate in the mitochondria of the ovum, channel through the genome and its products, and generate coupled self-organizing dissipative structures, producing phenotypes harmonized in the crucible of evolution, and emerging in *H. sapiens* as consciousness.

REFERENCES

Ardran, G. M. and F. H. Kemp 1952. The protection of the laryngeal airway during swallowing. *British Journal of Radiology* 25:406–16.

Ardran, G. M., F. H. Kemp, and J. Lind. 1958. A cineradiographic study of bottle feeding. *British Journal of Radiology* 31:11–22.

Courville, C. B. 1953. *Contributions to the study of cerebral anoxia.* Los Angeles: San Lucas Press.

Du Brul, E. L. 1958. *Evolution of the speech apparatus.* Springfield, IL: Charles C. Thomas.

Eccles, J. C. 1989. *Evolution of the brain: Creation of the self.* London: Routledge.

Farnsworth, D. W. 1940. High speed motion pictures of the human vocal folds. *Bell Laboratories Record* 18:203–08.

Fink, B. R. 1974. The thyroid cartilage as a spring. *Anesthesiology* 40:58–61.

Fink, B. R. 1975. *The human larynx: A functional study.* New York: Raven Press.

Fink, B. R., E. L. Frederickson, C. Gans, and S. E. Huggins. 1976. Evolution of laryngeal folding. *Annals of the New York Academy of Sciences* 280:650–59.

German, R. Z., and A. W. Crompton. 1994. Integration of swallowing and respiration in infant mammals. *Journal of Morphology* 220:348.

Goeppert, E. 1937. "Kehlkopf und Trachea." In *Handbuch der vergleichenden Anatomie der Wirbeltieren,* Vol. 3. edited by L. Bolk, E. Goeppert, E. Kallins, and W. Lubosch. Amsterdam: Asher (1967); Berlin: Urban and Schwarzenberg (1937).

Jordan, J. 1960. Quelques remarques sur la situation du larynx chez les lemuriens et les singes. *Acta Biologica Medica* (Gdansk) 4:39-51.

Larson, J. E., and S. W. Herring. 1994. Movement of the epiglottis in mammals. Submitted for publication.

Lehninger, A. L., D. L. Nelson, and C. M. Cox. 1993. *Principles of biochemistry,* 2nd ed. New York: Worth.

Lieberman, P. 1984. *The biology and evolution of language.* Cambridge, MA: Harvard University Press.

Meglitsch, P. A. and F. R. Schram. 1991. *Invertebrate zoology,* 3rd ed. New York: Oxford University Press.

Negus, V. E. 1929. *The mechanism of the larynx.* St. Louis: C.V. Mosby.

Negus, V. E. 1962. *The Comparative anatomy and physiology of the larynx.* New York: Hafner.

Prigogine, I., and I. Stengers. 1979. *La nouvelle alliance.* Paris: Gallimard.

Ramsey, G. H., J. S. Watson, R. Gramiak, and S. A. Weinberg. 1955. Cineflourographic analysis of the mechanism of swallowing. *Radiology* 64:498-518.

Saunders, J. B. de C. M., C. Davis, and E. R. Miller. 1951. The mechanism of deglutition (second stage) as revealed by cineradiography. *Annals of Otology, Rhinology and Laryngology* 60:897-916.

Warwick, R., and P. L. Williams. 1973. *Gray's anatomy.* 35th British ed. Philadelphia: W. B. Saunders.

Zubay, G. 1993. *Biochemistry,* 3rd ed. Dubuque, IA: Wm. C. Brown.

47 The Hierarchical Emergence of Consciousness

Alwyn C. Scott

INTRODUCTION

The wish to know who we are and where we come from is as old as mankind, arising—no doubt—with the emergence of self awareness as a special aspect of consciousness. Over the past few hundred years science has made substantial contributions to this quest, first of all by convincing us that our little planet is not the center of the universe with all the heavenly bodies revolving reverentially about, and later by showing that our species, *Homo sapiens*, was not specially created by God but evolved in a natural way from related forms of life.

Although such insights have disappointed some, science has been remarkably helpful in practical matters, providing denizens of the modern world with a cornucopia of tools and toys that are based on the laws of mechanics, electromagnetism, thermodynamics, quantum dynamics, or chemistry. From this success it is natural to expect—or at least to hope—that science might also provide an answer to the ultimate question posed above, and scientists often try to respond. Thus the story is told of a physicist who worked to develop a "theory of everything." After years of effort, he was successful. All knowledge—he had shown—could be expressed by the simple equation:

$$F = 0$$

Unfortunately he needed two hundred and forty-three large volumes to describe F. The present discussion is intended as a contribution to the same problem but is far less ambitious because there will be no attempt to stuff all knowledge into the Procrustean confines of a single equation.

Instead of assuming that a theory of everything exists—perhaps in some Platonic sphere—it seems appropriate to begin with a survey of what is known. To this end it is convenient to arrange current knowledge into the following hierarchy.

This diagram is not at all original; it is based upon observations of what those who know—or claim to know—write and talk about as a visit to any science library will confirm. Furthermore it is not complete. I would not quarrel with any reasonable additions or rearrangements; the aim is merely to suggest the hierarchical concept in a form that corresponds roughly to the organization of this paper.

<pre>
            culture
         consciousness
         phase sequence
                .
                .
                .
       assembly of assemblies
         assembly of neurons
          multiplex neuron
            nerve impulse
        axon-dendrite-synapse
   mitochondria-nucleus-cytoskeleton
     protein-membrane-nucleic acid
       phospholipid-ATP-amino acid
          inorganic chemistry
             atomic physics
</pre>

EMERGENT PHENOMENA

Upon examination of the hierarchy we see that new entities *emerge* from the dynamics of a particular level to provide a basis for the next higher level. From the perspective of modern *nonlinear dynamics,* with its interest in self-organization and pattern formation, this is to be expected (Scott 1970). Nonlinearity typically leads to the emergence of robust entities: eddies in a pool, tsunamis on the ocean, tornados in the atmosphere, and the Great Red Spot of the planet Jupiter are a few familiar examples. Within the realm of biology almost every dynamical process is nonlinear so emergent entities abound. In the present context, atoms form the building blocks for molecules, which combine to form proteins, which are components of biological cells, which compose organisms, and so on up the hierarchy to the heights—and depths—of human culture.

Linear systems are much easier to analyze than nonlinear systems since a complex cause can be expressed as a convenient sum of simple components. The combined effect is then the sum of the individual effects from each component of the total cause. For this reason linear systems have been favored by physical scientists, especially during the current century. In the nonlinear domain of biology, however, one bite stimulates the appetite while ten satisfy and twenty nauseate; a story told once can be amusing but told over again it becomes boring if not painful; one sperm will fertilize an egg, two can do no more. Nonlinear systems are more difficult to analyze because they are interesting. More things can happen—new *atomistic building blocks* emerge at each hierarchical level—and that is why the realm of biology is so rich.

For historical reasons, physical scientists tend to approach problems with the belief that the simplest of several competing theories is to be preferred. While this is a good rule for the study of cannon balls and chemistry, it is not necessarily appropriate for biology, and biologists often comment among themselves that one should not trust a biophysicist who has never got his or her hands wet in the laboratory. Among the specimen tanks and Petri dishes and test tubes one can seek out the most insignificant living species and then isolate the least interesting cell of that plant or animal. This cell will be—not in theory but in reality—wonderfully complex, to a degree not appreciated by many, if not most, physical scientists. Physicists must be careful not to decapitate biological reality with Ockham's razor.

As a simple example of the way that nonlinearity can lead to the emergence of a new dynamical entity, consider a large trampoline that is constructed from a rubber sheet. The *escape energy* of a child on the trampoline is the work that he or she must do to climb up to the edge. Suppose, to be specific, that the escape energy is equal to the square of the child's weight. Then the escape energy for two children (of weights W_a and W_b) standing together is:

$$\text{Escape energy} = (W_a + W_b)^2 = W_a^2 + W_b^2 + 2W_a W_b$$

Thus $2W_a W_b$ is the *extra* escape energy that appears because the two children are standing together on the trampoline. If they try to move apart, they must do $2W_a W_b$ units of work. As a pair, they are held together by this energy of attraction or binding energy. Thus the *pair of children* is a new atomistic entity that emerges because the trampoline is nonlinear.

A team of bicycle racers, a flock of geese, and a school of fish are extensions of this idea where the emergent team, flock, and school are held together by attractive energies that arise because the surrounding fluid (air or water) moves with them. But there are many variations on the theme. The atoms of a molecule of benzene are bound by a combination of electrostatic and chemical valence (quantum) energies; the membrane of a living cell organizes itself as a molecular bilayer in response to the shape and charge distribution of lipid molecules and the large dielectric constant of water; and the flame of a candle expresses a balance between the nonlinear localizing effect of combustion and the dispersive effect of thermal diffusion. At higher levels one can consider a burst of nervous activity, a memory, or a thought. It is, in fact, difficult to think of a biological *thing* that does not take advantage of nonlinearity to establish its integrity, its oneness.

As we survey the hierarchy of knowledge, we find that the appropriate dynamic laws (or laws of science) differ at each level. There is not one science, it seems, but a collection of them, each wedded to a particular level of the hierarchy. The biochemist pays little attention to atomic physics, the electrophysiologist has scant knowledge of psychology, and the chemist has no professional interest in ethnology. Although it is sometimes claimed that knowledge of the laws at one level helps to understand other levels, this is seldom so. At best the dynamical laws that emerge at one level may suggest a metaphor for those at another. There is a detailed theory for each level that a bona fide expert must master, and failure to do so is the mark of a dilettante, be he or she trained as a physiologist or as a philosopher.

One might ask why the hierarchy is not extended below atomic physics into the realms of nuclear dynamics and high-energy physics. Although such knowledge is interesting from the perspective of physical science, these studies have little if any relevance for atomic physics and, therefore, none at all for understanding the nature of human experience, and this understanding is a fundamental aim of the present discussion.

This point of view is reinforced by an evaluation of the reductionist approach to knowledge that is currently taking place in the physics community (Schweber 1993). In this debate, the reductionist position has been challenged by Philip Anderson, a distinguished condensed matter physicist, who has argued that:

> the reductionist hypothesis does not by any means imply a "constructionist" one: The ability to reduce everything to simple fundamental laws does not imply the ability to start from those laws and reconstruct the universe. In fact the more the elementary-particle physicists tell us about the nature of the fundamental laws, the less relevance they seem to have to the very real problems of the rest of science, much less to those of society. The constructionist hypothesis breaks down when confronted with the twin difficulties of scale and complexity. (Anderson 1972)

THE CELL ASSEMBLY

Above the nerve cell in our hierarchy is a level called "assembly of neurons." The emergent phenomenon at this level is the "cell assembly," a concept that Donald Hebb introduced into the theory of psychology in the late forties. His aim was to explain results of mammalian learning experiments that could not be understood in the context of either the Gestalt or the behaviorist frameworks that dominated mid-century American psychology (Hebb 1949). That this concept has stood the test of time is affirmed in a recent review of the cell assembly theory and of the research leading to its initial formulation (Hebb 1980).

Among the many billions of neurons in a human brain, Hebb supposed that some—through happenstance, practice, or both—become closely connected into a sort of "three dimensional fishnet" that can be widely dispersed throughout the brain. If a certain fraction of these interconnected neurons are induced to fire, they all fire and the assembly is said to be *ignited*. The term is apt. As the candle is a metaphor for pulse propagation along a nerve fiber, a pile of dried brush is a graphic model for a cell assembly. If enough twigs are lit, the entire pile begins to burn.

As a means for storing mental information, the cell assembly has two signal advantages over an individual neuron: first, like a hologram a cell assembly is robust to loss of information when some of its neurons are damaged or die because the information it carries is embodied in a pattern of interconnections that is distributed over many neurons, and second, a cell assembly can integrate concepts that are stored in different regions of the brain.

To introduce the basic idea, let us consider an analogous effect that occurs at the level of human culture. The residents of a large city—we imagine—have organized themselves into interest groups such as: alumni of the local university, the Methodist church, a motorcycle gang, the Junior League, a teen-age babysitting club, a chapter of the National Rifle Association (NRA), a committee for protecting the environment, and so on. Members of these *social assemblies* are interconnected in the sense that they share lists of addresses and telephone numbers, but any particular assembly is normally dormant. One morning the radio reports a proposed law that would require a longer waiting period for the purchase of a handgun. Two or three NRA members hear this news and almost immediately their entire assembly is ignited and begins firing messages at the state capitol. A week later half of the motorcycle gang is killed in a territorial dispute, but this is not the end of the club because the loss initiates a recruitment drive that soon brings the gang back to full strength.

Returning from this metaphor to the human brain, we note two important points. First a particular neuron may participate in many different assemblies—an NRA member might also be involved in protecting the environment—and accounting for this multiple participation is vital for determining the number of assemblies that can be organized in a brain site (Scott 1977b). Second, and even more important, Hebb's cell assembly shares two basic properties that characterize an individual neuron: (1) A *threshold* for stimulation below which the assembly remains quiescent, and (2) *All-or-nothing* activity.

Thus, several "assemblies of neurons" can become interconnected into an "assembly of assemblies" that shares the same two basic properties. Several of these, in turn, combine to form an "assembly of assemblies of assemblies" with the same properties; and so on up through the hierarchical structure of the brain.

To see how neatly Hebb's theoretical picture corresponds to the way our brains store information, consider the process of learning to read. First we learn to recognize lines and angles by organizing assemblies of neurons in the primary visual area of the cortex. Next these assemblies become interconnected to permit the perception of letters by the activation of assemblies of assemblies, or *second-order assemblies* in Hebb's (1980) terms. Then words by assemblies of assemblies of assemblies, or third-order assemblies, and so on up to the understanding of paragraphs, chapters, and the theme of a book. The dynamic complexity attained by the ideas perceived at these higher levels is difficult to imagine (Scott 1977a).

COMBINATORIAL BARRIERS

There is a powerful reason for the isolation between practitioners at different levels of the hierarchy. It is that the building blocks at a particular hierarchical level provide so many possible structures at the next higher level that these possibilities, although finite in number, cannot be listed. How can this be?

Although one tends to think that a finite number of items can, by definition, be put in a list, there is a practical sense in which this is not the case. A candidate for such a number is:

$$I = 10^{110}$$

because I is about equal to the atomic weight of the universe (or the mass of the universe measured in protons) times the age of the universe measured in picoseconds (10^{-12} seconds) or about the period of a molecular vibration). Walter Elsasser (1958, 1969) has proposed using the term *immense* to describe an integer that is equal to or greater than I. There is no way that a list of an immense number of items could be constructed because no conceivable computer could store such a list; and even if it were possible, there would never be enough time to inspect the list.

Immense sets abound. The number of chess games is immense, and this makes a match interesting; while the number of tick-tack-toe sequences is small enough to be listed by a twelve-year-old, so the game is dull. It is easy to show that the number of possible molecules is also immense as is the number of distinct proteins. There is an immense number of possible nerve cells and an immense number of primary assemblies that can be constructed from them. Similarly there is an immense number of second-order assemblies that can be constructed from those of first order, and an immense number of third-order assemblies can be constructed from those of second order, and so on (Scott 1977a). Near the top of the hierarchy there are immense numbers of possible ideas and songs and tools and human personalities and languages and cultural configurations. At the highest level of generality, there are an immense number of "possible worlds" so we can be certain that Doctor Pangloss failed to consider them all.

The immense number of structures that might emerge as we proceed from one level of the hierarchy of knowledge to the next precludes the possibility of predicting the character of the next higher level. It is necessary to look and see what is there, and chemists, biochemists, cytologists, neuroanatomists, electrophysiologists, psychologists, sociologists, and cultural anthropologists spend much of their time doing just this. Although atomic physics has something to say about chemistry, it cannot predict the molecules that we find interesting, just as the laws of biochemistry do not determine the proteins that are necessary to sustain the life of a tree or a triceratops. It is for this reason that each level of the hierarchy must be undertaken as an independent area of study. There is, in other words, a combinatorial barrier between adjacent levels, and our fabled physicist, it seems to me, had it wrong. Instead of trying to construct a theory of everything based upon the single equation $F = 0$ he should have considered many equations of the form:

$$F_j = 0$$

where $j = 2, 3, \ldots$ indicates a particular hierarchical level. This is the crux of the "reductionist-emergence" issue in the physics community (Anderson 1972, Schweber 1993). (It should be emphasized that this notation places no restrictions on the nature of the dynamics. If, for example, at level L of the hierarchy the dynamics is described by the set of differential equations: $dx/dt = f(x)$, where x is a vector, then the equation $F_L = 0$ can be written $dx/dt - f(x) = 0$. But this is *only* an example. If the dynamics can be described with mathematics, it can be expressed in the form of $F_j = 0$.)

Perhaps the diagram in Figure 47.1 will help to make things clear. Each of the shaded disks is intended to represent the domain of definition (or set of *atoms* on which the science is defined) for a particular level of the hierarchy. If the upper disk (labeled $F_1 = 0$) represents a particular human culture, then each point on it corresponds to an individual person. One of the people is represented by the next lower disk (labeled $F_2 = 0$), upon which each point might represent an aspect of consciousness, perhaps a complex cell assembly. And so on down through assemblies of assemblies, assemblies, neurons, axonal and dendritic branchings, membrane proteins, amino acids, molecular dynamics, to atomic physics. At each level of this diagram, the domain of definition (the structure of the shaded disk) cannot be specified *a priori* because there is an immense number of possible "atoms." At level j, averages over the dynamics at lower levels appear as parameters in the equation $F_j = 0$, and the variables at higher levels appear as boundary conditions. Thus causal influence is directed both upward and downward on the hierarchy.

CONSCIOUSNESS

Perhaps the most interesting event in the history of the universe—from the perspective of humankind—has been the evolution of consciousness: a phenomenon that is again becoming accepted as a subject for serious study.

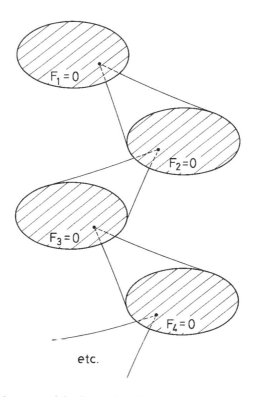

Figure 47.1 A diagram of the hierarchical organization of knowledge. Roughly speaking, each shaded disk corresponds to a branch of science.

An aim of this paper is to show how consciousness might fit into the hierarchical description of knowledge.

Physical scientists often approach this subject with reluctance because it seems necessary to choose among the following positions:

1. The *materialist* view that all of reality can be reduced to the implications of the "laws of physics and chemistry"; thus, whatever consciousness might be, it has no causal effectiveness in the real world, and
2. The *dualist* view that consciousness is "something more than" the activity of the brain, with which it interacts.

Neither of these positions feels comfortable. The outlook of the materialist is at variance with everyday experience, while the dualist seems to backslide into eighteenth century animism. The hierarchical perspective suggests a middle way between these unhappy choices.

The view presented here is that the brain itself is a hierarchical structure with several levels of dynamical activity (Scott 1978). From this perspective, consciousness emerges as a level of mental activity involving interactions among assemblies of assemblies of . . . of assemblies of neurons that are temporally organized into thought trains—the *phase sequence*—and the traditional dispute between monism and dualism becomes superseded by a perspective that honors the dynamical realities at all levels of the scientific hierarchy.

Psychic entities (such as feelings, dreams, fears, memories, ideas, habits, appetites, and so on) are enormously complicated things, and although some progress has been made in describing them—despite the baneful influence of behaviorism—one must admit that we are a long way from understanding how they might interact. Thus, natural science is not in a position to deny the possibility that consciousness emerges at a hierarchical level that enables it to monitor and guide (to some extent) the neurodynamics of the brain, even as it acts as an atom of human culture.

THE LOCUS OF CONSCIOUSNESS

In the hierarchical scheme presented above, consciousness was rather casually placed between human culture and the brain, but this says nothing about its physical location. Where should we look for it? From introspection, consciousness seems to me a little spotlight located somewhere behind the forehead that plays over the activities of the central nervous system, seeing a new relationship between hitherto disjoint ideas in one moment and feeling a sore knee or a twinge of guilt the next. Daniel Dennett tells us, however, that we should ignore such naive impressions. The "Cartesian Theater" is an illusion, he says, because cognition is distributed about the brain (Dennett 1991). Okay, but *how* is it distributed?

Some insight into the nature and locus of consciousness can be gleaned from a consideration of the phenomenon of life. In the context of a single-celled animal, life seems to emerge from the following hierarchy of scientific levels:

cytology
biochemistry
chemistry

The way that life might emerge from the biological hierarchy was suggested about fifteen years ago by Manfred Eigen and Peter Schuster in their book entitled *The Hypercycle: A Principle of Natural Self-Organization* (Eigen and Schuster 1979). Their basic concept is of an interrelated hierarchy of cyclic reaction networks (or a cycle of cycles of cycles) which they call a *hypercycle*. An example of a basic cycle is the citric acid cycle (which uses a molecule of oxaloacetate over and over again to extract the energy from acetic acid), but there are hundreds of these cycles in a simple organism. At the next level of the hierarchy, several such catalytic cycles are organized into an autocatalytic cycle, which manages its own reproduction. An example of this is the replication mechanism of single-stranded RNA. Several autocatalytic cycles are then organized at a higher level into a *catalytic hypercycle,* like a virus, which can evolve into more efficient structures. Thus, life has the hierarchical form:

catalytic cycle
↑↓
autocatalytic cycle
↑↓
basic catalytic cycle

In these nested and interacting cycles, energy, mass, and information are all currencies that are traded back and forth along the arrows.

Biomathematicians usually concentrate their attention on a single level of the scientific hierarchy, but hypercycles require a broader perspective. Instead of a description that is expressed as a single system of differential equations, a nested hierarchy of such mathematical descriptions is needed. Averages over the variables at one level may appear as parameters in the higher-level systems, and instantaneous values of the variables at a higher level establish the structures that are seen at lower levels. A number of biologists and applied mathematicians are beginning to consider life from the perspective of Eigen and Schuster (Baas 1994, Farmer, Kauffman, and Packard 1986, Nuño *et al.* 1993), and it is my belief that studies of the dynamics of such hierarchical structures will become increasingly important in the years ahead.

In one of the more striking of these efforts, Walter Fontana and Leo Buss have developed a minimal theory of biological organization that generates higher order structures from a simple representation of molecular interactions (Fontana and Buss 1994). This algebraic theory, called the *lambda-calculus,* was introduced in 1941 by Alonzo Church as a means to study the computability of mathematical algorithms (Church 1941). (In the briefest possible terms, it can be said that Church's lambda-calculus puts mathematical functions and their arguments on an equal theoretical basis.) Fontana and Buss show that the lambda-calculus provides a natural setting for considering the question: How do atoms organize themselves into molecular structures? In the context of this calculus, they find that the atoms in (the mathematical equivalent of) a well-stirred pot organize themselves into different hierarchical levels, where "Level 0 is defined by self-copying objects or simple ensembles of copying objects. Level 1 denotes a new object class, whose objects are self-maintaining organizations made of Level 0 objects, and Level 2 is defined by self-maintaining metaorganizations composed of Level 1 organizations." The corresponding diagram:

Level 2

↑↓

Level 1

↑↓

Level 0

is closely related to that of Eigen and Schuster for the structure of life.

Note carefully how the mathematics is being used here. It is not used to describe the dynamical details of the life processes but rather to study the manner in which a creature becomes organized. Can a similar approach be applied to the study of consciousness?

From the perspective of this paper (Scott 1995), an appropriate hierarchy for the dynamics of the mind is:

cultural configuration

↑↓

phase sequence

↑↓

complex assembly

↑↓

assembly of assemblies

↑↓

assembly of neurons

↑↓

multiplex neuron

↑↓

nerve membrane

↑↓

membrane protein

where the currency passed back and forth along the arrows is now information alone. Energy and mass are mere housekeeping details, of no more importance than yesterday's trash.

At the level of the nerve membrane, consciousness can be altered by chemicals that bind to membrane proteins and block ionic transport across the membrane. At the level of the neuron, consciousness is routinely switched off and on by anesthetic agents that change the actions of the synaptic contacts between cells. At higher levels, one is conscious of something called thought, which is stored in the myriad complex assemblies that have been pieced together throughout the years of learning. Thought, in turn, is formed by and interacts with the culture in which it develops. Up and down the hierarchy, from membrane ion channels to the ebb and flow of cultural interactions, an intense intercourse between the levels continues.

Just as life emerges from cycles of cycles of cycles of biochemical activity, consciousness seems to emerge from assemblies of assemblies of . . . of assemblies of neurons. This is what Erich Harth (1993, 1996) calls a "creative loop," and it is also termed a "hyperstructure" by Nils Baas (1996). Can such a hierarchy provide a solution to what David Chalmers (1996) calls the "hard problem" ? Can it explain the essence of consciousness?

Chalmers argues that an answer to this question requires the specification of an "extra ingredient" beyond mere mechanism. Traditionally this ingredient has been called the *soul,* although the behaviorists have dealt with the hard problem by denying it. From the perspective of natural science, both of these approaches are unacceptable. And what is left?

There are at least three suggestions for the extra ingredient. First, John Eccles (1994), Stuart Hameroff (1994), Roger Penrose (1994), and Henry Stapp (1993)—among others—suggest that consciousness may be embodied in a quantum mechanical wave function of some sort. Second, David Chalmers (1996) and Eugene Wigner (1969) feel that consciousness may be a new *primitive*, a fundamental property like the mass or electrical charge of an elementary particle. Finally, Donald Hebb (1949, 1980), John Searle (1992), Erich Harth (1993, 1996), Nils Baas (1996), and I—among others—suggest that consciousness may be a qualitatively new phenomenon that emerges from several levels of the mental hyperstructure in a nonreductive manner. Perhaps the third view was put best by Erwin Schrödinger, a founder of quantum mechanics, who did not see a quantum role in biology. Instead he wrote that the reason we cannot find our sentient self in our world picture is because (Schrödinger 1958):

It is itself that world picture.

If this position makes sense, it follows that research into the nature of consciousness must be truly interdisciplinary. In the study of life it is generally recognized that the joint efforts of chemists, biologists, biophysicists, biochemists, cytologists, physiologists, and physicians are all required in order to make progress. Similarly research on consciousness should meld the activities and insights of biochemists, cytologists, electrophysiologists, neuroscientists, engineers, computer scientists, physicians (particularly anesthesiologists, neurologists, and psychiatrists), psychologists, sociologists, and ethnologists. We must learn to listen to each other. Philosophers, physicists, and mathematicians may also have modest roles to play, but they should not seek to dominate the discussion.

Interdisciplinary research is often difficult to arrange because scientists have strong tribal instincts and tend to work on real biochemistry, real electrophysiology, real psychology, or real ethnology. This is of course entirely understandable. It takes much talent and years of hard work to become a first-rate biochemist, electrophysiologist, psychologist, or ethnologist. But to understand consciousness we must get beyond these parochial limits. We must set aside the outdated assumption that it is somehow more scientific to concentrate on a single layer of the hierarchy. Nowadays the really exciting science is interdisciplinary.

The new *Journal of Consciousness Studies* seems to be a step in this direction. The editors of this journal aim to:

> provoke serious, spirited debate by actively seeking opposing views, and are prepared to include a wide diversity of topics and approaches, from hard science to spiritual metaphysics; from the cultural imperialism of the "information superhighway" to deconstructionism.

It will be fun to watch this debate unfold.

REFERENCES

Anderson, P. A. 1972. More is different. *Science* 177:393–96.

Baas, N. A. 1994. "Emergence, hierarchies, and hyperstructures." In *Artificial life III*, edited by C. G. Langton. Reading, MA: Addison-Wesley.

Baas, N. A. 1996. "A framework for higher-order cognition and consciousness." in *Toward a science of consciousness*, edited by S. R. Hameroff, A. W. Kaszniak, and A. C. Scott. Cambridge, MA: MIT Press.

Chalmers, D. J. 1996. "Facing up to the problem of consciousness." In *Toward a science of consciousness*, edited by S. R. Hameroff, A. W. Kaszniak, and A. C. Scott. Cambridge, MA: MIT Press.

Church, A. 1941. *The calculi of lambda-conversion.* Princeton, NJ: Princeton University Press.

Dennett, D. C. 1991. *Consciousness explained.* Boston, MA: Little Brown.

Eccles, J. C. 1994. *How the self controls its brain.* Berlin: Springer-Verlag.

Eigen, M., and P. Schuster. 1979. *The hypercycle: A principle of natural self-organization.* Berlin: Springer-Verlag.

Elsasser, W. M. 1958. *Atom and organism: A new approach to theoretical biology.* Princeton, NJ: Princeton University Press.

Elsasser, W. M. 1969. Acausal phenomena in physics and biology: A case for reconstruction. *American Scientist* 4:502–16.

Farmer, J. D., S. A. Kauffman, and N. H. Packard. 1986. Autocatalytic replication of polymers. *Physica D* 22:50–67.

Fontana, W., and L. W. Buss. 1994. The arrival of the fittest: Toward a theory of biological organization. *Bull. Math. Biology* 56:1–64.

Hameroff, S. R. 1994. Quantum coherence in microtubules: An intra-neuronal substrate for emergent consciousness? *Journal of Consciousness Studies* 1:91–118.

Harth, E. 1993. *The creative loop: How the brain makes a mind.* Reading, MA: Addison-Wesley.

Harth, E. 1996. "Self-referent mechanisms as the neuronal basis of consciousness." In *Toward a science of consciousness: The first Tucson discussions and debates.* edited by S. R. Hameroff, A. W. Kaszniak, and A. C. Scott. Cambridge, MA: MIT Press.

Hebb, D. O. 1949. *The organization of behavior.* New York: John Wiley & Sons.

Hebb, D. O. 1980. *Essay on mind.* Hillsdale, NJ: Lawrence Erlbaum.

Nuño, J. C., M. A. Andrade, F. Morán, and F. Montero. 1993. A model of an autocatalytic network formed by error-prone self-replicative species. *Bull. Math. Biology.* 55:385–415.

Penrose, R. 1994. *Shadows of the mind.* Oxford, England: Oxford University Press.

Schweber, S. S. 1993. Physics, community and the crisis in physical theory. *Physics Today* 46:34–40.

Schrödinger, E. 1958. *Mind and matter.* Cambridge, England: Cambridge University Press.

Scott, A. C. 1970. *Active and nonlinear wave propagation in electronics.* New York: John Wiley & Sons. 1970. (Published in the USSR in 1977.)

Scott, A. C. 1977a. Neurodynamics (a critical survey). *J. Math. Psychology* 15:1–45.

Scott, A. C. 1977b. *Neurophysics.* New York: John Wiley & Sons.

Scott, A. C. 1978. Brain theory from a hierarchical perspective. *Brain Theory Newsletter* 3:66–9.

Scott, A. C. 1995 *Stairway to the mind.* New York: Copernicus (Springer-Verlag).

Searle, J. R. 1992. *The rediscovery of the mind.* Cambridge, MA: MIT Press.

Stapp, H. P. 1993. *Mind, matter, and quantum mechanics.* Berlin: Springer-Verlag.

Wigner, E. P. 1969. Are we machines? *Proc. Amer. Philosophical Soc.* 113:95–101.

X Phenomenology

Although this book is about consciousness, we have as yet read little of its direct experience. We have had theories of consciousness and discussions of its philosophical basis, we have been told about pathological states of consciousness and neural network models that might represent it. We have heard descriptions of cellular biology and quantum theory and nonlocal effects in space and time and the hierarchical ways that consciousness might be organized, but at this point the reader might well be asking: "What *is* consciousness?"

To begin to answer this question is the aim of the chapters in the present section. In response to those who might complain that the work presented here is "unscientific," we would point out that the first task of a scientist—the first step in application of the scientific method—is to observe carefully what needs to be explained and understood. Only when we know what we have before us, can we properly construct a theory of the mind.

The first chapter of the section is by Andrew Weil, a physician who has spent over two decades studying the effects of psychoactive drugs on the mind. In Chapter 48 he vividly recounts visions, shared hallucinations, and subjective phenomenology regarding reality itself—induced by naturally occurring plant and animal compounds. Although a considerable body of evidence is accumulating on what such chemicals do, Weil reports that we have no idea how they do it. The mental attitude of the subject (let us not say "patient") is clearly important to the effect, as is the cultural environment, but current knowledge of neural networks, microtubules, and quantum theory—in Weil's opinion—explains nothing. Theorists take note.

In Chapter 49 Brian Josephson and Tethys Carpenter come at the phenomenon of consciousness from an entirely different direction: the perception of music. They suggest that aesthetic beauty, including the appreciation of music, is a fundamental component of reality. Interestingly, Josephson and Carpenter see a relationship between the processes of life and those of aesthetics. That this discussion necessarily strays into the realm of aesthetics should not be unexpected. A century ago, George Santayana (1896/1955)—then a young colleague and former student of William James—undertook a careful study of aesthetics from the perspective of natural philosophy and pointed out the close relationship between the phenomenon of consciousness and the "ultimately irrational" sense of beauty.

Arthur Deikman is a psychiatrist who is concerned with the nature of spiritual experience, a subject of great interest to William James (1902/1929). Both emphasize the palpable reality of the mystical, so widely recognized by the poets and so memorably described by Martin Buber (1970) in recalling his perception of a linden tree:

> But it can also happen, if will and grace are joined, that as I contemplate the tree I am drawn into a relation and the tree ceases to be an It. The power of exclusiveness has seized me.

In Chapter 50 Deikman considers in some detail the differences between *instrumental consciousness*, which is familiar to scientists, and *receptive consciousness*, the mode of the poets (similar distinctions are made in chapters by Güven Güzeldere and Ian Marshall). In Deikman's view, a receptive mode is necessary in order to understand consciousness, but this does not mean that the instrumental mode must be ignored; both are important. As Buber put it, truly *seeing* the linden tree:

> ... does not require me to forego any of the modes of contemplation. There is nothing that I must not see in order to see, and there is no knowledge that I must forget. Rather is everything, picture and movement, species and instance, law and number included and inseparably fused.

Returning to the laboratory in Chapter 51, Richard Atkinson and Heath Earl describe experiments that show the positive effect of meditation on a perceptual task. Although the effect observed is relatively small in relation to the statistical uncertainty (a difference of 1.92 compared with an average standard deviation of 1.45), their results may be significant.

Finally in Chapter 52, José-Luis Diaz describes consciousness as "not a substance or an essence but a process that unfolds in time." He presents phenomenological features supporting a "Jamesian" *stream* of consciousness with several levels including: dreaming, being awake, the self-conscious state, and the state of ecstasy described above by Buber. Diaz's "continuously moving window of the present" depicts a link between consciousness and the physics of time. Some means for testing his model are suggested.

In review, Andrew Weil vividly recounts visions, shared hallucinations, and subjective phenomena regarding reality itself—induced by naturally occurring compounds. Brian Josephson and Tethys Carpenter suggest that aesthetic beauty, including appreciated music, is a fundamental component of reality, and Arthur Deikman considers modes of conscious thought best suited for spiritual experience. Richard Atkinson and Heath Earl report on studies of meditation effects, and José-Luis Diaz narrates a Jamesian stream of consciousness.

In future conferences of this sort, such careful reports of the phenomenology of consciousness should play a greater role.

REFERENCES

Buber, M. 1970. *I and thou.* New York: Scribner.

James, W. 1902/1929. *The varieties of religious experience.* New York: Random House.

Santayana, G. 1896 (1955). *The sense of beauty.* New York: Dover.

48 Pharmacology of Consciousness: A Narrative of Subjective Experience

Andrew Weil

This is a change-of-pace session, and I feel a certain responsibility in opening it. By training I am a botanist and a physician. I have only met four or five other physicians who majored in botany as undergraduates, and none of them make use of botanical training in their medical practices. I use a lot of it. As a practitioner of natural medicine, I recommend many plant-derived remedies to patients.

My present interests are healing and placebo responses, which I consider to be examples of healing responses mediated by the mind. Placebo responses are really the "meat" of medicine. Rather than worry about ruling them out, I think we should be finding out how to elicit them more of the time with treatments that are less invasive and less costly.

Some people who know me for my work in natural medicine are uneasy when they learn that my past work was research on psychotropic plants and drugs. I conducted the first controlled human experiments with marijuana in 1968, investigated psilocybin-containing mushrooms in North and South America, and studied a variety of South American hallucinogenic plants as well as medicinal plants. I did much of this work between 1971 and 1985, when I was a Research Associate in Ethnopharmacology at the Harvard Botanical Museum. Really, I cannot separate those two phases of my career, because much of my understanding of mind-body interactions and their relevance to healing derives from observations I made of people in altered states of consciousness induced by psychoactive drugs.

One of the most common questions I am asked in talking about psychoactive drugs—specifically about hallucinogenic or psychedelic drugs—is how they work. My answer is that we have not a clue. What we know about the pharmacology of these substances—for example, that they interact with serotonin receptors—tells us nothing about how they induce the complex, rich, and varied experiences that draw some people to them and make others run from them in terror. The purpose of my talk today is to draw your attention to the tremendous gulf that exists between what we understand neurochemically and neurobiologically what people actually experience of consciousness.

In the course of this talk I will refer to two ethnopharmacological puzzles that have recently been solved through research. One concerns an animal, the other a plant, both sources of unusual psychoactive drugs associated with very unusual experiences.

First let me tell you one brief anecdote from the time that I did controlled studies of marijuana. These were carried out at Boston University in 1968 (Weil, Zinberg, and Nelsen 1968).

Our basic intent was simply to show that you could study marijuana in human beings in a laboratory—no small feat for those days. There were great obstacles to obtaining marijuana legally, and to getting permission from all the Federal, state, and university bureaucracies. In some ways the universities proved most difficult. I was a senior student at Harvard Medical School at that time, and the only control that Harvard had over the project was through me; their human subjects committee threatened to withhold my medical degree if they did not like our experimental design. It was central to that design to use marijuana-naive subjects, because I was interested in finding out what marijuana did on its own to people who had no expectations of it as a result of prior experience. Harvard did not want me to use marijuana-naive subjects. They thought that if we introduced people to marijuana, the university was going to be sued when those people eventually turned into heroin addicts. After much wrangling, they decided that we could not use Harvard students as subjects but could use other people's students. (By the way, through all of this disagreement, my colleagues and I were having great trouble finding marijuana-naive subjects in the student population of Boston.)

Finally, the experiments commenced, and I was fascinated by the reactions of the marijuana-naive subjects, each of whom came in for three sessions, one with placebo cigarettes of ground-up male hemp stalks devoid of THC, and two with active marijuana cigarettes of different potencies. Although the experiments were double-blind, I could make good guesses as to who smoked marijuana and who did not because of the purely physical signs that people showed after consuming the cigarettes. Those who got THC had red eyes and increased heart rates, twenty or thirty points above baseline values. It was very impressive to see subject after subject with red eyes and increased heart rate sitting there saying, "Did I have a drug tonight? I can't believe that I had the real thing." Very interesting. Now this was a neutral laboratory setting, where no encouragement was given toward getting high, and this was what we found: people who were marijuana-naive, even though they showed all the physiological effects of the drug, had no experiences, no significant alterations of consciousness. I think that is a very important observation, one that demands explanation and one that suggests the kinds of problems that limit the use of pharmacology in trying to explain human experience.

In working over the years with various psychoactive substances, interviewing many users of them and reflecting on my own experiences with them, I have become a strong proponent of the view that drug experiences are created by three groups of factors: drug, set, and setting. Under "drug," I would include the pharmacological nature of the substance, dose, and route of administration, which can have enormous influence on the effects of psychoactive drugs. "Set" is that group of factors relating to people's expectations as to what will happen to them, both conscious and unconscious. A problem in analyzing set is that unconscious expectations are often more immediately determinative of experiences than conscious expectations. Unconscious expectations, by definition, are hard to elucidate. "Setting" refers to the physical environment in which a drug is taken as well as to the social and cultural environment.

These groups of factors interact to create an experience in a particular person at a particular time and place. Pharmacology alone, no matter how sophisticated, cannot explain these interactions by itself. The pharmacology of a drug is just one group of factors that interacts equally with the other two. In my experience, pharmacologists pay lip service to the concepts of set and setting, but then go right ahead and act as if all that matters is pharmacology. The hard fact is that under appropriately designed conditions of set and setting an experimentally administered dose of an amphetamine can cause a subject to fall asleep, while an experimentally administered dose of a barbiturate can cause a person to become alert and stimulated. Those responses are 180-degree reversals of pharmacological actions as we determine them in animals or in human subjects. The power of set and setting is vast; if you ignore these variables in trying to interpret responses to drugs, you do so at your peril. I tell you this as background for the experiences I will now describe.

I would like to read you a passage from a book that I came across in 1971. The book was *Wizard of the Upper Amazon*, and the author was Bruce Lamb, a man I subsequently came to know very well who died last year. He worked as a technical consultant for a large lumber company based in New York that some years ago was extracting tropical woods from the jungles of Peru. Bruce Lamb had managed large work crews of Peruvian laborers. One man he liked very much in the labor force seemed to command the respect of all other people. He had some sort of special charisma. As Lamb got to know him, he found that he was a famous curandero, a traditional healer. Over the years, they became good friends. The healer, whose name was Manuel Cùrdova, told his life story to Lamb, who eventually wrote it up in the book I am going to read a passage from. Manuel Cúrdova died at an old age some 15 years ago; his story continues to fascinate.

As a young boy, around the turn of the century, he was kidnapped by Amahuaca Indians in the northwest Amazon, whose territory was being invaded by rubber cutters. In dreams and visions, the Indians had seen that their way of life was about to be extinguished. They conceived the idea that if they got hold of the right person from the invading culture and trained him in their ways, he could save them. That is why they took the young Manuel Cùrdova to live with them. Their main method of training him was to give him repeated doses of the hallucinogenic beverage ayahuasca, which I think you heard about earlier in this conference from Fred Alan Wolf. Ayahuasca is the main hallucinogenic plant preparation of South America, used especially in the northwest Amazon. It is made differently in different regions, but the constant ingredient is a woody vine, *Banisteriopsis caapi*, whose active ingredient is harmine, one of the indole amine series of hallucinogenic drugs, the same series in which you find LSD and psilocybin. These drugs have structural similarities with serotonin and melatonin.

The bark of the *Banisteriopsis* vine is usually combined with leaves of other plants that contain a related psychedelic drug, dimethyltryptamine or DMT. When anthropologists and botanists first described the preparation of this beverage, they wrote that the DMT in it could not contribute to the activity because it is destroyed by monoamine oxidase in the gut and is only known to be active parenterally. Indians who make and drink ayahuasca say that they add the DMT-containing leaves to make the visions brighter. Their whole purpose in taking this beverage is to have visions, which they use in a very practical way to diagnose illness and determine correct courses of action through the changing circumstances of life. It turns out that harmine inactivates monoamine oxidase, so that when you combine these two plants, you get an orally active preparation of DMT. That is quite remarkable. If you ask anthropologists and botanists how Indians hit upon such a sophisticated pharmacological preparation, they tell you it was a matter of trial and error. If you are sitting in a lecture hall and a professor tells you that, you write it down in your notebook without questioning it. But I can tell you that if you are down in the rainforest looking at the profusion of plants, it is hard to imagine a shaman cooking up a new batch of Banisteriopsis (Lamb 1974) every day and saying, "Well, let's see. Today I will try this leaf." If you ask the shamans who make ayahuasca how they learned the technique, they also give you a very consistent answer. I have asked a lot of them, and they always reply that the *Banisteriopsis* vine showed them the other plants in visions. That is the answer you get from all the shamans, that they learned it in visions.

Cùrdova spent a number of years with this tribe until his relationship with them fell apart, and he fled back to his own culture. I would like to read you a brief description of one of his training sessions. At the beginning he took the beverage alone with the chief of the tribe, who was a master at arranging set and setting to manipulate the experience of the hallucinogenic potion. Only later, when Cùrdova was more familiar with ayahuasca did the chief allow him to join in group vision sessions.

> It was a select group of 12 that went to the secluded glade in the forest. It included some of the older men and several of the best hunters. The ritual and chants were similar to previous occasions, perhaps a little more elaborate. From the preparatory chants of the fragrant smoke and evocation of the spirit of the honi xuma (that is, of the vine that provides the drug), it was evident that Chief Xumu was attempting in this session to fix in my consciousness all the important or essential circumstances of their tribal life. There seemed to be an intense feeling of rapport among the group, all dedicated to the purpose of the old man.
>
> I was aware of the fragile hand that poured the magic fluid and passed the cups around to each. We drank in unison and settled into a quiet reverie of joint communion, savoring the fragrant smoke in the stillness of the silent forest A quiet chant held our conscious thoughts together as the potion took effect. A second cup was passed to intensify the reaction.

Color visions, indefinite in form, began to evolve into immense vistas of enchanting beauty. Soon, subtle but evocative chants led by the chief took control of the progression of our visions. Embellishments to both the chants and the visions came from the participants.

Soon, the procession of animals began, starting with the jungle cats. Some of these I had not seen before. There was a tawny puma, several varieties of the smaller spotted ocelot, then a giant rosetta-spotted jaguar. A murmur from the assembly indicated recognition. This tremendous animal shuffled along with head hanging down, mouth open and tongue lolling out. Hideous, large teeth filled the open mouth. An instant change of demeanor to vicious alertness caused a tremor through the circle of phantom viewers.

So, here are people having shared visions and reacting to the content of the visions in a consistent way.

Now, in hearing an account of an experience like this, the first question that comes up for me is, "Is it credible?" Do I believe it? How can I believe something so out of the ordinary unless I have had a similar experience or feel totally confident about the credibility of the people telling the story. I did not know Manuel Cùrdova, but I did know Bruce Lamb and had no reason to doubt his accuracy of reporting. Of course, he could have been fooled by his informant. In this case I have had an experience with a related substance that makes me willing to believe the story I just quoted.

The first of the two ethnopharmacological puzzles I promised to tell you about concerns a wonderful animal that is native to our desert here in Tucson, named *Bufo alvarius*, which used to be called the Colorado River toad and is now called the Sonoran Desert toad (Valdés, in press). It is an enormous toad: Large specimens are almost the size of footballs. At this time of year they are hibernating underground, but as the nights warm up in about another month, they will begin to come out. They congregate around lights at night, and eat every creature smaller than themselves. When the summer rains come, they mate, and their tadpoles develop quickly in standing water throughout the desert. Like all members of this genus *Bufo*, they have poisonous secretions. They have venom glands, large ones, behind the eyes and smaller ones on the legs that bathe the skin in toxins. *Bufo alvarius* has no predators. Many die under the tires of cars, but nothing eats them.

In the 1960s and early 1970s, the University of Texas Press published a monograph on the genus *Bufo* that included a chapter on the venoms of different species. It was noted in this obscure publication that *Bufo alvarius* was unique in having an unusual enzyme that converts bufotenin, a common toad toxin, into another analog of serotonin, a very potent psychedelic called 5-methoxy-dimethyltryptamine, or 5-MeO-DMT, a very close relative of the DMT in the leaves that make the visions brighter for ayahuasca drinkers. In fact, these two tryptamines often occur together in some South American plants that Indians prepare into hallucinogenic snuffs. *Bufo alvarius* venom contains up to 15 percent 5-MeO-DMT. I suspect that some people who read that monograph got the idea of trying the venom as a psychedelic, but I have no way to prove it. Some people think that knowledge of the psychoactivity of this toad came from native peoples who used

Figure 48.1 Sonoran desert toad (*Bufo alvarius*, or Colorado River Toad). Large glands (one is visible behind the right eye) produce venom which contains up to 15 percent 5-methoxy-dimethyltryptamine, a very potent psychedelic. This is the first known case of a hallucinogen from an animal source. (Photograph courtesy of Arizona Poison and Drug Information Center, University of Arizona Health Sciences Center.)

the venom as a hallucinogen in pre-Columbian times. In any case it is the first proved hallucinogen from an animal source.

Significant numbers of people here and elsewhere are now collecting these toads and milking their venom, which they dry and smoke. I am afraid this new interest represents a threat to our toads, especially since their habitats are also disappearing with urban growth. It is possible to milk the venom glands without harming the animals. The method is to squeeze a gland until it squirts a jet of venom, which can be collected on a plate. When the toad is released after this procedure, it goes right back to eating everything in sight, apparently undisturbed.

When I first heard about people inhaling the vapors of *Bufo alvarius* venom to alter consciousness, I was interested but skeptical because our toad has a strong reputation for being dangerously poisonous. It regularly poisons dogs that mouth it, often fatally. I know one dog owner who got a toad out of his dog's mouth within 10 seconds, but, nonetheless, after 30 minutes, the dog began salivating profusely, then had seizures and died in respiratory arrest. There is one report in the medical literature of a seven-year-old boy from southeastern Arizona who licked one of these animals and was brought to this institution in status epilepticus; he survived. Apparently, when you smoke the venom—and I say apparently because we do not yet have a chemical analysis to prove it—the powerful neurotoxicity is destroyed or largely destroyed, while the 5-MeO-DMT gets into your system quite efficiently. I know many people who use toad venom, some who have now smoked it hundreds of times, and they do not seem any the worse for wear.

When the vapors of 5-MeO-DMT are inhaled, they cause an almost instantaneous change in consciousness. Typically, a person will fall backwards and simply not be present for a few minutes. The experience is very brief, five to seven minutes, followed by a gradual return to ordinary awareness. Often people only remember the coming back; sometimes they recall a sense of having been dissolved, of having not existed for a period of time. Some people find the whole thing very scary. Others find the return, the reconstitution of the self, to be extremely pleasant. I think that high doses of *Bufo* venom would be quite frightening for most persons with little prior experience of drug-induced alterations of consciousness.

I have used the venom and synthetic 5-MeO-DMT just a few times, and I will relate one experience I had with it. I inhaled the vapors with a friend who had used it before, and the initial phase of the intoxication was—well, I just do not remember it. But in the period of reconstituting and becoming aware of myself and external reality, there was a distinct experience that lasted for a minute or two, in which we simply seemed to be present in each other's consciousness. When we were able to talk again, my friend told me he had had the same experience. So, while I have never seen simultaneous visions in a group, I have had a little taste of an experience that makes me think that such sharing of consciousness is possible, especially with training and under the skillful direction of someone who knew how to manipulate the altered state induced by a powerful hallucinogen.

Now, I would simply like to state my opinion that if, in fact, the experience I read you was real, I think we are as far as we can imagine from being able to explain that experience in terms of what we know about microtubules or neural networks. I am all for research on microtubules and neural networks. I also oppose the reduction of complex experiences that have immense meaning to those who experience them into any kind of mechanistic framework.

Now I will give you another example, having to do with the second ethnopharmacological puzzle, which concerns a plant that has long baffled researchers. Albert Hofmann, the Swiss pharmaceutical chemist who discovered LSD, and R. Gordon Wasson, the late New York banker who discovered the ongoing uses of hallucinogenic mushrooms in Mexico in the 1950s, teamed up to investigate the ethnopharmacology of three hallucinogens used traditionally by Mazatec Indians in the Mexican state of Oaxaca. Hofmann identified psilocybin from Wasson's mushroom specimens, then synthesized it. They then took the synthetic compound to María Sabina, the shaman who was Wasson's chief informant, in her little village in the mountains. She conducted a velada, a mushroom ceremony, using the synthetic drug from Switzerland. At the end, she pronounced it good and told them she was happy because she no longer had to depend on rains for her ceremonies—a great triumph of laboratory pharmacology.

Hofmann and Wasson then turned their attention to the Mexican morning glories, another hallucinogen that had been noted by early Spanish conquerors, but whose identity was unknown. They identified the chief species in use and found in it the closest analog that nature produces to LSD, a substance called LAE, the monoethylamide of lysergic acid.

Then they went after a third hallucinogenic plant known as ska María pastora, which means, the leaves of Mary the Shepherdess. The Indians told Wasson this substance was not as powerful as the other two but was important for training shamans, especially for the first level of training. Eventually, specimens of the plant were collected and identified as a mint, an unexpected discovery, because the mint family, the Lamiaceae, is not a plant group rich in alkaloids or toxins, and it had no known psychoactive species. María pastora was further identified as a member of the genus *Salvia*, the sages. You know culinary sage, and have probably seen many ornamental sages, none of which are pharmacologically active. This was a new species, soon named *Salvia divinorum* because of its association with shamanism and divination. But Hofmann, despite his previous triumphs, was unable to find an active constituent in it. After much fruitless investigation, he concluded that if there were an active constituent, it must be very unstable.

In the 1970s, several Mexican investigators worked on *Salvia divinorum*, and they too could find nothing significant in it. Over the years, the idea grew that if this plant had any psychoactivity at all, it must be of very low potency and contain something quite unstable. As it turns out, those suppositions were wrong. It is only very recently that the responsible constituents have been correctly identified, and they turn out to be extremely potent and powerful and also very unusual. The principal active component is a drug called salvinorin A. It is a diterpene, lacking nitrogen. Almost all major psychoactive drugs contain at least one nitrogen atom. The compounds in kava kava (*Piper methysticum*) from the Pacific are non-nitrogenous, as is THC in marijuana, but there are no exceptions among the true psychedelics or hallucinogens. All of these are alkaloids, which, by definition, contain nitrogen. Salvinorin A has a completely different chemical structure. It is also completely insoluble in water. So one problem that delayed its discovery is that if you simply administer the leaves of the plant orally, there may be no absorption of the active components and hence no psychoactivity. There is absorption if the leaves are first emulsified. Indians prepare *Salvia divinorum* by picking pairs of leaves and rolling them into a quid which they suck or chew, or by making an infusion in water with a great deal of agitation to produce a foam. They say that the strength of the preparation depends on the heaviness of the foam. Presumably, this is an emulsion that converts the salvinorin A into an absorbable form.

In the past two years, the psychedelic underground in this country has taken great interest in this plant, which is now in widespread cultivation here. It is not a controlled substance, grows readily, and produces many leaves. The yield of salvinorin A from the leaves is enormous, and the compound is easily purified. Some psychedelic explorers have found that you can also take this drug, or a concentrated extract of the leaves, in a manner similar to smoking free-base cocaine; that is, you can heat it in a small glass pipe and then inhale the vapors. When used in this way, salvinorin A turns out to be the most potent natural hallucinogen yet discovered, active in doses of 200–500 mcg, which is only slightly above the range of activity of LSD. Used in his way, I think there is little potential for abuse, because the experience is terrifying. I tried it once and would never do it again.

Recently, I reviewed a paper on *Salvia divinorum* that is about to be published in *The Journal of Psychoactive Drugs,* the best review I have seen. The author is L. J. Valdés, a chemist who has written other papers on the subject. I want to read you a quote from it. Valdés had the good fortune to work at length with a Mazatec shaman who taught him how to use the plant. He recorded this information in great detail and then did his chemical and pharmacological studies. Here is his description of his second experience. The shaman had first given him what he called a "beginner's dose" of the emulsion made from about twenty pairs of leaves. Now, on this occasion, about six months later, he took a larger dose. The shaman said that it was essential to take this in conditions of night and darkness, since only then would the visions develop fully. He also explained a necessary progression of visions that would ultimately lead to seeing religious figures and saints from whom there was much to learn. I won't describe the whole ceremony, which included many prayers and offerings. The shaman crushed the leaves in water, agitating the mixture vigorously, then offered it to Valdés. Then he spent an hour describing the journey that Valdés would take to heaven, the things he would see and the people he would meet. Unfortunately the village was extremely noisy that night, with much barking of dogs.

After about 15 minutes, we began to have visions. [I might say here that included in the "we" is an investigator who is attending this conference, José-Luis Díaz, and if any of you have questions about *Salvia divinorum,* he is one of the few experts on it.] This time I spoke mine out, alternating between English and Spanish, which helped to fix them in my mind. Díaz spoke first and mentioned flowers. I then saw eidetic images that evolved to plants and flowers. These later became giant fruits and seeds. At the same time, I felt that I was twisting inside my body as well as spinning around. I saw a burning cross with two horizontal rays. It stopped flaming and began to emit light. Suddenly, I began to feel very heavy.

After about fifty minutes, the shaman stopped the session, saying it was too noisy to have a meaningful experience; he suggested that Valdés and Díaz leave. Someone was there to drive them out of the village to the motel they were staying in. As they were leaving, the shaman told them the visions would come back and last throughout the night. He told them that with more experience they would begin to understand the use of the plant and the ways of healing and could study on their own. Valdés reports that during the return journey in the quiet darkness of the car, the imagery returned. "I saw the Virgin of Guadalupe. If the vision began to fade, I could will it back. We returned to the well-lit motel where there was music and noise. I thought that the experience was over, and things had returned to normal." They ate a light supper, showered, turned off the light and went to bed. It was about 11:30.

It was now some two-and-a-half hours after I first drank the potion. In the motel room, the imagery came back stronger than ever. Even though I didn't speak out, I saw pulsating purplish light that changed to an insect-like shape, perhaps a bee or a moth, and then into a pulsating sea anemone. It expanded into a desert full of prickly pear cacti and remained so for several minutes. During the first session and throughout the night, my visions had all appeared to be something like a cross between a silent movie picture and a cartoon. I felt myself to be an observer of these mute visions rather than being an actual part of them. Suddenly, however, I was in a broad meadow with brightly colored flowers. I had just crossed a stream by way of a small wooden bridge. Next to me was something that seemed to be the skeleton of a giant model airplane made of rainbow-colored inner tubing. The sky was bright blue, and I could see a woods in the distance. I found myself talking to a man in a shining white robe who was either shaking my hand or holding onto it. It was an amazing hallucination, as I truly believed that I was in the meadow. It was not like a dream. After a few moments, the desert landscape returned. I slowly went to sleep after an hour or so. I rose early the next morning, feeling no adverse effects.

The most amazing experience was that which happened to me on our return to the motel. Don Alejandro had described what the visionary journey would be like. I am sure that when I found myself in the meadow talking to the man in white, who fit the description of a saint, I was in the shaman's heaven. The hallucination was quite complete, being visual, aural, and tactile.

Again, I would say to you that I think that we are very, very far from being able to analyze or understand that kind of experience in terms of anything we know about the pharmacology of hallucinogenic drugs or processing by neural networks.

I want to end with a personal comment. I appreciate and respect what I would call the quality of mystery in the universe. The literal meaning of mystery is "that which is hidden." There are aspects of reality that are hidden from us, at least from our intellects. Yet mystery can be experienced even if it cannot be understood intellectually. I do not think that we can eliminate mystery from our experience of reality, nor is it the business of science to try do that. Science can approach mystery. But it seems that the brighter we illuminate reality with the light of science, the more we become aware of the extent of the surrounding darkness.

QUESTIONS AND COMMENTS

Q. . . . . Last year, eight million people in the United States spent over 1.5 billion dollars on Prozac and Prozac relatives for the largest psychological experiment in the history of this Earth. I would be very interested to hear your comments.

A. Let's see what happens twenty years from now—how Prozac is regarded then. Remember the famous aphorism, that a new remedy should be used as much as possible while it still has the power to heal. So, let's see what happens. As I said at the start, I am a great believer in placebo medicine. Placebo responses really are healing responses from within, triggered by belief. The ideal would be to elicit a healing response without doing anything to a patient. But you can elicit healing responses more reliably by giving treatments that both you and the patient believe in. I like to use the term "active placebo" for that kind of intervention. I think that all psychoactive drugs are active placebos. They definitely affect you physiologically, but what you do with that change in your physiology is a matter of factors outside the realm of pharmacology.

Q. Early in his career, Sol Snyder, the famous pharmacologist, studied a series of hallucinogenic drugs, all of which—correct me if I am wrong—have polyaromatic groups, indole rings, and so on. He studied the energy of the highest occupied molecular orbital. The highest occupied molecular orbital correlated perfectly with the potency of these drugs. The point is this: we have been talking in this conference about quantum theory and trying to relate it to activities of single quantum events like electrons and bosons and things of that sort. Your anecdotes about the mutual visions and shared consciousness can be explained in the context of quantum effects, non local effects. I think that this may be a link or clue to a role of quantum effects in these types of phenomena.

A. I think that the more consistency we can find between the results of neurobiological research and human experience the better. We want to try to bring these two realms together. If quantum physics provides a model to do that, so much the better. At the same time, I feel sure that if we solve one problem, there will be no shortage of others.

Q. I am assuming that you are aware of Terrence McKenna's speculations about the role of psychoactive plants in the evolution of human consciousness, which to a lot of us are wild speculations. Would you comment on them? Also, I know the original name for harmine was telepathine, and I am wondering if there have been any controlled studies in the West with it and what you think the possibilities for those are.

A. To the first question, I would say that the word "speculations" is perfectly appropriate, and I have no objection to "wild" either. McKenna's main theory is that ingestion of psychedelic plants, specifically mushrooms, was the chief evolutionary triggering factor that moved primates along to human consciousness. Anyone with a sense of animal behavior would agree, I think, that an individual member of any primate species that accidentally ingested a psychedelic drug would be at great risk to be eliminated from the gene pool. I think ingestion of these substances only becomes manageable with a level of civilization that can provide a secure setting in which the distortions of sensory awareness would not be dangerous.

As for the second question . . . I think the virtual absence of useful research on these substances in human beings is most unfortunate. At the least they are windows on both neurobiology and conscious experience, and made-to-order research tools. In the process of criminalizing disliked psychoactive drugs, our society has made it impossible for researchers to work with the psychedelics; they are as heavily stigmatized as heroin and PCP. All of them are in Schedule One of the Controlled Substances Act, which, for practical purposes, makes them unavailable for research. This is a great loss to the science of consciousness. Maybe there is some hope. Last year the FDA appointed a new director of the office in charge of psychoactive drugs. That official has indicated some willingness to change policies of the past and make it a little easier to do human research with these compounds. Very recently, just in the past year, human research has started with LSD and with MDMA. A psychiatrist at the University of New Mexico is studying DMT in human subjects. I hope more studies will get underway.

Q. My question is, could you describe the experience that made you never want to smoke the extract of *Salvia divinorum* again.

A. I was fairly naive at the time, because I had not read Valdés' paper, and all I knew of *Salvia divinorum* was that it was a great puzzle. Someone came to me with a sample of the concentrated extract and offered to show me how to inhale the vapors from a glass pipe. He said people were having "fantastic" experiences with it. Of course, when you inhale vapors, it is very difficult to quantify the dose you receive. My guess is that I inhaled more than 500 mcg of salvinorin A. I had an instantaneous sensation of falling backward, but not in ordinary space. Then I felt enveloped by some kind of higher-dimensional lattice and felt as if I were being smothered. My ego consciousness did not disappear as with 5-MeO-DMT. Rather, it was reduced to a tiny point that was imprisoned in something very dense and heavy. There was an overwhelming sense that reality had come to an end—that time, space, and reality had simply ended forever. There was also a recurring thought or image of the pipe. I knew I had done this to myself by using the pipe, and someone or something was laughing at me malevolently. It was a great cosmic joke at my expense. I had brought reality to an end, and that was it. I must say that I was quite amazed when reality began to come back. It took me a long time before I could talk. When I did, I was asked how much time I thought had elapsed since I inhaled the stuff. I thought it had been ten or twelve minutes; actually, it had been fifty. The experience disturbed me psychically and left me feeling disturbed for some time. Now that I know the plant can be taken as an oral emulsion, that seems a much wiser way of interacting with it. I have a specimen of the plant growing in a pot—not very happily because it is a cloud forest native and needs more humidity than exists in Tucson. When I walk by it, I look at it with considerable respect.

REFERENCES

Lamb, F. B. 1975. *Wizard of the upper amazon: The story of Manuel Cùrdova-Rios* Boston: Houghton Mifflin, pp. 32–3.

Valdés, L. J. in press. Salvia divinorum and the unique diterpene hallucinogen, salvinorin (divinorin) A. *Journal of Psychoactive Drugs.*

Weil, A. T., N. E. Zinberg, and J. M. Nelsen. 1968. Clinical and psychological effects of marijuana in man. *Science* 162:1234–42.

49 What Can Music Tell Us About the Nature of the Mind? A Platonic Model

Brian D. Josephson and Tethys Carpenter

Is the phenomenon of music to be understood in conventional biological terms, or is it instead an activity dependent upon subtler aspects of mind? Conventional explanations may be able to explain certain capacities in music (such as the ability to recognize and define particular categories of pattern, structure, or relationship), on the basis of the fact that possession of such abilities may confer selective advantages. What is more difficult to account for, using such arguments, is the specific forms that appear to be favored in music, and which appear to possess a curious generative capacity or "fertility" not possessed by arbitrary patterns of sound. Specifically, one often finds at the beginning of a piece of music a short and usually discrete unit (typical examples being the first theme of Mozart's Symphony No. 40, the opening bars of Beethoven's String Quartet Op. 95, and, on a larger scale, the *leitmotif* that begins Wagner's opera *Tristan und Isolde*), containing distinctive harmonic, melodic, and emotional patterning, which functions as the germ of elaborations in the course of the subsequent development. The fertility of such special forms or musical ideas is emphasized by the way a composer may develop an existing idea in a new way (for example, Schubert's use of the initial idea in Mozart's String Quintet in C major in his own String Quintet in C). These "musical ideas" resemble the memes of Dawkins (1989). It will be argued here that the specificity of these forms cannot be readily accounted for within conventional frameworks of explanation, and that better explanations are likely to be obtained by involving subtler aspects of mind than those normally taken into account.

The phenomenon of interest can be defined as the special effects on consciousness of the specific constituent patterns or ideas found in good music, the striking effects of the latter being in clear contrast to those of the forms created by mediocre composers or by mechanical procedures. Skilled composers do not produce innovations in a mechanical way; rather they appear to possess an intuitive ability to be aware of the creative potentials of particular patterns of sound even when considered in their most elementary forms, and then develop a composition from these "germs" (Schoenberg 1977, 1984).

At a superficial level, the specificity that has been discussed resembles that of a resonance, but this analogy does not appear to be a very helpful one since we are dealing with a highly nonlinear system; the true mechanism must be rather different in nature. We shall dismiss explanations based on conditioning, because the differences in style between an innovative composition and music that a listener has been exposed to previously account for a considerable part of its interest, and while conditioning can account for a listener's ability to process competently a new piece of music in a familiar style it cannot explain why particular innovations should have particularly powerful effects. Apart from this there are two main categories of possible explanation:

1. Genetic explanations. For each musical idea there is a corresponding gene coding for nervous system structure corresponding to selective sensitivity to that idea. Regarding this type of explanation, while some musical ideas may have correlates in the natural world, the majority do not, so that there would be no selective advantage in possessing such sensitivity. It also seems unlikely that the collection of sensitivities to musical ideas can be explained as accidental consequences of other adaptations. It seems to us that the only way of avoiding these problems would be to postulate that during the course of evolution there had been a species that used as a means of communication music very similar to the kind produced by human composers, and which had undergone a process of evolution that was the genetic equivalent of the human cultural evolution of music, by this means evolving genes corresponding to the musical sensitivities that have been postulated. This seems unlikely to have been the case.

2. "Theory of Everything" type explanations. The idea that there may be some universal formula or principle that distinguishes effective musical ideas from ineffective ones (in the same way as in the case of chemistry there is a universal formula, namely, Schrödinger's equation, that can distinguish between stable and unstable molecules). A related perceptual mechanism would provide the observed discrimination between good music and bad.

Attempts have been made by musical psychologists (such as Lerdahl and Jackendoff 1983, Narmour 1990) to discover such principles, but these attempts seem to us, in their present stage of development, grossly inadequate to the purpose, and to provide us with little illumination concerning the problems addressed here. There is in addition a further argument against "theory of everything" type explanations of musicality. This kind of explanation, in contrast to the other kinds of explanation discussed (namely, cultural and genetic explanations), allows essentially no scope for arbitrary factors to enter into the determination of the preferred forms. Given the apparently capricious nature of musical regularities, the kind of explanation that does allow arbitrary factors to enter seems considerably more plausible.

While none of the above arguments is conclusive, the difficulties we have noted provide some motivation for seeking alternatives. Elsewhere

(Josephson and Carpenter 1994) we have commented on the existence of interesting parallels (principally involving matters concerning information and regulation) between aesthetic processes and life processes. These parallels will now be developed further.

In the present context, the most basic parallel is that between effective musical pattern and gene, both being informational structures playing a key role in the activities of those structures that contain them (organism and musical mind). In the case of life, the genes help to determine the forms and activities of the structures that cause the genes to be replicated so that they survive. Particular gene structures generate particularly effective functional systems, and this very often entails high complexity since complex means are generally needed to produce simple results in an effective manner. Other contributions to complexity come from the complexity of the chemistry involved and the fact that genes often do not produce their effects in isolation. Further, in organisms there are clear means-end relationships related to the functionality of structures, by virtue of which the functional structures can be considered "significant." In contrast to the perspicuity of the processes involved at the functional level, the details at the structural level are complex, and related to function in a complex way, which generally has arbitrary aspects as a consequence of the way that nature operates opportunistically rather than logically.

We now observe the ways in which music possesses features paralleling those discussed in the case of life:

1. Effective musical structures are highly specific, as well as being (subjectively) functional;
2. While there is an overall logic behind the way that a given piece of music works, many of the details of form appear essentially arbitrary. The functional descriptions are considerably simpler than descriptions at the detailed level of the structure.

A further fact about music that is clarified by this picture is its perceived semantic aspect. Biological structures in general can be considered to have a semantic connected with means-end relationships. These semantic aspects can be divided into internal ones (related to direct maintenance of the organism independent of its environment) and external (maintenance dependent upon interactions with the environment). Correspondingly, in the case of music, some components appear to have external reference (that is to say they appear to relate in a general way to ordinary events in the world) (Meyer 1959), while others appear to be significant only in relation to the piece of music in which they appear.

These features of music could be understood if the mode of operation of mind were in general terms similar to that of life. According to this view, intelligence would be the product of a collection of adaptations capable of being specified by a coding system related to that of music. The fertility of particular musical patterns would reflect the operation of the specific adaptations specified by these patterns. Individual minds would make use of such adaptations in the same way as in ordinary biology individual organisms make use of genes. While the development of the organism, excluding mind, centers around the use that can be made of chemistry, the development of mind centers around the use that may be made of ideas and thought.

The question arises: Which is the mindsystem in which the processes we have been discussing occur? It cannot be the minds of individuals, since the preferences that the model is intended to explain are not those of individuals. Neither can it be the cultural mind (consisting of individuals communicating with each other musically) because, as discussed in connection with explanations based on conditioning, the selective response to innovations cannot be explained purely culturally. What remains is activity involving some kind of collective or universal mind. Our model thus entails a Platonic picture of the mind, where much of the intelligence of the individual is the consequence of preexisting ideas in some mindsphere. It follows that the study of music is at the same time the study of the quasigenetic aspects of this subtler realm of mind. Such studies may thus be able to inform us of aspects of mind not accessible to conventional studies that tend to focus on the more intellectual aspects of mind to the exclusion of its more intuitive ones. It may be worth pointing out here also that the idea that there is a fundamental connection between sound and form is an ancient one, dating back thousands of years in the Eastern philosophical tradition.

ACKNOWLEDGMENTS

We wish to thank Professor Robin Faichney and Dr. Marek Lees for helpful comments.

REFERENCES

Dawkins, R. 1989. *The selfish gene.* Oxford, England: Oxford University Press.

Josephson, B. D., and T. Carpenter. 1994. Music and mind—A theory of aesthetic dynamics: *On self-organization.* Springer series in Synergetics, Vol. 61, Heidelberg: Springer-Verlag, pp. 280–7.

Lerdahl, F., and R. Jackendoff. 1983. *A generative theory of tonal music.* Cambridge, MA: MIT Press.

Meyer, L. B. 1959. *Emotion and meaning in music.* Chicago: University of Chicago Press.

Narmour, E. 1990. *The analysis and cognition of basic melodic structures: the implication realization model.* Chicago: University of Chicago Press.

Schoenberg, A. 1977. *Fundamentals of Musical Composition.* Ch. III. London: Faber and Faber.

Schoenberg, A. 1984. "Folkloristic symphonies." In *Style and idea: Selected writings of Arnold Schoenberg,* edited by L. Stein. London: Faber and Faber, pp. 161-166.

50 Intention, Self, and Spiritual Experience: A Functional Model of Consciousness

Arthur J. Deikman

The spiritual traditions have been exploring consciousness for thousands of years. Their procedures and findings should be of particular interest to scientists endeavoring to build a comprehensive theory of consciousness. However, using their data for this purpose faces special difficulties. It was remarked earlier in the conference that "consciousness" has been a taboo word in science. How much more so have been the words "spiritual" and "mystic." These terms generally are used by scientists in a pejorative sense, as Steen Rasmussen did this morning when he said, "Nothing mystical about this," suggesting that, in contrast to mysticism, what he was talking about was understandable and real.[1]

Science's antipathy is understandable. "Spiritual" and "mystical" are labels applied to a wide range of behavior some of which is like a circus sideshow, ranging from people shaking tambourines and collecting money in airports to groups committing suicide at the behest of a "divinely inspired" leader. Furthermore, spiritual writings are usually wrapped in religious metaphors, and only recently has science freed itself from the tyranny of religious dogma. So, it is not surprising if a scientist shudders when someone gets up and starts using those taboo words.

The spiritual experiences to which I will be referring are not of the circus sideshow variety but the product of a disciplined system for perceptual development—as you discover when you read the mystical literature widely and thoroughly. There, mystics are explicit in cautioning that spiritual perception has nothing to do with sensations and images, with visions of angels, or emotional fits. What they are talking about is a different way of knowing, one that transcends sensory information. For example, Walter Hilton, an English mystic from the fourteenth century is quite explicit:

... visions of revelations by spirits ... do not constitute true contemplation. This applies equally to any other sensible experiences of seemingly spiritual origin, whether of sound, taste, smell or of warmth felt like a glowing fire in the breast. (Hilton 1953)

Philip Kapleau (1967) sums up the Zen point of view:

Other religions and sects place great store by the experiences which involve visions of God or hearing heavenly voices, performing miracles, receiving divine messages, or becoming purified through various rites ...

yet from the Zen point of view all are morbid states devoid of true religious significance and hence only *makyo* [disturbing illusions]. (Kapleau 1967)

There is a remarkable concordance among mystics from widely differing cultures and epochs as to the nature of spiritual consciousness and the means for attaining it. This consistency lends credence to their statements, impressing many people who have studied the mystical literature, not the least of which was William James (1902/1929). In his classic, *The Varieties of Religious Experience,* he wrote: There is about mystical utterance an eternal unanimity which ought to make a critic stop and think . . . the mystical classics have, as has been said, neither birthday nor native land (James 1902/1929).

Finally, mystics tend to be practical phenomenologists, rather than metaphysicians. Their injunction is: "Taste and you will know."

Here are some samples from the mystical literature dealing with typical themes: the existence of hidden dimensions, the nature of the true self, the importance of renunciation or nonattachment, and meditation.

In the mystical experience, reality is perceived in a radically different way. The following is from the Sufi tradition:

Another Dimension

The hidden world has its clouds and rain, but of a different kind. Its sky and sunshine are of a different kind. This is made apparent only to the refined ones—those not deceived by the seeming completeness of the ordinary world.

Unaware

You know nothing of yourself here and in this state. You are like the wax in the honeycomb: what does it know of fire or guttering? When it gets to the stage of the waxen candle and when light is emitted, then it knows.

Similarly, you will know that when you were alive you were dead, and only thought yourself alive (Reprinted by permission from *The way of the Sufi,* by Idries Shah (Octagon Press Ltd., London).

Mystics emphasize that human beings are confused about their essential self, as illustrated in the following passages from the Buddhist literature:

Since beginningless time, sentient beings have been lead astray by mistaking the nature of their mind to be the same as the nature of any other object . . . they thus lose their true and essential Mind.. . . (Goddard 1970)

Although good and evil are different, our self-nature is non-dual, and this non-dual nature is called the genuine nature. (Fung and Fung 1964)

Many statements refer to the importance of nonattachment, as in the following passage from the Upanishads:

When all of the desires in the heart fall away, then the mortal becomes immortal and here attains Brahman. When all the ties of the heart are severed here on Earth, then the mortal becomes immortal. This much alone is the teaching. (Nikhilananda 1949)

There is also considerable emphasis on meditation. The following quotation is from a twentieth-century Zen Master:

The most important thing is to forget all gaining ideas, all dualistic ideas. In other words, just practice zazen [meditation] in a certain posture. Do not think about anything, then eventually you will resume your own true nature, that is to say, your own true nature resumes itself. (Suzuki 1970)

All in all, mystics from markedly different cultural contexts and historical periods make similar assertions about spiritual experience. These can be summarized as follows:

1. Ordinary experience is narrow in scope and illusory in content, leading to a false belief about the nature of the self.
2. It is possible to know the transcendent reality that underlies appearances. This reality is characterized by unity, purpose, and positive value.
3. This higher, spiritual knowledge cannot be described in the language, concepts, and images of the ordinary world.
4. Attaining this experience requires the renunciation of self-centered aims and, depending on the circumstances, may be aided by the practice of contemplative meditation and service. A qualified teacher is necessary to prescribe practices and transmit nonverbal knowledge.

If we consider the possibility that these are valid assertions, referring to another way of perceiving the world and ourselves, the question arises as to how we might understand the function of the recommended practices and the consciousness that results from them.

A first step is to consider the way thought and perception develop in infants and young children. As the work of Piaget (1952), Gesell (1940), Spitz (1965), and Erikson (1951) indicate, our initial approach to the world is in terms of its object aspects. Thinking, linked to language, becomes object-based, and objects are linked to the body as template and model. Through repeated object-oriented actions, language, thought, vision, and muscle movement are coordinated in the learning process. Piaget, in particular, gives many examples showing the body to be the basic means for dealing with the world. Such experiences teach us the rules of thought; these are body rules—object rules. The concepts of space, time, and causality that we usually employ are the space, time, and causality that pertain to objects.

INSTRUMENTAL CONSCIOUSNESS

As adults, and especially in our role as scientists, we operate primarily in the object-mode of consciousness—the instrumental mode—derived from those early years of object learning. The mode is highly functional, it serves the intent of acting on the environment and enables us to survive as biological organisms. An example of this mode is what we experience when driving in heavy traffic. In accordance with our intention of acting on the environment, we have a self—the "survival self"—that is object-like, localized, and separate from others (Deikman, 1982).

"Self" is a very broad term that may refer to many different phenomena. For example, we experience a self that acts, "I am giving a lecture." There is an emotional self, "I feel happy"; a social self, "I am a psychiatrist"; and so on. I subsume all these usages under the term, the survival self, because they are all processes that have evolved to enable us to survive as individuals. The survival self relies on instrumental consciousness.

When our intention is to act on the world, we experience it as a collection of objects that operate in linear, causal fashion. Instrumental consciousness features focal attention, sharp perceptual and cognitive boundaries, logical thought, and reasoning. The formal (shape, size) dominates the sensual (color, texture). Thoughts concern the past and future.

RECEPTIVE CONSCIOUSNESS

There is a second mode, which we have from the beginning: receptive consciousness. It is the mode we need to use if we are going to enjoy a massage, listen to music, make love, enjoy a summer day, or experience communion with a loved one. The receptive mode serves the intent of taking in the environment.

A good example of this mode is our state of mind when soaking in a hot tub. Imagine that you've survived that drive through heavy traffic, made your plane, and arrived at your hotel. You are tired and frazzled and really want to unwind. So you draw a tub of hot water, slip into it, "Ah...." Your experience of self becomes relatively undifferentiated, nonlocalized, not distinct from the environment. A blurring or merging of boundaries takes place resulting in a more world-centered awareness. Attention is diffused. Paralogic, intuition, and fantasy take the place of discursive thought and the sensual dominates the formal. "Now" dominates over past and future. These two modes are compared in Figure 50.1.

INSTRUMENTAL CONSCIOUSNESS (EXAMPLE: DRIVING IN HEAVY TRAFFIC)	RECEPTIVE CONSCIOUSNESS (EXAMPLE: SOAKING IN A HOT TUB)
INTENT: To *ACT ON* the environment	**INTENT:** To *RECEIVE* the environment
SELF: Object-like, localized, separate from others Sharp boundaries Self-centered awareness	**SELF:** Undifferentiated, nonlocalized, not distinct from environment Blurring or merging of boundaries World-centered awareness
WORLD: Emphasis on objects, distinctions, and linear causality	**WORLD:** Emphasis on process, merging, and simultaneity
CONSCIOUSNESS: Focal attention Sharp perceptual and cognitive boundaries Logical thought, reasoning Formal dominates sensual Past/future	**CONSCIOUSNESS:** Diffuse attention Blurred boundaries Paralogic, intuition, fantasy Sensual dominates formal Now
COMMUNICATION: Language	**COMMUNICATION:** Music/art

Figure 50.1 Instrumental consciousness and receptive consciousness compared.

AN EXPERIMENT

In order to appreciate the critical role played by intention in shaping consciousness, try a brief experiment. Find another person to be your partner. Stand about an arm's distance away and look at each other's face. First, study your partner's face with the specific intent of analyzing it as if you were going to make a model of it. Spend a few moments doing that. Then stop and change your intention to one of receiving. Relax your gaze, be open to whatever your experience is of your partner's face. Just take it in.

Most people report they notice that shifting from one intention to the other results in a distinct difference in their perception. They use words like "richer" and "deeper" to describe the shift to the receptive mode. In that mode, the other person seems to gain dimension, presence. This exercise also will work with a tree or a flower, or you can try it the next time you visit a museum. In the latter case, you probably will enter the museum in the instrumental mode. If you are going to get anything out of the museum other than a stuffed sensation in the head, you will need to stop and shift to the receptive mode. Follow the same procedure given above, first analyze an individual artwork, then *allow* it to manifest itself, be open to it. Note the difference.[2]

At this point you might say, "Well, if the receptive mode is the key to spiritual development, all I need to get enlightened is to go out and buy a hot tub." Regrettably, it's not that simple. Decreasing the dominance of the instrumental mode is a necessary condition, but it's not sufficient. A further development of receptive consciousness is needed, one that depends on "forgetting the self." The self that is forgotten is the survival self; "forgetting" means to reduce markedly the dominance of consciousness by the survival self, permit a radically different sense of self and perception of our environment. Spiritual practices serve this function.

FORGETTING THE SELF BY SHIFTING INTENTION

Meditation emphasizes *allowing* (receptive mode intent), rather than *acting on* (instrumental mode intent). Typically, the meditator is instructed not to strive for anything and to relinquish analytic thought. The focus is on the present moment rather than the past or future (receptive time versus instrumental time). Thus, meditation procedures have a strong receptive intent that decreases instrumental activity.

Renunciation is a prominent feature of classical accounts of spiritual life. For most people, that aspect of spiritual practice is distinctly unappealing. When I first began reading the mystical literature and realized that meditation and renunciation were central exercises, I decided to do research on meditation, rather than renunciation. Like most people, I really didn't understand what the concept meant. Renunciation does not mean a drastic turning away from all the pleasures of the world. As a Zen master put it, "Renunciation is not giving up the things of this world; it is accepting that they go away" (Suzuki 1968). The essence of renunciation is nonattachment, a letting-go of self-centered aims, a relaxing of the grasping acquisitiveness that drives the instrumental mode of consciousness.

Along with meditation and renunciation, spiritual traditions emphasize service. Service provides a way of de-emphasizing the survival self in that it is *other-centered:* serving the task, rather than oneself. You may be familiar with the cynical argument that everything a person does is selfish, because even if you do something good for someone else, you get pleasure out of feeling that you have done a good deed. How does one escape that particular box? You can do so by serving the requirements of the task before you, *for its own sake.* At a superficial level, this might reflect only culturally conditioned standards of completeness or appropriateness. But there are deeper levels of "what is called for" that are harder to define. These are encountered when you respond to the real need of an individual, a community, or the world at large. The important point is that the service that mystics prescribe is not centered in some image or concept of oneself doing good, but in simply *doing what is called for.* That is how one gets beyond the limitations of self-gratification: service provides a goal that focuses on the needs of others, rather than the survival self.[3]

Through these spiritual practices, sense of self can be linked to a larger context, beyond that available to instrumental consciousness. Because it requires a shift in basic intent, there can be no cheating. To illustrate this point, imagine that you have been playing the stock market, but after reading some books on Buddhism you decide that enlightenment is a better deal and you will concentrate your efforts there. After all, the mystical spiritual literature seems to indicate that if you're enlightened you'll be blissful and won't be afraid of death. And so you fax a message to your brain computer. In the control room, a staff member reads the fax and hurries over to the boss: "Hey we ought to change the program; he wants enlightenment now." The boss replies, "No change needed; run the same program: acquisition."

Thus, it doesn't matter if you wear a yellow robe and sit in full lotus, intention and self will determine the consciousness you experience. To accomplish the change in perception with which the spiritual traditions are concerned, receptive intent needs to be combined with a profound shift away from the survival self.[4]

SELF AND PERCEPTION

I had a very vivid demonstration of the relationship between self and perception when I took part in a seven-day Zen meditation retreat. For most of the time, we meditated on a koan and once a day would sit and chant Japanese words printed in black on white paper. After a while, the words seemed to march across the page by themselves. They didn't seem to need me at all and the idea began to grow in my mind that the world didn't need me to be there for it to exist, so why not let my *self* vanish from it? The thought grew, "Let go! The world will still be here, let it exist without you.

See what it's like when you aren't here. Let go, let yourself disappear! Jump!" It felt as if I could do that and, finally, I did. I made some sort of mental leap and let go of my background sense of self—as if I disappeared. At the precise moment I did so, there was a dramatic change in my perception of the room. The other students sitting in their black robes were now awesome, beautiful Buddhas. When the meditation bell was struck, the sound rolled out like waves of silver. Everything was transfigured, archetypal, extra dimensional. It was an extraordinary change in perception that correlated exactly with the moment I let my survival self disappear. I don't know exactly how long the episode lasted; it faded as we walked from the meditation hall. The experience was a dramatic illustration of the linkage between the type of self that is dominant and the way in which the world is perceived.

$$C = f(I,S)$$

This formula sums up much of what I have been discussing. If the mystics are right, any theory of consciousness has to take into account both the intention that is operative and the type of self that is dominant.

Let us now make a closer comparison of the self associated with the instrumental mode and the self associated with the development of the receptive mode when "forgetting the self" has deepened and become pervasive. This latter self I am designating as the "spiritual" self.

Because the instrumental mode of consciousness emphasizes boundaries, form, and separation to aid in acquisition and control, the experience of self in that mode is object-like, discrete from its environment, vulnerable to isolation. The basic survival strategy of acquiring and defending underlies the traditional vices of greed and jealousy, and crimes such as murder.

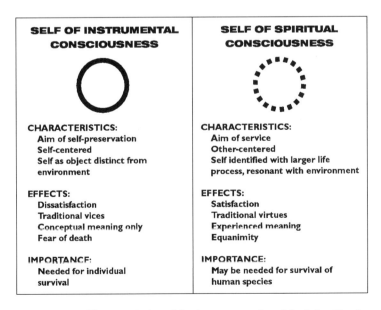

Figure 50.2 Characteristics of the instrumental and "spiritual" selves.

These can be considered functional for primitive biologic survival, but poor for satisfaction, meaning, and happiness.

However, reliance only on the instrumental mode results in dissatisfaction. The fundamental reason is that the acquisition of sensory pleasure is frustrated by the design of our central nervous system: pleasurable sensations habituate and fade; pain does not. Eventually, we can get bored with anything, and complain when the level to which we have adapted is reduced.[5] On the other hand, a toothache will continue to plague us without let up.

Furthermore, the sense that one's life is meaningless can result from dominance by the instrumental mode because meaning arises from the experience of relationship, of connection. Although the meaning of one's life can be conceptualized, the *experience* of meaning is something much deeper, better able to withstand the onset of existential despair. The instrumental separation of the self from its environment not only strips away meaning, but intensifies the fear of death. Death is the ultimate disaster for the isolated, acquisitive, object-self because death takes away everything—nothing is left.

In contrast, the self of spiritual consciousness is not experienced as separate from the environment but permeated by it. The sharp boundary of instrumental consciousness is muted, tapered, blended into the surrounding world. The self feels transparent, even invisible.

This analysis enables us to understand why service is emphasized in most spiritual disciplines. It is not a pious matter of doing good, but of providing a different experience of self. This larger self is not an occult phenomenon. All that may be required is to do volunteer work where you are needed. If you participate wholeheartedly, a subtle change in the experience of self occurs. You may only notice it if someone praises you for donating your time. Then, you are likely to reply, "Actually, I've received more than I've given." If you then are asked what you have been given, you may find it hard to articulate. The feeling is subtle, but definite. I suggest that you sense your connection to a larger process, and in some way feel part of it. Because you are engaged in serving something other than the survival self, a different perception of self becomes possible. This experience of connection is not possible for the self of instrumental consciousness.

The need—and resistance—to diminish survival self aims can be illustrated by imagining that you have come to a stream and want to drink. Knowing only one method of getting what you want—grasping—you try to grab the water. But when you open your fist there is nothing there. Frustrated, but not realizing the problem, you try again and again. A passerby says to you, "That won't work. If you want to drink the water you are going to have to cup your hands."

You might question her, "Oh, is that more spiritual and holy?"

"It has nothing to do with being holy," she replies. "It has to do with the nature of water. If you want to drink from the stream you have to cup your hands."

"But I don't want to cup my hands. It feels weak and I don't believe it will work."

"It's not a matter of what you believe, but of what you need to do."

"I'm not convinced."
"Suit yourself."

MEANING

By understanding the link between mode of consciousness and the experience of self, it is possible to help someone who feels their life lacks meaning. As a psychiatrist, I see this problem fairly often in people who are functioning well and not suffering from a specific emotional or mental illness. The theoretical models of psychoanalysis, neuroscience, and quantum physics provide little help in addressing this existential problem. However, with a functional model of consciousness in mind, I can inquire as to the extent to which the patient leads a self-centered life and then, if appropriate, indicate the consequences. By stressing the functional perspective, rather than the virtuous, it becomes possible to avoid the moralistic injunctions to which most of us have been exposed and see the need for serving others as a straightforward fact of human psychology.

UTILIZING BOTH MODES

If the needs of a task or of other persons become primary, then the lessened influence of the survival self permits a special state in which the two modes can be in harmonious, balanced interaction instead of alternating. When you have "surrendered" to the task, the ordinary sense of a discrete, controlling self is replaced by the experience of being both active and guided at the same time. For a writer or artist this is the special creative state they struggle to reach. But athletes, scientists, performers, and others also encounter it in the course of their work. For psychotherapists it is the "good hour" when the therapist acts in perfect synchrony with the patient's cues, following a subtle path that seems to carry as much as lead. That magical state is something most of us have experienced at one time or another but find hard to describe: "choiceless action," a state of "flow," "being in the zone." Reducing the influence of the survival self is the critical element that permits it to happen.

THE NEED OF SCIENCE FOR THE SPIRITUAL TRADITIONS

Just as the mode of consciousness that asks the question "What is the meaning of life?" is not the mode of consciousness that can hear the answer, the science that employs only instrumental consciousness is unlikely to develop an adequate theory of consciousness because of the way that mode shapes and limits perception. For this reason, we cannot develop an adequate understanding of consciousness using only the instrumental mode. We may need to turn our attention to the subtle, nonsensory perceptions with which the spiritual traditions have been concerned and adopt what Charles Tart (1972) has called "state specific science," in which the tools and plans for investigation are appropriate to the state of consciousness being investigated. This may entail the investigator having personal experience of that particular realm of consciousness. Even if the experience cannot be translated directly into object-logic and analytic language, it can provide a change in perspective and a framework within which a more comprehensive theory can be developed. Subtle perceptions may serve to orient us as to the question we should ask and the strategies we should adopt. I believe that the mind/body problem, for example, will not yield to one mode alone, but requires a pluralistic ontology.

GLOBAL NEED

From a larger perspective, more is at stake than a comprehensive theory. The problems that confront our planet—overpopulation, pollution, war—require for their solution the sense of connection to others that is provided by the development of receptive, other-centered consciousness. It would appear that this alternate mode of consciousness is needed for species survival, just as the instrumental is needed for individual survival. The mode that features connectedness—the traditional province of mystics—needs to be recognized as serving a vitally needed function. For this reason, the discoveries of the mystical traditions of the knowledge of how to bring about that shift in consciousness needs to be rescued from the never-never land of religion and made part of the overall framework of our science. There is no reason why we should not do so. Several hundred years ago the mystic Rumi stated, "New organs of perception come into being as a result of necessity. Therefore, O man, increase your necessity so that you may increase your perception" (Shah 1977).

We know now that infants of three months of age have about twice the neurons they will have as adults. A process of synaptic pruning takes place in which neurons that aren't used disappear. Thus, it seems that we may have a vast cognitive and perceptual potential that goes unutilized. Living in the sixteenth century, Rumi could not have invoked pyramidal neurons, synapses, dendritic spines, or microtubules. But today we can start to wonder whether we might be possessed at birth with neuronal circuits for the kind of knowing that mystics say is possible. Could the development of this potential be dependent in part on stimuli and requirements imposed by the culture in which we live? Can that perceptual potential be developed by specific exercises that make special demands on the organism? Would the development of that capacity have its own requirements involving diminished activity of competing circuits, such as those of the survival self? Considered in this light, the esoteric doctrines and practices of the mystical traditions cease to seem esoteric or irrational. They can be viewed as a sophisticated system for perceptual development discovered in ancient times and transmitted by symbols, artifacts, and metaphors indigenous to the particular culture in which this science was practiced. Thus, as strange and unscientific as mystical procedures may appear on the surface, especially when viewed apart from the culture in which they arose, we may find their effects reflected in neurophysiological changes associated with an increased range of function of the brain.

To develop that potential requires that the discoveries of the mystical traditions be translated into the concepts and language of a modern Western society so that they can be integrated with our science, and thus legitimized for our scientific culture as real, practical, functional. The result of such an integration may be a more effective approach to our problems as a society, based on a more comprehensive understanding of consciousness.

NOTES

1. I will be using the term "spiritual" to refer to the experiences reported in the mystical literature, not the theological interpretations associated with the various religions.

2. From this perspective, most museums are killers of aesthetic experience because they do not facilitate the shift to the receptive mode.

3. Service does not mean that instrumental mode functions cannot be utilized. They are, but in the service of other-centered goals.

4. Understanding the functional role of intention and self enables us to detect bogus spiritual teachers. If a teacher relies on threat, reward, or flattery to motivate followers, he or she is stimulating the survival self and, thereby, blocking spiritual development.

5. Almost everyone thinks they don't earn enough money. When asked how much would be enough, they specify a sum that is twice their income, regardless of their income level (Lapham 1988).

REFERENCES

Deikman, A. J. 1982. *The observing self: Mysticism and psychotherapy.* Boston: Beacon Press.

Erikson, E. 1951. *Childhood and society.* New York: Norton.

Fung, P., and G. Fung. 1964. *The sutra of the sixth patriarch on the pristine orthodox dharma.* San Francisco: Buddha's Universal Church.

Gesell, A. 1940. *The first year of life: A guide to the study of the pre-school child.* New York: Harper and Row.

Hilton, W. 1953. *The scale of perfection.* London: Burnes & Coates.

Goddard, D. 1970. *A Buddhist bible.* Boston: Beacon Press.

James, W. 1929. *The varieties of religious experience.* New York: Random House.

Kapleau, P. 1967. *The three pillars of Zen.* Boston: Beacon Press.

Lapham, L. 1988. *Money and class in America: Notes and observations on our civil religion.* New York: Weidenfeld & Nicolson.

Nikhilananda, S. 1949. *The Upanishads.* Vol. 1. New York: Bonanza Books.

Piaget, J. 1952. *The origins of intelligence in children.* New York: International Universities Press.

Shah, I. 1968. *The way of the Sufi.* London: Jonathan Cape.

Shah, I. 1977. *Tales of the dervishes.* London: Octagon Press.

Spitz, R. 1965. *The first year of life.* New York: International Universities Press.

Suzuki, S. 1970. *Zen mind, beginner's mind.* New York: Walker/Weatherhill.

Tart, C. 1972. States of consciousness and state-specific sciences. *Science* 186:1203–10.

51 Enhanced Vigilance in Guided Meditation: Implications of Altered Consciousness

Richard P. Atkinson and Heath Earl

Alterations in consciousness, such as those induced through meditation and the restricted environmental stimulation technique (REST), clearly involve modified levels of awareness and attention. These changes are so profound in meditation that some practitioners have characterized meditation-centering activities, such as concentration on breathing, as training in attention (Kapleau 1967), and the receptive, contemplative state achieved through these activities as passive awareness (Rajneesh 1972). The essential requirement of focused attention for proper meditation practice is evident from the use of such terms as onepointedness of mind (Ornstein 1986) and mindfulness (Kabat-Zinn 1994) to describe the desired states of consciousness to be achieved.

Likewise, Wallace and Fisher (1991) have suggested that REST is similar to concentrative meditation in that both require a reduction in sensory stimulation. Concentrative meditation accomplishes this reduction through focused attention upon a single unchanging or repetitive stimulus, while restricted environmental stimulation achieves the same effect through experimental manipulation of the environment with a REST chamber or flotation tank (Carrington 1986). In both conditions, there is a reduction in normal stimulus input processing, and a shift from a linear, detail-oriented cognitive processing strategy to a nonlinear, holistic-oriented one (Carrington 1977, Ornstein 1977, Suedfeld 1993). Through the process of turning off input processing for a time, an aftereffect of opening up or enhanced awareness is achieved in meditation and REST (Ornstein 1977, Suedfeld 1993).

The focused attention and increased awareness achieved through meditation and REST may be important determinants of enhanced perceptual performance, particularly on such vigilance tasks as signal detection. This assertion is supported by evidence that meditators display faster reaction times in response to visual tasks than do nonmeditators (Appelle and Oswald 1974), and discriminate more accurately between various sounds (Pirot 1978). Moreover, Atkinson and Sewell (1988) found that subjects in a flotation REST condition performed significantly better on a visual signal detection task (STD) than did subjects in relaxation and alertness conditions.

These perceptual enhancements may occur through sustained attention and awareness directed toward the object of the meditation. Deikman

(1966) refers to this phenomenon as deautomatization or dishabituation. Kasamatsu and Hirai (1966) found that advanced Zen meditators, but not nonmeditators, exhibited dishabituated brainwave activity to a clicking sound repeated every 15 seconds. These Zen masters appeared to sustain attention and awareness on a repetitive, monotonous stimulus. Similarly, perceptual vigilance tasks, such as visual signal detection, require one to deautomatize, to invest attention and awareness on the boring, repetitive task of discriminating a target stimulus from background stimuli (Atkinson and Sewell 1988).

As noted previously, flotation REST has been shown to enhance SDT performance (Atkinson and Sewell 1988). The present study extended these findings by assessing SDT performance before and after independent groups of subjects were exposed to guided meditation or chamber REST for one hour.

METHOD

Subjects

Subjects consisted of 52 volunteers (25 women, 27 men) from courses at Weber State University. Subjects were randomly assigned to chamber REST (N = 26) or guided meditation (N = 26) conditions.

Materials And Procedure

The experimental sessions were conducted inside a temperature-controlled (75°), lightproof, and sound-attenuated REST chamber with 7' × 7' × 7' interior dimensions. Subjects were seated on a comfortable reclining easy chair facing an Apple® 2e microcomputer and monitor and were informed that they would be completing a computer task involving extraction of a target stimulus from among similar background stimuli. Instructions for completing the task were generated on the monitor by an Apple® Basic "Signal Detection Task" program. Subjects were told to notify the experimenter before beginning Trial 1. The task instructions were presented as follows:

> Your environment has a large group of friends who are ordinarily so friendly that you welcome them. But be careful of enemies, for they will get you if you don't keep your eyes open and catch them. Your friends will be indicated on the screen by the letter F, and the enemy by the letter E. After each trial, the computer will ask you whether or not you saw an enemy. If the enemy was present, you should press P; if the enemy was absent, press A. Do not press return after your entry. Press the space bar to begin.

Subjects then completed 12 practice SDT trials to reduce the likelihood of obtaining posttest results due to practice effects, after which 36 SDT pretest trials were completed.

Following the pretest trials, subjects in the chamber REST condition were told to rest comfortably in a reclined position for 60 minutes. Instructions for the guided meditation group were identical except that subjects were required to listen through earphones for 60 minutes to a guided meditation audiotape, "Meditations on the Light." This audiotape includes breathing and relaxation exercises for approximately 5 minutes, followed by repeated suggestions for focused attention and meditation upon mentally created images of the moon during the remaining 55 minutes. The chamber light was then turned off and the chamber door was shut during exposure periods. After one hour of exposure, subjects were asked to move the chair to its full upright position to complete 36 posttest SDT trials on the computer.

For a given SDT trial, a sequence of Fs was presented at random positions one at a time on the monitor screen until there was a total of 60 letters. On 24 trials from a block of 36, one of the letters was an E instead of an F. The subjects' task was to detect the presence of the E. All letters faded from the screen within 30 seconds of trial onset, after which the following prompt appeared on the monitor: "If the enemy was present, press P; if the enemy was absent, press A." Subjects were required to make a response within 5 seconds. The next trial was then initiated by pressing the space bar.

Signal strength was manipulated by the timing of the target letter to be identified. On 12 strong signal trials, the E was presented within the first 15 seconds. On another 12 weak signal trials, the E was presented in the last 3 seconds of the trial. On the remaining 12 no signal trials, the E was not presented.

RESULTS

The mean number of prepost strong and weak signal hits and no signal false alarms for chamber REST and guided meditation conditions are found in Table 51.1. A Sensory Stimulation Level (Chamber REST, Guided Meditation) × Measurement Time (Pretest, Posttest) × Signal Strength (Strong Signal, Weak Signal, No Signal) analysis of variance (ANOVA) was performed on these data. Significant main effects were present for Time of

Table 51.1 Mean number of hits and false alarms on the pretest and posttest for chamber REST and guided meditation conditions.

Signal Strength	Pretest		Posttest	
	M	SD	M	SD
Strong Signal Hits				
Chamber REST	8.62	1.50	8.27	1.80
Guided Meditation	8.69	1.62	8.23	2.07
Weak Signal Hits				
Chamber REST	5.65	2.06	6.38	1.90
Guided Meditation	5.58	1.72	7.50	1.17
No Signal False Alarms				
Chamber REST	0.12	0.45	0.12	0.45
Guided Meditation	0.12	0.45	0.04	0.20

Measurement ($F(1,50) = 5.98$, $P<0.025$) and Signal Strength ($F(2,100) = 344.37$, $P<0.001$). A significant Time of Measurement × Signal Strength interaction effect was also present ($F(2,100) = 11.01$, $P<0.005$). Post-hoc comparisons showed no significant differences between chamber REST and guided meditation conditions on the strong, weak, and no signal pretest trials. The chamber REST and guided meditation conditions scored a significantly greater number of weak signal hits ($F(1,50) = 5.19$, $P<0.05$; $F(1,50) = 35.87$, $P<0.001$) from the pretest ($M = 5.65$; $M = 5.58$) to the posttest trials ($M = 6.38$; $M = 7.50$). The number of strong signal hits and no signal false alarms did not change significantly from pretest to posttest trials. The guided meditation condition ($M = 7.50$) scored a significantly greater number of weak signal hits ($F(1,50) = 6.50$, $P<0.025$) on the posttest trials than did the chamber REST condition ($M = 6.38$). The chamber REST and guided meditation conditions did not differ in their number of strong signal hits and no signal false alarms on the posttest trials.

DISCUSSION

The results of this study indicate that chamber REST and guided meditation significantly enhance perceptual vigilance for weak signals on a visual SDT from pretest to posttest, without producing a concomitant increase in false alarm rates. Moreover, these results suggest a significantly stronger perceptual vigilance-enhancing effect for guided meditation than for chamber REST on posttest weak stimulus signals. Overall, the results of this study support and extend a previous investigation that demonstrated enhanced perceptual vigilance on a SDT for flotation REST as compared with relaxation and alertness conditions (Atkinson and Sewell 1988).

One explanation for the enhanced perceptual vigilance found in guided meditation may be related to the nature of visual signal detection tasks. These tasks typically require subjects to identify a target stimulus from among a large number of similar background stimuli. Competent performance is aided by sustained and focused attention, as well as greater awareness of the nature and requirements of the task (Atkinson and Sewell 1988). As has already been noted, meditation is frequently described in terms of focused and sustained attention on external and internal stimuli and events (Kabat-Zinn 1994, Kapleau 1967, Ornstein 1986). Meditation involves the phenomenon of deautomatization (Deikman 1966), since proper meditation practice requires focused attention on automatic activities and processes. It is this ability of meditators to deautomatize that likely accounts, at least in part, for the enhanced vigilance demonstrated by the guided meditation group. Although the subjects in this group were not trained meditators, the audio-tape, "Meditations on the Light," provided repeated suggestions for focused attention upon breathing and relaxation, as well as suggestions for creation of and meditation upon mentally produced visual images of the moon.

As noted previously, meditation and REST share the commonality of functional sensory deprivation (Carrington 1986, Wallace and Fisher 1991). Ornstein (1986) has proposed that the turning off of input processing in meditation produces a shift from an active mode of cognitive processing, characterized by analytic, sequential, and discursive thought, to a receptive mode that is conceptualized as holistic, simultaneous, and intuitive in nature. Furthermore, Ornstein (1986) suggests that the shift to a receptive mode leads to an inward, quiescent state, which results in a primary aftereffect of enhanced awareness. Although Ornstein (1986) applies this theory to meditation, the restriction of input processing in REST should produce a similar shift from an active to a passive mode with its attendant aftereffect of enhanced awareness. In fact, Budzynski (1990) has proposed a related theory for REST in which the normally dominant cerebral hemisphere becomes deactivated when input processing is curtailed, allowing the nondominant hemisphere to assume a greater role, particularly with regard to nonlinear thinking and holistic perception. When considered in tandem, these theories seem to support a comparable shift in cognitive processing styles for meditation and REST, from an active or detail-oriented mode to a receptive or holistic one. In both instances, the shift has been proposed to be largely a function of the reduction or turning off of input processing.

The attentional and cognitive processing explanations are not mutually exclusive. The superior performance of the guided meditation group may be explicable in terms of the presence of both focused attention exercises and a holistic-oriented or receptive cognitive processing shift. Perceptual vigilance enhancement in chamber REST may be limited to the proposed cognitive processing shift, which may produce some increase in awareness but not at the same level as guided meditation.

REFERENCES

Appelle, S., and L. E. Oswald. 1974. Simple reaction time as a function of alertness and prior mental activity. *Perceptual and Motor Skills* 38:1263–68.

Atkinson, R. P., and M. M. Sewell. 1988. Enhancement of visual perception under conditions of short-term exposure to sensory isolation: A comparison of procedures for altering vigilance. *Perceptual and Motor Skills* 67:243–52.

Budzynski, T. H. 1990. "Hemispheric asymmetry and REST." In *Restricted environmental stimulation: Theoretical and empirical developments in flotation REST,* edited by P. Suedfeld, J. W. Turner, Jr., and T. H. Fine. New York: Springer-Verlag.

Carrington, P. 1977. *Freedom in meditation.* New York: Anchor Press/Doubleday.

Carrington, P. 1986. "Meditation as an access to altered states of consciousness." In *Handbook of states of consciousness,* edited by B. B. Wolman and M. Ullman. New York: Van Nostrand Rheinhold.

Deikman, A. J. 1966. Deautomatization and the mystic experience. *Psychiatry* 29:324–38.

Kabat-Zinn, J. 1994. *Wherever you go, there you are: Mindfulness meditation in everyday life.* New York: Hyperion Books.

Kapleau, P. 1967. *The three pillars of Zen.* Boston: Beacon Press.

Kasamatsu, A., and T. Hirai. 1966. An electroencephalographic study on the Zen meditation. *Folia Psychiatrica et Neurologia Japonica* 20:315–36.

Ornstein, R. E. 1977. *The psychology of consciousness,* 2nd ed. New York: Harcourt, Brace, Jovanovich.

Ornstein, R. E. 1986. *The psychology of consciousness* 3rd ed. New York: Penguin Books.

Pirot, M. 1978. "The effects of transcendental meditation technique upon auditory discrimination." In *Scientific research on the transcendental meditation program: Collected papers,* Vol. 1, edited by D. W. Orme-Johnson and J. T. Farrow. New York: MIU Press.

Rajneesh, G. S. 1972. *Dynamics of meditation.* Bombay: Life Awakening Movement Publications.

Suedfeld, P. 1993. "Stimulus and theoretical reductionism: What underlies REST effects?" In *Clinical and experimental restricted environmental stimulation: New developments and perspectives,* edited by A. F. Barabasz and M. Barabasz. New York: Springer-Verlag.

Wallace, B., and L. E. Fisher. 1991. *Consciousness and behavior,* 3rd ed. Needham, MA: Allyn and Bacon.

52 The Stream Revisited: A Process Model of Phenomenological Consciousness

José-Luis Díaz

INTRODUCTION

The scientific approach to consciousness requires a theory that makes the relationships between consciousness, behavior, and brain more transparent. In order to accomplish this task, it is necessary to develop robust theories concerning the nature of consciousness and to outline models of phenomenological consciousness that then can be related to specific neural events and behavioral processes. Even though it remains an incomplete and unsatisfactory definition, I will take the notion of phenomenological consciousness as equivalent to *awareness*, the third meaning of the word in the Oxford dictionary, and the common ground of the rest (see Natsoulas 1983); namely to the fact of a sentient individual having experiences such as sensations, feelings, thoughts, mental images, and intentions.

Heuristic models of phenomenological consciousness need to progressively fulfill two criteria. The first is a dynamic and patterned process architecture and the second is the inclusion of as many as possible of the main features of consciousness. At least seven features may be considered essential to the understanding of the nature of consciousness: temporality, processing activity, content, *qualia*, totality, attention, and levels. In the present work a Petri net diagram of phenomenological consciousness will be progressively developed by the successive definition and representation of these main features.

The model follows two guidelines and one assumption. The first guideline is William James' notion of a "stream of consciousness," a metaphor suitable for specification within the framework of a dynamic patterned process scheme. The second guideline appeals to the European Phenomenology aim to identify and describe the features of conscious phenomena but, in this case, without renouncing the need to deal with the fundamental question of the ontological nature of consciousness. In fact, in order to develop the process model it seems necessary to ascertain that conscious, neurophysiological, and behavioral processes are functionally and physiologically integrated, so that the dynamic models of each one of these would appear to be formally similar and capable of meaningful empirical and theoretical correlations. Such correlations would constitute psychophysical data, hypothesis and, eventually, laws. Thus, the present model assumes a dual aspect, neutral monist ontology according to which conscious phenomena are both brain and mental states of a peculiar informational class; namely, patterned processes that call for meaningful correlations but not for mutual reduction (Díaz 1989). The representation of consciousness as a dynamic process constitutes a model derived from this notion.

The process diagram will be introduced now by representing the aforementioned features of consciousness in seven stages of successive complexity.

THE SEVEN FEATURES

1. Temporality

Consciousness is not a substance or an essence but a process that unfolds in time. This is the core idea in William James' (1890) seminal concept of the "stream of consciousness." Similarly, Whitehead (1920) conceived consciousness as "duration," a peculiar type of natural event. More recently, Gennaro (1992) revived the Kantian theses that the most basic feature of consciousness is experiencing the world as having temporal duration, a feature that requires the operation of episodic memory. In turn we may say that time is that by means of which we apprehend the world.

In Figure 52.1 the temporal factor or unfolding of consciousness is represented as an arrow of time progressing (since Descartes) from left to right (1.1). The actual duration of awareness or unfolding of consciousness is conceived and represented in the diagram as a continuously moving "window" located in the forefront or tip of the arrow (1.2). Such "window of the present" has a brief duration and is functionally coupled with neurophysiological signals (Libet 1985), with working memory mechanisms (Tulving 1987), and with behavioral acts (Feldhutter, Schleidt, and Eibl-Eibesfeldt 1990).

The temporal aspect of consciousness (the subjective sensation of time), is far from straightforward. As it occurs with slow and fast motion in films, the events appear to develop at different velocities and the subjective speed is represented in the diagram by the width of the arrow (1.3).

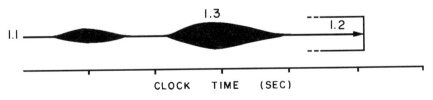

Figure 52.1 The arrow of subjective conscious time and the window of the present.

2. Processing Activity

Conscious experience not only occurs in time but also involves motion and work. Since the formulation of the ancient Buddhist psychology of the Abhidharma (Guenther 1957), mental contents can be considered to unfold as successions of factor-events that arise, undergo active processing, and vanish from awareness according to causal laws. Thus, in Figure 52.2, the time arrow of consciousness breaks into single arrows unfolding as a chain of mental factors (2.1).

Each arrow is represented as a trajectory defined by an initial point (the appearance of the event in the field of awareness), a course of active processing, and an arrowhead that signals the effect of the event in producing another and/or the point in time of its disappearance from consciousness (2.2). Indeed, many conscious events arise and disappear in the cognitive unconscious undercurrent or take place in the "fringe" (Magnan 1993).

3. Content

In contrast to the analytical philosophy notion of "intentionality" (meaning as reference and representation), the classical Phenomenology of Husserl and its development by Merleau-Ponty (Montero 1987) established that mental events are "phenomena" directly encountered in experience and, therefore, evident to the individual and grasped by insight. In accord with this notion, in the present model *content* refers to the specific mental factor-events comprising the unfolding of experience.

In order to represent content, the process diagram (Figure 52.3) depicts four broad classes of mental events: sensations (s), emotions (e), thoughts and mental images (t), and intentions (i). It would be possible to dissect each one of these classes in "families" (that is, the senses), "species" (each particular sensation), and even qualia (see below).

The causal relations among specific conscious mental events are nonlinear and complex. Thus, an event of one class may elicit another event of its

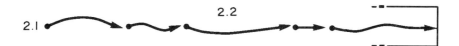

Figure 52.2 The processing activity of conscious events.

Figure 52.3 The contents of consciousness (s: sensations; t: thoughts, e: emotions, i: intentions).

own or other classes. Moreover, an event may arise and disappear from awareness by crossing the consciousness "threshold."

4. Qualia

The notion of qualia refers to the qualitative aspects of mental events (that is, the "raw feels" of sensations) applied either to their properties and/or to their affective tone. In a classic paper, Thomas Nagel (1974) alludes to this property when he skeptically wonders if it is possible to know what a sonar experience feels like to a bat.

Indeed, the qualitative properties of conscious events allow for the folk-psychological and academic textbook classification of sensations (seeing, hearing, feeling cold, and so on) and permit also the classification within a single sense (colors, textures, tones). On closer inspection, the specific properties comprising qualia define each single conscious event as personal (the red carpet as I see it as different from the same red carpet as you see it) and even unique (the red carpet as I see it now as different from the same red carpet as I saw it yesterday). In such an exhaustive description it is possible that content and qualia would coincide. Thus, the full specification (in the sense of describing the type and properties of a single conscious event) would include both features. Following the taxonomic metaphor used in the case of content, the description of specific qualia would be similar to the case of describing the unique peculiarities of a single specimen of a botanical species.

In Figure 52.4, qualia are represented by different arrow colors. It may be possible to formalize the relationship between color and qualia for each class of mental events, but this would require a more developed model and taxonomy.

5. Totality

The feature of totality refers to the unity of experience, to the fact that disparate contents such as sensations, emotions, or thoughts may appear simultaneously in the field of awareness integrating amalgams or occasions of consciousness. Such occasions arise from a coordination and mixture of events determined by a functional "binding" of neural processes and follow one another in the temporal continuum. Such functional connection and sequential processing endow consciousness with its peculiar holistic structure.

Figure 52.4 *Qualia*, the qualitative feeling of conscious contents.

Occasions of consciousness are represented in Figure 52.5 by the production of bi-dimensional strands of arrows, a procedure that allows for the simultaneous occurrence of more than one content, for complex causation links between events, and for the delimitation of "zones" or fields between vertical dotted lines (α–γ). In this case, the general classes of contents (sensations, emotions, thoughts, and intentions) are arranged in vertical rows. In the diagram, a particular strand of consciousness would constitute a "narrative" that could be "read" in terms of occurrences, transitions, and occasions.

At this point in the development of the diagram, it is worth mentioning that it conforms precisely to the characteristics of a Petri net diagram. These are occurrence-transition graphs developed by Carl Adam Petri of the University of Bonn in 1962 and applied by Anatol Holt to represent complex behavioral sequences (see Bateson 1991). Such behavioral sequences can properly be called "actograms."

The two-dimensional unfolding net can be said to correspond to the surface of the stream of consciousness seen from above. The last two features of consciousness will be included by expanding the frame or metaphor of the stream to include (as it happens in a river) margins (attention) and depth (levels).

6. Attention

Attention is one of the best known features of consciousness. The term refers to a system that permits the selection, from the internal and external stimuli, of those that require further processing. The selection of stimuli consists of preconscious priming mechanisms acting upon the stimulus where attention will focus (LaBerge 1990, Kinchla 1992).

There are two forms of attention (see Posner et al. 1987), one manifest in behavior (orientation reflexes, facial gestures, ocular movements, pupil diameter modifications) and the other covert (the capacity to localize and transfer the locus of attention without manifest behavioral expression). Moreover, there are several forms of covert attention that can be described using a common diagram of visual perception. Thus, in Figure 52.6 attention is represented (in comic strip fashion) by the dotted lines connecting the eye and the visual field (panel in the right). In terms of its scope or field

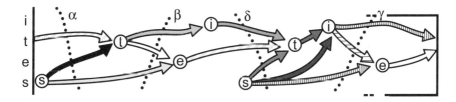

Figure 52.5 The unity or totality of conscious states represented as the bi-dimensional surface of a stream.

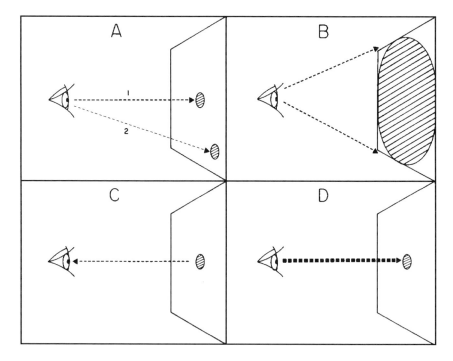

Figure 52.6 The forms of covert attention. Field: narrow (A) or panoramic (B); direction: aimed (A) or receptive (C); intensity: tenuous (A) or strong (D).

(the "zoom" property), attention can be either focal (A) or panoramic (B); concerning its direction (the "searchlight" property), it can be either aimed (A1, A2) or receptive (C); and concerning its intensity or effort, it can be either tenuous (A) or strong (D). The specific attention state varies continuously depending upon the moment-to-moment modifications in field, direction, and intensity.

The representation of these types of attention in the present time flow diagram requires the eye and the panel to disappear. The remaining virtual line which connects them would become an unfolding band defined by width (covering field) and shade (intensity). At the present moment it is not possible to adequately represent the three characteristics of attention. In any case, the covering field of unfolding attention can be incorporated in the Petri net developed in Figure 52.5 as the margin or boundary of the net (Figure 52.7). Using again the stream metaphor it can be said that the surface of the stream is defined by the successive occasions (with their dynamic contents and qualia), while attention constitutes the margins and shape of the stream itself.

There is one more feature of attention to schematize. It is the one concerning levels of operation, for it is well established that there are two levels of attention: one fast, passive, and automatic (functionally coupled to orientation reactions), and the other slow, active, and controlled (Kinchla 1992). It seems that automatic attention operates in a state of normal vigi-

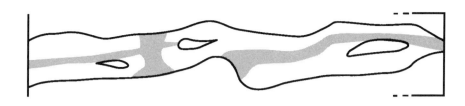

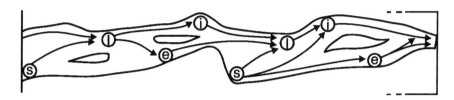

Figure 52.7 The shape of unfolding attention. Lower panel: the margins of the attentional state for the process shown in Figure 52. 5. Upper panel: Covering and intensity in the same process.

lance and that controlled attention occurs during self-consciousness. Thus, these characteristics of attention can actually be better understood and represented in the following section dealing with levels of consciousness.

7. Levels

In order to complete the description and modeling of consciousness, it is necessary to outline the existence of distinct levels of awareness. Such levels can be defined by specific features of attention, by the amount and type of processed information, by a hierarchy of information complexity, and by distinct thresholds between them.

The mere processing of cognitive information does not correspond with consciousness since there is wide documentation of episodes of perception, learning, elaboration, and skillful motor execution without awareness (Velmans 1991). The relationship between the vast and parallel unconscious cognitive processing and the selective and restricted conscious processing can be conceptualized as an "awareness threshold" represented in the present diagram (Figure 52.8) as a moving layer of surface "visibility" in the depth of the stream. Depending upon historical and adaptation determinants of the individual, some of the cognitive information items reach the threshold and undergo further processing until they disappear from awareness. In the diagram, the levels of consciousness would need to be established by variations in the depth of the awareness line within the stream. Whenever the line crosses the distinct thresholds between levels there would be a change in level of consciousness.

Four distinct levels of awareness have been defined through the ages: dream (except for lucid dreaming), awake (normal vigilance), self-consciousness, and ecstasy. The threshold between these levels can be defined with some precision, as it occurs with the point of awakening that establishes the difference between sleep and vigilance. For example, the reflective nature of self-consciousness (defined by Locke in 1690 as the capacity to be aware of one's own awareness), is characterized by controlled attention and by the state designated as *mindfulness* in which we are aware of the context and content of information. Thus, there is a distinct difference between this state and normal vigilance (or *mindlessness*) that operates with automatic attention (Langer 1992). Finally, the expanded states designed here by the rather peculiar noun "ecstasy" (related terms: "peak experiences," *samadhi, satori*, the "numinous," "mystical experiences") have been a subject of intense and prolonged interest, including William James' (1900) other classic, *The Varieties of Religious Experience* and the Dorpat School of Religious Psychology (Wulff 1985). There have been many attempts from traditional and academic psychology to construct definitions and "cartographies" of these states, especially in the early 1970s (Weil 1972, VanDusen 1972), but in the present work I would like merely to include them in the diagram for further analysis.

In the vertical process flow diagram (Figure 52.8), the levels of consciousness are represented by distinct levels in the depth of the layer of awareness. The diagram shows a vertical and longitudinal section of the flow as well as the thresholds or interfaces between the levels in the left. A hypothetical temporal progression of the actual band of awareness moves forward at different depths of the stream. Accordingly, some contents may be processed at different levels of awareness in consecutive moments. The model intends to show that distinct levels of consciousness differ in the amount and type of information processed, but this is an incomplete rendition since there are not only quantitative but qualitative differences between levels.

IMPLICATIONS OF THE MODEL

As with any model, the present diagram has advantages and limitations. The first advantage is didactic: the diagram correctly suggests the dynamic, everchanging nature of phenomenological consciousness and it defines and incorporates seven cardinal features of this elusive process. Moreover,

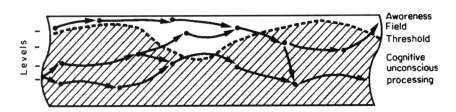

Figure 52.8 The levels of consciousness and cognitive unconscious processing represented as the depth flow of the stream.

it shows that the flow of consciousness is never uniform because of variations in the cross-sectional area (levels), in the changing width of the margins (attention), and in the intrinsic properties of the ongoing states (the dynamics of the contents). Another implication of interest is the fact that the Petri net diagram of conscious flow is formally identical to behavioral actograms. This similarity is in agreement with the multiple aspect ontology of the mind-body relation outlined above and it allows for possible empirical correlations between behavioral and conscious process.

In contrast with classical phenomenology and other theories of consciousness, but in agreement with Buddhist psychology (Guenther 1957, Goldstein 1983), with the ideas of Hume, and with modern trends in neuroscience (Klivington, 1989) or philosophy (see Lycan 1987, Dennett 1991), the present model does not require the inclusion of a "self" or an "ego." Consciousness constitutes a process defined by contents and states in continuous progression, and, in the concept of Irwin Lieb (1991), from one present moment to the next individuals are not unchanging entities: individuals are their actions. Even when a sense of a freedom of the will appears in the mind, it constitutes merely another content and a change of level. The elimination of an agent considerably facilitates the program of naturalization of consciousness.

One of the main problems of this or any other model of consciousness that intends to be testable (apart from didactic) is the possibility of empirically recording and analyzing a natural flow of awareness in terms of the model. At the present moment the self-recording of conscious flow in real time does not appear to be feasible even though some techniques of insight meditation that are based upon the labeling of experience (Goldstein 1983) could be eventually applied for this purpose with scientific tools. On the other hand, despite the spectacular advances in brain imaging techniques, the possibility of a *cerebroscope* or brain activity-recording device that could be reliably and specifically correlated with ongoing essential features of phenomenological consciousness seems remote despite the fact that a correlation with still unidentified neural processes should exist. In view of such limitations, it would seem that a possible application of the model would depend upon the development of a reliable "hermeneutic" (Jaspers 1913) or "heterophenomenological" (Dennett 1991) procedure. This would imply an actualization of introspection techniques in order to standarize verbal reports from trained subjects and obtain high interobserver agreements among evaluators of these reports. Indeed, several forms of this approach have been successfully used to study specific aspects and operations of consciousness for some time (Paivio 1975), and there are some analyses of consciousness in literary criticism (Cohn 1978) that could be useful.

Nevertheless, a minimally faithful rendition of ongoing conscious states would require a highly systematic procedure to analyze first-person accounts and third-person transcriptions of this material into occurrence-transition or Petri net diagrams. This seems a worthwhile and attainable program, and the present model intends to be of use in this direction.

ACKNOWLEDGMENTS

The present paper was produced with the aid of the National University of Mexico (a sabbatical fellowship to the author and a grant to the Interdisciplinary Cognitive Science Group, DGAPA-IN602491) and within the amiable hospitality of the Cognitive Science Program at the University of Arizona. The author thanks Merrill Garrett and Alfred Kazniak for their relevant comments to an earlier draft of the manuscript.

REFERENCES

Bateson, M. C. 1991. *Our own metaphor.* Washington, DC: Smithsonian Institution Press.

Cohn, D. 1978. *Transparent minds.* Princeton, NJ: Princeton University Press.

Dennett, D. C. 1991. *Consciousness explained.* New York: Little Brown.

Díaz, J. L. 1989. *Psicobiología y Conducta. Rutas de una Indagación.* Mexico City, Mexico: Fondo de Cultura Económica.

Feldhutter, I., M. Schleidt, and I. Eibl-Eibensfeldt. 1990. Moving in the beat of seconds. Analysis of the time structure of human action. *Ethology and Sociobiology* 11: 511-20.

Gennaro, R. 1992. Consciousness, self-consciousness and episodic memory. *Philosophical Psychology* 5:333-47.

Goldstein, J. 1983. *The experience of insight.* London and Berkeley: Shambala.

Guenther, H. V. 1957/1976. *Philosophy and psychology in the Abhidharma.* London and Berkeley: Shambala.

James, W. 1890/1950. *The principles of psychology.* New York: Dover.

James, W. 1900/1974. *The varieties of religious experience.* London: Fontana.

Jaspers, K. 1913/1973. *Psicopatología General.* Buenos Aires: Beta.

Kinchla, R. A. 1992. Attention. *Annual Review of Psychology* 43:711-42.

Klivington, K., ed. 1989. *The science of mind.* Cambridge, MA: MIT Press.

LaBerge, D. L. 1990. Attention. *Psychological Scientist* 1: 156-62.

Langer, E. 1992. Matters of mind: Mindfulness/mindlessness in perspective. *Consciousness and Cognition* 1:289-305.

Libet, B. 1985. Unconscious cerebral initiative and the role of conscious will in voluntary action. *Behavioral Brain Sciences* 8:529-66.

Lieb, I. C. 1991. *Past, present, and future.* Chicago: University of Chicago Press.

Lycan, W. 1987. *Consciousness.* Cambridge, MA: MIT Press.

Magnan, B. 1993. Taking phenomenology seriously: The "fringe" and its implications in cognitive research. *Consciousness and Cognition* 1.

Montero, F. 1987. *Retorno a la Fenomenología.* Barcelona: Anthropos.

Nagel, T. 1974. What is it like to be a bat? *Philosophical Review* 83: 435-51.

Natsoulas, T. 1983. Concepts of consciousness. *Journal of Mind and Behavior* 4:13-59.

Paivio, S. 1975. Neomentalism. *Canadian Journal of Psychology* 29:263-91.

Posner, M. I., A. W. Inhoff, F. J. Friederich, and A. Cohen. 1987. Isolating attentional systems: A cognitive anatomical analysis. *Psychobiology* 15:107-21.

Tulving, E. 1987. Multiple memory systems and consciousness. *Human Neurobiology* 6:67-80.

VanDusen, W. 1972. *The natural depth in man.* New York: Harper & Row.

Velmans, M. 1991. Is human information processing conscious? *Behavioral Brain Sciences* 14:651-68.

Weil, A. 1972. *The natural mind.* Boston: Houghton-Mifflin.

Whitehead, A. N. 1920/1971. *Concept of nature.* Cambridge, England: Cambridge University Press.

Wulff, D. M. 1985. Experimental introspection and the religious experience: The Dorpat School of Religious Psychology. *Journal of History of Behavioral Sciences* 21:131-50.

XI Overview

Looking back over the chapters in this book and remembering the formal lectures and poster sessions and informal discussions of the conference from which they sprang, one is impressed with two thoughts. First of all there is a remarkably diverse spectrum of views within the scientific community concerning the nature of the phenomenon called consciousness. Secondly—and as a result of this diverse spectrum—science does not have a collective opinion on the matter.

Nonetheless there has been some progress over the past decade or so. No longer is the "C-word" taboo in the august halls of science; it can now be mentioned in polite discourse without a loss of credibility. The subject is on the table, but the question remains: Where do we go from here?

Of course different people have diverse answers, but some lines of thought emerge from the preceding chapters. Surely it is important to gather more and better experimental data, and—as several of the chapters demonstrate—this is becoming available month by month and year by year in impressive quantities. But the *direction* of future scientific research on consciousness is yet to be clearly established. Is the traditional scientific paradigm of reductive materialism up to the task as some would have us believe? Or is some "new ingredient" needed? And if so, what? The chapters of the present section address this question from three different perspectives.

In Chapter 53 Ian Marshall considers the paradigms of the field known as *artificial intelligence* (which asserts that mind is not and cannot be more than some aspect of the computations that are carried out by the brain), and he then argues that higher mental processes—creativity, humor, spirituality, intention, and the like—cannot be explained by traditional computational paradigms. Perhaps analogous to "introspective consciousness," "higher-order thought" (for example, Chapter 2), and Penrose's "noncomputational processing," the mode which Marshall describes as "creative" and "evolving" requires, he suggests, quantum mechanics for its implementation.

A somewhat wider perspective on the options open to consciousness science is presented in Chapter 54 by Jean Burns, who defines three issues amenable to empirical testing. These are: (1) the ontological relationship between consciousness and the brain/physical world, (2) the physical characteristics associated with the mind/brain interface, and (3) whether consciousness can act on the brain independently of any brain processes. Burns provides a useful taxonomy on the ontological issue: *Physicalism* (including *reductionism* and *emergent physicalism*) is countered by *dualism, monistic idealism,* Bohm's *implicate order,* and comparable views. On the third issue, Burns observes that if we thoroughly understand brain functions, we can then determine whether or not some brain action (that is, "free will") occurs unrelated to previous conditions.

Appropriately, the most ambitious chapter of this section is the final one by Willis Harman. Is the conceptual framework of science sufficiently broad to encompass the phenomenon of consciousness, he asks, or must it be somehow enlarged to fit the facts of mental reality? Attempting an answer, he considers the degree to which science can claim to be objective and to what extent it is influenced by the culture in which it is immersed. Those who disagree might pause to consider the religious perspective from which modern science has emerged.

There is reason to suppose that the roots of our bias toward determinism lie deeper in our cultural history than many are accustomed to suppose. Indeed, it is possible that this bias may even *predate* modern scientific methods. In his analysis of thirteenth-century European philosophy, Henry Adams (1904) archly observed: "Saint Thomas did not allow the Deity the right to contradict himself, which is one of Man's chief pleasures." One wonders to what extent reductive science has merely replaced Thomas's God with the theory of everything.

The question of scientific objectivity becomes more compelling when one considers that doubts about the reductive paradigm are by no means new. William James (1890), Charles Sherrington (1951), Erwin Schrödinger (1944, 1958), Karl Popper and John Eccles (1977)—among others—have insisted that the reductive view is inadequate to describe reality. This is not a fringe group. They are among the most thoughtful and highly honored philosophers and scientists of the past century. How is it that their deeply held and vividly expressed views have been so widely ignored? Is it not that we *need* to see the world as better organized than the evidence suggests?

Take the matter of "downward causation" to which Harman gives some attention. Why should this be an issue in brain dynamics? As Erich Harth points out in Chapter 44, connections between higher and lower centers of the brain are reciprocal. They go both ways, up and down. The evidence (the *scientific* evidence) for downward causation was established decades ago by the celebrated Spanish histologist Ramon y Cajal, yet the discussion goes on. Why? The answer seems clear: If brains work like machines, they are easier to understand. The facts be damned!

REFERENCES

Adams, H. 1986, first published in 1904. *Mont Saint Michel and Chartres.* New York: Penguin Classics.

James, W. 1890. *Principles of psychology.* New York: Holt (republished in 1950 by Dover).

Popper, K. R., and J.C. Eccles. 1977. *The self and its brain.* Berlin: Springer-Verlag.

Schrödinger, E. 1944, reprinted 1967. *What is life?* Cambridge, England: Cambridge University Press.

Schrödinger, E. 1958, reprinted 1967. *Mind and matter.* Cambridge, England: Cambridge University Press.

Sherrington, C. S. 1951. *Man on his nature,* 2nd ed. Cambridge, England: Cambridge University Press.

53 Three Kinds of Thinking

I. N. Marshall

INTRODUCTION

There are clear mathematical models of two kinds of computation, serial and parallel processing. I shall argue that human thinking uses both of these and also a third process.

Mental arithmetic is done in a slow, step-by-step, rule-following fashion. Such serial processing resembles the style of a personal computer (Minsky 1972, Boden 1988). But recognizing a face is a another style of computation: the rapid combination of many pieces of data into a meaningful whole. This is the style of the more recently developed parallel processors. (Tank and Hopfield 1987, Rumelhart and McLelland 1986). "People do seem to have at least two modes of operation, one rapid, efficient and subconscious, the other slow, serial and conscious" (Norman 1987).

Serial and parallel processors have complementary strengths and weaknesses. Serial processors are good at precise, rational (algorithmic) tasks but poor at pattern recognition or learning complex skills such as driving. Parallel processors, which learn by association, not reason, lack language or explicit rule-following abilities. Both abilities are used by good chess players: immediate "intuitive" recognition and evaluation of the general type of position (parallel processing), plus a limited amount of detailed analysis (serial processing) (Seymour and Norwood 1993).

The human brain appears to have the "hardware" for both kinds of processor. Parallel processors consist of many richly interconnected elements somewhat resembling the local networks of neurons found throughout the brain. Serial processors require precise point-to-point wiring like telephone cables, resembling the brain's nerve tracts. Parallel processors learn slowly by repetition, whereas serial processors can store information presented only once. The brain has both types of memory. The one-off memory system involves the hippocampus, whereas the more primitive habit memory system is distributed throughout the brain. The two learning mechanisms have different biochemistries (Kandel and Hawkins 1992).

The dominant artificial intelligence view is that all human thinking is serial and/or parallel processing. But some writers have argued that this picture is incomplete. Phenomena it may leave out are:

1. Consciousness (many workers). Specifically, the unity of consciousness (Penrose 1989, Marshall 1989, Zohar 1990, Zohar and Marshall 1993).
2. Human agency, freedom, and responsibility (Lucas 1970).
3. Creative (nonalgorithmic) thinking (Penrose 1989).
4. Higher psychological processes bearing on truth, creativity, humor, spirituality (many workers).
5. Meanings, intentionality, semantics (Searle 1992).

The first three sets of phenomena have been linked by their proponents with a possible quantum mechanical aspect of brain function. These arguments will not be discussed here. This paper will focus on the final two sets of phenomena, linking them to a quantum physical model.

HIGHER MENTAL PROCESSES

Psychodynamic and religious systems concerned with human development have often postulated three cognitive processes (prepersonal, personal, and transpersonal), not two (Wilber 1983, Chapter 7). This fact may indicate that a third process, if it exists, is an evolutionary one. This area will be very briefly reviewed.

Freud described a primary process (the "id"), often unconscious, which worked by instinct and association. It could recognize patterns but not do serial calculations. This contrasted with a more rational secondary process (the "ego"), which developed gradually (Rycroft 1968). These two concepts, which have become part of our culture, correspond closely to parallel and serial processing. They have been confirmed by psychodynamic research (Peterson 1992).[1] Other related dichotomies are feeling-thinking, extraversion-introversion (Peterson 1992), and right brain–left brain (Ornstein 1972). The last correlation is unsurprising since in right-handed people the language area, required for many serial processing tasks, is in the left cortex.

But neither the ego nor the id can support genuine commitment to science, art, friends or morality. Freud did not regard the higher functions as mere expressions of instinct. (However, the Freudian superego did not manifest a third process, but only habit learning of parental rules.) Yet, during treatment or otherwise, people do evolve higher modes of function that "sublimate" sexual or aggressive instinctual energies (Rycroft 1968). The process of "insight" which leads to such evolution is not a purely rational ego-based matter, it is agreed, but psychoanalysis has remained vague at this crucial point.[2] It is difficult or impossible to give a mechanical model of this area.

Jung accepted the broad outlines of Freud's views on the ego and id, but he was more concerned than Freud with evolutionary processes, which he called "individuation" (Rycroft 1968). Many religious systems have similarly maintained that human evolution, perhaps aided by meditation, is possible. If so, it is not reducible to serial or parallel processing. One relevant kind of experience is that of contacting a timeless, changeless, realm, resembling Plato's forms of oneness, truth, beauty, and goodness. In a common metaphor, people may be "coming from" any of three perspectives: "head" (rational, wisdom), "heart" (creative, love), or "gut" (instinctive, power).

Basing her argument on empirical studies of child development and on cultural history, Donaldson (1992) distinguished four modes of thinking. The first three are closely related to the three kinds of thinking just outlined. A young baby's primary concern is with the here-and-now, the "point mode." After the age of six months it will often try to find and retrieve a dropped toy and in other ways behave as if objects had a continuing existence. This is the "line mode" of personal narratives. Later comes the "construct mode" whose concern is with the general patterns of art, science, or philosophy occurring anywhere in space-time. Her "transcendent mode," which refers to mystical and meditative experiences not apparently within space-time at all, is outside the scope of this paper. (But see Zohar and Marshall 1993, Chapter 10.)

Built in styles of thinking, one might expect, would be reflected in possible styles of human organization of the environment, including social groups. In fact one can find such analogies (Zohar and Marshall 1993). An example of parallel processing is the way people at a party independently move about until they find compatible subgroups. Such decentralized, trial-and-error behavior is common in the decision-making processes of informal small groups. Serial processing corresponds to bureaucratic rule-based forms of central organization. Both styles may co-exist in a group's life, as they do in the brain. But neither style describes the evolving culture of a relationship or group, which can develop new insights and values. This involves both shared experiences and explicit dialogue (Zohar and Marshall 1993).

The tentative pattern-matching of this section is summarized in Table 53.1. It is a background for the next section, which will focus on language and semantics.

A SEMANTIC THEORY

There are reductive and phenomenological traditions in semantics. Reductive approaches, favored by logical positivists, linguistic analysts, and most cognitive scientists, try to reduce meanings to something more tangible, such as truth conditions, linguistic behavior, or the syntax and observation statements of a formal language. But there has been great difficulty in implementing this program (Boden 1988), for example in mechanical translation. Several arguments, ranging from Gödel's Theorem (Penrose 1989) to Searle's Chinese Room (Searle 1992) have challenged the adequacy of a purely syntactic approach to meaning. Furthermore, there is no clear reductive account of the generation of new explicit concepts.

Table 53.1

Process	Parallel	Serial	Creative
	Associative (Feeling)	Rational (Thinking)	Creative
Language	As-if	Fixed	Evolving
Memory	Implicit, Repetitive	Explicit, one-off	?
	Content-Addressable	Name-Addressable	
	Distributed	Hippocampal	
Brain Structure	Neural Nets	Tracts and language area	Quantum Field?
Psychodynamic Process	Primary (Id)	Secondary (Ego)	Sublimation (Jung: Individuation)
Mode	Point	Line	Construct
Social Organization	Informal Small Groups	Bureaucracy	Pluralistic Dialogue

The phenomenological tradition, including Husserl and other continental philosophers, Johnson (1987), Gendlin (1981) and in some ways Searle (1992), preserves the richness and messiness of our experience of meaning in a nonreductive treatment. The difficulty here is to make any contact with computational science or brain physiology.

Functionalism, the doctrine that mental states can be characterized by their causal roles, has a foot in both camps (Smith and Jones 1986). "Hard functionalism" tries to eliminate all reference to mental states, leaving only a computer-like model. "Soft functionalism" retains the idea of mental states but tries to explore the causal interconnections of perception, beliefs, desires, and behavior. It is compatible with the theory that follows.

In this section, I will develop a phenomenological theory of meaning in a form that can be related to computation and to the brain. A parallel processor has no language or explicit rules, though its discriminative behavior is "as if" it did. (It can be "taught" to respond to one specific voice or face.) A serial processor has both language and program from the outset but these are fixed. Neither processor has a way of *creating* explicit language or program. So a starting point is the question: what must be added to a parallel processor to produce an explicit language? I will treat this semiformally.

In a parallel processor, let a,b,... represent its possible states. Each one might or might not be associated with a state of consciousness, depending on whether a separate consciousness-producing system is functioning and connected to that processor.

Let (a→b) be the proposition that state a often brings about state b. Such an association may be formed by repeated contiguity, or may be based on similarity or past explicit learning. The processor is assumed to be somewhat "noisy," so that a may evoke any of $(b_1, b_2, \ldots b_N)$, each with a defined probability.

Since any physical system whatsoever, ranging from a washing machine to the weather, can be represented in the (a,→) notation, we need a further notation to indicate consciousness. Let [a] be the conscious state associated with a when the consciousness system is on line. Its content, of course, is not "a" but things like "It is raining" or "Jane" or "angry."

Now a wide sense of meaning can be defined. If [a] → [b] in someone's parallel processor, then to him or her [a] connotes [b], as in "to me, summer means holidays and swimming." The first idea often evokes the second idea. Such associative meanings are far from logical definitions. They are often multiple, ambiguous, and dependent on the person and the context. Nevertheless, they are the basis of more sophisticated and explicit kinds of meaning.

A gesture or utterance conveys a new meaning to someone (that is, results in learning) if it changes the dynamics of their parallel processing systems—either by adding new states or by changing the strengths of the interconnections between existing states. The latter can happen in a diffuse, holistic manner, as with body language, metaphor, art, or charismatic speeches. The effect of music on one's mood may not be reducible to a simple set of observations and expectations, as objectivist theories of meaning require. It may change many things slightly. Associative meanings resemble fields as much as particles. This approach to meaning is empiricist, that is, based on experience. But it includes music and metaphor as well as the isolatable pointer readings that are the paradigm experiences of operationalism in science.

More precise meanings can be produced from networks of association by constraining the whole, like carving a statue, rather than by assembling logical atoms. If the associations to [a] and to [b] are each multiple-valued but overlap, then under certain common conditions what a joint presentation of [a] and [b] evokes will probably be restricted to what the two sets of meanings have in common.[3] Thus, new concepts or attitudes can be defined first by long narratives or widespread social trends before a short name makes them explicit. Spontaneous imagery is often a simultaneous response to several aspects of a situation. This accords with our experiences of children, animals, and cultures. All appear to have associative meanings beyond their linguistic capacity. Semantics in this wide sense precedes any fixed, conventional language. Nonverbal meanings, unconscious meanings, misunderstandings, and contradictions fit easily into the present framework, unlike an account of meanings based purely on logic and observation.

But associative meaning is only one kind of meaning, related to passive experience. There is also meaning in the sense of intention; for example, "We mean to repaint the house this year" or "I didn't mean it seriously." Conventional meanings or explicit meanings are agreements to follow given rules of interpretation; that is, joint, relatively permanent intentions. Early writing was semipictorial, later becoming conventional. Thus I suggest that meanings of all varieties can be derived from two roots: associative meanings and intentions.

Intention marks the difference between passively entertaining some mental content [a] and actively choosing it. The latter will be marked by an assertion symbol, that is, ⊢ [a]. If [a] is a proposition, ⊢ [a] asserts that proposition. If [a] is a possible action or decision, ⊢ [a] is to perform that action or make that decision. If [a] is a description, then ⊢ [a] is to have confidence that or presuppose that something answers to that description. If [a] is an experience, perhaps of pleasure or pain, ⊢ [a] is to draw attention to that experience, rather like adding an exclamation mark. This list is not exhaustive.

I want to outline a theory of part of this vast field: the evolution of explicit names from the inexplicit associative matrix. I believe that this happens in three stages.

Stage 0 is a stream of point-mode consciousness: [a] → [b] → [c] → ... without awareness of the causal links between states.

Stage 1 [a→b] will denote the explicit awareness that [a] often evokes [b] by association. One is aware not only of [a] and [b] but also of their connection. (Joint awareness is possible across a time interval of a few seconds, the specious present.) Such awareness is in the line mode.

Stage 2 [a→] denotes the connotations of [a], that is, whatever [a] evokes. The description may apply ambiguously to any of a set [b_1],..., [b_N]. The connection symbol → has now become a function which, applied to [a], yields the values [b_n]. The function → has been abstracted and made explicit.[4] Whereas [a] is a mental content, [a →] is a putative name or symbol to be interpreted.

Stage 3 ⊢ [a →] denotes intentional use of [a →] to evoke [b], either in communication or in one's own thinking. ⊢ [a →] is an explicit name, at least on that occasion, if the description [a →] has one and only one main value [b].

For example, at stage 1 one might be aware that shouting [a] was often followed by aggression [b]. Knowing this, if two people arguing begin to shout involuntarily, they each become aware of the other's potential aggression (stage 2). Deliberate shouting to warn or intimidate the other would be stage 3. The growling of dogs invites a similar analysis.

In the present account parallel processing plus the capacity for awareness [ ] and intention ⊢ leads to the development of explicit and growing languages. How might this system communicate with the brain's serial processing system? We cannot plausibly assume that there is a single serial computer using a fixed language. An innate "language of thought" containing in principle all the meanings to be used later does not seem feasible (see Boden 1988, Chapter 7). People constantly learn new meanings for themselves ranging from pizzas to symphonies to new faces. But to introduce new classes of objects to a serial computer we have to declare them explicitly.

Alternatively, could the single-serial processor have a fixed "machine language" but use a compiler program, like our personal computers? Then it would behave as if it spoke a higher-level language but in fact it would translate everything into its machine language and calculate in the latter. This would make it become much slower. But our immediate memory does not work like this. A person can repeat back about six or seven letters; but four or five short familiar words, that is, fifteen or twenty letters (Miller 1962). Serial calculation is in familiar words, not letters. Many other experiments point to the same conclusion. No compiler program can explain these results.

Hence, though our brains do serial processing, there is no single fixed serial processor or machine language. Apparently our brains create a sequence of serial processors, that is, syntactical rules tailored to the languages developed so far. This accords with the persistent intuition that one's thinking is not tied to any one formal language, and as such is not restricted by Gödel's Theorem on formal languages (Penrose 1989). The previous description of how explicit names could develop is part of the requisite explanation along these lines. Accounts of the development of predicates, variables, connectives, formation rules, and so on would also be needed, but will not be attempted here.[5]

What broad kinds of meanings is the human brain capable of creating? In a reflective person the units a, b, ... may include not only sense perceptions and their interrelations but also one's own intentions and states of self-awareness. In the present theory, not only quasiscientific meanings but all kinds of artistic, conceptual, religious, or archetypal productions are candidates. They may fall into certain broad classes if spontaneous self-organization, as in complexity theory (Briggs and Peat 1990), is a feature of brain processing. How far any such class of meanings reflects objective aspects of the world is another question that is outside the scope of this paper.

A PHYSICAL MODEL

Most nerve interconnections in the human cortex are multiple and local within 1–2 mm. We may regard the cortex as containing many small overlapping or interconnected parallel processors. At a higher level there seem to be parallel networks involving larger areas such as the primary visual cortex, and also precise interconnections reminiscent of serial processors.

The semantic theory just given requires a third type of processor capable of consciousness [] and intention ⊢. It has been argued elsewhere that the holistic properties of consciousness correspond to a macroscopic quantum system in the brain, a Bose-Einstein condensate (Marshall 1989, Zohar 1990, Zohar and Marshall 1993). Several candidates have been proposed (see also Jibu *et al.* 1994). A pilot experiment has suggested that such a quantum brain system may exist (Nunn, Clarke and Blott 1994). An outline is in Zohar and Marshall (1993).[6] Now I want to show that it is feasible that the same quantum system assumed capable of consciousness [] is also capable of intention ⊢.

If a conscious physical system were to do something intentionally, what would be visible to an observer? What is going on in someone's brain when she makes a decision? I will assume "property dualism" (Smith and Jones 1986). That is, awareness and intention are not reducible to the categories of current physics (mass, length, time, and charge); but they or their precursors are equally fundamental properties of the same entities. Hence, we cannot define the psychological properties in terms of the physical properties, but the two sets of properties must be compatible. (Compare the psychology and the physics of a conversation or a game of chess.) So we can formulate necessary conditions (for the psychology of intention to be compatible with the physics) but not sufficient conditions (for eliminative definition). I have argued earlier that intention in its various forms is always a choice between fully or dimly conscious alternatives. So we require of the physics that:

1. A set of alternatives $[a_1], \ldots, [a_N]$ is replaced by just one alternative $[a_n]$
2. This change is not determined by current physics (else, the "choice" was only an epiphenomenon).
3. Before the change, each $[a_n]$ was to some degree conscious (else, the choice was random and blind).
4. Any choice can be viewed as favoring the system's survival or evolution (else, the choice was unmotivated)

These four conditions for intentional choice can all be satisfied by a quantum mechanical process: the collapse of the wave function (in a Bose-Einstein condensate, which was assumed capable of consciousness). I shall have to be slightly more technical here.

A quantum system may exist in a "superposition" of several states, but on measurement it is found to be in some one definite state. There are several theories to explain this. (Bell 1988, Chapter 20). One kind of theory is that under certain specified circumstances the system spontaneously and indeterministically "collapses" from a superposition to a single state—for example, the G.R.W. theory (Bell 1988, Chapter 22), the Penrose theory (Penrose 1989), or a theory due to myself (Zohar and Marshall 1993, Chapter 7). Such a collapse process clearly satisfies requirements (1) and (2) above.

In the present quantum theory of consciousness, states of consciousness are excitations of a ground state, like ripples on a lake. I suggest also that states of clear consciousness correspond to possible results of a specific kind of measurement, for example, of energy or phase difference. (Technically, they would be eigenstates of a specific operator.) Then an unclear state of consciousness, as before making a decision or focusing one's attention, would be a superposition of several clear states, like hearing several conversations or musical instruments at once. Each possible choice would be dimly conscious (requirement 3). Finally, if increased objective coherence and subjective clarity can be regarded as "good" or "pleasurable," requirement 4 would be satisfied as well.

In this model, physical and mental properties belong to the same entities, but are answers to different questions. A game of chess, for example, is fully described from the physics point of view by listing all the actual moves. But moves that in terms of the physics are "random" can be understood further from the psychological perspective which asks why the players made them. If there is such a further level of explanation then the physical aspect is not random after all. But it may be pseudorandom, that is, showing no obvious pattern, like the digits of π.

How could such a quantum field interact with a parallel processor in the brain? There are two sorts of dynamics in the neurons. Neuron firing via the synapses, whose properties alter on learning (Kandel and Hawkins 1992), could support a parallel processor. The slower EEG potentials that are generated by (nonfiring) neurons and pervade the brain could conceivably contain a macroscopic quantum element, as the pilot experiment has suggested. (Nunn, Clarke, and Blott 1994). The two systems, discrete firings, and continuous EEG interact at every point.

This paper has been concerned with general outlines and has ignored many detailed issues. It sketched a new type of processing model in which there are three interacting kinds of brain computational process. The quantum process is physical but not an epiphenomenon; it has emergent dynamics. The model could be regarded as within the field of artificial intelligence, but in a much expanded sense.

NOTES

1. However, many of Freud's detailed hypotheses have been disconfirmed. For example, there is an "oral character type" as he described, but it appears unrelated to childhood feeding and weaning methods (Peterson 1992).

2. A good phenomenological account of insight in psychotherapy is in Gendlin (1981).

3. Small activations of a parallel processor are expected to give approximately linear input-output functions for each unit. Then, superposed patterns of activation scarcely interact. (Smolensky, P., Chapter 22 of Rumelhart and McLelland 1986, Vol. 2). This independence means that the probabilities of any shared outcomes are additive.

4. Technically, [a →] abbreviates (⊤ b) [a → b]. A description operator ⊤ is applied to [a → b] (Church 1956). But here, the result may have zero, one, or many values: "Any b such that a evokes b."

5. A first step in defining variables might be to regard a multiple-valued [a →] as a symbol for a variable whose possible values are the $[b_n]$.

6. In this experiment numbers were flashed on a screen about one per second, and the subjects were asked to press a button when certain numbers appeared. Meanwhile an EEG apparatus was randomly switched on and off, at times unknown to the subjects. EEG measurement significantly affected task performance—a result which is explicable only by a macroscopic quantum element in brain function. The result needs independent replication, of course.

REFERENCES

Bell, J. S. 1988. *Speakable and unspeakable in quantum mechanics.* New York: Cambridge University Press.

Boden, M. A. 1988. *Computer models of mind.* New York: Cambridge University Press.

Briggs, J., and F. D. Peat. 1990. *Turbulent mirror.* New York: Harper and Row.

Church, A. 1956. *Introduction to mathematical logic, Vol. I.* Princeton, NJ: Princeton University Press.

Donaldson, M. 1992. *Human minds.* London:Penguin Books.

Gendlin, E. T. 1981. *Focusing.* New York: Bantam Books.

Johnson, M. 1987. *The body in the mind.* Chicago: University of Chicago Press.

Jibu, M., S. Hagan, S. R. Hameroff, K. H. Pribram, and K. Yasue. 1994. Quantum optical coherence in cytoskeletal microtubules: Implications for Brain Function. *Biosystems.* 32:195–209.

Kandel, E. R., and R. D. Hawkins. 1992. The biological basis of learning and individuality. *Scientific American* Sept., pp. 78–86.

Lucas, J. R. 1970. *The freedom of the will.* Oxford, England: Oxford University Press.

Marshall, I. N. 1989. Consciousness and Bose-Einstein condensates. *New Ideas in Psychology* 7 (1):73–83.

Miller, G. A. 1962. *Psychology.* London: Penguin Books.

Minsky, M. 1972. *Computation.* London: Prentice-Hall International.

Norman, D. A. 1987. "Reflections on cognition and parallel distributed processing." In *Parallel distributed processing,* edited by D. E. Rumelhart and J. L. McLelland, Vol. 2. Cambridge, MA: MIT Press, pp. 531–46.

Nunn, C. M. H., C. J. S. Clarke, and B. H. Blott. 1994. Collapse of a quantum field may affect brain function. *Journal of Consciousness Studies* 1(1):127–39.

Ornstein, R. E. 1972. *The psychology of consciousness.* London: Penguin Books.

Penrose, R. 1989. *The emperor's new mind.* Oxford, England: Oxford University Press.

Peterson, C. 1992. *Personality.* 2nd ed. New York: Harcourt Brace Jovanovich.

Rumelhart, D. E., and J. L. McLelland. eds. 1986. *Parallel distributed processing,* 2 volumes. Cambridge, MA: MIT Press.

Rycroft, C. 1968. *A critical dictionary of psychoanalysis.* London: Thomas Nelson.

Searle, J. R. 1992. *The rediscovery of the mind.* Cambridge, MA: MIT Press.

Seymour, J., and D. Norwood. 1993. A Game for Life. *New Scientist* 139:23–6.

Smith, P., and O. R. Jones. 1986. *The philosophy of mind.* Cambridge, England: Cambridge University Press.

Tank, D. W., and J. J. Hopfield. 1987. Collective computation in neuron-like circuits. *Scientific American* 257(6).

Wilber, H. 1983. *Eye to eye.* New York: Anchor Books.

Zohar, D. 1990. *The quantum self.* New York: William Morrow.

Zohar, D., and I. N. Marshall. 1993. *The quantum society.* London: Bloomsbury.

54 The Possibility of Empirical Test of Hypotheses About Consciousness

Jean E. Burns

Recent models of consciousness have made a variety of hypotheses about the relationship of conscious experience to physical laws (Burns 1990, 1991). I discuss three issues that are addressed by many models and ask whether it is possible for hypotheses about these issues to be subject to empirical test. The term *consciousness* is used here to refer to the aggregate of all aspects of conscious experience, including those describable in terms of an information content, such as thoughts and sensory perceptions, and those which are not, such as awareness itself.

1. What is the ontological relationship between consciousness and the brain/physical world? *Physicalism* holds that all aspects of consciousness can be explained in terms of physical laws, and other hypotheses (*dualism, monistic idealism, implicate order,* and so on) claim that not all aspects of consciousness can be explained in this way.

The varieties of physicalism hypothesized in current literature are *reductionism* and *emergent physicalism*. Reductionism holds that all aspects of conscious experience can be explained in terms of physical laws. Emergent physicalism is similar to reductionism but holds that some aspects of consciousness (for example, awareness) are new properties of matter that emerge only in certain physical conditions (for example, complexity) and therefore cannot be traced to the laws describing ordinary matter.

Physical law ultimately is based on various symmetries and their associated conservation principles. Physicist SaulPaul Sirag, who has developed a unified field theory describing these basic principles, has pointed out that the mathematical space, which describes these principles intersects another mathematical space, which might represent the basic principles underlying consciousness (Sirag 1993a, 1993b). Because these principles are different from those of the physical world, he refers to his model as dualistic.

Models of monistic idealism (Goswami 1989, 1993) and the implicate order (Bohm 1982) propose that matter and individual mind arise out of consciousness, which is more fundamental than either of these. (These models differ in their hypotheses about the nature of the mind/brain interface.)

The hypothesis of reductionism is testable in that, if the brain is thoroughly understood and all aspects of consciousness can be traced back to underlying physical mechanisms in the brain, reductionism would then be demonstrated. However, there is no way to differentiate experimentally between the hypotheses of emergent physicalism, dualism, monistic idealism, and the like, because each one states that some aspect of consciousness cannot be traced back to known physical laws.

2. What physical characteristics are associated with the mind/brain interface? A variety of proposals have been made concerning the physical nature of the brain-mind interface (Burns 1990, 1993). (In physicalism, the interface is considered to be the cause of all attributes of consciousness. Nonphysicalist theories claim that the interface simply provides the means by which independently existing qualities of consciousness can be associated with the brain.)

Contemporary proposals about the interface include the following: Gregory Bateson (1979) and E. Roy John (1976) have each proposed that consciousness is associated with aggregations of matter, such as neurons in the brain, that are linked in sufficiently complex ways. (This postulate is frequently also made by researchers in artificial intelligence who are interested in consciousness.)

Goswami (1989, 1993), Penrose (1989), Stapp (1993), and Walker (1975) have each proposed that consciousness is associated with quantum mechanical processes in the brain.

Mathematician E. C. Zeeman (1976, 1977) has pointed out that sensory experience is probably associated with synchronous electrochemical oscillations in the brain, and this association has recently been confirmed by experimental work (Freeman 1991, Skarda and Freeman 1987). Crick and Koch (1990, Crick 1994) have suggested that the mind/brain interface is associated with various neural processes, including synchronous oscillations.

Synaptic transmission and much other activity of neurons is dependent upon the action of microtubules within the cell. Hameroff (1987) has pointed out that loss of consciousness is associated with loss of ability of microtubule proteins to change conformation, and has proposed that the interface is associated with these conformational changes.

The central problem in evaluating any interface hypothesis is knowing which entities are conscious and which are not. It is reasonable to assume that all humans who exhibit ordinary behavior are conscious. This criterion contains gray areas: During a petit mal seizure a person may continue actions previously initiated (and continue to play the piano, for instance), but not be conscious (Penfield 1975). Even so, if it appears that consciousness is present whenever the human brain manifests a particular physical condition, but never present when it is lacking, it is reasonable to suppose that this condition is associated with the mind/brain interface.

It is when we wish to know the complete set of conditions necessary to human consciousness, or whether nonhuman entities are conscious, that a problem arises. Suppose, for instance, Hameroff's proposal proves correct, that the mindbrain interface is associated with the ability of microtubule proteins to make changes in their conformation. Microtubules exist within all eukaryotic cells. Is a single cell conscious? Or are additional conditions, always present in human brains, necessary to consciousness, such as complexity?

We can only investigate the nature of the mind/brain interface in nonhuman entities if we know which of these are conscious. It is sometimes suggested that if a nonhuman entity can reproduce human behavior, it must be conscious (the Turing test). But this argument leads to an absurd conclusion: holographic images of humans perfectly replicate human behavior; therefore, such images must be conscious.

We have no present way to determine whether nonhuman entities are conscious. Therefore, we have no way to empirically investigate the full nature of the interface.

3. Can consciousness act on the brain independently of any brain process? All known processes of the physical world are either deterministic or exhibit quantum randomness. For that reason, it is sometimes suggested that free will does not exist, and that our experience of it is an illusion (Dennett 1984).

Nevertheless, a number of models of consciousness have postulated that free will can act to select between alternative brain programs, with this action being independent of any physical process (Burns 1991). Several researchers have shown that such action would contradict the second law of thermodynamics (Burns 1991). Therefore, a model that made such a postulate could not be reductionist. However, it could be based on emergent physicalism or any of the other ontologies.

If we thoroughly understood the workings of the brain, it might be possible to show that some brain action could not be traced back to previous physical conditions. Or alternatively, it might be shown that all brain action can be so traced. Until such understanding of the brain is attained—perhaps some centuries hence—it will not be possible to demonstrate empirically whether free will does or does not occur.

REFERENCES

Bateson, G. 1979. *Mind and nature: A necessary unity.* New York: Dutton.

Bohm, D. 1982. *Wholeness and the implicate order.* London: Routledge and Kegan Paul.

Burns, J. E. 1990. "Contemporary models of consciousness: Part I." *Journal of mind and behavior* 11:153–72.

Burns, J. E. 1991. "Contemporary models of consciousness: Part II." *Journal of Mind and Behavior* 12:407–20.

Burns, J. E. 1993. "Current hypotheses about the nature of the mind-brain relationship and their relationship to findings in parapsychology." In *Cultivating consciousness,* edited by K. R. Rao. New York: Praeger, pp. 139–48.

Crick, F. 1994. *The astonishing hypothesis: The scientific search for the soul.* New York: C. Scribners.

Crick, F., and C. Koch. 1990. "Towards a neurobiological theory of consciousness." *Seminars in the Neurosciences* 2:263–75.

Dennett, D. C. 1984. *Elbow room: The varieties of free will worth having.* Cambridge, MA: MIT Press.

Freeman, W. J. 1991. "The physiology of perception." *Scientific American* February, pp. 78–85.

Goswami, A. 1989. "The idealistic interpretation of quantum mechanics." *Physics Essays* 2:385–400.

Goswami, A. 1993. *The self-aware universe: How consciousness creates the material world.* New York: Tarcher/Putnam.

Hameroff, S. R. 1987. *Ultimate computing: biomolecular consciousness and nanotechnology.* New York: North-Holland.

John, E. R. 1976. "A model of consciousness." In *Consciousness and self-regulation,* Vol. 1, edited by G. E. Schwartz, and D. Shapiro. New York: Plenum. pp. 1–50.

Penfield, W. 1975. *Mystery of the mind.* Princeton, NJ: Princeton University Press.

Penrose, R. 1989. *The emperor's new mind.* New York: Oxford University Press.

Sirag, S.-P. 1993a. "Hyperspace reflections." In *Silver threads: 25 years of parapsychology research,* edited by B. Kane, J. Millay, and D. Brown. New York: Praeger, pp. 156-65.

Sirag, S.-P. 1993b. "Consciousness: A hyperspace view." Appendix to *Roots of Consciousness,* 2nd ed., by J. Mishlove. Tulsa, OK: Council Oak Books, pp. 327–65.

Skarda, C. A., and W. J. Freeman. 1987. "How brains make chaos in order to make sense of the world." *Behavioral and Brain Sciences* 10:161–95.

Stapp, H. P. 1993. *Mind, matter and quantum mechanics.* New York: Springer-Verlag.

Walker, E. H. 1975. "Foundations of paraphysical and parapsychological phenomena." In *Quantum Physics and Parapsychology,* edited by L. Oteri. New York: Parapsychology Foundation, pp. 1–53.

Zeeman, E. C. 1976. Catastrophe theory. *Scientific American* April, pp. 65–83.

Zeeman, E. C. 1977. *Catastrophe theory: Selected papers, 1972-1977,* Chapters 1,8,9. Reading, MA: Addison-Wesley.

55 Toward a Science of Consciousness: Addressing Two Central Questions

Willis W. Harman

Awareness of the need for an integrated science of consciousness is increasing rapidly. This conference has been an important first step. To make further progress we have to deal with two important questions. The first is one of strategy; the second of epistemology.

THE STRATEGY QUESTION: HOW BROADLY SHOULD CONSCIOUSNESS BE DEFINED?

Concentrating for now on conscious awareness alone would seem a reasonable strategy. Science has often progressed by focusing on the simplest and most tractable case first, and then proceeding to the more complex. Thus a straightforward approach would comprise an initial attempt to solve the puzzle of simple, everyday conscious awareness, leaving the more complex or recondite aspects of consciousness for later consideration.

However there is an alternate strategy that also has precedent in the history of science. Consider the origin of the evolutionary hypothesis. In the mid-nineteenth century there was much to be learned from studying separately the great variety of microorganisms, plants, and animals with which the planet is populated. But Charles Darwin boldly turned his attention to the synthesizing question: How can we understand all of these together? The result was the concept of evolution, around which practically all of biology is now organized.

There is an analogous situation in the study of consciousness. We seem to need some sort of conceptual framework within which to understand a broad range of phenomena and experiences including conscious awareness, intentionality (volition), unconscious processes, perception, imagery, states of consciousness, innovative problem solving, and so on. Or one might even think of extending the range to what some would be tempted to term "ridiculously broad"—to reports of awesome creative insight; out-of-body and near-death experiences; apparent memories of other lives; mystical experiences; and extraordinary psychic abilities. (This is not to suggest, of course, that all reports of anomalous phenomena or experience are to be believed, any more than everyone's sense perception is to be trusted. But when reports are persistent enough, even when the phenomena appear erratic, a certain face validity must be granted.) Thus the alternate strategy suggests itself, namely to concentrate on the question: What sorts of conceptual frameworks and organizing metaphors can be used to help us understand the many facets and dimensions of consciousness all considered together?

A Basic Epistemological Dilemma

If the latter course is to be taken, we must give attention to an intrinsic epistemological dilemma which is too little noted. In order to have confidence in the scientific view of reality, we have to answer the epistemological question: How do we know what we claim to know scientifically? Our view of reality is inevitably determined partly by that reality, and partly by the mental processes through which we arrived at the view we have. However, to know about those processes we need already a scientific study of the mind, for which we need a scientific epistemology—which, with a frustrating circularity, leads us back to the initial question.

This circularity implies that even if the results of generations of scientific inquiry appear to be convergent toward a particular picture of reality, a profound caution is advisable regarding how much faith is put in that picture. (This point has been made in another way by Thomas Kuhn [1970].) In any case, this dilemma suggests that as we search for the appropriate epistemology for a comprehensive science of consciousness, we need to pay particular attention to what is known about unconscious mental processes.

The Influence of the Unconscious in Constructing Science

Research on perception, hypnosis, repression, selective attention, mental imagery, sleep and dreams, memory and memory retrieval, acculturation, and so on, all suggests that the influence of the unconscious on how we experience ourselves and our environment is far greater than is typically taken into account. Science itself has never been thoroughly reassessed in the light of this recently discovered pervasive influence of the unconscious mind of the scientist.

The unconscious enters importantly into the construction of science in at least two ways: (1) the creative/intuitive mind (an aspect of the unconscious) is intimately involved with all the important conceptual advances in science; and (2) the contents and processes of the unconscious influence (individually and collectively) perceptions, "rational thinking," openness to challenging evidence, ability to contemplate alternative conceptual frameworks and metaphors, scientific interests and disinterests, scientific judgment—all to an indeterminate extent. The clear implication is that we must accept the presence of unconscious processes and contents, not as a minor perturbation, but as a potentially major factor in the construction of any society's construction of its particular form of science. (This consideration even puts into question whether or not the logical construction of a science from a rational epistemology may already be a culturally biased approach.)

Taking this influence of the unconscious into account, we now must reassess the vaunted "objectivity" of science. Research on perception makes it clear that science has been constructed on the basis of scientists' intersubjectively shared subjective experience, so that the difference between so-called "objective" data (for example, the read-out of a measuring instrument) and "subjective" data (an inner image) is one of degree only. (This point has been effectively made by Velmans 1993.) That is to say, science has always been based solely on the subjective experience of scientists, and the bias toward quantifiable "objective" data has been a preference, not an intrinsic quality, of scientific inquiry.

There is some precedent for taking into account unconscious processes and contents in the training of the researcher. In training to be a psychotherapist, the individual has to go through inner explorations similar to those anticipated in his/her future clients; with these experiences comes learning, and personal change. Similarly, in training to be a cultural anthropologist, the person must learn to experience being of another culture; this too brings personal change. In training for the bench, the future judge—at least ideally—goes through self-examination to uncover personality characteristics that could cause one to be unconsciously biased. In general, learning to be a faithful observer implies inner change. The scientist who would explore the topic of consciousness (in the broad sense we are considering here) must be willing to risk being transformed in the process of exploration.

Admissibility of a Consciousness Metaphor

One other consideration should be brought in before turning to the epistemological question. That is the permissibility of a consciousness metaphor. In science, as in ordinary life, we use metaphors to understand or communicate about the unfamiliar in terms of the familiar. Great mischief results, however, when these models and metaphors are mistakenly taken to be the "true" description of reality. When they are, people then feel driven to defend them, and to stamp out competing views. Many of the conflicts in the history of science (as in religion) have been battles between groups where each insists that their metaphors are "really" how reality is.

It is a peculiarity of modern science that some kinds of metaphors are allowed and others disallowed. It is perfectly acceptable to use metaphors that derive directly from our experience of the physical world (such as fundamental particles or acoustic waves), as well as metaphors representing what can be measured only in terms of its effects (such as gravitational, electromagnetic, or quantum fields). It has further become acceptable to use more holistic and nonquantifiable metaphors such as organism, personality, ecological community, Gaia, universe.

It is, however, taboo to use nonsensory "metaphors of mind"— metaphors that tap into images and experiences familiar from our own inner awareness. I am not allowed to say (scientifically) that some aspects of my experience of reality are reminiscent of my experience of my own mind—to observe, for example, that some aspects of animal behavior appear as though they were tapping into some supraindividual nonphysical mind, or as though there were in instinctual behavior and in evolution something like my experience in my own mind of purpose.

The implicit or explicit epistemological position of the "hard" scientist is that we know what we know through the empirical observation of quantifiable, replicable observations and interventions in the physical world. However, a less dogmatic attitude would hold that reality has many aspects, and is never fully captured in any model or metaphor. For example, admission of the consciousness metaphor would allow us to gain insight from a vantage point resembling philosophical idealism without having to commit to an ontological position that insists that ultimate reality is consciousness.

WHAT IS A SUITABLE EPISTEMOLOGY FOR RESEARCHING CONSCIOUSNESS?

Attempts to explore consciousness scientifically inevitably raise the question of whether the accepted epistemology of neurobiology, or quantum physics, or cognitive science is ultimately suited to the exploration of consciousness. The epistemology, or accepted set of "rules of evidence," of Western science has been extraordinarily effective in achieving the goals of prediction, control, and generation of manipulative technology. However, the study of consciousness is qualitatively different from the extant sciences.

Pushing the Frontiers of Present Science

Building upon the acknowledged success of the accepted scientific epistemology is the most obvious approach to use in "explaining" consciousness. However, there is good reason to suspect that it may in the end prove to be inadequate.

As an example of this approach, several quantum physicists have attempted to show the compatibility of quantum theory with philosophical idealism. (In this regard I would mention in particular Goswami 1993, Herbert 1993, and Jahn and Dunne 1987.) This compatibility is a fascinating conclusion. There would have seemed little reason to expect that the positivistic epistemology of Western science would lead to a result (quantum physics) that contradicts the initial premises and leads to totally different ontological assumptions. Once that has been shown, however, the implication is not that we may expect to learn more about consciousness by studying theoretical physics. Rather, it would seem to be that we must now seriously consider augmenting or replacing the dominant epistemology (that of prediction-and-control focused science) by an epistemology more appropriate to the exploration of consciousness—if you will, an epistemology of subjectivity.

Seeking an Epistemology of Subjectivity

Advocacy of an introspective or phenomenological epistemology is not a new development in science. In the past the proposal has been unequivocally rejected, for reasons that appeared adequate at the time. However, apparent cultural changes over the past few decades increase the likelihood that such an approach might now achieve gradual acceptance within the scientific community.

A recent effort to identify a suitable epistemology for the study of consciousness in the broadest sense resulted in the following nine proposed characteristics:

1. The epistemology will be "radically empirical" (in the sense urged by William James 1912) in that it will be phenomenological or experiential in a broad sense (that is, it will include subjective experience as primary data, rather than being essentially limited to physical-sense data) and it will address the totality of human experience (in other words, no reported phenomena will be written off because they "violate known scientific laws"). Thus, consciousness is not a "thing" to be studied by an observer who is somehow apart from it; consciousness involves the interaction of the observer and the observed, or if you like, the experience of observing.

2. It will aim at being objective in the sense of being open and free from hidden bias, while dealing with both "external" and "internal" (subjective) experience as origins of data.

3. It will insist on open inquiry and public (intersubjective) validation of knowledge; at the same time, it will recognize that these goals may, at any given time, be met only incompletely, particularly when seeking knowledge that includes deeper understanding of inner experience.

4. It will place emphasis upon the unity of experience. It will thus be congenial to a holistic view in which the parts are understood through the whole, while not excluding a reductionistic approach that seeks to understand the whole through the parts. Hence it will recognize the importance of subjective and cultural meanings in all human experience, including experiences—such as some religious or interpersonal experiences—that seem particularly rich in meaning even though they may be ineffable. In a holistic view, such meaningful experiences will not be explained away by reducing them to combinations of simpler experiences or to physiological/biochemical events. Rather, in a holistic approach, the meanings of experiences may be understood by discovering their interconnections with other meaningful experiences.

5. It will recognize that science deals with models and metaphors representing certain aspects of experienced reality, and that any model or

metaphor may be permissible if it is useful in helping to order knowledge, even though it may seem to conflict with another model which is also useful. (The classic example is the history of wave and particle models in physics.)

6. It will thus recognize the partial nature of all scientific concepts of causality. (For example, the "upward causation" of physiomotor action resulting from a brain state does not necessarily invalidate the "downward causation" implied in the subjective feeling of volition.) In other words, it will implicitly question the assumption that a nomothetic science—one characterized by inviolable "scientific laws"—can in the end adequately deal with causality.

7. It will be participatory in recognizing that understanding comes, not alone from being detached, objective, analytical, coldly clinical, but also from cooperating with or identifying with the observed, and experiencing it subjectively. This implies a real partnership between the researcher and the phenomenon, individual, or culture being researched; an attitude of "exploring together" and sharing understandings.

8. It will involve recognition of the inescapable role of the personal characteristics of the observer, including the processes and contents of the unconscious mind. The corollary follows, that to be a competent investigator, the researcher must be willing to risk being profoundly changed through the process of exploration.

9. Because of this potential transformation of observers, an epistemology that is accepted now may in time have to be replaced by another, more satisfactory by new criteria, for which it has laid the intellectual and experiential foundations.

CONSIDERING THE POSSIBILITY OF DIFFERENT ONTOLOGICAL ASSUMPTIONS

If indeed something like the above epistemology were to be adopted as the scientific community attempts to construct a true science of consciousness, it would seem that serious attention would have to be paid to the inner explorations that have gone on for thousands of years within the world's spiritual traditions. The distillation of these explorations is sometimes termed the "perennial philosophy," and there are ontological implications, which are examined in a recent paper by Ken Wilber (1993), entitled "The Great Chain of Being."

Based on some very sophisticated (if prescientific) exploration, this ancient view centers around the following proposition: "Reality, according to the perennial philosophy, is composed of different grades or levels, reaching from the lowest and most dense and least conscious to the highest and most subtle and most conscious. At one end of this continuum of being or spectrum of consciousness is what we in the West would call "matter" or the insentient and the nonconscious, and at the other end is "spirit" or "godhead" or the "superconscious" (which is also said to be the all-pervading ground of the entire sequence)... The central claim of the perennial philosophy is that men and women can grow and develop (or evolve) all the way up the hierarchy to Spirit itself, therein to realize a "supreme identity" with "Godhead."

A central understanding of this "perennial wisdom" is that the world of material things is somehow embedded in a living universe, which in turn is within a realm of consciousness, or Spirit. Similarly, a cell is within an organ, which is within a body, which is within a society... and so on. Things are not—cannot be—separate; everything is a part of this "great chain of being."

As Wilber observes, Western science became restricted to the matter end of the continuum only, and to "upward" causation only. With that restriction came a faith that in the end, a nomothetic science can adequately represent reality—a faith that phenomena are governed by inviolable, quantified "scientific laws." From that restriction came both the power of modern science (basically, to create manipulative technology) and the limitation of its epistemology. From it also stem all sorts of classical "problems"—the "mind-body problem," "action at a distance," "free will versus determinism," "science versus spirit," and so on.

This restriction of science to only a portion of "the great chain of being" was useful and justifiable for a particular period in history. The only mistake made was to become so impressed with the powers of prediction-and-control science that we were tempted to believe that that kind of science could lead us to an understanding of the whole: Fundamentally, there is no reason to suppose that reductionistic science can ever provide an adequate understanding of the whole.

What must be done now, according to Wilber, is to retain the open-minded scientific spirit, and the tradition of open, public validation of knowledge (that is, abjuring any scientific priesthood), but to open up the field of inquiry to the entire continuum and to downward as well as upward causation. Whether that will be soon done within science is a good question. However, because of the cultural shift that appears to be taking place, attaching increasing importance to the transcendental, there may be increasing public insistence that some such development take place in science if science is to retain its present position as the only generally accepted cognitive authority in the modern world.

CONCLUDING OBSERVATION

It would be a mistake to underestimate the confusion in our present situation. Any serious exploration of consciousness (specifically, one guided by the epistemology suggested above) reveals a discrepancy with the worldview implied by mainstream Western science, which presently prevails in all powerful modern institutions, from scientific and economic to educational and medical. The prestige of science is sufficiently impressive that "new paradigm" people often seek to authenticate their inner truth by reference to "the new physics," "quantum healing," "holographic theory," and "chaos theory." However, the incipient culture shift goes far more deeply into underlying metaphysical assumptions than do these "revolutions" within science. The emerging view implies a revolution of science, or at least of the way we think about science. It in no way denies the power of Western science for the purposes for which it was devised—prediction, control, and manipulation of the physical environment. It does hold in question the modern tendency to yield authority to the reductionistic scientific worldview as the reality by which we should guide our lives, individually and collectively.

We must not minimize the fundamental nature of the challenge implicit in consciousness research. Western science is about understanding cause. It is a tenet of modern society that that science can lead us toward the ultimate explanations for phenomena. However, the very conviction that a complete nomothetic science is possible—that everything can be ultimately explained through inviolable scientific laws—rules out consciousness (mind, spirit) as a causal reality. At the same time, everything in our personal experience affirms the importance of our ability to choose, and our deep inner guidance toward the better choice. This poses a fundamental dilemma. Either we must deny our own innate wisdom because "science knows better," or we have to face the fundamental inability of science in its present form (quantum physics and all) to give us an adequate cosmology to live by and to guide our society by.

It may well be that the last-described of the three basic approaches, namely that of re-basing science on a new set of ontological assumptions, is where the consciousness issue will end up eventually. However, for the time being this seems to have little likelihood of gaining very widespread acceptance. In the meantime we can urge adoption of the epistemology implied by the nine characteristics. On this basis, a serious and thoroughgoing exploration of all aspects of human consciousness can be mounted. The territory can be roughly charted, and areas of potentially fruitful research identified.

Meanwhile, if there is a continuation of the present cultural shift toward finding ultimate causality not in the outer, objective world, but in the subjective, the alternative posed by Wilber looks more and more feasible. As the exploration of consciousness proceeds, increasing numbers of scientists will undergo significant shifts in internalized assumptions and modes of perception. These changes in the "observing instrument" will tend to open up the possibility of a bolder epistemology.

REFERENCES

Goswami, A. 1993. *The self-aware universe.* New York: Putnam.

Herbert, N. 1993. *Elemental mind.* New York: D.P. Dutton.

Jahn, R., and B. Dunne. 1987. *Margins of reality: The role of consciousness in the physical world.* New York: Harcourt Brace Jovanovich.

James, W. 1912. *Essays in radical empiricism.* New York: Longmans, Green and Co. The present discussion is based on Taylor, 1994. "Radical empiricism and the conduct of research." In *New Metaphysical Foundations of Modern Science,* edited by W. Harman and J. Clark. Sausalito, CA: Institute of Noetic Sciences.

Kuhn, T. 1970. *The structure of scientific revolutions,* 2nd ed. Chicago: University of Chicago Press.

Velmans, M. M. 1993. "A reflexive science of consciousness" in *Experimental and theoretical studies of consciousness,* CIBA Foundation Symposium No. 174. Chichester, England: Wiley.

Wilber, K. 1993. The great chain of being. *Journal of Humanistic Psychology,* 33(3):52–65.

POSTSCRIPT

Science and Philosophy face a daunting chasm between reductive materialism and subjective experience. As we stand on the brink and ponder this task, we realize, as Andrew Weil concludes in Chapter 48:

...the brighter we illuminate reality with the light of science, the more we become aware of the surrounding darkness.

Contributors

Geoffrey L. Ahern
Department of Neurology and Psychology
University of Arizona
Tucson, Arizona 85721

Britt Anderson
Department of Neurology and the
  Alzheimer's Disease Center
University of Alabama at Birmingham
Birmingham, Alabama 35294-0007

Richard P. Atkinson
Department of Psychology
Weber State University
Ogden, Utah 84408-1202

Nils A. Baas
Department of Mathematical Sciences
University of Trondheim, NTH
N-7034 Trondheim, Norway

Mikael Bergenheim, PhD
Division of Work Physiology
National institute of Occupational Health
and
Department of Physiology
University of Umea
S-901 87 Umeå, Sweden

B. H. Blott
University of Southampton
Southampton S09 5NH
United Kingdom

John J. Boitano
Department of Psychology
Fairfield University
Fairfleld, Connecticut 06430

Jean E. Bums
Consciousness Research
San Leandro, California 94578

Tethys Carpenter
Department of Music
Royal Holloway and Bedford New College
Egham, Surrey TW2 OEX
United Kingdom

David J. Chalmers
Department of Philosophy
University of California, Santa Cruz
Santa Cruz, California 95064

Michael Conrad
Department of Computer Science
Wayne State University
Detroit, Michigan 48202

Gina DiTraglia Christenson
Department of Psychology
University of Arizona
Tucson, Arizona 85721

C. J. S. Clarke
University of Southampton
Southampton S09 5NH
United Kingdom

Randall C. Cork
Department of Anesthesiology
Louisiana State University Medical Center
New Orleans, Louisiana 70112

Thaddeus M. Cowan
Department of Psychology
Kansas State University
Manhattan, Kansas 66506

Alan Cowey
Department of Experimental Psychology
Oxford University
Oxford, OX 1 3 IB
United Kingdom

Richard J. Davidson
Department of Psychology
University of Wisconsin
Madison, Wisconsin 53792

José-Luis Díaz
Centro de Neurobiologia
Universidad Nacional Autónoma de
  México,
Instituto Mexicano de Psiquiatría
and
Cognitive Science Program
University of Arizona
Tucson, Arizona 85721

Arthur J. Deikman
Department of Psychiatry
University of California at San Francisco
San Francisco, California 94143-0844

Heath Earl
Department of Psychology
Weber State University
Ogden, Utah 84408-1202

Avshalom S. Elitzur
Department of Chemical Physics
The Weizmann Institute of Science
Rehovot, Israel 76100

B. Raymond Fink
Department of Anesthesiology
University of Washington
Seattle, Washington 98195

Owen Flanagan
Department of Philosophy
Duke University
Durham, North Carolina 27708

David Galin
Department of Psychiatry and Langley
    Porter Psychiatric Institute
University of California at San Francisco
San Francisco, California 94143-0844

Brittmarie Granlund
Division of Work Physiology
National Institute of Occupational
    Health
S-907 13 Umeå, Sweden

Güven Güzeldere
Stanford University
Center for the Study of Language and
    Information
Stanford, California 94305-4115

Scott Hagan
Department of Physics
McGill University
Montréal, Québec
H3A 2T8 Canada

Stuart Hameroff
Departments of Anesthesiology
    and Psychology
University of Arizona
Tucson, AZ 85724

Valerie Gray Hardcastle
Department of Philosophy
Virginia Polytechnic Institute and State
    University
Blacksburg, Virginia 24061-0126

Willis W. Harman
Institute of Noetic Sciences
Sausalito, California 94965

Erich Harth
Department of Physics
University of Syracuse
Syracuse, New York 13244-1130

Thomas Head
Department of Neurology
University of Alabama at Birmingham
Birmingham, Alabama 35294-0007

Polly Henninger
Boston Veterans Administration Medical
    Center
Boston, Massachusetts 02130

Marco Iacoboni
Department of Psychology
University of California, Los Angeles
Los Angeles, CA 90024-1563

E. M. Insinna
18, Allée des Frères Lumière
77600 Bussy St. Georges
France

Håkan Johansson
Division of Work Physiology
National Institute of Occupational
    Health
S-907 13 Umeå, Sweden

Mati Jibu
Department of Anesthesiology
Okayama University Medical School
and
Research Institute for Informatics and
    Science
Notre Dame Seishin University
Okayama 700
Japan

Brian D. Josephson
Cavendish Laboratory
Cambridge University
Cambridge CB3 OHE
United Kingdom

Alfred Kaszniak
Departments of Psychology, Psychiatry,
    and Neurology
University of Arizona
Tucson, Arizona 85721

John F. Kihlstrom
Department of Psychology
Yale University
New Haven, Connecticut 06520-8205

Christof Koch
Professor of Computation and Neural
    Systems
Division of Biology
California Institute of Technology
Pasadena, California 91125

Richard D. Lane
Departments of Psychiatry and
    Psychology
University of Arizona
Tucson, Arizona 85721

Benjamin Libet
Department of Physiology
University of California at San Francisco
San Francisco, California 94143-0444

Dyan Louria
College of Medicine
University of Arizona
Tucson, Arizona 85724

Victor Mark
Department of Neuroscience
University of North Dakota Medical
    Education Center
Fargo, North Dakota 58102

I. N. Marshall
57 Bainton Road
Oxford, OX2 7AG
United Kingdom

Douglas J. Matzke
1516 Copper Creek Drive
Plano, Texas 75075

C. M. H. Nunn
75/3 Northlands Rd.
Southampton SOI5 2LP
United Kingdom

Jonas Pedersen
Division of Work Physiology
National Institute of Occupational
    Health
S-907 13 Umeå, Sweden

Roger Penrose
Mathematical Institute
University of Oxford
Oxford, OX1 31B
United Kingdom

Karl H. Pribram
Center for Brain Research and
    Informational Sciences
Department of Psychology
Radford University
Radford, Virginia 24142

Jan Rayman
Department of Psychology
University of California, Los Angeles
Los Angeles, California 90024-1563

Eric M. Reiman
Positron Emission Tomography Center
Good Samaritan Regional Medical
    Center
Phoenix, Arizona 85006

M. V. Satarić
Faculty of Technical Sciences
Serbia, Yugoslavia

Gary E. Schwartz
Department of Psychology
University of Arizona
Tucson, Arizona 85721

Alwyn C. Scott
Department of Mathematics
University of Arizona
Tucson, Arizona 85721

D. Sept
Department of Physics
University of Alberta
Edmonton, AB T6G 2J1
Canada

Saul-Paul Sirag
25 Owosso Drive
Eugene, Oregon

Petra Stoerig
Institute of Medical Psychology
Ludwig-Maximilians-University
D-80336 Munich
Germany

Leopold Stubenberg
Department of Philosophy
University of Notre Dame
Notre Dame, Indiana 46556

John Taylor
Centre for Neural Networks
Department of Mathematics
Kings College
Strand, London
United Kingdom

Jeff Tollaksen
P.O. Box 500
Fallsburg, NY 12733

J.A. Tuszyński
Department of Physics
University of Alberta
Edmonton, AB T6G 231 Canada

D. Trpisová
Department of Physics
University of Alberta
Edmonton, AB T6G 2JI
Canada

Mario Varvoglis
Director, LRIP
Morsang sur Orge, 91390
France

Ron Wallace
Department of Sociology and
    Anthropology
University of Central Florida
Orlando, Florida 32816

Richard C. Watt
Department of Anesthesiology
University of Arizona
Health Sciences Center
Tucson, Arizona 85724

John G. Watterson
Faculty of Engineering and Applied
    Science
Griffith University - Gold Coast
Australia

Andrew Weil
Center for Integrative Medicine
University of Arizona
Tucson, Arizona 85724

James E, Whinnery
Naval Air Warfare Center
Warminster, Pennsylvania 18974-5000

Fred Alan Wolf
43 15th Avenue, Apt. 3
San Francisco, California 94118-2828

Andrew Wuensche
48 Esmond Road
London W4 1JQ
United Kingdom

Tokiko Yamanoue
Faculty of Engineering
Kyushu Institute of Technology
Tobata, Kitakyushu, 804
Japan

Kunio Yasue
Research Institute for Informatics and
    Science
Notre Dame Seishin University
Okayama 700
Japan

Eran Zaidel
Department of Psychology
University of California, Los Angeles
Los Angeles, California 90024-1563

Danah Zohar
57 Bainton Road
Oxford OX2 7AG
United Kingdom

Index

Abelson, R., 365
Abhidharma, Buddhist psychology of, 715
A Cognitive Theory of Consciousness, 11
Activation-Synthesis Hypothesis, 75
activities of daily living (ADL) scales, 230
Acuna, C., 149
Adams, H., 726
adaptation, 635
Adelson, E. H., 263
aerial combat maneuvering (ACM), 169
Aertsen, A., 56, 61
Agarwal, G. S., 500
age-associated memory impairment (AAMI), 234
agnosia, 274
  associative, 130, 134
  sensory-specific, 148
Aharonov, Y., 525, 532, 541, 551–52, 559–60, 563–64
Ahern, Geoffrey, 245, 293
Ahumada, A., 154
Aicardi, J., 190
Akins, K., 22
Alavi, F., 361
Albano, J. E., 263
Albert, D. Z., 451–52, 466, 556, 559–60, 563
Alberts, W. W., 338, 346, 359, 361, 373
Albrecht, D. G., 254
Albright, T. D., 265
Alexander, C., 574
Alexander, I., 386
Allard, T., 118
Allison, A. C., 503
Allman, J., 263
all-or-nothing activity, 663
all-or-nothing response, 609
Allport, A., 12–13
Alopex process, 623–26
Alpert, N. M., 318, 618, 622
altered consciousness, 707–12

altitude-induced hypoxia and acceleration, 178
Alzheimer's Disease (AD)
  microtubule dysfunction and, 514
  self-awareness of deficit in patients with, 227–42
aminergic neurons, 78
amino acids, 400
Amit, D., 360
Amitai, Y., 56
amnesia, 97, 274, 296
Amos, L. A., 408, 513
Amsel, A., 156
amygdala, 115
  involvement in emotion, 316
  projection from thalamus to, 313
amygdalectomy, 150–51
Anandan, J., 551–52, 558, 564, 585
anatomic protection of nervous system, 178–79
Andermann, E., 190
Andermann, F., 190, 313
Anderson, Britt, 166
Anderson, J. R., 422
Anderson, P. A., 662, 664
Anderson, R. M., 156
Anderson, S. W., 230, 234–35
Andrade, M. A., 667
Andreu, J. M., 514
Andrews, G., 219
anesthesia
  awareness of, 295
  brain processes during, 319
  illuminating consciousness through understanding, 425
  implicit memory during, 295–302
  investigation of brain dynamics and, 321
anesthetic bindings, 425–34
Anninos, P. A., 614
anosognosia, 228–29

Anstis, S. M., 51
antedating, 339–42
anthropology, 69
antirealism, 42
Antrobus, J., 85
anxiety, 131, 317
Aoki, C., 514
aphasia, 223
a posteriori necessity, 18
Appelle, S., 707
archetypes, 598–99, 603–04
Ardran, G. M., 651
Arhem, P., 261
Aristotle, 34
Arkin, A., 85
Armstrong, David M., 2, 29, 32–36, 38, 42
Arnold, V. I., 581, 582–84, 586
Arshansky, R. I., 548
artificial consciousness, 640
artificial intelligence, 56, 725, 729
artificial life, 393
artificial neural networks, 383–84
Artin, E., 108
Aserinsky, E., 85
Ashby, Ross, 350, 383
Aspect, A., 443, 494–95
assembly of neurons, 609
association, 59
associative agnosia, 130, 134
Atema, J., 515
Athenstaedt, H., 407, 513
Atkin, A., 264
Atkinson, Richard P., 674, 707–12
atomic nuclei, granular arrays of, 522–23
atoms
 building blocks of, 660
 Newtonian vs. quantum view of, 444
 Rydberg, 419
 wave/particle duality of, 507
atonic seizures, 189
attended versus preattended contents, 123
attention
 artificial, in oscillatory neural network, 377–82
 consciousness and, 717–19
 focus of, 6, 125–26
 as functional aspect of awareness, 377
 process of, 142
 theory of, 76
attitudes, as determinants of behavior, 101
attractors, archetypes and, 603
Auchus, A. P., 231
audition, sense of, 106
Augustine, Saint, 67, 83

Ausiello, D. A., 515
automata, as basis for self-awareness, 452
autonomic processes, 102, 225, 519
Avila, J., 514
awareness
 all-or-nothing character of, 344
 attention as functional aspect of, 377
 awakeness as level of, 719
 categories related to self-monitoring, 131–32
 derived from self-monitoring, 132
 as direct availability of global control, 20
 distinction between basic capacity for and contents of, 439
 emotional, 133
 feature, 129–30
 as first-person subjective experience, 135
 four distinct levels of, 719
 James's model of, 122–23, 127
 Mangan's rehabilitation of fringe, 126
 mechanisms of, 22
 passive, 707
 phenomenological consciousness as, 713
 primary visual cortex and, 253
 relationship between neuronal activity and, 247
 as self-reported ratings of perceptions, 279
 spiritual, 292
 spotlight metaphor for, 125–26
 theory of levels of, 279–93
 topic summaries of, 130–31
 under anesthesia, 295
 varieties of, 129–34
 variety of types of, 136
 visual, 11
 without awareness," 280
 See also conscious awareness
axoneme, 422
ayahuasca, 679–80
Aydede, Murat, 38
Ayer, A. J., 47, 490
Azzopardi, P., 265

Baars, B. J., 11, 22, 124–25, 131, 136, 372, 507
Baas, Nils A., 610, 614–15, 633–48, 667, 669
Baas, P. W., 514, 528
Babinksi, M. J., 228
Bachevalier, J., 274
"back pathways," 253
Baddeley, A., 361, 363, 373
Bagshaw, M. H., 150–51
Baird, B., 58
Bamber, D., 110

Bandura, A., 95
Banisteriopsis vine, 681
Barbaro, N. M., 190, 343
Barbas, H., 251
Barbur, J. L., 253, 263, 265
Bard, P. A., 314
Bargh, J. A., 96, 100–101
Barnett, M., 409
Barnhardt, T. M., 96
Baron, R., 157
Barrett, T. W., 158
Barron, R., 158
basal forebrain, 77
basal ganglia, 148–49, 252
basin of attraction, 350, 385, 387–88, 609
b-aspects of consciousness, 68
Bateson, Gregory, 24, 740
Bateson, M. C., 717
Baudry, M., 421
Bauer, R. M., 56, 223, 264, 274
Baumgartner, G., 249
Beach, F. A., 150
Beatty, W. W., 229
Beaudry, A., 647
Bechtold, G., 517
Beck, F., 436, 562
"becoming," notion of, 545
Beek, B., 614
behavior
  attitudes as determinants of, 101
  consequences as outcomes of, 153
  delayed experience versus quick
    responses of, 345
  deliberate control of, 6
  implicit perception in evaluation and,
    98–99
  missing element in physical account of,
    545–47
  species specific, 150
Behr, S. E., 297
Bekenstein, 572
beliefs, cognitive states and, 32–33
Bell, I. R., 279–81
Bell, J. S., 494–95, 736
Bell's theorem, 552
Belyi, B. I., 205, 218
Bem, D. J., 292, 591, 593
Benioff, P., 519, 531
Bennett, H. L., 296
Bensimon, G., 514
Benson, D. F., 191, 194, 229
Benton, J. S., 228
Berg, G., 229–30
Bergenheim, Mikael, 244, 303–10

Berger, D. H., 144
Berger, Hans, 321
Berger, R., 591
Bergson, 544
Berkeley, 34
Bernard, Claud, 425
Berren, Melissa, 102
Besso, Michele, 547
Biederman, I., 58
Biersack, H. J., 191
Bigot, D., 514
binding, 10, 59
binding problem, 2, 436, 611
  neurobiological oscillations and, 51–61
  as problematic feature of consciousness,
    508
  time and the, 52–61
bindings, anesthetic, 427
biochemistry, 667
bioenergetic foundations of conscious-
    ness, 649–57
biological naturalism, 46
biological systems, hierarchical struc-
    tures in, 635
biological theory of consciousness, 113–19
biology
  quantum theory and, 496
  subneural, 393–434
bio-psychokinesis, 591
Birbaumer, N., 252
Birch, J. R., 497
Birks, C., 514
Bisiach, E., 121, 228–29, 259
Bitstring Physics, 572
Black, N., 271
blackout, 175
Blackwell, S. T., 253
Blake, R., 254
Blakeslee, S., 52
Blamire, A., 334
Bleecker, D., 580
blindness, 263
blindsight, 223
  phenomenal vision lost in, 273
  testing in monkeys, 265–73
  victims of, 118
blindsmell, 279, 283–90
blindtouch, 285
Block, N., 37, 87, 259
Block Universe model, 545
Blott, B. H., 245, 331–36, 446, 449, 735, 737
Blum, L., 644
Blumstein, S., 225
Bock, 89

Boden, M. A., 729, 731, 734
Bodis-Wollner, I., 264
Boeijinga, P. H., 369
Bogen, J. E., 123, 191, 193–94, 197, 205, 207, 216, 219
Bogen, P. J., 203
Boghossian, P. A., 42
Bohm, David, 334, 442, 476, 541, 726, 739
Bohr, Niels, 552, 597, 599, 606
Bohr, W. A. 602
Boitano, John J., 90–91, 113–19, 639
Boltzmann machines, 422
Bolz, J., 56, 61
Bond, R. N., 100–01
Boolean network, 383–87
Borges, J. L., 436
Boring, Gary, 158
Bose-Einstein condensates, 436, 516, 552, 735
  model for biological, 447
  as providers of quantum unity, 445
Bottini, G., 223
bottlenecks, Newtonian and von Neumann, 570
Bowden, P., 386
Bowen, D. M., 228
Bower, J., 58
"boxes-systems" approach, 357
Boyd, I. A., 307
Braddick, O. J., 249
brain
  as classical system, 495
  as complex system, 615
  connectionist views of mind and, 393–434
  convergence zone in, 612
  distinguishing unconscious from conscious mental functions in, 342–46
  induction of consciousness in the ischemic, 169–87
  mapping as feature of organization of, 114
  of mammals drawn on same scale, 654
  olfaction consciousness and the, 279–93
  orchestrated reduction of quantum coherence in microtubules of, 507–40
  percolation network model of, 482–84
  quantum superpositions of different levels of, 562
  representation of time in, 303
  sensory systems of the, 142
  states, 35
  stem, 115, 252
  synchronous behavior of cells in, 405

use of phosphoryl chemical energy by, 649
Brandon, Robert, 82, 86
Brandt, S., 263
Brassel, F., 191
Braud, W., 592, 594
Braun, J., 254–55
Bray, D., 422
Brecht, B., 106–08
Breiter, H., 334–35
Brentano, Franz, 29–30
Briggs, J., 735
Broad, C. D., 581, 589–96
Broca's area, 116–17
Brody, B. A., 149, 156
Brooks, C. V., 153
Brown, A. S., 124, 134
Brown, Burnell R., 433, 534
Brown, T. H., 97, 421
Brown, V., 358
Brown, W., 219
Buber, Martin, 674
Buchsbaum, G., 253
Bucy, P. C., 150–51, 154
Buddhist psychology, 721, 715
Budzynski, T. H., 711
Bufo alvarius, 683
Buhman, 56
Bullier, J., 265
Buonanno, F. S., 318, 618, 622
Burian, R., 86
Burke, S., 365
Burnham, C. A., 110
Burns Jean E., 725–26, 739–41
Burns, R. B., 524
burst manifold, 146
Burton, R. R., 175
Buss, Leo, 667

Cahn, R. N., 580
Cajal, Ramon y, 726
calculus, lambda, 667–68
Cammarota, J. P., 172, 175
Campbell, C., 300
Campbell, D., 110
Campbell, F. W., 158
Canavan, A. G. M., 252
Cantiello, H. F., 515
Cantor, N., 94
Cantrell, Dave, 534
Capec, M., 544
Capra, F., 159
carbon macromolecules, 472
Carden, D., 514

Cariani, P., 645
Carlier, M. F., 409, 515
Carlton, E. H., 147
Carpenter, Tethys, 674, 693
Carpenter, M. B., 191, 307
Carrington, P., 707, 711
Carruthers, Peter, 29
Carter, M., 358
Cartesianism, 46
Cartesian Theater, 619, 666
Carver, C. S., 134
catalytic hypercycle, 667
catastrophe theory, 581
Cauller, L. J., 265
causality, 601, 606
CBF. *See* cerebral blood flow (CBF)
cell
  activity of single, 54
  assembly of, 662–63
  assembly theory, 639
  eukaryotic, 512
  pyramidal, 56
  self-assembly paradigm of, 481
  synchronized activity of living, 397
cellular automata, in microtubules, 517–18
cephalic nervous system (CPNS), 171–74, 176
cerebral blood flow (CBF), 311
cerebral commissurotomy, 203
cerebral hemispheres, 190
cerebral representation, in "time-on" theory, 344
cerebral time requirement, 337–39
cerebroscope, 721
cerebrovascular disease, 231
Chabris, C. F., 318, 618, 622
Chaffin, R., 365
Chalmers, David J., 1, 3, 5–28, 38, 292, 353, 436, 507, 669
chaos theory, 750
chaotic dynamics, 14
charge-transfer complexes, 419
Chase, D., 407
Chaves, F., 190
Cheek, D. B., 296
chemistry, 667
Chen, D. F., 61
Chen, J. C., 482
Cheng, K., 114
Chernat, R., 514
Chiarello, C., 200
Chinese Room, 731
cholinergic neurons, 76, 78
Chou, K.-C., 515

Chow, K. L., 154
Christenson, Gina DiTraglia, 166, 237–38
Chung, S., 61
chunking, 126
Church, A., 667, 737
Churchland, Paul, 1, 29, 32–36, 38, 121–40
Churchland, Patricia, 82
Chwirot, W. B., 446
Cianci, C., 514
Cicero, 67
cingulate gyrus, 115
circadian cycle, 317
clairvoyance, 589
Claparede case, 362
Clark, A., 13, 22
Clark, S. A., 118
Clarke, C. J. S., 245, 331–36, 446, 449, 735, 737
claustrum, 252
Clegg, J. S., 401, 527
Clive, 118
Clore, G. l., 134
coconscious processing, 96
cognition
  accessibility of, 11
  framework for higher-order consciousness and, 633–48
  function of, 7
  modeling, 8
  processes as syntactic and semantic, 638
cognitive science, 89–164
  conception of brain in, 495
  natural method and, 69
cognitive states, beliefs as paradigmatic of, 32–33
cognitive system, integration of information by, 6
Cohen, A., 717
Cohen, D. S., 359
Cohen, J., 83
coherence of motion, 61
coherent pumped phonons, 515–17
Cohn D., 721
collapse criterion, 518–27
collapse fraction, 528–30
collective unconscious, 598, 605–06
colliculus, 252
Collins, A., 134
color experience, projectivist account of, 42
color qualia, 41
color sensations, 20
coma, brain processes during, 319
commissurotomy
  inkblot testing for subjects of, 203–21

surgical, 197
"common sense," 612
communication
  microstructure of, 158
  conflicting behavior in split-brain patient, 189–96
comparison net, 357
compatibilism, simple, 44
competing topic summary awareness, 131
competitive relational mind model, 367
competitive system, 363
complementarity, 597, 606
complexity
  as criteria for consciousness, 494–95
  as hyperstructure, 638
computation, consciousness as, 569–71
confidence, 153
connectionist models of consciousness, 493–495
Conners, B. W., 56
Conrad, Michael, 397, 437, 469–92, 531
conscious access, theories of, 225
conscious experience, 7. *See also* experience
  biological roots and social usages of, 141–64
  forms of, 141
  opportunity for modulation of, 346
  synchronization of sensory information to, 303–10
Conscious Mind, The, 27
consciousness
  addressing two central questions toward a science of, 743–51
  artificial, 640
  binding problem of, 436, 611
  bioenergetic foundations of, 649–57
  Bose-Einstein condensates and, 439–50
  brain and living tissues underlying, 397
  components in manifestation of, 105
  as computation, 569–71
  conglomerate nature of, 259
  connectionist models of, 493–95
  contents of, 148–53, 715
  criteria for, 493–96
  degrees of, 640
  dissociations in, 364
  distinction between self-consciousness and, 485
  double-aspect theories of, 439–40
  "double-tiered" theories of, 32
  dual hemisphere, 193
  Edelman's biological theory of, 113–19
  efference and the extension of, 105–11
  emergence and, 633–35

epistemology and, 744, 746–48
esoteric descriptions of, 159
evolution of, 665
excitations of, 448
first-person point of view of, 68
fluctuon model of, 470–72
framework for higher-order cognition and, 633–48
frontal cortex role in, 290
functional explanation of, 7
functional model of, 695–706
global workspace theory of, 11
"hard problem" of, 507
hierarchical emergence of, 659–72
higher-order, 30–32, 116–17
holistic natures of, 436
how broadly should it be defined?, 743
how does brain distinguish unconscious from, 342–44
hyperstructure and, 636
implications of, 707–12
inducing, 169
induction in the ischemic brain of, 169–87
instrumental, 674, 697–98
instrumental vs. spiritual, 701
interdisciplinary research into, 669–70
"intermediate level" theory of, 12
introspective, 2, 29, 34, 273
introspective link principle and, 29–41
intuitive, 158
James's "stream of," 135
locus of, 666
loss of, as protective mechanism, 177–81
macromolecules in, 477–80
main features of, 470
metaphor for, 745
microtubules as biological substrate for, 407–17
model of time as created by, 562
as multilevel process, 640
multiple drafts model of, 12
neural correlates of, 243
neural Darwinism model of, 12
neural membrane as "master computer" of, 423
neurobiological theory of, 10
as neurologic state, 182–85
neuronal basis of, 262–65
new computational paradigm for, 569–78
new insights from quantum theory on, 551–68
nonreductive explanations of, 16
nonreductive theory of, 19
objective, 148, 697–98

olfaction and, 279–93
neural time factors in, 337–47
PET to improve understanding of, 316
pharmacology of, 677–90
phenomenal, 7, 273
physical model of, 439, 735–37
physics of, 439
possible physical model of, 353–76
possibility of empirical test of hypotheses about, 739–42
primary, 115, 227
privacy of, 484–85
problems of, 5–28, 41
psi and research on, 589–96
quantum brain structures relevant to, 512
quantum theories of, 14, 440
receptive, 674, 698
reconciling physics with, 546
reductive methods to address, 5
reduction of quantum coherence in brain MTs as model for, 507–40
reflective, 227
self-identification hierarchy and, 462
self-referent mechanisms as neuronal basis of, 611–32
semantic memory content of, 369
semantic theory of, 731–35
sensory performance in absence of, 285
seven successively complex stages of, 714–20
spiritual aspects of, 157, 696
states of, 142–43
structure of, 20
subcellular quantum optical coherence for, 493–506
subjective aspects of, 68
subjective unity and, 621
substrate or neural correlate of, 22
temporal dynamics of, elementary integration units and, 564
theory of, mathematical strategy for, 579–88
time and, 543–50
as touchstone for philosophical and neuroscientific research, 259
towards the neuronal substrate of visual, 247–57
toward a theory of, 18–26
two-vector theory of, 561–63
understanding through anesthesia, 425
unity of, 367, 443
visual perception and phenomenal, 259–78
as wakefulness, 317

what are life and, 640–41
what is function of, 273–75
who has? 260–61
Consciousness and Contemporary Science, 121
conscious observation, collapse of, 435
conscious processes, transition from pre-consciousnes/subconscious to, 508
conscious recall, 223
conscious states, mutual exclusiveness of, 143
conscious thought, collapse criterion and, 518–27
consequences, as outcomes of behavior, 153
consistency frameworks, observation and, 572–73
constructive naturalism, dreaming and, 68–70
content, consciousness and, 715–16
content-addressable memory, 384
Contingencies of Reinforcement, 260
contour maps, hydrophobic, 427
control, integration and, 8
controlled processing, 102
conventionalism, 470, 489
convergence zone in brain, 612
Conway, J. H., 580, 582, 586
Conway, L., 575
Cooper, A., 401
Cooper, Greg, 82
Cooper, L. A., 318, 618
Copenhagen interpretation of quantum theory, 436, 507, 510, 599
Corbetta, M., 317
Cork, Randall C., 295–302
Corkin, S., 118, 274
Cormack, R. H., 254
corporeal reality, 158
corpus callosum, 190
Correa, D. D., 236
cortex, 115, 359
　40–Hz oscillations in, 52
　far frontal, 153–57
　function of, 81
　limbic medial frontal-cingulate, 155
　microprocessing in, 143–48
　neurons in layers of, 262
　occipital-parietal, 134
　perifissural, 148
　peripheral visual pathways and, 617
　primary visual, 252
　somatosensory, 337
　synaptodendritic network in, 147
Cosgrove, G. R., 191

Costa de Beauregard, O., 594
Cotterill, R. M, 58
Coulter, L., 230
Couture, Lawrence, 102
covariation, 155
Cowan, J. D., 359
Cowan, M. W., 189
Cowan, Thaddeus M., 90–91, 105–11
Cowey, Alan, 134, 243, 250, 259–78
Cox, C. M., 650
Coxeter, H. S. M., 580, 582
CPNS. *See* cephalic nervous system (CPNS)
Crane, T., 43
Creative Loop: How the Brain Makes a Mind, The, 609
creativity, 443, 730
Crick, Francis, 10, 13, 22, 27, 52, 84, 86, 125, 243, 247–57, 331, 381, 495, 507, 529, 616, 619–20, 740
Crompton, A. W., 651
Cronin-Golomb, A., 205, 207–08, 217, 219
Crook, T. H., 232–34
Crowley, Aleister, 106–08
Csermely, T. J., 614
Culbertson, J. T., 585
Cummings, J. L., 229
Cummins, R., 34
curandero, 679
Cùrdova, Manuel, 679–81
current topic summary awareness, 130
Cvitanovic, P., 604
cytology, 667
cytoplasmic transitions, 401
cytoskeleton
  components of, 513
  microtubules and, 407, 512–15
Czeki, S., 250

Daffner, K. R., 191
Dager, S. R., 229–30
Dalibard, J., 494–95
Damasio, A. R., 51, 97, 130, 274, 315–16
Damasio, H., 97, 315–16
DíAmato, S., 563
Dance of the Wu Li Masters, 159
Daneri, A., 555
Danielcyzk, W., 228
Darley, J. M., 94
Darwin, Charles, 159, 275, 743
Darwinism
  neural, 12, 113–14
  quantum mechanics and, 423
Dasheiff, R. M., 264

Daugman, J. G., 147
Davey, M. R., 307
David, A., 334
Davidson, Richard, 245
Davies, P., 105–11
Davis, C., 651
Davis, H. S., 296
Davison, A. N., 228
Dawkins, R., 638, 691
Dayhoff, J. E., 390, 482, 514
death
  neurologic state of, 183
  as ultimate disaster for object-self, 702
DeBettignies, B. H., 230–31
De Brabander, M., 513
Deecke, L., 155, 342, 520, 531
definite, meanings of, 127
De Haan, E., 264
Deikman, Arthur J., 674, 695–708, 710
déjà vu, 152
Delattre, D. L., 359, 361, 373
deLeon, M. J., 230–31, 234, 238
Del Giudice, E., 446, 448, 501, 527
Delis, D. D., 205
Dement, W., 85, 86
dementia, independent living skills impairment in, 231
Democritus, 42
dendrodendritic synapses, 358
Dennett, Daniel C., 1, 12–13, 29, 37, 67, 83, 259–60, 303, 307–09, 372, 496, 619–20, 627, 666, 721, 740
Derr, P., 591
DesAutels, Peggy, 27
Descartes, Rene, 42, 67, 160, 443
descriptive realism, 44, 48
Design for a Brain, 383
Desimone, R., 252–53
Desmond, J. E., 58
d'Espagnat, B., 583
Deutsch, D., 449, 452, 512, 519, 531, 564
DeValois, K. K., 147
DeValois, R., 147, 254
Díaz, José-Luis, 674, 685, 714
Diaz-Nido, J., 514
Dicke, R. H., 501
Dikkes, P., 118
Dikman, Ziya V., 279–81, 292
Dimattia, B. V., 152
Diósi, L., 435, 509, 511–12, 519
Dirac, P. A. M., 580
dissociation, 186, 195, 364
distal events, 106–08
Dive, D., 319

DMT, 680, 688
Dobmeyer, S. S. G. L., 317
Doglia, S., 446, 448, 501, 527
domains, 157
Domich, L., 358
Donaldson, M, 731
Donoghue, J. P., 61
Dorpat School of Religious Psychology, 720
dorsal system, 155
double-aspect theory, 1, 24–26, 439–40
double-tiered theories of consciousness, 32–33
Douglas, R. J., 152–53
Dowling, J., 358
dreams, 247, 674
  of Amahuaca Indians, 679
  arising of images in, 464
  deconstructing, 67–88
  double aspect model of, 70–74
  four philosophical problems about, 67–68
  functionality of, 67
  identification of mentation during, 85
  invented functions of, 81
  as level of awareness, 719
  lucid, 79, 465
  natural functions and, 78–81
  premonitory, 599
  quantum mechanics of, 451–68
Dreskin, M., 231
Dretske, F., 32, 37–38
Dreyfus, Hubert L., 243, 570
Driver, M. V., 192
drugs
  hallucinogenic, 430, 677
  psychotropic, 677
dual consciousness, 189–96
dualism, 1, 18, 42, 439, 569, 620, 666, 726, 739
DuBois-Reymond, E., 260
Dubrul, E. L., 655
DU-Evidenz, 261
Duke, D. W., 323
Dukkbecjm, N., 574
Dulac, O., 190
Dunn, D., 38
Dunne, B., 746
Durwen, H. F., 191
Dustin, P., 407, 513
Dwayne-Miller, R. J., 517
Dyer, D. C., 430
dynamic universality, 483
Dynkin, E. B., 581–82

Earl, Heath, 674, 707–12
earth, gravitational field of, 177
easy problems of consciousness, 1, 5–6
Eccles, John C., 344, 436, 529, 562, 607, 654, 669, 726
Eckhorn, 10, 54–55, 60–61
ecstasy, 719
Edelman, Gerald, M., 12–13, 58, 90–91, 135, 259, 331, 372, 507, 614, 639
Edelman's biological theory of consciousness, 113–19
Edinger, L., 194
efference, extension of consciousness and, 105–11
Efron, R., 205, 249
ego, 730
Ehresmann, A. C., 639, 645
Eibl-Eibesfeldt, I., 714
Eigen, Manfred, 646, 667–68
eigenstates, 736
  "choice" of, 510
  of mass distribution, 525
eigenvalues, quantum theory and, 580
Einstein, Albert, 452, 547, 552, 609
  theory of gravity, 541, 579
  on time, 557
EIU. *See* Elementary Integration Units (EIU), 564
Ekman, P., 133
Elbert, T., 252
electrochemical synaptodendritic states, 142
electroencephalogram (EEG)
  activity, 332
  changes in, 174
  registration, of subliminal olfaction, 279–91
  use in experiment for measuring consciousness, 448
  viewing electrical act of individual neurons with, 452
electromagnetism, 16
electrons
  as elaborate logical structures composed of events, 47
  non-Born-Oppenheimer, 478
Elementary Integration Units (EIU), 564
Elger, C. E., 193
eliminativism, physicalism without, 43
Elitzur, Avshalom C., 524, 541–50, 557
Ellenberg, L., 219
Ellman, S. J., 85
Elsasser, Walter, M., 664
Elton, Matthew, 27

emergence, 633–35, 641–45
  deducible, 639, 642
  definition of, 635
  as function of observation, 637
  hierarchical, 659–72
  phenomena of, 660–62
emergent physicalism, 726, 739
Emmett, Kathleen, 85
emotion
  awareness of, 133
  brain regions involved in normal human, 313–16
  domain of goals and, 133–34
  involvement of thalamus in, 315
  responses, 99
  states, 455–58
  structural properties of, 21
energy, registration of subtle information and, 292
Engel, A. K., 53, 56–57, 61, 118
Engelborghs, Y., 514
environmental entanglement, 526–30
environmental stimuli, 6
epicritic processing, 148
epilepsy
  absence, 189
  seizures in, 189
  use of EEG in diagnosis of, 321
epiphenomenalism, 546
episode
  boundaries of, 150–51
  as context, 152
episodic consciousness, 150–53
episodic memory
  content of, 369–70
  net, 357
epistemology, consciousness and, 744, 746–48
Epstein, J. A., 154
Erikson, E., 697
Ermentrrout, G. B., 359
escape energy, 661
Eslinger, P. J., 229
eukaryotic cells, 512
evaluation, implicit perception in behavior and, 98–99
Evans, D. R., 231
events
  as basic "building blocks" of the universe, 47
  proximal and distal, 106–08
everything, theory of, 572
evolution
  biology, 69
  towards higher-order structure, 635–36
excitation, threshold for, 609
Exner, J. E., 205, 216
experience
  awareness as first-person subjective, 135
  conceptual gap between physical process and, 18
  delayed versus quick behavioral responses, 345
  extension across the senses, 105–06
  fringe, 123–24, 126
  as fundamental, 17
  Kantian distinction between precondition for experience and, 474
  near-death, 183
  new approach to explain, 9
  oceanic, 157
  out-of-body, 157
  problem of, 6, 12–13
  as sensitive to details of social situation, 93
  states of, 7
  structure of subjective, 121–40
  subjective character of, 41
  subjective timing of, 339–42, 345
  tip-of-the-tongue, 124, 128
  variety of conscious, 141–64
experiential selection, 113
experimental neuroscience, 243–347
explanatory gap, 9
explicit memory, 295, 317–18
explicit recognition, 274
extracorporeal reality, difference between corporeal reality and, 158
extra ingredient, soul as, 669
eye
  "inner," 618
  peripheral visual pathways and, 617

Faichney, Robin, 694
"fallacy of the homunculus," 250
familiarity, notion of, 368
familiarization, 152
Farah, M. J., 22, 264
Far East, mystical traditions of, 157
Farmer, J. D., 322, 667
Farnsworth, D. W., 655
Farthing, G. W., 227
Fazio, R. H., 94
fear, autonomic expression of, 313
Featherstone, 503
feature awareness, James's nucleus as, 129–30
Feher, E. P., 232–34

Feinstein, Bertram, 303, 309, 337–39, 345–46, 359–61, 373, 520, 525, 532
Feinstein, D. I., 86
Felbain-Keramdias, S. L., 143
Feldhutter, I., 714
Felleman, D. J., 248, 250, 252
Fenwick, P., 373
fermions, vacuum, 470–72
Fernandez, Mercedes, 279–81, 292
Ferrari, D. C., 591–93
Ferris, S. H., 229–31, 234, 238
Ferster, D., 56
Festinger, L., 110
Fetz, E. E., 61
Feynman, R. P., 519, 531
Field, D. J., 525
fighter aviation, +Gz-stress and, 169–71
Fink, Raymond B., 610, 649–57
Finkel, L. H., 58
Fiorani, M. Jr., 249
first-order structures, 634
Fischer, B., 194
Fischer, K. H., 410
Fisher, L. E., 707, 711
Fivush, R., 365
Flament, D., 61
Flanagan, F., 151
Flanagan, Owen J., 1–3, 37–38, 52, 67–88, 122, 134, 227, 259, 275
Flavell, J. H., 227
Fleschig, P., 148
Flohr, H., 12–13, 265
Flory, P. G., 401
fluctuon model, 470–72
focus of attention, 125–26
Fodor, Jerry, 260, 275
Fontana, Walter, 667
forebrain, 115
form, connection between sound and, 694
Forster, E. M., 174
Fortier, P. A., 61
Foulkes, D., 85
Ford, A., 219
Gramiak, R., 651
Frackowiak, R. S. J., 253, 265
Franchi, Stefano, 38
Franck, G., 319
Franco, G., 319
Frank, H. S., 399
Frank, R., 315–16
Franks, F., 500
Franks, N. P., 425, 503
Frederiks, J. A. M., 229
free association tests, 297

free-floating anxiety, 131
Freeman. R. D., 56
Freeman, W. J., 373, 507, 614–15, 740
"free will," nondeterministic, 508
Freidenberg, D. l., 231
Frenkel, A., 435, 511
Freud, Sigmund, 2, 123, 159, 346, 606, 730
Frey, S., 265
Friederich, F. J., 717
Friedrich, P., 514
Fries, W., 252
Friley, K., 300
fringe experience, 123–24, 126
Frith, C. D., 131
Fröhlich, Herbert, 430, 446–48, 482, 496, 515, 641
frontal cortex, role in consciousness, 290
frontal eye fields (FEF), 365
Fujita, I., 114
Fuller, J. L., 151
Fulton, J. F., 151
function, 8
functionalism, 732
Fung, G., 696
Fung, P., 696
Fuster, J. M., 248, 251–52, 612

Gaal, G., 61
Gabor, Dennis, 158
Gabora, Liane, 27
Gage, Phineas, 316
Gaia, 745
Galaburda, A., 315–16
Galanter, A., 154, 157
Galileo, 42, 159
Galin, David, 90, 121–40
Galkin, T. W., 252
Gall, Frances, 141
gamma range oscillations, 55–56
Ganellen, R. J., 231
Ganong, A. H., 421
Gans, C., 653
"ganzfeld" studies, 591
Gao, B., 514
garden-of-Eden states, 387–88
Gardner, 89
Garrett, Merrill, 722
Gattas, R., 249
Gazzaniga, M. S., 167, 207, 218
gedanken device, 481
Gelade, G., 58
gelation, osmosis and, 400
gel state, 402
Geminiani, G., 228–29

Genberg, L., 517
Gendlin, E. T., 732, 737
general social-interaction cycle, 93–95
Gennaro, R., 714
Genzel, L., 517
George, H. J., 525
Georgopoulos, A., 149
Gerfen, C. R., 251
Gerler, D., 124, 133
German, R. Z., 651
Geroch, R., 512
Gesell, A., 697
Gestalt psychology, 123
Ghirardi, G. C., 435, 511–12, 519
Ghose, G. M, 56
Giannini, J. A., 296
Gilbert, J. M., 525
Gill, M., 143
Gillies, B., 433
Gilmore, R., 584
Girrard, P., 265
Gizzi, M. S., 263
Gleason, C. A., 342–43
Gleick, J., 604
glial memory cells, brief description of, 451
Glisky, Elizabeth, 102
Glisky, Martha, 102
global blackboard model, 372
global control
  awareness as direct availability for, 20
  nature of, 359
global gate, 357–58
globally accessible information, 11
global mapping, 114–15
global workspace
  as neural correlate of consciousness, 22
  theory of consciousness, 11
globus pallidus, 252
G-LOC syndrome, 172–77
  event sequence of, 181
  near death experiences and, 185
  as protective mechanism, 179
Gloor, P., 313
Glymour, Clark, 83
goals, emotion and the domain of, 133–34
Goddard, D., 696
Goddard, P., 580
Gödel, K., 644
Gödel's theorem, 644, 731, 735
Godfrey-Smith, P., 86
godhead, 749
Goeppert, E., 651, 653
Golavan, A., 366
Goldman, L., 296

Goldman, P. S., 51
Goldman-Rakic, P., 51
Goldstein, F. C., 231, 233
Goldstein, J., 721
Goldstone, J., 499
Goldstone bosons, 497
Goldstone modes, 499
Goodale, M. A., 381
Goode, R., 118
Goodglass, H., 131, 205
Gordon, M., 230–31, 234, 238
Gorelick, P. B., 231
Gorse, D., 373
Goswami, A., 436, 564, 586, 739–40, 746
Gould, S. J., 83
Goutières, F., 190
Grabowski, T., 315–16
Grace, A. A., 56
Graff, D., 514
grain problem, 45–46
Grand Unified Theory, 572
Grangier, P., 443
Granlund, Brittmarie, 244, 303–10
Grassberger, P., 322–23, 327
Grassi, R., 435, 511–12, 519
Grattan, L. M., 229
Graves, R. E., 131, 205, 236
gravitational field, earth's, 177
gravity theories, 579–80
Gray, C. M., 56–57, 60, 118, 507
Greco-Vigorito, C., 365
Green, J. P., 231, 233, 430
Green, M., 584
Green, R. C., 231, 233
Greene, P. H., 350
Greengard, P., 514
Greenwald, A. G., 96, 101
Grey-Walter, W., 520, 531
Gross, C. G., 265
Gross, D., 584
Grossberg, S., 373
Grotzinger, B., 342, 520, 531
Grueninger, W. E., 152
Grundler, W., 517
Gruneberg, M. M., 133
Guenther, H. V., 715, 721
Guitton, D., 263
Gupta, S., 231
Gur, R. C., 207
Gusakova, V., 366
Gustafson, J. W., 143
Gustafson, L., 228–29
Güzeldere, Güven, 2–3, 29–41, 674
+Gz-stress, fighter aviation and, 169–71

Haan, E. D., 223
Hagan, Scott, 404, 437, 482, 493–506, 527, 530, 735
Hahan, E. L., 503
Haken, H., 414
Hall, Z. W., 421
Halligan, P., 223
hallucinogenic drugs, 430, 677
Halpain, S., 514
Halsey, M. J., 426, 503
Haltiner, A., 239
Hameroff, Stuart R., 14, 260, 321, 389–90, 394, 404, 407, 422, 425, 430, 432, 437, 448, 452–53, 466, 482, 507–40, 613, 619, 633, 640–41, 669, 735, 740
Hamilton, R. J., 175
Hamilton, S. E., 318, 618, 622
Hao Bai-lin, B. L., 604
Hardcastle, Valerie Gray, 2–3, 51–61
Hardin, C. L., 13, 22
hard problem of consciousness, 1, 5–7, 507, 669
Hardy, B., 420
Harman, Willis W., 726, 743–51
Hart, T., 124, 133
Harth, Erich, 609–32, 669, 726
Hartle, J. B., 512
Hartog, J., 154
Hartree-Fock type self-consistent field, 478
Hasted, J. B., 497
Hawking, S., 571–72
Hawkins, R. D., 737
Hayne, H., 365
Head, Henry, 148
Head, J., 525
healing
  placebo responses and, 677
  quantum, 750
Hearst, E., 151
Heaton, J., 300
Hebb, Donald O., 507, 529, 609–10, 614, 639, 662–63, 669
Hebden, M. V., 296
Heilman, K. M., 192, 232, 617
Heisenberg, 435, 442
Heisenberg equations of motion, 500
Hellige, J. B., 195
Helmstaedter, C., 193
hemianesthesia, 223
Henninger, K. M., 219
Henninger, Polly, 166, 203–21
Henninger inkblot, 209
Herbert, N., 585–86

Herman, J. H., 85
Hermann, D. J., 365
Hermans, H. J. M., 149
Herring, S. W., 651
Hersh, N. A., 149
heterophenomenological procedure, 721
Hier, D. B., 231
hierarchical emergence of consciousness, 659–72
hierarchies, 636, 668
higher level perceptual phenomena, 54
higher mental processes, 730–31
higher-order
  cognition and consciousness, 633–48
  consciousness (HOC), 116–17
  perception (HOP) theories of consciousness, 32–36
  representation (HOR) theories of consciousness, 31–32
  structure principle, 636
  thought theories, 37
Hilbert Space, 572
Hiley, B., 541
Hilgard, E. R., 143, 357, 364, 372
Hill, C. S., 18
hill-climbing algorithm, 623
Hilton, Walter, 695
Hinton, G. E., 125–26
Hinton, H. C., 579
hippocampus, 114–15, 155, 247, 316
Hirai, T., 708
Hirokawa, N., 511
Hirst, W., 89, 131
History of Experimental Psychology, 158
Hitch, G., 361, 373
Ho, M. W., 641
Hoare, R. D., 192
Hobbes, 42
Hobson, Allan, 71, 82, 84
Hobson, J. A., 73, 75, 77, 83, 86–87
HOC. *See* higher-order consciousness (HOC)
Hofmann, Albert, 683–84
Hofstadter, D. R., 331
holism
  of brain states, 552
  of quantum mechanics, 552–53
Holmes, P., 58
holographic theory, 750
Holt, Anatol, 717
Homo sapiens, consciousness in, 469, 459
Homskaya, E. D., 153–54
Honorton, C., 292, 591–92, 593
Hopfield, John J., 86, 350, 384, 609, 614, 729

Horio, T., 407
Horn, D., 58
Horne, P. V., 373
Horowitz, S., 313
Horwitz, L. P., 548–49
Hosobuchi, Y., 343
Hotani, H., 407
Howell, P., 97
Hubel, D. H., 51, 114, 639
Huber, S. J., 231
Huberman, B. A., 58
Huggins, S. E., 653
Hughes, p., 106–08
human creativity, 443
human language, 150
human "real intelligence," 571
Hume, 47, 721
Humel, J. E., 58
humor, listener autonomic activity and, 225
Humphrey, D., 231
Humphrey, N., 12–13
Humphreys, J. E., 580
Hunt, S. C., 156
Hunt, S. P., 514
Huntington's disease, 229
Hurt, C. A., 297
Husain, S. K., 497
Husserl, Phenomenology of, 715
Huttenlocher, P. R., 189
Hyde, T. S., 229–30
hydrophobic contour maps, 427
hypercycle, 667
Hypercycle: A Principle of Natural Self-Organization, The, 667
hyperstructure, 610, 615, 635–38, 646–47
  complexity as, 638
  consciousness and, 636
  as result of higher-order structure principle, 636
  schematic example of, 637
hypnosis, 143, 364
hypothalamus, 77, 316, 115
hypoxia
  altitude-induced acceleration and, 178
  ischemia, 118

Iacoboni, Marco, 166, 193, 197
id, 730
identity theory, 546
imagery
  mental, 622
  neural basis of, 618
imagination, 442

immortality, 696
immune system, sleep and the, 87
implicate order, 726, 739
implicit knowledge, 123
implicit memory, 295–302, 317–18
implicit perception, in evaluation and behavior, 98–99
implicit recognition, 274
impression formation, 99–100
Inbody, S. B., 232–34
independent living skills (ILS), impairment in dementia of, 231
information
  as basic ingredient of universe, 1
  double-aspect theory of, 24–26
  globally accessible, 11
  integration of, 6
  no central cortical exchange of, 51
  physical, 24
  processing of, 407–17
Ingber, D. E., 515
Inhoff, A. W., 717
"inner eye," 618
inputs, neuronal, 262
Insinna, Ezio M., 390, 607, 542
instinct, as species-shared propensity, 150
instrumental consciousness, 674, 697–98, 701
integration, control and, 8
intelligence
  artificial, 725
  social, 94
  tying qualia to, 476
intention, 695–706
intentional fallacy, 35–36
intentionality, philosophical understanding of, 35–36
interdisciplinary research into consciousness, 669–70
interhemispheric dissociation, 195
intermediate level theory, 12
internal access, 8
internal mental images, 21
intracortical feedback pathways, 56
introspection, 31
introspective consciousness, 2, 29, 34, 273
introspective link principle, 29–41
inverse problem, 622
ischemic brain, induction of consciousness in the, 169–87
isoflurane, memory with pure, 297–300
Iwai, E., 265

Jackendoff, R., 13, 122, 126, 129, 134–36, 692
Jackson, F., 18
Jacoby, L. L., 356, 362
Jagadeesh, B., 56
Jagust, W. J., 230
Jahn, R. G., 574–75, 746
jamais vu, 152
James, William, 1–2, 29–30, 47, 96, 122–35, 136, 141, 275, 344, 356–57, 370, 372, 552, 674, 696, 713–14, 720, 726, 747
James's model of awareness, 122–23, 127
Janowsky, J. S., 229
Jarvik, L. F., 514
Jaspers, K., 721
Jeannerod, M., 265
Jenkins, W. M., 118
Jensen, A. R., 339
Jensen, C. G., 524
Jensen, K. S., 407, 515, 517–18
Jerison, H. J., 421, 422
Jervey, J. P., 248
Jibu, J., 482
Jibu, Mari, 404, 437, 448, 493–506, 527, 530, 735
Johansson, Håkan, 244, 303–10
John, E. Roy, 740
Johnson, M. K., 97, 131, 133, 732
Johnson-Laird, P. N., 134
Jolles, J., 232
Jona-Lasinio, G., 499
Jones, D. E., 100
Jones, O. R., 732, 736
Jordan, J., 654
Jorgensen, K., 419
Josephson, Brian D., 674, 693
Josza, R., 519, 531
Jouvet, M., 85
Joyce, G., 647
Jozsa, R., 564
Julez, B., 158
Jung, C. G. , 2, 125–26, 597, 602–04, 606–07, 730
Jung's theory of synchronicity, 542

Kaada, B. R., 154
Kabat-Zinn, J., 707, 710
Kafatos, M., 613
Kafka, Franz, 371
Kahneman, D., 125–26
Kaivarainen, A. I., 403, 531
Kaku, M., 572
Kamback, M. C., 156
Kammen, D., 58
Kamppinen, 89
Kanazirska, M., 515
Kandel, E. R., 51, 729, 737
Kanemoto, Ansel, 328
Kanerva, P., 571, 575
Kaneshiro, E. S., 422
Kang, S., 430
Kaplan, E., 207
Kapleau, Philip, 695–96, 707, 710
Karamporsala, H., 407
Karampurwala, H., 515, 517–18
Károlyházy, F., 435, 511
Karplus, M., 515
Kasamatsu, A., 708
Kaszniak, Alfred W., 166, 229–30, 233–34, 236–38, 722
Kauffman, Stuart A., 383, 667
Kaufmann, L. H., 582
Keilmann, F., 517
Keller, E. F., 86
Kelley, C. M., 356, 362
Kelso, S. R., 421
Kemp, F. H., 651
Kempen, H. J. G., 149
Kesner, R. P., 152
Keus, H., 517
Keysar, B., 58
Kihlstrom, John F., 89, 94, 96, 101, 131–32, 250, 296–97, 299–300, 315
Killheffer, R., 569
Kim, H., 524
Kim, J. K., 44, 61, 97
Kinchla, R. A., 717–18
kinesthesis, 106
King, S. M., 265
Kinney, H. C., 118
Kinnunen, P. K., 420
Kinsbourne, M., 156, 194, 216, 219, 303, 307–09
Kiper, D. C., 254
Kirschner, M., 514
Kitcher, P., 79, 86
Klassonde, M. C., 158
Kleitman, N., 85
Klima, H., 504
Kline, John P., 279–81, 293
Klinger, M. R., 101
Klivington, K., 721
Kluft, R. P., 193
Klug, A., 408, 513
Klüver, H., 150–51, 265
Knight, R. T., 205
knowledge
  hierarchical organization of, 665
  implicit, 123

not accessible to conscious recall, 223
  tacit, 123
  by triangulation, 69
Ko, A. Y., 497
Koch, Christof, 10, 13, 22, 27, 52, 58, 84, 125, 243, 247–57, 331, 381, 443, 449, 495, 507, 520, 529, 619, 740
Koffka, K., 123
Kolb, F. C., 254, 255
König, P., 53, 56–57, 118
Konow, A., 154
Korein, J., 118
Kornhuber, H. H., 155, 342, 520, 531
Korsakoff's syndrome, 97
Koruga, Djuro, 390, 433
Kosslyn, S. M., 318, 618, 622
Kostant, B., 582
Kreiter, A. K., 56, 61
Kremer, F., 517
Kripke, S., 18
Kronheimer, P. B., 582, 584
Kruse, W., 61
Kryger, M. H., 86
Krystosek, A., 515
Kudo, T., 514
Kuhn, Thomas, 744
Kulics, A. T., 265
Kupfermann, I., 191
Kurthen, M., 191, 193

Laandwehr, R. S., 124
La Berge, D. L., 358, 717
La Berge, S., 465
Lahoz-Beltra, R., 390, 482, 535
Lamb, Bruce, 679–80
Lamb, M. R., 205
lambda-calculus, 667–68
Land, E. H., 253
Landauer, R., 570
Land effect, 253–54
Landis, T., 131, 205
Lane, Richard, 245
Lange, D. G., 517
Langer, E., 720
Langton, C. G., 393, 640
language
  acquisition of, 116–17
  comprehension in severe "sensory aphasic," 223–25
  human, 150
  theories of comprehension of, 225
Lapham, L., 704
Laplacian net, 359
Larrabee, G. J., 234, 239

Larson, J. E., 651
Lashley, Karl, 158
Lassonde, M., 265
lateral geniculate (G) body, 72
lateral limits technique, 208
Laurikainen, W. A., 602
Laverty, S. G., 110
Law, Heather, 102
Law of Minimization of Mystery, 14
Lawrence, T., 593
Laxer, K. D., 190
Lazarus, R., 133
Ledoux, Joseph, 313–14
Lee, C., 61, 248
Lee, J. C., 525
Lee, J. R., 158
Lees, Marek, 694
left hemisphere, 153–57, 204, 217
left visual field (LVF), 197
Legéndy, C. R., 52–53
Lehninger, A. L., 650
Leibniz, 42
Leonesio, R. J., 124
Lepor, E. F., 265
Lerdahl, F., 692
Lesser, 383, 386–87
levels of awareness
  "awareness without awareness" and, 290–91
  consciousness and, 719–20
  theory of, 279–93
Levin, G., 359, 361, 373
Levin, H. S., 239
Levine, J., 9, 18
Levinson, B. W., 296
Levitt, J. B., 254
Levy, J., 205–06, 217–18
Lewis, D., 18
Lewis, R. T., 206
Lewontin, R. C., 83
Li, K. H., 504
Liberman, E. A., 482
Libet, Benjamin, 22, 133, 244–45, 303, 309, 334–35, 359–61, 373, 520, 525, 529, 532, 565, 714
Lieb, Irwin, 721
Lieb, W. R., 425, 503
Lieberman, P., 651
light
  in G-LOC experiences, 186
  physical theory of, 44–45
  speed of, 572–73
Light, L. L., 296
limbic system, 115

forebrain and, 150–53
  medial frontal-cingulate cortex, 155
  Papez circuit and, 359
Lind, J., 651
Lindahl, B. I. B., 261
Lindsay, R. D., 614
Linke, D. B., 191, 193
liquid crystal neural membrane, 419
Liu, T. J., 101
Livingston, J., 190
Livingstone, M. S., 51
Llinás, R. R., 56, 84–86, 359
Lloyd, B. B., 129
Lloyd, E., 86
Loar, B., 18
locality, 494–95, 620
Locke, J., 2, 29, 32–36, 38, 42
Lockwood, M., 436, 547
Loebel, J. P., 229–30
Logan, G. D., 102
Logothetis, N. K., 248, 263
Lohman, A. H. M., 369
Loinger, A., 555
Lombardi, W. J., 100–01
Long, M., 155
Lopes da Silva, F. H., 369
Lopez-Barneo, J., 514
Lorentz covariant mechanism, 565
Lorenz, K., 261
loss of consciousness, 177–81
Louria, Dyan, 394, 514, 530
LSD, 683–84, 688
Lucas, J., 730
lucid dreams, 79, 465, 719
Lucky, R., 571
Lukacs, B., 435, 511
Lumer, E., 58
Luria, A. R., 153–54
Lycan, William, 29, 32–38, 721
Lynch, B., 421
Lynch, J. C., 149
Lynch, Ron, 433
Lyons, W., 31, 38

MacCuish, J., 219
MacKay, D. M., 201
MacKay, V., 201
MacLeod, Don, 254
McCall, S. L., 503
McCammon, J. A., 515
McCann, J. J., 253
McCarthy, K., 564
McCarthy, M., 230–31, 234, 238
McCarthy, R. A., 149

McClelland, J. L., 125–26
McCloskey, D. I., 345
McCormick, D. A., 85–96
McCulloch, W. S., 349–50
McDermott, J. J., 123
McGinn, Colin, 16, 69, 83, 259
McGlone, J. 231–33
McGlynn, S. M., 228–29, 237–38
McGoveran, D., 572
Mach, 47
Mach's principle, 471
Mack, L. L., 192
McKay, J., 583
McKenna, Terrence, 687
McLachlan, D. R., 235–36, 238
McLelland, J. L., 729, 737
McLendon, G., 517
McMoneagle, J., 575
McNaughton, B. L., 367
Macon, J. B., 307
macromolecules, 472, 477–80
macroscopic ordered state, 499
Maggiora, M. G., 515
magnetic resonance imaging (MRI), 35, 189
Mahurin, R. K., 230–34
makyo, 696
Malcolm, Norman, 67, 83
Maljkovic, V., 618, 622
Mallany, H. M., 497
Malmo, R. B., 156
Mamelak, A. N., 190
mammals, brains drawn on same scale of, 654
Mandelkow, E. M., 408, 514
Mandler, G., 125–26
Mangan, B., 122, 124–31, 136, 370, 372
Mangone, C. A., 231
manifolds
  burst, 146
  receptive field, 145
Mann, D. M. A., 228
mapping
  as feature of brain organization, 114
  global, 114–15
MAPs. *See* microtubule associated proteins
Maquet, P., 319
Marcel, A. J., 121, 259, 357, 364, 372
Margolus, N., 571
Margulis, L. L., 407
marijuana, 677–78
Mark, Victor, 165–66, 189–96
Marsh, Gail, 82, 89
Marshall, D., 736

Marshall, Ian N., 331, 334, 432, 436, 446, 449, 495, 516, 530, 674, 729–38
Marshall, J., 118, 223, 263
Marsolek, Chad, 102
Martial, 97
Mascarenhas, S., 513
materialists, 569, 666
Matsumoto, G., 482
Matsuno, K., 477
Matsuyama, S. S., 514
matter
  microstructure of, 158
  philosophical positions on the relation between mind and, 440
Matthews, G. B., 83
Mattson, K. E., 229
Matzke, Douglas J., 541–42, 569–78
Maudlin, T., 569
Maunsell, J. H. R., 250, 263
Maxwell, 16–17
  vector analysis by, 579–80
  equations of, 472
  Kinetic Theory of Gases, 398
Mayes, A., 223
MDMA, 688
Mead, Carver, 575
meaning, as existential problem, 703
mechanisms
  of awareness, 22
  that perform function, 8
medicine, 165–242
meditation, 157, 465
  enhanced vigilance in guided, 707–12
  renunciation and, 699–700
  Zen, 710
Meglitsch, P. A., 650
Meier, C. H., 599–603
Meissner effect, in superconductivity, 527
Meissner's corpuscles, 307
Melki, R., 409, 515
Mellor, D. H., 43
memetic selection, 70
memory
  age-related decline in, 230
  "content-addressable," 384
  emergence of, 383–92
  far from equilibrium, 389
  implicit, during anesthesia, 295–302
  implicit and explicit, 317–18
  with pure isoflurane, 297
  social judgment as implicit, 97
  "value-category," 116
  working, 361
Memory Observation Questionnaire, 232

mental functions, neural time factors in conscious and unconscious, 337–47
mental images, internal, 21
mental processes
  higher, 730–31
  similarities between quantum processes and, 442
mental representation, 34–36
mental space, 494
mental states, 356
  as brain states, 35
  higher-level representing of lower-level, 29
  meta-level, 31–32
  reportability of, 6
mentation
  NREM and REM sleep and, 71, 81
  identification of dream, 85
Merikle, P. M., 101
Merkel's receptors, 307
Merle, J., 216
MerleauPonty, 715
Merrill, C. R., 430
Merzenich, M., 118
Mesulam, M. M., 191, 194
meta-analysis, 589–90
metacognition, 227
meta-level mental state, 31–32
Metamemory in Adulthood (MIA) questionnaire, 236
metamemory processes, in AD, 236
"Metamorphosis, The," 371
metaphysical theory, neutral monism as, 48
Metcalfe, J., 227
Metzinger, T., 38
Meunier, M. J., 274
Meyer, H. H., 425
Meyer, L. B., 693
Meyer-Overton correlation, 426
Michalak, R., 623, 628
Michel, F., 285
microtubule associated proteins (MAPs), 421, 509–10, 514, 518, 524
microtubules (MTs), 407, 439, 497, 740
  automaton simulation of, 518
  cellular automata in, 517–18
  cytoskeleton and, 512–15
  information processing, 408, 515–18
  interconnection of, 509
  isolating quantum coherence from, 526–30
  neuronal, illus., 511
  orchestrated reduction of quantum coherence in brain, 507–40

quantum coherence in, 526
  schematic graph of quantum coherence in, 530
  structure from x-ray crystallography, illus., 513
  wave function self-collapse in, 518–30
  as modulators of membrane receptor sensitivity, 524
MID. *See* multi-infarct dementia (MID)
Miezin, F. M., 263, 317
Milani, M., 446, 448, 501, 527
Milberg, W., 225
Mileusnic, R., 514
Miller, B. A., 157
Miller, E. R., 651
Miller, G. A., 154, 735
Miller, J. G., 126, 280
Milner, B., 156, 274
Milstein, J., 157
Minami, I., 114
Minch, E., 633
mind
  connectionist views of brain and, 393–434
  crossdisciplinary physics as basis for, 571–72
  hierarchy for dynamics of, 668
  nonphysical sciences for the study of, 569–70
  philosophical positions on the relation between matter and, 440
  philosophy of, 1–88
mind-body problem, 569
  relativity theory and, 543–550
mindfulness, 720
"Mind of God," 436
mindlessness, 720
"mind's eye," 36
Minina, S. V., 482
Minsky, M., 167, 496, 619, 729
Mirsen, T., 231
Mirsky, A. F., 150–51
Mishkin, M., 148, 250, 252, 274
Mishra, R. K., 482
Misner, C. W., 477
Mitchison, G., 86
Mitchison, T., 514
Mjakotina, O. L., 482
modeling
  neurophysiological and cognitive, 8
  what it is like to be, 353–76
moderator variables, psi research and, 593
modern functionalist view, 260
Moffit, A., 464
Mojtahedi, S., 190

molecular biology, 393
monism, 439, 666
monistic idealism, 436, 726, 739
Monroe, R., 575
Monson, N., 229
Monte-Carlo simulations, 413, 416
Montero, F., 667, 715
Montoro, R. J., 514
Moody, R., 575
Moore, J. W., 58
Morán, F., 667
Moravec, H., 243
Morgan, L. C., 645
Moriyasu, K., 580
Morledge, D. M., 342–43
Morse, M., 157
Morton, William, 295
Moscovitch, M., 235–36, 238
motor control, 305
Mountcastle, V. B., 149, 614
Mouritsen, O. G., 419
Movshan, J. A., 254
Movshon, A. J., 263
MRI. *See* magnetic resonance imaging (MRI)
MT. *See* microtubules
multi-infarct dementia (MID), 228, 231
multiple drafts model, 12 303, 308–09
multiple personality disorder, 193
multiple universes, 435
Münsterberg, H., 110
Murray, E. A., 274
Murray, J., 359
Murthy, V. N., 61
music, nature of mind and, 691–94
Myers, J. J., 207, 217, 219
Myerson, J., 263
Mylrea, Kenneth, 328
myoclonic jerking, 174
mysticism, 157, 695–97, 720

n. medialis dorsalis, 154
Nadeau, R., 613
Nagel, Thomas, 6, 353, 368, 371, 716
Nagl, W., 504
Nambu, Y., 499
Narens, L., 124, 133
Narmour, E., 692
narrative consciousness, 153–57
National Protein Data Base, 427
Natsoulas, T., 290, 713
naturalism, biological, 46
naturalistic dualism, 18
naturalist thesis, 260

natural method, 2
near-death experiences (NDEs), 183
Neary, D., 228
Nebes, R. D., 204–05
negative kinetic energy, 560–61
Negus, V. E., 652–55
Neisser, U., 123, 125
Nelkin, N., 259, 271
Nelsen, J. M., 678
Nelson, D. K., 207
Nelson, D. L., 650
Nelson, R., 592
Nelson, T. O., 124, 133
neocortex, 247
neostriatum, 252
nerons, synaptic transmissions of, 740
nerve cell, 419, 662
nervous system
 anatomy and physiology protect, 178
 neurologic states and substates of, 182
 visual awareness and neuronal activity in, 247
net
 comparison, 357
 episodic memory, 357
 Laplacian, 359
 semantic, 357
Neubauer, C., 517
neural correlate of consciousness (NCC), 22, 243, 247
neural Darwinism, 12, 113–14
neural groups, oscillation of relevant, 10
neurally delayed experience, 345
neural membrane
 as "master computer" of consciousness, 423
 microcomputational evolution in, 422
 quantum computation in, 419–24
neural net, intermediary, 261
neural networks, 349–92, 494
 artificial "attention" in oscillatory, 377–82
 how does memory emerge in simple artificial?, 383
 psi and, 593
 theory of, 349
neural states, 452–53
neural systems, dynamics of, 613–15
neural time factors, 337–47
neuroanatomical correlates of consciousness, 317
neurobiological theory of consciousness, 10
neurologic states, thermodynamic considerations of, 181–83

neuronal basis of consciousness, self-referent mechanisms as, 611–32
neuronal microtubules, illus., 511
neuronal substrate of visual consciousness, towards the, 247–57
neurons
 aminergic and cholinergic, 78
 assembly of, 609, 662
 behavior of, 54
 cholinergic, 76
 cortical or subcortical, 262
 discharge properties of, 262
 McCulloch-Pitts, 349
 morphology of, 262
 number in infants vs. adults, 704
 output, 626
 schematic of central region of, 510
neurophysiological modeling, 8
neurophysiology, conception of brain in, 495
neuroscience
 experimental, 243–347
 natural method and, 69
 problems with solutions of, 58
neurotransmitters, 262
neutral monism
 as metaphysical theory, 48
 qualia and, 46–48
Newcombe, F., 264
Newell, A., 7, 273
Newer, M., 158
new physics, 750
Newsome, W. T., 250, 263
Newton, universal gravity theory of, 579
Newtonian bottleneck, 570
Newtonian physics, 440–41
Nichols, D. E., 430
Nicol, E., 497
Niedenthal, P. M., 99–100
Nielsen, C., 58
night terrors, 71
Nikhilananda, S., 696
Nilsson, L., 228–29
Nip, M. L. A., 513
Nisbett, R., 38
Nishimura, T., 514
Nobili, R., 451–52, 465
nonalgorithmic processing, 14
nonbasic principles, Chalmers', 19
non-Born-Oppenheimer electrons, 478
nonconscious "contexts," 131
nondeterministic "free will," 508
nonlinear dynamics, 14, 660
nonlocality

of quantum mechanics, 552–53
in synchronicity and quantum physics, 606
quantum coherence, 436
of space and time, 541–608
nonphysical sciences, for study of mind, 569–70
nonreductive explanations of consciousness, 16–18
nonreductive physicalism, 44
nonreductive theory of consciousness, 19
non-REM (NREM) sleep, 71, 74
norepinephrine, 76
Norman, D. A., 133, 135–36, 373, 729
normative duality in split brain, 197–202
Northen, B., 228
Norwood, D., 729
Noton, D., 365–66
Noyes, P., 572
nuclei, granular arrays of atomic, 522–23
nucleus reticularis thalami (NRT), 349, 358
"numinous" experiences, 720
Nunn, C. M. H., 245, 331–36, 446, 449, 735, 737
Nunn, J. F., 503
Nuño, J. C., 667
Nuwer, M., 157

Oakley, D. A., 421
Oakson, G., 358
Oatley, K., 134
objective consciousness, 148
objective reduction (OR), 518–30
objective threshold, 101
object-mode of consciousness, 697–98
Obs, 635–36, 646
obsession, 85
occipital-parietal cortex, lesions of, 134
oceanic experiences, 157
olfaction
  consciousness the brain and, 279–93
  EEG registration of subliminal, 279–91
Olive, D., 580
Olivier, A., 313
Ono, H., 110
Onsager, L., 515
"on-track-ness," feeling of, 136
Oppenheimer, S., 231
orchestrated reduction, 507–40
"order of arrival" theory, 303
ordered water, 527
organizational invariance, principle of, 22
Organization of Behavior, The, 609
Ornstein, R. E., 574, 707, 710–11, 730

Ortony, A., 134
Oscar-Berman, M., 154
oscillations
  40–Hz, 22, 52, 84
  binding problem and neurobiological, 51–61
  gamma range, 55
  phase locked pattern of, 55
  synchronous 40–Hz, 118
  in thalamic input, 56
oscillatory neural networks (ONN), 377
osmosis, gelation and, 400
osmotic mechanism, 404
Oswald, L. E., 707
outer-plexiform layer (OPL), 358
out-of-body experiences, 157, 186
output neurons, 626
Overton, E., 425
O-water studies, 319
Oxbury, J. M., 192, 265
Oxbury, S. M., 192, 265

Packard, N. H., 667
Paillard, J. F., 285
Paivio, S., 721
Palmer, R. G., 86
Palmini, A., 190
Pandya, A. S., 623–24, 626, 629
Panigrahy, A., 118
Pantaloni, D., 409, 515
Papanicolaou, A. C., 133
Papex limbic circuit, 359
parallel processors, serial and, 729
parapsychology, 599, 589
parallelism, 546
Paré, D., 84–86
Park, J., 61
Pasik, P., 265
Pasik, T., 265
p-aspects of consciousness, 68
Passerini, D., 223
passive awareness, 707
Pastor, R., 420
Pauli, Wolfgang, 597, 599–604, 606
Paus, T., 277
PCP, 688
p-dreams, 80
"peak experiences," 720
Pearl, D. K., 303, 309, 339, 342–43, 345, 360, 373, 520, 525, 532
Pearle, P., 511–12, 519
Peat, F. D., 584, 735
Pedersen, Jonas, 244, 303–10
Peirce, C. S., 149

Penfield, W., 148, 309, 740
Pennington, D. S., 158
Penrose, Roger, 14, 245, 259–60, 331, 334, 432, 435–37, 443, 449, 470–72, 507–41, 549, 572, 580, 583, 586, 613, 619, 640, 644, 669, 725, 730–31, 735–36, 740
perception
  awareness as self-reported ratings of, 279
  causal account of, 47
  common sense understanding of, 31
  as excitation of consciousness, 448
  implicit, 98–99
  "of the mental," 29
  as organized in layers, 272
  self and, 700–03
  sensory aspects of, 142
  vision as paradigmatic of, 32–33
percepts, 47
percolation
  collapse of quantum parallelism and, 469–92
  network model of, 482–84
Perenin, M.-T., 265
"perennial philosophy," 748
Perez, S., 219
perifissural cortex, 148
peripheral vision, loss of, 175
Perry, John, 38
personal/extrapersonal distinction, 148–49
personal goals, emotion and, 133
Pertile, G., 614
PET. *See* positron emission tomography (PET)
Peterhans, E., 249
Petersen, S. E., 76, 86, 252, 315, 317
Peterson, C., 730, 737
Peterson, J. M., 279–81
petit mal, 189
Petri, Carl Adam, 717
Petrides, M., 156
PGO waves, 72–73, 80
phase-locked oscillations, 55–57
phase sequence, 666
Phelan, A. M., 517
Phelps, E. A., 218
phenomena, functionally definable, 8
phenomenal consciousness, 7, 259–78
phenomenalism, demarcation between neutral monism and, 47
phenomenological consciousness, process model of, 713–24
phenomenology, 69, 673–723
Philipps University, 54

philosophical relativity, principle of, 487–90
phonons, coherent pumped, 515–17
phosphergy, 649–55
phosphoryl chemical energy, 649
photic blink response, 263
phylogenetic data, 610
physical information, 24
physicalism, 726, 739
  objections to noneliminative, 43
  science vs., 42–43
  without eliminativism, 43
physical model of consciousness, 735–37
physical theory, 17
physics
  crossdisciplinary, as basis for mind, 571–72
  ignoring of time's passage in, 544
  theories that seeks to reconcile consciousness with, 546
Piaget, J., 697
Pickersgill, M. J., 254
Pick's disease, 229
"picture theory" of mental representation, 35–36
Pietromonaco, P., 100
Pietsch, P., 575
Pihlaja, D. J., 514
Piper methysticum, 684
Pirot, M., 707
Pirozzolo, F. J., 230–34
Pitts, Walter, 349–50
pixel, pressure, 398
placebo responses, healing and, 677
planet, problems confronting our, 704
planning, prefrontal brain areas and, 250
Plato, 67, 541, 579
platonic model, to understand music, 691–94
Plourde, G., 216
Plum, F., 194
Podlachikova, L., 366
Poglitsch, A., 517
Pohl, W. G., 149
Poincaré, H., 470, 487
"point of view," determination of, 369
Poirrier, R., 319
Polak, Ernest, 293
Poland, J., 43–44
Polanyi, M., 123
Poletti, C. E., 307
Pollen, D. A., 147, 158
Polson, J. S., 321, 425
Polster, M. R., 318

Ponds, R. W., 232
pons, 252
Ponsot, G., 190
pontine brainstem, 77
Popescu, S., 551–52, 564
Popp, F. A., 504, 641
Poppel, E., 564–65
Popper, Karl R., 261, 275, 726
positron emission tomography (PET), 311–13, 618
Posner, J. B., 194
Posner, M. I., 58, 76, 86, 89, 315, 717
posterior cerebral convexity, 148
posteromedial lateral suprasylvian area (PMLS), 61
posthypnotic amnesia, 296
Prat, A. G., 515
precognition, 589
preconscious processing, 102
  in impression formation, 99–100
  limits on, 100–02
  transition from conscious to, 508
prefrontal brain areas, planning and, 250
premonitory dream or vision, 599
pressure pixel, 398
Presti, E., 58
Pribram, Karl H., 90–91, 141–64, 404, 446, 452, 482, 527, 530, 617, 735
Price, T., 229
Prigatano, G. P., 131, 228
Prigogine, I., 547, 656
primary consciousness, 115, 227
Primary Repertoire, 113
primary visual cortex, 252–53
prime, subliminal, 101
primitive, 669
principle of scientific realism (PSR), 43
Principles of Psychology, 1, 122, 124
Prinzmetal, W., 58
Pritchard, W. S., 323
privacy, varieties of consciousness and, 484–85
problem
  inverse, 622
  of locality, 620
  mind-body or mind-brain, 569
Procaccia, I., 322, 327
processing activity, consciousness and, 715
projective technique, Rorschach Inkblot Test as first, 205
proprioceptive states or desires, 31
prosopagnosics, 223
Prosperi, G. M., 555
protein
  conformation of, 515
  hydrophobic pockets of, 425–34
  structure of, 400
protons, as elaborate logical structures composed of events, 47
proximal events, distal events and, 106–08
Prozac, 687
psi, 589–96
psychedelic drugs, 430
psychic entities, 666
psychic healing, 591
psychic phenomena, 589–96
psychoïd factors, 604
psychokinesis (PK), 589
psycholinguistics, 101
psychological unconscious, 96
psychology, natural method and, 69
psychophysical theory, 17, 19
psychotropic plants and drugs, 677
Ptito, A., 265
Ptito, M., 158, 265, 277
Puck, T. T., 515
Puthoff, H., 575
Putnam, F. W., 193
Pylyshyn, Z., 52
pyramidal cells, 56

qualia, 2, 7, 247, 713
  absent and inverted, 24
  defined, 41
  irreducibility of, 259
  model of choice and, 469–92
  neutral monism and, 46–48
  place in the world of science for, 41–49
  as qualitative feeling of conscious contents, 716
  as qualities of subjective experience, 473–76
  science and the elimination of, 42
  as tenable referents, 473–76
  tying intelligence to, 476
qualitative consciousness, 41
Quant, M., 591
quantum brain dynamics (QBD), 499
quantum brain structures, consciousness-relevant, 512
quantum coherence
  in microtubules, 526
  orchestrated reduction of in brain microtubules, 507–40
  schematic graph of proposed, 529–30
quantum computation, 419–24
quantum field, brain function and collapse of, 321–36

Quantum Gravity, 572
quantum healing, 750
quantum mechanics (QM), 14, 551, 634
  Copenhagen interpretation of, 599
  Darwinian theory and, 423
  of dreams and the emergence of self-awareness, 451–68
  holism and nonlocality of, 552–53
  two-vector formalism of, 558–65
quantum optical coherence, subcellular model for, 497–503
quantum parallelism, percolation and collapse of, 469–92
quantum phenomena, subneuron level, 448
quantum physical neural states, self-reflective, 452–53
quantum physics, differences between Newtonian physics and, 441
quantum self-interrogation, 455–58
quantum superposed states, 507. *See also* wave functions
quantum superposition, of brain levels, 562
quantum systems
  collapse from superposition to single state, 736
  synchronicity and nonlocal information in, 597–608
quantum theory, 435–540
  biology and, 496
  of consciousness, 14, 440
  Copenhagen interpretation of, 436, 507, 510
  new insights on time consciousness and reality from, 551–68
  observable eigenvalues with, 580
Quesney, L. F., 313
Quillian, M. R., 123
Quinlan, Karen Ann, 118
Quinn, P. C., 365

Raab, E., 264
Rabin, A. I., 205
Radin, D., 592–93
Rajneesh, G. S., 707
Ramachandran, V. S., 51, 207
Ramsey, 651
random Boolean networks, 383–87
  architecture of, 386
  basin of attraction field of, 385
  biological model for, 389
  learning algorithms for, 390–91
rapid eye movement (REM) sleep, 70–81, 312, 366

Rasmussen, Steen, 390, 407, 482, 515, 517–18, 695
Rasmussen, T., 191, 309
Rauch, S. L., 318, 618, 622
raw sensation, 41
Rayleigh-Bénard convection cells, 605
Rayman, Jan, 166, 193
"real intelligence," human, 571
realism, descriptive, 44, 48
reality
  constantly changing, 543
  contrasting modes of organizing, 203–21
  difference between corporeal and extracorporeal, 158
  new insights from quantum theory on, 551–68
Rebhund, L. I., 524
receptive consciousness, 674, 698
receptive field manifolds, 145
reciprocal connections, 113
reciprocal determinism, 95
recognition, implicit and explicit, 274
reduction, 435. *See also* wave function, collapse of
reductionism, 726, 739
reductive explanation, 8
reductive methods, 5, 15
reductive physicalism, 43
Reed, B. R., 230
reentrant connections, 113
reference, self, 622
reflection space, 580
reflective consciousness, 227
reflexive behavior, 261
Regard, M., 205–06
Reichmann, K., 191
Reiman, Eric M., 245, 318
Reingold, E. M., 101
Reisberg, B., 229–31, 234, 238
Reisch, Gregor, 612
Reitbock, H. J., 54, 56, 61
Relational Theory of Mind, 360
relative recency, 156
relativity, philosophical, 487–90
relativity theory
  dismissal of simultaneity by, 544
  ignoring of time's passage by, 547
  mind-body problem and, 543–50
religious experience, variety of, 141
REM. *See* rapid eye movements (REM) sleep
renunciation, meditation and, 699–700
reportability, 8
Rescher, Nicholas, 48

resemblance theory, 34–36
Restak, R. M., 118
restricted environmental stimulation technique (REST), 707
retina, 253
Reuter, B. M., 191
Revuonso, A., 89
Rhine, J. B., 601
Rhodes, Paul A., 27, 56
Ribary, U., 84, 359
Ricciardi, L. M., 496
Richard, L., 517
Riddock, G., 263
Rigamonti, D., 229
right hemisphere, 204, 217, 232
  basal ganglia and, 148–49
  tumors in, 205
"rightness," feeling of, 136
right visual field (RVF), 197
Rimini, A., 435, 511–12, 519
Rindler, W., 572–73
Risse, G., 97
Rittenhouse, C. D., 87
Robertson, L. C., 205
Robinson, D. L., 252
Robitaille, Y., 190
Robson, J. G., 158
Rocha-Miranda, C. E., 249
Rockstroh, B., 252
Roffwarg, H. P., 73, 85
Roger, G., 443, 494–95
Rohrer, W. H., 248
Rohrlich, D., 559
Roland, P. E., 254
Roos, L., 154
Roots of Consciousness, The, 582
Rorschach, Hermann, 205
Rorschach Inkblot Test, 205–19
Rosa, M. G. P., 249
Rosch, E., 129, 635
Rose, J. E., 148
Rose, S. P., 514
Rosen, D., 497
Rosen, N., 549
Rosenberg, Gregg, 27
Rosenquist, A. C., 263
Rosenthal, David, 29–30, 38
Rosenthal, R., 591
Rosenwald, Larry, 85
Rossi, F., 451, 464
Rosvold, A. E., 150–51
Rosvold, H. E., 51
Roth, L. E., 514
Roth, T., 86

Rothblatt, L., 617
Rothi, L. J., 192
Rovee-Collier, C., 365
Rowlands, S., 446
Rozendaal, N., 232
Rozin, P., 207
Rubens, A., 130, 135
Rubin, N., 191
Rueckert, L., 218
Ruff, M. N., 149
Rulving, E., 250
Rumelhart, D. E., 729, 737
Rumi, 704
Rumnelhart, D. E., 125–26
Rusconi, M., 223
Rush, A., 365
Russek, L. G., 293
Russell, Bertrand, 2, 45, 47–48, 349, 547
Russell, J. W., 151
Russell, R. W., 151
Ryback, I., 366
Rycroft, C., 730
Rydberg atoms, 419
Rydberg wave packets, 420
Ryder, L. H., 504

Sabina, María, 683
Sabo, S., 219
Sacks, O., 118, 130
Sadzot, B., 319
Sagi, D., 58
Sahraie, A., 263
Saint-Cyr, J. A., 252
Sakai, H., 482
Sakata, H., 149
Salam, A., 499
Salin, P-A., 265
Salmon, E., 319
Salvia divinorum, 684, 688
Samsonovich, A. V., 482
Sanders, M. D., 118, 263
Sanderson, R. E., 110
Sanes, J. N., 61
Sanford, David, 82
Sanger-Brown, 150–51
Santayana, G., 673
Sarnat, H. B., 189
Satarić, M. V., 394, 407–17, 513, 515
Satir, P., 527
Saunders, J. B. deC. M., 651
Schacter, D. L., 96, 228–29, 235–36, 238, 250, 274, 295, 297, 299, 317–18
Schacter, D. S., 123, 131, 134
Schäefer, E. A., 150–51

Schall, J. D., 248, 263
Schank, R., 365
Schanze, T., 61
Schechter, E., 591, 593
Scheier, M. F., 134
Schein, S. J., 253
Schiebel, A. B., 358
Schiffer, M., 572
Schiffer, S., 44
Schill, K., 564
schizophrenia, 449
Schleidt, M., 714
Schlitz, M., 592, 594
Schmidt, H., 58, 594
Schneck, M. K., 229
Schneider, W., 377, 507
Schoenberg, A., 691
Schomer, D. L., 191
Schooler, J., 83
Schram, F. R., 650
Schramm, J., 193
Schreiber, H., 155
Schrödinger, Erwin, 393, 350, 435, 511, 562, 669, 726
Schrödinger wave equation, 331, 441, 465, 510, 600, 692
Schultz, D. H., 197, 207
Schuster, Peter, 646, 667–68
Schwartz, C., 56, 61
Schwartz, Gary E., 245, 279–93
Schwartz, J. H., 51
Schwarz, J. H., 580
Schweber, S. S., 662, 664
Schwender, D., 565
science
  answers to ultimate question by, 659
  antipathy to mystical tradition, 695
  cognitive, 89–164
  conventionalist philosophy of, 470
  and the elimination of qualia, 42
  influence of unconscious in constructing, 744
  nonphysical, for study of mind, 569–70
  vs. physicalism, 42–43
  pushing the frontiers of present, 746
  spiritual traditions and, 703–04
  "state specific," 703
  three interpretations of, 41–46
  Western, as restricted to matter and "upward" causation, 749
scientific instrumentalism, 48
Scott, Alwyn C., 349, 433, 482, 529, 534, 610, 614–15, 621, 633, 639, 659–72

Searle, John R., 1, 22–23, 37, 46, 259, 353, 367–68, 621, 669, 730–32
Secondary Repertoire, 113
second-order assemblies, 663
second-order structure, 634
segmentation, binding association and, 59
Seidenberg, M., 239
seizures, epileptic, 189
Sejnowski, T. J., 53, 85–86
selection mechanism, 635
self
  concept of, 136–37
  forgetting by shifting intention, 699–700
  of instrumental vs. spiritual consciousness, 701
  intention, spiritual experience and, 695–706
  perception and, 700–03
  survival, 697–98
self-awareness, emergence of, 451–68
self-concept, 132
self-consciousness, 273
  distinction between consciousness and, 485
  as level of awareness, 719
self-identification, formation of hierarchy of, 462
self-image, self-inquiry hierarchy for ordering, 453
self-induced transparency, 502
self-interrogation, quantum, 455–58
self-monitor, 131–32
self/not self boundary, 458
self-reference, 622
self reflection
  mathematical structure of, 453–55
  quantum physical neural states and, 452–53
Sellars, Wilfrid, 43–45
semantic capability, 486–87, 638
semantic memory, of consciousness, 369
semantic net, 357
semantic processing, 102
semantic theory, 731–35
senses
  extension of experience across the various, 105–06
  five, 367
  systems of the brain for, 142
sensory aphasic, language comprehension by, 223–25
sensory experience, projection of, 148
sensory information, synchronization of conscious experience to, 303–10

sensory performance, in absence of conscious awareness, 285
sensory-specific agnosias, 148
sentience, 41
Sept, D., 394, 407–17
septum, 115
Serafetinides, E. A., 192
Sereno, M. I., 51
Sergent, J., 193
serial processors, parallel and, 729
serotonin, 76, 430
Setterlund, M. B., 100
Sewell, M. M., 707–08, 710
Seymour, J., 729
Shadows of the Mind, 509, 511, 541
Shah, Idries, 696, 704
Shah, M. V., 296
Shallice, T., 22, 133, 136, 363, 373
shaman, 683, 685
Shames, Victor, 102
Shannon, C. E., 24, 148
Shaw, J. C., 273
Sheldrake, R., 574
Sheng He, 254
Shepard, R. N., 618
Shepherd, G. M., 389
Sherk, H., 252
Sherrington, Charles, 393–94, 726
Sherrington, S., 393–94
Shertsova, N., 366
Shimamura, A. P., 227, 229, 296
Shimony, A., 553
Shipp, S., 250
Shishidura, M., 645
Shklovsky-Kordy, N. E., 482
Shor, P., 571
Shore, Peter, 574
Shub, Mm., 644
Shulgin, A. T., 430
Sidtis, J. J., 193
Siekevitz, P., 514
signal detection task (STD), 707
Silva, L. R., 56
similar motor areas (SMA), 366
Simon, H. A., 273, 638
Simons, D. J., 146
Sims, N. R., 228
simultaneity, in synchronicity and quantum physics, 606
Singer, G., 151
Singer, W., 53, 56–57, 60–61, 118, 503, 507
Singh, A., 296
Sirag, Saul-Paul, 541–42, 579–88, 739
Sirockman, B. E., 233

Skarda, C. A., 740
skin-induced sensation, 340
Skinner, B. F., 260
skin potential tracings, 224
sleep
　difference between wakefulness and, 6
　dreams as spandrels of, 67–88
　EEG patterns in diagnosing, 321
　mentation during NREM and REM, 71
　O-water studies to investigate, 319
　rapid eye movement (REM), 366
　REM, 312
　slow-wave, 366
Sloane, N. J. A., 580, 582, 586
Slowdowy, P., 583
slow-wave sleep (SWS), 366. *See also* non-REM sleep
Smale, S., 644
smell, sense of, 105–06
Smith, Brian C., 38
Smith, C. W., 446, 448
Smith, Lloyd, 293
Smith, P., 732, 736
Smith, S. A., 517
Smith, W. S., 61
Smolensky, P., 125–27
Smythies, J. R., 581
Snowden, J. S., 228
Snyder, S. H., 430, 687
social assemblies, 662
social intelligence, 94
social interaction
　general cycle of, 93–95
　unconscious processes in, 93–104
social judgment, as implicit memory, 97
social psychology
　cardinal principle of, 93
　natural method and, 69
sol-gel states, 527
Solymosi, L., 193
somatosensory (SI) cortex, electrical stimulation of, 337
Sorenson, L., 218
Sosa, E., 43
Soso, M. J., 264
soul, 292, 669
sound, connection between form and, 694
Space, Hilbert, 572
space
　nonlocal time and, 541–608
　reflection, 580
space-time, geometry of, 541
space-time dynamics, theory of, 548
Sparks, D. L., 248, 252

spatial continuity, 61
species-shared behavior patterns, 150
speed of light, 572–73
Sperling, G., 110
Sperry, Roger W., 135,165, 204–07, 216–17, 219, 633, 639
Spinoza, 490
Spinelli, 156, 158
spirit, 292
  as realm of consciousness, 749
  vital, 15
spiritual aspects of consciousness, 157
"spiritual awareness," 292
spiritual consciousness
  nature of, 696
  self of, 701
spiritual experience, intention self and, 695–706
spiritual traditions, science and, 703–04
Spitz, R., 697
split brain
  normative duality in, 197–202
  operation, 203
  patient, conflicting communicative behavior in, 189–96
  procedure, 190
  syndrome, 274
spontaneous symmetry breaking, 498–500
Sporns, O., 58
spotlight metaphor for awareness, 125–26
Sprague, J. M., 263
Springfield, Christina, 328
Spurtzheim, G., 141
Squire, L. R., 229
Squires, E., 512
Srull, T. K., 99
Stahlkopf, Deborah, 82
Stairway to the Mind, 621
Stapp, Henry P., 436, 509, 541, 549, 552–53, 563, 613, 669, 740
Stark, L., 365–66
states of consciousness, 142–43
state specific science, 703
statistical thermodynamics, 401
statistical causality, quantum physics and, 601
STD. *See* signal detection task
Stefan, H., 191
Stein, D. l., 416
Stelmack, G., 285
Stengers, I., 656
Steriade, M., 56, 85, 358, 617
Stern-Gerlach magnet, 553
Sterzi, R., 223

Stevenson, J. A. F., 151, 157
Stickgold, R., 75, 86, 87
Stillings, 89
Stoerig, Petra, 134, 243, 250, 259–78
Stone, M., 151
Stow, J. L., 515
Stoy, J. E., 644
stream of consciousness, 135, 344, 674, 713
Strochi, P., 525
structural coherence, principle of, 20–22
structures
  of awareness, 20
  of consciousness, 20
  as criteria for consciousness, 493–94
  first- and second-order, 634
  hierarchical, 635
Stryker, M., 52
Stuart, C. I., 497
Stubenberg, Leopold, 2–3, 41–49, 436
Stuss, D. T., 194, 229
subcellular quantum optical coherence, 493–506
subconsciousness, as neurologic state, 182–85
subconscious processing, 96
subjective experience
  order of arrival theory and, 305
  pharmacology of consciousness and narrative of, 677–90
  structure of, 121–40
subjective referral backwards in time, 339–42
subjective threshold, 101
subjectivity, seeking epistemology of, 747–48
subliminal olfaction, EEG registration of, 279–91
subneural biology, 393–434
substantia nigra, 252
substrate of consciousness, 22
Sudilovsky, A., 234
Suedfeld, P., 707
superconductivity, Meissner effect in, 527
superconscious, 749
superego, 730
"superposition catastrophe," 52–61
superradiance, 500–01
superstring theories, 579
supervisory attentional system (SAS), 373
supplementary eye field (SEF), 365
"survival self," 697–98
Sutin, J., 307
Suzuki, S., 697, 700
Swenberg, C. E., 514

Sykes, R. N., 133
Sykes, W. S., 295
synaptic function, mechanism for regulating, 514
synaptodendritic domains, 157
synaptodendritic processing, 143
synchronicity
  causality and, 601
  Jung's theory of, 597–99
  nonlocal information in quantum systems and, 597–608
synchronized neuronal assemblies, 55
synchronous 40–Hz oscillations, 118
syntactics, cognitive processes as, 638
Szent-Gyorgyi, A., 393

tacit knowledge, 123
tactile senses, 106
Tada, K., 514
Takahashi, Y., 497
Takeda, M., 514
Tampieri, D., 190
Tanaka, K., 56, 114
Tank, D. W., 729
Tao of Physics, 159
Taravath, S., 190
Tart, Charles, 703
taste, sense of, 105–06
Tataryn, D. J., 96
Taylor, J. G., 110, 349, 356, 358, 360–64, 368
Taylor, J. H., 147, 158
Taylor, J. L., 345
Taylor, M. A., 239
TCA. *See* temporal central availability
telepathy, 589
temporal binding, 495
temporal central availability (TCA), 564
temporal cortex, 115
temporality, consciousness and, 714
temporal order
  subjective experience of, 303
  subjective judging of, 305–06
  See also time
Teuber, H.-L., 274
thalamic input, oscillations in, 56
thalamus, 115, 357, 359
  hyperpolarized in slow-wave sleep, 366
  intralaminar nuclei of, 252
  involvement in emotion, 315
  lateral geniculate (G) body of, 72
  medial portion of, 154
  peripheral visual pathways and, 617
  sensory relay station in, 313

theory of everything, 572, 579–80, 659, 692, 726
Theory of Neuronal Group Selection (TNGS), 113
thermodynamic description of loss of consciousness, 181
thermodynamics, statistical, 401
Theurkauf, W. E., 514
thinking, three kinds of, 729–38
"third eye," 36
TH-NRT-C complex, 363, 366
Thom, Rene, 581
Thomas, W., 386
Thompson, E., 635
Thompson, V. M., 318
Thompson, W. L., 618, 622
Thorell, L. G., 254
Thorne, K. S., 477
thought
  as excitation of consciousness, 448
  as sensitive to details of social situation, 93
  stream of, 136
threshold for excitation, 609
threshold for stimulation, 663
Thurston, N., 214, 219
Tien, H. T., 515
Tillson, P., 514
Timasheff, S. N., 409, 515
time
  binding problem and, 52–61
  consciousness and, 543–50
  does it pass?, 543–45
  as illusion, 557
  mental and biological functions on different scales of, 53
  model of, as created consciousness, 562
  neural factors of, 337–47
  new insights from quantum theory on, 551–68
  nonlocal space and, 541–608
  representation in brain of, 303
  subjective referral backwards in, 339–42
  subjective sensation of, 714
"time-on" theory, 344–46
tip-of-the-tongue experience, 124, 128, 130, 390
TNGS. *See* Theory of Neuronal Group Selection
To, L., 407
Tobias, B. A., 131
Toffoli, T., 571
Tollaksen, Jeff, 524, 532, 541–42
Tomaiuolo, F., 277

Tomkins, S., 133
Tonomi, G., 58
Total, M. E., 100–101
totality, consciousness and, 716–17
Tranel, D., 97, 230, 234–35, 274
transcendental consciouness, 157–59
Transcendental Meditation, 465
Treisman, A., 51, 125–26, 379
Trevarthen, C., 194, 205–06, 217
triangulation, knowledge by, 69
Triesman, A., 58
Trope, I., 207
Trosset, M. W., 230, 233–34
Trpisová, B., 394, 407–17, 513
Tsotsos, J. K., 449
tubulin, 409
tubulin states, schematic model of, 516
Tulving, E., 235–36, 238, 714
tumors, use of EEG in diagnosis of, 321
tunnel vision, 175, 186
Turing machine, 493–94
Tuszyński, J. A., 394, 407–17, 513, 515
Tyler, L., 225
Tzanakou, E., 623, 628

Uleman, J. S., 96
Umezawa, H., 496–98
unawareness, operational definition of, 234
unconscious
  collective, 598, 605–06
  influence in constructing science of, 744
  psychological, 96
unconscious mental functions
  how does brain distinguish conscious from, 342–44
  neural time factors in, 337–47
  speed of, 346
unconsciousness
  inducing, 169
  as neurologic state, 182–85
  self-identification hierarchy and, 462
Ungerleider, L. G., 250, 252
unity of consciousness, 443
universe, events as basic "building blocks" of, 47
unified field theory, 580
Unnikrishnan, K. P., 623–24, 626, 629
Unruh, W. G., 572
"unus mundus," 598
Upanishads, 696
Usher, M., 58

vacuum fermions, 470–72

vague, meanings of, 127
Vaidman, L., 525, 532, 541, 551–52, 556, 559–60, 564
Vaidyanath, R., 515, 517–18
Valdés, L. J., 681, 685
Valdiserri, Michael, 102
Valdiserri, Susan, 102
Valenstein, T. E., 617
Vallar, G., 223
Vallee, R. B., 514
"value-category" memory, 116
Vanbremeersch, J. P., 639, 645
Vander Stoep, A., 156
van der Waals forces, 426, 428–29
Van-Dusen, W., 720
Van Essen, D. C., 248, 250, 252
Van Heerden, P. J., 158
Van Loon, R. J. P., 149
Varela, F. J., 635
Varieties of Religious Experience, The, 124, 141, 696, 720
Varvoglis, Mario, 542, 589–96
Vassilev, P., 515
Vavdyanatu, R., 407
Vecker, A., 318
vector analysis, 580
vekada, 683
Velleman, D. J., 42
Velmans, M. M., 132, 135, 719, 745
Venable, R., 420
Venecia, D., 157
Verhey, F. R., 232
vertebrates, examples of O logistics in, 652
veto theory, 261
Villemure, J. G., 265
Virtanen, J. A., 420
viscerautonomic effects, 154
vision
  loss of peripheral, 175
  as paradigmatic of perception, 32–33
  sense of, 106
  tunnel, 175
visions
  Amahuaca Indiansí, 679
  of God, 695
  premonitory, 599
visual awareness, 11
  biological usefulness of, 250
  neural correlates of, 247
visual consciousness, 247–57
visual cortical hierarchy, 251
visual cortical system, 247
visual data, processing of, 51
visual deficits, hierarchy of, 264

visual dreams, phenomenal, 272
visual fields
  complex geometry of, 20
  right and left, 197
visual imagination, 247
visual perception, phenomenal consciousness and, 259–78
visual system, lesions of, 264
vital spirit, 15
Vitiello, Giuseppe, 446, 448, 501
Vogel, J. P., 197
Vogel, P. J., 203, 207
Vogel-Bogen series, 203, 206–07
Volkenstein, M. V., 478
voluntary act, initiation of, 342
von Bekesy, G., 309
von der Heydt, R., 249
von der Malsburg, C., 52–54, 56, 377, 507
von Franz, M. L., 603–04
von Frenckell, R., 319
von Helmholtz, H., 110
von Neuman, J., 435, 511, 517, 553, 562, 580, 594
von Steinbuchel, N., 564
Vrba, E., 83

Wada, J., 191
Wahl, Sharon, 27
wakefulness
  difference between sleep and, 6
  parts of brain controlling states of, 77
Walker, E. H., 436, 586, 740
Walker, J. A., 190
Wall, P., 151
Wallace, B., 394, 707, 711
Wallace, Ron, 394, 419–24
Wallace, S. F., 263
Wang, N., 515
Wappenschmidt, J., 191
Warrington, E. K., 118, 149, 216, 263
Warwick, R., 656
Washburn, M. F., 110
Wasson, R. Gordon, 683–84
water
  clusters as pixels of life, 397–405
  ordered, 527
  role in origin and maintenance of life, 398–401
Watson, J. D. G., 253, 265
Watson, R. T., 232, 617, 651
Watt, C., 593
Watt, D., 373
Watt, Richard C., 245, 321, 407, 425, 430, 482, 509, 517

Watterson, John G., 393–94, 527
wave/particle duality of atoms, 507
wave function, 470–72, 435, 507
  collapse of, 510–12, 599–603
  meaning of , 558–60
  Schrödinger, 331
  self-collapse in microtubules, 518–30
  See also reduction
wave packets, Rydberg, 420
Way of the Sufi, The, 696
Weaver, W. , 148
Weber, T., 511, 519
Weil, Andrew T., 430, 677–90, 720, 753
Weinberg, S. A., 499, 651
Weintraub, D., 105–11
Weise, S. B., 318
Weisenberg, R. C., 514
Weiskrantz, L., 118, 134, 250, 263, 265, 273, 285
Wellman, H. M., 227
Wells, Horace, 295
Wen, W. Y., 399
Werbos, D. J., 482
Wernicke's area, 116–17, 356
West-Eberhard, M. J., 86
Western culture, esoteric tradition in, 157
Western science, as restricted to matter and "upward" causation, 749
Western thought, "basic limiting principles" of, 589
Wheeler, John A., 25, 435, 477, 520, 552, 557, 571, 575, 600, 602
Whinnery, C. C. M., 175
Whinnery, James E., 165, 175, 181
Whitehead, Alfred North, 273, 349, 544, 714
Wiesel, T. N., 114, 639
Wigner, E. P., 580, 594, 600, 602, 669
Wilber, H., 730–31
Wilber, Ken, 748–50
Wilkes, K. V., 12–13, 259
Williams, P. L., 656
Wilson, M. A., 58, 367
Wilson, T., 38
Wilson, W. H., 151
Wimsatt, W. C., 121, 135, 136
Wise, S. B., 618, 622
Witter, M. P., 369
Wittgenstein, L., 487
Wizard of the Upper Amazon, 679
WM/Semantic modules, 364
Wolf, Fred Alan, 2, 436–37, 451, 679
Wolfram, S., 384
Wolkstein, M., 264

Wong, R., 614–15
Wood, 156
Woolsey, C. N., 148
word completion tests, 297
word paradigms, 296
working memory, 361
Wright, E. W., 303, 309, 338–39, 342, 345–46, 359–61, 373, 520, 525, 532
Wright, J. J., 149
Wright, K. P., 279–81
Wuensche, Andrew, 350, 383–92
Wulf, 503
Wulff, D. M., 720
Wurtz, R. H., 263
Wyer, R. S., 99

Yamanoue, Tokiko, 350, 377–82
Yame, S., 56
Yarbus, A. L., 365
Yarom, Y., 56
Yasue, Kunio, 404, 437, 448, 482, 493–506, 527, 530, 735
Yates, J. L., 206
Yates, P. O., 228
Yoga, 157
Young, A. W., 223, 264
Young, M. P., 56, 250
Yu, W., 514, 528
Yukie, M., 265
Yun, L. S., 318

Zaidel, D., 204–05
Zaidel, Eran, 131, 166, 193–94, 197
Zakula, R. B., 412, 515
Zeeman, E. C., 740
Zeh, H. D., 548
Zeigarnik, B., 142
Zeki, S. M., 51, 253, 265
Zen philosophy, 157, 695–96, 699, 708
Zhang, C.-T., 515
Zhang, Y., 420
Zinberg, N. E., 678
Zohar, Danah, 436–37, 439–50, 730–31, 735–36
Zubay, G., 650
Zukav, Gary, 159